EWBANK'S HYDRAULICS & MECHANICS

A

DESCRIPTIVE AND HISTORICAL ACCOUNT

OF

HYDRAULIC AND OTHER MACHINES

FOR

RAISING WATER,

Ancient and Modern:

WITH OBSERVATIONS ON VARIOUS SUBJECTS

CONNECTED WITH THE

MECHANIC ARTS:

INCLUDING THE PROGRESSIVE DEVELOPMENT OF

THE STEAM ENGINE:

DESCRIPTIONS OF EVERY VARIETY OF BELLOWS, PISTON, AND ROTARY PUMPS—FIRE ENGINES—WATER RAMS—PRESSURE ENGINES—AIR MACHINES—EOLIPILES, &C. REMARKS ON ANCIENT WELLS—AIR BEDS—COG WHEELS—BLOWPIPES—BELLOWS OF VARIOUS PEOPLE—MAGIC GOBLETS—STEAM IDOLS, AND OTHER MACHINERY OF ANCIENT TEMPLES. TO WHICH ARE ADDED EXPERIMENTS ON BLOWING AND SPOUTING TUBES, AND OTHER ORIGINAL DEVICES—NATURE'S MODES AND MACHINERY FOR RAISING WATER. HISTORICAL NOTICES RESPECTING SIPHONS, FOUNTAINS, WATER ORGANS, CLEPSYDRÆ, PIPES, VALVES, COCKS, &C.

IN FIVE BOOKS.

ILLUSTRATED BY NEARLY THREE HUNDRED ENGRAVINGS.

FOURTH EDITION,

REVISED AND CORRECTED—TO WHICH IS ADDED, A SUPPLEMENT

BY THOMAS EWBANK.

It is a cruel mortification in searching for what is instructive in the history of past times, to find the exploits of conquerors who have desolated the earth, and the freaks of tyrants who have rendered nations unhappy, are recorded with minute and often disgusting accuracy—while the discovery of useful arts, and the progress of the most beneficial branches of commerce, are passed over in silence, and suffered to sink into oblivion.
Robertson's India.

NEW YORK:
SOLD BY BANGS, PLATT, & CO.,
204 BROADWAY.

1850.

Entered according to the Act of Congress, in the year 1842, by THOMAS EWBANK, in the Clerk's Office of the Southern District of New-York.

Dill, Stereotyper, 128 Fulton st.

PREFACE.

CIRCUMSTANCES having led me, in early life, to take an interest in practical hydraulics, I became anxious to obtain an account of all the contrivances employed by different people to raise water—whether for domestic, agricultural, mining, manufacturing, or other purposes; and great was the disappointment I felt on learning that no book containing the information I sought had ever been published. This was the case between thirty and forty years ago; and, notwithstanding the numerous journals and other works devoted to the useful arts, it is in a great measure the case still. No one publication, so far as my knowledge extends, has ever been devoted to the great variety of devices which the human intellect has developed for raising liquids. That such a work is wanted by a large class of mechanics, if not by others, can hardly be questioned; and it is somewhat surprising that it was never undertaken.

It appears from La Hire's Preface to Mariotte's Treatise on the Motion of Fluids, that the latter philosopher often expressed a determination to write "on the different pumps and other engines which are in use, or which have been proposed," but unfortunately he did not live to carry his design into effect. The celebrated work of Belidor, from its extent, and the variety of subjects embraced and illustrated, stands at the head of modern works on hydraulic devices; but of the four large volumes, a small part only is devoted to machines for raising water, and many such are not noticed at all: besides, the cost of the work and the language in which it is written will always prevent it from becoming a popular one with American or English machinists.

Having in the course of several years collected memoranda and procured most of the works quoted in the following pages, I have attempted to prepare a popular volume on the subject—something like the one I formerly longed for—feeling persuaded that it will be as acceptable to mechanics under circumstances similar to those to which I have alluded as it would then have been to myself. Every individual device for raising water has, of course, not been described, for that would have been impossible; but every class or species will be found noticed, with such examples of each as will enable the general reader to comprehend the principle and action of all. In addition to which, inventors of hydraulic machines can here see what has been accomplished, and thus avoid wasting their energies on things previously known.

PREFACE.

In a work of this kind little that is new can be expected; I have not, however, servilely copied any author, but have written the whole as if little had been written before. I have sought for information wherever I could find it; and with this view have perused more volumes than it would be prudent to name. A few gleanings which modern writers have passed over have been picked up—two or three ancient devices have been snatched from oblivion, as the atmospheric sprinkling pot and the philosophical bellows, and some erroneous opinions have been corrected; that, for example, respecting the origin of the safety valve. There is little room for the charge of arrogance in claiming this much, since it is all I have to claim and it is nothing but what a little industry in any one else would have realized. Several devices of my own have also been introduced which must speak for themselves. On referring to old works that are expensive or of rare occurrence I have generally quoted the very words of the writers, under the impression that some of these works will not long be met with at all. For the convenience of perusal the work is broken into chapters, and as much miscellaneous matter has been introduced, an index is added. The general arrangement and division of the subject will be found at the close of the first chapter.

In tracing the progress of any one of the primitive arts, it is difficult to avoid reference to others. They are all so connected that none can be perfectly isolated. I have therefore introduced such notices of inventions and inventors as seemed useful to be known : facts which appeared interesting to the writer as a mechanic, he supposed would not be wholly without interest in the opinion of his brethren. In this, I am aware, it is easy to to be mistaken; for it is a common error to imagine that things which are interesting to ourselves must be equally so to others. As, however, all those devices that contribute to the conveniences of life will ever possess an intrinsic value, the hope is indulged that the following account of several important ones, although it may present little attraction to general readers, will at least be found useful to those for whom it is more especially designed. It certainly is not what I could wish, but it is the best I could produce. I am sensible that it has many imperfections, and there are doubtless many more which have not been perceived. That I have often been diverted from the subjects embraced in the title-page is true; and as the whole was written at long intervals, even of years, a want of order and connection may be perceived in some parts, and obscurity felt in others. All that I can offer to diminish the severity of criticism, is freely to admit there is much room for it.

In noticing various hydraulic devices, I have endeavored to award honor to whomsoever it was due : to say nothing of the ancients, with whom most of them originated, it may here be observed that the Germans were the earliest cultivators of practical hydraulics in modern times. The Dutch (part of that people) contributed to extend a knowledge of their inventions. It was a Dutchman who constructed the famous machinery at Marli, and England was indebted to another for her first water-works at London Bridge. The simplest pump-box or piston known, the inverted cone of leather, is of German origin, and so is the tube-pump of Muschenbroek. Hose for fire-engines, both of leather and canvas, was invented by Dutchmen. They carried the chain-pump of China to their settlements in India, and also to Europe. Van Braam brought it to the U. States. A German invented the air-pump, and the first high pressure steam-engine figured in books was by another. As regards hydraulic machinery, the Dutch have been to the moderns, in some degree, what the Egyptians were to the ancients—their teachers. The physical geography of Holland and Egypt necessarily led the inhabitants of both countries to cultivate to the utmost extent the art of raising water. Wind-mills for draining water off land first occur (in modern days) in Holland. It is indeed the constant employment of this element—wind—that preserves the Dutch from destruction by another; for, as a nation, they are in much the same predicament they formerly put unruly felons in, viz : confining each in a close vault with a pump, and then admitting a stream of water that required his unceasing efforts to pump out, to prevent himself from drowning.

The French have contributed the neatest machine known ; the ram of Montgolfier—theirs is the double pump of La Hire, and the frictionless piston of Gosset—La Faye improved the old tympanum of Asia—Papin was one of the authors of the steam-engine, and Le Demour devised the centrifugal pump. Rotary pumps and the reintroduction of air-vessels and fire-engines rest between Germany and France. Drawn leaden pipes were projected by Dalesme. The English revived the plunger pump and stuffing-box of Moreland, and furnished the expanding metallic pistons of Cartwright and Barton—the steam-engines of Worcester and Savery, Newcomen and Watt—the pneumatic apparatus of Brown, and motive engines of Cecil and others—Whitehurst was the first to apply the principle of the ram, and the quicksilver pump was invented by Hawkins—Hales invented the milling of sheet lead, and the first drawn pipes were made by Wilkinson. Switzerland contributed the spiral pump of Wirtz—America has furnished the riveted hose of Sellers and Pennock, the motive machine of Morey, and high pressure engines of Evans ; and both have given numerous modifications of every hydraulic device. The Italians have preserved many ancient devices, and to them the discoveries of Gallileo and Torricelli respecting atmospheric pressure are due. Porta has given the first figure of a device for raising water by steam, and Venturi's experiments have extended their claims.

Remarks have occasionally been introduced on the importance of the mechanic arts and the real dignity attached to their profession, notwithstanding the degraded state in which operatives have ever been held by those who have lived on their ingenuity and become enriched by their skill. But this state of things we believe is passing away, and the time is not distant when such men, instead of being deemed, as under the old regime, virtual serfs, will exert an influence in society commensurate with their contributions to its welfare. And where, it may be asked, is there a comfort, or convenience, or luxury of life, which they do not create or assist to furnish, from the bread that sustains the body to the volume that informs the mind ?

Few classes have a more honorable career before them than intelligent mechanics. Certainly none have better opportunities of associating their names with those of the best of their species. Science and the arts open the paths to true glory ; and greater triumphs remain to be achieved in both than the world has yet witnessed. Human toil has not been dispensed with, but it certainly will be superseded, in a great measure, if not altogether, by forces derived from inanimate nature. A great part of the globe is yet a desert, inhabited by beasts of prey, or by men more savage than they ; whereas the Creator designs the whole to be a garden and peopled with happy intelligences, as in the first Eden. It is much too common to seek ephemeral distinction on the troubled sea of politics or party ; but of the thousands who launch their barks upon it, how few ever reach the haven of their wishes ! The greater part are soon engulphed in oblivion, while not a few, exhausted by useless struggles, are bereft of their energies and quickly sink in despair—but no fame is more certain or more durable than that which arises from useful inventions. Whitney and Whittemore, Evans and Fulton, will be remembered as long as cotton gins, carding machines, steam-engines, and steam-boats are known on these continents, and when contemporary politicians are wholly forgotten—in fact most of these are so already. The name of Watt will be known while that of every warrior and monarch and statesman of his day has perished; and so it ought to be, for with few exceptions, he contributed more to the happiness of his species than have such men from the beginning of time. No one is now interested in learning any thing respecting the sanguinary Bull of Burgundy and his wily antagonist, the eleventh Louis of France, whose contests kept for years the European world in an uproar; and the latter, not content with murdering his species by wholesale, in his old age slew infants that he might acquire new vigor by bathing in their blood: but as long as time endures, the world will revere the names of their contemporaries—Gottenburg, Koster, Faust, and Schœffer and their associates in printing and type-founding.

Science and the arts are renovating the constitution of society. The destiny of nations cannot be much longer held by political gamblers, wealthy dolts, titled buffoons, and royal puppets; these no longer sustained by factitious aids must descend to their own level. Theories of governments will not be opposed to nature and carried out in violation of her laws; but practical science will be the ruling principle; and practical philosophers will be, as God designed they should be, the master spirits of the world. The history and progress of the useful arts will soon become a subject of general study. Historians will hereafter trace in them the rise and fall of nations; for power and preëminence will depend upon new discoveries in and applications of science. Battles will soon be fought by engineers instead of generals, and by mechanism in place of men. But battles, we trust, will hereafter be few; for if ever men were called upon by that which is dear to them and their race—by that which is calculated to rouse the purest feelings and exterminate the worst ones, it is to denounce that spirit of military glory which encourages and induces offensive wars. Take away all the false glare and pomp of wars, and tyranny will expire—for it would have nothing to support it. Put war in its true light, and no well regulated mind would ever embrace it as a profession.

To poets and writers of romance, the annals of mechanism present unexplored sources of materials. They are mines of the richest ores—fields teeming with the choicest fruits and flowers. Here are to be found incidents as agreeable and exciting in their natures, and as important in their effects as anything that can be realized by the imagination alone; such too, as present nothing to offend the finest taste, or conflict with the purest morals. When novelists have worn out the common ground; (and they seem already to have done so,) when mere sentiment grows flat, and the exhibition of the passions becomes stale; when politics, history and love are exhausted—works founded on the origin, progress, and maturity of the useful arts will both charm the imagination and improve the judgment of readers. Does an author wish to introduce characters who have left permanent impressions of their genius upon the world? Where can he find them in such variety as in the race of *inventors?* Is he desirous of enriching his pages with singular coincidences, curious facts, surprising results —to fascinate his readers, and cause them to anticipate the end of his pages with regret? Let him detail the circumstances that led to the conception, and accompanied the improvement of those inventions and discoveries that have elevated civilized man above the savage.

Is such a writer desirous, for instance, to entertain the sex? He could

PREFACE. vii

hardly do it more effectually than by writing a volume on the labors of primitive spinsters, ere the distaff was adopted, or the spindle (the original fly-wheel) was invented; by detailing the circumstances that gave birth to those implements, with the trials, observations, customs and anecdotes connected with their introduction and their uses—imagining the congratulations that were poured upon the artist who wove the first web in a loom, and the praises bestowed upon the author of that machine and the shuttle—recalling the times and scenes when groups of laughing females were hastening to examine the first colored mantles; and recording the bursts of admiration which dropped from them (in all the force of oriental hyperbole) upon witnessing the processes by which purple and scarlet and crimson and green, &c. were produced—recounting the methods by which the art of dyeing wrought a revolution in costume, and how it became one of the great sources of wealth to Babylon and Tyre—referring to the gratification which the invention of needles and pins, of thimbles and combs, conferred on ancient dames; and noticing the influence of these in improving the dress and deportment of women—describing the trials of artists before they succeeded in perfecting these instruments, and so on, until every addition to domestic dwellings, to household furniture, and to dress be reviewed—until every thing which a modern lady possesses over an Indian's squaw be brought forward and described, with all the known facts and circumstances associated with its history and application; —and thus form a series of essays on the arts, in which every line would be poetry, and every incident new.

A new species of drama might here take its rise; one possessing equal attractions and exhibiting equally interesting pictures of human life, as any thing which writers of comedy or tragedy have yet produced. Here are characters and customs of every variety, age, and nation—incidents and adventure in the greatest profusion—the extremes of misery and bliss, of poverty and wealth, of suffering virtue and unrequited toil, and their opposites. Here the humblest individuals have, by industry and ingenuity, risen from obscurity and astonished the world. Mechanics have become kings like the old potter of Sicily, (Agathocles,) Aurelius the blacksmith of Rome, and Leitz the tinker who founded the caliph dynasty of the Soffarites. Kings have left their thrones to become workmen in brass and silver, wood and iron; as Demetrius at his lathe, Æropus making lamps and tables, Charles V. in his watchmaker's shop; and if some bizarre examples are wanted, there is still to be seen the mantua-making apartment of Ferdinand VII. with specimens of his work.

A play might be founded on the fairs held at Delos, (the Pittsburg of of the old Greeks,) where merchants (observes Pliny) assembled from all parts of the world to purchase hardware and bronze. An island whose artists were ennobled for the beauty and finish of their works in the metals, and who particularly excelled in brazen feet for chairs, tables, and bedsteads, and in statues and other large works in brass. Then there was the workmen of Ægina, who beat all others in fabricating branches and and sockets of candelabra; while those of Tarentum produced the best pedestals or shafts. In connection with which, there is the singular story of the Lady Gegania, who, after giving 50,000 sesterces for a bronze candlestick, adopted its ill-favored and hump-backed maker for her companion and heir.

How rich in interest would a dramatic scene be if laid in an antediluvian smith's shop! (Forges have always been places of resort.) To notice the characters of the visitants, listen to their remarks, examine the instruments fabricated by the artist, his materials, fuel, bellows, and other tools!

PREFACE.

There is not a more interesting scene in all the Iliad than the description of Vulcan at work. But if such a distance of time is too remote, there is the forge of *Kawah*, the blacksmith of Ispahan, he whose *apron* was for centuries the banner of the Persian empire. The forge of Aurelius also, where he made the sword by which he was while emperor slain.

A scene might open in the barber's shop of Alexandria, in which the boy Ctesibius used to play, and where the first scintillations of his genius broke out; while his subsequent speculations, his private essays and public experiments, some of which were probably exhibited before the reigning Ptolemies, might be brought into view—his pupil, Heron, and other philosophers and literati might also be included in the plot. Of the connection of barbers with important events there is no end—there was the tatling artist of Midas, the spruce hair-dresser of Julian the emperor, the inquisitive one that saved Cæsar's life by listening to the conversation of assassins—the history of the silver shaving vessel with which the benevolent father of Marc Anthony relieved the pecuniary distresses of a friend—there was the wicked Oliver Dain; and the ancestor of Tunstall, the famous Bishop of Durham, was barber to William the Conqueror: hence the bishop's coat of arms contained *three combs*.

Who would not go to see a representation of the impostures of the heathen priesthood? Men who in the darkest times applied some of the finest principles of science to the purposes of delusion! With what emotions should we enter their secret recesses in the temples!—places where their chemical processes were matured, their automaton figures and other mechanical apparatus conceived and fabricated, and where experiments were made before the miracles were consummated in public. But it is impossible to enumerate a tithe of the subjects and incidents for the drama that might be derived from the history of the arts: they are more numerous than the mechanical professions—more diversified than articles of traffic or implements of trades. The plots, too, might be rendered as complicated, and their denouement as agreeable or disagreeable as could be desired: and what is better than all, in such plays the moral, intellectual and inventive faculties of an audience would be excited and improved—science would pervade every piece, and her professors would be the principal performers.

THOS. EWBANK.

New-York, December, 1841.

CONTENTS.

BOOK I.
PRIMITIVE AND ANCIENT DEVICES FOR RAISING WATER.

CHAPTER I.
The subject of raising water interesting to philosophers and mechanics—Led to the invention of the steam engine—Connected with the present advanced state of the arts—Origin of the useful arts lost—Their history neglected by the ancients—First inventors the greatest benefactors—Memorials of them perished, while accounts of warriors and their acts pervade and pollute the pages of history—A record of the origin and early progress of the arts more useful and interesting than all the works of historians extant—The history of a single tool (as that of a hammer) invaluable—In the general wreck of the arts of the ancients, most of their devices for raising water preserved—Cause of this—Hydraulic machines of very remote origin—Few invented by the Greeks and Romans—Arrangement and division of the subject   -   -   -   -   -   1

CHAPTER II.
Water—Its importance in the economy of nature—Forms part of all substances—Food of all animals—Great physical changes effected by it—Earliest source of inanimate motive power—Its distribution over the earth not uniform—Sufferings of the orientals from want of water—A knowledge of this necessary to understand their writers—Political ingenuity of Mahomet—Water a prominent feature in the paradise of the Asiatics—Camels often slain by travelers, to obtain water from their stomachs—Cost of a draught of such water—Hydraulic machine referred to in Ecclesiastes—The useful arts originated in Asia—Primitive modes of procuring water—Using the hand as a cup—Traditions respecting Adam—Scythian tradition—Palladium—Observations on the primitive state of man, and the origin of the arts   -   -   -   -   -   -   -   -   9

CHAPTER III.
Origin of vessels for containing water—The calabash the first one—It has always been used—Found by Columbus in the cabins of Americans—Inhabitants of New Zealand, Java, Sumatra, and of the Pacific Islands employ it—Principal vessel of the Africans—Curious remark of Pliny respecting it—Common among the ancient Mexicans, Romans and Egyptians—Offered by the latter people on their altars—The model after which vessels of capacity were originally formed—Its figure still preserved in several—Ancient American vessels copied from it—Peruvian bottles—Gurgulets—The form of the calabash the first one in the vases and goblets of the ancients—Extract from Persius' satires—Ancient vessels for heating water modeled after it—Pipkin—Saucepan—Anecdote of a Roman dictator—The common cast-iron cauldron of great antiquity: similar in shape to those used in Egypt in the time of Rameses—Often referred to in the Bible and in the Iliad—Grecian, Roman, Celtic, Chinese and Peruvian cauldrons—Expertness of Chinese tinkers—Crœsus and the Delphic oracle—Uniformity in the figure of cauldrons—Cause of this—Superiority of their form over straight-sided boilers—Brazen cauldrons highly prized—Water pots of the Hindoos—Women drawing water—Anecdote of Darius and a young female of Sardis—Dexterity of oriental women in balancing water pots—Origin of the canopus—Ingenuity and fraud of an Egyptian priest—Ecclesiastical deceptions in the middle ages   -   -   -   -   -   -   -   14

CHAPTER IV.
On wells—Water one of the first objects of ancient husbandmen—Lot—Wells before the deluge—Digging them through rock subsequent to the use of metals—Art of digging them carried to great perfection by the Asiatics—Modern methods of making them in loose soils derived from the East—Wells often the nuclei of cities—Private wells common of old—Public wells infested by banditti—Wells numerous in Greece—Introduced there by Danaus—Facts connected with them in the mythologic ages—Persian ambassadors to Athens and Lacedemon thrown into wells—Phenician, Carthagenian and Roman wells extant—Cæsar and Pompey's knowledge of making wells enabled them to conquer—City of Pompeii discovered by digging a well—Wells in China, Persia, Palestine, India and Turkey—Cisterns of Solomon—Sufferings of travelers from thirst—Affecting account from Leo Africanus—Mr. Bruce in Abyssinia—Dr. Ryers in Gombroon—Hindoos praying for water—Caravan of 2000 persons and 1800 camels perished in the African desert—Crusaders   24

CHAPTER V
Subject of Wells continued—Wells worshiped—River Ganges—Sacred well at Benares—Oaths taken at wells—Tradition of the rabbins—Altars erected near them—Invoked—Ceremonies with regard to water in Egypt, Greece, Peru, Mexico, Rome, and Judea—Temples erected over wells—The

B

CONTENTS.

fountain of Apollo—Well Zem Zem—Prophet Joel—Temple of Isis—Mahommedan mosques—Hindoo temples—Woden's well—Wells in Chinese temples—Pliny—Celts—Gauls—Modern superstitions with regard to water and wells—Hindoos—Algerines—Nineveh—Greeks—Tombs of saints near wells—Superstitions of the Persians—Anglo-Saxons—Hindoos—Scotch—English—St. Genevieve's well—St. Winifred's well—House and well 'warming' - - - - 33

CHAPTER VI.

Wells continued—Depth of ancient wells—In Hindostan—Well of Tyre—Carthagenian wells—Wells in Greece, Herculaneum and Pompeii—Wells without curbs—Ancient laws to prevent accidents from persons and animals falling into them—Sagacity and revenge of an elephant—Hylas—Archelaus of Macedon—Thracian soldier and a lady at Thebes—Wooden covers—Wells in Judea—Reasons for not placing curbs round wells—Scythians—Arabs—Aquilius—Abraham—Hezekiah—David—Mardonius—Moses and the people of Edom—Burckhardt in Petra—Woman of Bahurim—Persian tradition—Ali the fourth caliph—Covering wells with large stones—Mahommedan tradition—Themistocles—Edicts of Greek emperors—Well at Heliopolis—Juvenal—Roman and Grecian curbs of marble—Capitals of ancient columns converted into curbs for wells - - 37

CHAPTER VII.

Wells concluded—Description of Jacob's well—Of Zemzem in Mecca—Of Joseph's well at Cairo—Reflections on wells—Oldest monuments extant—Wells at Elim—Bethlehem—Cos—Scyros—Heliopolis—Persepolis—Jerusalem—Troy—Ephesus—Tadmor—Mizra—Sarcophagi employed as watering troughs—Stone coffin of Richard III used as one—Ancient American wells—Indicate the existence in past times of a more refined people than the present red men—Their examination desirable—Might furnish (like the wells at Athens) important data of former ages - - - 44

CHAPTER VIII.

Ancient methods of raising water from wells: Inclined planes—Stairs within wells: in Mesopotamia—Abyssinia—Hindostan—Persia—Judea—Greece—Thrace—England. Cord and bucket: used at Jacob's well—By the patriarchs—Mahomet—In Palestine—India—Alexandria—Arabian vizier drawing water—Gaza—Herculaneum and Pompeii—Wells within the houses of the latter city—Aleppo—Tyre—Carthage—Cleanthes the 'well drawer' of Athens, and successor of Zeno—Democritus—Plautus—Asclepiades and Menedemus—Cistern pole—Roman cisterns and cement—Ancient modes of purifying water - - - - - - - - - - - - 51

CHAPTER IX.

The pulley its origin unknown—Used in the erection of ancient buildings and in ships—Ancient one found in Egypt—Probably first used to raise water—Not extensively used in ancient Grecian wells: cause of this—Used in Mecca and Japan—Led to the employment of animals to raise water—Simple mode of adapting them to this purpose in the east. Pulley and two buckets: used by the Anglo-Saxons, Normans, &c—Italian mode of raising water to upper floors—Desagulier's mode—Self-acting or gaining and losing buckets—Marquis of Worcester—Heron of Alexandria—Robert Fludd—Lever bucket engine—Bucket of Bologna—Materials of ancient buckets - - - 58

CHAPTER X.

The windlass: its origin unknown—Employed in raising water from wells, and ore from mines—Chinese windlass—Other inventions of that people, as table forks, winnowing machines, &c. &c. Fusee: its application to raise water from wells—Its inventor not known. Wheel and pinion—Anglo-Saxon crane—Drum attached to the windlass roller, and turned by a rope: used in Birmah, England, &c. Tread wheels: used by the ancients—Moved by men and various animals—Jacks—Horizontal tread-wheels—Common wheel or capstan. Observations on the introduction of table forks into Europe - - - - - - - - - - - 68

CHAPTER XI.

Agriculture gave rise to numerous devices for raising water—Curious definition of Egyptian husbandry—Irrigation always practiced in the east—Great fertility of watered land—The construction of the lakes and canals of Egypt and China subsequent to the use of hydraulic machines—Phenomenon in ancient Thebes—Similarity of the early histories of the Egyptians and Chinese—Mythology based on agriculture and irrigation: both inculcated as a part of religion—Asiatic tanks—Watering land with the yoke and pots—An employment of the Israelites in Egypt—Hindoo water bearer—Curious shaped vessels—Aquarius, 'the water pourer,' an emblem of irrigation—Connection of astronomy with agriculture—Swinging baskets of Egypt, China and Hindostan. Arts and customs of the ancient Egyptians - - - - - - - - - 79

CHAPTER XII.

Gutters: single do.—double do.—Jantu of Hindostan: ingenious mode of working it—Referred to in Deuteronomy—Other Asiatic machines moved in a similar manner—its antiquity. Combination of levers and gutters—Swinging or pendulum machine—Rocking gutters—Dutch scoop—Flash wheel 86

CHAPTER XIII.

The swape: used in modern and ancient Egypt—Represented in sculptures at Thebes—Alluded to by Herodotus and Marcellus—Described by Pliny—Picotah of India: agility of the Hindoos in working it. Chinese swape—Similar to the machines employed in erecting the pyramids—The swape seen in paradise by Mahomet—Figure of one near the city of Magnesia—Anglo-Saxon swape—Formerly used in English manufactories—Figures from the Nuremburg Chronicle, Munster's Cosmography, and Besson's Theatre des Instrumens. The swape common in North and South America—Examples of its use in watering gardens—Figures of it the oldest representations of any hydraulic machine—Mechanical speculations of ecclesiastics: Wilkins's projects for aerial navigation—Mechanical and theological pursuits combined in the middle ages—Gerbert—Dunstan—Bishops famous as castle architects—Androides—Roode of grace—Shrine of Becket—Speaking images—Chemical deceptions—Illuminated manuscripts - - - - - 94

CONTENTS.   xi

CHAPTER XIV.

Wheels for raising water—Machines described by Vitruvius—Tympanum—De La Faye's improvement.—Scoop wheel—Chinese noria—Roman do.—Egyptian do.—Noria with pots—Supposed origin of toothed wheels—Substitute for wheels and pinions—Persian wheel: common in Syria—Large ones at Hamath—Various modes of propelling the noria by men and animals—Early employment of the latter to raise water. Antiquity of the noria—Supposed to be the 'wheel of fortune'—An appropriate emblem of abundance in Egypt—Sphinx—Lions' heads—Vases—Cornucopia—Ancient emblems of irrigation—Medea: inventress of vapor baths—Ctesibius—Metallic and glass mirrors—Barbers - - - - - - - - - - - - 109

CHAPTER XV.

The chain of pots—Its origin—Used in Joseph's well at Cairo—Numerous in Egypt—Attempt of Belzoni to supersede it and the noria—Chain of pots of the Romans, Hindoos, Japanese and Europeans—Described by Agricola—Spanish one—Modern one—Applications of it to other purposes than raising water—Employed as a first mover and substitute for overshot wheels—Francini's machine—Antiquity of the chain of pots—Often confounded with the noria by ancient and modern authors—Introduced into Greece by Danaus—Opinions of modern writers on its antiquity—Referred to by Solomon—Babylonian engine that raised the water of the Euphrates to supply the hanging gardens—Rope pump—Hydraulic belt - - - - - - - - 122

CHAPTER XVI.

The screw—An original device—Various modes of constructing it—Roman screw—Often re-invented—Introduced into England from Germany—Combination of several to raise water to great elevations—Marquis of Worcester's proposition relating to it exemplified by M. Pattu—Ascent of water in it formerly considered inexplicable—Its history—Not invented by Archimedes—Supposed to have been in early use in Egypt—Vitruvius silent respecting its author—Conon its inventor or re-inventor—This philosopher employed as a first mover for his flattery of Ptolemy and Berenice—Dinocrates the architect—Suspension of metallic substances without support—The screw not attributed to Archimedes till after his death—Inventions often given to others than their authors—Screws used as ship pumps by the Greeks—Flatterers like Conon too often found among men of science—Dedications of European writers often blasphemous—Hereditary titles and distinctions—Their acceptance unworthy of philosophers—Evil influence of scientific men in accepting them—Their denunciation a proof of the wisdom and virtue of the framers of the U. S. constitution—Their extinction in Europe desirable—Plato, Solon, and Socrates—George III—George IV—James Watt—Arago—Description of the 'Syracusan,' a ship built by Archimedes, in which the screw pump was used - - 137

CHAPTER XVII.

The chain pump—Not mentioned by Vitruvius—Its supposed origin—Resemblance between it and the common pump—Not used by the Hindoos, Egyptians, Greeks or Romans—Derived from China—Description of the Chinese pump and the various modes of propelling it—Chain pump from Agricola—Paternoster pumps—Chain pump of Besson—Old French pump from Belidor—Superiority of the Chinese pump—Carried by the Spaniards and Dutch to their Asiatic possessions—Best mode of making and using it—Wooden chains—Chain pump in British ships of war—Dampier—Modern improvements—Dutch pump—Cole's pump and experiments—Notice of chain pumps in the American navy—Description of those in the U. S. ship Independence—Chinese pump introduced into America by Van Braam—Employed in South America—Recently introduced into Egypt—Used as a substitute for water wheels—Peculiar feature in Chinese ship building—Its advantages - - 148

CHAPTER XVIII.

On the hydraulic works of the ancient inhabitants of America: population of Anahuac—Ferocity of the Spanish invaders—Subject of ancient hydraulic works interesting—Aqueducts of the Toltecs—Ancient Mexican wells—Houses supplied with water by pipes—Palace of Motezuma—Perfection of Mexican works in metals—Cortez—Market in ancient Mexico—Hydraulic works—Fountains and jets d'eau—Noria and other machines—Palenque: its aqueducts, hieroglyphics, &c.—Wells in ancient and modern Yucatan—Relics of former ages, and traditions of the Indians. Hydraulic works of the Peruvians: Customs relating to water—Humanity of the early incas—Aqueducts and reservoirs—Resemblance of Peruvian and Egyptian customs—Garcilasso—Civilization in Peru before the times of the incas—Giants—Wells—Stupendous aqueducts and other monuments—Atabalipa—Pulleys—Cisterns of gold and silver in the houses of the incas—Temples and gardens supplied by pipes—Temple at Cusco: its water-works and utensils—Embroidered cloth—Manco Capac - - - - - - - - - - - - - - - 159

BOOK II.
MACHINES FO RAISING WATER BY THE PRESSURE OF THE TMOSPHERE.

CHAPTER I.

On machines that raise water by atmospheric pressure—Principle of their action formerly unknown—Suction a chimera—Ascent of water in pumps incomprehensible without a knowledge of atmospheric pressure—Phenomena in the organization, habits and motions of animals—Rotation of the atmosphere with the earth—Air tangible—Compressible—Expansible—Elastic—Air beds—Ancient beds and bedsteads—Weight of air—Its pressure—Examples—American Indians and the air pump—Boa constrictor—Swallowing oysters—Shooting bullets by the rarefaction of air—Boy's sucker—Suspension of flies against gravity—Lizards—Frogs—Walrus—Connection between all departments of knowledge—Sucking fish—Remora—Lampreys—Dampier—Christopher Columbus at St. Domingo—Ferdinand Columbus—Ancient fable—Sudden expansion of air bursting the bladders of fish—Pressure of the atmosphere on liquids - - - - - - - - - 173

CHAPTER II.

Discovery of atmospheric pressure—Circumstances which led to it—Galileo—Torricelli—Beautiful experiment of the latter—Controversy respecting the results—Pascal—His demonstration of the

CONTENTS.

cause of the ascent of water in pumps—Invention of the air pump—Barometer and its various applications—Intensity of atmospheric pressure different at different parts of the earth—A knowledge of this necessary to pump makers—The limits to which water may be raised in atmospheric pumps known to ancient pump makers - - - - - - - - - - 187

CHAPTER III.

Ancient experiments on air—Various applications of it—Siphons used in ancient Egypt—Primitive experiments with vessels inverted in water—Suspension of liquids in them—Ancient atmospheric sprinkling pot—Watering gardens with it—Probably referred to by St. Paul, and also by Shakespeare—Glass sprinkling vessel and wine taster from Pompeii—Religious uses of sprinkling pots among the ancient heathen—Figure of one from Montfaucon—Vestals—Miracle of Tutia carrying water in a sieve described and explained—Modern liquor taster and dropping tubes—Trick performed with various liquids by a Chinese juggler—Various frauds of the ancients with liquids—Divining cups - - - - - - - - - - - - 191

CHAPTER IV.

Suction: impossible to raise liquids by that which is so called—Action of the muscles of the thorax and abdomen in *sucking* explained—Two kinds of suction—Why the term is continued—Sucking poison from wounds—Cupping and cupping horns—Ingenuity of a raven—Sucking tubes original atmospheric pumps—The sanguisuchello—Peruvian mode of taking tea by sucking it through tubes—Reflections on it—New application of such tubes suggested—Explanation of an ambiguous proverbial expression - - - - - - - - - - - - 201

CHAPTER V.

On bellows pumps: great variety in the forms and materials of machines to raise water—Simple bellows pump—Ancient German pump—French pump—Gosset's frictionless pump: subsequently re-invented—Martin's pump—Robison's bag pump—Disadvantages of bellows pumps—Natural pumps in men, quadrupeds, insects, birds, &c.—Reflections on them. Ancient vases figured in this chapter - - - - - - - - - - - 205

CHAPTER VI.

The atmospheric pump supposed by some persons to be of modern origin—Injustice towards the ancients—Their knowledge of hydrodynamics—Absurdity of an alledged proof of their ignorance of a simple principle of hydrostatics—Common cylindrical pump—Its antiquity—Anciently known under the name of a siphon—The *antlia* of the Greeks—Used as a ship pump by the Romans—Bilge pump—Portable pumps—Wooden pumps always used in ships—Description of some in the U. S. Navy—Ingenuity of sailors—Singular mode of making wooden pumps, from Dampier—Old draining pump—Pumps in public and private wells—In mines—Pump from Agricola, with figures of various boxes—Double pump formerly used in the mines of Germany, from Fludd's works—The wooden pump not improved by the moderns—Its use confined chiefly to civilized states - 211

CHAPTER VII.

Metallic pumps—Of more extended application than those of wood—Description of one—Devices to prevent water in them from freezing—Wells being closed, no obstacle in raising water from them—Application of the atmospheric pump to draw water from great distances as well as depth—Singular circumstance attending the trial of a Spanish pump in Seville—Excitement produced by it—Water raised to great elevations by atmospheric pressure when mixed with air—Deceptions practiced on this principle—Device to raise water fifty feet by atmospheric pressure—Modifications of the pumps innumerable—Pumps with two pistons—French marine pump—Curved pump—Muschenbroeck's pump—Centrifugal pump—West's pump—Jorge's improvement—Original centrifugal pump—Ancient buckets figured in this chapter - - - - - - 221

BOOK III.

MACHINES FOR RAISING WATER BY COMPRESSURE INDEPENDENTLY OF ATMOSPHERIC INFLUENCE.

CHAPTER I.

Definition of machines described in this Book—Forcing pumps—Analogy between them and bellows—History of the bellows that of the pump—Forcing pumps are water bellows—The bellows of antediluvian origin—Tubal Cain—Anacharsis—Vulcan in his forge—Egyptian, Hindoo, and Peruvian blowing tubes—Primitive bellows of goldsmiths in Barbary—Similar instruments employed to eject liquids—Devices to obtain a continuous blast—Double bellows of the Foulah blacksmiths without valves—Simple Asiatic bellows—Domestic bellows of modern Egypt—Double bellows of the ancient Egyptians. Bellows blowers in the middle ages—Lantern bellows common over all the east—Specimens from Agricola—Used by negroes in the interior of Africa—Modern Egyptian blacksmiths' bellows—Vulcan's bellows—Various kinds of Roman bellows—Bellows of Grecian blacksmiths referred to in a prediction of the Delphic oracle—Application of lantern bellows as forcing pumps—Sucking and forcing bellows pumps—Modern domestic bellows of ancient origin—Used to raise water—Common blacksmiths' bellows employed as forcing pumps—Ventilation of mines 231

CHAPTER II.

Piston bellows: used in water organs—Engraved on a medal of Valentinian—Used in Asia and Africa. Bellows of Madagascar. Chinese bellows: account of two in the Philadelphia museum—Remarks on a knowledge of the pump among the ancient Chinese—Chinese bellows similar in their construction to the water forcer of Ctesibius, the double acting pump of La Hire, the cylindrical steam engine, and condensing and exhausting air pumps. Double acting bellows of Madagascar—Alledged ignorance of the old Peruvian and Mexican smiths of bellows: their constant use of blowing tubes no proof of this—Examples from Asiatic gold and silver smiths—Balsas—Sarbacans—Mexican Vulcan. Natural bellows pumps: blowing apparatus of the whale—Elephant—

CONTENTS. xiii

Rise and descent of marine animals—Jaculator fish—Llama—Spurting snake—Lamprey—Bees—The heart of man and animals—Every human being a living pump: wonders of its mechanism, and of the duration of its motions and materials—Advantages of studying the mechanism of animals - - - - - - - - - - - - - 244

CHAPTER III.

Forcing pumps with solid pistons: the syringe: its uses, materials and antiquity—Employed by the Hindoos in religious festivals—Figured on an old coat of arms—Simple garden pump—Single valve forcing pump—Common forcing pump—Stomach pump—Forcing pump with air vessel—Machine of Ctesibius: its description by Vitruvius—Remarks on its origin—Errors of the ancients respecting the authors of several inventions—Claims of Ctesibius to the pump limited—Air vessel probably invented by him—Compressed air a prominent feature in all his inventions—Air vessels—In Heron's fountain—Apparently referred to by Pliny—Air gun of Ctesibius—The hookah - - 259

CHAPTER IV.

Forcing pumps continued: La Hire's double acting pump—Plunger pump: invented by Moreland; the most valuable of modern improvements on the pump—Application of it to other purposes than raising water—Frictionless plunger pump—Quicksilver pumps—Application of the principle of Bramah's press by bees in forcing honey into their cells. Forcing pumps with hollow pistons: employed in French water works—Specimen from the works at Notre Dame—Lifting pump from Agricola—Modern lifting pumps—Extract from an old pump-maker's circular—Lifting pumps with two pistons—Combination of hollow and solid pistons—Trevethick's pump—Perkins's pump - 271

CHAPTER V.

Rotary or rotatory pumps: uniformity in efforts made to improve machines—Prevailing custom to convert rectilinear and reciprocating movements into circular ones—Epigram of Antipater—Ancient opinion respecting circular motions—Advantages of rotary motions exemplified in various machines—Operations of spinning and weaving; historical anecdotes respecting them—Rotary pump from Serviere—Interesting inventions of his—Classification of rotary pumps—Eve's steam engine and pump—Another class of rotary pumps—Rotary pump of the 16th century—Pump with sliding butment—Trotter's engine and pump—French rotary pump—Bramah and Dickenson's pump—Rotary pumps with pistons in the form of vanes—Centrifugal pump—Defects of rotary pumps—Reciprocating rotary pumps: a French one—An English one—Defects of these pumps - 281

CHAPTER VI.

Application of pumps in modern water works: first used by the Germans—Water works at Augsburgh and Bremen—Singular android in the latter city—Old water works at Toledo—At London Bridge—Other London works moved by horses, water, wind and steam—Water engine at Exeter—Water works erected on Pont Neuf and Pont Notre Dame at Paris—Celebrated works at Marli—Error of Rannequin in making them unnecessarily complex. American water works: a history of them desirable—Introduction of pumps into wells in New-York city—Extracts from the minutes of the Common Council previous to the war of independence—Public water works proposed and commenced in 1774—Treasury notes issued to meet the expense—Copy of one—Manhattan Company—Water works at Fairmount, Philadelphia - - - - - - 293

CHAPTER VII.

Fire engines: probably used in Babylon and Tyre—Employed by ancient warriors—Other devices of theirs—Fire engines referred to by Apollodorus—These probably equal in effect to ours: Spiritalia of Heron: fire engine described in it—Pumps used to promote conflagrations—Greek fire a liquid projected by pumps—Fires and wars commonly united—Generals the greatest incendiaries—Saying of Crates respecting them—Fire pumps the forerunners of guns—Use of engines in Rome—Mentioned in a letter of Pliny to Trajan, and by Seneca, Hesychius, and Isidore. Roman firemen—Frequency of fires noticed by Juvenal—Detestable practice of Crassus—Portable engines in Roman houses—Modern engines derived from the Spiritalia—Forgotten in the middle ages—Superstitions with regard to fires—Fires attributed to demons—Consecrated bells employed as substitutes for water and fire engines—Extracts from the Paris ritual, Wynken de Worde, Barnaby Googe and Peter Martyr respecting them—Emblematic device of an old duke of Milan—Firemen's apparatus from Agricola—Syringes used in London to quench fires in the 17th century—Still employed in Constantinople—Anecdote of the Capudan Pacha—Syringe engine from Besson—German engines of the 16th century—Pump engine from Decaus—Pump engines in London—Extracts from the minutes of the London Common Council respecting engines and squirts in 1667—Experiment of Maurice mentioned by Stow the historian—Extract from 'a history of the first inventers' 302

CHAPTER VIII.

Fire engines continued: engines by Hautsch—Nuremberg—Fire engines at Strasbourg and Ypres—Coupling screws—Old engine with air chamber—Canvas and leather hose and Dutch engines—Engines of Perier and Leopold—Old English engines—Newsham's engines—Modern French engine—Air chambers—Table of the height of jets—Modes of working fire engines—Engines worked by steam. Fire engines in America: regulations respecting fires in New Amsterdam—Proclamations of Governor Stuyvesant—Extracts from old minutes of the Common Council—First fire engines—Philadelphia and New-York engines—Riveted hose—Steam fire engines now being constructed. Devices to extinguish fire without engines—Water bombs—Protecting buildings from fire—Fire escapes—Couvre feu—Curfew bells—Measuring time with candles—Ancient laws respecting fires and incendiaries—The dress in which Roman incendiaries were burnt retained in the auto da fe - 323

CHAPTER IX.

Pressure engines: of limited application—Are modifications of gaining and losing buckets and pumps—Two kinds of pressure engines—Piston pressure engine described by Fludd—Pressure engine from Belidor—Another by Westgarth—Motive pressure engines—These exhibit a novel mode of employing water as a motive agent—Variety of applications of a piston and cylinder—Causes of the ancients being ignorant of the steam engine—Secret of making improvements in the arts—Fulton, Eli Whitney, and Arkwright—Pressure engines might have been anticipated, and valuable lessons in science may be derived from a disordered pump—Archimedes—Heron's fountain—Portable ones recommended in flower gardens and drawing rooms in hot weather—Their invention gave rise to a new class of hydraulic engines—Pressure engine at Chemnitz—Another modification of Heron's fountain—Spiral pump of Wirtz - - - - - - 352

BOOK IV.
MACHINES FOR RAISING WATER (CHIEFLY OF MODERN ORIGIN) INCLUDING EARLY APPLICATIONS OF STEAM FOR THAT PURPOSE.

CHAPTER I.

Devices of the lower animals—Some animals aware that force is increased by the space through which a body moves—Birds drop shell fish from great elevations to break the shells—Death of Æschylus—Combats between the males of sheep and goats—Military ram of the ancients—Water rams—Waves—Momentum acquired by running water—Examples—Whitehurst's machine—Hydraulic ram of Montgolfier—'Canne hydraulique' and its modifications - - - - 365

CHAPTER II.

Machines for raising water by fire: air machines—Ancient weather glasses—Dilatation of air by heat and condensation by cold—Ancient Egyptian air machines—Statue of Memnon—Statues of Serapis and the Bird of Memnon—Decaus's and Kircher's machinery to account for the sounds of the Theban idol—Remarks on the statue of Memnon—Machine for raising water by the sun's heat, from Heron—Similar machines in the 16th century—Air machines by Porta and Decaus—Distilling by the sun's heat—Musical air machines by Drebble and Decaus—Air machines acted on by ordinary fire—Modifications of them employed in ancient altars—Bronze altars—Tricks performed by the heathen priests with fire—Others by heated air and vapor—Bellows employed in ancient altars—Tricks performed at altars mentioned by Heron—Altar that feeds itself with flame from Heron—Ingenuity displayed by ancient priests—Secrets of the temples—The Spiritalia—Sketch of its contents—Curious lustral vase - - - - - - - - - - 374

CHAPTER III.

On steam: miserable condition of the great portion of the human race in past times—Brighter prospects for posterity—Inorganic motive forces—Wonders of steam—Its beneficial influence on man's future destiny—Will supersede nearly all human drudgery—Progress of the arts—Cause why steam was not formerly employed—Pots boiling over and primitive experiments by females—Steam an agent in working prodigies—Priests familiar with steam—Sacrifices boiled—Seething bones—Earthquakes—Anthemius and Zeno—Hot baths at Rome—Ball supported on a jet of steam, from the Spiritalia—Heron's whirling eolipile—Steam engines on the same principle—Eolipiles described by Vitruvius—Their various uses—Heraldic device—Eolipiles from Rivius—Cupelo furnace and eolipile from Erckers—Similar applications of steam revived and patented—Eolipiles of the human form—Ancient tenures—Jack of Hilton—Puster a steam deity of the ancient Germans—Ingenuity of the priests in constructing and working it—Supposed allusions to eolipilic idols in the Bible—Employed in ancient wars to project streams of liquid fire—Draft of chimneys improved, perfumes dispersed, and music produced by eolipiles—Eolipiles the germ of modern steam engines 388

CHAPTER IV.

Employment of steam in former times—Claims of various people to the steam engine—Application of steam as a motive agent perceived by Roger Bacon—Other modern inventions and discoveries known to him—Spanish steam ship in 1543—Official documents relating to it—Remarks on these—Antiquity of paddle wheels as propellers—Project of the author for propelling vessels—Experiments on steam in the 16th century—Jerome Cardan—Vacuum formed by the condensation of steam known to the alchymists—Experiments from Fludd—Others from Porta—Expansive force of steam illustrated by old authors—Interesting example of raising water by steam from Porta—Mathesius, Canini and Besson—Device for raising hot water from Decaus—Invention of the steam engine claimed by Arago for France—Nothing new in the apparatus of Decaus nor in the principle of its operation—Hot springs—Geysers—Boilers with tubular spouts—Eolipiles—Observations on Decaus—Writings of Porta—Claims of Arago in behalf of Decaus untenable—Instances of hot water raised by steam in the arts—Manufacture of soap—Discovery of iodine—Ancient soap makers—Soap vats in Pompeii—Manipulations of ancient mechanics—Loss of ancient writings—Large sums anciently expended on soap—Logic of Omar - - - - - - 402

CHAPTER V.

Few inventions formerly recorded—Lord Bacon—His project for draining mines—Thomas Bushell—Ice produced by hydraulic machines—Eolipiles—Branca's application of the blast of one to produce motion—Its inutility—Curious extract from Wilkins—Ramseye's patent for raising water by fire—Manufacture of nitre—Figure illustrating the application of steam, from an old English work—Kircher's device for raising water by steam—John Bate—Antiquity of boys' kites in England—Discovery of atmospheric pressure—Engine of motion—Anecdotes of Oliver Evans and John Fitch—Elasticity and condensation of steam—Steam engines modifications of guns—A moving piston the essential feature in both—Classification of modern steam engines—Guerricke's apparatus—The same adopted in steam engines—Guerricke one of the authors of the steam engine - - 416

CHAPTER VI.

Reasons of old inventors for concealing their discoveries—Century of Inventions—Marquis of Worcester—His Inventions matured before the civil wars—Several revived since his death—Problems in the 'Century' in older authors—Bird roasting itself—Imprisoning chair—Portable fortifications—Flying—Diving—Drebble's sub-marine ship—The 68th problem—This remarkably explicit—The device consisted of one boiler and two receivers—The receivers charged by atmospheric pressure—Three and four-way cocks—An hydraulic machine of Worcester mentioned by Cosmo

CONTENTS.

de Medicis—Worcester's machine superior to preceding ones, and similar to Savery's—Piston steam engine also made by him—Copy of the last three problems in the Century—Ingenious mode of stating them—Forcing pumps worked by steam engines intended—Ancient riddle—Steam boat invented by Worcester—Projectors despised in his time—Patentees caricatured in a public procession—Neglect of Worcester—His death—Persecution of his widow—Worcester one of the greatest mechanicians of any age or nation—Glauber - - - - - - - 427

CHAPTER VII.

Hautefeuille, Huyghens and Hooke—Moreland—His table of cylinders—His pumps worked by a cylindrical high-pressure steam engine—He made no claim to a steam engine in England—Simple device by which he probably worked his plunger pumps—Inventions of his at Vauxhall—Anecdote of him from Evelyn's Diary—Early steam projectors courtiers—Ridiculous origin of some honors—Edict of Nantes—Papin—Digesters—Safety valve—Papin's plan to transmit power through pipes by means of air—Cause of its failure—Another plan by compressed air—Papin's experiments to move a piston by gunpowder and by steam—The latter abandoned by him—The safety valve improved, not invented by Papin—Mercurial safety valves—Water lute—Steam machine of Papin for raising water and imparting motion to machinery - - - - - - - 441

CHAPTER VIII.

Experimenters contemporary with Papin—Savery—This engineer publishes his inventions—His project for propelling vessels—Ridicules the surveyor of the navy for opposing it—His first experiments on steam made in a tavern—Account of them by Desaguliers and Switzer—Savery's first engine—Its operation—Engine with a single receiver—Savery's improved engine described—Gauge cocks—Excellent features of his improved engine—Its various parts connected by coupling screws—Had no safety valve—Rejected by miners on account of the danger from the boilers exploding—Solder melted by steam—Opinions respecting the origin of Savery's engine—It bears no relation to the piston engine—Modifications of Savery's engine by Desaguliers, Leopold, Blakey and others—Rivatz—Engines by Gensanne—De Moura—De Rigny—Francois and others—Amonton's fire mill—Newcomen and Cawley—Their engine superior to Savery's—Newcomen acquainted with the previous experiments of Papin—Circumstances favorable to the introduction of Newcomen's engine—Description of it—Condensation by injection discovered by chance—Chains and sectors—Savery's claim to a share in Newcomen's patent an unjust one—Merits of Newcomen and Cawley 453

CHAPTER IX.

General adoption of Newcomen and Cawley's engine—Leopold's machine—Steam applied as a mover of general machinery—Wooden and granite boilers—Generating steam by the heat of the sun—Floats—Greenhouses and dwellings heated by steam—Cooking by steam—Explosive engines—Vapor engines—English, French, and American motive engines—Woisard's air machine—Vapor of mercury—Liquefied gases—Decomposition and recomposition of water - - - - 468

BOOK V.

NOVEL DEVICES FOR RAISING WATER, WITH AN ACCOUNT OF SIPHONS, COCKS, VALVES, CLEPSYDRÆ, &c. &c.

CHAPTER I.

Subjects treated in the fifth book—Lateral communication of motion—This observed by the ancients—Wind at the Falls of Niagara—The trombe described—Natural trombes—Tasting hot liquids—Waterspouts—Various operations of the human mouth—Currents of water—Gulf Stream—Large rivers—Adventures of a bottle—Experiments of Venturi—Expenditure of water from various formed ajutages—Contracted vein—Cause of increased discharge from conical tubes—Sale of a water power—Regulation of the ancient Romans to prevent an excess of water from being drawn by pipes from the aqueducts - - - - - - - - - - 475

CHAPTER II.

Water raised by currents of air—Fall of the barometer during storms—Hurricanes commence at the leeward—Damage done by storms not always by the impulse of the wind—Vacuum produced by storms of wind—Draft of chimneys—Currents of wind in houses—Fire grates and parabolic jambs—Experiments with a sheet of paper—Experiments with currents of air through tubes variously connected—Effect of conical ajutages to blowing tubes—Application of these tubes to increase the draft of chimneys, and to ventilate wells, mines and ships - - - - - - 481

CHAPTER III.

Vacuum by currents of steam—Various modes of applying them in blowing tubes—Experiments-Effects of conical ajutages—Results of slight changes in the position of vacuum tubes within blowing ones—Double blowing tube—Experiments with it—Raising water by currents of steam—Ventilation of mines—Experimental apparatus for concentrating sirups in vacuo—Drawing air through liquids to promote their evaporation—Remarks on the origin of obtaining a vacuum by currents of steam - - - - - - - - - - - - - - 489

CHAPTER IV.

Spouting tubes—Water easily disturbed—Force economically transmitted by the oscillation of liquids—Experiments on the ascent of water in differently shaped tubes—Application of one form to siphons—Movement given to spouting tubes—These produce a jet both by their ascent and descent—Experiments with plain conical tubes—Spouting tubes with air pipes attached—Experiments with various sized tubes—Observations respecting their movements—Advantages arising from inertia—Modes of communicating motion to spouting tubes—Purposes for which they are applicable—The souffleur - - - - - - - - - - - - 497

CONTENTS.

CHAPTER V.

Nature's devices for raising water—Their influence—More common than other natural operations—The globe a self-moving hydraulic engine—Streams flowing on its surface—Others ejected from its bowels—Subterranean cisterns, tubes and siphons—Intermitting springs—Natural rams and pressure engines—Eruption of water on the coast of Italy—Water raised in vapor—Clouds—Water raised by steam—Geysers—Earthquakes—Vegetation—Advantages of studying it—Erroneous views of future happiness—Circulation of sap—This fluid wonderfully varied in its effects and movements—Pitcher plant and Peruvian canes—Trees of Australia—Endosmosis—Waterspouts—Ascent of liquids by capillary attraction—Tenacity and other properties of liquids—Ascent of liquids up inclined planes—Liquid drops—Their uniform diffusion when not counteracted by gravity—Their form and size—Soft and hard soldering—Ascent of water in capillary tubes limited only by its volume—Cohesion of liquids—Ascent of water through sand and rags—Rise of oil in lamp wicks and through the pores of boxwood - - - - - - - - - 505

CHAPTER VI.

Siphons—Mode of charging them—Principle on which their action depends—Cohesion of liquids—Siphons act in vacuo—Variety of siphons—Their antiquity—Of eastern origin—Portrayed in the tombs at Thebes—Mixed wines—Siphons in ancient Egyptian kitchens—Probably used at the feast at Cana—Their application by old jugglers—Siphons from Heron's Spiritalia—Tricks with liquids of different specific gravities—Fresh water dipped from the surface of the sea—Figures of Tantalus's cups—Tricks of old publicans—Magic pitcher—Goblet for unwelcome visitors—Tartar necromancy with cups—Roman baths—Siphons used by the ancients for tasting wine—Siphons, A. D. 1511—Figures of modern siphons—Sucking tube—Valve siphon—Tin plate—Wirtemburg siphon—Argand's siphon—Chemists' siphons—Siphons by the author—Water conveyed over extensive grounds by siphons—Limit of the application of siphons known to ancient plumbers—Error of Porta and other writers respecting siphons—Decaus—Siphons for discharging liquids at the bend -Ram siphon - - - - - - - - - - - - - 514

CHAPTER VII.

Fountains: variety of their forms, ornaments and accompaniments—Landscape gardeners—Curious fountains from Decaus—Fountains in old Rome—Water issuing from statues—Fountains in Pompeii—Automaton trumpeter—Fountains by John of Bologna and M. Angelo—Old fountains in Nuremberg, Augsburg and Brussels—Shakespeare, Drayton and Spencer quoted—Fountains of Alcinous—The younger Pliny's account of fountains in the gardens of his Tuscan villa—Eating in gardens—Alluded to in Solomon's Song—Cato the Censor—Singular fountains in Italy—Fountains described by Marco Paulo and other old writers—Predilection for artificial trees in fountains—Perfumed and musical fountains—Fountains within public and private buildings—Enormous cost of perfumed waters at Roman feasts—Lucan quoted—Introduction of fountains into modern theatres and churches recommended—Fountains in the apartments of eastern princes—Water conveyed through pipes by the ancients into fields for the use of their cattle—Three and four-way cocks - 532

CHAPTER VIII.

Clepsydræ and hydraulic organs: Time measured by the sun—Obelisks—Dial in Syracuse—Time measured in the night by slow matches, candles, &c.—Modes of announcing the hours—"Jack of the clock"—Clepsydræ—Their curious origin in Egypt—Their variety—Used by the Siamese, Hindoos, Chinese, &c.—Ancient hourglasses—Indexes to water clocks—Sand clocks in China—Musical clock of Plato—Clock carried in triumph by Pompey—Clepsydra of Ctesibius—Clock presented to Charles V—Modern clepsydræ—Hourglasses in coffins—Dial of the Peruvians. Hydraulic organs: imperfectly described by Heron and Vitruvius—Plato, Archimedes, Plutarch, Pliny, Suetonius, St. Jerome—Organs sent from Constantinople to Pepin—Water organs of Louis Debonnaire—A woman expired in ecstasies while hearing one play—Organs made by monks—Old Regal 542

CHAPTER IX.

Sheet lead: Lead early known—Roman pig lead—Ancient uses of lead—Leaden and iron coffins—Casting sheet lead—Solder—Leaden books—Roofs covered with lead—Invention of rolled lead—Lead sheathing. Leaden pipes: of great antiquity—Made from sheet lead by the Romans—Ordinance of Justinian—Leaden pipes in Spain in the 9th century—Damascus—Leather pipes—Modern iron pipes—Invention of cast leaden pipes—Another plan in France—Joints united without solder—Invention of drawn leaden pipes—Burr's mode of making leaden pipes—Antiquity of window lead—Water injured by passing through leaden pipes—Tinned pipes. Valves: their antiquity and variety—Nuremberg engineers. Cocks: of great variety and materials in ancient times—Horapollo—Cocks attached to the laver of brass and the brazen sea—Also to golden and silver cisterns in the temple at Delphi—Found in Japanese baths—Figure of an ancient bronze cock—Superior in its construction to modern ones—Cock from a Roman fountain—Numbers found at Pompeii—Silver pipes and cocks in Roman baths—Golden and silver pipes and cocks in Peruvian baths—Sliding cocks by the author. Water closets: of ancient date—Common in the east. Traps for drains, &c. - - - - - - - - - - - - - - - 550

APPENDIX.

John Bate—Phocion—Well worship—Wells with stairs—Tourne-broche—Raising water by a screw—Perpetual motions—Chain pumps in ships—Sprinkling pots—Old frictionless pump—Water power—Vulcan's trip-hammers—Eolipiles—Blowpipe—Philosophical bellows—Charging eolipiles—Eolipilic idols referred to in the Bible—Palladium—Laban's images—Expansive force of steam—Steam and air—Windmills—Imprisoning chairs—Eolipilic war-machines - - - - 565

INDEX - - - - - - - - - - - - 575

A DESCRIPTIVE

AND

HISTORICAL ACCOUNT

OF

HYDRAULIC AND OTHER MACHINES

FOR RAISING WATER.

BOOK I.
PRIMITIVE AND ANCIENT DEVICES FOR RAISING WATER.

CHAPTER I.

The subject of raising water, interesting to Philosophers and Mechanics—Led to the invention of the Steam Engine—Connected with the present advanced state of the Arts—Origin of the useful arts lost— Their history neglected by the Ancients—First Inventors the greatest benefactors—Memorials of them perished, while accounts of warriors and their acts pervade and pollute the pages of history—A record of the origin and early progress of the arts more useful and interesting than all the works of historians extant—The history of a single tool, (as that of a hammer,) invaluable—In the general wreck of the arts of the ancients, most of their devices for raising water preserved—Cause of this—Hydraulic machines of very remote origin—Few invented by the Greeks and Romans—Arrangement and division of the subject.

ALTHOUGH the subject of this work may present nothing very alluring to the general reader, it is not destitute of interest to the philosopher and intelligent mechanic. The art of raising water has ever been closely connected with the progress of man in civilization, so much so, indeed, that the state of this art, among a people, may be taken as an index of their position on the scale of refinement. It is also an art, which, from its importance called forth the ingenuity of man in the infancy of society ; nor is it improbable, that it originated some of the simple machines, or *mechanic powers* themselves.
It was a favorite subject of research with eminent mathematicians and engineers of old ; and the labors of their successors in modern days, have been rewarded with the most valuable machine which the arts ever presented to man—the STEAM ENGINE—for it was "raising of water" that exercised the ingenuity of DECAUS and WORCESTER, MORELAND and PAPIN, SAVARY and NEWCOMEN; and those illustrious men, whose suc-

cessive labors developed and matured that "semi-omnipotent engine," which "driveth up water by fire." A machine that has already greatly changed and immeasurably improved the state of civil society; and one which, in conjunction with the PRINTING PRESS, is destined to renovate both the political and moral world. The subject is therefore, intimately connected with the present advanced state of the arts; and the amazing progress made in them during the last two centuries, may be attributed in some degree to its cultivation.

The origin and early history of this art, (and of all others of primitive times) are irrecoverably lost. Tradition has scarcely preserved a single anecdote or circumstance relating to those meritorious men, with whom any of the useful arts originated; and when in process of time, HISTORY took her station in the temple of science, her professors deemed it beneath her dignity, to record the actions and lives of men, who were merely inventors of machines, or improvers of the useful arts; thus nearly all knowledge of those to whom the world is under the highest of obligations, has perished forever.

The SCHOLAR mourns, and the ANTIQUARY weeps over the wreck of ancient learning and art—the PHILOSOPHER regrets that sufficient of both has not been preserved to elucidate several interesting discoveries, which history has mentioned; nor to prove that those principles of science, upon which the action of some old machines depended, were understood; and the MECHANIC inquires in vain for the processes by which his predecessors in remote ages, worked the hardest granite without iron, transported it in masses that astound us, and used them in the erection of stupendous buildings, apparently with the facility that modern workmen lay bricks, or raise the lintels of doors. The machines by which they were elevated are as unknown as the individuals who directed their movements. We are almost as ignorant of their modes of working the metals, of their alloys which rivalled steel in hardness, of their furnaces, crucibles, and moulds; the details of forming the ennobling statue, or the more useful skillet or cauldron. Did the ancients laminate metal between rollers, and draw wire through plates, as we do? or, was it extended by hammers, as some specimens of both seem to show?[a] On these and a thousand other subjects, much uncertainty prevails. Unfortunately learned men of old, deemed it a part of wisdom, to conceal from the vulgar, all discoveries in science. With this view, they wrapped them in mystical figures, that the people might not apprehend them. The custom was at one time so general, that philosophers refused to leave any thing in writing, explanatory of their researches.

Whenever we attempt to penetrate that obscurity which conceals from our view, the works of the ancients, we are led to regret, that some of their MECHANICS did not undertake, for the sake of posterity and their own fame, to write a history and description of their machines and manufactures.

We know that philosophers, generally, would not condescend to perform such a task, or stoop to acquire the requisite information, for they deemed it discreditable to apply their energies and learning, to the elucidation of such subjects. (Few could boast with Hippias—who was master of the liberal and *mechanical* arts—the ring on his finger, the tunic, cloak,

---

[a] "And they did *beat the gold* into thin plates, and *cut* it into wires." Exod. xxxix, 3. These plates, were probably similar to those made by the ancient goldsmiths of Mexico, which were "three quarters of a yard long, foure fingers broad, and as thicke as parchment." Purchas' Pilgrimage, 984. "Silver *spread into plates*, is brought from Tarshish, and gold from Uphaz." Jer. x, 9.

and shoes which he wore, were the work of his own hands.) Plato inveighed with great indignation against Archytas and Eudoxus, for having debased and corrupted the excellency of geometry, by mechanical solutions, causing her to descend, as he said, from incorporeal and intellectual to sensible things; and obliging her to make use of matter, which requires manual labor, and is the object of servile trades.[a]

To the prevalence of such unphilosophical notions amongst the learned men of old, may be attributed, the irretrievable loss of information respecting the prominent mechanics of the early ages, those
. . . . " Searching wits,
Who graced their age with new invented arts." *Virgil, En.* vi, 900.

Their works, their inventions, and their names, are buried beneath the waves of oblivion; whilst the light and worthless memorials of heroes, falsely so called, have floated on the surface, and history has become polluted with tainted descriptions of men, who, without having added an atom to the wealth, or to the happiness of society, have been permitted to riot on the fruit of other men's labors; to wade in the blood of their species, and to be heralded as the honorable of the earth! And still, as in former times, humanity shudders, at these monsters being held up, as they impiously are, to the admiration of the world, and even by some christians too, as examples for our children.

" We may reasonably hope," says Mr. Davies in his popular work on the Chinese, " that the science and civilization which have already greatly enlarged the bounds of our knowledge of foreign countries, may, by diminishing the vulgar admiration of such pests and scourges of the human race, as military conquerors have usually proved, advance and facilitate the peaceful intercourse of the most remote countries with each other, and thereby increase *the general stock of knowledge and happiness* among mankind." Vol. 1, 18.

"Of what utility to us at this day, is either Nimrod, Cyrus, or Alexander, or their successors, who have astonished mankind from time to time ? With all their magnificence and vast designs, they are returned into nothing with regard to us. They are dispersed like vapors, and have vanished like phantoms. But the INVENTORS of the ARTS and SCIENCES labored for ALL AGES. We still enjoy the fruits of their application and industry—they have procured for us, *all the conveniencies of life*—they have converted all nature to our uses. Yet, all our admiration turns generally on the side of those heroes in blood, while we scarce take any notice of what we owe to the INVENTORS OF THE ARTS." *Rollin's Introduction to the Arts and Sciences of the Ancients.*

Who that consults history, only for that which is *useful*, would not prefer to peruse a journal of the daily manipulations of the laborers and mechanics who furnished clothing, arms, culinary utensils, and food for the armies of old—to the most eloquent descriptions of their generals, or their battles ? And as it is now with respect to accounts of such transactions in past ages—so will it be in future with regard to similar ones of modern times. Narrations of political convulsions, recitals of battles, and of honors conferred on statesmen and heroes, while dripping with human gore, will hereafter be unnoticed, or will be read with horror and disgust, while DISCOVERIES IN SCIENCE and DESCRIPTIONS OF USEFUL MACHINES, will be all in all.

It is pleasing to anticipate that day, which the present extensive and extending diffusion of knowledge is about to usher in, when despotism

---
[a] Plutarch's Life of Marcellus.

shall no longer hold the GREAT MASS of our species, in a state of unnatural ignorance, and of physical degradation, beneath that of the beasts which perish; but when the mechanics of the world, the creators of its wealth, shall exercise that influence in society to which their labors entitle them.

If we judged correctly of human character, we should admit that the mechanic who made the chair in which Xerxes sat, when he reviewed his mighty host, or witnessed the sea fight at Salamis, was a more useful member of society than that great king:—and, that the artisans who constructed the drinking vessels of Mardonius, and the brass mangers in which his horses were fed, were really more worthy of posthumous fame, than that general, or the monarch he served: and, if it be more virtuous, more praiseworthy, to alleviate human sufferings than to cause or increase them; then that old mechanician, who, when Marcus Sergius lost his hand in the Punic war, furnished him with an *iron* one, was an incomparably better man, than that or any other mere warrior: and so was he, who, according to Herodotus, constructed an artificial foot for Hegisostratus.[a]

Notwithstanding the opinion of Plato—we believe a description of the WORKSHOPS of DÆDALUS, and of TALUS his nephew; those of THEODORUS of Samos and of GLAUCUS of Chios, (the alleged inventor of the inlaying of metals;) an account of the process of making the famous Lesbian and Dodonean cauldrons,[b] and of the method by which those celebrated paintings in glass, were executed, fragments of which have come down to us, and which have puzzled, and still continue to puzzle, both our artists and our chemists; (the figures in which, of the most minute and exquisite finish, pass entirely and uniformly through the glass;)[c] if to these were added, the particulars of a working jeweller's shop of Persepolis and of Troy; of a lapidary's and an engraver's of Memphis; of a cutler's and upholsterer's of Damascus; and of a cabinet maker's and brazier's of Rome; together with those of a Sidonian or Athenian ship yard—such a record would have been more truly useful, and more really *interesting*, than almost all that ancient philosophers ever wrote, or poets ever sung.

A description of the FOUNDRIES and FORGES of India and of Egypt; of Babylon and Byzantium; of Sidon, and Carthage and Tyre; would have imparted to us a more accurate and extensive knowledge of the ancients, of their manners and customs, their intelligence and progress in science, than all the works of their historians extant; and would have been of infinitely greater service to mankind.

Had a narrative been preserved, of all the circumstances which led to the invention and early applications of the LEVER, the SCREW, the WEDGE, PULLEY, WHEEL and AXLE, &c.; and of those which contributed to the discovery and working of the metals, the use and management of fire, agriculture, spinning of thread, matting of felt, weaving of cloth, &c. it would have been the most perfect history of our species—the most valuable of earthly legacies. Though such a work might have been deemed of trifling import by philosophers of old, with what intense interest would it have been perused by scientific men in modern times! and what pure delight its examination would have imparted to every inquisitive and intelligent mind!

Such a record, would not only have filled the mighty chasm in the early history of the world, but would have had an important influence in pro-

---

[a] Herod. ix, 37.   [b] Eneid, iii, 595, and v, 350.   Herod. iv, 61.   [c] Ed. Encyc. Art. Glass.

moting the best interests of our race. It would have embraced incidents respecting man's early wants, and his rude efforts to supply them; particulars respecting eminent individuals, and the origin of antediluvian discoveries and inventions, &c. of such thrilling interest, as no modern novelist could equal, nor the most fertile imagination surpass.

It would have included a detail of those eventful experiments in which iron was first cast into cauldrons, forged into hatchets, and drawn into wire; with an account of the individuals, by whose ingenuity and perseverance, these invaluable operations, were, for the first time on this planet, successfully performed. Finally, it would have convinced us, that these men were the *true* HEROES of old, the genuine benefactors of their species, whose labors were for the benefit of *all* ages, and *all* people ; and an account of whose lives (not those of robbers,) should have occupied the pages of history, and whose names should have been embalmed in everlasting remembrance.

A chronological account of a few *mechanical implements*, would have afforded a clearer insight into the state of society in remote times, than any writings now subsisting. Nay, if we could realize a complete history of a *single tool*, as a hammer, a saw, a chisel, a hatchet, an auger, or a loom, it would form a more comprehensive history of the world, than has ever been, or perhaps ever will be written. Take for example a *hammer;* what a multitude of interesting circumstances are inseparably connected with its development and early uses ! circumstances, which, if we were in possession of, would explain almost all that is dark and mysterious respecting our ancient progenitors. A history of this implement would embrace the origin and general progress of all the useful arts ; and would elucidate the civil and scientific acquirements of man, in every age. It would open to our view, the public and private economy of the ancients; introduce us into the interior of their workshops, their dwellings and their temples; it would illustrate their manners, politics, religion, superstition, &c. In tracing the various purposes to which it was applied, we should become acquainted with all the material transactions in the lives of some ancient individuals from their birth to their death ; and also, with the circumstances which led to the rise and fall of empires. Like the celebrated "History of a Guinea," it would open to our inspection all the minutiæ in private and public life.

How infinitely various, are the materials, sizes, forms, and uses of the hammer ? and how indicative are they all of the state of society and manners ? At first, a club; then a rude mallet of wood ; next, the head formed of stone, and bound to the handle by withes, or by the sinews of animals ; afterward, the heads formed of metal. These, before iron or steel was known, were often of copper and even of gold; and subsequently, those of the latter material were faced, like some ancient chisels, with the more scarce and expensive iron.[a]

Ancient hammers varied as now in size, from the huge sledge of the Cyclops, to the portable one, with which Vulcan chased the more delicate work on the shield of Achilles,—from the maul, by which masses of ore were separated from their beds in the mines, to the diminutive ones, which Myrmecides of Miletus, and Theodorus of Samos, used to fabricate carriages and horses of metal, which were so minute as to be covered by the

---

[a] "It appears that in the tangible remains of smelting furnaces, found in Siberia, that gold hammers, knives, chisels, &c. have been discovered, *the edges of which* were skilfully *tipped with iron;* showing the scarcity of the ore, the difficulty of manufacturing it, and the plenty and apparently trifling value of the other." Scientific Tracts, Boston, 1833. Vol. iii, 411.

wings of a fly. Its *figure* has always varied with its uses, and none but modern workers in the metals can realize the endless variety of its shapes, which the ancient smiths required, to fabricate the wonderfully diversified articles of their manuafcture: from the massive brazen altars and chariots, to the chased goblets, and invaluable tripods or vases, for the possession of which, whole cities contended.

The history of the hammer in its widest range, would let us into the secrets of the statuaries and stone cutters of old: we should learn the process of making those metallic compounds, and working them into tools, with which the Egyptian mechanics sculptured those indurate columns that resist the best tempered steel of modern days. It would introduce us to the ancient chariot makers, cutlers and armorers; and would teach us how to make and temper the blades of Damascus; as well as those which were forged in the extensive manufactory of the father of Demosthenes. It would make us familiar with the arts of the ancient carpenters, coiners, coopers and jewellers. We should learn from it, the process of forging dies and striking money in the temple of Juno Moneta; of making the bodkins and pins for the head dresses of Greek and Roman ladies; while at the religious festivals, we should behold other forms of this implement in use, to knock down victims for sacrifice by the altars.

Finally, a perfect history of the hammer, would not only have made us acquainted with the origin and progress of the useful arts, among the primeval inhabitants of *this* hemisphere; but would have solved the great problems respecting their connection with, and migration from the eastern world.

But although we justly deplore the want of information relating to the arts in general of the remote ancients; it is probable that few of their devices for raising water have been wholly lost. If there was one art of more importance than another to the early inhabitants of CENTRAL ASIA and the VALLEY OF THE NILE, it was that of raising water for agricultural purposes. Not merely their general welfare, but their very existence depended upon the artificial irrigation of the land; hence their ingenuity was early directed to the construction of *machines* for this purpose; and they were stimulated in devising them, by the most powerful of all inducements. That machines must have been *indispensable* in past, as in present times, is evident from the climates and physical constitution of those countries. Their importance therefore, and universal use, have been the means of their preservation. Nor is it probable that any of them were ever lost in the numerous political convulsions of old. These seldom affected the pursuits of agriculture, and never changed the long established modes of cultivation; besides, hydraulic apparatus, from their utility, were as necessary to the conquerors as the conquered.[a]

Perhaps in no department of the useful arts, has less change taken place than in Asiatic and Egyptian agriculture. It is the same now, that it was thousands of years ago. The implements of husbandry, modes of irrigation, and devices for raising water are similar to those in use, when Ninus and Nebuchadnezzar, Sesostris, Solomon, and Cyrus flourished. And it would appear that the same *uniformity* in these machines prevailed over *all* the east, in ancient as in modern times: a fact accounted for, by the great and constant intercourse between continental and neighboring nations; the practice of warriors, of transporting the inhabitants and especially the mechanics and works of art, into other lands; and also from the great importance and universal use of artificial irrigation.

---

[a]Battles were sometimes fought in one field, while laborers were cultivating unmolested the land of an adjoining one.

Chap. 1.] *Hydraulic machines of the Ancients not lost.* 7

Every part of the eastern world has often had its inhabitants torn from it by war, and their places occupied by others. This practice of conquerors was sometimes modified, as respected the peasantry of a subdued country, but it appears that from very remote ages, *mechanics* were invariably carried off. The Phenicians, in a war with the Jews, deprived them of every man who could forge iron.[a] "There was no smith found throughout all the land of Israel; for the Philistines said, lest the Hebrews make swords and spears." SHALMANEZER, when he took Samaria, carried the people "away out of their own land to Assyria, and the king of Assyria, brought men from Babylon, and from Cuthah, and from Ava, and from Hamath, and from Sepharvaim, instead of the children of Israel; and they possessed Samaria, and dwelt in the cities thereof."[b] When Nebuchadnezzar took Jerusalem, he carried off, with the treasure of the temple, "all the craftsmen and smiths." Jeremiah says he carried away the "carpenters and smiths, and brought them to Babylon." Diodorus says, the palaces of Persepolis and Susa were built by mechanics that Cambyses carried from Egypt.[c] Ancient history is full of similar examples. Alexander practised it to a great extent. After his death, there was found among his tablets, a resolution to build several cities, some in Europe and some in Asia; and his design was to people those in Asia with Europeans, and those in Europe with Asiatics.[d] In this manner some of the most useful arts, necessarily became common to all the nations of old; and their perpetuity in some degrree secured, especially such as related to the tillage and irrigation of the soil.

We are inclined to believe that the hydraulic machines of the Assyrians, Babylonians, Persians and Egyptians, have all, or nearly all, come down to us. Most of them have been continued in uninterrupted use in those countries to the present times; while others have reached us through the Greeks and Romans, Saracens and Moors; or, have been obtained in modern days from China and Hindostan.

It is remarkable that almost all machines for raising water, originated with the *older* nations of the world; neither the Greeks, (if the screw of Conon be excepted, and even it was invented IN EGYPT,) nor the Romans, added a single one to the ancient stock; nor is this surprising; for with few exceptions, those in use at the present day, are either identical with, or but modifications of those of the ancients.

It is alleged that Archytas of Tarentum, 400, B. C. invented *"hydraulic machines,"* but no account of them has reached our times, nor do we know that they were designed to *raise* water. They consisted probably, in the application of the windlass or crane, (the latter it is said he invent ed) to move machines for this purpose. Had any important or useful machine for raising water, been devised by him, it would have been continued in use; and would certainly have been noticed by Vitruvius, who was acquainted with his inventions, and who mentions him several times in his work. 1 b. chap. 1., and 9 b. chap. 3.[e]

We have arranged the machines described in this work in five classes; to each of which, a separate BOOK is devoted. A few chapters of the first book, are occupied with remarks on WATER; on the ORIGIN OF VES-

---

[a]1 Sam. chap. xiii, 19, 22. [b]2 Kings chap. xvii, 23, 24. [c]Goguet, Tom. iii, 13.
[d]Diodorus Siculus, quoted by Robertson. India page 191. See Wilkinson's Ancient Egyptians, 1 vol. 206.
[e]Archytas made an automaton pigeon of wood which would fly. It was this probably, which gave the idea to the modern mechanician of Nuremburgh, who constructed an eagle, which flew towards Charles V. on his entrance into that city

SELS for containing it; on WELLS and FOUNTAINS, and customs connected with them, &c.

Some persons are apt to suppose the term HYDRAULIC *machines*, comprises every device for *raising* water; but such is not the fact. Apparatus *propelled* by it, as tide mills, &c. are hydraulic machines; these do not raise the liquid at all; while on the contrary, all those for elevating it, which are comprised in the second class, are pneumatic or hydro-pneumatic machines, their action depending on the pressure and elasticity of the atmosphere.

The *first* Class includes those, by which the liquid is elevated in movable vessels, by mechanical force applied to the latter.

Water raised in a bucket, suspended to a cord, and elevated by the hand, or by a windlass; the common pole and bucket, used daily in our rain water cisterns; the sweep or lever so common among our farmers, are examples of this class; so are the various wheels, as the tympanum, noria, chain of pots; and also the chain pump, and its modifications. This Class embraces all the principal machines used in the ancient world; and the greater part of modern hydraulic machinery is derived from it.

The *second* Class comprises such as raise water through tubes, by means of the elasticity and pressure, or weight of the atmosphere; as *sucking* pumps, so named; siphons, syringes, &c.

The aplication of these machines, unlike those of the first class is limited, because the atmosphere is only sufficient to support a column of water of from thirty to thirty five feet in perpendicular height; and in elevated countries, (as Mexico) much less. Numerous modifications of these machines have been made in modern times, but the pump itself is of ancient origin.

Those which act by *compression* are described in the *third* Class. The liquid being first admitted into a close vessel, is then forcibly expelled through an aperture made for the purpose. In some machines this is effected by a solid body impinging on the surface of the liquid; as the piston of a pump: in others, the weight of a column of water, is used to accomplish the same purpose.

Syringes, fire engines, pumps which are constructed on the same principle as the common bellows, are examples of the former; and the famous machine at Chemnitz in Hungary, Heron'sfountain, pressure engines, of the latter. Nor can the original invention of these be claimed by the moderns. Like the preceding, they were first developed by the energy of ancient intellects.

Fourth Class. There is however another class, which embraces several machines, which are supposed to be exclusively of modern origin; and some of them are by far the most interesting and philosophical of all. Such as the Bèlier hydraulique, or ram of Montgolfier; the centrifugal pump; the fire engine, so named because it raised water "by the help of fire;" that is, the original steam engine, or machine of Worcester, Moreland, Savary and Papin.

In the *fifth* Class, we have noticed such modern devices, as are either practically useful, or interesting from their novelty, or the principles upon which they act. An account of siphons is comprised in this class. Remarks on natural modes of raising water. Observations on cocks, pipes, valves, &c; and some general reflections are added.

CHAPTER II.

WATER—Its importance in the economy of nature—Forms part of all substances—Food of all animals—Great physical changes effected by it—Earliest source of inanimate motive power—Its distribution over the earth not uniform—Sufferings of the orientals from want of water—A knowledge of this necessary to understand their writers—Political ingenuity of Mahomet—Water a prominent feature in the paradise of the Asiatics—Camels often slain by travellers, to obtain water from their stomachs—Cost of a draught of such water—Hydraulic machine referred to in Ecclesiastes—The useful arts originated in Asia—Primitive modes of procuring water—Using the hand as a cup—Traditions respecting Adam—Scythian tradition—Palladium—Observations on the primitive state of man and the origin of the arts.

WATER is, in many respects, the most important substance known to man: it is more extensively diffused throughout nature than almost any other. It covers the greater part of the earth's surface, and is found to pervade its interior wherever excavations are made. It enters into every or nearly every combination of matter, and was supposed by some ancient philosophers, to be the origin of *all* matter; the primordial element; of which every object in nature was formed. The mineral kingdom, with its variegated substances and chrystalizations; the infinitely diversified and enchanting productions of the vegetable world; and every living being in animated nature, were supposed to be so many modifications of this aqueous fluid. According to VITRUVIUS, the Egyptian priests taught, that "all things consist of water;"[a] and Egypt was doubtless the source whence Thales and others derived the doctrine. PLINY, says "this one element seemeth to rule and command all the rest."[b] And it was remarked by PINDAR—

"Of all things, water is the best."

Modern science has shown that it is not a simple substance, but is composed of at least two others; neither of which, it is possible, is elementary.

Water not only forms part of the bodies of all animals,[c] but it constitutes the greatest portion of their food. Every comfort of civilized or savage life depends more or less upon it; and life itself cannot be sustained without it. If there were no rains or fertilizing dews, vegetation would cease, and every animated being would perish. Even terrestrial animals may be considered as existing in water, for the atmosphere in which we live and move, is an immense aerial reservoir of it, and one more capacious than all the seas on the face of the earth.

Water is also the prominent agent, by which those great physical and chemical changes are effected, which the earth is continually undergoing; and the stupendous effects produced by it, through the long series of past ages, have given rise, in modern times, to some of the most interesting departments of physical science.

The mechanical effects produced by it, render it of the highest importance in the arts. It was the earliest source of inanimate motive power; and has contributed more than all other agents to the amelioration of man's condition. By its inertia in a running stream, and by its gravity in a falling one, it has superseded much human toil; and has administered to our wants, our pleasures and our profits; and by its expansion into the aeriform

---

[a] Proem to b. viii. [b] Nat. Hist. xxxi, 1.
[c] A human corpse which weighed an hundred and sixty pounds—when the moisture was evaporated, weighed but twelve.

state, it appears to be destined, (through the steam engine) to accomplish the greatest moral and physical changes, which the intellectual inhabitants of this planet have ever experienced, since our species became its denizens.

The distribution of water is not uniform over the earth's surface, nor yet under its crust. While in some countries, natural fountains, capacious rivers, and frequent rains, present abundant sources for all the purposes of human life; in others, it is extremely scarce, and procured only with difficulty, and constant labor. This has ever been the case in various parts of Asia, and also in Egypt and other parts of Africa, where rain seldom falls. It is only from a knowledge of this fact and of the temperature and debilitating influences of eastern climates, that we are enabled to appreciate the peculiar force and beauty of numerous allusions to water, which pervade all the writings of eastern authors, both sacred and profane. Nor without this knowledge could we understand many of the peculiar customs of the people of the east.

MAHOMET well knew that his followers, living under the scorching rays of the sun, their flesh shrivelled with the desiccating influences of the air, and "dried up with thirst," could only be moved to embrace his doctrines by such promises as he made them, of "springs of living waters," "security in shades," "amidst gardens" and "fountains pouring forth plenty of water."[a] Nor could his ingenuity have devised a more *appropriate* punishment, than that with which he threatened unbelieving Arabs in hell. They were to have no mitigation of their torments; no cessation of them, except at certain intervals, when they were to take copious draughts of "*filthy* and *boiling* water."[b] It was universally believed by the ancients, that the MANES of their deceased friends experienced a suspension of punishment in the infernal regions, while partaking of the provisions which their relatives placed on their graves. The Arabian legislator improved upon the tradition.

The orientals have always considered water, either figuratively or literally, as one of the principal enjoyments of a future state. Gardens, shades, and fountains, are the prominent objects in their paradise. In the Revelations we are told "the Lamb shall lead them, (the righteous,) unto living fountains of waters." Chap. vii, 17.—"A pure river of water of life." Chap. xxii, 1. The book which contains an account of the religion and philosophy of the Hindoos, is named ANBERTKEND, signifying, "the cistern of the waters of life."[c]

Inhabitants of temperate climates, seldom or never experience that excruciating thirst implied in such expressions as "the soul panting for water;" nor that extremity of despair when, under such suffering, the exhausted traveller arrives at a place "where no water is." Under these circumstances, the orientals have often been compelled to slay their camels, for the sake of the water they might find in their stomachs; and a sum exceeding five hundred dollars, has been given for a single draught of it.

It is necessary to experience something like this, in order fully to comprehend the importance of the Savior's precept, respecting the giving "a cup of cold water," and to know the real *value* of such a gift. We should then see that sources of this liquid are to the orientals, literally "*fountains of life*," and "*wells of salvation.*" And when we become acquainted with their methods of *raising* water, we shall perceive how singularly apposite are those illustrations, which the author of Ecclesiastes has drawn from "the pitcher broken at the fountain;" and from "the wheel broken at the cistern." Chap. xii, 6.

---

[a] SALE's Koran, chaps. 55, 76, 83. [b] Koran, chaps. 14, 22, 37. [c] Million of Facts, p. 253.

In attempting to discover the origin, and to trace the progress of the art of raising water, we must have recourse to ASIA, the birthplace of the arts and sciences; from whence, as from a centre, they have become extended to the circumference of the earth. It was *there* the original families of our race dwelt, and the inventive faculties of man were first developed. It was from the ancient inhabitants of that continent that much of the knowledge, nearly all the arts, and not a few of the machines which we possess at this day, were derived.

That man at the first imitated the lower animals in quenching his thirst at the running stream, there can be no doubt. It was natural, and because it was so, his descendants have always been found, when under similar circumstances, to follow his example. The inhabitants of New Holland, and other savages quench their thirst in this manner, (i. e. by laying down.) The Indians of California were observed by Shelvock in 1719, to pursue the same method. " When they want to drink they go to the river."[a]

The heathen deities, who in general were distinguished men and women, that were idolized after death, are represented as practising this and similar primeval customs. Thus Ovid describes LATONA on a journey, and languishing with thirst, she arrives at a brook,

.... And *kneeling* on the brink
Stooped at the fresh repast, prepared to drink,
But was hindered by the rabble race. *Metam.* vi. 500.

When circumstances rendered it difficult to reach the liquid with the mouth, then " the *hollow of the hand*" was used to transfer it.

Gideon's soldiers pursued both modes in allaying their thirst;[b] and it was the practice of the *last*, which Diogenes witnessed in a boy at Athens, which induced that philosopher to throw away his jug, as an implement no longer necessary.

Virgil represents *Eneas* practising it:

Then water in his hollow palm he took
From Tyber's flood. *En.* viii, 95. *Dryden*.

And *Turnus*, in the absence of a suitable vessel, made libations in the same way. The practice was common.

.... As by the brook he stood,
He scooped the water from the chrystal flood;
Then with his hands the drops to heaven he throws,
And loads the powers above with offered vows.

" At sunrise, the Bramins take water out of a tank with the hollow of their hands, which they throw sometimes behind and sometimes before them, invoking Brama."[c]

Herodotus, describing the Nasamones, an ancient people of northern Africa, observes, " when they pledge their word, they drink alternately from each other's hands;" (b. iv, 172.) a custom still retained among their descendants. It is, according to Dr. Shaw, " the only ceremony that is used by the Algerines in their marriages." (Travels, p. 303.)

A Hindoo, says Mr. Ward, " drinks out of a brass cup or takes up liquids in the balls of his hands." (View of the Hindoos, p. 130.) This mode of drinking may appear to us constrained and awkward; but in warm climates, the flexibility of the human body, and custom, make the performance of it easy and not ungraceful.

" I drank repeatedly as I walked along, wherever the pebbles at the bottom gleamed clearest—just deep enough to use *one's hand* as a cup."

---

[a] Voyages round the World, ii, 231. Lon. 1774 [b] Judges, vii, 5, 6.
[c] Sonnerat, Voyage to the East Indies and China. i. 161. Calcutta, 1789.

(Lord Lindsay's Travels, letter 7, Arabia.) Another English traveller noticed women in India use "their hands as ladles to fill their pitchers."

Some writers suppose that Adam, at the beginning of his existence, was not subject to such inconvenient modes of supplying his natural wants. They will have it, that he possessed the knowledge of a philosopher, and was equally expert as a modern mechanic, in applying it to the practical purposes of life. It need scarcely be remarked, that this is imaginary: we might as well credit the visionary tales of the rabbis, or digest the equally authentic accounts of Mahomedan writers. According to these, Adam must have been a *blacksmith*, for he brought down *from paradise* with him, five things made of *iron*; an *anvil*, a *pair* of *tongs*, two *hammers*, a large and a small one, and a *needle!* Analogous to this is the affirmation of the Scythians, mentioned by Herodotus,[a] that there fell from heaven into the Scythian district, four things made of gold; a *plough*, a *yoke*, an *axe*, and a *goblet*. The palladium of Troy, it was said, also, fell down from heaven. It was a small statue of Pallas, holding a *distaff* and *spindle*.[b]

We believe there is no authority in the bible, either for the superiority of Adam's knowledge, or of the circumstances in which he was placed: on the contrary, Moses represents him and his immediate descendants, in that rude state, in which all the original and distinct tribes of men have been found at one time or another; living on the spontaneous productions of the earth, on fruits and roots; ignorant of the existence and use of the metals, (and there could be no civilization where these were unknown;) naked and insensible of the advantages of clothing: in process of time, using a slight covering of leaves, or other vegetable productions, and subsequently applying the skins of animals to the same purpose; then constructing huts or dwellings of the leaves and branches of trees; attaining the knowledge of, and use of fire; and making slight attempts to cultivate the earth; for slight indeed they must have been, in the infancy of the human race, before animal power was applied to agricultural labor, or the implements of husbandry were known. Of these last, rude implements formed of sticks, might have been, and probably were used, as they have been by rude people in all ages. Virgil's description of the aborigines of Italy, previous to the reign of Saturn, is merely a poetic version of traditions of man in primeval times:

> Nor laws they knew, nor manners, nor the care
> Of lab'ring oxen, nor the shining share, (the plough.)
> Nor arts of gain, nor what they gained to spare.
> Their exercise the chase : *the running flood*
> *Supplied their thirst :* the *trees* supplied *their food*.
> Then Saturn came. *En.* viii, 420.

Vitruvius says, " In ancient times, men, like wild beasts, lived in forests, caves, and groves, feeding on wild food; and that they acquired the art of producing fire, from observing it evolved from the branches of trees, when violently rubbed against each other, during tempestuous winds."[c]

Similar traditions of their ancestors were preserved by all the ancient nations, and some of their religious ceremonies were based upon them. Thus at the Plynteria, a festival of the Greeks in honor of Minerva, it was customary to carry in the procession a cluster of *figs*, which intimated the progress of civilization among the first inhabitants of the earth, as figs served them for food, after they had acquired a disrelish for *acorns*. The Arcadians eat apples till the Lacedemonians warred with them.[d]

The oak was revered because it afforded man in the first ages, both food and drink, by its *acorns* and *honey*, (bees frequently making their hives

---

[a] iv, 5. [b] These and similar traditions of other people, indicate the *extreme antiquity* of the implements named. The ancients were as ignorant of their origin as we are.
[c] ii, 1. [d] Plutarch in Alcibiades and Coriolanus.

upon it,) and from this circumstance probably, was it made "sacred to Jupiter." The elder Pliny, in the proem to his 16th book, speaks of trees which bear mast, which says he, "ministered the *first* food unto our forefathers." Thus Ovid in his description of the golden age:—

> The teeming earth, yet guiltless of the plough,
> And unprovoked, did fruitful stores allow :
> Content with food which nature freely bred,
> On wildings and on strawberries they fed ;
> Cornels and bramble-berries gave the rest,
> And falling acorns furnished out the feast. *Metam.* ii, 135.

In the ancient histories of the Chinese, it is recorded of their remote ancestors, that they were entirely naked and lived in caves; their food wild herbs and fruits, and the raw flesh of animals; until the art of obtaining fire by the rubbing of two sticks together was discovered, and husbandry introduced.

There are persons however, who suppose it dishonoring the Creator, to imagine that ADAM, the immediate work of his hands, and the intellectual and moral head of the human family, should at any period of his existence have been destitute of many of those resources which the Indians of our continent, and other savages possess; although it is evident, that some time must have elapsed before he could realize, (if he ever did,) all the conveniences which even they enjoy.

There is nothing unreasonable or unscriptural in supposing that *all* the primitive arts originated in man's immediate wants. Indeed, they could not have been introduced in any other way, for it is preposterous to suppose the Creator would directly reveal an art to man, the utility of which he could not perceive, and the exercise of which his wants did not require.

Nor could any art have been *preserved* in the early ages, except it furnished conveniences which could not otherwise be procured. On no other consideration could the early inhabitants of the world have been induced to practice it. But when success attended the exercise of their ingenuity in devising means to supply their natural and artificial wants, the simple arts would be gradually introduced, and their progress and perpetuity secured by *practice* and by that alone.

This appears to have been the opinion of the ancients:

> Jove willed that man, by *long experience* taught,
> Should various arts invent by *gradual thought.* *Geor.* i, 150.

CHAPTER III.

ORIGIN OF VESSELS for containing water—The Calabash the first one—It has always been used—Found by Columbus in the cabins of Americans—Inhabitants of New Zealand, Java, Sumatra, and of the Pacific Islands employ it—Principal vessel of the Africans—Curious remark of Pliny respecting it—Common among the ancient Mexicans, Romans and Egyptians—Offered by the latter people on their altars—The model after which vessels of capacity were originally formed—Its figure still preserved in several—Ancient American vessels copied from it—Peruvian bottles—Gurgulets—The form of the Calabash prevailed in the vases and goblets of the ancients—Extract from Persius' Satires—Ancient vessels for heating water modeled after it---Pipkin---Sauce-pan---Anecdote of a Roman Dictator---The common cast iron cauldron, of great antiquity; similar in shape to those used in Egypt, in the time of Rameses---Often referred to in the Bible and in the Iliad---Grecian, Roman, Celtic, Chinese, and Peruvian cauldrons---Expertness of Chinese tinkers---Crœsus and the Delphic oracle---Uniformity in the figure of cauldrons---Cause of this---Superiority of their form over straight sided boilers---Brazen cauldrons highly prized---WATER POTS of the Hindoos---Women drawing water---Anecdote of Darius and a young female of Sardis---Dexterity of oriental women in balancing water pots---Origin of the Canopus---Ingenuity and fraud of an Egyptian priest---Ecclesiastical deceptions in the middle ages.

WATER being equally necessary as more solid food, man would early be impelled by his appetite, to procure it in larger quantities than were required to allay his thirst upon a single occasion; and, also the means by which he might convey it with him, in his wanderings, and to his family. It is not improbable that this was the first of man's natural wants which required the exercise of his inventive faculties to supply. The luxuriance of the vegetable region, in which all agree that he was placed, furnished in abundance the means that he sought; and which his natural sagacity would lead him, almost instinctively, to adopt. The CALABASH or GOURD, was probably the first vessel used by man for collecting and containing water: and although we have no direct proof of this, there is evidence, (that may be deemed equally conclusive,) in the general fact—that man, in the infancy of the arts, has always, when under *similar circumstances*, adopted the *same means*, to accomplish the *same objects*. Of this, proofs innumerable, might be adduced from the history of the old world, particularly with regard to the uses and application of natural productions; and when at the close of the fifteenth century, Columbus opened the way to a *new* world, having in his search after one continent discovered another (of which neither he, nor his contemporaries ever dreamt, and which in extent exceeded all that his visions ever portrayed;) he found the CALABASH the principal vessel in use among the inhabitants, both for containing and transporting water.

The calabashes of the Indians, (says Washington Irving,) served all the purposes of glass and earthenware, supplying them with all sorts of domestic utensils. They are produced on stately trees, of the size of elms.[a] The New Zealanders possessed *no other* vessel for holding liquids; and the same remark is applicable at the present day to numerous savage tribes. Osbeck, in his Voyage to China, remarks, that the Javanese sold to European ships, among other necessaries, " bottles of gourds filled with water, as it is made up for their own use."[b]

When Kotzebue was at Owhyhee, Tamaahmaah the king, although he

---
[a] Irving's Colum. 1, 105, and Penny Mag. for 1834, p. 416. [b] i, 150.

Chap. 3] *The Calabash.* 15

possessed elegant European table utensils, used at dinner, a gourd containing taro-dough, into which he dipped his fingers, and conveyed it by them to his mouth, observing to the Russian navigator, " this is the custom in my country and I will not depart from it."[a] This conduct of Tamaahmaah, resembled that of Motezuma. Solis observes, that he had " cups of gold and salvers of the same," but that he sometimes drank out of cocoas and natural shells.[b]

When Kotzebue revisited the Radack Islands, " he carried to them seeds of gourds for valuable vessels," as well as others of which the fruit is eaten.[c]

" There is a gourd more esteemed by the inhabitants of Johanna for the large shell, than for the meat. It will hold a pailful. Its figure is like a man's head, and therefore called a calabash."[d]

The people of Sumatra drink out of the fruit called *labu*, resembling the calabash of the West Indies: a hole being made in the side of the neck and another one at the top for vent. In drinking they generally hold the vessel at a distance above their mouths, (like the ancient Greeks and Romans) and catch the stream as it falls; the liquid descending to the stomach without the action of swallowing.[e]

The Japanese have a tradition that the first man owed his being to a calabash.[f]

Capt. Harris, in his " Wild sports of Southern Africa" (chap. xvii.) in describing the residence of the king of Kapaue, observes, " the furniture consisted exclusively of calabashes of beer, ranged round the wall." And again in chap. xx:—" a few melons, rather deserving the name of vegetables, were the only fruit we met with; and these I presume are nurtured chiefly for the gourd, which becomes their calabash or water flagon."

Clavigero says, " the drinking vessels of the ancient Mexicans, were made of a fruit similar to gourds."[g]

For such purposes, the calabash has ever been used wherever it was known, and will continue to be so, as long as it grows and man lives.

The elder Pliny, in speaking of the cultivation of gourds, a species of which were used as food by the Romans, observes, " *of late* they have been used in baths and hot houses for pots and pitchers;" but he adds, that they were used in *ancient times* to contain wine, " in place of rundlets and barrels." From him we learn that the ancients had discovered the means of controlling their forms at pleasure. He says, *long* gourds are produced from seeds taken from the *neck*; while those from the middle produce *round* or *spherical* ones, and those from the sides, bring forth such as are *short and thick*.[h]

Among the offerings which the Egyptians placed on their altars, was the gourd. An undeniable proof of its value in their estimation; for nothing was ever offered by the ancients to their gods, which was not highly esteemed by themselves.[i] The consecration of this primeval vessel, in common with other objects of ancient sacrifice, doubtless originated in its universal use in the early ages; and most likely gave rise to the subsequent practice of dedicating cups and goblets, of gold, silver, and sometimes of precious stones.

As the gourd or calabash was not only the first vessel used to collect and convey water, but one apparently designed by the Creator for these purposes, a figure of it is here given.

[a] Voyage Discov. Lon. 1821. i, 313, and ii, 193. [b] Conquest Mexico, Lon. 1724. iii, 83.
[c] iii, 175. [d] A New Account of East India and Persia, by Dr. Fryer. Lon. 1698. 17
[e] Marden's Sumat. 61. [f] Montanus' Japan. 275. [g] Hist. of Mexico. Lon. 1837. i. 438.
[h] Nat. Hist. xix, 5. [i] Wilkinson i, 276.

No. 1.

This interesting production of nature is entitled to particular notice, because, it is, in all probability, the *original model* of the earliest artificial vessels of capacity; the pattern from which they were formed. It is impossible to glance at the figure without recognizing its striking resemblance to our jugs, flasks, jars, demijohns, &c. Indeed when man first began to make vessels of clay, he had no other pattern to guide him in their formation but this, one with which he had been so long familiar, and the figure of which experience had taught him was so well adapted to his wants. Independent of other advantages of this form, it is the best to impart *strength* to fragile materials.

That the long necked vases of the ancients were modeled after it, is obvious. Many of them differ nothing from it in form, except in the addition of a handle and base. The oldest vessels figured in the GRANDE DESCRIPTION OF EGYPT, by the Savans of France, and in Mr. Wilkinson's late work on the ancient Egyptians, are fac-similes of it. The same remark applies to those of the Hindoos and Chinese.

No. 2. Ancient Vases.

The first three on the left are of earthenware from Thebes, from Wilkinson's second volume, p. 345, 354. " Golden ewers" of a similar form were used by the rich Egyptians for containing water, to wash the hands and feet of their guests. (page 202.) The next is Etruscan, from the " History of the ancient people of Italy." Florence 1832. Plate 82. The adjoining one is a Chinese vase, from " Designs of Chinese Buildings, Furniture," &c. Lon. 1757. The last is from Egypt. Similar shaped vessels of the Greeks, Romans, and other people might easily be produced. See Salt's Voyage to Abyssinia, page 408, and *Grande Description*, E. M. Vol. 2. Plates I, I, and F, F. In the Hamilton Collection of Vases, examples may be found. In the splendid volume of plates to D'Agincourt's Storia Dell' Arte, the figure of the gourd may be seen to have prevailed in artificial vessels in the fourth, fifth, and up to the twelfth centuries.

Numerous vessels from the tombs of the INCAS, are identical in figure with the calabash; while others, retaining its general feature, have the bellied part worked into resemblances of the human face. As several old Peruvian bottles exhibit a peculiar and useful feature, we have inserted (figure 3,) a representation of one, in the possession of J. R. Chilton, M. D. of this city. An opening is formed in the inner side of the handle which communicates with the interior of the vessel, by a smaller one made through the side, as shown in the section. By this device air is admitted, and a person can either drink from, or pour out the contents, with-

Peruvian Vessels.

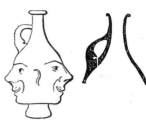

No. 3. Peruvian Bottle.

out experiencing that disagreeable gurgling which accompanies the emptying of a modern bottle. The openings are so arranged as to form a very *shrill whistle*—by blowing into the mouth of the vessel, a sound is produced, equal to that from a boatswain's call on board a man of war.

These vessels have been noticed by most travelers in South America. They are sometimes found double—two being connected at the bottom with only one discharging orifice. Some are of silver. *Frezier*, among others, gives a figure of one resembling two gourds united. It "consists of two bottles joined together, each about six inches high, having a hole (tube) of communication at the bottom. One of them is open, and the other has on its orifice a little animal, like a monkey, eating a cod of some sort; under which is a hole which makes a whistling when water is poured out of the mouth of the other bottle, or when that within is but shaken; because the air being pressed along the surface of both bottles, is forced out at that little hole in a violent manner."[a]

These whistles are so constructed, as to play either when the air is *drawn in* through them, or *forced out*. Perhaps the water organs of the ancients, were *originally* little more than an assemblage of similar vessels. M. Frezier thought the smallest of these bottles were designed expressly to produce music; if so, they are (we suppose) the only water instruments extant.[b]

The large earthen vessels used by the water carriers of Mexico, strictly resemble the gourd. Saturday Mag. vol. vi, 128.

The "gurgulets" of the Persians, Hindoos, and Egyptians of the present day, are rather larger, but of the same shape as the Florence flask, i. e. of the gourd. They are formed " of a porous earth, and are so called, from the sound made when water is poured out of them to be drunk, as

---

[a] A Voyage to the South Sea, &c. in 1712, '13, '14. Lon. 1717. 274.

[b] The following extract from a late newspaper affords additional information respecting these vessels in remote ages:

"*The Peruvian Pompeii.*—We recently gave a description of an ancient subterranean city, destroyed by an earthquake, or some other sudden convulsion of nature, lately discovered near the port of Guarmey, in Truxillo, on the coast of Peru. The only account of it which appears to have as yet been received in the United States, was brought by Capt. Ray of Nantucket, who a few weeks since returned from the South Seas in the ship Logan, and who, having visited the spot whilst the inhabitants of Guarmey were excavating the buried streets and buildings, obtained several interesting relics of its ancient but unknown population. The Portland Orion describes some of these, of which we did not find any mention in the Nantucket Inquirer from whom we derived our former information, and they are of a character which may possibly afford the diligent antiquary some clue to the age and origin of the people to whom they belonged. They are two grotesquely shaped earthen vessels, somewhat rudely yet ingeniously constructed, of a species of clay, colored or burnt nearly black. One of these, which is capable of holding about a pint, is shaped somewhat like a quail, with a spout two inches long, rising from the centre of the back, from which also a handle extends to the side.

The other is a double vessel, connected at the centre, and also at the top, by a handle reaching from the spout or nozzle of one vessel to the upper part of the other—the latter not being perforated but wrought into the likeness of a very unprepossessing human countenance. At the back of what may be considered the head of this face, is a small hole, so contrived that on blowing into the mouth of the vessel a shrill note is produced, similar to that of a boatswain's call. From the activity with which the excavations were proceeding when Capt. Ray left the place, it may be hoped that discoveries will be made which will greatly add to the antiquarian history of this continent."

the Indians do without touching it with their lips."[a] The bottles of the Negroes of Africa, are made of woven grass of the same shape. Earthen gurgulets for cooling liquids are made in this city.

The gourd was not merely imitated by primitive potters and braziers, but when the arts were at their zenith, its figure predominated in the most elaborate of vases. The preceding remarks show, that the *forms* of many of our ordinary vessels of capacity, did not originate in caprice or by chance, but are derived from nature; that the pattern which man has copied, was furnished him by his Maker; and that with all his ingenuity, he has never been able to supersede it. PERSIUS in his third Satire, alludes to the transition from primitive earthenware and brazen vessels to those which luxury had introduced in his days :

> Now gold hath banished Numa's simple vase,
> And the plain brass of Saturn's frugal days.—
> Now do we see to precious goblets turn,
> The Tuscan pitcher, and the vestal urn. *Drummond*, 105.

VESSELS FOR HEATING WATER.

Although not strictly connected with the subject, we may observe that the gourd is probably the original vessel for HEATING *water, cooking, &c.* In these and other applications, the neck is sometimes used as a handle, and an opening made into the body by removing a portion of it, (see illustration No. 4,) its exterior being kept moistened by water while on the fire, as still practised by some people, while others apply a coating of clay to protect it from the effects of flame.

In some parts where the calabash or gourd is not cultivated, cocoa shells are used in the same manner. KOTZEBUE found the Radack Islanders thus heating liquids. "On my return, I fell in with a company sitting round a fire and boiling something in cocoa shells."[b] A primitive *Sumatran* vessel for boiling rice is the bamboo, which is still used—by the time the rice is dressed, the vessel is nearly destroyed by the fire.[c] When in process of time, vessels for heating water were formed wholly of clay, they were fashioned after the gourd. Figures of ancient *saucepans* both of metal and fictile ware, greatly resemble it, and so do some of those of modern times. The common earthenware PIPKIN is an example.

This useful implement has come down from very remote ages, and apparently with slight alteration in its figure. (See figure in No. 4.) In some parts of Europe, its form approaches still nearer to that of the gourd.

No.   Gourd, Cauldron, and Pipkin.

It is used over all the eastern world. Dampier observed in Tonquin, "women sitting in the streets with a pipkin, over a small fire full of chau," or tea, which they thus prepared and sold.[d] Fosbroke enumerating the household utensils represented in Egyptian sculptures, remarks, " we meet too with vessels of the precise form of modern saucepans."[e] An interesting circumstance is recorded in Roman history in connection with one of these vessels. Marcus Curius Dentatus, who was three times Consul, was as

---

[a] Fryer's India and Persia, p. 47.
[b] Voyage Discov. ii, 109, and iii, 152, and Fryer's India, 7.
[c] Marsden's Sumatra, 60. [d] Dampier's Voyage. Lon. 1705. ii, 31 [e] For. Topog. 83.

remarkable for his frugality as his patriotism. During the time that he swayed the destinies of his country, the ambassadors of the Samnites visited him at his cottage, and found him boiling vegetables in an earthen pot or pipkin; they attempted to bribe him with large presents; but he characteristically replied, "I prefer my earthen pots to all your vessels of gold and of silver." To this Juvenal alludes, when contrasting the frugality of former times with the luxury of his contemporaries :

> When with the herbs[a] he gathered, CURIUS stood
> And seethed his pottage o'er the flaming wood ;
> That simple mess, an old Dictator's treat,
> The highway laborer now would scorn to eat. *Sat.* xi, 105.

The common cast iron *bellied* kettle or CAULDRON, furnishes another proof of the forms of culinary vessels having undergone little or no change, while passing through so many ages: its shape is precisely the same as that of the SITULA or POT, sculptured on the obelisk of Heliopolis, (See its figure in No. 4, and Dr. Shaw's Travels, 402, 413.) Others with ears and feet, are delineated in the Theban sculptures. In the tomb of Rameses the Third, is a graphic representation of an Egyptian kitchen, showing the processes of slaying the animals—cutting the joints—preparing ingredients for seasoning—*boiling* the meat—stirring the fire—making and baking bread, &c. &c. The cauldrons of various sizes are similar in shape to ours. Wilkinson's An. Egyp. ii, 351, 383, 385. There is reason to believe that boilers of this form were common to all the nations of the ancient world; that the 'pottage' by which Jacob defrauded Esau of his birthright; and the 'savoury meat,' which Rebecca cooked for Isaac, were prepared in them. To one of these, Job referred; "out of his nostrils goeth smoke, as out of a seething pot or cauldron." xli, 20. And Elisha also, when he said to his servant. "Set on the great pot, and seethe pottage for the sons of the prophets." 2 Kings, iv, 38. It is often mentioned by Homer, in whose writings it forms a conspicuous object:

> And soon the flames encompassing around its *ample belly.*
> *Iliad*, xviii, 427. Cowper

Such were the boilers of Argos, (respecting which arose the saying, " a *cook* from Elis—a *cauldron* from Argos—*tapestry* from Corinth, &c.) and of the Spartans, in which they prepared their famous 'black broth.' A figure of a Roman cauldron, in which the priests boiled their portion of the sacrifice, is given by Misson, in the first volume of his Travels, plate 4. It has a bail, three studs or feet, and is of a spherical shape resembling ours, but ornamented with figures round its sides.

The same shaped boilers were common among the Gauls, who probably derived the knowledge of making them from the Phenicians. The art of *tinning* culinary vessels, which they are said to have invented, (Pliny, Nat. Hist. xxxiv, 17,) was most likely obtained from the same source.[b] The Celtiberi are said to have been expert workers of iron. Their "most ancient iron pot," had ears and feet, and was shaped like those of the Egyptians. (See its figure in 'Scottish Gael.' p. 316. The cast iron cauldrons of the Chinese are also examples. These are made very thin; and what is singular, their mechanics have the art of soldering them when cracked, with portions of the same metal, by means of a blow-pipe and small furnace.[c] They are the principal article of furniture in the dwell-

---

[a] Plutarch, says they were *turnips.*
[b] Pliny, b. xii, 1, says, the Gauls were first induced to invade Rome, by one of their countrymen, a *smith*, who had long worked in that city. He carried home, figs, raisins, oil and wine, which "set the teeth of his countrymen watering." *Holland's Trans.*
[c] " During our short stay this morning in the village of *Fan-koun*, I had an oppor

ings of the poor. The kettles of the Chinese says Mr. Bell, (who lodged one day in a cook's house near Pekin,) "are indeed very thin, and made of cast iron, being extremely smooth both within and without." Fuel is scarce and they used bellows to heat them.[a] These we have no reason to suppose have undergone any change from the remotest times, and they are in all probability of the same form as the celebrated cauldrons of antiquity. That those of the Scythians, the ancient Tartars and Chinese, were similar to those of the Greeks, is asserted by Herodotus. "As Scythia is barren of wood, they have the following contrivance to dress the flesh of the victim: having flayed the animal, they strip the flesh from the bones; and if they have them at hand, they throw it into *certain pots made in Scythia*, and RESEMBLING THE LESBIAN CAULDRONS, though somewhat larger." Herod. iv, 61.

The boilers of the ancient Mexicans and Peruvians, had the same general form. See plate 31 of Frezier's Voyage to the South Sea, in 1712, '13, '14. As these people had not the use of iron, their vessels were of earthenware, copper and its alloys, silver, and even of gold. In the temple ple at Cusco, "were boyling pots and other vessels of gold." Two enormous cauldrons were carried by the conquerors to Spain, "each sufficient wherein to boyle a cow." (Purchas' Pilgrimage, 1061, and 1073.) The negroes of Africa, made theirs of the same shape. (Generale Histoire, tom. v, Planche 88.) *Large* cauldrons were common of old; they are frequently mentioned by Homer, Herodotus, &c. and in the Bible. Mahomet, in the 34th chapter of the Koran, speaks of large cauldrons belonging to David. Some of those represented at Thebes, appear sufficiently capacious to contain the cooks that attend them. Crœsus boiled together a tortoise and a lamb in a large brasen cauldron, which had a *cover* of the same metal; hence the reply of the Delphic oracle, to the demand of his ambassadors to be informed what Crœsus was at that moment doing:

> E'en now the odors to my sense that rise
> A tortoise boiling with a lamb supplies,
> Where *brass* below, and *brass* above it lies.   *Herod.* i, 47.

The question naturally arises—why such uniformity in the figure of this utensil? and what has induced people in distant times and countries to make it resemble a portion of a hollow sphere or spheroid, instead of forming it with plane sides and bottom? It is clear there was some controlling reason for this—else why should the fanciful Greek and Roman artists, have permitted it to retain its primitive form, while all other household implements, as lamps, vases, drinking vessels, and tripods, &c. were moulded by them into endless shapes. Brasen cauldrons we know were highly prized. They were sometimes polished, and their sides richly ornamented, but still their general form was the same as those of more ancient people. In this respect, both Greeks and Romans left them as they found them. The reason is obvious. When a liquid is heated in a cylindrical or other vessel having perpendicular sides, it easily '*boils over*;' but when the sides incline *inwards* at the *top*, as in these cauldrons; it cannot well be thrown out by ebullition alone; for the heated waves as they

---

tunity of seeing a tinker execute what I believe is unknown in Europe. He mended and soldered frying-pans of cast iron, that were cracked and full of holes, and restored them to their primitive state, so that they became as serviceable as ever. He even took so little pains to effect this, and succeeded so speedily, as to excite my astonishment." Van Braam's Journal of the Dutch embassy to China, 1794—5. Lon. 1798. ii, 78, and Chinese Repository, Canton, 1838. iv, 38.

[a] Travels from Petersburgh to diverse parts of Asia. Lon 1764. i, 312.

Chap. 3.] *Cauldrons.* 21

rise are directed towards the centre, where their force is expended against each other. Dyers, brewers, distillers, &c. are well aware of this fact. The remote ancients had therefore observed the inefficiency of straight sided boilers, and applied a simple and beautiful remedy; one whch was possibly suggested by the previous use of natural vessels, as the gourd, &c. This is no mean proof of their sagacity, and of the early progress of the arts of founding and moulding. From the extreme antiquity of these cauldrons, it is not improbable that their form is similar to the pattern, which Tubal-Cain himself used, and which he taught his pupils to imitate. Similar vessels are found in the workshops of Vulcan. See plate 20, Painting, in D'Agincourt's Storia Dell'Arte, Prato, 1827. Brasen cauldrons were formerly considered suitable presents for kings—rewards of valor—prizes in the games, &c. Of the gifts offered by Agamemnon to to appease the wrath of Achilles, were—

> Seven tripods, never sullied yet by fire;
> Of gold, ten talents; TWENTY CAULDRONS bright."
> *Iliad*, ix, 150. *Cowper.*

They were among the goods which Priam took to redeem the body of Hector.

> He also took ten talents forth of gold,
> All weighed; two splendid tripods; CAULDRONS four;
> And after these a cup of matchless worth. *Ib.* xxiv. 294.

The prizes at the funeral games on the death of Patrocles, were—

> 'Capacious CAULDRONS, tripods bright.'

In the 17th century, they were considered suitable presents to a Persian Emir—" At length he came, and was presented by the caravan-Bashi with a piece of satin, half a piece of scarlet cloth, and two large copper cauldrons." Tavernier's Trav. Lon. 1678. 61.

These unobtrusive vessels are now used without exciting a thought of their worth, or of the ingenuity of those to whom we are indebted for them; although they have contributed infinitely more to the real comfort and innocent gratification of man, than all the splendid VASES that were ever made. These have always had their admirers and historians. Volumes embellished with costly illustrations, have been written on their forms, materials, ages and authors; but no modern Hamilton, has entered the kitchen to record and illustrate the origin, improvement, modifications and various uses of the cauldron. This vessel, like a despised but necessary attendant, has been the inseparable companion of man in his progress from barbarism to refinement, and has administered to his necessities at every stage: yet it has ever been disregarded, while literary *cuisiniers* have expatiated in numerous treatises on the virtues of meats prepared in it. Endless are the essays on sauces, but the history of the more useful sauce-pan is yet to be written. An account of this vessel and of the cauldron, would place in a very novel and instructive light, the domestic manners of the world; and an examination of the various modes of *heating* the latter, would bring to view many excellent devices for economizing fuel.[a]

VASES used by oriental women to convey water from public wells and fountains for domestic purposes, are often referred to, by sacred and profane authors. Figure No. 5, represents a female of Hindostan, bearing

---

[a] See the ancient Peruvian furnace in Frezier's Voyage to the South Seas, by which *three* cauldrons were heated by a very small pot of lama's dung, or of the plant *icho*; wh'ch were used for want of other fuel.

No. 5.

one, the shape of which, closely resembles the gourd with the neck removed. This is their general form throughout the east. The Hindoos, have them of copper or brass, as well as of earthenware, but they are all shaped alike. This is not a little singular, because a deviation from a globular to a cylindrical form, would enable their mechanics to make *those of metal* at much less expense. They therefore adhere to the primitive model, because of its superiority over others, or from that adhesion to ancient customs which forms so prominent a feature in Asiatic character. In the early ages it was the universal custom for young women to draw water. The daughters of princes and chief men, were not exempt from it. Isis and Osiris are sometimes represented with water vessels on their heads. There are several interesting examples in the Old Testament. Homer, as might be expected, frequently introduces females thus occupied. When Nestor entertained Telemachus, he bade

· · · · · The handmaids for the feast prepare,
The seats to range, the fragrant wood to bring,
And limpid waters from the living spring. *Odys.* iii, 544. *Pope.*

And again at Ithaca;

· · · · · With duteous haste a bevy fair,
Of twenty virgins to the spring repair:
    *   *   *   *
Soon from the fount, with each a brimming urn,
(Eumæus in their train) the maids return. *Ib.* xx, 193 and 202.

Fountains and wells became the ordinary places of assembly for young people—especially, "at the time of the evening, the time that women go out to draw water." Gen. xxiv, 11. Several of the Patriarchs first beheld their future wives on these occasions; and were doubtless as much captivated by their industry and benevolent dispositions in relieving the wants of strangers and travelers, as by their personal charms. It was

· · · · · · · · Beside a chrystal spring—

that Ulysses met the daughter of Antiphates. Travelers have often noticed the singular tact with which Asiatic women balance several of these water pots on their heads without once touching them with their hands. "The finest dames of the Gentoos disdained not to carry water on their heads, with sometimes *two* or *three* earthen pots *over one another*, for household service; the like do all the women of the Gentiles." Fryer's Trav. 117. At one of their religious festivals, Hindoo women, "have a custom of dancing with several pots of water on their heads, placed one above another." Sonnerat, i, 150.

A very pleasing instance of female dexterity in carrying water, is recorded by Herodotus, v, 12. As Darius, king of Persia, was sitting publicly in one of the streets of Sardis, he observed a young woman of great elegance and beauty, bearing a vessel on her head, leading a horse by a bridle fastened round her arm, and at the same time spinning some thread. Darius viewed her as she passed, with attentive curiosity, observing that her employments were not those of a Persian, Lydian, nor indeed of any Asiatic female; prompted by what he had seen, he sent some of his attendants to observe what she did with the horse. They accordingly followed her—When she came to the river, she gave the horse some water and then filled her pitcher: having done this, she returned by the way she came, with the pitcher of water on her head, the horse fastened by a bridle to her arm, and as before, employed in spinning.

Industrious labor is an ornament to every young woman—indeed neither the symmetry of her person, nor the vigor of her mind, can be perfectly developed without it. The fine forms and glowing health of the women of old, were chiefly owing to their temperate modes of living, their industrious habits, and the exercise they took in the open air.

No. 6. A Canopus.

A circumstance recorded in the history of the Egyptians, accounts for the peculiar form of one of their favorite vessels, the *Canopus;* the annexed figure of which, is taken from the 'History of the ancient people of Italy,' plate 27. It was named after one of their deities, who became famous on account of a victory which he obtained over the Chaldean deity, FIRE;—the story of which exhibits no small degree of ingenuity in a priest, and it affords a fair specimen of the miracles by which people were deluded in remote times. The Chaldeans boasted, as they justly might, of the unlimited power of their god, and they carried him about to combat with those of other provinces, all which he easily overcame and destroyed, for none of their images were able to resist the force of *fire !*—At length a shrewd priest of Canopus, devised this artifice and challenged the Chaldeans to a trial. He took an earthen jar, in the bottom and sides of which he drilled a great number of small holes;—these he stopt up with wax, and then filled the jar with water: he secured the head of an old image upon it, and having painted and sufficiently disguised it, brought it forth as the god Canopus! In the conflict with the Chaldean Deity the wax was soon melted by the latter, when the water rushed out of the holes, and quickly extinguished the flames. Univ. Hist. i, 206. In memory of this victory, vessels resembling the figure of the god used on this occasion became common. Dr. Shaw gives the figure of one which he brought with him from Egypt. Trav. 425. See Montfaucon, tom. ii, liv. i, cap. 18. A figure of one throwing out water from numerous holes on every side is also given. Tom. ii, liv. iii.

A somewhat similar case of superstition in the middle ages, is quoted by Bayle from Baronius; being a trial of the virtue in the bones of two saints; or rather a contest of priestly skill. *St. Martin's* relics being carried over all France came to Auxerre, and were deposited in the church of *St. Germain,* where they wrought several miracles. The priests of the latter considered him as great a saint as the former; they therefore demanded one half of the receipts, " which were considerable;" but Martin's priests contended that it was his relics that performed all the miracles, and therefore *all* the gifts belonged to them. To prove this, they proposed that a sick person should be put between the shrines of the saints, to ascertain which performed the cure. They therefore laid a *leper* between them, and he was healed on that side which was next to St. Martin's bones, and not on the other ! the sick man then very naturally turned his other side, and was instantly healed on that also! Cardinal Baronius in commenting on this result, seriously observes, that St. Germain was as great a saint as St. Martin, but that as the latter had done him the *favor of a visit,* he suspended the influence he had with God, to do his *guest* the greatest honor ! The custom of having patron saints or gods was universal among the ancient heathen; and the same system was carried by half pagan christians of the dark ages to an incredible extent. Ecclesiastics peddled the country, like itinerant jugglers, with sacks of bones and other relics from the charnel house—the pretended virtues of which, they sold to the deluded multitude as in the above instance.

CHAPTER IV.

ON WELLS—Water one of the first objects of ancient husbandmen—Lot—Wells before the deluge—Digging them through rock subsequent to the use of metals—Art of digging them carried to great perfection by the Asiatics—Modern methods of making them in loose soils derived from the East—Wells often the nuclei of cities—Private wells common of old—Public wells infested by Banditti—Wells numerous in Greece—Introduced there by Danaus—Facts connected with them in the mythologic ages—Persian ambassadors to Athens and Lacedemon thrown into wells—Phenician, Carthagenian and Roman wells extant—Cæsar and Pompey's knowledge of making wells enabled them to conquer—City of Pompeii discovered by digging a well—Wells in China, Persia, Palestine, India, and Turkey—Cisterns of Solomon—Sufferings of travelers from thirst—Affecting account from Leo Africanus—Mr. Bruce in Abyssinia—Dr. Ryers in Gombroon—Hindoos praying for water—Caravan of 2000 persons and 1800 camels perished in the African desert—Crusaders.

As the human family multiplied, its members necessarily kept extending themselves more and more from their first abode; and in searching for suitable locations the prospect of obtaining water would necessarily exert a controlling influence in their decisions. An example of this, in later times, is given by Moses in the case of Abraham and Lot. The land was too much crowded by their families and flocks, "so that they could not dwell together," and when they had concluded to separate, Lot selected the plain of Jordan, because "it was *well watered* every where." Gen. xiii, 10. In the figurative language of the East, "Lot lifted up his eyes and beheld all the plain of Jordan;" in plain English, he went and carefully examined it. When thus extending themselves, the early inhabitants of the world, would frequently meet with locations every way adapted to their wants with the single exception of water; circumstances, which necessarily must have excited their ingenuity in devising means to obtain it.

At what period of mans' history he first had recourse to wells, we have no account; nor of the circumstances which led him to *penetrate the earth*, in search of water. Wells, we have no doubt, are of antediluvian origin, and the knowledge of them, like that of the primitive arts, has been preserved by uninterrupted use from the period of their first discovery. At first, they were probably nothing more than shallow cavities dug in moist places; and their depth occasionally increased, in order to contain the *surface* water that might drain into them within certain intervals of time; a mode of obtaining it still practised among barbarous people. The wells of Latakoo, described by Mr. Campbell, in his "Travels in South Africa," were of this description. They were but two feet deep and were emptied every morning. The people of New Holland, the most wretched and ignorant of our species, had similar excavations, at which *Dampier*, when on the coast in 1688, obtained a supply for his ships. He says, "we filled our barrels with water at wells which had been dug by the natives." Burney's Voy. iv, 260. Wells are also connected with the superstitions of the New Zealanders; and the Radack Islanders, when discovered by Kotzebue, had pits or square wells, which they had dug for water. Kotzebue's Voy. ii, 28, 66, and iii, 145, 223. The fresh water which Columbus found in the huts belonging to the Indians of Cuba, was probably obtained from similar wells; but which the Spaniards, who found none but salt water, were unable to discover. *Personal Nar. of Colum.* 67. *Boston,* 1827.

These simple excavations would naturally be multiplied and their dimensions enlarged as far as the limited means of man, in the early ages, would permit, and his increasing wants require. But when the discovery of the metals took place, (in the seventh generation from the first pair, according to both Moses and Sanchoniathon,) the *depth* of wells would no longer be arrested by rocks, nor their construction limited to locations where these did not occur. From very ancient wells which still remain, it is certain, that at a time long anterior to the commencement of history, the knowledge of procuring water by means of them, was well understood, perhaps, equally so as at present. On this supposition only, can we reconcile the selection of locations for them composed *wholly of rock.* Some of the oldest wells known are dug entirely through that material, and to a prodigious depth.

Man's ingenuity was, perhaps, first exercised in procuring water; and it is not improbable, that the art of constructing wells was more rapidly carried to perfection than any other. The physical character of central Asia, its climate, universal deficiency of water, its swarms of inhabitants, and their pastoral, and agricultural pursuits, would necessarily contribute to this result. The Abbe Fleury, in his "Manners of the Ancient Israelites," justly observes, "their numerous herds of cattle necessarily induced them to set a very high value on their wells and cisterns; and more especially as they occupied a country where there was no river but Jordan, and where rain seldom fell." Chap. iii. In no other part of the world, even in modern times, has more science been evinced, or mechanical skill displayed in penetrating the earth, than is exhibited in some of the ancient wells of the east; and it is to their authors, that WE are indebted for the only known method of sinking wells of great depth, through loose soils and quicksands, viz: by first constructing a curb, (of stone, brick, &c.) which settles as the excavation is deepened, and thereby resists the pressure of the surrounding soil.

Wells are mentioned by Moses, as in *common use* among the ancient Canaanites; some of which at that remote age adjoined roads, for the benefit of travelers and the public at large. Indeed, all people who have had recourse to wells, have consecrated some of them to the convenience of strangers and travelers. The first wells were probably all of this description. Most of those mentioned in history were certainly such. At one of these, Hagar rested and refreshed herself, when she fled from the ill treatment of Sarah. And it was "by the way" of this well, that Isaac was going when he first met with Rebecca. And we learn from Gen. xxv, 11, that he subsequently took up his abode near it; a custom by which wells frequently became nuclei of ancient cities. Jacob's well is an example, if really dug by him. When that patriarch and his family drank of its waters, few dwellings were near it; (Gen. xxiii, 19;) but, before the time of Alexander, these had so far increased, as with the ancient Shalem, to form the capital city of Samaria. And 600 years before Alexander's conquest of Judea, Jeroboam when he governed the ten tribes had a palace in the vicinity of this well. Josephus, Antiq. viii, 3. "Tadmor in the wilderness," or Palmyra, one of the most splendid cities of the old world, was built by Solomon (2 Chron. viii, 4,) in the Syrian desert, and its location determined according to Josephus, (Antiq. viii, 6,) "because at that place only there are springs and pits (wells) of water." Pliny makes the same remark, and speaks of its "abundance of water." Nat. His. v, 25. Bonnini, in his 'Syracuse Antichi,' remarks that most of the Sicilian cities took their names from the fountains they were near, or the rivers they bordered upon. The deep well in the Cumean Sybil's cave, gave its name *Lilybe*, both to the cape and town near it. Breval's Remarks on Europe, 19 and 39. The same may be said of other European cities. BATH in England derived its name from the springs near it. It was named Caer-Badon, or the place of baths, before the Roman invasion. The city of WELLS, also, was named after the wells of water near it, especially the one now known as St. Andrew's Well. Lewis's Topographical Dictionary. Many others might be named.

Private wells were, however, very common in ancient times. Abraham and Isaac constructed several for the use of their own families and flocks. David's spies were secreted in the well of a private house. "Water out of thine own cistern and running waters out of thine own well," is the language of Proverbs, v, 15; and in the 2d Book of Kings, xviii, 31, we read of "every one drinking water out his of own cistern;" or *pit* as it is in the margin; a term often used by eastern writers, synonymously with well. In the plans of private houses at Karnac, it appears that the ancient Egyptians arranged their houses and court yards *(Grande Description,* tom. iii, Planche xvi,) in a manner very similar to those of the Romans, as seen at Pompeii, and like these, each house was generally furnished with a round well and an oblong cistern. Lardner's Arts of the Greeks and Romans, i, 44. "If I knew a man incurably thankless," says Seneca, "I would yet be so kind as to put him on his way, to let him light a candle at mine, or draw water *at my well.*" Seneca on Benefits; L'Estrange's Trans. The story of Apono, an Italian philosopher, and reputed magician, of the 13th century, indicates that almost every house had a well. He, however, had not one, or it was dry, and his neighbor having refused to let his maid draw water from his well, Apono, it was said, by his magic caused it through revenge to be carried off by devils. *Bayle.*

Numerous wells of extreme antiquity are still to be seen in Egypt. *Van Sleb* notices several. Besides those in some of the pyramids, there are others which are probably as old as those structures. Mr. Wilkinson mentions one *near* the pyramids of Geezer. An. Egyp. vol. iii. Among the ruins of Nineveh, a city whose foundations were laid by Ashur, the son of an antediluvian, is a remarkable well, which supplies the peasants of the vicinity with water, and who attribute to it many virtues.[a] Captain Rich named it *Thisbe's* Well. The immediate successors of that Pharaoh who patronized Joseph erected stations to command the wells, (which were previously in use, and probably had been for ages,) at *Wadee Jasous*, and these same wells still supply the port of Philoteras or Ænnum, on the Red Sea, with water, as they did four thousand years ago.[b]

The building of stations to protect wells was common in ancient times, on account of robbers laying in wait near them. There is an allusion to this in Judges, "They are delivered from the noise of archers in the places of drawing water." Chap. v, 11. It was at the public fountains that the Pelasgi attacked the Athenian women. Near the ruins of an Egyptian Temple at Wady El Mecah, is an enclosure, in the centre of which is a well. "All round the well there is a platform or gallery raised six feet. on which a guard of soldiers might walk all round. In the upper part

---

[a]Narrative of a residence in Koordistan, and on the site of ancient Nineveh, by C. J. Rich, Lon 1836. Vol. ii, 26 and 34. [b]An. Egyp. Vol i, 46.

of the wall are holes for discharging arrows." *Fosbrokes' For. Top.* 322. The custom of guarding the roads, especially in the vicinity of tanks and wells, is still common. Fryer in his Travels in India, noticed it. "We found them in arms, not suffering their women to stir out of the town unguarded, to fetch water." Page 126, 222. In Shaw's Travels in Mauritania, he noticed a beautiful rill of water, which flowed into a basin of Roman workmanship, named '*Shrub we Krub*,' *i. e.* "drink and be off," on account of the danger of meeting assassins in its vicinity. Sandys speaks of the "wells of fear." Travels, p. 140.

In ancient Greece, wells were very numerous. The inhabitants of Attica were supplied with water principally from them. Vitruvius remarks, that the other water which they had, was of bad quality. B. viii, Chap. 3. Plutarch has preserved some of the laws of Solon respecting wells. By these it was enacted that all persons who lived within four furlongs of a *public* well, had liberty to use it; but when the distance was greater, they were to dig one for themselves; and they were required to dig at least six feet from their neighbor's ground. Life of Solon. According to Pliny, Danaus sunk the first wells in Greece. Nat. His. vii, 56. Plutarch, in his life of Cimon, says the Athenians taught the rest of the Greeks " to sow bread corn, to avail themselves of the use of wells, and of the benefit of fire " From the connection in which wells are here mentioned, it is evident, that in the opinion of the ancient Greeks, they were among the first of man's inventions; and hence the antiquity of devices to raise water from them. In the mythologic ages, the labor of raising water out of deep wells was imposed as a punishment on the daughters of Danaus, for the murder of their husbands. The daughters of Phaedon (who was put to death by the thirty tyrants) threw themselves into a well, preferring death to dishonor. The body of Chrysippus, son of Pelops, was disposed of in the same way, after being murdered by his brothers, or his step-mother. When Darius sent two heralds to demand earth and water of the Athenians, (the giving of which was an acknowledgment of subjection,) they threw one of them into a *ditch*, and the other into a WELL, telling them in mockery to take what they came for. Plutarch. And Herodotus informs us, that the Lacedemonians treated the Persian ambassadors, who were sent to them on the same errand, in precisely the same manner. Herod. b. viii. 133. These brutal acts led to the invasion of Greece by Xerxes.

Shortly after Alexander's death, Perdiccas and Roxana murdered Statira and her sisters, and had their bodies thrown into a well. Hence, wells were probably common in Babylon as well as in Nineveh; for this was most likely a private one; a public one would scarcely have been selected, where concealment was required. Sir R. K. Porter, in his Travels in Georgia, Persia, Armenia, and ancient Babylon, Vol. i. 698, speaks of the remains of an ancient and "amazing deep well," near Shiraz. Remains of Phenician and Carthagenian wells are still to be seen. Near the ancient *Barca*, Della Cella discovered "wells of great depth, some of which still afford most excellent water."[a] At *Arar*, are others, some of which are excavated through rocks of sandstone. At Arzew, the ancient *Arsenaria*, Dr. Shaw observed a number of wells, "which from the masonry appear to be as old as the city."[b] The celebrated fountain of the sun of the ancients, near the temple of Jupiter Ammon, according to Belzoni, is a WELL sixty feet deep, and eight feet square. (In this case and

---

[a] Russel's Barbary States. [b] Trav. p. 29.

numerous others, the terms "well" and "fountain," are synonymous. "The following is among the first observations of Sir William Gell, after landing on the Troad; "we past many wells on the road, a proof that the country was once more populous than at present.[a] The inhabitants of Ithaca, the birth place of Ulysses and Telemachus, and the scene of some of the principal events recorded in the poetry of Homer, still draw their supplies of water, as in former times, from wells.[b] And as in other places, a tower was anciently erected to guard one of these wells, and protect the inhabitants while drawing water from it.[c]

The ancient Egyptians irrigated the borders of the desert above the reach of the inundations of the Nile, *from* WELLS, which they dug for that purpose.[d] The Chinese also use wells to water their land.

As it regards the antiquity and importance of wells, it has been observed that the earliest account on record of the *purchase* of land, 23 Gen. was subsequent to that of a well, Gen. xxi, 30.

Roman wells are found in every country which that people conquered. Their armies had constant recourse to them, when other sources of water failed, or were cut off by their enemies. Paulus Emilius, Pompey, and Cæsar, often preserved their troops from destruction by having recourse to them. This was strikingly illustrated by Cæsar when besieged in Alexandria; the water in the cisterns having been spoiled by the Egyptians. It was Pompey's superior knowledge in thus obtaining water, which enabled him to overthrow Mithridates, by retaining possession of an important post, which the latter abandoned for want of water. Thus the destinies of these manslayers and their armies, frequently depended on the wells which they made.

The city of Rome, previous to the time of Appius Claudius Cæcus, who first conveyed water to it by an aqueduct, A. U. C. 411, was supplied chiefly from fountains and wells, several of which are preserved to this day. (At Chartres in France, a Roman well is still known as the 'Saints' Well,' on account of martyrs drowned in it by the Romans.)

In noticing the wells of ancient Italy, we may refer to a circumstance, which although trivial in itself, led to the most surprising discovery that had ever taken place on this globe, and one which in the interest it has excited is unexampled. In the early part of the eighteenth century, 1711, an Italian peasant while *digging a* WELL near his cottage, found some fragments of colored marble. These attracting attention, led to further excavation, when a statue of Hercules was disinterred, and shortly afterwards a mutilated one of Cleopatra. These specimens of ancient art, were found at a considerable depth below the surface, and in a place which subsequently proved to be a temple situated in the centre of the ancient city of Herculaneum! This city was overwhelmed with ashes and lava, during an eruption of Vesuvius, A. D. 79, being the same in which the elder Pliny perished, who was suffocated with sulphurous vapors, like Lot's wife in a similar calamity. Herculaneum therefore had been buried 1630 years! and while every memorial of it was lost, and even the site unknown, it was thus suddenly, by a resurrection then unparalleled in the annals of the world, brought again to light; and streets, temples, houses, statues, paintings, jewellery, professional implements, kitchen utensils, and other articles connected with ancient domestic life, were to be seen arranged, as when their owners were actively mov-

---

[a] Top. of Troy, Lon. 1804, p. 5.   [b] Ed. Encyc. Art. Ithaca.
[c] Lard. Arts of the Greeks and Rom. Vol. i, 136.   [d] Wilk. Vol. i, 220.

ing among them. Even the skeletons of some of the inhabitants were found; one, near the threshold of his door, with a bag of money in his hand, and apparently in the act of escaping. The light which this important discovery reflected upon numerous subjects connected with the ancients, has greatly eclipsed all previous sources of information; and as regards some of the arts of the Romans, the information thus obtained, may be considered almost as full and satisfactory, as if one of their mechanics had risen from the dead and described them.

Among the early discoveries made in this city of Hercules, (it having been founded by, or in honor of him, 1250, B. C.) not the least interesting is one of its public wells; which having been covered by an arch and surrounded by a curb, the ashes were excluded. Phil. Trans. xlvii, 151. This well was found in a high state of preservation—it still contains excellent water, and is in the same condition as when the last females retired from it, bearing vases of its water to their dwellings, and probably on the evening that preceded the calamity, which drove them from it for ever.

Forty years after the discovery of Herculaneum, another city overwhelmed at the same time, was "destined to be the partner of its disinterment, as well as of its burial." This was Pompeii, the very name of which had been almost forgotten. As it lay at a greater distance from Vesuvius than Herculaneum, the stream of lava never reached it. It was inhumed by showers of ashes, pumice and stones, which formed a bed of variable depth from twelve to twenty feet, and which is easily removed; whereas the former city was entombed in ashes and lava to the depth of from seventy to a hundred feet. With the exception of the upper stories of the houses, which were either consumed by red hot stones ejected from the volcano, or crushed by the weight of the matter collected on their roofs, we behold in Pompeii a flourishing city nearly in the state in which it existed eighteen centuries ago! The buildings unaltered by newer fashions; the paintings undimmed by the leaden touch of time; household furniture left in the confusion of use; articles even of intrinsic value abandoned in the hurry of escape, yet safe from the robber, or scattered about as they fell from the trembling hand which could not stoop or pause for the most valuable possessions; and in some instances the bones of the inhabitants, bearing sad testimony to the suddenness and completeness of the calamity which overwhelmed them. Pompeii, i, 5. Lib. Entertaining Knowledge. In the prison, skeletons of unfortunate men were discovered, their leg bones being enclosed in shackles, and are so preserved in the museum at Portici.

I noticed, says M. Simond, a striking memorial of this mighty eruption, in the Forum opposite to the temple of Jupiter; a new altar of white marble exquisitely beautiful, and apparently just out of the hands of the sculptor, had been erected there; an enclosure was building all around; the mortar just dashed against the side of the wall, was but half spread out; you saw the long sliding stroke of the trowel about to return and obliterate its own track—but it never did return; the hand of the workman was suddenly arrested; and, after the lapse of 1800 years, the whole looks so fresh, that you would almost swear the mason was only gone to his dinner, and about to come back immediately to finish his work! We can scarcely conceive it possible for an event connected with the arts of former ages, ever to happen in future times, equal in interest to the resurrection of these Roman towns, unless it be the reappearance of the Phenician cities of the plain.

From the facility of removing the materials at Pompeii, much greater advances have been made in uncovering the buildings and clearing the streets, than will probably ever be accomplished in Herculaneum. As might have been expected, several *wells* have been found, besides *rainwater cisterns* and *fountains* in great numbers. The latter were so common, that scarcely a street has been found without one; and every house was provided with one or more of the former.

During the excavations immediately previous to the publication of Sir Wm. Gell's splendid work, 'Pompeiana,' in 1832, a very fine well was discovered near the gate of the Pantheon, 116 feet in depth and containing 15 feet of water![a]

That wells were numerous in Asia and the east generally, we can readily believe, when we learn that some of the most fertile districts, could neither be cultivated nor inhabited without them. Not less than fifty thousand wells were counted in *one* district of Hindostan, when taken possession of by the British; several of which are of very high antiquity. In China, wells are numerous, and often of large dimensions, and even lined with marble. In Pekin they are very common, some of the deepest wells of the world are in this country. M. Arago, (in his Essay on Artesian Wells,) observes that the Chinese have sunk them to the enormous depth of eighteen hundred feet! "Dig a well before you are thirsty," is one of their ancient proverbs. The scarcity of water over all Persia has been noticed by every traveler in that country. In general the inhabitants depend entirely on wells, the water of which is commonly bad. Fryer, xxxv, 67.

To provide water for the thirsty has always been esteemed in the east, one of the most excellent of moral duties, hence benevolent princes and rich men, have, from the remotest ages, consecrated a portion of their wealth to the construction of wells, tanks, fountains, &c. for public use. It is recorded as one of the glories of Uzziah's reign, that he "digged many wells." Over all Persia, there are numerous cisterns built for public use by the rich. Fryer, 225. "Another work of charity among the Hindoos" observes Mr. Ward, "is the digging of pools, to supply the thirsty traveler with water. The cutting of these, and building flights of steps, in order to descend into them, is in many cases very expensive; 4,000 rupees, (2,000 dollars,) are frequently expended on one." At the ceremony of setting it apart for public use, a Brahmin, in the name of the donor, exclaims, "I offer this pond of water to quench the thirst of mankind," after which the owner cannot appropriate it to his own use. Hist. Hindoos, 374.

Ferose, one of the monarchs of India, in the fourteenth century, "built fifty sluices" (to irrigate the land,) and "one hundred and fifty wells." One of the objects, which the fakirs, or mendicant philosophers of India, have frequently in view, in collecting alms, is to 'dig a well,' and thereby atone for some particular sin. Other devotees stand in the roads with vessels of water, and give drink to thirsty travelers from the same motives. Among the supposed causes of Job's affliction, adduced by Eliphaz, was, "thou hast not given water to the weary to drink," xxii, 7: a most horrible accusation in such a country as Syria, and one which that righteous man denied with the awful imprecation, "then let mine arm fall from my shoulder blade, and mine arm be broken from the bone." xxxi, 22.

"The sun was setting," says Mr. Emerson, "as we descended the last chain, and with the departure of daylight, our tortures commenced, as it was too dark to see any of the fountains charitably erected by the Turks near the road."[b] Large legacies are sometimes left by pious Turks for the

---

[a] Pompeiana, Preface.   [a] Letters from the Egean, Let. 5.

erection of fountains, who believe they can do no act more acceptable to God.[a] This mode of expending their wealth, at the same time that it conferred real and lasting benefits on the public, was the surest way of transmitting to posterity the names of the donors. The pools of Solomon, might have preserved his name from oblivion had nothing else respecting him been known. These noble structures, in a land where every other work of art has been hurried to destruction, remain almost as perfect as when they were constructed, and Jerusalem is still supplied with water from them, by an earthen pipe about ten inches in diameter. "These reservoirs are really worthy of Solomon; I had formed no conception of their magnificence; they are three in number, the smallest between four, and five hundred feet in length." The waters are discharged from one into another, and conveyed from the lowest to the city. "I descended into the third and largest; it is lined with plaister like the Indian chunan, and hanging terraces run all round it." Lindsay's Trav. Let. 9.

According to the moral doctrines of the Chinese, "to repair a road, make a bridge, or dig a well," will atone for many sins. Davis' China, ii, 89. The Hindoos, says Sonnerat, believe the digging of tanks on the highways, renders the gods propitious to them; and he adds, "Is not this the best manner of honoring the deity, as it contributes to the natural good of his creatures?" Vol. i, 94.

SUFFERINGS OF TRAVELERS FROM THIRST.

The extreme sufferings which orientals have been, and are still called to endure from the want of water, have been noticed by all modern travelers, from Rubriques and Marco Paulo, to Burckhardt and Niebuhr. Wells in some routes, are a hundred miles apart, and are sometimes found empty; hence travelers have often been obliged to slay their camels for the water these animals retain in their stomachs. *Leo Africanus* noticed two marble monuments in his travels; upon one of which was an epitaph, recording the manner in which those who slept beneath them had met their doom. One was a rich merchant, the other a water carrier, who furnished caravans with water and provisions. On reaching this spot, scorched by the sun and their entrails tortured by the most excruciating thirst; there remained but a small quantity of water between them. The rich man, whose thirst now made him regard his gold as dirt, purchased a single cup of it for ten thousand ducats; but that which possibly might have been sufficient to save the life of one of them, being divided between both, served only to prolong their sufferings for a moment, and they both sunk into that sleep from which there is no waking upon earth. Lives of Travelers, by St. John.

Mr. Bruce, when in Abyssinia, obtained water from the stomachs of camels, which his companions slew for that purpose. Sometimes the mouths and tongues of travelers, from want of this precious liquid, become dry and hard like those of parrots; but these are not the only people who suffer from thirst. During the long continuance of a drought which prevailed over all Judea in Ahab's reign, every class of people suffered. 1 Kings, xvii and xviii. And such droughts are not uncommon. "The poor and needy seek water, and there is none, and their tongue faileth for thirst," (Isa. xli, 17,) in modern times as when the prophet wrote, and not the poor alone, for "the honorable men are famished," and, as well as the multitude, are "dried up with thirst." Isa. v, 13

---

[a] Com. Porter's Letters from Constantinople, i, 101.

Mechanics in cities were not exempt. " The smith with the tongs, both worketh in the coals and fashioneth it with hammers, and worketh it with the strength of his arms, is hungry and his strength faileth, he *drinketh no water and is faint.*" Isa. xliv, 12.

Dr. Ryers, who lived in the city of Gombroon, on the Persian Gulf, when describing the heat of the climate and the deficiency and bad quality of the water, observes that the heat made " the mountains gape, the rocks cleft in sunder, the waters stagnate, to which the birds with hanging wings repair to quench their thirst; for want of which the herds do low, the camels cry, the barren earth opens wide for drink; and all things appear calamitous for want of kindly moisture; in lieu of which hot blasts of wind and showers of sand infest the purer air, and drive not only us, but birds and beasts to seek remote dwellings, or else to perish here;" and after removing to a village some miles distant, "for the sake of water," by a metaphor, that will appear to some persons as bordering on blasphemy, he says, " it was as welcome to our parched throats, as a drop of that cool liquid, to the importunate *Dives.*" Fryer, p. 418. Under similar circumstances, the Hindoos, night and day run through the streets, carrying boards with earth on their heads, and loudly repeating after the Brahmins, a prayer, signifying " God give us water." Even in Greece and Rome, where water was in comparative abundance, agricultural laborers considered the Frog an object of envy, inasmuch as it had always enough to drink in the most sultry weather. Lard. Arts Greeks and Rom. Vol. ii, 20. The ignorant and clamorous Israelites, enraged with thirst, abused Moses, and were ready to stone him, because they had no water.

One of the most appalling facts that is recorded of suffering from thirst occurred in 1805. A caravan proceeding from Timboctoo to Talifet, was disappointed in not finding water at the usual watering places; when, horrible to relate, all the persons belonging to it, two thousand in number, besides eighteen hundred camels, *perished by thirst!* Occurrences like this, account for the vast quantities of human and other bones, which are found heaped together in various parts of the desert. Wonders of the World, p. 246. While the crusaders besieged Jerusalem, great numbers perished of thirst, for the Turks had filled the wells in the vicinity. Memorials of their sufferings may yet be found in the heraldic bearings of their descendants. The charge of a foraging party 'for water,' we are told, "was an office of distinction;" hence, some of the commanders on these occasions, subsequently adopted *water buckets* in their coats of arms, as emblems of their labors in Palestine. 'Water Bougettes,' formed part of the arms of Sir Humphrey Bouchier, who was slain at the battle of Barnet, in 1471. Moules' Ant. of Westminster Abbey.

CHAPTER V.

Subject of WELLS continued—Wells worshipped—River Ganges—Sacred well at Benares—Oaths taken at Wells—Tradition of the Rabbins-Altars erected near them—Invoked-Ceremonies with regard to water in Egypt, Greece, Peru, Mexico, Rome, and Judea---Temples erected over wells---The fountain of Apollo---Well Zem Zem---Prophet Joel---Temple of Isis---Mahommedan Mosques---Hindoo temples ---Woden's well---Wells in Chinese temples---Pliny---Celts --Gauls---Modern superstitions with regard to water and wells---Hindoos---Algerines---Nineveh---Greeks---Tombs of saints near wells---Superstitions of the Persians---Anglo Saxons---Hindoos---Scotch---English---St. Genevieve's well---St. Winifred's well---House and well 'warming.'

In the early ages *water* was reverenced as the substance of which all things in the universe were supposed to be made, and the vivifying principle that animated the whole; hence, rivers, fountains, and wells, were worshipped and religious feasts and ceremonies instituted in honor of them, or of the spirits which were believed to preside over them. Almost all nations retain relics of this superstition, while in some it is practised to a lamentable extent. Asia exhibits the humiliating spectacle of millions of her people degraded by it, as in former ages. Shoals of pilgrims are constantly in motion over all Hindoston, on their way to the 'sacred Ganges;' their tracks stained with the blood and covered with the bones of thousands that perish on the road. With these people, it is deemed a virtue even to *think* of this river; while to bathe in its waters washes away all sin, and to expire on its brink, or be suffocated in it, is the climax of human felicity. The holy WELL in the city of Benares is visited by devotees from all parts of India; to it they offer rice, &c. as to their idols.

From this sacred character of water, it very early became a custom, in order to render obligations inviolable, to take oaths, conclude treaties, make bargains, &c. at wells. We learn that when Jacob was on his way to Egypt, he came to the "well of the oath," and offered sacrifices to God. Josephus, Ant. ii, 7. At the same well, his grandfather Abraham concluded a treaty with Abimelech, which was accompanied with ceremonies and oaths. Gen. xxi. At the celebrated *Puteal Libonis*, at Rome, oaths were publicly administered every morning; a representation of this well is on the reverse of a medal of Libo. Encyc. Ant. 412. It was believed that the "oaths of the Gods" was also by water. Univer. His. Vol. iv, 17. The Rabbins have a tradition that their kings were always anointed by the side of a fountain. Solomon was carried by order of David to the 'fountain of Gihon,' and there proclaimed king. Joseph. Ant. vii, 14.

The ancient Cuthites, says Mr. Bryant, and the Persians after them, had a great veneration for fountains and streams. ALTARS were erected in the vicinity of wells and fountains, and religious ceremonies performed around them. Thus Ulysses:

> Beside a fountain's sacred brink, we raised
> Our verdant altars, and the victims blazed. *Iliad* ii, 368.

"Wherever a spring rises, or a river flows," says Seneca, "there we should build altars and offer sacrifices," and a thousand years before Seneca lived, the author of the 68th Psalm spoke of worshipping God from the "fountains of Israel." The Syracusans held great festivals every year at the fountains of Aretnusa, and they sacrificed black bulls to Pluto at the fountain of Cyane. Wells were sometimes dedicated to particular deities, as the oracular fountain mentioned by Pausanias, near the sea at Patra, which still remains nearly as he described it; and having been re-dedicated to a christian saint, "is still a sacred WELL." Divination by water, was practised at this well. A mirror was suspended by a thread, having its polished surface upwards, and while floating on the water, presages were drawn from the images reflected.

Polynices, in Œdipus Coloneus, swears "by our native fountains and our kindred gods." Antigone, when about to be sacrificed, appeals to the "fountains of Dirce, and the grove of Thebe." Ajax before he slew himself, called on the sun, the soil of Salamis, and "ye fountains and rivers here." Trag. of Sophocles lit. trans. 1837.

"At Peneus' fount Aristeus stood and bowed with woe,
Breathed his deep murmurs to the nymph below: *Georgics* L. iv, 365.
Cyrene! thou whom these fair springs revere."

The fountain of Aponeus, (now Albano) the birth place of Livy, was an oracular one. That of Pirene at Corinth, was sacred to the muses. Eneas invoked "living fountains" among other "Ethereal Gods." And old Latinus

" Sought the shades renowned for prophecy,
Which near Albuneas' sulphureous fountain lie." En. vii, 124.

Cicero says, the Roman priests and augurs, in their prayers, called on the names of rivers, brooks, and springs.

Vessels of water were carried by the Egyptian priests in their sacred processions, to denote the great blessings derived from it, and that it was the beginning of all things. Vitruvius says they were accustomed to place a vase of it in their temples with great devotion, and prostrating themselves on the earth, returned thanks to the divine goodness for its protection. Book viii, Proem. In the celebration of the Eleusinian mysteries, those who entered the temple, washed their hands in holy water, and on the ninth and last day of the festival, vessels of water were offered with great ceremonies, and accompanied with mystical expressions to the Gods. Those who were initiated were prohibited from ever sitting on the cover of a well. Sojourners among the Greeks carried in the religious processions, small vessels formed in the shape of boats; and their daughters *water pots* with umbrellas. Rob. Ant. Greece. Plutarch says, "*fishes* were not eaten of old, from reverence of springs."

Among the ancient Peruvians, certain Indians were appointed to sacrifice "to fountains, springs, and rivers." Pur. Pil. 1076. Holy water was placed near the altars of the Mexicans. Ibid, 987. Tlaloc was their God of water; on fulfilling particular vows they bathed in the sacred pond Tezcapan. The water of the fountain Toxpalatl was drank only at the most solemn feasts: no one was allowed to taste it at any other time. Clavigero, Lon. 1786, vol. i, 251 and 265. The *Fontinalia* of the Romans, were religious festivals, held in October, in honor of the Nymphs of wells and fountains; part of the ceremonies consisted in throwing nosegays into fountains, and decorating the curbs of wells with wreaths of flowers.

The Jews had a religious festival in connection with water, the origin of which is not clearly ascertained. It was kept on the last day of the feast of tabernacles, when they drew water with great ceremony from the pool of Siloah and conveyed it to the temple.[a] It is supposed, the Sa-

---

[a] Uni.-Hist. i, 607.

vior alludes to this practice, when on "the last day, that great day of the feast, he stood and cried, saying, if any man thirst, let him come unto me, and drink. He that believeth on me, as the scripture hath said, out of his belly shall flow rivers of living waters." John, vii, 37. One of the five solemn festivals of the people of Pegu, is 'the feast of water,' during which, ' the king, nobles and all the people throw water upon one another.' Ovington's Voy. to Surat. 1689. 597. The superstitious veneration for wells, induced the ancients to erect temples near, and sometimes over them; as the fountain of Apollo, near the temple of Jupiter Ammon; the well Zemzem in the temple of Mecca, &c. In accordance with this prevailing custom, we find the prophet Joel speaks of a fountain which should come forth *out* of the *house of the Lord*, and water the valley. iii, 18. And when Jeroboam built a temple, that the ten tribes might not be obliged to go to Jerusalem to worship, and there be seduced from him, Josephus tells us, that he built it by the fountains of the lesser Jordan. Antiq. viii, cap. 8. In the temple of Isis, at Pompeii, the 'sacred well' has been found. Pompeii, i, 277, 279.

The ancient custom of enclosing wells in religious edifices was adopted by both Christians and Mahommedans. Among the latter it is still continued, and it is not altogether abandoned by the former.

"This afternoon," says Fryer, speaking of one of the mosques in India, "their *sanctum sanctorum* was open, the priest entering in barefoot, and prostrating himself on one of the mats spread on the floor, whither I must not have gone, could his authority have kept me out. The walls were white and clean but plain, only the commandments wrote in Arabic at the west end, were hung over a table in an arched place, where the priest expounds, on an ascent of seven steps, railed at top with stone very handsomely. Underneath are fine cool vaults, and stone stairs to descend to a *deep tank*."

As it was formerly death to a christian who entered a mosque, we shall add a more recent instance. In 1831, Mr. St. John disguised himself, like Burckhardt, in the costume of a native, and visited the mosques of Cairo. In that of Sultan Hassan, he observes, "ascending a long flight of steps, and passing under a magnificent doorway, we entered the vestibule, and proceeded towards the *most sacred* portion of the edifice, where, on stepping over a small railing, it was necessary to take off our babooshes, or red Turkish shoes. Here we beheld a spacious square court, paved with marble of various colors, fancifully arranged, with a beautiful octagonal *marble fountain* in the centre." Egypt and Mohammed Ali, ii, 338. It is the same in Persia. Tavern. Trav. Lon. 1678. 29. The temples of India says Sonnerat, have a sacred tank, deified by the Brahmins The figures of gods are sometimes thrown 'into a tank or well.' Voy. i, 111, 132. In old times, churches were removed from other buildings, and were surrounded with courts, in the centre of which there were fountains, where people washed before going to prayers. Moreri Dic. In one of the old churches at Upsal, is an ancient well, that had formerly been famous 'for its miraculous cures.' Woden's well is still shown in the same city. It was in the vicinity of the old temple of that great northern deity. De la Mortraye's Trav. ii, 262. Van Braam noticed a well in one of the large temples of China. Journ. ii, 224. 'Sacred springs,' are mentioned by Juvenal. 3 Sat. 30. Pliny speaks of fountains and wells of water as very 'wholesome and proper for the cure of many diseases;' to which, he says, there is ascribed some divine power, insomuch that they give names to sundry gods and goddesses, xxxi, 2. The Celts venerated lakes, rivers, and fountains, into which they threw gold.

The Britons and Picts did the same. Scot. Gael, 258. Mezeray, in his History of France, when speaking of the church in the third and fourth centuries, remarks, 'Hitherto very few of the French had received the light of the gospel; they yet adored trees, *fountains*, serpents, and birds.' i, 4. In the eighth century, the council of Soissons condemned a heretic, who built oratories and set up crosses near fountains, &c. Ib. 113.

Ancient superstitions with regard to water are still practised more or less over a great part of the world. At the first new moon in October, the Hindoos hold a great celebration to their Deities. "The next moon, their women flock to the *sacred wells*." Fryer, 110. Many of the ceremonies performed in old times by women in honor of wells and fountains, are yet practised in some of the Grecian islands. There the females still dance round the wells, the ancient *Callichorus*, accompanied with songs in honor of Ceres. *Dr. Clarke.* "I have just returned this morning," (says Mr. Campbell in his Letters from the South, Phila. Ed. 1836, 102,) "from witnessing a superstitious ceremony, which, though unwarranted by the Koran, is practised by all Mahometans here, [Algiers] black, brown, and white, nay by the Jews also. It consists in sacrificing the life of some eatable animal to one of the devils who inhabit certain fountains near Algiers. The victims were fowls, they were dipped in the *sacred* sea, as Homer calls it, after which the high priest took them to a neighboring fountain, and having waved his knife thrice around the head of an old woman, who sat squatting beside it, cut their throats," &c.

The custom was probably a common one in ancient Nineveh; for once a year the peasants assemble and sacrifice a *sheep* at Thisbe's well, with music and other festivities. The Greeks are so much attached to grottoes and wells, that "there is scarcely one in all Greece and the islands, which is not consecrated to the Virgin, who seems to have succeeded the ancient nymphs in the guardianship of these places.[a]

The supposed sanctity of wells also led to the custom of interring the bodies of saints or holy persons near them; thus in all parts of Egypt, the tombs of saints are found in the vicinity of those places, "where the wandering dervishes stop to pray, and less pious travelers to quench their thirst." Some, says Fryer, are buried with " their heels upwards, like Diogenes."

Worship of wells, like many other superstitions of Pagan origin, was early incorporated with the ceremonies of the christian church, and carried to an idolatrous excess. A schism took place in Persia among the Armenians, in the tenth century; one party was accused of 'despising the holy well of Vagarsciebat.' In Europe it was at one time universal. In England, in the reigns of Canute and Edgar, edicts were issued prohibiting well worship. When Hereward the Saxon hero, held the marshes of Ely against the Norman conqueror, he said he heard his hostess conversing with a witch at midnight! he arose silently from his bed, and followed them into the garden, to a 'fountain of water,' and there he 'heard them holding converse with the spirit of the fountain.' From a collection of Anglo Saxon remains, the following example is taken. "If any one observe lots or divinations, or keep his wake, [watch] at any *wells*, or at any other created things, except at God's church, let him fast three years; the first one on bread and water," &c. In a Saxon homily against witchcraft and magic, in the library of the University of Cambridge, it is said, "some men are so blind, that they bring their offerings to immovable *rocks*, and also to *trees* and to *wells*, as witches do teach." [b]The Hindoos still wor-

---

[a] Rich's Nar. of a Residence in Koordistan. ii, 42. [b] Foreign Quarterly. July, 1838.

Chap. 6.] *Depth of Wells.* 37

ship stones, trees, and water, and make offerings to them.[a] In a manuscript written in the early part of the fifteenth century, there is a humorous song, in which there is an allusion to this superstition. It begins thus:

'The last tyme I the *wel* woke
Sir John caght me with a croke,
He made me swere be bel and boke
I shuld not tel.'

Even so late as the seventeenth century, people in Scotland were in the habit of visiting wells, at which they performed numerous acts of superstition. Shaw, in his History of the Province of Moray, says that 'heathen customs were much practised among the people,' and among them, he instances their 'performing pilgrimages to wells,' and 'building chapels to fountains.'[b] At the present time in some parts of England, remains of well worship are preserved, in the custom of performing annual processions to them, decorating them with wreaths and chaplets of flowers, singing of hymns, and even reading a portion of the gospel as part of the ceremonies.

These same customs gave rise to the numerous *holy* wells, which formerly abounded throughout the old world, and the memory of many of which is still preserved in names of towns. In the church of *Nanterre*, near Paris, the birth place of Saint Genevieve, is a well, by the water of which, this patroness of the Parisians miraculously restored her blind mother and many others to sight! Breval's Eu. 307. Saint Winifred's well in Flintshire, Eng. from its sacred character gave name to the town of Holywell. Mr. Pennant says, the custom of visiting this well in pilgrimage, and offering up devotions there, was not in his time entirely laid aside: "in the summer, a few are to be seen in the water, in deep devotion up to their chin for hours, sending up their prayers, or performing a number of evolutions round the polygonal well." Even so late as 1804, a Roman catholic bishop of Wolverhampton, took much pains to persuade the world, that an ignorant proselyte of his, named Winifred White was miraculously cured at this well of various chronic diseases!

The custom of 'house-warming' is very ancient; the *same ceremonies*, were formerly performed on the completion of new wells.

CHAPTER VI.

Wells continued: Depth of ancient wells—In Hindostan—Well of Tyre—Carthagenian wells—Wells in Greece, Herculaneum and Pompeii—Wells without curbs—Ancient laws to prevent accidents from persons and animals falling into them—Sagacity and revenge of an elephant—Hylas—Archelaus of Macedon—Thracian soldier and a lady at Thebes—Wooden covers—Wells in Judea—Reasons for not placing curbs round wells—Scythians—Arabs—Aquilius—Abraham—Hezekiah—David—Mardonius—Moses and the people of Edom—Burckhardt in Petra—Woman of Bahurim—Persian tradition—Ali, the fourth Caliph—Covering wells with large stones—Mahommedan tradition—Themistocles—Edicts of Greek emperors—Well at Heliopolis—Juvenal—Roman and Grecian curbs of marble—Capitals of ancient columns converted into curbs for wells.

A knowledge of the depth and other circumstances, relating to some ancient wells, is necessary to a due investigation of the various methods of raising water from them. We cannot indeed form a correct judgment of the latter, without some acquaintance with the former.

---

[a] Ward's Hindoos, 342, 352.   [b] Hone's Every Day Book, ii, 636, 685. Fosbroke, 684.

The wells of Asia are generally of great depth, and of course were so in former times. In Guzzerat, they are from eighty to a hundred feet; in the adjoining province of Mulwah, they are frequently three hundred feet. In Ajmeer, they are from one to two hundred feet. Mr. Elphinstone in his mission to Cabaul observes, 'the wells are often three hundred feet deep; one was three hundred and forty five;' and with this enormous depth, some are only three feet in diameter. The famous well of ancient Tyre, 'whose merchants were princes, and whose traffickers were the honorable of the earth,' is, according to some travelers, without a bottom; but La Roque, is said by Volney, to have found it at the depth of 'six and thirty fathom.'

Shalmanezer besieged this city of *mechanics* for five years, without being able to take it; at last he cut off the waters of this well, when the inhabitants dug others within the city; after which they held out against Nebuchadnezzar, and the whole power of the Babylonian empire for *thirteen* years; being the longest siege on record, except that of Ashdod. Jos. Antiq. ix, 14. Ancient Carthagenian wells of great depth have been already mentioned. Dr. Shaw (Trav. 135,) observes of a tribe of the Kabyles, ' their country is very dry, they have no fountains or rivulets, and in order to obtain water, they dig wells ' to the depth of from one to two hundred fathom.' Jacob's well is a hundred and nine feet, and Joseph's well at Cairo, near three hundred feet deep. The well Zemzem at Mecca, is two hundred and ten feet. 'Exceeding deep wells' in Surat, are mentioned by Toreen, in Osbeck's Voyage to China. That the wells of Attica were generally deep, is obvious from a provision in Solon's law respecting them, by which a person, after digging to the depth of sixty feet without obtaining water, was allowed to fill a six gallon vessel twice a day at his neighbor's well. The frequency of not meeting with water at that depth, evidently gave rise to this provision.[a] The wells of Herculaneum and Pompeii, were probably all of considerable depth, if we judge from those that have been discovered.

WELLS WITHOUT CURBS.

Another feature in ancient—particularly Asiatic—wells, was, they were often *without curbs* or parapets built round them; hence animals often fell into them and were killed. A very ancient law enacted, that, 'if a man shall open or dig a pit, [a well] and not cover it; and an ox or an ass fall therein, the owner of the pit shall make it good, and give money to the owner of them, and the dead beasts shall be his.' Exo. xxi, 33, 34. This was probably an old Phenician and Egyptian law which the Israelites adopted from its obvious utility. Josephus' account of it is more explicit: 'let those that dig a WELL or a pit, be careful to lay planks over them, and so keep them shut up, not to hinder persons from drawing water, but that there may be no danger of falling into them.' Antiq. iv, 8. Numerous examples of the utility of such a law might be produced from oriental histories. Benaiah, one of the three famous warriors of David, who broke through the hosts of the Philistines and drew water for him out of the well of Bethlehem, 'slew a lion in the midst of a pit in the time of snow.' Sam. xxiii, 20 : from Josephus, this appears to have been one of the ordinary wells of the country, which having no curb, had been left open, and the 'lion slipped and fell into it.' Antiq. vii, 12.

'On our way back to the town, we saw a poor ass dying in a pit, into

---

[a] Plutarch's Life of Solon.

Chap. 6.] *Wells without Curbs.* 39

which he had fallen with his legs tied, that being the practice of the Arabs when they send out these animals to feed.'[a] The custom of the Arabs in this respect has probably, like many others, undergone no change. It explains the necessity of the law in Exodus, as quoted above.

As two elephant drivers, each on his elephant, one of which was remarkably large and powerful, and the other small and weak, were approaching a well, the latter carried at the end of his proboscis a bucket by which to raise the water. The larger animal instigated by his driver, (who was not provided with one,) seized and easily wrested it from the weaker elephant, which, though unable to resent the insult, obviously felt it. At length, watching his opportunity when the other was standing amid the crowd with his side to the well, he retired backwards in a very quiet and unsuspicious manner, and then rushing forward with all his might, drove his head against the side of the robber, and fairly pushed him into the well—the surface of the water in which, was twenty feet below the level of the ground.

But animals were not the only sufferers:—There are passages in ancient authors which indicate the loss of human life both accidentally and by design, in consequence of the absence of curbs to wells. Thus Hylas who accompanied Hercules on the Argonautic expedition, went ashore to draw water from a well or fountain, and he fell in and was drowned. Virgil represents the companions of Hylas after missing him, as spreading themselves along the coast and loudly repeating his name:

> And Hylas, whom his messmates loud deplore,
> While Hylas! Hylas! rings from all the shore.
> *Ec.* vi, 48. *Wrangham.*

Archelaus of Macedon, a contemporary of Socrates, ascended the throne by the most horrid crimes. Among others whom he murdered, was his own brother, a boy only seven years old. He threw his body into a well, and endeavored to make his mother believe that the child fell in, 'as he was running after a goose.' *Bayle.*

When Alexander, like a demon, destroyed the city of Thebes, (the capital of one of the States of Greece,) and murdered six thousand of its inhabitants, a party of Thracian soldiers belonging to his army demolished the house of Timoclea, a lady of distinguished virtue and honor. The soldiers carried off the booty, and their captain having violated the lady, asked her, if she had not concealed some of her treasures: she told him she had, and taking him alone with her into the garden, she showed him a well, into which she said she had thrown every thing of value. Now we are told, that as he stooped down to look into the well, this high spirited and much injured lady pushed him in, and killed him with stones.[b]

From these accounts, it appears that wells belonging to private houses in ancient Greece, were sometimes without curbs, although they probably had portable or *wooden* covers. That these were common, is evident from a passage already quoted from Josephus; and the remains of one have been discovered in Pompeii.[c] The private well mentioned in 2 Sam. xvii, 18, had no curb. Indeed it is evident from the New Testament, that the ancient custom of leaving the upper surface of wells level with the ground, prevailed among the Jews, through the whole of their history, from their independence as a nation, to their final overthrow by Titus. ' What man among you having one sheep, if it *fall into* a *pit* on the sabbath day, will

---

[a] St. John's Egypt, i, 354. [b] Plutarch's Life of Alexander. [c] Pompeii, ii, 204.

he not lay hold of it and lift it out?' Matt. xii, 11. And again in Luke, 'which of you shall have an ass, or an ox fallen into a *pit*, and will not straightway pull him out on the sabbath day.' xiv, 5.

In these passages, which are parallel to those quoted from Exodus and Josephus, the word 'pit' is synonymous with 'well.' In Antiq. vii, 12. 'the well of Bethlehem,' is called a 'pit.' Wells without curbs are met with in Judea and the east generally, at the present time, although they are not so numerous as formerly. Mr. Stephens, in his 'Incidents of Travel,' observed on the road to Gaza, 'two remarkable wells of the very best Roman workmanship, about fifty feet deep, lined with large hard stones, as firm and perfect as on the day on which they were laid; the uppermost layer on the top of the well, 'was on a *level* with the pavement.[a] In some illustrations of the Book of Genesis, executed in the fourth or fifth century, one represents the interview between Rebecca and Eliezer; the well is square, and the curb but a few inches high.

REASONS FOR NOT PLACING CURBS ROUND THE MOUTHS OF WELLS.

The motives which induced the ancients to leave their wells without curbs were various:

1. That they might be more readily *concealed*. This was a universal custom in times of war. When Darius invaded Scythia, the inhabitants did not attempt an open resistance, but covered up their wells and springs and retired. Herod. iv, 120. Mr. Elphinstone, in his mission to Cabaul, says, the people 'have a mode of covering their wells with boards, heaped with sand, that effectually conceals them from an enemy.' Diodorus Siculus, remarked the same of the Bedouin Arabs, eighteen centuries ago, and they still practise it. Travelers in the Lybian desert are often six and seven days without water, and frequently perish for want of it; 'the drifting sand having covered the *marks* of the wells.'[b] Wells, when thus concealed 'can only be found by persons whose profession it is to pilot caravans across this ocean of sand, and the sagacity with which these men perform their duty is wonderful;' like pilots at sea with nothing but the stars to direct them.

2. To prevent them from being poisoned or filled up, both of which frequently occurred. The Roman General Aquilius conquered the cities of the kingdom of Pergamus, one by one, by poisoning the waters. This horrid crime has always prevailed. In 1320, many Jews were burnt in France, while others were massacred by the infuriate people, under the belief that they had poisoned the wells and fountains of Paris. The Earl of Savoy was poisoned in this manner in 1384, and the practice was common in the fifteenth century.[c] Some of the wells belonging to Abraham, were stopped up by the inhabitants. 'And Isaac digged again the wells of water, which they had digged in the days of Abraham his father, for the Philistines had stopped them, after the death of Abraham.' Gen. xxvi, 18. 'We walked on some distance to a well, which we found *full of sand*; Hussein scooped it out with his hands, when the water rose and all of us drank.' Lindsay's Trav. Let. 7. When the Assyrians under Senacherib, invaded Judea in the eighth century, B. C. 'Hezekiah took counsel with his princes and mighty men, to stop the waters of the fountains which were *without* the city· and they stopped all the fountains, saying, why should the king of Assyria come here and find much water?' 2 Kings, iii, 19. 25.

---

[a] Vol. ii, 101, and Lindsay's Trav. Let. 9. [b] Ogilvy's Africa, 281.
[c] Mezeray's France. Lon. 1683. pp. 349, 408, 414.

The custom of leaving the principal supply of water without the walls of the more ancient cities, is remarkable; and the reason for it has not yet been satisfactorily explained. The water which supplied Alba Longa, lay in a very deep glen, and was therefore scarcely defensible; but the springs of the Scamander at Troy, of Enneacrunus at Athens; of Dirce at Thebes, and innumerable others, prove that such instances were common.[a] When David waged war against the Ammonites, his success, according to Josephus, was chiefly owing to his general cutting off their waters, and especially those of a particular well. Antiq. vii, 1. Mardonius stopped up the Gargaphian fountain, which supplied the Grecian army with water, an act which brought on in its vicinity, the famous battle of Platea, in which he was slain, and the power of Persia in Greece finally prostrated. A remarkable instance of the labor and perseverance of ancient soldiers, in cutting off a well or fountain from besieged places, is given by Cæsar in his Commentaries on the War in Gaul. viii, 33.

3. To prevent the water from being *stolen;* which could scarcely have been prevented at wells with curbs, for they could not then have been concealed. We must bear in mind that the extreme scarcity of water in the east, required a vigilant and parsimonious care of it; and hence continual quarrels arose from attempts to purloin it, or to take it by force. ' And the herdsmen of Gerar did strive with Isaac's herdsmen, saying, the water is ours.' Gen. xxvi, 20. This kind of strife, says Dr. Richardson, between the different villagers, still exists, as it did in the days of Abraham and Lot. It was customary for shepherds to seize on the wells before others came, lest there should not be sufficient water for all their flocks, and it was at an occurrence of this kind, that Moses first became acquainted with Zipporah and her sisters. Jos. Antiq. ii, 11. " Nearly six hours beyond the ruined town of Kournou, and two beyond the dry bed of a small stream called El Gerara, [the brook of *Gerar?*] we were surprised at finding two large and deep wells, beautifully built of hewn stone. The uppermost course, and about a dozen troughs for watering cattle disposed round them, of a coarse white marble; they were evidently coeval with the Romans. Quite a patriarchal scene presented itself as we drew near to the wells; the Bedouins were watering their flocks; two men at each well letting down the skins and pulling them up again, with almost ferocious haste, and with quick savage shouts." Lindsay's Trav. Let. 9.

The scarcity of water in those countries has from the remotest times made it an object of *merchandise.*—" Ye shall also buy water of them for money that ye may drink." Deut. ii, 6, 28. And Jeremiah—" we have drunken our water for money." Lam. v, 4. See Ezekiel, iv, 16, 17. This value of water may be perceived in the negotiation of Moses with the king of Edom, for a passage through that country. He pledged himself that his countrymen would not injure the fields or the vineyards; "neither," says he, "will we drink of the waters of the wells;" and in a subsequent proposition, he adds, " if I and my cattle drink of thy waters, then I will *pay* for it." Num. xx, 17, 19. It is we think evident from the text, that the great quantities of water which such a host would require, was the principal objection urged by the people of Edom; they were afraid, and very naturally too, that a million of souls might drain all their wells while passing through the land, a calamity that might prove fatal to themselves. Brooks and rivers, were dried up by the army of Xerxes as he advanced towards Greece.

It may be observed here, that when in 1811, Burckhardt discovered

---

[a] Gell's Topography of Rome, i, 34.

Petra, the long lost capital of Edom, an intense interest was excited among the learned men of Europe, and several hastened to behold the most extraordinary city of the world; a city excavated out of the rocks, whose origin goes back to the times of Esau, the 'father of Edom,' and which had for more than a thousand years, been completely lost to the civilized world. But the natives swore, as in the times of Moses, they should not enter their country, nor *drink* of their *water*, and they threatened to shoot them like dogs, if they attempted it. It was with much difficulty and danger, that Burckhardt at length succeeded in obtaining a glimpse of this singular city. He was disguised as an Arab, and passed under the name of Sheik Ibrahim. The difficulty and danger of a visit to Petra, is now however in a great measure removed by the present Pasha, Mahommed Ali.

From the custom of concealing many ancient wells, we learn the important fact, that machines for raising the water could not have been *attached* to, or *permanently* placed near them. As these, as well as curbs or parapets projecting above the ground, would have betrayed to enemies and strangers their location. When the woman at Bahurim secreted David's spies in the well belonging to her house, and " spread a covering over the well's mouth, and spread ground corn thereon;" 2 Sam. xvii, 19, her device could not have succeeded, if a curb had enclosed its mouth, or if any permanent machine had been erected to raise the water from it; as these would have indicated the well to the soldiers of Absalom, who would certainly have examined it, because wells were frequently used as hiding places in those days. There is a tradition in Persia that one of the Armenian patriarchs, was concealed several years in a well, during the persecution of the Christians under Dioclesian and Maximinian; and was 'privately relieved by the daily charity of a poor godly woman.' Fryer, 271.

When Ali the fourth Caliph of the Arabians, marched with ninety thousand men into Syria, the army was in want of water. An old hermit, whose cell was near the camp, was applied to; he said he knew but of one cistern, which might contain two or three buckets of water. The Caliph replied that the *ancient patriarchs* had dug wells in that neighborhood. The hermit said there was a tradition of a well whose mouth was closed by a stone of an enormous size, but no person knew where it was. Ali caused his men to dig in a spot which he pointed out, and not far from the surface, the mouth of the well was found.[a]

Where wells were too well known to be concealed, as those in the neighborhood of towns, villages, &c. they were sometimes secured by large stones placed over them, which required the combined strength of several persons to remove. 'A *great* stone was upon the well's mouth; and they rolled the stone from the well's mouth and watered the sheep, and put the stone again upon the well's mouth.' Gen. xxix, 2, 3. The Mahommedans have a tradition that the well at which Moses watered the flocks of his father-in-law, was covered by a stone which required several men to remove it. It is indeed obvious large stones only could have been used, for small ones could not extend across the wells, which were frequently of large diameter. Jacob's well is nine feet across, and some were larger The curb round the well Zemzem at Mecca, is *ten feet* in diameter. " Another time we passed an ancient well," says Lindsay, Let. 10, " in an excursion from Jerusalem to Jericho and the Dead Sea, its mouth sealed with a large stone, with a hole in the centre, through which

---

[a] Martigny's History of the Arabians, ii, 49.

Chap. 6.]   *Roman and Grecian Curbs.*   43

we threw a pebble, but there was no water, and we should have been sorry had there been any, for our *united strength* could not have removed the seal."

Notwithstanding the precautions used, shepherds were often detected in fraudulently watering their flocks at their neighbors' wells, to prevent which, *locks* were used to secure the covers. These continued to be used till recent times. M. Chardin noticed them in several parts of Asia. The wells at Suez, according to Niebuhr, are surrounded by a strong wall to keep out the Arabs, and entered by a door 'fastened with enormous clamps of iron.' In Greece as in Asia, those were *fined* who stole water. When Themistocles during his banishment was in Sardis, he observed in the temple of Cybele a female figure of brass, called '*Hydrophorus*' or Water Bearer, which he himself had caused to be made and dedicated out of the *fines* of such as had *stolen* the water, or diverted the stream.[a] One of the Greek emperors of Constantinople issued an edict A. D. 404, imposing a fine of a *pound* of gold for every *ounce* of water surreptitiously taken from the reservoirs.[b] And a more ancient ruler remarked that '*stolen waters* are sweet.' Proverbs, ix, 17. The ancient Peruvians had a similar law.

Curbs or parapets were generally placed round the mouths of wells in the *cities* of Greece and Rome, as appears from many of them preserved to the present time, as well as those discovered in Pompeii and Herculaneum. The celebrated mosaic pavement at Preneste, contains the representation of an ancient well; by some authors supposed to be the famous fountain of Heliopolis. Montfaucon and Dr. Shaw have given a figure of it. The curb is represented as built of brick or cut stone. Curbs were generally massive cylinders of marble and mostly formed of *one block*, but sometimes of two, cramped together with iron. Their exterior resembled round altars. Those of the Greeks were ornamented with highly wrought sculptures and were about twenty inches high. Roman curbs were generally plain, but one has been found in the street of the Mercuries at Pompeii, beautifully ornamented with triglyphs. To these curbs Juvenal appears to allude :

<blockquote>
Oh! how much more devoutly should we cling
To thoughts that hover round the sacred spring,
Were it still margined with its native green,
And not a marble near the spot were seen. Sat. iii, 30 *Badham.*
</blockquote>

That Roman wells were generally protected by curbs, appears also from a remark of the elder Pliny : " at Gades the fountain next to the temple of Hercules, is enclosed about like a well." B. ii, 97. Dr. Shaw mentions several Roman wells with corridors round, and cupolas over them, in various parts of Mauritania. Trav. 237. Mr. Dodwell describes the rich curb of a Corinthian well, ten figures of divinities being carved on it. Such decorations he says were common to the sacred wells of Greece.

In various parts of Asia and Egypt, the finest columns have been broken and hollowed out to serve as curbs to wells; and in some instances, the *capitals* of splendid shafts may be seen appropriated to the same purpose. Although such scenes are anything but pleasant to the enlightened traveler, the preservation of valuable fragments of antiquity has been secured by these and similar applications of them. They certainly are less subject to destruction, as curbs of wells, than when employed, like the fine Corinthian capital of Parian marble, which Dr. Shaw observed at Arzew, ' as a block for a blacksmith's *anvil*.' Trav. 29, 30.

---

[a] Plutarch's Life of Themistocles.   [b] Hydraulia, p. 232. Lon. 1835.

CHAPTER VII.

Wells concluded: Description of Jacob's well—Of Zemzem in Mecca—Of Joseph's well at Cairo—Reflections on wells—Oldest monuments extant—Wells at Elim—Bethlehem—Cos—Scyros—Heliopolis—Persepolis—Jerusalem—Troy—Ephesus—Tadmor—Mizra—Sarcophagi employed as watering troughs—Stone coffin of Richard III used as one—Ancient American wells—Indicate the existence in past times of a more refined people than the present red men—Their examination desirable—Might furnish (like the wells at Athens,) important data of former ages.

A description of some celebrated wells may here be inserted, as we shall have occasion to refer to them hereafter. *Jacob's well*, is one of the most ancient and interesting. Through a period of thirty-five centuries it has been used by that patriarch's descendants, and distinguished by his name. This well is, as every reader of scripture knows, near Sychar, the ancient *Shechem*, on the road to Jerusalem, and has been visited by pilgrims in all ages. Long before the christian era, it was greatly revered, and subsequently it has been celebrated on account of the interview which the Savior had with the woman of Samaria near it. Its location according to Dr. Clarke is so distinctly marked by the Evangelist, and so little liable to uncertainty from the circumstances of the well itself, and the features of the country, that if no tradition existed for its identity, the site of it could hardly be mistaken.

The date of its construction may, for aught that is known to the contrary, extend far beyond the times of Jacob; for we are not informed that it was digged by him. As it is on land which he purchased for a residence, "of the sons of Hamor the father of Shechem," and was in the vicinity of a Canaanitish town; it *may* have been constructed by the former owners of the soil, and probably was so. The woman of Samaria when conversing with the Savior respecting it, asks 'Art thou greater than our father Jacob who *gave* us the well, and drank thereof himself, his children and cattle?" John, iv, 12. She does not say he dug it. This famous well is *one hundred and five feet deep*, and *nine feet in diameter*, and when Maundrell visited it, it contained fifteen feet of water. Its great antiquity will not appear very extraordinary, if we reflect that it is bored through the solid rock, and therefore could not be destroyed, except by an earthquake or some other convulsion of nature; indeed wells of this description, are the most durable of all man's labors, and may, for aught we know, last as long as the world itself.

The well *Zemzem* at Mecca, may be regarded as another very ancient one. It is considered by Mahometans one of the three holiest things in the world, and as the source whence the great progenitor of the Arabs was refreshed when he and his mother left his father's house. " She saw a well of water, and she went and filled the bottle with water and gave the lad to drink." Gen. xxi, 19. This well, the Caaba and the black stone,[a] were connected with the idolatry of the ancient Arabs, centuries before the time of Mahomet. The Caaba is said to have been built by Abraham and Ishmael, and it is certain that their names have been connected with it from the remotest ages. Diodorus Siculus, mentions it as

---

[a] This stone like those of the Hindoos and the one mentioned in Acts, xix, "fell down from heaven" and is probably a meteorite.

being held in great veneration by the Arabs in his time. [50, B. C.] The ceremonies still performed, of "encircling the Caaba seven times, kissing the black stone, and drinking of the water of the well Zemzem," by the pilgrims, were practices of the ancient idolaters, and which Mahomet, as an adroit politician, incorporated into his system, when unable to repress them. The conduct of the pilgrims when approaching this well and drinking of its water, has direct reference to that of Hagar, and to her feelings when searching for water to preserve the life of her expiring son.

If we reflect on the infinite value of wells in Syria—on the jealous care with which they have always been preserved—that while they afforded good water, they could never be lost—that Mecca is one of the most ancient cities of the world, the supposed Mesa of the scriptures, Gen. x, 30, —and that this well is the only one in the city, whose waters can be drunk:—we cannot but admit the possibility at least, that it is the identical one, as the Arabs contend, of whose waters, Ishmael and his mother partook.

We are not aware that any modern author has had an opportunity of closely examining it; it being death for a christian to enter the Caaba. Burckhardt visited the temple in the disguise of a pilgrim, but we believe he had not an opportunity to ascertain any particulars relating to its depth, &c. Purchas, quoting Barthema, who visited Mecca in 1503, says it is "three score and ten yards deepe," [210 feet,] "thereat stand sixe or eight men, appointed to draw water for the people, who after their sevenfold ceremonie come to the brinke," &c. Pil. p. 306. In Crichton's History of Arabia, Ed. 1833. Vol. ii, 218, this well is said to be fifty-six feet to the surface of the water. The curb is of fine white marble, five feet high, and seven feet eight inches in its interior diameter. In the 317th year of the Hegira, the Karmatians slew seventeen thousand pilgrims within the circumference of the Caaba, and filled this famous well with the dead bodies;—they also carried off the Black Stone.

JOSEPH'S WELL.—The most remarkable well ever made by man, is Joseph's well at Cairo. Its magnitude, and the skill displayed in its construction, which is perfectly unique, have never been surpassed. All travelers have spoken of it with admiration.

This stupendous well is an oblong square, twenty-four feet by eighteen; being sufficiently capacious to admit within its mouth a moderate sized house. It is excavated (of these dimensions,) through solid rock to the depth of one hundred and sixty-five feet where it is enlarged into a capacious chamber, in the bottom of which is formed a basin or reservoir, to receive the water raised from *below*, (for this chamber is not the bottom of the well.) On one side of the reservoir another shaft is continued, one hundred and thirty feet lower, where it emerges through the rock into a bed of gravel, in which the water is found. The whole depth, being two hundred and ninety-seven feet. The lower shaft is not in the same vertical line with the upper one, nor is it so large, being fifteen feet by nine. As the water is first raised into the basin, by means of machinery propelled by horses or oxen *within the chamber*, it may be asked, how are these animals conveyed to that depth in this tremendous pit, and by what means do they ascend? It is the solution of this problem that renders Joseph's well so peculiarly interesting, and which indicates an advanced state of the arts, at the period of its construction.

A spiral passage-way is cut through the rock, from the surface of the ground to the chamber, independent of the well, round which it winds with so gentle a descent, that persons sometimes ride up or down upon

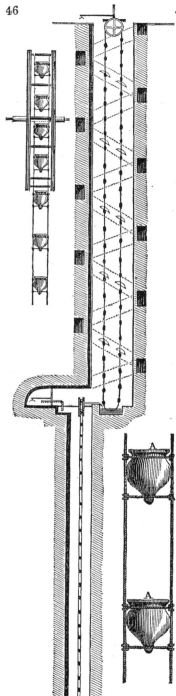

asses or mules. It is six feet four inches wide, and seven feet two inches high. Between it and the interior of the well, a wall of rock is left, to prevent persons falling into, or even looking down it, (which in some cases would be equally fatal,) except through certain openings or windows, by means of which, it is faintly lighted from the interior of the well: by this passage the animals descend, which drive the machinery that raises the water from the lower shaft into the reservoir or basin, from which it is again elevated by similar machinery, and other oxen on the surface of the ground. See figure. In the lower shaft, a path is also cut down to the water, but as no partition is left between it and the well, it is extremely perilous for strangers to descend.

The square openings represented on each side of the upper shaft, are sections of the spiral passage, and the zigzag lines indicate its direction. The wheels at the top carry endless ropes, the lower parts of which reach down to the water; to these, earthenware vases are secured by ligatures, see A, A, at equal distances through the whole of their length, so that when the machinery is moved, these vessels ascend full of water on one side of the wheels, discharge it into troughs as they pass over them and descend in an inverted position on the other. For a further description of this apparatus, see the chapter on the *chain of pots*.

This celebrated production of former times, it will be perceived, resembles an enormous hollow screw, the centre of which forms the well, and the threads, a winding stair-case round it. To erect of granite a flight of "geometrical" or "well stairs," two or three hundred feet high, on the surface of the ground, would require extraordinary skill; although in the execution, every aid from rules, measures, and the light of day, would guide the workmen at every step; but to begin such a work at the top, and construct it *downwards* by excavation alone, in the dark bowels of the earth, is a more arduous undertaking, especially as deviations from the correct lines could not be remedied; yet in Joseph's well, the partition of rock between the pit and the passage-way, and the uniform inclination of the latter, seem to have been ascertained with equal precision, as if the whole had been constructed of cut stone on the surface. Was the pit, or the passage, formed first; or were they simultaneously carried on, and the excavated masses from both borne up the latter? The extreme thinness of the partition wall, excited the astonishment of M. Jomard, whose account of the well is inserted in the second volume of Memoirs in Napoleon's great Work on Egypt, part 2nd, p. 691. It is, according to him, but sixteen centimetres thick, [about six inches!] He justly remarks that it must have required singular care to leave and preserve so small a portion while excavating the rock from both sides of it. It would seem no stronger in proportion, than sheets of pasteboard placed on edge, to support one end of the stairs of a modern built house, for it should be borne in mind, that the massive roof of the spiral passage *next the well*, has nothing but this film of rock to support it, or to prevent such portions from falling, as are loosened by fissures, or such, as from changes in the direction of the strata, are not firmly united to the general mass. But this is not all: thin and insufficient as it may seem, the bold designer has pierced it through its whole extent with semicircular openings, to admit light from the well: those on one side are shown in the figure.

Opinions respecting the date of this well are exceedingly various. Pococke thought it was built by a vizier named Joseph, eight hundred years ago; other authorities more generally attribute it to Saladin, the intrepid defender of his country against the hordes of European savages, who, under the name of crusaders, spread rapine and carnage through his land. His name was Yussef, [Joseph.] By the common people of Egypt, it has long been ascribed to the patriarch of that name, and their traditions are often well-founded; of which we shall give an example in the account of the *Swape*. Van Sleb, who visited Egypt several times in the 17th century, says, some of the people in his time, thought it was digged by spirits, and he adds, " I am almost inclined to believe it, for I cannot conceive how man can compass so wonderful a work."[a] This mode of accounting for ancient works is common among ignorant people, and may be considered as proof of their great antiquity. Dr. Robertson, in speaking of ancient monuments in India, remarks that they are of such high antiquity, that as the natives cannot, either from history or tradition, give any information concerning the time in which they were executed, they universally ascribe the formation of them to superior beings.[b] Some writers believe this well to have been the work of a more scientific people than any of the comparatively modern possessors of Egypt—in other words, they think it the production of the same people that built the pyramids and the unrivalled monuments of Thebes, Dendarah and Ebsamboul. Lastly, Cairo is supposed by others, to occupy the site of Egyptian Babylon, and this well is considered by them, one of the remains of that ancient city. Amidst this variety of opinions respecting its origin, it is certain, that it is every way worthy of the ancient mechanics of Egypt; and in its magnitude exhibits one of the prominent features which characterize all their known productions.

Why was this celebrated well made oblong? Its designer had certainly his reasons for it. May not this form have been intended to enlighten more perfectly the interior, by sooner receiving and retaining longer the rays of the sun? To what point of the compass its longest

---

[a] The present state of Egypt, by F. Van Sleb. Lon. 1678. p. 248.   [b] India, Appendix

sides coincide, has not, that we are aware of, been recorded. Should they prove to be in the direction of the rising and setting sun, the reason suggested, may possibly be the true one.

In Ogilvy's Africa, it is remarked that at the last city to the south of Egypt, " is a deep well, into whose bosom the sun shines at noon, while he passes to and again through the northern signs." p. 99. This is the same well that Strabo mentions at *Syene*, which marked the summer solstice—the day was known, when the style of the sun dial cast no shade at noon, and the vertical sun darted his rays to the bottom of the well. It was at Syene, that Eratosthenes, 220 B. C. made the first attempt to measure the circumference of the earth—and to the same city, the poet Juvenal was banished.

REFLECTIONS ON ANCIENT WELLS.

Before leaving this part of the subject, it may be remarked that ancient wells are of very high interest, inasmuch as many of them are the only memorials, that have come down to us, of the early inhabitants of the world; and they differ from almost all other monuments of man in former times; not only in their origin, design, and duration, but above all, in their UTILITY. In this respect, no barren monument, of whatever magnitude or material, which ambition, vanity, or power, has erected, at the expense of the labor and lives of the oppressed, can ever be compared with them. Such monuments are, with few exceptions, proofs of a people's sufferings; and were generally erected to the basest of our species: whereas ancient wells have, through the long series of past ages, continually alleviated human woe; and have furnished man with one of nature's best gifts without the least alloy.

It would almost appear, as if the divine Being had established a law, by which works of pure beneficence and real utility should endure almost for ever; while those of mere magnificence, however elaborately constructed, should in time pass away. The temple of Solomon—his golden house, ivory palaces, and splendid gardens are wholly gone; but the plain cisterns, which he built to supply his people with water, remain almost as perfect as ever. Thus the pride of man is punished by a law, to which the most favored of mortals formed no exception.

An additional interest is attached to several wells and fountains of the old world, from the frequent allusion to them in the Scriptures, and by the classical writers of Greece and Rome. In addition to those already named, the following may be noticed. When the Israelites left Egypt, " they came to *Elim*, where were twelve wells of water, and three score and ten palm trees." Now the Grove of Elim yet flourishes; and its fountains have neither increased nor diminished, since the Israelites encamped by them.[a] Modern travelers in Palestine often allay their thirst at the well which belonged to the birth place of David, the " Well of Bethlehem," whose waters he so greatly preferred to all others. The inhabitants of *Cos*, drink of the same spring which Hippocrates used twenty-three hundred years ago; and their traditions still connect it with his name. The nymphs of *Scyros*, another island in the Egean, in the early ages assembled at a certain fountain to draw water for domestic uses. This fountain, says Dr. Clarke, exists in its original state; and is still the same rendezvous as formerly, of love or of gallantry, of gossip-

---

[a] We are aware that Dr. Shaw—Travels, p. 350—observed but *nine* wells. He says, at that time, three of them were filled up with sand; but the whole were to be seen a short time previous to his visiting them, and we believe since.

ping and tale telling. Young women may be seen coming from it in groups, and singing, with vases on their heads, precisely as represented on ancient marbles. It was at Scyros where young Achilles was concealed to prevent his going to the Trojan war. He was placed among, and habited like, the daughters of Lycomedes; but Ulysses adroitly discovered him, by offering for sale, in the disguise of a pedler, a fine suit of armor, among trinkets for women.

Heliopolis, the city of the Sun, the ON of Genesis, of which Joseph's father-in-law was governor and priest, and whose inhabitants, according to Herodotus, (ii. 3.) were the most ingenious of all the Egyptians, and where the philosophers of Greece assembled to acquire "the wisdom of Egypt," was famous for its fountain of excellent water:—this fountain, with a solitary obelisk, is all that remains to point out the place where that splendid city stood.

Aqueducts, fountains, cisterns and wells, are in numerous instances the only remains of some of the most celebrated cities of the ancient world. Of Heliopolis, Syene and Babylon in Egypt; of Tyre, Sidon, Palmyra, Nineveh, Carthage, Utica, Barca, and many others; and when, in the course of future ages, the remaining portals and columns of Persepolis are entirely decayed, and its sculptures crumbled to dust: its cisterns and and aqueduct (both hewn out of the rock) will serve to excite the curiosity of future antiquaries, when every other monument of the city to which they belonged has perished. The features of nature, says Dr. Clarke, continue the same, though works of art may be done away: the 'beautiful gate' of the Jerusalem temple is no more, but Siloah's Fountain still flows, and Kedron yet murmurs in the Valley of Jehoshaphat. According to Chateaubriand, the Pool of Bethesda, a reservoir, one hundred and fifty feet by forty, constructed of large stones cramped with iron, and lined with flints embedded in cement, is the only specimen remaining of the ancient architecture of that city.

Ephesus, too, is no more; and the temple of Diana, that according to Pliny was 220 years in building, and upon which was lavished the talent and treasure of the east; the pride of all Asia, and one of the wonders of the world, has vanished; while the fountains which furnished the citizens with water, remain as fresh and perfect as ever. And as a tremendous satire on all human grandeur, it may be remarked, that a few solitary marble sarcophagi, which once enclosed the mighty dead of Ephesus, have been preserved—but as *watering troughs for cattle!*[a] Cisterns have been discovered in the oldest citadels of Greece. The *fountains of Bounarbashi* are perhaps the *only* objects remaining, that can be relied on, in locating the palace of Priam and the site of ancient Troy. And the well near the outer walls of the temple of the sun at Palmyra, will, in all probability, furnish men with water, when other relics of Tadmor in the wilderness have disappeared.[b]

To conclude, a great number of the wells of the ancient world still supply man with water, although their history generally, is lost in the night of time.

---

[a] Mr. Addison, in his journey southward from Damascus, says the fountain at *Nazera*, in Gallilee, "trickles from a spout into a marble trough, which appears to have been an ancient sarcophagus." And close by the well at *Mizra*, he observed fragments of another, which had been used for a similar purpose. We may add, that *Speed*, the old English historian, remarks that the stone coffin of *Richard* 3d, " is now made a drinking trough for horses at a common Inn." Edition of 1615, p. 737.

[b] Lord Lindsay's Letters, (10.)—Phil. Trans. Lowthorp's Abridg. iii, 490.

ANCIENT AMERICAN WELLS.

As wells are among the most ancient of man's labors, that are extant in the old world, might we not expect to find some on these continents, relics of those races, who, in the unknown depths of time, are supposed to have cultivated the arts of civilization here? We might: and true it is that among the proofs that a populous and much more enlightened people than the Indians have ever been, were at one time the possessors of America, *ancient wells* have been adduced. "From the highest point of the Ohio, says Mr. T. Flint, to where I am now writing (St. Charles on the Missouri) and far up the upper Mississippi and Missouri, the more the country is explored and peopled, and the more its surface is penetrated, not only are there more mounds brought to view, but more incontestible marks of a numerous population. WELLS, ARTIFICIALLY WALLED, different structures of convenience or defence, *have been found in such numbers, as no longer to excite curiosity."*

But *American* antiquities were so novel, so unlooked for, and so insulated from those of the old world, that learned men were greatly perplexed at their appearance; and at a loss to account for their origin. This is still, in a great measure, the case. A mystery, hitherto impenetrable, hangs over the primeval inhabitants of these continents. Who they were, and whence they came, are problems that have hitherto defied all the researches of antiquarians. Nothing, perhaps, but the increasing occupation of the soil, and excavations which civilization induces, will eventually determine the question, whether these antiquities are to be attributed to European settlers of the sixteenth century; to the enterprising Scandinavians, the *North Men*, who, centuries before the voyages of Columbus and the Cabots, visited the shores of New England, New York and the Jerseys; or whether some of them did not belong to an indigenous or Cuthite race, who inhabited those prolific regions, in times when the mastodon and mammoth and megalonix were yet in the land.

No one can reflect on the myriads of our species who have occupied this half of the globe—perhaps from times anterior to the flood—without longing to know something of their history; of their physical and intellectual condition; their languages, manners and arts; of the revolutions through which they passed; and especially of those circumstances which caused them to disappear before the progenitors of the present red men. The subject is one of the most interesting that ever exercised the human mind. It is calculated to excite the most thrilling sensations, and we have often expressed our surprise, that one of the most obvious and promising sources of information has never been sufficiently investigated: we allude to ancient wells, a *close examination of which*, might lead to discoveries equally interesting, and far more important, than those which resulted from a similar examination of Grecian wells. Dr. Clarke says, that "Vases of Terra Cotta, of the *highest antiquity*, have been found in cleansing the wells of Athens."[a]

Some persons may perhaps suppose the old wells in the western parts of this continent, to be the work of Indians; but these people have never been known to make any thing like a regular well. Mr. Catlin, the artist,

---

[a] A Roman well was discovered in the seventeenth century, near the great road which leads to Carlisle, in England. Instead of being walled up with stone, it was lined with large casks or hogsheads, six feet deep, and made of pine. The well was covered with oak plank nine inches thick. In it were found *urns, drinking cups, sandals and shoes*, the soles of which were stitched and *nailed*. Phil. Trans. Lowthorp's Abridg. iii, 431.

who spent eight years among those on the upper waters of the Mississippi and Missouri, and another gentleman who had long been east of the Rocky mountains, among the Flat Heads, and other tribes towards the Pacific, both inform us that the wild and untutored Indians never have recourse to wells. They in fact have no need of them, as their villages are invariably located on the borders or vicinity of rivers. In some cases of suffering from thirst while traveling, they, in common with other savages, sometimes scrape a hole in sand or wet soil, to obtain a temporary supply.

CHAPTER VIII.

Ancient methods of raising water from wells: Inclined planes—Stairs within wells: In Mesopotamia—Abyssinia—Hindostan—Persia—Judea—Greece—Thrace—England—Cord and bucket: Used at Jacob's well—by the patriarchs—Mahomet—In Palestine—India—Alexandria—Arabian Vizier drawing water—Gaza—Herculaneum and Pompeii—Wells within the houses of the latter city—Aleppo—Tyre—Carthage—Cleanthes the 'Well Drawer' of Athens, and successor of Zeno—Democritus—Plautus—Asclepiades and Menedemus—Cistern pole—Roman cisterns and cement—Ancient modes of purifying water.

WE are now to examine the modes practised by the ancients, in obtaining water from wells. When the first simple excavations became so far deepened, that the water could no longer be reached by a vessel in the hand, some mode of readily procuring it under such circumstances would soon be devised. In all cases of moderate depth, the most simple and efficient, was to form an inclined plane or passage, from the surface of the ground to that of the water; a device by which the principal advatages of an open spring on the surface were retained, and one by which domestic animals could procure water for themselves without the aid or attendance of man. There is reason to believe that this was one of the primitive methods of obtaining the liquid, when it was but a short distance below the surface of the ground; and was most likely imperceptibly introduced by the gradual deepening of, or enlarging the cavities of natural springs, or artificial excavations.

But when in process of time, these became too deep for exterior passages of this kind to be convenient or practicable, the wells themselves were enlarged, and stairs or steps for descending to the water, constructed *within* them. The circumstances recorded in Genesis, xxiv, induce us to believe that the well at which Eliezer, the steward of Abraham, met Rebecca, was one of these. When the former arrived at Nahor, he made his camels "to kneel down without the city by a well of water, at the time of the evening that women go out to draw water: and Rebecca came out with her pitcher upon her shoulder—and she went *down* to the well, and filled her pitcher and *came up*." Had any machine been attached to this well, to raise its water, or had a vessel suspended to a cord been used, she could have had no occasion to descend. It therefore appears that the liquid was obtained by immersing the pitcher in it, and in order to do this, the persons 'went down' to the water. That this well was not deep, may be inferred from the fact that Rebecca drew water sufficient to quench the thirst of *ten* camels, for it is said, she supplied them, "till they had done drinking;" a task which no young female could have accomplished in the time implied in the text, if this well had been even moderately deep, and one which under all circumstances was a laborious performance; for these animals take a prodigious quantity of water at a time, sufficient to last them from ten to twenty days. Eliezer might well wonder at the ingenuous and benevolent disposition of Rebecca, and every reader of the account is equally surprised at his insensibilty, in permitting her to perform the labor unaided by himself or his attendants.

Wells with stairs by which to descend to the water, are still common. The inhabitants of Arkeko in Abyssinia, are supplied with water from six wells, which are twenty feet deep and fifteen in diameter. The water is collected and *carried up a broken ascent* by men, women and children.[a] Fryer in his Travels in India, p. 410, speaks of "deep wells many fathom under ground with *stately stone stairs*." Joseph's well in Egypt is another example of stairs both within and without. Bishop Heber observed one in Benares, with a tower over it, and a "steep flight of steps for descending to the water." Forrest, in his Tour along the Ganges and the Jumna, says, "near the village of Futtehpore, is a large well, ninety feet in circumference, with a broad stone staircase to descend to the water, which might be about thirty feet." Mr. Forbes, in his Oriental Memoirs, remarks that "many of the Guzzerat wells, have steps leading down to the water; while others have not." In a preceding page, we quoted a passage from Ward's History of the Hindoos to the same effect. Tavernier, speaking of the scarcity of water in Persia, says, of wells they have a great many, and he describes one with steps down to the water.[b] "We passed a large and well built tank, with *two flights of steps* descending into it, at the opposite angles, possibly the pool of Hebron, where David hanged the murderers of Ishbosheth."[c] The fountain of Siloam is reached by a descent of *thirty* steps cut in the solid rock.

The small quantity of water furnished by some wells, rendered a descent to it desirable, and hence it was often collected as fast as it appeared, by women who often waited for that purpose. "That which pleased me most of all," says Fryer, p. 126, "was a sudden surprise, when they brought me to the wrong side of a pretty square tank or well, with a wall of stone breast high; when expecting to find it covered with water, looking down five fathom deep, I saw a *clutter* of women, very handsome, *waiting* the distilling of the water from its dewy sides, which they catch in jars. It is cut out of a black marble rock, up almost to the top, with broad steps *to go down*. Mr. Addison in his 'Journey Southward from Damascus,' says, "at the fountain near D'jenneen, the women used their *hands* as ladles to fill their pitchers. This scarcity of water, and the practice of scooping it up in small quantities, are referred to, by both sacred and profane authors. "They came to the pits and found no water, they returned with their vessels empty." Jer. xiv, 3. "There shall not be found of it a sherd, [a potter's vessel,] to take fire from the hearth, or to take water out of the pit,"—that is, to scoop it up when too shallow to immerse a vase or pitcher in it. Isaiah, iii, 14. St. Peter speaks of wells 'without water,' and Hosea, of 'fountains dried up.'

The water nymphs lament their empty urns." *Ovid, Met.* ii, 278.

The inhabitants of Libya, where the wells often contain little water, "draw it out in little buckets, made of the shank bones of the camel.[d]"

Wells with stairs are not only of very remote origin, but they appear to have been used by *all* the nations of antiquity. They were common among the Greeks and Romans.[e] The well mentioned by Pausanias, of

---

[a] Ed. Encyc. Art. Arkeko. [b] Persian Trav. 157. [c] Lindsay's Trav. Let. 9.
[d] Ogilvy's Africa, 306. [e] Lardner's Arts of the Greeks and Romans. i, 138.

which we have spoken in a previous chapter, has steps which lead down to the water.ᵃ The well for the purification of worshippers, in the temple of Isis, in Pompeii, has a descent by steps to the water.ᵇ The wells of Thrace, had generally a covered flight of steps.ᶜ Ancient wells of similar construction are still to be seen in various parts of Europe. There is one near Hempstead, Eng. for the protection of which, an act of parliament was passed in the reign of Henry VIII.

Such wells, probably gave rise to the beautiful circular stairs so common in old towers, and still known, as ' well stairs.'

In Galveston, (Texas,) and other parts of America, where there are no springs, cisterns are sunk in the sand between hillocks, into which the surface water drains, and steps are formed to lead down to it.

CORD AND BUCKET.

No.   Modern Greek female drawing water.   No. 9. From a manuscript of the 12th century.

However old and numerous wells with stairs within them may be, most of the ancient ones were constructed without them ; hence the necessity of some mode of raising the water. From the earliest ages, a vessel *suspended to a cord*, has been used by all nations—a device more simple and more extensively employed than any other, and one which was undoubtedly the germ of the most useful hydraulic machines of the ancients, as the chain of pots, chain pump, &c. That a cord and bucket were used to raise water from Jacob's well, nineteen centuries ago, is evident from the account of the interview, which the Savior had with the woman of Samaria at it. " Then cometh he to a city of Samaria, called Sichar ; now Jacob's well was there, and Jesus being wearied sat on the well ; and there cometh a woman of Samaria to draw water ; Jesus saith unto her, give me to drink." Had any machine been attached to this well at that time, by which a traveler or stranger could raise it, he could have procured it for himself ; and as he was thirsty, he probably would have done so, without waiting for any one to draw it for him ; but the reason why he did not, is subsequently explained by the woman herself ; who, in replying to one of his remarks, the meaning of which she misapprehended, said " Sir, thou hast nothing to *draw with*, and the well is deep." This well, as already remarked, is one hundred and five feet deep. Hence at that period every one carried the means of raising the water with him. No. 9. of the illustrations, is a representation of the woman of Samaria drawing water. It is from a Greek illuminated manuscript of the 12th century, from D'Agincourt's Storia Dell'Arte.

It is still the general practice in the east, for any one, who goes to

---

ᵃ For. Top. 196.   ᵇ Pompeii, i, 277.   ᶜ Hydraulia, 166.

draw water, to carry a vessel and cord with him, a custom which without doubt, has prevailed there since the patriarchal ages. This was the opinion of Mahomet, whose testimony on such a subject is unexceptionable. He was an Arab—a people who pride themselves on the preservation of the customs of their celebrated ancestors, Abraham, Ishmael, and Job. In his account of Joseph's deliverance from the pit, into which his brethren had cast him, (and which many commentators believe was a well, which at the time contained little or no water,) he says: "Certain travelers came, and sent one to draw water, (who went to the well in which Joseph was,) and *he let down his bucket,*" &c. Koran, chap. xii. This account is perfectly consistent with that of Moses. Josephus, also, seems to have believed it to be a well: "Reubel took the lad and tied him to a cord, and let him down gently into the pit, *for it had no water in it.*" Antiq. B. ii. 3.

At 3 o'clock, (says Mr. Addison in his "*Journey Southward from Damascus,*") we rode to a well (in approaching Cana of Galilee) in a field, where an Arab was watering his goats. There was a long stone trough by the side of the well, and this was filled with water by means of a leathern bucket attached to a rope, which the Arab *carried about with him,* for the convenience of himself and his herds. It was just such a scene as that described in Genesis: "And behold a well in the field, and lo, there were three flocks of sheep lying by it, for out of that well they watered their flocks, and a great stone was upon the well's mouth." Among the ruins of Mizra, in the great plain of Jezreel, the same traveler observes: "Surprised at the desolate aspect of the spot, I rode with my servant to a well a few yards distant, where two solitary men were watering their goats, by means of a *leathern bucket attached to a rope;* and dismounting, I sat on the stone at the well's mouth." Mr. Forbes, after a residence of many years in Asia, said he "did not recollect *any* wells furnished with buckets and ropes for the convenience of strangers; most travelers are therefore *provided with them;* and halcarras and religious pilgrims frequently carry a small brass pot affixed to a long string for this purpose."

In ancient Alexandria, where the arts were cultivated and science flourished to an extent perhaps unequaled in any older city, water was drawn up from the cisterns, with which every house was provided, with the simple cord and bucket. This city was supplied with water from the Nile: it was admitted into vaulted reservoirs or cisterns, which were constructed at the time the foundations of the city were laid by Alexander. They were sufficiently capacious to contain water for a whole year, being filled only at the annual inundation of the river, through a canal made for the purpose. Apertures or well openings, through which the water was raised from these reservoirs, are still to be seen. "Whole lines of ancient streets are traceable," (says Lord Lindsay, Travels, Letter 2.) "*by the wells* recurring *every six or seven yards:* by which the contiguous houses, long since crumbled away, drew water from the vast cisterns with which the whole city was undermined."

"Every house," says Rollin, "had an opening into its cistern, like the mouth of a well, through which the water was taken up either in buckets or pitchers." It may be said, this last quotation is not conclusive, since it does not indicate the manner in which the bucket was elevated—by a windlass? a pulley? or by the hand alone? We have satisfactory evidence that it was by the latter. The pavement of the old city is from ten to thirty feet below the surface of the modern streets, and excavations are frequently made by the Pasha's workmen, for the stones of the old pavement and of the buildings. In this manner the marble mouths of the vaulted reservoirs or cisterns are frequently brought to light; St. John's Egypt, vol. i. 8 : and they *invariably* exhibit *traces of the ropes* used for raising the water. Grooves are found *worn in them*, (by the ropes) to the depth of two inches, and such grooves are often numerous in each curb or mouth. Dry wells are built over some of these, and continued to the level of the present streets. Through them the inhabitants still draw water from the ancient reservoirs; and in the same manner as it was raised from them when the Ptolemies ruled over the land. A person in raising the bucket, stands at a short distance from the curb or mouth, and pulls the rope horizontally, or nearly so, towards him. In this way, the rope *rubs against the top* and *inside* of the curb, and in time wears deep grooves in it, such as are found in the ancient ones just mentioned. Sometimes, in order to avoid the friction, and consequent loss of power and wear of the ropes, the person drawing would stand on the edge of the curb, so as to keep the cord clear; but the practice is too perilous ever to have been general. It is, however, practised occasionally by the Hindoos.

El Makin, the Arabian historian, says that *Moclach*, the Vizier of Rhadi, who was deprived of his right hand and his tongue, and was confined in a lower room of the palace, where was a well; and having no person to attend him, he drew water for himself, pulling the rope with his left hand, and stopping it with his teeth, till the bucket came within his reach." This was in the tenth century. Martigny's History of the Arabians, vol. iv. 7. The wells on the road to Gaza, noticed by Mr. Stephens, had their upper surfaces formed of marble, which he observes had many grooves cut in it, " apparently being worn by the long continued use of ropes in drawing water." Incidents of Travel, vol. ii. 102.

That the same mode of raising it was adopted in the *public* wells of the ancient cities of Greece and Rome, is evident from those of Herculaneum and Pompeii; and from discoveries made in the latter city, it is obvious that it was practised in obtaining water from the wells and cisterns of *private houses*. This is a very interesting fact in connection with our subject, as it shows conclusively that the pump, if used at all by the Romans in their private houses, it was only to a very limited extent. In 1834, besides theatres, baths, temples and other public buildings, eighty houses had been disinterred. These were found to be almost uniformly provided with cisterns, built under ground and cemented, for the collection of rain-water. Each of these has an opening, enclosed in a curb, through which the water was drawn up. These are generally formed of a white calcareous stone, *on which are to be seen deep channels*, (Pompeii, vol. i. 88,) like those on the mouths of the Alexandrian cisterns, and produced from the same cause—the friction of the ropes used in drawing the water. The hypæthrum, says Sir William Gell, in his description of the house of the Dioscuri, in this case served as a compluvium ; receiving the water which fell from the roof, and transmitting it to a reservoir below, to which there is a marble mouth or puteal, exhibiting the traces of long use, in the furrows *worn by the ropes*, by which the water was drawn up. Pompeiana, vol. ii, 27.

The great variety of buildings to which wells and cisterns having their curbs *thus worn* were attached, show that this mode of raising water was nearly universal in Pompeii. The simple cord and bucket was equally used in the palace of the quæstor, and the humble dwelling of the private citizen. It was by them, the priests drew water for the uses of the temples, and mechanics for various purposes in the arts. Bakers thus raised water for their kneading-troughs, from cisterns or wells under the floor of their shops. *Three* bakers' shops, at least, have been found, and all of them in a tolerable state of preservation : their mills, ovens, kneading troughs, flour, loaves of bread, (with their quality, or the bakers' names stamped on them,) leaven, vessels for containing water, and their reservoirs of the latter, &c. have been discovered, so as to leave almost nothing wanting to perfect our knowledge of this art among the Romans. It is probable that wells were not infrequent in the *interior* of the houses in Pompeii, for another one was discovered in the house of a medical man, as presumed from *chirurgical instruments* found in it.[a]

The custom of Roman bakers having wells or cisterns within their houses, continued to modern times. When the Royal Academy of Sciences of France, undertook in the last century, the noble task of publishing a detailed account of all the useful arts, with a view to their universal diffusion and perpetuity—the baker is represented drawing water from a well, under the floor of his shop, and in a manner analagous to that practised by his predecessors of Pompeii.[b] London bakers also had wells in their cellars, for the same purpose, and probably still have them to some extent.

The inhabitants of the city of Aleppo, the metropolis of Syria, drew water from their cisterns or subterraneous reservoirs, and also from their wells, with which 'almost every house'[c] was provided, with a cord and bucket, in the same manner as the Egyptians of Alexandria ; and so do the inhabitants of *Soor*, which occupies the site of ancient Tyre, a town which contained in 1816, according to Mr. Buckingham, eight hundred stone built houses, most of which, he observes, had wells. Ancient Carthage was built like Alexandria, upon cisterns—a common practice of old. The modern inhabitants of Arzew, the ancient Arsenaria, as observed by Dr. Shaw in his Travels, dwell in the old cisterns, as in so many hovels; the water from which, was doubtless drawn in former times, by the simple cord and bucket—the *universal implements* still used throughout Egypt, Palestine, Syria, Asia Minor, Persia, Hindostan, and generally through all the east. This primeval device for raising water, has been used in all ages, and will doubtless continue to be so used, to the end of time.

An interesting circumstance is recorded, respecting an individual, who, from his occupation in ancient Athens, was named the ' Well-Drawer,' which may here be noticed. This was Cleanthes, a native of Lydia, who went to Athens as a wrestler, about 300 B. C. and acquiring a taste for philosophy there, determined to place himself under the tuition of some eminent philosopher, although he possessed no more than four *drachmæ*, or sixty-two cents ! He became a disciple of Zeno, and that he might have leisure to attend the schools of philosophy in the day-time, he *drew water* by night, as a common laborer in the public gardens. For several years he was so very poor, that he wrote the heads of his master's lectures, on bones and shells, for want of money to buy better materials : at last, some Athenian citizens observing, that though he appeared strong and healthy, he had no visible means of subsistence, summoned him before the Areopagus, according to a law borrowed from the Egyptians, to give an account of his manner of living. Upon this, he produced the gardener for whom he drew water, and a woman for whom he ground meal, as witnesses to prove that he subsisted by the labor of his hands. The judges, we are

---

[a] Lardner's Arts, &c. i, 268. [b] Descriptions des Arts et Metiers. Paris, 1761. Art. du Boulanger. Planche 5. [c] Russel's History of Aleppo, p. 7.

told, were so much struck with admiration of his conduct, that they ordered ten *minæ*, [one hundred and sixty dollars] to be paid him out of the public treasury.

The conduct of Cleanthes explains the secret of the great celebrity of many ancient philosophers, and shows the *only* means by which eminence in any department of human knowledge can be acquired: viz. by *industry* and *perseverance*. Besides his poverty, which of itself was sufficient to paralyze the efforts of most men, he was so singularly dull in apprehension, that his fellow disciples used to call him the *ass;* but resolution and application raised him above them all, made him a complete master of the stoic philosophy, and qualified him as successor of the illustrious Zeno. Democritus beautifully expressed the same sentiment, by representing Truth as hid in the bottom of a well; to intimate the difficulty with which she is found.

Analogous to the conduct of Cleanthes, was that of *Plautus*, the poet, who being reduced from competence to the meanest poverty, hired himself to a baker as a common laborer, and while employed in grinding corn, exercised his mind in study. The same may be remarked of Asclepiades and Menedemus, two Grecian philosophers, who were both so poor, that at one period, they hired themselves as *bricklayer's laborers*, and were employed in carrying mortar to the tops of buildings. Asclepiades, was not ashamed to be seen thus engaged, but his companion "hid himself if he saw any one passing by." *Athenæus*, says they were at one time summoned, like Cleanthes, before the Areopagites, to account for their manner of living—when they requested a miller to be sent for, who testified that "they came *every night* to his mill, where they labored and gained *two* drachmæ."

No. 8, in the last engraving, represents a modern Greek female drawing water. It is from a sketch of Capo D'Istrias' house. See the Westminster Review for September, 1838.

CISTERN POLE.

No. 10. Cistern Pole.

This simple implement, may be thought too insignificant to deserve a particular notice, but as it is extensively used in our rain-water cisterns, and is no modern device, we are unwilling to pass it. It was known to the Romans. Pliny expressly mentions it, when speaking of various modes of watering gardens. He says water is drawn from a well or tank, "by plain poles, hooks and buckets," B. xix, 4; and that it was a *domestic* implement in old times as at present, in raising water from cisterns, is proved by the discovery of some of the *hooks* at Pompeii. Lard. Arts, &c. i, 205. Having mentioned the rain water cisterns of the Romans, it may be observed, that they were as common in Pompeii as they are in this city, every house having been furnished with one.

As Pliny's account of these cisterns may be useful to some mechanics, especially masons, we shall make no apology for inserting it. "The walls were lined with strong cement, formed of five parts of sharp sand, and two of quicklime mixed with flints; the bottom being paved with the same, and well beaten with an iron rammer." B. xxxvi, 23. Holland's Trans. The composition of this cement, differs from that which Dr. Shaw says has been used in modern times in the east; and which he thinks is the same as that of the ancients. He says the cisterns which were built by *Sultan ben Eglib,* in several parts of the kingdom of Tunis, are equal in solidity with the famous ones at Carthage, continuing to this day (unless where they been designedly broken,) as firm and compact, as if they were just finished. The composition is made in this manner: they take two parts of wood ashes, three of lime, and one of fine sand, which after being well sifted and mixed together, they beat for three days and nights incessantly with wooden mallets, sprinkling them alternately and at proper times, with a little oil and water, till they become of a due consistence. This composition is chiefly used in their arches, cisterns and terraces. But the pipes of their aqueducts, are joined by beating tow and lime together, with oil only, without any mixture of water. Both these compositions quickly assume the hardness of stone, and suffer no water to pervade them. Trav. 286.

If the Romans wished to have water perfectly pure, they made two and sometimes three cisterns, at different levels; so that the water successively deposited the impurities with which it might be charged. From this, we see that the recent introduction of *two* cisterns for the same purpose, in some of our best houses, is a pretty old contrivance. It in fact dates far beyond the Roman era. The famous cisterns of Solomon are examples of it. Rain-water was frequently boiled by the Romans before they used it. Pliny xxxi, 3. This was also an ancient practice among older nations. Herodotus, says the water of the Choaspes, which was drunk by the Persian kings, was previously boiled, and kept in vessels of silver. B. i, 188.

CHAPTER IX.

The Pulley: Its origin unknown—Used in the erection of ancient buildings and in ships—Ancient one found in Egypt—Probably first used to raise water—Not extensively used in ancient Grecian wells: Cause of this—Used in Mecca and Japan—Led to the employment of animals to raise water—Simple mode of adapting them to this purpose, in the east. Pulley and two buckets: Used by the Anglo Saxons, Normans, &c.—Italian mode of raising water to upper floors—Desagulier's mode—Self-acting, or gaining and losing buckets—Marquis of Worcester—Heron of Alexandria—Robert Fludd—Lever bucket engine —Bucket of Bologna—Materials of ancient buckets.

PULLEY AND SINGLE BUCKET.

WE now come to the period when some of the simple machines, or mechanical powers, as they are improperly named, were applied to raise water. *When* this first took place, is unknown : That it was at an early stage in the progress of the arts, few persons will doubt; but the time is as uncertain, as that of the invention of those admirable contrivances for transmitting and modifying forces. It was among the devices by which the famous structures of antiquity were raised; and Egyptian engineers under the Pharaohs, were undoubtedly acquainted with all the combinations of it now known. Had Vitruvius neither described it, nor mentioned its applications, a circumstance which occurred at the close of Cleopatra's life, would have sufficiently proved its general use, in the erection of elevated buildings under the Ptolemies. The Egyptian queen,

Chap. 9.] Pulley and Single Bucket. 59

to avoid falling into the hands of Octavius,*took refuge in a very high tower, accessible only from above. Into this, she and her two maids, drew up Antony, (who had given himself a fatal wound,) by means of ropes and pullies, which happened to be there, for the purpose of raising stones to the top of the building. But the pulley was an essential requisite in the sailing vessels of Egypt, India and China, in the remotest ages. Neither trading ships, nor the war fleets of Sesostris, or previous warriors, could have traversed the Indian ocean without this appendage to raise and lower the sails, or quickly to regulate their movements by halliards. The ancient Egyptians, says Mr. Wilkinson, " were not ignorant of the pulley." The remains of one have actually been disinterred, and are now preserved in the museum of Leyden. The sides are of *athul* or tamarisk wood, the roller of fir: part of the rope made of *leef* or fibres of the date tree, was found at the same time. This relic of former times, is supposed to have been used in drawing water from a well. Its date is uncertain.

There are reasons which render it probable that the *single* pulley, was devised to raise water and earth from wells, and probability is all that can ever be attained with regard to its origin. But may not the pulley have been known *before* wells? We think not, and for the following reasons: 1. Most barbarous people have been found in possession of some of the latter, but not of the former; and in the infancy of the arts, man has in all ages, had recourse to the same expedients, and in the *same order*. 2. Wells are not only of the highest antiquity, but they are the only known works of man in early times, in which the pulley could have been required or applied. 3. The importance of water in those parts of Asia where the former generations of men dwelt, must have urged them at an early period to facilitate by the pulley, the labor of raising it. That it preceded the invention of ships, and the erection of lofty buildings of stone, is all but certain; but for what purpose, except for raising water, the pulley could have previously been required, it would be difficult to divine. It seems to have been the first addition made to those primitive implements, the cord and bucket; and when once adopted, it naturally led, as we shall find in the sequel, to the most valuable machine which the ancients employed. By it the friction of the rope in rubbing against the curb, and the consequent loss of a portion of the power expended in raising the water, were avoided, and by it also a beneficial change in the direction of the power, was attained: instead of being exerted in an ascending direction, as in Nos. 8 and 9, it is applied more conveniently and efficiently in a descending one, as in the figure.

N 1. Pulley and Bucket.

Notwithstanding the obvious advantages of using the pulley, it would appear that it was not extensively used in the public wells of the ancients, except in those from which the water was raised by oxen. No example of its use has occurred in the wells of Herculaneum or Pompeii. Nor does it appear to have been employed to any great extent by the Greeks; for with them, a vessel by which to draw water, was as necessary a utensil to their mendicants, as to the modern pilgrims and fakirs of Asia. The poorest of beggars, Aristophanes' *Telepheus*, had a staff, a broken cup, and a bucket, although it leaked. This custom therefore of carrying a vessel, and cord to draw water, shows that no permanent one was attached to their public wells, which would have been the case had the pulley been used. If such had been the custom, neither the mendicant Telepheus, nor Diogenes the philosopher, would have carried about with them, vessels for the purpose.

It is not easy to account for the partial rejection of the pulley by the Greeks in raising water, when its introduction would have materially diminished human labor. It certainly did not arise from ignorance of its advantages, as their constant application of it to other purposes, attests; and there is reason to believe, they adopted it to some extent in raising water from the holds of their ships, in common with the maritime people of Asia. It was indeed used in some of their wells,[a] but only to a limited extent. The principal reason for not employing it in public wells, was probably this—With it, a single person only could draw water at a time, while without it, numbers could lower and raise their vessels simultaneously, without interfering with each other. In the former case, altercations would be frequent and unavoidable; and the inconvenience of numbers of people waiting for water in warm climates a serious evil. The rich, and those who had servants would always procure it, while the poor and such as had no leisure, would obtain it with difficulty. The *large diameter* of their wells and those of other nations, it would seem, was solely designed to accommodate several people at the same time. These reasons it is admitted, do not apply to the *private* wells and cisterns of the Greeks and Romans, in which the pulley might have been used; but those people followed the practice of older nations, and from the great number of their slaves, (who drew the water) they had no inducement or disposition to lessen their labor.

A bucket suspended over a pulley, is still extensively used in raising water from wells throughout the world. The Arabians use it at the well Zemzem; the mouth of which, is "surrounded by a brim of fine white marble five feet high, and ten feet in diameter; upon this the persons stand, who draw water in leathern buckets, attached to *pulleys*, an iron railing being so placed as to prevent their falling in."[b]

Apparatus precisely similar to the figure in No. 11, are used by the Japanese and other Asiatics. Montanus' Japan. 294.

The pulley has but recently given place to pumps, in workshops and dwellings, and in these only to a limited extent—being confined chiefly to a few cities in the United States and Europe. In France and England, it was a common appendage to wells in the interior of houses, during the last century; and in such cases it is still extensively used throughout Spain, Portugal and other parts of Europe. It is very common in this country, and also in South America.

But the grand advantage of the pulley in the early ages was this;—by it the vertical direction in which men exerted their strength, could be directly changed into a horizontal one, by which change, *animals* could be employed in place of men. The wells of Asia, frequently varying from two to three, and even four hundred feet in depth, obviously required more than one person to raise the contents of an ordinary sized vessel: and where numbers of people depended on such wells, not merely to supply their domestic wants, but for the purposes of irrigation, the substitution of animals in place of men, to raise water, became a matter almost of necessity, and was certainly adopted at a very early period. In employing an *ox* for this purpose, the simplest way, and one which deviated the least from their accustomed method, was merely to attach the end of the rope to the yoke, after passing it over a pulley fixed sufficiently

---

[a] Lardner's Arts, &c i, 138.   [b] Crichton's Arabia. ii, 219.

Chap. 9]   *Application of Animals to Raise Water.*   61

high above the mouth of the well, and then driving the animal in a direct line from it, and to a distance equal to its depth, when the bucket charged with the liquid would be raised from the bottom. This, the most direct and efficient, was, (it is believed,) the identical mode adopted, and like other devices of the ancients, it is still continued by their descendants in Africa and Asia. Its value in the estimation of the moderns, may be learned from the fact, that it is adopted in this and other cities for raising coals, &c. from the holds of ships; for which and similar purposes, it has been in use for ages in Europe. It has also been used to work pumps, the further end of the rope being attached to a heavy piston working in a very long chamber or cylinder.

No. 12. Ancient and Modern method of raising water in Asia.

This was probably one of the first operations, and certainly one of the most obvious, where human labor was superseded by that of animals, and in accomplishing it, the pulley itself was perhaps discovered. This mode is common in Egypt, Arabia, India—through all Hindostan, and various other parts of the east. Mr. Elphinstone mentions a large well under the walls of the fort at Bikaneer, from fifteen to twenty-two feet in diameter, and three hundred feet deep. In this well *four* large buckets are used, each thus drawn up by a *pair* of oxen, and *all* worked at the same time. When any one of them was let down, "its striking the water, made a noise like a great gun." But simple as this mode of raising water by animals is, it is capable of an improvement equally simple, though not perhaps obvious to general readers. It was not however left to modern mechanicians to discover, but is one among hundreds of ancient devices, whose origin is lost in the remoteness of time. It is this—Instead of the animal receding from the well on level ground, it is made to descend an *inclined plane*, so that the *weight* of its body contributes towards raising the load. This is characteristic of Asiatic devices. At a very early period, the principle of *combining* the *weight* of men and animals with their *muscular energy*, in propelling machines, was adopted. We shall meet with other examples of it.

PULLEY AND TWO BUCKETS.

The addition of another bucket, so as to have one at each end of the rope, was the next step in the progress of improvement; and although so simple a device may appear too obvious to have remained long unperceived, and one which required no stretch of intellect to accomplish, it was one of no small importance, since it effected what is seldom witnessed in practical mechanics—a saving both of time and labor. Thus, by it, the empty vessel descended and became filled, as the other was elevated, (without the expenditure of any additional time and labor to lower it, as with the single bucket,) while its weight in descending, contributed towards raising the charged one.

These advantages were not the only results of the simple addition of another bucket; though they were probably all that were anticipated by the author at the time. It really imparted a *new feature* to the apparatus, and one which naturally led to the development of that great machine, in which terminated all the improvements of the older mechanics on the primitive cord and bucket—and to which, modern ingenuity has added—nothing—viz: THE ENDLESS CHAIN OF POTS—indeed nothing more was then wanting, but to unite the two ends of the rope together, and attach a number of vessels to it, at equal distances from each other, through the whole of its length, and the machine just named was all but complete.

No. 13. Ancient.   No. 14. Modern.
[From sepulchral monuments.]

The Anglo Saxons used two buckets hooped with iron, one at each end of a chain which passed over a pulley.[a] And in the old Norman castles, water was raised by the same means. In one of the *keeps or towers*, still remaining, which was built by Gundulph, bishop of Rochester, in the reigns of the Conqueror and William Rufus, the mode of elevating the water is obvious. "For water, there was a well in the very *middle* of the partition *wall:* it was also made to go through the whole wall, from the bottom of the tower up to the very leads, (i. e. the roof) and on every floor were small arches in the wall, forming a communication between the pipe of the wall, and the several apartments, so that by a pulley, water was communicated every where." And in Newcastle, a similar tower exhibits the same device for obtaining the water: "a remarkable pillar from which arches branched out very beautifully on each side, inclosed a pipe, (that is, the continuation of the well,) which conducted water from the well."[b] It appears to have been, in the middle ages, the uniform practice to enclose wells within the walls of towers, that in case of sieges, the water might not be cut off. It was the same in early Rome: the capitol was supplied by a deep well at the foot of the Tarpeian Rock, into which buckets were lowered through an artificial groove or passage made in the rock.[c] The double bucket is still used in inns in Spain. See a figure in Sat. Mag. Vol. vii, 58.

A simple mode is practised in Italy, by which a person in the upper story of a house, and at some distance from the well or cistern, (which is generally in the court yard,) raises water without being obliged to descend. One end of a strong iron rod or wire, is fixed to the house above the window of an upper landing or passage, and the other end in the ground,

---

[a] Encyc. Antiq. 524.   [b] Ibid, 82.   [c] Gell's Topography of Rome, ii, 203.

Chap. 9.] *Raising Water to Upper Floors.* 63

on the farther side of the well and in a line with its centre as in No. 15. A ring which slides easily over the wire is secured to the handle of the bucket, to which a cord is also attached and passes over a pulley fixed above the window. Thus when the cord is slackened, the bucket descends along the wire into the water, and when filled is drawn up by a person at the window. (Kitchens in the houses of Italy, like those of London and Paris are often on the upper floors.) " This mode of raising water to the upper stories of houses is practised in Venice and some other towns in Italy."[a] We are not acquainted with the origin of this device. From the circumstance of the ancient, (as well as the modern) inhabitants of Asia, Greece, Italy, &c. having had jets d'eau and tanks of water in the centre of their court-yards, it is possible that this mode of raising water to the upper floors of dwellings, may be of ancient date. It was in use in the 16th century, and is described in Serviere's collection, from which the figure is taken.[b] In the same work are devices for raising water in buckets to the tops of buildings ', ulleys, ropes, &c. moved by water wheels.

No. 15. Italian mode of raising water to the upper floors of a house.

Of modern devices for raising water with the pulley and bucket, the most efficient is said to be that of Dr. Desaguliers. After passing the rope over a pulley, he suspended to its end a frame of wood on which a man could stand—the bucket at the other end was made heavier than this frame, and therefore descended of itself. The length of the rope was such, that when the bucket was at the bottom, the frame was level with the place to which the water was to be raised. As soon as the bucket was filled with water, for the admission of which a hole was made in its bottom, and covered by a flap or valve, a man, whose weight exceeded, (with the frame) that of the bucket and water, stepped upon the frame, and sunk down with it to the bottom, and consequently raised the bucket of water to the required height, when a hook caught in a hasp at the side of the bucket, turned it over, and discharged its contents into the reservoir. As soon as the bucket was empty, the man at the bottom stepped off the frame and ran up a flight of stairs made for the purpose, to the place whence he descended; and in the mean time, the bucket being heavier than the frame, descended to the water, and was again raised by the same process.

Such a device is well enough for philosophical experiment, but is certainly not adapted for practical purposes. Simple as it may appear, there are requisites necessary to its efficient application, which in common practice are unattainable.

---

[a] Cadell's Journey into Carniola, Italy, &c. Edinburgh, 1820, i, 481.
[b] Recueil D'Ouvrages Curieux de Mathematique et de Mechanique, ou Description du Cabinet de M. Grollier de Serviere, avec des figures en taille douce, par M. Grollier de Serviere, son petit fils. A Lyon, 1719. The elder Serviere died in the 17th century.

SELF-ACTING, OR GAINING AND LOSING BUCKETS.

In the latter part of the 16th, or beginning of the 17th century, a machine for raising water, was in use in Italy, which is entitled to particular notice, on account of its being alleged to be the first one of the kind which was *self-acting;* and in that respect, was the forerunner of the motive 'Fire Engine' itself. It appears to have been first described by Schottus in his *Technia Curiosa.* According to Moxon, his description was taken from one in actual operation "at a nobleman's house at Basil." (Mech. Pow. 107.) But Belidor, says the first one who put such a thing in execution, was Gironimo Finugio, at Rome in 1616; although Schottus had long before *contrived* an engine for this purpose. Moxon has given a figure and description of one, but without naming the source from whence he obtained it: he says it was "*made at Rome*, in the convent of St. Maria de Victoria : the lesser bucket did contain more than a whole urn of water, (at Rome they say *un barile*,) but before, while they used lesser buckets, the engine wanted success." It would seem that it was to one of these 'Roman Engines,' that the Marquis of Worcester referred, in the 21st proposition of his Century of Inventions: "How to raise water constantly with two buckets only, day and night, *without any other force than its own motion,* using not so much as any force, wheel or sucker, nor more pulleys than one, on which the cord or chain rolleth, with a bucket fastened at each end. This I confess I have seen and learned of the great mathematician Claudius, his studies at *Rome*, he having made a present thereof unto a cardinal, and I desire not to own any other men's inventions, but if I set down any, to nominate likewise the inventor."

The machine described by Moxon, is encumbered with too many appendages for popular illustration—its essential parts will be understood by the accompanying diagram, from Hachette's Traité Elémentaire des Machines, Paris, 1819. Over a pulley S, are suspended two vessels A and B, of unequal dimensions. The smaller one B, is made heavier than A when both are empty, but lighter when they are filled. It is required to raise by them, part of the water from the spring or reservoir E, into the cistern Z. As the smaller bucket B, by its superior gravity, descends into E, (a flap or valve in its bottom admitting the water,) it consequently raises A into the position represented in the figure. A pipe F, then conveys water from the reservoir into A, the orifice or bore of which pipe, is so proportioned, that both vessels are filled *simultaneously.* The larger bucket then preponderates, descending to O, and B at the same time rising to the upper edge of Z, when the projecting pins O O, catch against others on the lower sides of the buckets, and overturn them at the same moment. The bails or handles are attached by swivels to the sides, a little above the

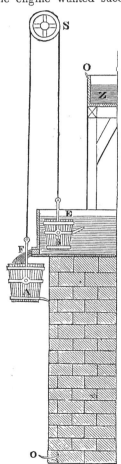

No. 16. Gaining and Losing Buckets.

centre of gravity. As soon as both vessels are emptied, B again preponderates, and the operation is repeated without any attendance, as long as there is water in E and the apparatus continues in order.

In Moxon's machine, the vessels were filled by two separate tubes of unequal bore; the orifices being covered by valves, to prevent the escape of water while the buckets were in motion; these valves were opened and closed by means of cords attached to the buckets. The efflux through F in the figure, may easily be stopped as soon as A begins to descend, by the action of either bucket on the end of a lever attached to a valve, or by other obvious contrivances. The water discharged from A, runs to waste through some channel provided for the purpose. These machines are of limited application, since they require a fall for the descent of A, equal to the elevation to which the liquid is raised in B. They may however be modified to suit locations where a less descent only can be obtained. Thus, by connecting the rope of B to the periphery of a large wheel, while that of A is united to a smaller one on the same axis, water may be raised higher than the larger bucket falls, but the quantity raised will of course be proportionally diminished.

In Serviere's Collection, a Gaining and Losing Bucket Machine is described. Another one was invented in 1725, by George Gerves an English carpenter, who probably was not aware that he had been anticipated by continental mechanics upwards of a century before. He erected one in Buckinghamshire, which was much approved of by Sir Isaac Newton, Beighton, Desaguliers, Switzer, and others. Mr. Beighton who drew up a description of it, observes that it was so free from friction, that "it is likely to continue an age without repair;" and Dr. Desaguliers on inserting an account of it in his Experimental Philosophy, vol. ii, 461, says, "this engine has not been out of order since it was first set up, about fifteen years ago." Notwithstanding these favorable testimonials, it has fallen into disuse. It was much too complex and cumbersome, and of too limited application ever to become popular.

The principle of self-action in all these machines is no modern discovery, for it was described by Heron of Alexandria, who applied it to the opening and closing the doors of a temple, and to other purposes. The motive bucket when filled, descended and communicated by a secret cord the movement required, and when its contents were discharged (by a siphon similar to the one figured in the Clepsydra of Ctesibius, in our fifth book,) it was again raised by a weight at the other end of the cord, like the bucket, in the last figure. See De Naturæ Simia seu technica macrocosmi historia, by Robert Fludd, (the English Rosicrucian.) Oppenheim, 1618, pp. 478 and 489, where several similar contrivances are figured—hence the device is much older than has been supposed. Perhaps the best modification of the 'Gaining and Losing Bucket', is Francini's, a description of which may be seen in our account of the Endless Chain of Pots.

A *lever* machine described by Dr. Desaguliers may here be noticed. "A A, (No. 17,) are two spouts running from a gutter or spring of water, into the two buckets D and E. D containing about thirty gallons and being called the *losing bucket*, and E the *gaining* bucket, containing less than a quarter part of D, as for example six gallons. D E, is a lever or beam movable about the axis or centre C, which is supported by the pieces F F, between which the bucket D can descend when the contrary bucket E is raised up, D C, is to C E, as one is to four. G L is an upright piece, through the top of which the lever K I moves about the centre L, sometimes resting on the prop H, and sometimes raised from it by the pressure of the arm C E on the end I. The bucket D when empty, has its mouth upwards, being suspended as above mentioned. The end D with its bucket is also lighter than the end with the bucket E, when both are empty. By reason of the different bore of the spouts, D is filled almost as soon as E, and immediately preponderating, sinks down to D, and thereby raises the contrary end of the lever and its bucket up to the cistern M, into which it discharges its water; but immediately the bucket D becoming full, pours out its water, and the end of the lever E comes down again into its horizontal situation, and striking upon the end I of the loaded lever I K, raises the weight K, by which means the force of its blow is broken. If the distance A B or fall of the water be about six feet, this machine will raise the water into the cistern M twenty-four feet high. Such a machine is very simple and may be made in any proportion according to the fall of the water, the quantity allowed to be wasted, and the height to which the water must be raised."

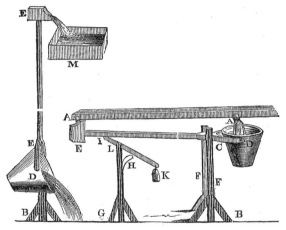

No. 17. Gaining and Losing Buckets.

"Some years ago," Dr. Desaguliers continues, "a gentleman showed me a model of such an engine varying something from this, but so contrived as to stop the running of the water at A A, when the lever D E began to move. He told me he had set up an engine in *Ireland*, which raised about half a hogshead of water in a minute, forty feet high, and did not cost forty shillings a year to keep it in repair, and that it was not very expensive to set up at first." Experimental Philosophy, vol. i, 78.

There is a singular historical fact connected with the use of buckets to raise water from wells, which will serve to conclude this part of the subject. Every person knows, that war between nations has often arisen from the most trifling causes; when thousands of human beings, alike ignorant and innocent of its origin, hired by its authors, armed with murderous weapons and incessantly exercised in the use of them, are marshaled into the presence of a similar host; when both being stimulated by inflaming addresses, and often excited by ardent spirits, destroy each other like infuriated tigers! Then after one party is overcome, the other glorying in the slaughter, hail their leader a *hero*, and not infrequently do that, which fiends would shudder to think of—viz. return thanks to the benign *Savior* of men, for having enabled them thus to de-

stroy their species; and to produce an amount of misery, as evinced in the shrieks of the wounded—the agonies of the dying—the unutterable pangs of widows, and the untold sufferings of orphans—that would suffice to draw tears from demons! And all this for what? Why, at one time, according to Tasso, and it is degrading to our nature to repeat it, because some thieves of Modena *stole a bucket belonging to a public well of Bologna!* This fatal bucket is still preserved in the cathedral of Modena—a memorial of a sanguinary war, and of the evils attending the most horrible of all human delusions, *military glory.*

"In the year 1005, some soldiers of the commonwealth of Modena ran away with a bucket from a public well, belonging to the State of Bologna. This implement might be worth a shilling; but it produced a bloody quarrel which was worked up into a bloody war. Henry, the king of Sardinia, for the Emperor Henry the second, assisted the Modenese to keep possession of the bucket; and in one of the battles he was made prisoner. His father, the Emperor, offered a chain of gold that would encircle Bologna, which is seven miles in compass, for his son's ransom, but in vain. After twenty-two years imprisonment, and his father being dead, he pined away and died. His monument is still extant in the church of the Dominicans. This fatal bucket is still exhibited in the tower of the cathedral of Modena, enclosed in an iron cage."

MATERIALS OF BUCKETS.—Neptune and Andromache watered horses with metallic ones. Both Greeks and Romans had them of wood, metal and leather. Sometimes wooden ones were hooped with *brass.* One of these was found in a Roman barrow in England. The ancient British had them without hoops and cut out of solid timber. The Anglo Saxons made them of staves as at present. Those of the old Egyptians were of metal, wood, skins or leather, and probably of earthenware. See figures in 11th and 13th chapters. We have given figures of some metallic ones discovered in Pompeii, in Book II. The bucket of Bologna is formed of staves and bound with iron hoops.[a]

The old error that 'water has no weight in water,' arose from not perceiving the weight of a bucket, until it was raised out of the liquid in which it was plunged.

Although poetry is foreign to the design of this work, and cold water is not remarkably inspiring, nor a bucket a very poetical object, yet the following beautiful lines of S. Woodworth, on 'The Bucket,' are as refreshing in the midst of a dry discussion, as a draught of the sparkling liquid to a weary traveller of the desert.[b]

> That moss-covered vessel I hail as a treasure;
> For often at noon, when returned from the field,
> I found it the source of an exquisite pleasure,
> The purest and sweetest that nature can yield.
>
> How ardent I seized it, with hands that were glowing,
> And quick to the white-pebbled bottom it fell:
> Then soon, with the emblem of truth overflowing,
> And dripping with coolness, it rose from the well.
>
> How sweet from the green mossy brim to receive it,
> As poised on the curb it inclined to my lips!
> Not a full blushing goblet could tempt me to leave it,
> Though filled with the nectar that Jupiter sips.

---

[a] Misson's Travels, iii, 327, and Keysler's Travels, iii, 138.
[b] They have been erroneously attributed to the British Poet Wordsworth.

CHAPTER X.

The Windlass: Its origin unknown—Employed in raising water from wells, and ore from mines—Chinese windlass—Other inventions of that people, as table forks, winnowing machines, &c. &c. Fusee: Its application to raise water from wells—Its inventor not known. Wheel and pinion—Anglo-Saxon crane—Drum attached to the windlass roller, and turned by a rope: Used in Birmah, England, &c. Tread wheels: Used by the Ancients—Moved by men and various animals—Jacks—Horizontal tread wheels—Common wheel or capstan. Observations on the introduction of table forks into Europe.

THE WINDLASS.

ALTHOUGH it may never be known to whom the world is indebted for the *windlass*, there are circumstances which point to the construction of wells and raising of water from them, as among the first uses to which it, as well as the pulley, was applied. The windlass possesses an important advantage over the *single* pulley in lifting weights, or overcoming any resistance; since the intensity of the force transmitted through it, can be modified, either by varying the length of the crank, or the circumference of the roller on which the rope is coiled. Sometimes a single vessel and rope, but more frequently two, are employed, as in the figure, No. 18.

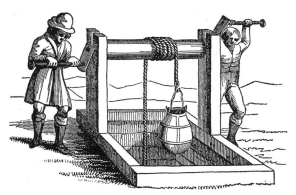

No. 18. Windlass. From Kircher's Mundus Subterraneus.

The buckets are suspended from opposite sides of the roller, the rope winding round it in different directions, so that, as one ascends, the other descends. Pliny, in his Natural History, xix, 4, mentions this machine as used by the Romans for raising water; and in the 36th book, cap. xv, when speaking of a canal for draining the marsh *Fucinus*, part of which passed through a mountain, he says the water which flowed in upon the workmen was raised up " with device of engines and *windles*." As there was not any apparatus attached to the public wells in Greek and Roman cities, or if so, to a very limited extent, it is probable the windlass was chiefly used in the country, where its application to deep wells was perhaps as common as it is in other parts of the world at the present time.

It has always been used in raising ore and water from mines. Agricola has given several figures of it as employed in those of Hungary, where

Chap. 10.] Chinese Windlass.

it has probably been in uninterrupted use since the Roman era.[a] Sometimes it was placed on one side of the well, and at a short distance from it, the ropes passing through pulleys that were suspended over its mouth. By this arrangement water may be raised to any required height *above* the windlass; an advantage in some cases very desirable. Belidor has given a similar figure, and observes that such machines were extensively used in the Low Countries.[b] Sometimes a series of pulleys were combined with it. In an old work, we have seen the windlass attached to a large tub in which water or coal was raised, so that one or more persons might ascend and descend, without the aid of others on the surface of the ground; the ropes being passed through a block above the mouth of the pit.[c] It is very probable that these applications of it were known to the Greeks and Romans. Switzer, in his 'Hydrostatics,' says, the ancients used the windlass for raising water, and that all their machines of a similar construction were classed under the general name of *Budromia*.

There is a very peculiar and exceedingly ingenious modification of the windlass, which may here be noticed, and for which we are indebted to the Chinese. It furnishes the means of increasing mechanical energy to almost any extent, and as it is used by them to raise water from some of those prodigiously deep wells already noticed, (p. 30,) a figure of it, (No. 19.) is inserted. The roller consists of two parts of unequal diameters, to the extremities of which, the ends of the rope are fastened on opposite sides, so as to wind round both parts in different directions. As the load to be raised is suspended to a pulley, (See fig.) every turn of the roller raises a portion of the rope equal to the circumference of the thicker part, but at the same time lets down a portion equal to that of the smaller; consequently the weight is raised at each turn, through a space equal only to half the difference between the circumferences of the two parts of the roller. The action of this machine is therefore slow, but the mechanical advantages are proportionably great.[d]

No. 19. Chinese Windlass. No. 20. Fusee Windlass.

This is the neatest and most simple modification of the wheel and axle, that human ingenuity has devised, and is a proof that the principles of mechanical science were well understood in remote ages; for every me-

---

[a] De Re Metallica, Basil. 1657. p. 118, 119, 160.
[b] Architecture Hydraulique, tom. 2, p. 333.
[c] Besson's Théatre des Instrvmens Mathematiques et Mechaniques. A Lyon, 1579.
[d] 'The Chinese,' by J. F. Davis, vol. ii, 286.

chanician, we think will admit, that mechanical tact and ingenuity, unaided by scientific knowledge, could never have devised it. It exhibits a species of originality so unique, so simple and efficient, that evidently shows it to have been the conception of no common mind. At what time it was first taken to Europe, we have not the means of ascertaining. It has but recently, i. e. comparatively so, been described in books. We are not aware of its having been noticed in any, previous to the last half century. It appears to have been introduced like several other standard machines from the same source, so gradually, that the precise period of its first arrival cannot easily be determined. Considering the long period, during which European nations have maintained an intercourse with the Chinese, the recent introduction of this machine may appear singular; but very little is yet known of that people, although an intimate acquaintance with their arts, would probably enrich us with treasures, more valuable than their teas and their porcelain.

There is a large debt of gratitude due to the Chinese, which has never been sufficiently acknowledged. It is to them, we are indebted for some of the most important discoveries connected with the present state of the arts and sciences. From them was derived the chief of all arts, PRINTING, and even movable types, and that invaluable acquisition, the mariner's compass; peculiar stoves,[a] chain-bridges, spectacles, silver forks,[b] India ink,[c] chain-pump, winnowing machine,[d] besides many others; and to correct a popular error, which attributes to our fellow citizens of Connecticut, the invention of 'wooden hams,' it may as well be remarked, that these are also of Chinese origin. Le Comte, says they are so adroitly constructed, that numerous buyers are constantly deceived; and frequently it is not till one is boiled and ready to be eaten, that it is dicovered to be "nothing but a large piece of wood under a hog's skin." But if China has produced specimens of dishonest ingenuity, she has, in the *tread-mill*, furnished one of the greatest terrors to evil doers.

A large *Fusee* is sometimes used in place of the cylindrical roller of a windlass, especially in wells of great depth. When a bucket is at the bottom, and the weight of a long rope or chain has to be overcome in addition to that of the water, it is accomplished more easily by winding up

---

[a] "These *stoves* are extremely convenient, and deserve to be made known universally in our country. Some of our company took such stoves with them to Gottenburgh, as models for those who might want to know their construction." Osbeck's Voyage to China, vol. i, 322.

[b] "The use of silver forks with us, by some of our spruce gallants taken up of late, came from China to Italy, and from thence to England." Heylin's Cosmography, Lon. 1670. p. 865.

[c] The secret of making it, was brought by a Dutch supercargo to Göttingen in 1756, and there divulged. Lon. Mag. for 1756. p. 403.

[d] This was also brought first to Holland in the beginning of the 18th century, whence it soon spread over Europe. It was carried to Scotland in 1710. Walter Scott, has incorporated in one of his novels, an historical fact relating to the superstition of his countrymen respecting it. When first introduced, the religious feelings of some were greatly shocked at an invention, by which artificial whirlwinds were produced in calm weather, when, as they supposed, it was the will of God for the air to remain still. As they considered it a moral duty to wait patiently for a natural wind, they looked upon the use of this machine, as rebellion against heaven, and an attempt to take the government of the world out of the Creator's hands! Constant readers of the Bible, the more superstitious of the Covenanters imagined it was a cunning device of the Wicked One, the 'Prince of the power of the Air,' and therefore one of those works, which Christians are called to guard against and renounce! It was introduced into America in 1761, as a "Dutch machine for winnowing grain." The first one, was made in Massachusetts, "by the directions of a gentleman in the Jersies," during the same year. Lon. Mag. for 1761. p. 273. Davis' Chinese, vol. ii, 361.

Chap. 10.] *Wheel and Pinion.* 71

the rope on the small end of the fusee; and as the length diminishes, it coils round the larger part. (See No. 20, which is however inaccurately drawn—as the bucket is at the top of the well, it should have been represented as suspended from the large end of the fusee.) The value of a device like this, will be appreciated when the great depth of some wells is considered, and the consequent additional weight of the chains. In the fortress of Dresden is a well, eighteen hundred feet deep; at Spangenburgh one of sixty toises; at Homberg, one of eighty; at Augustburgh, is a well at which half an hour is required to raise the bucket; and at Nuremburgh another, sixteen hundred feet deep. In all these, the water is raised by chains, and the weight of the last one is stated to be upwards of a ton: Misson, (vol. i, 116,) says three thousand pounds.

It is to be regretted that the name of the inventor of the fusee, and the date of its origin, are alike unknown. It forms an essential part in the mechanism of ordinary watches; for without it they would not be correct measurers of time. Every person knows that the moving power in a clock is a weight, and that the various movements are regulated by a pendulum; but neither weights nor pendulums are suited to portable clocks, or watches; hence a spiral spring is adopted as the first mover in the latter, and when coiled up, as it is by the act of 'winding up' a watch, the force which it exerts, imparts motion to the train of wheels; but as this force gradually diminishes as the spring unwinds, the velocity of the train would diminish also, if some mode of equalizing the effect of this varying force was not adopted: It is the fusee which does this, by receiving the energy of the spring when at its maximum, on its *smaller* end; and as this energy diminishes, it acts on the *larger* parts, as on the ends of levers, which lengthen in the same ratio as the force that moves them is diminished.

No. 21. Windlass with cog-wheel and pinion.

In another modification of the windlass, a *cog-wheel* is fixed to one end of the roller, and moved by a *pinion* that is secured on a separate shaft, and turned by a crank, as in the figure. By proportioning the diameter of the wheel and that of the pinion, (or the number of teeth on each) according to the power employed; a bucket and its contents may be raised from any depth, since a diminution in the velocity of the wheel from a smaller pinion, is accompanied with an increase of the energy transmitted to the roller and *vice versa.*

The Greeks and Romans employed the wheel and pinion in several of their war engines, and in various other machinery. Part of a cog-wheel was discovered in Pompeii. They probably were also employed, as in No. 21, to raise water from deep wells, a purpose for which they have been long used in Europe. See Belidor, tom. ii, liv. 4. From some experiments made by Mr. Robertson Buchanan, it was ascertained that the labor of a man in working a pump, turning a winch, ringing a bell, and rowing a boat, might be represented respectively by the numbers, 100, 167, 227, and 248; hence it appears that the effect of a man's labor in turning a windlass, is fifty per cent. more than in working a pump in the ordinary way by l er

No. 22. Anglo Saxon Crane.

As a man cannot, with effect, apply his strength conveniently to a crank that describes a circle exceeding three or four feet in diameter, another ancient contrivance enabled him to transmit it through a series of revolving levers, inserted into one or both ends of the roller; and which extended to a greater distance from the centre than the crank, as in the copper-plate printing press, the steering wheel of ships and steam vessels, and numerous other apparatus employed in the arts. It was formerly used to raise water in buckets from mines and wells, and even to work pumps: (cams being secured to the roller, raised the piston rods in a manner similar to the common stamping mills.) Agricola has figured it as applied to both purposes. De Re Metallica, 118, 129, 141. No. 22, is an example of its application by the Anglo Saxons, from Strutt's Antiquities

No. 23. Drum attached to a Windlass.

"There cannot be a more expeditious way to raise water from a deep well, than to make a large wheel, [drum] at the end of the *winlace*, that may be two or three times the diameter of the winlace, on which a smaller and longer rope may be wound, than that which raises the bucket, so that when the bucket is in the well, the same rope is all wound on the greater wheel, [drum] the end whereof may be taken on the shoulder, and the man may walk or run forwards, till the bucket be drawn up. The bucket may have a round hole in the midst of the bottom with a cover fitted to it, like the sucker of a pump, that when the bucket rests on the water, the hole may open and the bucket fill." Dictionarium Rusticum, Lon. 1704. See No. 23.

This is one of the modes of raising heavy weights, described by Vitruvius, in Book X of his Architecture, and is so figured in some of the old editions, that of Barbaro for example. Venice 1567. It appears to have been adopted to raise water from the deep wells of Asia in ancient times, and is still continued in use there. In Sym's Embassy to Ava, there is a notice of the Petroleum Wells, the oil from which is universally employed throughout the Birman empire. One which he examined was four feet square, and thirty-seven fathoms, [222 feet] deep. The water and oil "were drawn up in an iron pot, fastened to a rope passed over a wooden cylinder, which revolves on an axis supported by two upright posts. When the pot is filled, two men take the rope by the end, and run down a declivity, which is cut in the ground to a distance equivalent to the depth of the well; thus when they reach the end of their track, the pot is raised to its proper elevation."[a] The contents, water and oil, are then discharged into a cistern, and the water is afterwards drawn off through a hole in the bottom. A ratchet wheel and click to detain the bucket when elevated, would enable a single person to work this machine, or the bucket might be suspended to its bail by swivels, and overturned at the top by a catch, as in No. 16.

No. 24. Tread Wheel.

Another mode of communicating motion to the roller, is by means of a *tread wheel*, attached like the drum in figure 23, to one end of it. In this, a man or an animal walks or rather climbs up one side, somewhat like a squirrel in its cage, and by his weight turns the wheel, and raises the water, as represented in No. 24.

This appears to have been a common mode of applying human effort among the ancients. Some of their cranes for raising columns and other

---

[a] Embassy to Ava, Lon. 1800, vol. iii, 236. See also an account of these wells, and modes of raising their contents, in vol. ix, of Tilloch's Phil. Mag. p. 226

heavy weights, were moved by tread wheels, (Vitruvius, x, 4.) A figure of one is preserved in a bas-relief, in the wall of the market place at Capua.[a] Like other ancient devices for raising water it has been continued in use in Europe since Roman times, and is described by most of the old writers on Hydraulics. Agricola figures it as used in the mines of Germany. "To raise water from a deepe well," says an old English writer, "some use a large wheele for man or beast to walk in." At *Nice*, two men raise the water from a deep well, by walking in one of these wheels. It is the 'Kentish fashion' according to Fosbroke, the wheels being propelled both by men and asses. The Anglo Saxons and Normans also used them for drawing water.

Whether the Greeks and Romans employed *animals* in tread wheels, we know not, but the practice is very old, and from the obvious advantage of quadrupeds over biped man in climbing ascents, it is probable that they were so employed by the ancients. Oxen, horses, mules, asses, dogs, goats and bears, have all been used to propel these wheels, and to raise water by them. At Spangenburgh, water was raised from the well mentioned in a previous chapter, by an *ass*. In the Isle of Wight, Eng. one was thus engaged for the extraordinary period of forty years, in raising water from a well two hundred feet deep, which was supposed to have been dug by the Romans. Long practice had taught the animal to know exactly how many revolutions were required to raise the bucket; when by a backward movement he would instantly stop the wheel. *Goats* are remarkable for scaling precipices, and therefore seem well adapted for this kind of labor. In Europe, they are employed to raise both water and ore from mines, by tread wheels. In the Chapter on the Chain-Pump, we have inserted a cut from Agricola, representing them thus engaged, But of the larger animals, if there is one better adapted than another, from its conformation and habits, it is the *Bear;* and it is not a little singular, that the *Goths* actually employed that animal in such wheels to raise water.[b]

It is probable that the Chinese have from remote times employed various animals in them; this we infer from a remark of one of their writers, quoted by Dr. Milne. In exhorting husbands to instruct their wives, he encourages them in the *arduous* task by reminding them, that "even monkeys may be taught to play antics—*dogs* may be taught to *tread in a mill*—*cats* may be taught to *run round a cylinder*, and parrots may be taught to recite verses;" and hence he concludes it possible to teach women something![c] Many ancient customs relating to the taming and using of animals are still practised in China. The old Egyptians, had baboons and monkeys trained to gather fruit from trees and precipices inaccessible to men.[d] The Chinese employ them for the same purpose.[e] Mark Antony, Nero, and others, in imitation of more ancient warriors, were sometimes drawn in chariots by lions—Chinese charlatans, both ride on, and are drawn by them, and by tigers also. (Nieuhoff's Embassy.)

Before *smoke* jacks were invented, joints of meat while roasting, were often turned by *dogs* running in tread-wheels. They were of a peculiar breed, (now nearly extinct) long backed, having short legs, and from

---

[a] Fosbroke Antiq. 257, and Winkleman's Arts of the Ancients. Paris, Ed. tom. ii, planche 13.
[b] Olaus Magnus, quoted by Fosbroke. Encyc. Antiq. 71. [c] Downing's Stranger in China, vol. ii, 172. [d] Wilkinson, vol. ii, 150. [e] Breton's China.

their occupation, were named *turnspits*. The mode of teaching them, was more summary than humane. The animal was put into a wheel, the sides of which were closed, and a burning coal thrown in behind him; hence he could not stop climbing without having his legs burned. As might be supposed, they were by no means attached to their profession; of which the following incident has been adduced as a proof: In a certain city, having agreeably to custom, attended their owners to church, the lesson for the day, happened to be that chapter of Ezekiel wherein the self-moving chariots are described. When the minister first pronounced the word '*wheel*,' they all pricked up their ears in alarm—at the second mention of it, they set up a doleful howl; and when the awful word was uttered a third time, every one it is said, made the best of his way out of the church.

But the most singular animal formerly used in thus turning the spit, was a *bird*, and of a species too which furnished more victims for the roast than any other, viz. the *goose!* Moxon, observes that although dogs were commonly used, "geese are better, for they will bear their labor longer, so that if there be need, they will continue their labor twelve hours."[a] A singular illustration of man's power over the lower animals, in thus compelling one to cook another of its own species for his use.

The old jack, consisting of three toothed wheels and a weight, was used as early as 1444. The *smoke* jack was known in the following century, if not before, for it was described by Cardan, and afterwards (in 1571) by Bartolomeo Scappi, cook to Pope Pius V. in a book on culinary operations. In 1601, a 'jack maker' was a regular trade in Europe, and the ingenuity of the manufacturers was then often displayed in decorating them with moving puppets, as in some ancient clocks, and in the organs, &c. of street musicians. Bishop Wilkins, (Mathematical Magic, B. ii, cap. 3) speaks of "jacks no bigger than a walnut to turn any joint of meat."

The *name* of these machines, and a certain vulgar phrase, not yet quite obsolete, are all that is left to recal to mind, a class of domestics, whose occupation, like Othello's, is gone. These were men whose duty it was to turn the spit, and who answered to the familiar cognomen of 'Jack,' formerly a common name for a man-servant, and now applied to designate numerous instruments that supply his place. Seated at one side of a huge fire, his duty was to turn the roast by a crank attached to one end of the spit. See a figure of a French *tourne-broche* in the exercise of his vocation, in 'Hone's Every Day Book.' vol. ii, 1057. The office was far from being a sinecure, since no slight labor was required to move the large joints of olden times, the whole of a sheep or an ox being frequently roasted at once; hence in some kitchens built in the 13th century, it was particularly directed, that each should be provided with furnaces sufficiently large to roast two or even three oxen. It was from the custom of these artists, of surreptitiously helping themselves to small pieces of the roast, while in the performance of their duty, the unclassical expression 'licking the fingers,' came, and verifying a Turkish proverb, 'he that watches the kettle, is sure to have some of the soup.' The phrase however, was not then so obnoxious to good taste, nor the act to good manners, as now; for table-forks were generally unknown, and moderate sized joints were handed round on the spit, so that every one at table, separated by a knife a slice to his taste, and conveyed it by his fingers

---

[a] Mechanick Powers, Lon. 1696. p. 72.

to his plate, and thence to his mouth. Hence the advice of *Ovid*, for neither Greeks nor Romans used table-forks:

"Your meat genteelly with your fingers raise ;
And—as in eating there's a certain grace,
Beware, with greasy hands, lest you besmear your face.

A German writer in the middle of the 16th century, in suggesting the whirling Eolipile as a turnspit, remarks, "it eats nothing, and gives withal an assurance to those partaking of the feast, whose suspicious natures nurse queasy appetites, that the haunch has not been pawed by the turnspit, in the absence of the housewife's eyes, for the pleasure of licking his unclean fingers." This evil propensity of human turnspits, however, eventually led to their dismissal, and to the employment of another species, which, if not better disposed to resist the same temptations, had less opportunities afforded of falling into them. These were the canine laborers already noticed.

No. 25. Horizontal Tread-Wheel, from Agricola.

Horizontal tread-wheels for raising water are described by Agricola, from whose work, *De Re Metallica*, we have copied the figure. Two men on opposite sides of a horizontal bar, against which they lean, push with their feet the bars of the wheel on which they tread, behind them. Similar wheels, inclined to the horizon were also used. For another kind of tread-wheel, see chapters 14, and 17. On the Noria and Chain-pump.

In all the preceding machines the roller is used in a *horizontal* position; but at some unknown period of past ages, another modification was devised, one, by which the power could be applied at any distance from the centre. Instead of placing the roller as before, over the well's mouth, it was removed a short distance from it, and secured in a vertical position, by which it was converted into the wheel or capstan. One or more horizontal bars were attached to it, of a length adapted to the power employed, whether of men or animals; and an alternating rotary movement imparted to it, as in the common wheel or capstan, represented in the next figure, No. 26. It appears from Belidor, (Tom. ii, 333) that machines of this kind, and worked by *men* were common in Europe previous to, and at the time he wrote. Sometimes the shaft was placed in the edge of the well, so that the person that moved it walked round the latter, and thus occupied less space.

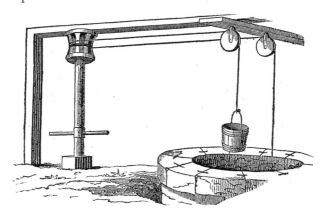

No. 26. Common Wheel or Capstan.

Circumstances, highly illustrative of European manners during the early part of the 17th and preceding centuries, are associated with the introduction of *table-forks*. They were partially known in Italy in the 11th century, for in a letter of Peter Damiani, who died in 1073, mention is made of a lady from Constantinople, who was married to the Doge of Venice, and who among other strange customs, required *rain-water* to wash herself, and was so fastidious respecting her food, as to use a fork, and a golden one too, to take her meat, which was previously cut into small pieces by her servant. Lon. Quart. Review, vol. 58. (April, 1837.) They are mentioned, (probably as curiosities) in a charter of Ferdinand I. of Spain, 1101, and in the wardrobe accounts of Edward I. of England, 'a pair of knives with sheathes of silver enameled, and a forke of chrystal,' are specified. Fosbroke, Ency. Ant. Forks were common in Italy in the 15th century, although nearly unknown in France and England in the following one. At the close of the 16th, they are noticed as a luxury in France, and *lately introduced*. Henry the Fourth's fork is still preserved —it has two prongs, and is of steel. So late as 1641, they were not universal in Paris. In a representation of a great feast held by the cobblers in that year, and attended by musicians, &c.—no forks are on the table—the carver holds what appears to be a leg of mutton with one hand, while with the other he cuts a slice off, for a lady seated next to him. Hone's Every Day Book, vol. ii, 1055.

They were not used in England till about the same time, a period much later than might have been supposed. In 1611, an Englishman was ridiculed for using one. This was Coryatt the eccentric traveler. "I observed," he says, "a custom in all those Italian cities and townes through the which I passed, that is not used in any other country that I saw in my travels; neither do I think that any other nation of Christendome doth use it, but only Italy. The Italian and most strangers that are commorant, [dwelling] in Italy do alwaies at their meales *use a little forke*, when they cut their meate. For while with their knife, which they hold in one hand, they cut the meate out of the dish, they fasten their forke which they hold in the other hand, upon the same dish. So that whatsoever he be, that sitting in the company of any others at meale, should unadvisedly touch the dish of meate with his fingers, from which all at the table doe cut, he will give occasion of offense unto the company, as having transgressed the lawes of good manners; insomuch that for his error, he shall be at least brow-beaten, if not reprehended in words. This forme of feeding, I understand, is generally used in all places in Italy, their forkes being for the most part made of yron or steele, and some of silver; but these are used only by gentlemen. The reason of this their curiosity is, because the Italian cannot by any means have his dish touched with fingers, seeing all men's fingers are not alike cleane—hereupon I myself thought good to imitate the Italian fashion, by this *forked* cutting of meate, not only while I was in Italy, but also in Germany, and oftentimes in England since I came home : being once quipped for that frequent using of my forke, by a certain learned gentleman, a familiar friend of mine, one Master Laurence Whitaker, who in his merry humour doubted not to call me *Furcifer*, only for using a forke at feeding."

In this extract, we have a view of Italian gentlemen 'feeding' in the beginning of the 17th century, and in the following one, we obtain an insight into British manners during the middle of it. Forty years after the publication of Coryatt's Travels, a Manual of Cookery appeared, containing the following instructions to British ladies, when at table. "A gentlewoman being at table, abroad or at home, must observe to keep her body straighte, and lean not by any means upon her elbowes—nor by ravenous gesture discover a voracious appetite. Talke not when you have meate in your mouthe; and do not smacke like a pig—nor eat spoonemeat so hot that the tears stand in your eyes. It is very uncourtly to drinke so large a draught that your breath is almost gone, and you are forced to blow strongly to recover yourselfe. Throwing down your liquor as into a funnel, is an action fitter for a juggler than a gentlewoman. In carving at your own table, distribute the best pieces first, and it will appeare very decent and comely to *use a forke ; so touch no piece of meat without it.*" This elegant extract is from ' *The Accomplished Lady's Rich Closet of Rarities.*' London, 1653. Neither knives nor forks were used at the tables of the Egyptians. A representation of a feasting party is sculptured on a tomb near the pyramids, a copy of which is inserted in Vol. II, of Wilkinson's interesting work. One gentleman holds a small joint of meat in his hand, two are eating fish which they retain in their fingers, while another is separating the wing of a goose with the same implements.

CHAPTER XI.

Agriculture gave rise to numerous devices for raising water—Curious definition of Egyptian husbandry—Irrigation always practised in the east—Great fertility of watered land—The construction of the lakes and canals of Egypt and China, subsequent to the use of hydraulic machines—Phenomenon in ancient Thebes—Similarity of the early histories of the Egyptians and Chinese—Mythology based on agriculture and irrigation: Both inculcated as a part of religion—Asiatic tanks—Watering land with the yoke and pots—An employment of the Israelites in Egypt—Hindoo Water Bearer—Curious shaped vessels—Aquarius, 'the Water Pourer,' an emblem of irrigation—Connection of astronomy with agriculture—Swinging baskets of Egypt, China and Hindostan. Arts and customs of the ancient Egyptians.

THE last three chapters include most of the methods adopted by the ancients to raise water for domestic purposes. There is, however, another class of machines of equal merit and importance, which probably had their origin in agriculture, i. e. in the irrigation of land. Persons who live in temperate climates, where water generally abounds, can scarcely realize the importance of *artificial irrigation* to the people of Asia and other parts of the earth. It was this, which chiefly contributed to support those swarms of human beings, who anciently dwelt on the plains of the Euphrates, the Ganges, the Nile, and other large rivers. In Egypt alone, the existence of millions of our species has in all times depended wholly upon it, and hence the antiquity of machines to raise water among that people. The definition of oriental agriculture is all but incomprehensible to an uninformed American or European—it is said to consist chiefly, "in having suitable machines for raising water," a definition sufficiently descriptive of the profession of our firemen, but few people would ever suppose it explanatory of that of a farmer. It is however literally true. Irrigation is everything—the whole system of husbandry is included in it; and no greater proof of its value need be given, than the fact of machines employed to raise water for that purpose in Egypt, being taxed.

The agricultural pursuits of man, must at a very early period have convinced him of the value of water in increasing the fruitfulness of the soil: he could not but observe the fertilizing effects of rain, and the rich vegetation consequent on the periodical inundations of rivers; nor on the other hand, could he possibly have remained ignorant of the sterility consequent on long continued droughts: hence nature taught man the art of irrigating land, and confirmed him in the practice of it, by the benefits it invariably produced. In some countries the soil was thus rendered so exceedingly fruitful as to exceed credibility. Herodotus, when speaking of Babylonia, which was chiefly watered by artificial irrigation, (for the Assyrians he observes, 'had but little rain,') says, it was the most fruitful of all the countries he had visited. Corn, he said never produced less than two hundred-fold, and sometimes three hundred; and after reciting some other examples, he remarks, that those persons who had not seen the country, would deem his account of it a violation of probability—in other words, a traveler's tale. Clio, 193. Five hundred years afterwards, the elder Pliny speaking of the same country, observes, "there is not a territory in all the east comparable to it in fertility;" while in another part of his work, he refers to the *cause* of its fruitfulness—he says, the principal care required, was, "to keep the ground well watered." Nat. Hist. vi, 26, and xviii, 17.

Mr. St. John, mentions a species of Indian corn growing in the fields of Egypt, prodigiously prolific. On one ear, three thousand grains were reckoned! and a lady, who frequently made the experiment in the Thebaid, constantly found between eighteen hundred and two thousand. Egypt and Mahommed Ali, vol. i, 143. Another proof of the value of irrigation is given by Herodotus. When speaking of that part of Egypt near Memphis, he observes, that the people enjoyed the fruits of the earth with the smallest labor. "They have no occasion for the process *nor the instruments* of agriculture, usual and necessary in other countries." This remark of the historian, has been ridiculed by some authors; but its truth has been verified by recent travelers.[a]

The advantages of artificial irrigation have not only been known from the earliest ages, but some of the most stupendous works which the intellect of man ever called into existence, were designed for that purpose: works so ancient as to perplex our chronologists, and so vast as to incline some historians to class them among natural formations. Ancient writers unite in asserting that *Lake Mœris* was 'the work of men's hands,' and constructed by a king of that name; its prodigious extent, however, has led some modern authors to question its alleged origin, although artificial works still extant, equal it in the amount of labor required; as the *Wall* of China, the Pyramids, and other works of ancient Egypt. Sir William Chambers, when comparing the works of the *remote* ancients with those of Greece, observes that the city of Babylon would have covered all Attica; that a greater number of men were employed in building it, than there were inhabitants of Greece; that more materials were consumed in a single Egyptian Pyramid, than in all the public structures of Athens; and that Lake Mareotis could have deluged the Peloponnesus, and ruined all Greece. But incredible as the accounts of Lakes Mœris and Mareotis may appear, these works did not surpass, if they equaled, another example of Egyptian engineering, which had *previously* been executed. This was the removal of the Nile itself! In the reign of *Menes*, (the first, or one of the first sovereigns) it swept along the Libyan chain of mountains, that is, on *one side* of the valley that constitutes Egypt; and in order to render it equally beneficial to both sides, a new channel was formed through the centre of the valley, into which it was directed: an undertaking which indicates a high degree of scientific knowledge at that early period.

Before the lakes and canals of Egypt or China could have been undertaken, the inhabitants must have been long under a regular government, and one which could command the resources of a settled people, and of a people too, who from experience could appreciate the value of such works for the purpose of irrigation, as well as the inefficiency of previous devices for the same object; that is, of *machines* for raising the water: for if it be supposed the construction of canals to convey, of reservoirs to contain, and of locks and sluices to distribute water, *preceded* the use of MACHINES for raising it—it would be admitting that men in ignorant times had the ability to conceive, and the skill to execute the most extensive and perfect works that civil engineering ever produced—to have formed lakes like oceans, and conveyed rivers through deserts, ere they well knew how to raise water in a bucket, or transmit it through a pipe or a gutter. The fact is, ages must necessarily have elapsed before such works could have been dreamt of, and more before they could have been accomplished. Individuals would naturally have recourse to rivers in their immediate vicinity, from which (the Nile for example) they must long have toiled in

---

[a] St. John, vol. i, 181. Lindsay, Let. 5.

raising water, before they would ever think of procuring it from other parts of the same stream, at distances varying from ten to a hundred miles, or consent to labor for its conveyance over such extensive spaces.

How extremely ancient, then, must hydraulic machinery be in Egypt, when such works as we have named, were executed in times that transpired long before the commencement of history—times that have been considered as extending back to the infancy of the world! But if, as is generally supposed, civilization and the arts *descended* the Nile, then machines for raising water must have been employed on the *upper* borders of that river before Lower Egypt was peopled at all. In Nubia and Abyssinia they have at all times been indispensable, in consequence of the elevation of the banks and the absence of rain; while in Middle and Lower Egypt, during the intervals of the annual overflow, they were also the only resource; and at no time even there, could those grounds which lay beyond the reach, or above the inundation, be irrigated without them. The surface of the river during 'low Nile' is, in parts of Nubia, 30 feet below the banks, in Middle Egypt 20 feet, and diminishes to the Delta.[a] In other countries, *rain*-water was (and still is) collected into reservoirs on the highest places, and distributed as required; but nothing of this kind could ever have taken place in the land of the Pharaohs. We are informed the inhabitants of ancient Thebes were once thrown into great consternation by a phenomenon which they looked upon as an omen of fearful import—it was simply a *shower of rain!* This circumstance explains the opinion of the old Egyptians, in supposing the Greeks subject to famines, as their harvests depended on rain. Herod. ii, 13.

A striking result derived from an examination of Egyptian history, observes Mr. Wilkinson, is the conviction, that the earliest times into which we are able to penetrate, *civilized communities* already existed, and society possessed all the features of later ages. The most remote period to which we can see, opens with a nation possessing all the *arts* of civilized life *already matured*. The pyramids of Memphis were erected within three hundred years of the Deluge; and the Tombs of Beni-Hassan, hewn and painted with subjects describing the arts and manners of a refined people, about six hundred years after that event. From these paintings, &c. we learn that the manufacture of linen, of cabinet work, of glass, of elaborate works in gold and silver, bronze, &c. and numerous games,[b] &c. were then, as now, in vogue. The style of architecture was grand and chaste, for the fluted columns of Beni-Hassan "are of a character calling to mind the purity of the Doric." Indeed, modern science and the researches of travelers are daily adducing facts which set at defiance our ordinary chronological theories, but which appear to strengthen the opinion of those commentators of the scriptures, who consider the Deluge of Noah to have been, like that of Deucalion, *local*, not universal: a doctrine inconsistent with the postdiluvian origin of the mechanic arts generally, and of simple machines for raising water in particular.

There is a striking resemblance between the early history of Egypt and that of China, with regard to the origin of, and esteem for agriculture, religious ceremonies connected with it, artificial irrigation, modes of working the soil, and implements used. This art among both people was coeval with their existence as nations, and doubtless extends back, as ancient writers assert, to periods anterior to Noah's flood.[c] SHINNUNG, the 'divine

---
[a] Wilk. An. Egypt, i, 9.
[b] Lifting weights, wrestling, single stick, and bull-fights, games of ball, throwing knives odd and even, draughts, dice, and thimble-rig, are all represented on the monuments.
[c] Le Comte's, China, 118, Lon. 1738.

husbandman' resembled Osiris. He began to reign 2832, B. C. or nearly five hundred years before the deluge; and it is in imitation of this antediluvian monarch, that the emperors at the present day, plough a portion of land with their own hands, and sow different kinds of grain, once a year. Such it appears was also the practice of Egyptian monarchs, for emblematical representations of them breaking up the soil with a hoe, are found among the sculptures. Osiris instructed them in agriculture, and taught them the value and practice of *irrigation*. He confined the Nile within banks, and formed sluices to water the land. On these accounts was he idolized after death, and worshipped as a god; and the ox, which he taught them to employ in the cultivation of the soil, was selected to personate him. In accordance with this custom, the Israelites made and worshipped the calf (or ox) in the wilderness; and Jereboam to gratify the ignorance or superstition of his countrymen, placed golden statues of them in Dan and Bethel; or, perhaps from policy, introducing the worship of Apis, as a compliment to Shishak, with whom he had found a refuge from the wrath of Solomon. Similar to the worship of Apis, is the great festival of the Chinese when the sun reaches the 15° of Aquarius, during which a figure of a *cow* is carried in the processions. The feast of Lanterns of this people, is also like that of *lamps* anciently held over all Egypt. Herod. ii, 62. The celebrated Yu was, like some of his predecessors, raised to the throne of China, on account of his agricultural labors. We are informed that he drained a great portion of the land; and that he wrote several books on the cultivation of the soil, and on *irrigation*. He flourished 2205, B. C. or about 140 years after the flood. Seventeen years after his death, it is admitted that husbandry became *systematically organized*, which necessarily included a settled plan of artificial irrigation. A writer in the Chinese Repository observes, "in these early ages, the fundamental maxims of the science of husbandry were established, which so far as we can learn, *have been practised to the present day*." Vol. iii, 121. Egyptian husbandry we are told consisted chiefly in having *proper machines* for raising water, and small canals judiciously disposed to distribute it over their fields. In both these respects it resembled that of China at the present day; a portable machine to raise water, being as necessary an implement to a Chinese peasant as a spade is to one of ours.

The *mythology* of the whole ancient world seems not only to have been intimately associated with agriculture, but appears to have been based upon it. To the invention of the plough, and the irrigation of land, all the mysteries of Ceres may be referred. The great importance of agriculture in furnishing food to man, induced legislators at an early period to devise means to promote it. This they accomplished by connecting it with the worship of the gods; and by classing the labors of husbandry among the most essential of religious duties. This system seems to have been universal. It was incorporated by the Roman lawgivers with the institutions of that people. Plutarch, in his life of Numa, expressly states that some of the laws were designed to recommend agriculture "as a part of religion." See also Pliny, xvii. 2. The *sacred books* of almost all the ancient nations placed irrigation, digging of tanks and wells, among those acts most acceptable to the gods; and hence the Zendavesta of the Persians, and the Shaster of the Hindoos, distinctly inculcate the conveying of water to barren land, as one of the precepts of religion.

As no country depended more upon agriculture for the physical and political existence of its inhabitants, than Egypt; so nowhere was religious veneration for it carried to a greater extent. We are informed their priests and statesmen used "all their influence in advancing the prosperity and *splendor* of agriculture;" and as Osiris made his own plough, both kings and priests wore sceptres of the form of that implement. So sensible were they of the blessings derived from agriculture, that they not only enforced the worship of Osiris and Isis for its discovery and introduction, but caused even the animals (oxen) that assisted in the cultivation of their lands to be honored with religious rites. It cannot therefore be supposed, that a people, who thus consecrated almost every thing connected with husbandry, should neglect that in which it chiefly consisted, viz: IRRIGATION; for this was the real cause of the amazing fruitfulness of their soil, and of their individual and national prosperity. Thus Isis is often found represented with a bucket, as an emblem of irrigation, and of the fecundity of the Nile.[a]

From the foregoing remarks the great antiquity of machines for raising water may be inferred; for artificial irrigation certainly gave birth to the most valuable of our hydraulic engines, if it be not indeed the great parent of them all.

The ancient practice of constructing large tanks, to collect water for irrigation, is still followed in various parts of the east, and their dimensions render our reservoirs comparatively insignificant. In the Carnatic, some are *eight miles in length,* and *three* in breadth. In Bengal, they frequently cover a hundred acres, and are lined with stone. Knox, in his Historical Relation of Ceylon, (Lon. 1681,) says, the natives formed them of two and three fathoms deep, some of which were in length 'above a mile.' (p. 9.) To these reservoirs, and the difficulty of making them sufficiently tight to hold the water, there is an allusion in Jeremiah ii, 13. Le Comte, mentions them. When limited quantities of water were required at a distance, at places situated higher than its source, it was often carried in vessels suspended from a yoke, and borne across the shoulders. This mode is still practised. In the plains of Damar in Arabia, water is drawn from deep wells, and thus carried to the fields. Dr. Shaw observes of the country of the Benni Mezzah, (in ancient Mauritania) it has no rivulets, "but is supplied *altogether* with well water." "Persia," says Fryer, "is chiefly beholden to wholesome springs of living water to quench the thirst of plants as well as living creatures." This method is pursued by the Hindoos, Japanese, Chinese, Javanese, Tartars, &c. in watering those terraces which they construct and cultivate on the sides of mountains.

No. 27. Egyptians watering land with Pots and Yoke.

"The Chinese," says Le Comte, "have everywhere in Xensi and Xansi, for want of rain, certain pits from twenty to a hundred and twenty feet deep, from which they draw water by an incredible toil to irrigate their land." That the mode of carrying water to the fields by the yoke, was practised in Egypt in ancient times, appears from the figures, No. 27, copied from sculptures at Beni Hassan, the oldest monuments extant in that country. One of these wooden yokes, was found at

---

[a] See Shaw's Trav. 403, 412; Univer. Hist. Vol. i, 205; and our remarks on the Noria, in Chap. XIV.

Thebes, with leather straps and bronze *buckles*, and is now preserved in England. The yoke is about three feet seven inches long, and the straps, which are double, about sixteen inches. Some yokes had from four to eight straps, according to the purposes for which they were intended. Wilk. An. Egypt, vol. ii, 138.

As watering the land has always been the staple employment in Egypt, there can be little doubt, that the Israelites were employed in this service. We are informed, "their lives were made bitter unto them by hard bondage, in mortar and in brick, and in *all manner of service in the field.*" And in Deuteronomy, xi, 10, irrigating the land, is expressly mentioned as one of their labors. In Leviticus, they are reminded that it was God who brought them out of Egypt, and delivered them from slavery, who broke the bands of their yoke and made them go upright; alluding to the stooping posture consequent on the long continued use of the yoke; and in case of disobedience, they were threatened with subsequent thraldom, to "serve their enemies in hunger and nakedness, and with a yoke on their necks;" a literal description of them when thus employed in watering the lands of their oppressors. A passage in the 81st Psalm, appears to refer expressly to their deliverance from this, and another laborious method of watering the soil. "I removed his shoulder from the burden, his hands were delivered from the pots." The severity of this labor, may be inferred from that of Chinese peasants, who carrying burdens like the Egyptians, have deep impressions worn on their shoulders by the yoke. Osbeck's Voy. i, 252.

It was a common custom of old to employ slaves and prisoners of war, in watering and working the land. Herodotus, i, 66, observes of the Lacedemonians, that after the death of Lycurgus, they invaded the Tegeans, and carried with them a quantity of fetters to bind their enemies; but they were themselves defeated, and loaded with their own fetters, were employed in the fields of the Tegeans: and Joshua, in accordance with this custom, made the captive Gibeonites 'hewers of wood and drawers of water.' Isaiah alludes to the same, lx, 5: *strangers shall stand and feed your flocks,* and the sons of the *alien* shall be your *ploughmen* and your *vine dressers.*

No. 23. Hindoo Water Carrier.

This figure represents a modern laborer of Hindostan, and it will serve also to represent those of China and other Asiatics, who carry water to their gardens and fields in precisely the same way. It will be perceived that the *form* of the vessels is similar to that of the old Egyptian Pots in the preceding figure, and that both of them serve to corroborate our views respecting the origin of the forms of these, and other domestic vessels of capacity.

There is another mode of carrying water, which was anciently practised, but of which we do not remember to have seen any particular notice. It is represented in Plates 50, and 75. Pitt. Storia Dell'Arte. A conical vessel bent like the horn of an ox, is borne on the shoulder, the large and open end projecting in front, so that the bearer could discharge any part of its contents, by inclining it, which, in one figure is effected by means of a cord going round it, and one end held in the hand. These vessels are figured as large as the bodies of

Chap. 11.]   *Egyptian Mental.*   85

those who carry them, and appear to have been formed of staves, and hooped.

As an evidence of the antiquity of watering land with pots, we may refer to one of the constellations, to "Aquarius" or the "Water Pourer," a figure which was adopted as an expressive emblem of that season, when rains descended, and the lands were irrigated by nature alone. Although it may possibly be true, as some authors suppose, that some of the present signs of the Zodiac were substituted for more ancient ones at some period of time, posterior to the Argonautic expedition—(see Goguet's Dissertation on the names and figures of the constellations—Origine des Loix, Tom. ii.) there are others, and among them the "Water Pourer," which are for any thing known to the contrary, *original figures*, adopted by the first cultivators of astronomical knowledge, i. e. by the antediluvian sons of Seth, who, according to Josephus, were the "inventors of that peculiar sort of wisdom which is concerned with the heavenly bodies and their order;" (Ant. B. i, chap. 2.) and which signs were continued by their successors, the Chaldeans, in the first ages after the flood, and have remained unaltered to our days. The extreme antiquity of astronomy, and its *connection with agriculture*, are undoubted, of which we shall meet with other examples besides the one just given. This connection was the source of the great mass of *symbolical imagery* which pervades the history, mythology, and almost every thing connected with the remote ancients; most of which is so perplexing to decipher, and the greater part of which has defeated all attempts of the moderns satisfactorily to explain. In the time of Job, who is supposed to have lived before Moses, the constellations were well known. "Canst thou bind the sweet influences of Pleiades, or loose the bands of Orion," xxxviii, 31. "Which maketh Arcturus, Orion and Pleiades, and the chambers of the south." ix, 9. Indeed, M. Bailey and others have admitted that the astronomy of Chaldea, India, and Egypt, is but the wreck of a great system of astronomical science, which was carried to a high degree of perfection in the early ages of the world.

When water is only required to be raised two or three feet from a tank or river, a vessel suspended by four cords, and worked by two men, is very extensively used in the east. In Egypt it is named the "Méntal," the figure of which is copied from *Grande Description*.

No. 29. Egyptian Méntal.

A small trench is dug on the edge of the river, on the borders of which two men stand opposite each other. They hold in each hand a cord, the ends of which are attached to a *basket* of palm leaves covered with leather. After launching it into the water, they lean backwards so as to be half seated on small mounds of earth raised for the purpose, by which the weight of the body assists in raising the load, as it is swung towards the gutter or basin formed on the bank to receive it. The movements of the men are regulated by chanting, a custom of great antiquity, and adopted in all kinds of manual labor where more than one person were engaged.

Sonnerat has figured and described (Vol. ii, 132,) a similar contrivance of the Hindoos. " They use a basket for watering, which is made impenetrable with cow dung and clay; it is suspended by four cords; two men hold a cord in each hand, draw up the water, and empty it in balancing the basket." Mr. Ward says this machine is commonly used in the south of Bengal to water the land. Hist. Hindoos, 92. Travelers in China have noticed it in use there. " Where the elevation of the bank over which water is to be lifted is trifling, they sometimes adopt the following simple method. A light water-tight basket or bucket is held suspended on ropes between two men, who, by alternately tightening and relaxing the ropes by which they hold it between them, give a certain swinging motion to the bucket, which first fills it with water, and then empties it by a jerk on the higher level; the elastic spring which is in the bend of the ropes, serving to diminish the labor." Davis' China, Vol. ii, 358. Chinese Repos. Vol. iii, 125. Sir George Staunton also described it, with an engraving; by which it appears the Chinese do not use a bank of earth or any other prop, like the Egyptians, to support them in their labor. Osbeck has noticed a peculiar feature in working these baskets. He says Chinese laborers TWIST the cords as they lower the vessel, and when it is raised, the *untwisting* of them, overturns it and discharges the contents. This mode of raising water in China, was noticed by Gamelli, in 1695, although not particularly described by him: he says " the Chinese draw up water in a basket, two men working at the rope."

Of all employments in ancient and modern Egypt, this may be considered the most laborious and degrading. The wretched peasants, naked or nearly so, may be seen daily, from one end of Egypt to the other, in the exercise of this severe labor. " I have seen them," says Volney, " pass whole days thus drawing water from the Nile, exposed naked to the sun, which would kill us." To this mode of raising water there is probably an allusion in the latter clause of the passage already quoted from the 81st Psalm: " His hands were delivered from the pots," or " *baskets*," as the word is sometimes translated, and is so in this instance in the margin of the common English version. Indeed, it was peculiarly appropriate that a Psalm, written as this was, to celebrate the deliverance of the Israelites from Egyptian bondage, should allude to some of the severest tasks imposed upon them while under it. Raising of water to irrigate the land was emphatically " THE LABOR OF EGYPT," from which they were freed.

Some additional remarks to those on page 81, respecting other arts and customs delineated on Egyptian monuments, may interest some readers.

Salting fish seems to have been a regular profession in ancient Egypt, and by processes similar to those now in use; although it was not till the 15th century that the art was known in modern Europe, when William Bukkum, a Dutchman, who died in 1447, " found out the art of salting, smoking, and preserving herrings." It is also not a little singular that the Egyptians had a religious rite, in which, as in modern *Lent*, every person ate fish. They used the spear, hook and line; drag, seine and other nets. Part of a net, with leads to sink it, has been found at Thebes. Wealthy individuals had private fish-ponds, in which they angled. They hunted with dogs; and also with the lion, which was tamed for that purpose. The noose or lasso, and various traps, were common. Cattle were branded with the names of their owners. In taking birds, they had decoys and nets, like modern fowlers. Beer was an Egyptian beverage, and onions a favorite esculent—these were as superior in taste to ours, as in the elegance of the bunches in which they were tied. At feasts they had music and dancing, castanets, and even the *pirouette* of Italian and French *artistes*. They had '*grace*' at meals; and wore wreaths of flowers and nosegays. Essences in bottles and ointments, the odor of some of which remains. The ladies wore necklaces formed of beads of gold, glass, and of precious stones, and even of *imitation* stones. In dress they had cotton and linen cloths: some of the latter were so fine as to be compared to *woven air*, through which the person was distinctly seen; and the former of patterns similar to those of modern calicos. Ezekiel speaks of "fine linen with broidered work from Egypt;" and in Exodus it is often mentioned. They had tissues of silver and gold, and cloth formed wholly of the latter. In furniture, carpets and rugs: one of the latter was found at Thebes, having figures of a boy and a goose wrought on it. Toilet boxes inlaid with various colored woods, and ornamented with ivory and golden studs. Sofas, chairs, stools and ottomans, all imitated in modern articles. Bedsteads enclosed in mosqueto nets; and pillows, the latter of wood, the material of which they were formerly made in Europe. Inlaid works of gold, silver, and bronze. Vases of elegant forms and elaborate workmanship: great numbers of these are represented among the varieties of *tribute* carried by *foreigners* to Thothmes III, in whose reign the Israelites left Egypt. Door-hinges and bolts of bronze, similar to the modern; scale-beams, enameling. Gold-beating and gilding. Gold and silver wire; some specimens are flattened with the hammer, others are believed to have been drawn. Vessels with spouts like those of our tea-kettles: one of the best proofs of skill in working sheet metal.

Glass blowers are represented at work, and vessels identical with our demijohns and Florence flasks have been found, and both protected with reed or wicker work—besides, pocket bottles covered with leather, and other vessels of glass, cut, cast and blown. Goldsmiths in their shops are shown, with bellows, blow-pipes, crucibles and furnaces; golden baskets of open work; solder, hard and soft, the latter an alloy of tin and lead. Stone cutting; the form of the mallet the same as ours. Chisels of bronze; one found, is nine and a quarter inches long, and weighs one pound twelve ounces—its form resembles those now in use. Wheelwrights and carriage makers at work; from which it is ascertained that the bent or improved carriage pole of modern days, was in use upwards of three thousand years ago. Carpenters' and cabinet makers' shops, are represented; from which and from specimens of work extant, we learn that *dovetailing* and doweling, glue and *veneering* were *common*. Adzes, saws, hatchets, drills and bows, were all of bronze. Models of boats. The leather cutter's knife had a semicircular blade, and was identical with the modern one. Shoe and sandal makers had straight and *bent* awls; the latter was supposed to have been a modern invention—the bristle at the end of a thread does not seem to have been used, as one person is seen drawing the thread through a hole with his teeth. Lastly, Egyptian ladies wore their hair plaited and curled; they had mirrors, needles, pins, and jewelry in great abundance; they had fans and combs; one of the latter has teeth larger on one side than on the other, and the centre is carved and was probably inlaid. Their children had dolls and other toys; and the gentlemen used walking canes and wore wigs, which were common.

CHAPTER XII.

Gutters: Single do.—Double do.—Jantu of Hindostan: Ingenious mode of working it—Referred to in Deuteronomy—Other Asiatic machines moved in a similar mannner—Its Antiquity. Combination of levers and gutters—Swinging or Pendulum Machine—Rocking gutters—Dutch Scoop—Flash Wheel.

MOST of the machines hitherto noticed, raise water by means of flexible cords or chains, and are generally applicable to wells of great depth. We now enter upon the examination of another variety, which, with one exception, (the chain of pots) are composed of inflexible materials, and raise water to limited heights only. Another important distinction between them is this—In preceding machines, the 'mechanical powers' are distinct from the hydraulic apparatus, i. e. the wheels, pulleys, windlass, capstan, &c. form no essential part of the machines proper for raising the water, but are merely employed to transmit motion to them; whereas those we are now about to describe, are made in the *form* of levers, wheels, &c. and are propelled as such. The following figure, represents one of the earliest specimens.

No. 30. Single Gutter.

It is simply a trough or gutter, the open end of which rests on the bank, over which the water is to be elevated; the other end being closed is plunged into the liquid, and then raised till its contents are discharged. It forms what is called a lever of the second order, the load being between the fulcrum and the power.

No. 31. Double Gutter.

This figure represents an improvement, being a double gutter, or two of the former united and placed across a trough or reservoir designed to receive the water. A partition is formed in the centre, and two openings made through the bottom on each of its sides, through which the water that is raised escapes. The machine is worked by one or more men, who alternately plunge the ends into the water, and consequently produce a continuous discharge. Sometimes, openings are made in the bottom next the laborers, and covered by flaps, to admit the water without the necessity of wholly immersing those ends. Machines of this kind are described by Belidor, but he has not indicated their origin. From their simplicity, they probably date from remote antiquity. They are obviously, modifications of the *Jantu* of Hindostan and other parts of Asia, and were perhaps carried to Europe, (if not known there before) among other oriental devices, soon after a communication with that country was opened by the Cape of Good Hope.

THE JANTU.

The jantu is a machine extensively used in Bengal and other parts of India, to raise water for the irrigation of land, and is thus described by by Mr. Ward, in his History of the Hindoos. "It consists of a hollow trough of wood, about fifteen feet long, six inches wide, and ten inches deep, and is placed on a horizontal beam lying on bamboos fixed in the bank of a pond or river. One end of the trough rests upon the bank, where a gutter is prepared to carry off the water, and the other end is dipped in the water, by a man standing on a stage, *plunging it in with his foot*. A long bamboo with a large weight of earth at the farther end of it, is fastened to the end of the jantu near the river, and passing over the gallows before mentioned, poises up the jantu full of water, and causes it to empty itself into the gutter. This machine raises water *three feet*, but by placing a series of them one above another, it may be raised to *any height*, the water being discharged into small reservoirs, sufficiently deep to admit the jantu above, to be plunged low enough to fill it." Mr. Ward observes, that water is thus conveyed over rising ground to the distance of a *mile* and more. In some parts of Bengal, they have different methods of raising water, "*but the principle is the same.*"

There is in this apparently rude machine, a more perfect application of mechanical science, than would appear to a general observer. As the object of the long bamboo lever is to overcome the weight of the water, it might be asked, why not load the end of the jantu itself, which is next the bank sufficiently for that purpose, and thereby avoid the use of this additional lever, which renders the apparatus more complex, and apparently unnecessarily so? A little reflection will develope the reasons that led to its introduction, and will at the same time furnish another proof of oriental ingenuity. As the position of the jantu is nearly horizontal when it discharges the water, if the end were loaded as proposed, it would descend on the bank with an increasing velocity; for the weight would be at the end of a lever which virtually lengthened as it approached the horizontal position; and this effect would be still further augmented by the resistance of the water diminishing as the jantu rose, that is, by its flowing towards the centre—the consequence would be, that the violent concussions, when thus brought in contact with the bank, would speedily shake it to pieces. Now this result is ingeniously avoided by the lever and its weight. Thus, when the laborer has plunged the end of the jantu next him into the water, this lever (as we suppose, for we have not seen a figure of it) is placed, so as to be nearly in a *horizontal position*, by which its maximum force is exerted at the precise time when it is required, i. e. when the jantu is at its lowest position and full of water; and as the latter ascends, the loaded end of the lever descends, and its force diminishing, brings the end of the jantu gradually to rest. A somewhat similar effect might be produced, by making the load on the lever descend into the water, especially if its specific gravity varied but little from that fluid. Traits like this, which are often found in ancient devices, are no mean proofs of skill in the older mechanicians; and as professors of the fine arts, discover the works of masters by certain characteristic touches, and by the general effect of a painting or sculpture—so professors of the useful arts may point to features like the above, as proofs that they bear the impress of the master mechanics of old.

At what period in the early history of our species this class of machines was first devised, can only be conjectured; they are evidently of very high antiquity; this is inferable not only from their simplicity, extensive use over all Asia—where it may be said, machines for raising water have never changed—but also from the mode of working them, *by the feet.* Every one acquainted with the bible, knows that numerous operations were thus performed. The juice of grapes was expressed by men treading them; and the tombs of Egypt contain sculptures representing this and other operations. Mortar was mixed and clay prepared for the potter by the feet. The Chinese work their mangles by the feet; and both they and modern Egyptians, and Hindoos, move a variety of other machines by the same means: among these are several for raising water, as the Picotah of Hindostan, (described in the next chapter,) the chain pump of China, and we may here remark, that *all* the machines for raising water described by Vitruvius, with one exception, were propelled by the feet, or as expressed in the English translation, by the "treading of men." It is not at all improbable, that to the JANTU, Moses alluded when describing to his countrymen the land to which he was leading them: "A land of hills and valleys," that "drinketh water of the rain of heaven," where they should not be employed, as in Egypt, where rain was generally unknown, in the perpetual labor of raising it to irrigate the soil: "For the land whither thou goest in to possess it, is not as the land of Egypt from whence ye came out, where thou sowedst thy seed, and *wateredst it with thy foot.*" Deut. xi, 10. Some authors suppose this passage refers to the oriental custom of opening and closing the small channels for water, that intersect the fields; but this trifling labor would scarcely have been mentioned by Moses, as constituting an *important distinction* between the two countries. It was in fact common to both. It is much more probable that he referred to the severe and incessant toil of *raising* water, to which they had been subject in Egypt, and which would be in a great degree superseded in Canaan by the "rain of heaven." He could not possibly have pointed out to them, a more encouraging feature of the country to which they were migrating.

A very interesting proof that the Egyptians in the time of Moses did propel machines by the feet, has recently been brought to light. In one of the tombs at Thebes, which bears the name of THOTHMES III. there is a sculptured representation of some Egyptian *bellows* which were thus worked. We shall have occasion to refer to them when we come to inquire into the history of the pump, in the third book. This mode of transmitting human energy appears to have been quite a favorite one in ancient times; for the purpose of illustration we will describe one which is identical with the Jantu; and is moreover one of the *most common* implements connected with ancient and modern agriculture in the east: "The PEDAL," says Mr. Ward, "is a rough piece of wood, generally the trunk of a tree, balanced on a pivot, with a head something like a mallet; it is used to separate rice from the husk, to pound brick dust for buildings, &c. A person stands at the further end, and with his feet presses it down, which raises up the head, after which he lets it fall on the rice or brick.

Combination of Gutters and Levers.

"One of these pedals is set up at *almost every house* in country places." This primitive implement is also in general use in the agricultural districts of CHINA. "The next thing," says a writer in the Chinese Repository, Vol. iii, 233, "is to divest the grain of the husk; this is done by pounding it in stone mortars; two of these are placed in the ground together, and have corresponding pestles of wood or stone attached to long levers. A laborer by alternately *stepping upon each lever* pounds the grain, &c." Paper mills of the Chinese, by which the shreds of bamboo and the farina of rice are reduced to a pulp, are precisely the same, and worked by men treading on levers as in the jantu.[a] And we may add, that the paste of which Macaroni is made, is kneaded by a similar implement, and which the Romans probably received from the east.

Hence it appears that the jantu is merely one of a class of machines of similar construction and moved in the same manner; and as the pedal of the Hindoos is supposed to be as old as their agriculture, the jantu may certainly be considered equally ancient, for it is the more important machine of the two. They both, however, appear to have had a common origin; and to have come down together through the long vista of past ages, without the slightest alteration. The fact of the jantu being still used in India proves its antiquity, for it is well known that the Hindoos retain the same customs and peculiarities that distinguished their ancestors thousands of years ago. "A country," says Dr. Robertson, "where the customs, manners, and even dress of the people, are almost as permanent and invariable as the face of nature itself." This attachment to ancient customs exists with singular force in regard to every thing connected with their agriculture. Like the Chinese and some other people of the east, nothing can induce them to deviate from the practice of their forefathers, either as it regards their implements or modes of cultivation. And when we bear in mind, that the Hindoos were among the earliest of civilized people; that it was their arts and their science which enlightened the people, who, in the early ages dwelt in the valley of the Nile; we can readily admit that the jantu was used, in the time of Moses, and that to it he alluded in the passage already quoted; but, be this as it may, it may safely be considered as a fair specimen of primeval ingenuity in applying human effort, as well as in raising water; and in both respects is entitled to the lengthened notice we have given it.

These machines when worked by the feet raise water only about three feet, but where the elevation is greater, they have been moved by the hands, by means of ropes and a double lever, as in the next figure; the open ends being attached by pins to the edge of the reservoir. In this manner water may be raised five or six feet at a single lift, according to the length of the gutter.

Contrivances of the kind were formerly used in Europe; and, as in the eastern world, series of them were sometimes employed to raise water to great elevations, to the top of buildings, &c. They are figured and described in Serviere's collection. A number of cisterns are placed at equal distances above each other from the ground to the roof. In these, gutters are arranged as in the figure; the lowermost raises water into the first, into which others dip and convey it to the next one, and so to the highest. In some, the gutters are worked by a combination of levers; in others, by ropes passing over pulleys at the highest part of the building and united to a crank that is attached to a water wheel or other first mover

---

[a] Breton's China, Vol. ii, 39, and Vol. iv, 27.

Various forms of the gutters are figured, (the heads of some like large bowls,) as well as modes of working them. See figure No. 32.

No. 32. Combination of Levers and Gutters.

There is another modification of the jantu, by which water may be raised to great elevations. A number of gutters, open at both ends, are permanently connected to, and over each other, in a zigzag direction, so that while one end of the lowest dips in the water, its other end inclines upwards at an angle proportioned to the length of the gutter and the motion to be given to it, and is united to the lower end of the next one, which also inclines upwards, but in an opposite direction, and is united to the next, and so on, the length of each diminishing as it approaches the top, as in the following figure.

No. 33. Pendulum or Swinging Gutters.

In the bottom of each, an opening is made and covered by a flap or valve to prevent the water, when once past through, from returning. All

Chap. 12.]   Dutch Scoop.   93

the gutters are secured to a frame of wood which is suspended on a pin secured to a beam, so that by pulling the cords alternately the whole may be made to oscillate like a pendulum. Thus, when pulled to one side, one of the lowest gutters dips into water, and scoops up a portion of it, to facilitate which the end is curved; and as it rises, the liquid runs along to the farther end, and passing through the valve is retained till the motion is reversed, when it flows down to the next gutter, and passing through its valve, is again continued in the same manner to the next; entering at every oscillation the gutter above, till it reaches the highest; and from which it is discharged into a reservoir, over which the last one is made to project. A double set of gutters, as shown in the figure, was sometimes attached to the same frame, so that a continuous stream could be discharged into the reservoir. Machines like the above are more ingenious than useful. They do not appear to have ever been extensively used, although they are to be found in the works of several old writers on hydraulics. The one represented by the figure is described by Belidor as the invention of M. Morel, who raised water by it 15 or 16 feet. Similar machines were known in the preceding century. A pendulum for raising water is described at page 95, of the first volume of machines approved by the French Academy, and at page 205, is a "hydraulic machine" by A. De Courdemoy, similar to the one we have copied; except that square tubes were used instead of open gutters; they were also of equal length, and attached to a rectangular frame, but were suspended and worked in the same manner as No. 33.

A different mode of working these machines, was devised by an English engineer. Instead of suspending the frame like a pendulum, he made the lower part terminate in *rockers* like those of a cradle; these resting on a smooth horizontal plane, a slight impulse put the whole in motion. The lowest gutters at each oscillation dipped into the water, and raised a portion, as in the preceding figure.

No. 34. Dutch Scoop.

Among other simple devices, is the Dutch scoop, frequently used by that people in raising water over low dykes. It is a kind of box-shovel suspended by cords from a triangular frame, and worked as represented in the figure. By a sweeping movement, an expert laborer will throw up at each stroke, a quantity of water equal to the capacity of the shovel, although from its form, such a quantity could not be retained in it.

The Flash Wheel, is another contrivance to raise large quantities of water over moderate heights, being extensively used in draining wet lands, particularly the fens of England. It is made just like the wheel of a steamboat, and when put in rapid motion, generally by a windmill, it pushes the water up an inclined shute, which is so curved, that the paddles may sweep close to it, and consequently drive the liquid before them. The 'back water' thrown up by the paddle wheels of steam vessels is raised in a somewhat similar manner.

CHAPTER XIII.

THE SWAPE: Used in modern and ancient Egypt—Represented in sculptures at Thebes—Alluded to by Herodotus and Marcellus—Described by Pliny—Picotah of India; agility of the Hindoos in working it. Chinese Swape—Similar to the machines employed in erecting the pyramids—The Swape, seen in Paradise by Mahomet—Figure of one near the city of Magnesia—Anglo Saxon Swape—Formerly used in English manufactories—Figures from the Nuremburgh Chronicle, Munster's Cosmography, and Besson's Theatre des Instrumens. The Swape common in North and South America—Examples of its use in watering gardens—Figures of it, the oldest representations of any hydraulic machine—Mechanical speculations of Ecclesiastics: Wilkins' projects for aerial navigation—Mechanical and theological pursuits combined in the middle ages—Gerbert—Dunstan—Bishops famous as Castle architects—Androides—Roode of grace—Shrine of Becket—Speaking images—Chemical deceptions—Illuminated manuscripts.

OF machines for raising water, the *Swape* has been more extensively used in all ages, and by all nations, than any other. Like most implements for the same purpose, its application is confined within certain limits; but these are such as to render it of general utility. The méntal or swinging basket, and the jantu, raise the liquid from two to three feet only at a lift, while the swape elevates it from five to fifteen, and in some cases still higher. It is not, however, well adapted for greater elevations; a circumstance which accounts for its not having been much used in the *wells* of ancient cities—their depth rendered it inapplicable, as the generality of ours do at this day. In Egypt, this machine is named the *Shadoof*, and in no country has it been more extensively employed. In modern days, more persons are there engaged in raising water by it and the méntal, than are to be found in any other class of Egyptian laborers. They raise the liquid at each lift, about seven feet, and where it is required higher, series of swapes are placed at proper distances above each other, in a similar manner as the Hindoos arrange the jantu, and as shown in the figure, (No. 35.) The lowermost laborer empties his vessel into a cavity or basin formed in the rock, or in soil rendered impervious to water, three or four feet above him, and into which the next one plunges his bucket, who raises it into another, and so on till it reaches the required elevation. M. Jomard,[a] says it is not uncommon to see from thirty to fifty shadoofs at one place, raising water one above another. At Esne, he saw twenty-seven Arabs on one tier of stages, working fourteen

---

[a] Grande Description. E. M. Tom. ii. Memoirs, Part 2, p. 780.

Ancient Egyptian Swape.

double swapes, i. e. two on each frame, the bucket of one descending as the other rises. They were relieved every hour, so that fifty-four men were required to keep the machines constantly in motion. The overseer or task-master measured the time by the sun, and sometimes by a simple clepsydra or water-clock.

No. 35. Modern Egyptians using the Swape.

It is impossible to pass up the Nile in certain states of the river, without being surprised at the myriads of these levers, and at their unceasing movements; for by relays of men, they are often worked without intermission, both night and day. In Upper Egypt especially, where from the elevation of the banks they are more necessary, and of course more numerous, the spectacle is animating in a high degree, and cannot but recall to reflecting minds similar scenes in the very same places in past ages, when the population was greatly more dense than at present, and the country furnished grain for surrounding nations. In some parts, the banks appear alive with men raising water by swapes and the effect is rendered still more impressive by the songs and measured chantings of the laborers, and the incessant groans and creakings of the machines themselves. To the ancient custom of singing *while raising water*, there is an evident allusion in Isaiah, xii, 3. Therefore with joy shall ye draw water out of the wells of salvation.

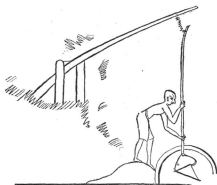

The Arabs have a tradition that the shadoof was used in the times of the Pharaohs, and a proof that such was the fact, has recently been furnished by Mr. Wilkinson. (Vol. ii. 5.) who found the remains of one in an ancient tomb at Thebes; in addition to which they are represented in sculptures which date from 1532 to 1550 B. C. a period extending beyond the Exodus. No. 36 represents it as used at that remote period for the irrigation of land.

No. 36. An Egyptian in the time of Moses raising water by the swape. From sculptures at Thebes.

It appears to have formed one of a series, designed to raise water over the elevation feebly portrayed in the back ground, in precisely the same way that is now common in Egypt and in the east, and as shown in No. 35. The remark of a traveler that a Chinese seemed to him "an antediluvian renewed," might with equal propriety, be applied to a modern Arab raising water by this implement from the Nile; and the figure, No. 36, might be taken as a probably correct representation of an antediluvian laborer engaged in the same employment. On comparing the last two cuts, the former having been sketched by Mr. Wilkinson, from life, but three years ago; and the latter copied from sculptures that have been executed upwards of three thousand years, we see at once, that the swape has undergone as little change in Egypt, since the times of the Pharaohs, as the *costume*, if such it may be called, of the laborers themselves; in other words both remain the same. The discovery of this implement among the sculptures of ancient Egypt tends to corroborate our views respecting the antiquity of other machines for the same purpose, and which like it are still in common use in the east. It also admonishes us not to reject as improbable or fabulous, current oriental traditions; since they are, as in the case of this machine, often, if not generally, founded in truth.

The swape seems to be alluded to by Herodotus, vi, 119, as used in Persia in his time. He observes that Darius, the father of Xerxes, sent some captives to a certain distance from Susa, and forty furlongs from a well, the contents of which were "drawn up with an engine, to which a kind of bucket is suspended, made of half a skin; it is then poured into one cistern and afterwards removed into a second." This appears to have been the shadoof of the Egyptians, as figured in No. 35, to which there is probably a reference also in Clio, 193, where he says the Assyrians irrigated their lands from the Euphrates "by *manual labor* and by *hydraulic engines*." Aristotle mentions the swape as in common use among the Greeks.[a] Dr. Clarke says some of the wells of Greece were not deep, and pulleys were not used, only buckets with ropes of twisted herbs, and sometimes the water was raised by a 'huge lever, great stones being a counterpoise to the other end.' A circumstance connected with the overthrow of the Syracusans, and the death of Archimedes, in which the swape is referred to, may here be noticed. When the Roman vessels, at the siege of Syracuse were grappled by hooks and elevated in the air, by levers that projected over the walls of the city, their resemblance to vessels of water raised by the swape, was so striking, that Marcellus was wont to say, "Archimedes used his ships to *draw water with*."[b] This remark of the Roman general clearly shows that the swape was very familiar to him and to his countrymen. But we are not left to circumstances like this to infer its use among the Romans. Pliny expressly mentions it among machines for raising water. As the passage is highly interesting, and as we shall have occasion to refer to it hereafter, it may as well be inserted here. It is in the fourth Chapter of the Nineteenth Book, "On Gardens:" "above all things there should be water at command, (if possible a river or brook running through it, but if neither can be obtained,) then they are to be watered with pit water, fed with springs; either *drawn up* by plain poles, hooks, and buckets; or *forced* by pumps and such like, going with the strength of wind enclosed, or else *weighed up* with *swipes and cranes*." Holland's Trans.

The Swape is extensively used over all Hindostan. "The peasants, morning and evening draw water out of wells by buffaloes or oxen, or

---

[a] Bishop Wilkins on the lever.   [b] Plutarch's life of Marcellus, Wrangham's notes.

Chap. 13.]   *The Picotah.*   97

else by a thwart post, poised with a sufficient weight at the extremity laid over one fixed in the earth; the water is drawn by a bucket of goat's skin."[a] In some districts, the Hindoos have a mode of working the Swape, which, so far as we know, is peculiar to themselves. In Patna it is common, and the machine when thus propelled, is named the *Picotah*.

No. 37. Picotah of Hindostan.

"Near the well or tank, a piece of wood is fixed, forked at the top; in this fork another piece of wood is fixed to form a swape, which is formed by a peg, and steps cut out at the bottom, that the person who works the machine may easily get up and down. Commonly, the lower part of the swape is the trunk of a tree; to the upper end is fixed a pole, at the end of which hangs a leather bucket. A man gets up the steps to the top of the swape, and supports himself by a bamboo screen erected by the sides of the machine." He plunges the bucket into the water, and draws it up by his weight; while another person stands ready to empty it. In the volume of plates to the Paris edition, 1806, of Sonnerat's Voyages, the machine is represented rather different from the above. The laborer alternately steps on and off the swape, from a ladder or stage of bamboos erected on one side of it See plate 23, Sonnerat.

The apparatus and mode of working it, is more fully described in the following extract from 'Shoberl's Hindostan in Miniature.' "By the side of the well a forked piece of wood, or even a stone, eight or ten feet high is fixed upright. In the fork, is fastened by means of a peg, a beam three times as long, which gradually tapers, and is furnished with steps like those of a ladder. To the extremity of this long beam, which is capable of moving up and down, is attached a pole, to the end of which is suspended a large leather bucket. The other end being the heaviest, when the machine is left to itself, the bucket hangs in the air at the height of twenty feet; but to make it descend, one man, and sometimes two, mount to the middle of the beam, and as they approach the bucket, it sinks to the bottom of the well, and fills itself with water. The men then move back to the opposite end, the bucket is raised, and another man empties it into a basin. This operation is performed with such celerity that the water *never ceases running*, and you *can scarcely see the man moving* along his beam; yet he is sometimes at the height of twenty feet, at others, touching the ground; and such is his confidence, that he laughs, sings, smokes, and eats in this apparently ticklish situation." Vol. v, p. 22, 24. This mode of applying human effort, was eurly adopted in the working of pumps—a piston rod being attached to each end of the vibrating beam. Dr. Lardner, has inserted a figure of it in his popular

---

[a] Fryer's Travels in India, 187.

treatise on Pneumatics. It is figured in most of the old authors, and was most likely, copied from the Picotah, and other oriental machines, which have been propelled in a similar manner from very remote times. See Gregory's Mechan. Vol. ii, 312. Ed. 1815.

The Swape is one of the ancient and modern implements of China, where it is used, as in Egypt and India, for the irrigation of land. It is frequently made to turn in a socket, (or the post itself moves round,) in addition to the ordinary vibratory motion. In several situations, this is a decided improvement, as the vessel of water when raised above the edge of a tank or river, can, if desirable, be swung round to any part of the circle which it describes. Sir George Staunton, has given a figure of it, which Mr. Davis has copied into his popular work on the Chinese. When thus constructed, it is according to Goguet, (Tom. iii, Origine des Loix,) identical with the engines mentioned by Herodotus, B. ii, 125, as employed in the erection of the Egyptian pyramids; these, he supposes were portable swapes, or levers of the first order, with a rotary movement like those of the Chinese. A number of these being placed on the lowest tier of stones which formed the basis of the pyramids, were used to raise those which form the second tier; after which, other swapes were placed on the latter and materials raised by them for the third range, and in like manner to the top. This was the *process* which Herodotus says was adopted. M. Goguet, supposes that two swapes were employed in raising every stone, one at each end, and that the levers were depressed by a number of men laying hold of short ropes attached to them for that purpose. This mode appears to accord with the meagre description of the machines used in the erection of the pyramids, which the father of history has given.

It has already been observed, that the engines employed by Archimedes to destroy the Roman ships in the harbor of Syracuse, were so analagous to the swape, as to elicit from Marcellus, an observation to that effect. In fact, machines similar to it, were used by ancient engineers both for attacking and defending cities. Vegetius, says they were used to raise soldiers to the tops of walls, &c. In the oldest translation of his work, (Erffurt, 1511,) there is a figure of it, which is identical with the Chinese swape, and with that which Goguet supposes was used by the old engineers of Egypt. Barbaro, in his edition of Vitruvius, also figures it. In Rollin's 'Arts and Sciences of the Ancients,' are several examples and figures of it, applied to the purposes of war; and among others to the destruction of the Roman vessels before Syracuse.

A story in the 'Hegiat al Megiales' shews how common it was in Arabia in the seventh and preceding centuries. Mahomet in one of his visions of paradise, "saw a machine much used in the Levant for drawing water out of wells, called by the Latins *Tollens*, and consisting of a long lever fixed on a post, [i.e. the swape.] Enquiring to whom it belonged, he was told it was Abougehel's, (the bitterest enemy to him and his religion.) Surprised at this, he exclaimed, ' what has Abougehel to do with paradise, he is never to enter there !' Shortly after, he understood the drift of the vision, for the son of his enemy became a Mussulman, upon which he exclaimed ' Abougehel was *the swape*, by which God drew up his son from the bottom of the pit of infidelity.'" It is used by the Japanese; and as figured by Montanus, the bucket is raised by pulling down the opposite end of the lever by means of cords attached to it.

In Fisher's " Constantinople, and the Scenery of the Seven Churches of Asia." Lon. 1839, is a beautiful view of the city of Magnesia near Mount Sipylus, in Asia Minor, a city founded by *Tantalus*, whose fabled punish-

Chap. 13.]   *Anglo Saxon Swape.*   99

ment has rendered his name notorious. In the foreground is represented the following figure of the swape, a machine which the writer observes, " forms a conspicuous object *in every landscape* in the east. One is seen erected in *every garden,* and as irrigation is constantly required in an arid soil, it is always in motion, and its dull and drowsy creaking is the sound incessantly heard by all travelers."

In this figure we behold not merely a sketch of modern Asiatic manners; but one, which as regards raising of water; the machine by which it is effected; animals around it; costume of the individuals; and portraiture of rural life,—has remained unchanged from times that reach back to the infancy of our race, and of which history has preserved no records.

[For this interesting cut, and for No. 35 also, I am indebted to my friend WILLIAM EVERDELL, Esq. who, besides ther contributions to this work, undertook the task, to him a novel one, of engraving them.]

No. 38.   Swape in Asia Minor.

The swape has probably been in continual use in Great Britain, from the period of its subjugation by the Romans, if not before. It is there known under the various names of ' *Swape,*' ' *Sweep,*' and in old authors, ' *Swipe.*' A figure of it, as used by the Anglo-Saxons, is here inserted, from Vol. i, of the 'Pictorial History of England,' copied from an ancient manuscript in the British Museum. The costume of the female, her masculine figure, the shingled well, and form of the vase or pitcher, are interesting, as indicative of manners and customs, &c. of former ages. The arm of the lever to which the bucket is suspended, appears extremely short, but this is to be attributed to its defective representation. The following summary of ancient British devices is from Fosbroke's Encyclopedia of Antiquities. " The Anglo-Saxons had a wheel for drawing water

No. 39.   Anglo Saxon Swape.

from wells. They were common annexations to houses. Rings were fixed to the chains of wells. We find a beam on a pivot, with a weight at one end for raising water. Wheels and coverings. A lever, the fulcrum of which was a kind of gallows over the well. Two buckets one at each end of a chain adapted to a versatile engine called volgolus. Buckets with iron hoops, and drawing water from deep wells as a punishment." The swape appears to have been the principal machine in England for raising water till quite recent times. In the 17th century it was used *in manufactories,* and is not yet, perhaps, wholly superseded by the pump. Bishop Wilkins, in speaking of the lever and its application by Archimedes in destroying the Roman fleet, says, "it was of the same form with that which is *commonly used* by *brewers and dyers* for the drawing of water. It consists of two posts, the one fastened perpendicularly in the ground, the other being jointed on cross to the top of it." Mathemat. Magic. B. i, Chaps. 4 and 12. This was published in 1638. In 1736, Mr. Ainsworth published his celebrated Latin Dictionary, and under the word *Rachâmus,* ' a truckle or pulley used in drawing up water ;' he adds, " perhaps not unlike the sweep *our brewers use :*" hence at that time, it continued to be used for raising water and transferring liquids in English breweries and similar establishments, as remarked by Wilkins one hundred years before.

No. 40. Swape. From the Nuremburgh Chronicle. A. D. 1493.

In Germany it was frequently, and still is, a prominent object in country towns and villages, as well as in farm yards. In the former it was frequently erected on, or at the end of bridges for the purpose of raising water from rivers and brooks. In the famous Nuremburgh Chronicle it is frequently figured. From a variety of different forms, we have selected No. 40, as a specimen.

In the Cosmography of Sebastian Munster, 1550, it is represented at page 729, as employed for raising water to supply, by means of pipes, a neighboring town. Agricola, in his De Re Metallica, has also figured it. pp. 443 and 458.

No. 41. Swape from S. Munster's Cosmography. 1550.

The Swape was very common in France and the neighboring nations on the European continent, in the last and preceding centuries. It is named *bascule* in France. The old *Dictionnaire de Trevoux,* says :

Les bascules les plus simple, sont celles qui ne consistent qu'en une pièce de bois soutenue d'une autre par le milieu ou autrement, comme d'un essieu, pour être plus au moins en équilibre. Lorsqu'on pese sur un des bouts l'autre hausse. Ces sortes de bascules sont les plus communes; on s'en sert pour élever des eaux. The last sentence is believed to be applicable to every part of Europe at the present time, perhaps equally so as at any former period.

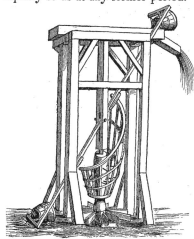

No. 42. Swape from Besson. 1568.

We subjoin a description of one proposed by James Besson, a French mechanician, 270 years ago, by which two buckets, one at each end, may be used. As the vibration of the beam is ingeniously effected by a continuous rotary movement, a figure of it will be acceptable to the intelligent mechanic.

The lever is suspended at the centre of its length, on a pin which passes through the lower part of the perpendicular post, the upper end of which is firmly secured to the frame and cross piece. A perpendicular shaft is placed immediately under the lower end of the post and in the same vertical line with it. The upper journal of the shaft enters and turns in the end of the post, while its lower one is received into a cavity in the block upon which it rests. This shaft forms the axis of an inverted cone of frame-work—a section of which, resembling an hyperbolic curve, acts as it revolves on the under side of the swape, and imparts to it the required movement. To lessen the friction, two long rollers are fixed to its under side, and upon these only does the curved edge of the cone act. The shaft may of course be turned by any motive power. In the figure, a horizontal water wheel is attached to the shaft, with oblique paddles which receive the impulse of the stream in which they are placed. This device may serve as an example of mechanical tact and resource in the early part of the 16th century, when practical mechanics began to be cultivated as a science.

The swape is commonly used by the farmers on this continent, in the vicinity of whose dwellings it may be seen, more or less, from the St. Lawrence to the Mississippi. In some of the states, it still bears the old English name of the 'sweep' as in Virginia—in others it is named the 'balance pole.' It prevails in Mexico, Central America, Peru, Chili, and generally throughout the southern continent. There is some uncertainty respecting its having been known here before the arrival of Europeans in the 16th century. See remarks on Ancient American Machines in the last chapter of this book.

The swape appears to have been used in all times, for watering *gardens* in the east, as already observed of Asia Minor, it is there seen erected in almost every one. No. 43 represents it employed in the gardens of Egypt during the sojourn of the Israelites in that country. The tree and plant are uniform hieroglyphical representations of gardens.

The labourer discharges the contents of his bucket into a wooden trough or gutter, by which the water is conveyed to the plants; a mode still followed through all the east. To this application of the swape there is probably a reference in the prediction of Balaam, delivered one hundred years *after* these figures were sculptured, 'he shall pour water out of his buckets, and his seed shall be in many waters, (Numb. xxiv, 7,) an intimation that the Israelites should possess a country, where this great desideratum should be in comparative abundance, a land "watered as a garden of herbs." The figure may serve also to illustrate the 'gutters and watering troughs' in which Jacob watered the flocks of Laban, his father in law. Gen. xxx, 38.

No. 43. Egyptian Shadoof employed in watering a garden. 1550, B. C.

The luxuriance of vegetation in an eastern garden, (when properly watered,) the richness of its scenery, the beauty of its foliage and flowers, form one of the most enchanting prospects in nature; hence it became the most favorite, as it was the most natural, metaphor of human felicity. When the prophets promised prosperity, it was in such language as the following: "Thou shalt be like a watered garden, and like a spring of water whose waters fail not," and "their soul shall be as a watered garden." On the contrary, when the wicked were denounced, "ye shall be as an oak whose leaf fadeth, and as a garden that hath no water." The same metaphor is a frequent one in ancient poets, and in most instances the use of the swape is implied. Thus Homer:

> As when autumnal Boreas sweeps the sky.
> And instant blows the *watered gardens* dry.

And Ovid

> As in a *watered garden's* blooming walk. *Met.* x, 277.

Pliny mentions it expressly for the same purpose, and to it Juvenal seems to allude in his third satire:

> There from the *shallow* well your hand shall pour
> The stream it loves on every opening flower.

This use of the swape is not now confined to the gardens of fallen Asia, Egypt and Greece, but it is employed by the most enlightened nations; and in London and in Paris, as in Athens and Alexandria, Memphis and Thebes, this primitive implement has not been superseded. In Breton's China, Lon. 1834, the Chinese swape is described; and the author observes, "it is similar to those which are seen in the *market gardens* round *London:*" and in a more recent work, 'Scripture illustrated from Egyptian Antiquities,' the author speaking of the Egyptian swape, says, it is the same as used in the *gardens of Brentford*.

Of the swape, it may be remarked, that the most ancient portraiture extant, of any hydraulic *machine*, is a sculptured representation of IT, between three and four thousand years old, and even at that remote period it was in all probability a very old affair, and in common use. These sculptures moreover prove, that it has remained in Egypt unaltered in its form, dimensions, mode and material of its construction and methods of using it, during *at least* thirty-four centuries! and this, notwithstanding the political convulsions to which that country has ever been subject, since its conquest by Cambyses; its inhabitants having been successively under the Persian, Grecian, Roman, Saracenic, and Turkish yoke, thus literally fulfilling a prophecy of Ezekiel, that, "there shall be no longer a prince of the land of Egypt,"—a descendant of its ancient kings; yet through all these mighty revolutions that have swept over it like the fatal Simoon, and destroyed every vital principle of its ancient grandeur, this simple machine has past through them all unchanged, and is still applied by the inhabitants to the same purposes, and in precisely the same way, for which it was used by their more enlightened progenitors.

We have seen it used by the Greeks and Romans, and we find it still in the possession of their descendants, wherever they dwell, as well as among those of more ancient people, the Hindoos, Arabs, and Chinese. And although we may be unable to keep it constantly in view in Europe, in those ages which immediately followed the fall of the Roman power, when the ferocious tyranny of the Saracens established a despotism over the mind as well as the body; and by the characteristic zeal of OMAR, entailed ignorance on the future, by consuming the very sources of knowledge under the baths of Alexandria; yet, when in the 15th century, the human intellect began to shake off the lethargy, which during the long night of the dark ages had paralyzed its energies, and *printing* was introduced—that mighty art which is ordained to sway the destinies of our race forever—among the *earliest* of printed books, with illustrations, this interesting implement may be found portrayed in *vignettes*, in *views of cities*, and of *rural life;* tangible proofs of its universal use throughout Europe at that time, as well as during the preceding ages.

Having referred in this and in a preceding chapter, to the 'Mathematical Magic' of Wilkins, we subjoin some remarks on the mechanical speculations of that and other old church dignitaries. [These remarks were at first designed for a note, but have been too far extended to be inserted as one.] The former was certainly one of the most ingenious and imaginative of mechanics that ever was made a bishop of, and not a few have worn the mitre. 'The Right Reverend Father in God, John, Lord Bishop of Chester,' (like friars Bacon and Bungey, the Jesuit Kircher, the Abbe Mical, and a host of others,) excelled equally in mechanical and theological science; and at one period of his researches in the former, seemed almost in danger of rendering the latter superfluous: viz. by developing a plan of conveying men to other worlds by machinery! See his Tract on on the 'Discovery of a New World in the Moon, and the possibility of a passage thither.' Lon. 1638. After removing with a facility truly delightful, those objections to such a 'passage' as arise from the 'extreme coldness and thinness of the etherial air,' 'the natural heaviness of a man's body,' and 'the vast distance of that place from us,' and the consequent necessity of rest and provisions during so long a journey, there being, as he observes, 'no inns to entertain passengers, nor any castles in the air to receive poor pilgrims'—he proposes three modes of accomplishing the object. 1. By the application of *wings* to the body; 'as angels are pictured, as Mercury and Dædalus are feigned, and as has been attempted by divers, particularly by a Turk in Constantinople, as Busbequius relates. 2. By means of *birds*, for as he quaintly says, "If there be such a great *ruck* in Madagascar, as Marcus Polus the *Venetian* mentions, the feathers in whose wings are twelve feet long, which can scoop up a horse and his rider, or an elephant, as our kites do a mouse; why then, 'tis but teaching one of these to carry a man, and he may ride up thither, as Ganymede did, upon an eagle." 3. Or, "if neither of these ways will serve, yet I do *seriously*, and upon *good ground*, affirm it possible to make a *flying chariot*, in which a man may sit, and give such a motion unto it as shall convey him through the air; and this perhaps might be made large enough to carry diverse men at the same time, together with food for their *viaticum*, and *commodities for traffic*." The construction of such a chariot, he says, was 'no difficult matter, if a man had leisure to show more particularly the means of composing it.' It is to be regretted that he did not pretermit some of his labors for that purpose, especially as his project was not merely to skim along the surface of this planet, like modern aeronauts, or ancient navigators creeping along shores—but like another Columbus, to launch out into the unknown regions of space, in search of other worlds.

Had Wilkins been a countryman as well as a contemporary of Galileo, his aerial flights would have been confined to a dungeon, and the wings of his genius would have been effectually clipped with Roman shears. Indeed we must admit that he was the greater sinner of the two! for Galileo merely taught the absurd doctrine of the sun's stability, and that the earth moved round it, in opposition to the evidence of his senses, to the doctrines of the church, and in flat contradiction of those passages in the Bible, which Bellarmine adduced as proofs indubitable, that the sun 'rises *up*' in the east every morning, and 'goes *down*' in the west every night, and that the earth is established and 'cannot be moved.' Whereas the heretical bishop, endeavored to open a way by which men could visit other worlds when they pleased, and that too, without consulting, or so much as saying 'by your leave,' to the successors of St. Peter!

The earliest English aeronaut was Elmer, a monk of the 11th century. He adapted wings to his hands and feet, and took his flight from a lofty tower. He sustained himself in the air for the space of a furlong, but his career, (like that of Dante in the fifteenth century) terminated unfortunately, for by some derangement of his machinery he fell, and both his legs were broken. Dante, after several successful experiments, fell on the roof of a church and broke his thigh.

It is a singular fact in the history of the arts, that mechanical skill was in former times intimately connected with *theological* pursuits, and that some of the cleverest workmen were ecclesiastics, and of the highest grades too; witness Gerbert, Dunstan, Albertus, and many others. The first was a French mechanician of the 10th century, whose researches led him at that early period, to experiment on steam, and on its application to produce music. He was successively archbishop of Rheims and Ravenna, and in 999 took his seat in St. Peter's chair, and was announced to the world as Pope Sylvester II. It may now appear strange that monks and friars, abbots, bishops, archbishops and popes, should have been among the chief cultivators of, and most expert manipulators in the arts, and that to them we are greatly indebted for their preservation through the dark ages; but, in those times, it was so far from being considered derogatory in ecclesiastics to work at 'a trade,' that those who did not, were accounted unworthy members of the church; hence monks were cooks, carpenters, bakers, farmers, turners, founders, smiths, painters, carvers, copyists, &c.; all had some occupation, besides the study of their peculiar duties. "In that famous colledg, our monasterie of *Bangor*, in which there were 2100 christian philosophers, that served for the profit of the people in Christ, living by the labor of their hands, according to St. Paul's doctrine."[a] This was in the 5th century, when *Pelagius* belonged to the same monastery. In the 7th, "almost all monks were addicted to manual arts," and according to St. Benedict, such only as lived by their own labor, "were truly monks."[b] "*They made and sold their wares to strangers*, for the use [benefit] of their monasterie, yet somewhat cheaper than others sold."[c] Many of these men naturally became expert workmen, especially in the metals—a branch of the arts that seems to have been a favorite one with them; hence, the best gold and silver smiths of the times were often found in cloisters; and the rich 'boles, cups, chalices, basens, lavatories of silver and gold, and other precious furniture' of the churches, were made by the priests themselves:—It may be a question, whether they were not right in thus combining mental and physical employments; as a compound being, manual labor seems necessary to the full development of man's intellect, and to its healthy and vigorous exercise. Dunstan, Archbishop of Canterbury in the 10th century, was skilled in metallurgical operations—he was a working jeweler, and a brass founder. Two large bells for the church at Abingdon were cast by him. He is said to have been the inventor of the Eolian harp, an instrument whose spontaneous music induced the people at that dark age, to consider him a conjurer—hence the old lines—

> St. Dunstan's harp, fast by the wall,
> Upon a pin did hang—a;
> The harp itself, with ty and all,
> *Untouched by hand* did twang—a.

The genius of some led them to cultivate *architecture*. Cathedrals and other buildings yet extant, attest their skill. Of celebrated architects in the 11th century, were *Mauritius*, bishop of London, and *Gundulphus*, bishop of Rochester. The latter visited the Holy Land previous to the crusades, and is said to have been one of the greatest builders, and the most eminent castle architect of his age. In the Towers of London and Rochester, he left specimens of his art. At page 62, we referred to the remains of a castle built by him, and to his mode of protecting the well, and raising the water to the different floors. In the 12th century, these reverend artists were numerous. In England, were Roger, bishop of Salisbury, and Ernulf, his successor—Alexander of Lincoln—Henry De Blois of Winchester, and Roger, of York; all of whom left remarkable proofs of their proficiency as builders. In France, 'in sundry times the ecclesiastics performed *carving*, *smelting*, *painting*, and *mosaic*.' Leo, bishop of Tours in the 6th century, 'was a great artist, especially in *carpentry*.' St. Eloy was at first a *sadler*, then a *goldsmith*, and at last bishop of Noyan; he built a monastery near Limoges, but he was most noted for *shrines* of gold, silver, and precious stones. He died in 668. The church of Notre Dame des Unes, in Flanders, was begun by Pierre, the 7th abbot, and completed in 1262, by Theodoric. 'The whole church was built by the monks themselves, assisted by the lay brothers and their servants.'[d]

Luther was accustomed to *turning*, and kept a lathe in his house, 'in order to gain his livelihood by his hands, if the word of God failed to support him.'

---

[a] Monastichon Britannicum, Lon. 1655. p. 40.   [b] Ib. 268.   [c] Ib. 301
[d] Ed. Encyc. Art. Civil Architecture.

Those in whom the 'organ of constructiveness,' or invention was prominent, produced among other curious machinery, *speaking heads, images of saints, &c.* These, it is believed, were imitations of similar contrivances in heathen temples. The statue of Serapis moved its eyes and lips. The bird of Memnon flapped his wings, and uttered sounds.[a] It is to be regretted that no detailed descriptions of these, and of such, as were used in European churches previous to the reformation, have been preserved. An account of the ingenious frauds of antiquity would be as valuable to a mechanician as it would be interesting to a philosopher. It would in all probability develope mechanical combinations both novel and useful; and would include all the mechanism of modern androides; and most of the deceptions to be derived from *natural* magic.

A famous image known as the *Roode of Grace*, is often mentioned by English historians. A few scattered notices of it are worth inserting. Speed in his history of Great Britain, (page 790,) says "it was by divers vices [devices] made to *bow down*, and to *lift up itselfe*, to *shake*, and to *stir both head, hands, and feet*, to *rowle its eyes, moove the lips*, and to *bend the brows.*" It was destroyed in Henry VIII's reign, being "broken and pulled in pieces, so likewise the images of our Lady of Walsingham and Ipswich, set and besprinkled with jewels, and gemmes, with divers others both of England and Wales, were brought to London, and burnt at Chelsea, before the Lord Crumwell." In the life of the last named individual some further particulars of it are given, and which explain the mode of operation. "Within the *Roode of Grace*, a man stood inclosed with an *hundred wyers*, wherewith he made the image *roll his eyes, nod his head, hang the lip, move and shake his jaws*; according as the value of the gift offered, pleased or displeased the priest; if it were a small piece of silver, he would *hang the lip*, if it were a good piece of gold, his *chaps* would *go merrily,*" &c. Cromwell discovering the cheat, caused the image "with all his engines to be openly showed at Paul's Cross, and there to be torn in pieces by the people." Clarke's Lives, Lon. 1675. It would have been a dangerous practice to have employed intelligent 'lay craftsmen' in making machines like this, or to have engaged them in 'pulling the wires.' The shrine of *Becket* showed great proficiency in some of the arts. It "did abound with more than princely riches, its meanest part was pure gold, garnished with many precious stones, as *Erasmus* that saw it, hath written; whereof the chiefest was a rich gemme of *France*, offered by king *Lewis*, who asked and obtained (you may be sure, he buying it so deare) that no passenger betwixt Dover and Whitesand should perish by shipwracke." The bones of Becket were laid in a splendid tomb. "The timber work of his shrine was covered with plates of gold, *damasked* and *embossed* with *wires of gold*, garnished with *broches, images, angels, precious stones, and great orient pearles*; all these defaced filled two chests, and were for price, of an unestimable value." A catalogue of the miracles wrought at his shrine filled two folio volumes![b]

---

[a] See Kircher's Musurgia Universalis, Rome, 1650, Tom ii, p. 413, for an ingenious figure of such an automaton.

[b] Accounts kept by Churchwardens previous to the reformation often exhibit curious information in relation to the repairing, replacing, and clothing of images, and to the sale of damaged or worn out ones, as appears by the following extracts from 'A boake of the stuffe in the cheyrche of Holbeche sowld by Cheyrchewardyns of the same, according to the injunctyons of the Kynges Magyste, A. D. 1447.' *The Trinity* with the Tabernacle, sold for two shilling and fourpence. The Tabernacle of *Nicholas* and *James* for six shillings and eight pence. "All the Apostyls *coats* and *other raggs,*" for eight shillings and four pence. And in 1547, " XX score and X hund, of *latyn*, at ii. s. and xi. d. the score." This item probably consisted of brazen utensils, images, &c. sold for their value as old metal. Stukely's Antiquities. London 1770, page 21.

Other devices, less complex than the Roode of Grace, but when adroitly managed, equally effective and imposing, consisted in the application of secret tubes, through which sound might be conveyed from a person at a distance. Sometimes the accomplice was concealed in the pedestal, or in the statue itself, or in the vicinity. "The craftinesse of the inchanters, (observes Peter Martyr,) led them to erect images against walles, and gave answer through holes bored in them; wherefore the people were marvellouslie amazed when they supposed the images spake. There were dailie woonders wrought at the images whereby the sillie people were in sundriewise seduced."[a] It was by a trick of this kind, that *Dunstan* confounded his adversaries in an important discussion—the crucifix hanging in the church opened its mouth and decided the question in his favor. Numerous examples of more recent times might be given. We add one from Keysler's Trav. Vol. i, 148: A monk having made a hole through a wall, behind an image of the Virgin, 'placed a concealed tube from it to his cell; and through it caused the image to utter whatever he wished the people to believe.' By such tubes figures of the Virgin have repeatedly declared her wishes, saluted her worshippers, and returned their compliments. It was by the same device that several statues of heathen deities performed prodigies; that of Jupiter for example, which burst forth into loud fits of laughter. Misson's Trav. Vol. ii, 412.

Within ancient temples, says Fosbroke, was a dark interior, answering to the choir of modern cathedrals, the *Penetrale*, into which the people were not permitted to enter. When the time of sacrifice arrived, the priest opened the doors that the people might see the altar and victim; for only the priests and privileged persons entered into the *cella*, i. e. into interior. Some temples admitted light only at the door, for darkness was deemed a most powerful aid to superstition. "The penetrale of the temple of Isis, at Pompeii is a small pavilion, raised upon steps, under which is a vault, that may have served for oracular impositions. A shrine of this kind is still open for inspection at Argos. In its original state it had been a temple; the further part where the altar was, being an excavation of the rock, and the front and roof constructed of baked tiles. The altar yet remains and part of the fictile superstructure, but the most remarkable thing is a secret subterraneous passage terminating *behind the altar* its entrance being at a considerable *distance*, towards the right of a person facing the altar, and so cunningly contrived as to have a small aperture, easily concealed, and level with the surface of the rock. This was barely large enough to admit the entrance of a single person, who could creep along to the back of the altar, where being hid by some colossal statue, or other screen, the sound of his voice would produce a most imposing effect among the listening votaries." Antiq. 33. It is a curious fact that conjurers and chiefs among American Indians, were found to practice similar cheats. In St. Domingo, some Spaniards having abruptly entered the cabin of a cacique, they were astonished to hear an idol apparently speaking (in the Indian tongue) with great volubility. Suspecting the nature of the imposture, they broke the image, and discovered a concealed tube, which proceeded from it to a distant corner, where an Indian was hid under some leaves. It was this man, speaking through the tube, that made the idol utter, whatever he wished the hearers to believe. The Cacique prayed the Spaniards to keep the trick secret, as it was by it, that he secured tribute and kept his people in subjection.[b]

---

[a] Common Places, Part ii, Chap. v. Lon. 1583.
[b] Histoire Gènèrale. La Haye. 1763. Tom. 18, p. 229.

Another device adopted by ecclesiastics, for subduing the turbulent passions of their ignorant people, and exciting in them feelings of respect for the church, was by making images of the Virgin and of Christ, to *weep*, and sometimes to *sweat* blood, &c. These effects being, of course, represented as the result of their impenitence. 'The fathers of Monte Vaccino made the wooden crucifix sweat that was fastened to the wall of their church; through which they had a passage for the water to run into the body of the crucifix, wherein they had drilled several pores, so that it passed through in little drops.' De La Mortraye's Trav. Vol. i, 23. This was a staple trick of heathen priests; hence Statius, in his *Thebaid*, B. ix, v. 906, represents the statue of Diana weeping.

> For tears descended from the sculptured stone.

And Lucan,

> The face of grief each marble statue wears,
> And Parian gods and heroes stand in tears.

In the temple of the great Syrian goddess at Hierapolis, were idols that could 'move, *sweat* and deliver oracles as if alive.'[a] Among ancient *chemical* deceptions, the liquefaction of St. Januarius' blood, is still performed; and once a year, all Naples is in suspense till the miracle is accomplished. We shall have occasion to notice other ingenious ancient devices for the same purposes of delusion, in the fourth Book, when speaking on the application of steam to raise water.

Although the monks present lamentable examples of misdirected talents and misapplied time, their labors tended to the general progress of refinement and learning. We may regret that unworthy spirits among them abused the superstitions of the times to their own advantage—imitating the statesmen and priests of antiquity, in making the oracles declare what they wished; still, they were the only lights of the dark ages, and even their introduction of images of saints, &c. in place of the pagan idols, contributed in the end to the overthrow of idolatry, and was perhaps the only condition on which the barbarous people, could be induced to give up their ancient deities. 'It can hardlie be credited,' says Peter Martyr, 'with how greate labor and difficultie, man could be brought from the worshipping of images.'

Another class devoted themselves to writing and copying, that is, to the art of multiplying books; and their industry and skill have never been, and in all human probability, never will be surpassed. The beauty, uniformity and effect of their pages, are equal to those of any printed volume. The richness of the illuminated letters, the fertility of imagination displayed in their endlessly variegated forms, the brightness of the colors and gilding, and the minuteness of finish, can only be appreciated by those who have had opportunities of examining them. We have seen some in which the illustrations equalled the finest paintings in miniature.[b] In a literary and useful point of view, the labors of these men are above all praise. They were the channels through which many valuable works of the ancients have been preserved and transmitted to us. And as regards the arts, both ornamental and useful, the monks were at one time almost their only cultivators.

---

[a] Univer. Hist, i, 373.   [b] In the Library of John Allan, Esq. of this city.

CHAPTER XIV.

WHEELS for raising water—Machines described by Vitruvius—Tympanum—De La Faye's improvement—Scoop Wheel—Chinese Noria—Roman do.—Egyptian do.—Noria with Pots—Supposed origin of Toothed Wheels—Substitute for wheels and pinions—Persian Wheel: Common in Syria—Large ones at Hamath—Various modes of propelling the Noria by men and animals—Early employment of the latter to raise water. Antiquity of the Noria—Supposed to be the 'Wheel of Fortune'—An appropriate emblem of abundance in Egypt—Sphinx—Lions' Heads—Vases—Cornucopia—Ancient emblems of irrigation—Medea: Inventress of Vapor Baths—Ctesibius—Metallic and glass mirrors—Barbers.

HAVING examined such devices for raising water, as from their simplicity have been generally unnoticed in treatises on hydraulic machines, we proceed to others more complex; and first, to such as revolve round the centres from which they are suspended, and which have a continuous instead of an alternating motion. Although differing in these respects and in their form, from the jantu or vibrating gutter and the swape, they will be found essentially the same; their change of figure being more apparent than real, and merely consequent on the new movement imparted to them. As these machines are obviously of later date than the preceding, it may perhaps be supposed, that the period of their introduction might be ascertained; but so it is, that with scarcely an exception, the time when, place where, and the persons by whom, they were invented, are absolutely unknown.

Although allusions to machines for raising water are found in several of their authors, it does not appear, that any general account or comprehensive treatise of them, was ever written by the ancients. If such a work was executed, it has perished in the general wreck of ancient records. About the beginning of the Christian era, a Roman architect and *engineer*, published a treatise on those professions, in which he inserted a brief description of some hydraulic engines. This is the only ancient work extant which treats professedly of them; and the whole that relates to them might be included in two pages of this volume.

The machines described by VITRUVIUS, for it is to him we allude, are the *Tympanum, Noria, Chain of Pots,* the *Screw,* and the *Machine of Ctesibius* or *Pump.* He has not mentioned the jantu, swape, the cord and bucket, with the various modes of using the latter; probably, because he considered these too simple in their construction to be properly classed among hydraulic machinery; he therefore passed by them, and modern authors have generally followed his example. Notwithstanding the omission of these, there are circumstances which render it probable that his account, brief as it is, includes all the *principal* machines that were used by the nations of the old world, if we except China. He wrote at a period the most favorable for acquiring and transmitting to posterity, a perfect knowledge of the mechanic arts of the ancient civilized nations; for he flourished during the last scenes of the mighty drama, when Rome had become the arbitress of the world, and the enlightened nations of the east—their wealth, learning, arts and artisans, were prostrate at her feet; so that if we were to suppose, absurd as it would be, that the previous intercourse of the Romans with Asia Minor, Egypt, Carthage and Greece, had not made them familiar with the arts of those countries, nothing could have prevented them from possessing such knowledge when they became Roman provinces—

hence we infer, that if there had been in use in any of those countries, (for some centuries previous to or during the life time of Vitruvius, and he was an old man when he published his work;) any efficient machine for raising water, different from those he has described, it would have been known to the Romans, and would have been noticed by him. Moreover, he was evidently familiar with the inventions of the mechanicians of former ages and frequently refers to them; and as all the machines described by him, were of foreign origin, and most of them of such high antiquity as to reach back to ages anterior to the birth of Romulus and the foundation of Rome; we have no reason to suppose that any important one has escaped him: to which we may add, if any useful machine for raising water had originated with his countrymen, he would scarcely have failed to record the fact.

The tympanum consists of a series of gutters united at their open ends to a horizontal shaft, which is made hollow at one end and placed a little higher than where the water is to be elevated; the gutters are arranged as radii, and are of sufficient length to extend from the shaft to a short distance below the surface of the water, as represented in the annexed diagram.

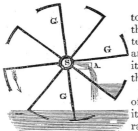

No. 44. Tympanum.

S, the shaft; G, G, the gutters; A, a trough to take away the water. The arrow indicates the direction in which the wheel turns; each gutter, as it revolves scoops up a portion of water and elevates it, till by the inclination to the axle, it flows towards the latter, and is discharged through one end of it.

Were the machine made as thus represented, i.e. of separate gutters and not connected to each other it could not be durable, as the weight of water raised at the end of each would have a tendency to break them at their junction with the shaft. The ancients therefore made two strong disks of plank well jointed together, of the diameter of the intended wheel, these they secured on a shaft, at a distance from each other, proportionate to the quantity of water required to be raised. Any number of plank partitions (Vitruvius says eight) were then inserted in the direction of radii between these disks, and were well secured to them, and made tight by caulking and pitch. The spaces between them, at the circumference of the wheel, were also closed, with the exception of an opening left for the admission of water to each; and where each partition joined the shaft, a hollow channel was formed in the latter, parallel to the axis, through which the water was discharged into a trough or gutter placed immediately under it. The tympanum is obviously a modification of the jantu of India, or rather it is a number of them combined, and having a revolving instead of a vibratory movement. It is the first machine described by Vitruvius; of which he observes, "it does not raise the water high, but it discharges a great quantity in a short time." B. x, Cap. 9. From its resemblance to a drum or tabor, it was named by the Romans *Tympanum*.

The prominent defect of the tympanum arises from the water being always at the extremity of a radius of the wheel, by which its resistance increases as it ascends to a level with the axis; being raised at the end of levers which virtually lengthen till the water is discharged from them. There is no reason to suppose, that this defect if perceived at all, by ancient mechanicians, was ever remedied by them; to most persons, the idea would never occur, that so simple a machine *could* be essentially improv

Chap. 14.]   *The Tympanum and Scoop Wheel.*   111

ed, and its having been described as represented in the last figure by a Roman philosopher and engineer; it was most likely used as thus constructed, through the remote ages of antiquity, to the early part of the last century, when a member of the Royal Academy of Sciences, of France, M. De La Faye, developed by geometrical reasoning, a beautiful and truly philosophical improvement. It is described by Belidor, (Tom. ii, 385,

No. 45. Tympanum improved by La Faye.

387,) together with the process of reasoning that led to it "When the circumference of a circle is developed; a curve is described, (the involute) of which all the radii are so many tangents to the circle; and are likewise all respectively perpendicular to the several points of the curve described, which has for its greatest radius, a line equal to the periphery of the circle evolved. Hence, having an axle whose circumference a little exceeds the height which the water is proposed to be elevated, let the circumference of the axle be evolved, and make a curved canal, whose curvature shall coincide throughout exactly with that of the involute just formed; if the further extremity of this canal be made to enter the water that is to be elevated, and the other extremity abut upon the shaft which is turned; then in the course of rotation, the water will rise in a VERTICAL DIRECTION, tangential to the shaft, and perpendicular to the canal, in whatever position it may be." See No. 45.

No. 46. Scoop Wheel.

The above figure from Belidor, is composed of four tubes only, but it is frequently contructed with double the number. Instead of tubes, curved partitions between the closed sides of the wheel are oftener used, as in the SCOOP WHEEL—which consists of a number of semicircular partitions, extending from the axle to the circumference of a large flat cylinder. As it revolves in the direction of the arrows, the extremities of the partitions dip into the water, and scoop it up, and as they ascend, discharge it into a trough placed under one end of the shaft, which is hollowed into as many compartments as there are partitions or scoops. Wheels of this description, and propelled by steam, are extensively used to drain the fens of Lincolnshire.

THE NORIA OR EGYPTIAN WHEEL.

The tympanum has been described as an assemblage of gutters, and the Noria may be considered as a number of revolving swapes. It consists of a series of poles united like the arms of a wheel to a horizontal shaft. To the extremity of each, a vessel is attached, which fills as it dips into the water, and is discharged into a reservoir or gutter at the upper part of the circle which it describes. See No. 47. Hence, the former raises water only through half a diameter, while this elevates it through a whole one. The idea of thus connecting a number of poles with their buckets, must have early occurred to the agricultural machinists of Asia. The advantages of such an arrangement being equally obvious as in the tympanum. The means that naturally suggested themselves, of strengthening a number of poles thus arranged, gradually brought these machines into the form of wheels. Sometimes, a rude ring was formed, to which the exterior ends were secured; at others, disks of plank were adopted, and the vessels were attached either to the sides or rim, and sometimes to both.

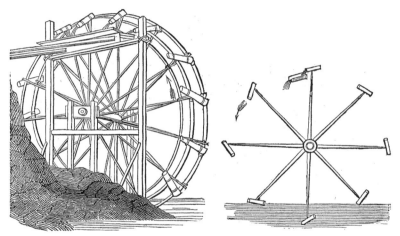

No. 48. Chinese Noria.   No. 47. Noria.

The Chinese make the noria, in what would seem to have been its primitive form; and with an admirable degree of economy, simplicity, and skill. With the exception of the axle and two posts to support it, the whole is of bamboo, and not a nail used in its construction. Even the vessels, are often joints of the same, being generally about four feet long and two or three inches in diameter. They are attached to the poles by ligatures at such an angle, as to fill nearly when in the water, and to discharge their contents when at, or near the top. See No. 48.

The periphery of the wheel is composed of three rings of unequal diameter, and so arranged, as to form a frustrum of a cone. The smallest one, to which the open ends of the tubes are attached, being next the bank over which the water is conveyed. By this arrangement, their contents are necessarily discharged into the gutter as they pass the end of it. When employed to raise water from running streams, they are propelled by the current in the usual way—the paddles being formed of woven bamboo. The sizes of these wheels, vary from twenty to seventy feet in

Chap. 14.] *Egyptian Noria.* 113

diameter. According to Staunton, some raise over three hundred tons of water in twenty-four hours. A writer in the Chinese Repository, mentions others which raise a hundred and fifty tons to the height of forty feet, during the same time. They combine strength and lightness in a remarkable degree.[a]

The mode of constructing and moving the noria by the Romans, is thus described by Vitruvius: "When water is to be raised higher, than by the tympanum, a wheel is made round an axis, of such a magnitude, as the height to which the water is to be raised requires. Around the extremity of the side of the wheel, square buckets cemented with pitch and wax are fixed; so that when the wheel is turned by the walking of men, the filled buckets being raised to the top, and turning again toward the bottom, discharge of themselves what they have brought into the reservoir." B. x, Cap. 9. Newton's Trans. As the drawings made by Vitruvius himself, and annexed to his work are all lost, his translators do not always agree respecting the precise form of the machines described by him. Newton has figured the noria as a large drum, to one side of which square boxes or buckets are secured. These buckets are closed on all sides, with the exception of an opening to admit and discharge the water. Perault has placed them on the paddles or floats of an undershot wheel, like Barbaro, except that the latter makes the bottom of the boxes or buckets serve at the same time as paddles to receive the impulse of the stream. Rivius, in his German Translation, (Nuremburgh 1548,) has given one figure resembling an *overshot wheel* with the motion reversed, a form in which it is still sometimes made; in another, it is similar to the *noria of Egypt* at the present day, a modification of it, probably of great antiquity.

No. 49. Egyptian Noria.

Instead of pots or other vessels secured to the arms by ligatures, or buckets attached to the sides of a wheel, as described by Vitruvius, the periphery of the wheel itself is made hollow, and is divided into a number of cells, or compartments, which answer the same purpose as separate ves-

---

[a] Van Braam's Journal, i, 172. Ellis's Journal of Amherst's Embassy, 280. Chinese Repository, iii, 125.

sels. The figure No. 49, is taken from the *Grande Description* of Egypt. Plate 3, Tom. 2, E. M. It was sketched from one near Rosetta, which raised the water nine feet. The liquid enters through openings in the rim, and is discharged from those on the sides. The arrow shows the direction in which it moves. The section of part of the rim, will render the internal construction obvious. Mr. P. S. Girard, author of the Memoir on the Agriculture of the Egyptians, says they are extensively used in the Delta, the *cog* wheels being very rudely formed.

The tympanum may be considered as a wheel with *hollow spokes*, while the noria, as above constructed, is one with *hollow felloes*, a term by which it is designated in French authors: 'Roue a jante creuses,' a name very expressive, and one which, in the absence of information respecting the construction of this machine, might enable a mechanic to make it.

In various parts of Asia, Greece, Turkey, Spain, &c. Earthenware jars or pots, are secured to the rim or side of the wheel, as in No. 50. Every farm and garden in Catalonia, says Arthur Young, has such a machine to raise water for the purpose of irrigating the soil. They are propelled by horses, oxen, mules, and sometimes by men. In Spain, the noria has remained unaltered from remote times. It is there still moved by means of a device which probably gave rise to toothed wheels.

No. 50. Noria with Pots.

In the axle of the noria are inserted two, (and sometimes four) strong sticks which cross each other at right angles, forming arms or spokes. The part of the shaft in which these are fixed, extends nearly to the centre of the path, round which the animal walks; and contiguous to it, is the vertical shaft to which the yoke or beam is attached: the bottom of this shaft has spokes inserted into it similar to the former, and which take hold of them in succession, and thereby keep the wheel or noria in rotation. See No. 50. This rude contrivance is common through all the east, and is in all probability identical with those of the early ages; in other words, the primitive substitute of the modern cog wheel.

In Besson's 'Theatre Des Instrumens' is an ingenious device by which a horizontal shaft with four spokes, as in the last figure, can impart motion to a vertical one, at any distance from the centre, and thereby answer the purpose of a number of wheels and pinions in modifying the velocity of the machinery, according to the work it has to perform, or to an increase or diminution of the motive force employed. On the horizontal shaft, (which is turned by a crank,) is a sliding socket to which the spokes are secured. The vertical shaft has also a similar socket, which is raised and lowered by means of a screw, and to it, arms and spokes are well secured. These are arranged in the form of a flat cone; so that by adjusting the sockets, the spokes in the horizontal shaft can be made to take hold on those which form the cone round the vertical one at any part, from its apex to its base.

Two prominent defects have been pointed out in the noria. First, part of the water escapes after being raised nearly to the required elevation. Second, a large portion is raised *higher* than the reservoir placed to receive it, into which it is discharged after the vessels begin to descend. (See No. 49, in which they are very conspicuous.) Consequently, part of the power expended in moving this wheel, produces no useful effect. These imperfections, however, did not escape the notice of ancient mechanicians, for to obviate them, the PERSIAN WHEEL was devised, and so named from its having been invented or extensively used in that country.

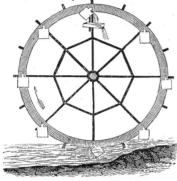

No. 51. Persian Wheel.

The vessels in which the water is raised, instead of being fastened to the rim, or forming part of it, as in the preceding figures, are suspended from pins, on which they turn, and thereby retain a vertical position through their entire ascent; and when at the top are inverted by their lower part coming in contact with a pin or roller attached to the edge of the gutter or reservoir, as represented in the figure. By this arrangement no water escapes in rising, nor is it elevated any 'higher' than the edge of the reservoir; hence the defects in the noria are avoided. Persian wheels it is believed, have been used in Europe ever since the Romans ruled over it, if not before. The greatest work in France according to Arthur Young, for the artificial irrigation of land, was a series of them in Languedoc, which raised the water thirty feet. In a Dutch translation of Virgil's Georgics in 1682, they are represented with huge buckets like barrels, suspended from both sides of the rim. They are common in Switzerland and the Tyrol. Travels in Poland by D'Ulanski, page 241. They were extensively used in England one hundred and fifty years ago. See Dict. Rusticum. Lon. 1704. We are not aware of their being much employed, if at all, in the United States.

They are common in various parts of Asia. "The water wheels still used in Syria," says Mr. Barrow, "differ only from those of China, by having loose buckets suspended at the circumference, instead of fixed tubes."[a] Dr. Russel, in his 'Natural History of Aleppo,' (p. 20,) says the inhabitants make use of large quantities of water, "which they raise with the Persian wheel," from the river. Perhaps the most interesting specimens of these machines extant, are to be found in another and very ancient city of Syria; in *Hamath* on the Orontes, so named after its founder, one of the sons of Canaan. "Two days journey below Homs, (says Volney) is Hamath, celebrated in Syria for its water works. The wheels are the largest in the country, being thirty-two feet in diameter." The city is built on both sides of the river, and is supplied with water from it by means of them, the buckets of which empty themselves into stone aqueducts, supported on lofty arches on a level with the ground on which the city stands. They are propelled by the current. Burckhardt observed about a dozen of them, the largest he says, "is called Naoura el Mahommeyde,

---

[a] Embassy to China, Lon. 1806. p. 540.

and is at least seventy feet."[a] They are, he remarks, the greatest curiosity which a modern traveler can find in the city. Their enormous magnitude will be apparent, if we consider that the loftiest class of buildings in this city, (N. York,) those of six stories, seldom exceed sixty feet. If therefore, the largest of the Persian wheels at Hamath, were placed on the pavement, with its side towards a range of such buildings, it would occupy a space nearly equal to the fronts of three of them, and would extend several feet over the roofs of the highest—and twelve of them would occupy a street, one sixth of a mile in length.

The construction of the water works of *Hamath* have probably remained unaltered in their general design, from very remote times. The peculiar location of this city, the rapidity of the river, (named El Ausi, the *swift*,) and its consequent adaptation to propel *undershot wheels*, which we know, were used in such works by the ancients, render it probable that the present mode of raising water, is much the same as when this city flourished under Solomon; and when the Romans under Aurelius, overthrew the queen of Palmyra and her army, in its immediate vicinity; and from the great antiquity of the noria, its extensive use over all Asia in former ages, and its peculiar adaptation to Hamath, and the tenacious adherence of the orientals to the devices of their forefathers; we infer that the machines which Burckhardt beheld with admiration, raising the water of the Orontes, were similar to others in use at the same city, when the spies of Moses, searched the land, 'from the wilderness of Zin unto Rehob, as men come to *Hamath*.'[b] These wheels may be cited as another proof of the preservation, (by continual use) of hydraulic machines, while every other memorial of the people by whom they were originally used, has long since disappeared.

MODES OF PROPELLING THE NORIA.—The tympanum, noria, chain of pots, and even the screw, were often turned, according to Vitruvius, by the 'treading' or 'walking of men,' i. e. except when employed to raise water from rapid streams, in which case they were propelled, he says, by the current acting on float boards or paddles, as in common under-shot wheels. There is a difference of opinion among his translators respecting the mode by which *men* moved these machines. *Rivius*, the translator of the German edition of 1548, seems to have thought that they walked round an upright shaft, (as in figs. 26 and 53,)which they turned by horizontal bars, and by means of cog wheels communicated the required motion. He has also represented the noria as moved by men turning a crank; a mode of propelling it that is figured in the first German edition of Vegetius, (1511.) *Barbaro*, (1567,) represents the tympanum as moved by a crank; the noria by a current of water; and the chain of pots, by a tread wheel, like the one figured in No. 24. *Perault*, also, in his figure of

---

[a] Travels in Syria, and the Holy Land. Lon. 1822, p. 146.
[b] There are several interesting circumstances recorded respecting Hamath. This city and Damascus were frequently subject to the Jews. The 'land of Hamath,' was particularly fatal to them and their kings. Zedekiah was there taken, and his sons and nobles slain in his presence; his own eyes were then put out, and he was carried a captive to Babylon, where he died. Jer. xxxix, 5. Pharaoh Necho there put Jehoahaz, another of their kings in bonds, whence he was taken a prisoner to Egypt, and confined till his death. 2 Kings xxiii, 34. Among the most interesting discoveries of modern times, connected with the ancient history of this people, are sculptured representations at Thebes, of the Jews captured by Shishak, with the hieroglyphical inscription, 'Jouhouda Melec,' king of the Jews. From the discoveries of Young and Champollion, the precision with which the dates are determined, is wonderful; 'many of the sculptures have the dates inscribed to the day and the month.' The figure of the Jewish king, is supposed to be a correct portrait, for we are told in those of the Egyptian monarchs, " the likenesses are always exactly preserved."

Chap. 14.]   *Early employment of Animals to raise Water.*   117

the tympanum, places the men in a similar one, and this interpretation of the text has been generally followed. It is corroborated by other ancient authors, and by Vitruvius himself, in Book x, cap. 4, where he speaks of a wheel to raise weights, 'by the walking of men therein,' that is, the common walking crane. *Philo*, who was contemporary with Vitruvius, or flourished shortly after him, mentions a wheel for raising water, which was turned by the motion of men's feet, 'by their ascending successively the several steps that are within it.' Tread wheels are mentioned also by Suetonius, and Strabo speaks of some for raising the water of the Nile, which were moved by a hundred and fifty slaves. Mr. Newton, the English translator, supposed the men walked on the *outside* of the wheel, like the modern tread mill. It is very probable that this mode was in use among the ancients, for it is common in Persia and other oriental countries, particularly China, where it is undoubtedly of great antiquity. Barbaro has figured the *screw*, as propelled by men pulling down spokes on the periphery of a wheel attached to it, or by treading on them.

About eighteen years ago, a person in this city, (N. York,) took out a patent for employing *animals* to propel such wheels. A horse was placed near the top and yoked to a horizontal beam fixed behind, and against which he drew. In January, 1795, a Mr. Eckhardt obtained a patent in England for 'A Method of applying Animals to Machinery in general.' His plan was to employ cattle and all other bulky animals to walk on the top of large wheels; he also proposed a flexible floor, like an endless chain, which passed over two wheels, and formed an inclined plane on which animals walked, and to increase the effect, they drew a loaded cart behind them.[a] Sixty years before this, viz. in 1734, Mr. W. Churchman exhibited before the Royal Society, a model of 'A new Engine for raising Water, in which Horses and other Animals draw without any loss of power.' This engine was a series of pumps worked by a large tread wheel, on the top of which horses were made to draw against a beam to which they were yoked. He also proposed to employ horses at the same time within the wheel.[b] But the contrivance was even then an old one, for in Agricola, a horse is figured imparting motion to bellows by walking *upon* a tread wheel.[c]

There is a passage in the second chapter of the Koran, which throws some light on the early employment of animals in raising water. Among the ancients, it was a prevailing custom when they sacrificed an ox, or a heifer, to select such as had never been broken to labor: hence the direction of the Sibyl to Eneas.

Seven bullocks *yet unyoked*, for Phœbus choose,
And for Diana, seven unspotted ewes.

The Israelites also, were instructed to offer " a red heifer without spot wherein is no blemish, and upon which *never came yoke*." " An heifer which hath *not been wrought with*, and which hath not drawn in the yoke." One which, according to Mahomet, was " not broken to plough the earth, or WATER *the field*." Now this interpretation is not only consistent with the text of Moses, but is exceedingly probable, for the Arabs have undoubtedly preserved with their independence and ancient habits, traditions of numerous transactions referred to in the Pentateuch, the particulars of which are not recorded; besides it indicates, what indeed might have been inferred: viz. that the principal employment of animals in the early ages, was to plough and *irrigate* the soil. But when in process of time,

---

[a] Repertory of Arts. Lon. 1795. Vol. ii.   [b] Phil. Trans. Abridged by Martyn, viii, 321.
[c] De Re Metallica, 169

human population became dense, then animal labor was in some degree superseded by that of man. The extensive employment of the latter, appears to have been a prominent feature in the political economy of ancient Egypt, just as it is in modern China. As the country teemed with inhabitants, the extensive use of animal labor would not only have interfered with the means of the great mass of the former in obtaining a living, but would have required too large a portion of the land to raise food merely for the latter.

The antiquity of the noria may be inferred from its name of " *Egyptian* wheel," the only one by which it was known in some countries. It is to be found if we mistake not, among the symbols of ancient mythology. In elucidating one of the religious precepts of Numa, which required persons when worshipping in the temples, *to turn round ;* Plutarch observes, that this change of posture may have an enigmatical meaning, "like the EGYPTIAN WHEELS, admonishing us of the instability of every thing human, and preparing us to acquiesce and rest satisfied with whatever turns and changes the divine being may allot." Life of Numa. This figurative application of the noria, is obviously used by Plutarch as a common and consequently a long established symbol of the mutability of human affairs ; and, as the sentiment which he illustrates by it, is precisely the *same* as that which the *wheel of the goddess of fortune* was designed to point out, the " instability of fortune," and of which it was the emblem, we conclude that the "wheel of fortune," was a *water* wheel, and no other than the NORIA ; and that to it, the Grecian philosopher in the above passage referred. The selection of an Egyptian wheel to denote the mutability of human affairs, indicates the origin not only of Plutarch's similitude, but also that of the fable of the goddess. Egypt was the source whence the Greeks obtained not only their arts and science, but also their mythology, with its deities, heroes and its mysterious system of symbolical imagery ; and if the Egyptians were not the *inventors* of the system of representing and concealing things by symbols, they certainly carried it to a greater extent than any other people, and at a period long before the Greeks had emerged from barbarism, or an Egyptian colony had settled in their country.

Although we are not aware that the *wheel of fortune* had any other signification, yet, as the same goddess presided over RICHES AND ABUNDANCE—a more expressive emblem of these in EGYPT could not have been devised. Agriculture was the grand source of wealth in that country, and *it* depended almost entirely upon artificial irrigation, for except during the annual inundation of the Nile, water was raised for that purpose by machines, and among these, the noria was one of the most prominent, and probably one of the most ancient. Egypt without irrigation would have been a dreary waste, and like its neighboring deserts uninhabited by man; but by means of it, the soil became so exceedingly fertile that Egypt became "the garden of the east,"—the " hot bed of nature," and the " granary of the world." It was artificial irrigation which, under the Pharaohs, produced food for seventeen millions of inhabitants, and in the reign of Rameses or Sesostris, a surplus sufficient for thirty-three millions more ; and even under the Grecian yoke, when its ancient glory had long departed, the prodigious quantities of grain, which it produced, enabled Ptolemy Philadelphus to amass treasure equal to nine hundred and fifty millions of dollars. There was therefore a peculiar propriety, whether designed or not, in the goddess of " prosperity," " riches," and " abundance," being accompanied with the noria or Egyptian wheel, *the* implement which contributed so greatly to produce them. The manner in which this deity was sometimes represented, appears to have had direct reference to agriculture and irrigation. She was seated on rocks, (emblems of sterility?) the wheel by her side and a river at her feet, (to signify irrigation?) and she held wheat ears, and flowers in her hand. But whether the ancient Egyptians adopted the noria or not, as the emblem of wealth and irrigation, one of their most favorite symbols has direct reference to the latter, and indirectly to the former: viz. the SPHINX; figures of which have been found among the ruins, from one end of the country to the other. This figure consists, as is universally known, of the the head and breasts of a woman, united to the body of a lion, and was symbolical of the annual overflow of the Nile, which occurred when the sun passed through the zodiacal signs, *Leo* and *Virgo*—hence the combination of these signs in the Sphinx, as an emblem of that general irrigation of the land once a year, upon which their prosperity so greatly depended. This was the origin of passing streams of water through the mouths of figures of lions, and sometimes, though more rarely, of virgins, as in the figures below—which are taken from Rivius' translation of Vitruvius.

No. 52. Orifices of Pipes, &c. symbolical of Irrigation.

The analogy between the form and ornaments of an object and its uses, seems to have always been kept in view by the ancients; although, from our imperfect knowledge of them, it is difficult and sometimes impossible to perceive it. That they displayed unrivaled skill in some of their designs and decorations is universally admitted. There is certainly no natural analogy between a lion and a fountain, and no obvious propriety in making water to flow out of the mouths of figures of these animals; on the contrary, they appear to be very inappropriate; but when we learn that the lion as an astronomical symbol, was intimately associated with a great natural hydraulic operation, of the first importance to the welfare of the Egyptians, we perceive at once their reasons for transferring figures of it to artificial discharges of the liquid, and hence the orifices of cocks, pipes, and spouts of gutters, fountains, &c. were decorated as above. In some ancient *fountains*, figures of virgins, as nymphs of springs, leaned upon urns of running water. In others, VASES *overturned,* (with figures of Aquarius, Oceanus, &c.) a beautiful device. Lions' heads for spouts are very common in Pompeii.

There is another ancient emblem, and one that is universally admired, which may here be noticed, as its origin is associated with *artificial irrigation*—the CORNUCOPIA, or 'Horn of Abundance.' This elegant symbol is probably of Egyptian origin, for ISIS was sometimes represented with it, and *Isis*, in the Egyptian language, signified the 'cause of abundance.' We have already seen that irrigation was and still is, the principal source of plenty in Egypt; and *water* in the scriptures is repeatedly used in the same sense. To understand the allegory, it must be borne in mind that *rivers* were anciently compared to *bulls;* the reasons for which at this remote period, are not very obvious; perhaps among others, from the noise of rapid streams, bearing some resemblance at a distance, to the lowing and bellowing of these animals; and the *branches of rivers* were compared to their *horns;* thus, the small branch of the Bosphorus, which forms the harbor of Constantinople, still retains its ancient name of the 'Golden Horn;' and in some of our dictionaries, 'winding streams' is given as one of the definitions of *horns.* The bull which is common on some Greek coins is supposed to have been the symbol of a river, perhaps from the overflow of some, when the sun passed through the zodiacal sign *Taurus.* According to the Greek version, one of the branches of the river *Achelous* in Epirus, was diverted or broken off by Hercules, to irrigate some parched land in its vicinity. This, like other labors of that hero, was allegorized by representing him engaged in conflict with a bull, (Achelous) whom he overcame, and *broke off one of his horns;* and this horn being filled with fruits and flowers, was emblematical of the *subsequent fertility of the soil.* Ovid describes the contest, when that hero

> . . . . 'twixt rage and scorn,
> From his maimed front, he tore the stubborn horn,
> This, heap'd with flowers and fruits, the Naiads bear,
> Sacred to plenty, and the bounteous year.
>   *   *   *   *
> But Achelous in his oozy bed
> Deep hides his brow deform'd, and rustic head,
> No real wound the victor's triumph show'd,
> But his lost honors griev'd the watery god. *Met.* ix.

Thus river gods were sometimes represented with a cornucopia in one hand, and the other resting on a vase of flowing water.

Another interesting allegory of the ancients has reference to water: the fable of MEDEA, who it was said, by *boiling* old people, made them young again, referred to warm or vapor baths, which she invented, and into which she infused fragrant herbs—in other words, the 'patent medicated vapor baths' of the present day. She also possessed the art of changing the color of the hair. When therefore, by her fomentations, persons appeared more active and improved in health, and their grey hairs changed into ringlets of jet, the belief in her magic powers became irresistible— and when at length, her apparatus, i. e. the *cauldrons, wood and fire, &c.* were discovered, (which she had sedulously concealed,) it was supposed that her patients were in reality *boiled.* From Ovid, it seems she had the modern *sulphur* bath also, and used it in the cure of Æson, the father of her husband Jason:

> . . . . the sleeping sire,
> She lustrates thrice with sulphur, water, fire.
>   *   *   *   *
> His feeble frame resumes a youthful air,
> A glossy brown, his hoary beard and hair.
> The meagre paleness from his aspect fled
> And in its room sprang up a florid red. *Met.* vii.

This lady was the great patroness of herb and steam doctors of old; and may be considered the ancient representative of modern manufacturers of specifics, which, as they allege, (and often truly) remove all diseases. The fable of her slaying her own children in the presence of Jason, is easily explained by her administering to them the *wrong medicine,* or too large a dose of the right one; the latter was certainly the case with old Pelias who expired under it.

Having noticed in this chapter the supposed origin of cog-wheels, we may as well introduce here an ancient mechanic, to whom we shall have occasion hereafter to allude; one, whose name is intimately associated with the most valuable machines for raising water, and with several important improvements in the mechanic arts. As the earliest distinct notice of cog-wheels is in the description of one of his machines, (see the clepsydra, page 547,) we may as well introduce him to the reader at this part of our subject, although we have not yet in the progress of our work, arrived at the period at which he flourished.

During the reign of Ptolemy Philadelphus over Egypt, an Egyptian barber pursued his vocation in the city of Alexandria. Like all professors of that ancient mystery, he possessed besides the inferior apparatus, the *two* most essential implements of all: a razor and a looking glass, or mirror, probably a metallic one. This mirror, we are informed, was suspended from the ceiling of his shop, and balanced by a weight, which moved in a concealed case in one corner of the room. Thus, when a customer had undergone the usual purifying operations, he drew down the mirror, that he might witness the improvement which the artist had wrought on his outer man; and, like Otho,

In the Speculum survey his charms.   *Juv. Sat.* ii.

after which he returned it to its former position for the use of the next customer.[a] It would seem that the case in which the weight moved was enclosed at the bottom, or pretty accurately made, for as the weight moved in it, and displaced the air, a certain sound was produced, either

---

[a] Metallic mirrors furnish one of the best proofs of skill in working the metals in the remotest times, for their antiquity extends beyond all records. In the first pages of history they are mentioned as in common use. The brazen laver of the Tabernacle, was made of the mirrors of the Israelitish women, which they carried with them out of Egypt. From some found at Thebes, as well as representations of others in the sculptures and paintings, we see at once that these 'looking glasses,' (as they are called in Exodus,) were similar to those of Greek and Roman ladies: viz. round or oval plates of metal, from three to six inches in diameter, and having handles of wood, stone and metal highly ornamented and of various forms, according to the taste of the wearer. Some have been found in Egypt with the lustre partially preserved. They are composed of an alloy of copper, and antimony or tin, and lead; and appear to have been carried about the person, secured to, or suspended from the girdle, as pincushions and scissors were formerly worn and are so still by some antiquated ladies. The Greeks and Romans had them also of silver and of *steel.* Some of the latter were found in Herculaneum. Plutarch mentions mirrors enclosed in very rich frames. Among the articles of the toilet found in Pompeii, are ear-rings, golden and *common pins,* and several metallic mirrors. One is round and eight inches in diameter, the other an oblong square. They had them with plane surfaces, and also convex and concave. Seneca says his countrywomen had them also, equal in length and breadth to a full grown person, superbly decorated with gold and silver, and precious stones. Their luxury in this article, seems to have been excessive, for the cost of one often exceeded a moderate fortune. The dowry which the Senate gave the daughter of Scipio, according to Seneca, would not purchase in his time, a mirror for the daughter of a freedman. The Anglo-Saxon dames had portable metallic mirrors, and wore them suspended from the waist. It is not a litle singular that the ancient Peruvians had them also, formed of silver, copper and its alloys, and also of obsidian stone. They had them plane, convex, and concave. Had not the art of making these mirrors been revived in the speculums of reflecting telescopes, their lustre could hardly have been appreciated; and they would probably have been considered as indifferent substitutes for the modern looking-glass. These last are supposed to have been manufactured in ancient Tyre, and of a black colored glass. Fluid lead or tin was afterwards used. It was poured on the plates while they were hot from the fire, and being suffered to cool, formed a back which reflected the image. Looking-glasses of this description were made in Venice, in the 13th century. It was not till about the 16th, that the present mode of coating the back with quicksilver and tin foil was introduced. The inventor is not known. Venus was sometimes represented with a speculum in one hand, and the astronomical symbol of the planet Venus is the figure of one. There is a chemical examination of an ancient speculum in the 17th volume of Tilloch's Phil. Mag.

Barbers flourished in the mythologic ages, for Apollo having prolonged the ears of Midas to a length resembling those of a certain animal, the latter it is said, endeavored to hide his disgrace by his hair, but found it impossible to conceal it from his *barber* Bronze razors were anciently common.

by its expulsion through some small orifice, or by its escape between the sides of the case and the weight. This sound had probably remained unnoticed like the ordinary creaking of a door, perhaps for years, until one day as the barber's son was amusing himself in his father's shop, his attention was arrested by it. This boy's subsequent reflections induced him to investigate its cause; and from this simple circumstance, he was led eventually either to invent, or greatly to improve the hydraulic organ, a musical instrument of great celebrity in ancient times. His ingenuity and industry were so conspicuous, that he was named ' The Delighter in Works of Art.' His studies in various branches of natural philosophy, were rewarded it is said, with the discovery of the pump, air-gun, fire-engine, &c. He also greatly improved the clepsydra or water-clock, in the construction of which he introduced *toothed* wheels, and even *jeweled holes*. Vitruvius, ix, 9. These ancient time-keepers, were therefore the origin of modern clocks and watches. Now this barber's son is the individual we wish to introduce to the reader, as CTESIBIUS OF ALEXANDRIA, one of the most eminent mathematicians and mechanicians of antiquity—one, whose claims upon our esteem, are not surpassed by those of any other individual, ancient or modern.

It will be perceived that the simple, the trivial sound produced by the descent of the weight in his father's shop, was to him, what the fall of the apple was to Newton, and the vibration of the lamp or chandelier in the church at Pisa, to Galileo. The circumstance presents another to the numerous proofs which might be adduced, that inquiries into the causes of the most trifling or insignificant of physical effects, are sure to lead, directly or indirectly, to important results—while to young men especially, it holds out the greatest encouragement to occupy their leisure in useful researches. It shows, that however unpropitious their circumstances may be, they may by industrious application, become distinguished in science, and may add their names to those of Ctesibius and Franklin, and many others—immortal examples of the moral grandeur of irrepressible perseverance in the midst of difficulties.

CHAPTER XV.

THE CHAIN OF POTS—Its origin—Used in Joseph's well at Cairo—Numerous in Egypt—Attempt of Belzoni to supersede it and the noria—Chain of pots of the Romans, Hindoos, Japanese, and Europeans—Described by Agricola—Spanish one—Modern one—Applications of it to other purposes than raising water—Employed as a first mover and substitute for overshot wheels—Francini's machine—Antiquity of the chain of pots—Often confounded with the noria by ancient and modern authors—Introduced into Greece by Danaus—Opinions of modern writers on its antiquity—Referred to by Solomon—Babylonian engine that raised the water of the Euphrates to supply the hanging gardens—Rope pump—Hydraulic Belt.

THE tympanum and noria in all their modifications, have been considered as originating in the gutter or jantu, and the swape; while the machine we are now to examine is evidently derived from the primitive cord and bucket. The first improvement of the latter was the introduction of a pulley (No. 11) over which the cord was directed—the next was the addition of another vessel, so as to have one at each end of the rope, (Nos. 13 and 14) and the last and most important consisted in uniting the ends of the rope, and securing to it a number of vessels at equal distances through the whole of its length—and the CHAIN OF POTS, was the result.

The general construction of this machine will appear from an examination of those which are employed to raise water from Joseph's well at Cairo, represented at page 46. Above the mouth of each shaft a vertical wheel is placed; over which two endless ropes pass and are suspended from it. These are kept parallel to, and at a short distance from each other, by rungs secured to them at regular intervals, so that when thus united, they form an endless ladder of ropes. The rungs are sometimes of wood, but more frequently of cord like the shrouds of a ship, and the whole is of such a length that the lowest part hangs two or three feet below the surface of the water that is to be raised. Between the rungs, earthenware vases (of the form figured No. 7) are secured by cords round the neck, and also round a knob formed on the bottom for that purpose. See A, A, in the figure. As the axis of the two wheels are at right angles to each other, two separate views of the chains are represented. In the lower pit, both ropes of one half of the chain is seen; while in the upper, the whole length of one is in view. The vases or pots are so arranged that in passing over the wheel, they fall in between the spokes which connect the two sides of the latter together, as shown in the section; and when they reach the top, their contents are discharged into a trough. [In some machines the trough passes under one rim which is made to project for that purpose; in others, it is placed below the wheel and between the chains.] There are in the upper pit, one hundred and thirty-eight pots and the distance from each other is about two feet seven inches. The contents of each are twenty cubic inches. The wheels that carry the chains are six feet and a half in diameter. They are put in motion by cog wheels (on the opposite end of their axles) working into others that are attached to the perpendicular shafts to which the blindfolded animals are yoked.

The chain of pots in Egypt is named the *Sakia*. Its superiority over the noria and tympanum, &c. in being adapted to raise water from every depth, has caused it to be more extensively employed for artificial irrigation than any other Egyptian machine—hence it is to be seen in operation, all along the borders of the Nile, from its mouth up to the first cataract. In Upper Egypt, and Nubia, they are so exceedingly numerous as to occur every hundred yards; and in some cases they are not forty yards apart. Their numbers and utility have rendered them a source of revenue, for we are informed that each sakia is taxed twenty dollars per annum, while the swape is assessed at half that amount. They are also common in Abyssinia. They were noticed there by Poncet in 1698. When *Sandys* was in Egypt, A. D. 1611, the great number of sakias did not escape his observation: " Upon the banks all along are infinite numbers of deepe and spacious vaults into which they doe let the river, drawing up the water into higher cesterns, with wheeles set round with pitchers, and turned about by buffaloes." Travels, page 118.

An attempt was made some years ago by an enterprising European to supersede the employment of these machines in Egypt, which on account of the interesting circumstances connected with it may here be noticed. In the latter part of the last century an intelligent young man of Padua was designed by his parents for a monk, and was sent to Rome to receive an appropriate education. His inclination however led him to prefer the study of natural philosophy to that of theology, and particularly hydraulics. Upon the invasion of Italy and capture of Rome by the French, he wandered over various parts of Europe, supporting himself by publicly per forming feats of agility and strength, and by scientific exhibitions. After roving thus for fifteen years, he determined to visit Egypt, under the belief that he would make his fortune there by introducing machinery on the principle of the pump, as substitutes for the noria and chain of pots, &c. In June 1815, he landed at Alexandria, and after some delay was introduced to Mahommed Ali, (the present Pasha,) who approved of his project, and in whose gardens at Soubra, three miles from Cairo, he constructed his machine. But no sooner was it completed and put in operation than he discovered in the Turkish and Arabic cultivators an unconquerable opposition to its introduction. Indeed this result might have been anticipated and, if we are not mistaken, they were right in preferring their own simple apparatus to an elaborate machine, of the principle of whose action they were utterly ignorant. Their rejection of it was looked upon as another example of superstitious adherence to the imperfect mechanism of former ages; but under all the circumstances, it was, we believe, an evidence of the correctness of their judgment. Thus disappointed, his brightest hopes blasted, and his pecuniary resources all but exhausted—for he received no renumeration, either for the loss of his time or his money—he, with an energy of character deserving all praise, determined to make the best of his misfortunes. He therefore turned his attention to that subject which necessarily occurs to every intelligent stranger in Egypt—its *antiquities*—and while the British Museum remains, and the colossal head of young Memnon is preserved, the name of BELZONI will be remembered and respected.

From the following description of the chain of pots by Vitruvius, it appears that the Romans made it of more durable materials than either the ancient or modern people of Asia. "But if a place of still greater height (than could be reached by the noria) is to be supplied; on the same axis of a wheel, a *double chain of iron* is wound and let down to the level of the bottom; having *brass buckets*, each containing a congius (seven pints) hanging thereto, so that upon the turning of the wheel, the chain revolving round the axis raises the buckets to the top; which when drawn upon the axis, become inverted and pour into the reservoir the water they have brought." Book x, Cap. 9, Newton's Trans. As no reference is made to the form of the vessels, by Vitruvius, we find them represented by translators in a variety of shapes, as cylinders, cubes, truncated cones, pyramids, as well as portions of, and combinations of them all. Some are left open at the top, and both with and without projecting lips in front, by which to shoot the contents over the edge of the reservoir as they pass the wheel or drum. Others are closed,

No. 53. Roman Chain of Pots.

and admit and discharge the water through an orifice or short tube as represented. (No. 53.) From the separate figure of one of the vessels it will be seen that the tubes are placed at the upper corner, and consequently retain the water till the vessels ascend the drum, when it is discharged as represented. Provision should be made for the escape of air from these vessels, as they enter the water, and also for its admission on the discharge of the liquid above. The wheel or drum which carries the chain is, in this figure, solid, and cut into a hexagonal form to prevent it from slipping.

There is also in old authors a great diversity in the construction of the chains, and also in their number. Some understand by the term '*double chain*,' merely a simple one doubled and its ends united; i. e. one whose length is equal to double the space through which the water is to be elevated by it. Others suppose two separate ones intended and placed parallel to each other, the vessels being connected to them as in the figure. Others again, and among them Barbaro, figure two sets of chains and pots carried by the same wheel. He has also made them pass under pulleys in the water, a useless device, except when the chains are employed in an inclined position.

The chain of pots is mentioned by most oriental travelers, although described by few. In *Terry's* voyage to India in 1615, speaking of the tanks and wells of the Hindoos, he observes, "they usually cover those wells with a building over head, and with oxen draw water out of them, which riseth up in many small buckets, whereof some are always going down, others continually coming up and emptying themselves in troughs or little rills, made to receive and convey the water, whither they please." p. 187. To the same machine *Fryer* refers, when speaking of the different modes of raising water from *deep wells*. It is drawn up, he says, by oxen "with huge leathern buckets or pots around a wheel." p. 410. And again at Surat, it is drawn up "in leathern bags upon wheels." p. 104. Had not *wells* been mentioned in connection with these extracts from Fryer, we might have supposed it was the *noria* to which he alluded. Tavernier mentions it in the same way as applied to draw water from wells in Persia. p. 143. When required to raise it from rivers, they were, as in the case of the Persian wheels on the Orontes, propelled by the current when it was sufficiently rapid for the purpose. "As for the Euphrates, (observes Tavernier,) certain it is that the great number of mills built upon it, to convey water to the neighboring grounds, have not only rendered it unnavigable, but made it very dangerous." Lucan in the 3d book of his Pharsalia alludes to this extensive diversion of the water for agricultural purposes, in his time.

> But soon Euphrates' parting waves divide,
> Covering, like fruitful Nile, the country wide.

These mills are probably similar to those referred to by Montanus in his account of *Japan*, p. 296. The city of Jonda, he observes was defended by a strong castle, which was "continually supplied with fresh water by two mills." It is a pity they were not described.

The chain of pots was used by all the celebrated nations of antiquity and it still is employed more or less over all Asia and Europe. Previous to the 16th century, it constituted the 'water works' for supplying European cities, and was often driven by windmills—as it still is in Holland. It seems to be the *ne plus ultra* of hydraulic engines among half civilized nations, while those only which are enlightened, have the pump. Even the materials of which it was made by different people of old, may be considered as emblematical of their national characters. The inhabitants of Egypt, central and southern Asia, employed light and fragile materials;

the ropes were fibres of the palm tree, and the vessels of earthenware; while the Romans made the chains of iron and the vessels of brass. The former people were soft, effeminate, and easily subdued; the latter stern and inflexible—an iron race.

It is described by Agricola as employed in the German mines. De Re Metallica, pp. 131, 132, 133. The chains and vessels are represented of various forms, and the latter both of iron and wood, and propelled by tread and water wheels.[a] In Besson's 'Theatre,' A. D. 1579, it is figured as worked by a pendulum and cog wheels—the teeth being continued over half the peripheries only.[b]

In Spain it has remained in continual use since the conquest and occupation of that country by the Romans; and was perhaps previously introduced by the *Phenicians*, a people, to whom Spain was early indebted for many valuable acquisitions. It was employed there by the Moors in the middle ages, under whom the inhabitants enjoyed a degree of prosperity and civilization unexampled during any subsequent period of their history. The arts and manufactures were carried to great perfection, so much so, that in the twelfth century, while the rest of Europe was in comparative barbarism, the tissues of Grenada and Andalusia were highly prized at Constantinople, and throughout the eastern empire. To the Moors of Spain, Europe was greatly indebted for the introduction and dissemination of many of the arts of the east; among others they introduced the Asiatic system of agriculture, with its inseparable adjunct *artificial irrigation*. We are told they divided the lands into small fields, which were kept constantly under tillage; and "they conveyed water to the highest and driest spots"

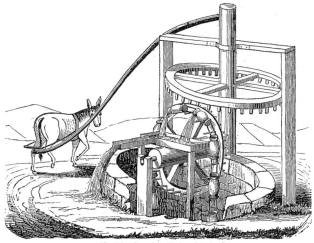

No. 54. Spanish Chain of Pots.

The chain of pots in Spain, is in the form and material of its vessels

---

[a] The De Re 'Metallica' of George Agricola is invaluable for its account of the hydraulic engines employed in the mines of Germany in the 16th and preceding centuries; being doubtless similar to those used by the Romans in some of the same mines. and continued uninterruptedly in use. The first edition of this work was published in 1546, others in 1556—1558—1561—1621—and 1657, all at Basil. Brunet's 'Manuel Du Libraire et De L'Amateur de Livres.' Paris, 1820. It is a copy of the last edition we make use of. The author was born in 1494, and died in 1555.

[b] See also Kircher's Mundus Subterraneus. Tom. ii. pp. 195, 228.

Chap. 15] Modern Chain of Pots.

and the imperfect substitutes for cog wheels, identical with those of Egypt and Asia, and may be considered a fair representative of this machine as used in the agricultural districts of the ancient world.

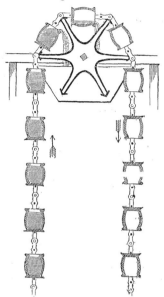

No. 55. Modern Chain of Pots.

No. 55. represents a section of a modern machine. The wheel is placed in or over a cistern designed to receive the water. Buckets are secured to the chain between the joints of the latter, and the wheel as it revolves, receives the centre of these joints on the ends of its arms, which are suitably shaped for the purpose. The buckets therefore fall in between the arms of the wheel and become inverted in passing over it as in the figure.

The chain of pots has been applied to a great variety of purposes. It has been employed for ages, in *cleansing docks, deepening harbors*, &c. The vessels being made of iron and formed like wide scoops, are made to pass under pulleys attached to the bottom of a moveable frame, which is raised and lowered, to suit the varying depth of the channel. Besson also proposed it to *raise mortar*, &c. to the top of city walls, fortifications, &c. and wherever large quantities were required; an application of it that is worthy of the notice of extensive builders, for the time consumed and exertion expended by a laborer, in ascending a long ladder or flight of stairs to deposit a modern hodful of mortar, and returning through the same space, is hardly consistent with the spirit of economy and useful research that characterizes the age. The amount of force consumed in bearing his own body *twice* over the space, independently of the load, would in a well regulated device of this kind produce an equal result. Oliver Evans introduced the chain of pots into his mills, for the purpose of transmitting flour and grain to the different floors.

It has been adopted as a substitute for water wheels. As the noria, when its motion is reversed by the admission of water into its buckets at the upper part of the periphery, is converted into an overshot wheel—so the chain of pots, has in a similar manner, been made to transmit power and communicate motion to other machines. In locations where there is a small supply of water, but which falls from a considerable height, it becomes a valuable substitute for the overshot wheel, as a first mover. It is remarkable that this obvious application of it should not have occurred to European mechanicians previous to the 17th century. It was designed by M. Francini, and by the direction of *Colbert*, the illustrious and patriotic minister of Louis XIV, one was erected in 1668, in one of the public gardens at Paris. A natural spring in this garden supplied water for the plants. It was received into a large basin, and to prevent its overflowing, the surplus or waste water was discharged by a gutter into a well, at the bottom of which it disappeared in the soil. M. Francini took advantage of this fall of waste water in the well, and made it the means of raising a portion of the spring water sufficiently high to form a jet d'eau.

He erected a chain of pots, E, B, No. 56, which reached from the bottom of the well to such a height above its mouth as the water to form the jet was required to be raised. From the upper wheel or drum, another chain of pots, D, C, was suspended and carried round by it, the lower end dipping into the water to be raised from the spring A. By this arrangement the weight of the water in *descending* the well in the buckets of the first chain, raised a smaller portion (allowing for friction) through the same space by the second one—and a proportionable quantity still higher. A spout conveyed the water into the buckets of the driving or motive chain as shown at B. These buckets were made of brass, and wide at the top, the better to receive water from the spring; and also that when one was filled, the surplus might fall down its sides into the next one below, and from that to the third one, and so on, that none might be lost by spilling over. The buckets of the other chain were of the same form and material, but instead of being open like the former, they were closed on all sides, the water being received into them at A, and discharged from them at *m*, through short necks or tubes, *e*, *s*, which are upwards when the buckets ascend, being connected to the smaller part of the latter. A pipe from the upper cistern *m*, conveyed the water to form the jet. The arrow indicates the direction in which both chains move. The vessels on the chain E, B, *below* B, descending into the well, (the bottom of which is not shown,) full—while those shown at D, C, are empty.

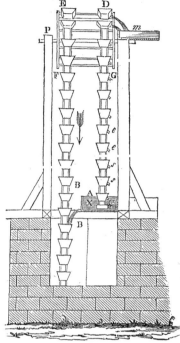

No. 56. Francini's Machine.

The chain of Pots has been employed to work pumps in mines, to propel thrashing machines, &c. &c.[a]

There is much confusion in the notices of the chain of pots by ancient authors, from their referring to it without discrimination as a '*wheel*,' and thus confounding it with the tympanum and noria, and that modification of the latter, known as the Persian wheel. From the circumstance of its having been propelled in the same manner as these, viz: by oxen in the usual way, (through the medium of cog wheels,) or by men walking upon or within a wheel, &c. it has from custom, inadvertence, or from a superficial knowledge of its distinctive features, been classed with them. It was of it that Strabo spoke, " which by wheels and pulleys raised the water of the Nile to the top of a very high hill; and which, instead of being moved by oxen, was propelled by one hundred and fifty slaves." And when Julius Cæsar was beseiged in Alexandria by the Egyptians,

---

[a] See Vol. i, of machines approved by the French Academy. Desaguliers' Philos Vol. ii   Edinburgh Encyc. Vol. x, 896.

the chain of pots was included among the "wheels and other engines," by which the latter raised water from the sea and discharged it into the cisterns that supplied Cæsar's army with fresh water. It was most likely among the "hydraulic engines," which Herodotus observes the Babylonians had, to raise water from the Euphrates to irrigate their lands. These 'engines' were certainly similar to those of India, Egypt, Greece, and other neighboring countries; for if they had been of novel construction, or peculiar to Chaldea, he would scarcely have failed to notice so important a fact, if he even omitted (as he has) to describe them.

The same lack of discrimination is obvious in almost all the accounts of modern as of ancient authors, respecting this machine. When they speak of wheels for raising water, it is as difficult to ascertain those to which they allude, as it is in the parallel passages of Philo and Diodorus, Strabo and Cæsar. Thus *Tavernier* in his passage down the Tigris to Bagdat, remarks, "all the day long, we saw nothing upon either side of the river, but pitiful huts, made of the branches of palm trees, where live certain poor people that turn the *wheels*, by means whereof they water the neighboring ground." Sometimes the chain of pots is mentioned by travelers as the *Persian wheel*, and popular extracts from their works tend greatly to perplex enquirers into its history. When we met with statements from SHAW's travels, that the Persian wheel was extensively used on the banks of the Nile, through all Egypt, they were so much at variance with the testimony of other oriental travelers, and so foreign to our impressions respecting the use of that machine in Egypt, that we had immediate reference to his work; when the apparent discrepancy was explained. He describes and figures the chain of pots (sakia) as the *Persian wheel*.[a] Norden commits the same error: "they likewise employ the Persian wheel with ropes of pitchers, which is turned by oxen."[b] *Twiss* also describes the Spanish chain of pots as the Persian wheel, and which he observes is used "all over Portugal, Spain, and the Levant."[c]

Other travelers speak of it as the *Noria*. Mr. Jacob, in his "Travels in the South of Spain," Lon. 1811, page 152, says the Spaniards "use a mill of Arabic origin, from which our chain pump is evidently derived; it is called a noria. A vertical wheel over a well has a series of earthen jars fastened together by cords of Esparto, which descend into the water and fill themselves by the motion of the wheel. The vertical wheel is put in motion by a horizontal one, which is turned by a cow. No machine can be more simple." In the *Grande Description* of Egypt, it is designated '*Roue a pots*,'[d] instead of naming it from the chain, its peculiar and distinguishing feature. It certainly has nothing in common with the noria, except the pots or vessels in which the water is raised; and these in the latter, are suspended from the arms of inflexible levers, and ascend in the arc of a circle; while in both these circumstances, and others might be named, there is no resemblance whatever between them. The chain of pots is generally named by French authors, '*Chapelet*,' from its resemblance to the string of beads, which Roman catholics, Mahometans, Budhists, &c. (like the Pagans of old) use in repeating their prayers. This appellation is sufficiently discriminating, and is appropriate; certainly more so than *Roue a pots*, since it serves to separate this machine from every species of *wheels*, and to preserve a distinction between two very different classes of hydraulic engines.

The chain of pots seems always to have been used to raise water from

---
[a] Travels, page 337. [b] Travels in Egypt and Nubia, Vol. i, 56.
[c] Travels through Portugal and Spain in 1772, and '73. Lon. 1780, page 329.
[d] Tom. 2. Memoirs, E. M. Plate 5.

Joseph's well. If the location of this well, its peculiar construction, division into two distinct shafts, the chamber between them for the animals which propel the machinery, the passage for their ascent and descent, and its enormous depth, be maturely considered, it will appear, we think, that no other machine could at any time have been used, or *intended* by its constructors to have been used in raising its water; if therefore this celebrated well be, as supposed, a work of the ancient Egyptians, or a relic of Babylon, then the endless chain of pots may safely be regarded as coeval with the foundation of that ancient city, if not, as it probably is, much more ancient.

It was probably the 'pump,' which according to tradition, DANAUS introduced into Greece, a thousand years before the building of Babylon by the Persians. During the time the Israelites were in Egypt, this prince, in consequence of domestic quarrels, left it with his family and friends, and sailed for Greece. They landed on the coast of Peloponessus or the Morea, and were hospitably entertained at Argos, where they settled. It is said, the Greeks did not at that time possess the knowledge of obtaining water from wells; the companions of Danaus having been the first to dig them, and to introduce PUMPS. Pliny vii. 56. If the inhabitants of Greece were ignorant of wells, previous to the arrival of these strangers, they could certainly have had no occasion for pumps; and it was natural for the Egyptians, when they dug wells, to introduce their own country methods of obtaining water from them.

As the word 'pumps,' is not however to be understood in the restricted sense in which it is at present used, the question occurs, what kind of machines were these? 1. They must have been simple in their construction, for otherwise they would have been ill adapted to a rude and uncultivated people, and such the Greeks were while ignorant of wells. 2. They must have been of general application to the *wells* of Greece. 3. They were such as, from their great utility, were continued in use through subsequent ages, for they were highly prized, and the memory of their introduction preserved. 4. They were such as were previously used in Egypt. Now, of all ancient devices for raising water, to which the term '*pump*' could with any propriety be applied, the chain of pots is the only one that fulfils the conditions premised. It is evident that the *jantu* and its modifications are wholly inapplicable to raise water from wells; and the *tympanum* and *noria* are equally so. The *swape* is not adapted to *deep* wells and those of Greece were generally such: yet as it is admirably adapted to raise water from small depths, and was so used by the ancient Greeks; it is probable that it was also introduced by Danaus; as we know that it was in common use in Egypt, in his time. (See figures 36 and 43.) It must however have been of extremely limited application to wells on account of their depth. (See page 38.) The modern inhabitants of Egypt raise water with it only about *seven* feet; and from the figures just referred to, it is obvious that in the time of Danaus, it was raised no higher by it. but if its application was even extended in Greece, to elevate water from twice that depth, its employment in wells must have been comparatively trifling.

It could not have been the *chain pump*, for it does not appear, that either the Greeks or Romans were acquainted with that machine. Vitruvius is silent respecting it. Nor can we suppose any thing like the *atmospheric* or *forcing pump* intended—even, if it could be proved that both were then known. They are too complex to have been at all suited to the Greeks at that remote age. Indeed they are altogether worthless to a rude people, who would be unable to keep them in order, or to detect the causes of their ceasing to act. But that the 'pumps' of Danaus were some kind of bucket machines, like the chain of pots, is inferable from the account of his daughters' punishment. They were condemned to draw water from deep wells, and would of course, use the machines their fa ther introduced. Now we are told that the vessels in which they raised the liquid leaked so much, that the water escaped from them ere it reached the surface—hence their *endless* punishment. The witty remark of Bion implies the same thing. A person speaking of the severe punishment of these young women, in perpetually drawing water in vessels full of holes, he remarked, "I should consider them much more to be pitied were they condemned to draw water in vessels without holes." Hence, we infer that the Egyptian *sakia* or CHAIN OF POTS, was the 'pump' introduced by Danaus, and that to it tradition refers. It was the *only one* to which, from its construction, and adaptation to *every depth*, the name of 'pump' could have been applied—while from its simplicity and efficiency, it was a gift of no ordinary value to the Greeks; and the introduction of it into their country was worthy of being preserved from oblivion.

It is believed to have been in uninterrupted use there since the age of Danaus; although history may not have preserved any record or representation of so early an employment of it. It is still used on the continent and in the islands, as well as throughout Syria and Asia Minor. At SMYRNA it is as common as a pump with us. In "Voyage Pittoresque de la Grece," Paris, 1782, Plate 49, contains a drawing, and page 9 a description of one in a garden at SCIO, the ancient *Chios*, and capital city of the island of the same name. It is similar to the one represented in No. 54, and is doubtless identical with those employed in the same cities, when Homer was born near the former, and when he kept a school in the latter.

On the antiquity of this and preceding machines, we add the opinions of recent writers. "A traveler standing on the edge of either the Libyan or Arabian desert, and overlooking Egypt, would behold before him one of the most magnificent prospects ever presented to human eyes. He would survey a deep valley, bright with vegetation, and teeming with a depressed but laborious population engaged in the various labors of agriculture. He would see opposite to him another eternal rampart, which, with the one he stands upon, shuts in this valley, and between them a mighty river, flowing in a winding course from the foot of one chain to the other, furnishing lateral canals, whence the water is elevated by wheels and buckets of the rudest structures, worked sometimes by men and sometimes by cattle, *and no doubt identical with the process in use in the days of Sesostris.*"[a] "These methods" (of raising water to irrigate the land,) "are not the invention of the modern Egyptians, *but have been used from time immemorial* without receiving the smallest *improvement.*"[b] "Even the creaking sound of the water wheels, as the blindfolded oxen went round and round, and of the tiny cascades splashing from the string of earthen pots into the trough which received and distributed the water to the wooden canals; were not disagreeable to my ears, since they called up before the imagination, the *primitive ages* of mankind, the rude contrivances of the *early kings of Egypt*, for the advancement of agriculture, which have *undergone little change or improvement up to the present hour.*"[c]

Like every other machine that has yet been named, the date of its ori-

---

[a] North American Review, Jan. 1839. p. 185-6. [b] History of the Operations of the French and British Armies in Egypt. Newcastle 1809. Vol. i, page 92.
[c] St. John, "Egypt and Mohammed Ali." Vol. i, 10.

gin is unknown. From its simplicity, its obvious derivation from the primitive cord and bucket, its employment over all Asia and Egypt at the present time, and its extensive use in the ancient world; there can be no question of its great antiquity. Vitruvius is alike silent respecting the origin of this, as of the noria and tympanum, and doubtless for the same reason—their origin extended too far into the abyss of past ages to be discovered. It is singular that the ancients, who attributed almost every agricultural and domestic implement to one or other of their deities, should not have derived the equally important machines for raising water from a similar source. The origin of the *plough* they gave to Osiris, of the *harrow* to Occator, the *rake* to Sarritor, the *scythe* to Saturn, the *sickle* to Ceres, the *flail* to Triptolemus, &c.; and as they attributed the art of *manuring* ground to a god, they surely ought to have given the invention of machines to irrigate it to another.

To the chain of pots, there is an allusion in the beautiful description of the decay and death of the human body, in the 12th chapter of Ecclesiastes: "Or ever the silver cord be loosed, or the golden bowl be broken; or the pitcher be broken at the fountain, or the wheel broken at the cistern." In the east, the chain is almost uniformly made of cord or rope; and the former part of the passage appears to refer to the ends, which are spliced or tied together, becoming loosened, when the vessels would necessarily be broken, for the whole would fall to the bottom; an occurrence which is not uncommon. The term *silver* cord, is expressive of its *whiteness*, the result of its constant exposure to water and the bleaching effect of the sun's rays: and *golden* bowl refers to the *red* earthenware pots or vases, in which the water is raised. Both pots and cords streaming with water, and glittering in the sun, presented to the vivid imaginations of the orientals, striking resemblances to burnished gold and silver. The circulation of the stream of life in man, (his blood)[a] its interruption in disease and old age, his energies failing, and the mechanism of his frame wearing out, and at last ceasing forever to move; are forcibly illustrated by the endless or circulating cord of this machine; its raising living waters and dispersing them through various channels, as so many streams of life, until its vessels, the pitchers, become broken, and the flow of the stream interrupted, and the wheel, upon which its movements depended, becoming deranged, broken, and destroyed.

That the pots or vases are frequently broken, we learn from numerous travelers. In the account of Joseph's well, in the *Grande Description*, it is said to be necessary for a man to be in constant attendance, to keep the animals which move it from stopping, and *to replace the pitchers that are broken*. And that the wheels were often deranged is more than probable, when we consider how exceedingly rude and imperfect is their construction over all the east. The surprise of travelers has often been elicited by their continuing to work at all, while exhibiting every symptom of derangement and decay. "The water wheels, pots, ropes, &c." says Mr. St. John, "had an extremely antique and dilapidated appearance; and, if much used, would undoubtedly fall to pieces."[b] We are told that a more striking picture of rude and imperfect mechanism could scarcely be conceived; and it is not improbable, that the 'Egyptian Wheel' as an emblem of *instability*, had reference to its defective construction and con-

---

[a] That the circulation of the blood was known to the ancients, see Dutens' 'Inquiry into the origin of the discoveries attributed to the Moderns' Lon. 1769, pp. 210, 222
[b] Egypt and Mohammed Ali, i, 126, 127.

stant liability to derangement, as much so as to its rotary movement. Nor is it likely that they were much superior at any time in Judea, for the Jews never cultivated the arts to any extent. The mechanics among them when they left Egypt were probably more numerous and expert than during any subsequent period of their history. In the eleventh century B. C. when Saul began to reign, there was not a blacksmith in the land, or one that could forge iron; they had been carried off by the Philistines; and although David at his death left numerous artificers, when his son built the temple and his own palace, he obtained mechanics from TYRE.

It is moreover possible that the plaints and moanings incident to old age, 'when the grasshopper shall be a burden and desire shall fail,' were also intended to be pointed out by the perpetual creaking of these rickety machines, as indicative of approaching dissolution. The harsh noise they make has been noticed by several travelers. St. John speaks of the creaking sound of the water wheels; and Stephens, in his 'Incidents of Travel,' observes, " it was moonlight, and the creaking of the water wheels on the banks, (of the Nile) sounded like the moaning spirit of an ancient Egyptian."

ON THE ENGINE THAT RAISED WATER FROM THE EUPHRATES TO SUPPLY THE HANGING GARDENS AT BABYLON.

There is a machine noticed by ancient authors, which probably belongs to this part of our subject, and it is by far the most interesting hydraulic engine mentioned in history. Some circumstances connected with it, are also worthy of notice. It was constructed and used in the most ancient and most splendid city of the postdiluvian world; a city which according to tradition existed like Joppa, before the deluge: viz. BABYLON—a city generally allowed to have been founded by the builders of Babel; subsequently enlarged by *Nimrod;* extended and beautified by *Semiramis;* and which reached its acme of unrivaled splendor under Nebuchadnezzar.

The engine which raised the water of the Euphrates to the top of the walls of this city, to supply the pensile or hanging gardens, greatly exceeded in the perpendicular height to which the water was elevated by it, the most famous hydraulic machinery of modern ages; and like most of the works of the remote ancients, it appears to have borne the impress of those mighty intellects, who never suffered any physical impediment to interfere with the accomplishment of their designs; and many of whose works almost induce us to believe that men 'were giants in those days.' The walls of Babylon, according to Herodotus, i, 178, were 350 feet high! Diodorus Siculus and others make them much less; but the descriptions of them by the latter, it is alleged, were applicable only, after the Persians under Darius Hystaspes retook the city upon its revolt, and demolished, or rather reduced their height to about 50 cubits; whereas the father of history gives their original elevation, and incredible as it may appear, his statement is believed to be correct. He is the oldest author who has described them; and he *visited* Babylon within one hundred and twenty years of Nebuchadnezzar's death; and four hundred before Diodorus flourished. He has recorded the impressions which at that time, the city made on his mind, in the following words, " its internal beauty and magnificence exceed whatever has come within my knowledge;" and Herodotus, it must be remembered, was well acquainted with the splendid cities of Egypt and the east. Had not the pyramids of Geezer, the temples and tombs of Thebes and Karnac, the artificial lakes and canals of Egypt, the wall of China, the caves of Ellora and Elephanta, &c.

come down to our times; descriptions of them by ancient authors, would have been deemed extravagant or fabulous, and their dimensions reduced to assimilate them with the works of modern times: so strongly are we inclined to depreciate the labors of the ancients, whenever they greatly excel our own. According to Berosus, who is quoted by Josephus, Antiq. x, 11, it was Nebuchadnezzar who constructed these gardens, so that the prophet Daniel must have witnessed their erection, and also that of the hydraulic engine; for he was a young man when taken a captive to Babylon in the beginning of Nebuchadnezzar's reign, and he continued there till the death of that monarch and of his successor. Amytis, the wife of Nebuchadnezzar, was a Mede, and as Babylon was situated on an extensive plain, she very sensibly felt the loss of the hills and woods of her native land. To supply this loss in some degree, these famous gardens, in which large forest trees were cultivated, were constructed. They extended in terraces formed one above another to the top of the city walls, and to supply them with the necessary moisture, the engine in question was erected.[a]

As no account of the nature of this machine has been preserved, we are left to conjecture the principle upon which it was constructed, from the only datum afforded, viz: the height to which it raised the water. We can easily conceive how water could have been supplied to the uppermost of these gardens by a *series* of machines, as now practised in the east to carry water over the highest elevations—but this is always mentioned as a *single* engine, not a series of them. Had its location been determined, that circumstance alone, would have aided materially in the investigation; but we do not certainly know whether it was placed on the highest terrace—on a level with the Euphrates—or at some intermediate elevation. The authors of the Universal History remark, "upon the *uppermost* of these terraces was a reservoir, supplied by a certain engine, from whence the gardens on the other terraces were supplied." They do not say where the engine itself was located. Rollin places it on the highest part of the gardens: "In the upper terrace there was an engine or kind of pump by which the water was drawn up."

The statement of an engine having been erected at the top is probably correct, for we are not aware that the ancients at that period possessed any machine which, like the forcing pump, projected water *above itself*. Ancient machines, (and every one which we have yet examined, is an example,) did not raise water higher than their own level. But if sucking and forcing pumps were then known and used in Babylon, a period however, anterior to that of their alleged invention, of at least 500 years, still if this engine was placed on the uppermost terrace, both would have been wholly inapplicable. If therefore we incline to the opinion that this engine was a modification of one of those ancient machines, which we have already examined; we are not led to this conclusion by supposing the state of the arts in Babylon at the period of its construction, to have been too crude and imperfect to admit of more complex or philosophical

---

[a] Paintings found in Pompeii, represent Villas of two stories having trees planted on their roofs. These kind of gardens were probably not very uncommon in ancient times in the east, though none perhaps ever equaled those of Babylon. They have been continued to modern times in Asia. Tavernier, when in Bagnagar (the modern Hyderabad) the capital of Golconda, found the roofs of the large courts of the palace terraced and containing gardens, in which were trees of such immense size "that it is a thing of great wonder how those arches should bear so vast a burden." The origin of these and of the city was similar to that of the Babylonian gardens. The King at the importunity of *Nagar*, one of his wives, founded the city and named it after her *Bagnagar*—i. e, " the gardens of Nagar."

apparatus—on the contrary, we know that the Babylonians carried many of the arts to the highest degree of refinement. "They were great contrivers," in this respect, and "fell short of no one nation under the sun, so far from it, that they in a great measure showed the way to every nation besides." Univer. His. Vol. i, 933. Besides, it is certainly more philosophical to suppose this famous engine to have been a modification of some machine, which we have reason to believe was used in Chaldea at that time, and capable of producing the results ascribed to the Babylonian engine, than of any other of which that people possibly knew nothing.

Of all ancient machines, the CHAIN OF POTS was certainly the best adapted for the purpose, and if we mistake not, the only one that could, with any regard to permanency and effect, have been adopted. It stands, and justly so, at the head of all ancient engines for raising water through great elevations; and it may be doubted whether any machine could now be produced *better adapted* for the hanging gardens of Babylon—either in the economy and simplicity of its construction; durability and effect; or be less liable to derangement, less expensive, or more difficult for ordinary people to repair. The project of raising water through a perpendicular elevation, exceeding three hundred feet, in numerous vessels attached to an endless chain, would probably startle most of our mechanicians; and some might suppose that the *weight* of so long a chain, if made of *iron*, would overcome the tenacity of the metal; but almost all the works of the remote ancients partook of the same bold features. *Magnitude* in some of their MACHINES, is as surprising as in other departments of their labors. Their engineers seem to have carried it to an extent that in modern days, would be considered as verging on the limits of the natural properties of materials.

That the chain of pots was the standard machine for raising water *in quantities from great depths* would appear from Vitruvius, since it is the only one adapted for that purpose which he has described, except the "machine of Ctesibius;" and as he professes to give an account of the " various machines for raising water," and his profession as a civil engineer would necessarily render him familiar with the *best of them*, it is clear that he was ignorant of any other having been in previous use. That the engine at Babylon was no other than the chain of pots, may be inferred from the employment of the latter in Joseph's well, where it raises water to an elevation nearly equal to that ascribed to the former; and if the subject were of sufficient interest, we think a connection might be traced between them, if Joseph's well be, as supposed, a relic of Egyptian Babylon. Both Egypt and Chaldea were subject to the same monarch at the time that city was built. Twenty two or three years only had elapsed after Nebuchadnezzar's death when Cyrus took Babylon, and with it the empire; and nine years after he was succeeded by his son Cambyses, who when in Egypt, it is alleged, founded a city on the site of modern Cairo, and named it after old Babylon. Cambyses reigned seven years and five months. If, therefore, the Babylonian machine was superior to the 'chain of pots,' (and it must have been, if it differed at all from the latter, for otherwise it would not have been selected,) then it would, we think, as a matter of course, have been adopted also in Joseph's well, in which the water was required to be elevated to about the same height as in the hanging gardens. Besides, if it possessed peculiar advantages, it would certainly have been *preserved in use*, as well as the chain of pots, for the wealth, comfort, and even existence of the people of the east, have at all times depended too much upon such machines to suffer any valuable one to be lost.

But was the chain of this machine formed of metal, or of ropes? Of the latter we have no doubt. They are generally made of flax or fibres of the palm tree at the present day over all the east. In great elevations, chains of rope possess important advantages over those of metal, in their superior lightness, being free from corrosion, and the facility of repairing them. But by far the most interesting problem connected with the Babylonian engine is, was the water of the Euphrates raised by it to the highest terrace at a SINGLE LIFT? If we had not been informed of one reservoir only, on the upper terrace " from whence the gardens on the others were watered," we should have supposed the water really raised as in Joseph's well, i. e. by *two*, or even more separate chains; and as it is, we cannot believe that so ingenious a people as the Babylonians would raise the whole of the water which the gardens required to the uppermost terrace, when the greatest portion of it was not wanted half so high. As the size of the terraces diminished as they approached the top of the walls, it is probable that full two thirds of the water was consumed within one hundred feet of the ground. We therefore conclude that this famous engine was composed of at least two, and probably more, separate chains of pots; and even then, it might with as much propriety, be noticed by ancient authors as a *single* machine, as that at Cairo still is, by all modern travelers. Winkelman says, the famous gardens at Babylon had canals, some of " which were supplied by pumps and other engines." And Kircher in his *Turris Babel*, 1679, represents fountains and *jets d'eau* on every terrace.

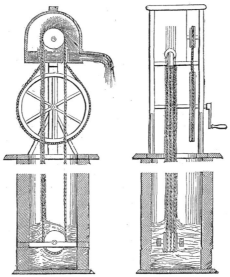

No. 57. Rope Pump.

There is another device that belongs to this chapter. Every person knows, that where water is dispersed over extended surfaces, and of too limited depth to allow the use of a vessel to scoop it up, various substances are employed to absorb it, as sponge and woolen rags, and from which it is separated by pressure. A housemaid, when washing a floor, thus collects in a cloth the liquid dispersed in the purifying process; and by wringing returns it to the vessel. The process is substantially the same as that adopted to raise water in Vera's Rope Pump. See No. 57.

This machine consists of one or more endless ropes, formed of loosely spun wool or horse hair, and stretched on two pulleys like the endless chain of pots. These pulleys have grooves formed on their surfaces for the reception of the ropes. One of them is placed over the mouth of a well, and the other suspended in or secured to the bottom. A rapid motion is communicated to the upper pulley, by a multiplying wheel, and the ascending side of each rope then carries up the water absorbed by it; and which is separated from it when passing over the upper pulley, partly by centrifugal force, and partly by being squeezed in the deep groove, or by passing through a tube as shown in the figure. In the beginning of the motion, the column of water adhering to the rope, is always less than when it has been worked for some time, and continues to increase till the surrounding air partakes of its motion. By the utmost efforts of a man, nine gallons of water were raised by one of these machines from a well, ninety-five feet deep, in one minute. Adam's Philos. Vol. iii, 494.

The HYDRAULIC BELT is a similar contrivance. It is an endless double band of woolen cloth, passing over two rollers, as in figure 57. It is driven with a velocity of not less than a thousand feet per minute; when the water contained between the two surfaces is carried up and discharged as it passes over the upper roller, by the pressure of the band. Some machines of this kind are stated to have produced an effect equal to seventy-five per cent. of the power expended, while that of ordinary pumps seldom exceeds sixty per cent. See Lon. Mechan. Mag. Vol. xxix, page 431.

CHAPTER XVI.

THE SCREW—An original device—Various modes of constructing it—Roman Screw—Often re-invented—Introduced into England from Germany—Combination of several to raise water to great elevations—Marquis of Worcester's proposition relating to it, exemplified by M. Pattu—Ascent of water in it formerly considered inexplicable—Its history—Not invented by Archimedes—Supposed to have been in early use in Egypt—Vitruvius silent respecting its author—Conon its inventor or re-inventor—This philosopher famous for his flattery of Ptolemy and Berenice—Dinocrates the architect—Suspension of metallic substances without support—The screw not attributed to Archimedes till after his death—Inventions often given to others than their authors—Screws used as ship pumps by the Greeks—Flatterers like Conon too often found among men of science—Dedications of European writers often blasphemous—Hereditary titles and distinctions—Their acceptance unworthy of philosophers—Evil influence of scientific men in accepting them—Their denunciation a proof of the wisdom and virtue of the framers of the U. S. Constitution—Their extinction in Europe desirable—Plato, Solon, and Socrates—George III—George IV—James Watt—Arago—Description of the 'Syracusan,' a ship built by Archimedes, in which the Screw Pump was used.

THE COCHLEON or EGYPTIAN SCREW, the machine next described by Vitruvius, is, in every respect, the most original one of which he has given an account. Unlike the preceding, which appear to have been in a great measure deduced from each other, it forms a species of itself; and whoever was its inventor, he has left in it a proof of his genius, and a lasting monument of his skill. If it be not the earliest hydraulic engine that was composed of *tubes*, or in the construction of which they were introduced, it certainly is the oldest one known of that description; and in its mode of operation it differs essentially from all other ancient tube machines; in the latter the tubes merely serve as conduits for the ascending water, and as such are at rest; while in the screw it is the tubes themselves in motion that raises the liquid.

This machine has been constructed in a variety of ways. Sometimes by winding, in the manner of a screw, one or more flexible tubes (generally of lead or strong leather) round a cylinder of wood or iron. 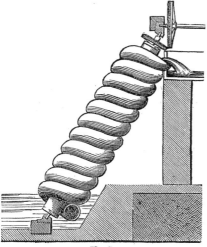 This cylinder is sustained by gudgeons in such a position, that at whatever angle with the horizon it is used, the plane of the helix must always be inclined to its axis at a greater angle; otherwise no water could be raised by it any more than by turning it in the wrong direction. The lower end being immersed in water, the liquid enters the tube and is gradually raised by each revolution until it is discharged above. These machines are commonly used at an inclination to the horizon of about 45°, although they sometimes are placed at 60°. See the figure.

58. Screw.

Instead of tubes wound round a cylinder, large grooves were sometimes formed in the latter and covered by boards or sheets of metal, closely nailed to the surfaces between the grooves—so that the latter might be considered as tubes sunk into the cylinder, instead of being folded round its exterior.

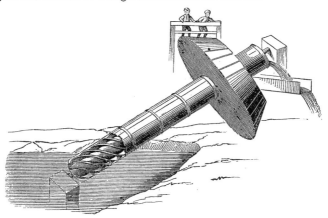

No. 59. Roman Screw.

Another mode was to make the threads of plank, arranged as a helix ound a solid cylinder, which was fitted with journals, and made to revolve in a *fixed* hollow cylinder of the same length; the edges or extremities of the threads rubbing against the sides of the latter, and consequently producing the same effect as No. 58. This modification of the cochleon is known as the German Snail. It has this advantage, that it may be worked in an open channel, or half a cylinder instead of a whole one, since it is only the lower half of the latter, that is essential to the the operation of raising water. Machines of this kind of large dimensions have long been employed by the Dutch, and are generally driven by windmills. But the outer cylinder is more generally fixed to the edges of the helix, and turned with it. It was made in this manner by the ancient Romans; the outer cylinder or case was of plank, well jointed together, and nailed to the edges of the screw, and the whole cemented with pitch, and bound together by iron hoops. It was moved like the noria, &c. " by the walking of men." Vitruvius, B. x, Chap. 11. See No. 59.

The screw as represented in the preceding figures, has never been lost to the world since its invention, although it has long been unknown in that country in which it was devised—Egypt. It appears early in printed books. In the first German edition of Vegetius, (1511) it is figured, and nearly in a vertical position. A laborer with a feather in his cap, and a sword at his side, is *seated* across the top of the frame, and turns it by a crank.[a]

Like almost every other hydraulic engine, the screw has often been re-invented. Cardan mentions a blacksmith of Milan, who imagining himself its original inventor, " for joy, ran out of his wits," and the writer recollects when a boy, hearing of an ingenious shoemaker in much the same predicament. It appears to have been, like other machines for the same purpose, introduced into England from Germany. " The Hollanders, (says Switzer,) have long ago, as some books that I have seen of theirs of fortification intimate, us'd them in draining their morassy and fenny ground, *from whence they have been brought into England;* and used in the fens of Lincolnshire, Cambridgeshire and other low countries. Those of the smallest kind that are worked by men have only an iron handle, as a grindstone has; but the largest that are wrought by horses, have a wheel like the cog-wheel of a horse mill. This engine, (he continues,) which takes hold of the water, as a cork screw does a cork, will throw up water as fast as an overshot wheel, whereby in a short time, an infinite number of water may be thrown up; and I remember when the foundation of the stately bridge of *Blenheim* was laid, we had some of them used with great success; and they are also used in the New River Works, about *Newbury, Berkshire,* and said to be the contrivance of a common soldier, who brought the invention out of Flanders." Hydrostatics, 296, 298.

When employed to raise water to great elevations, a series of two, three, or more, one above another, have been employed; the lower one discharging its contents into a basin, in which the inferior end of the next above is immersed, the whole being connected by cog wheels. Thus an old author observes, " you may raise water to any height in a narrow place, viz. within a tower to the top thereof, as we have known done at *Augusta,* in Germany; to wit, if the spiral pipes be multiplied, so that the water being raised by the lower spiral, and being poured out into some receptacle or cistern; hence, it may be raised higher again by another spiral, and so successively by more spirals, as high as you please, all which spirals may be moved by one power, viz. by the water of a river underneath, or by another animated power." Moxon.

It was one of the objects of the Marquis of Worcester, and his 'unparalleled workman, Caspar Kaltoff,' to avoid the necessity of thus combin-

---

[a] Whether *sitting* was the usual position of European laborers and mechanics when at work, in the middle ages, we know not; but Cambden has a remark which intimates that all *English* mechanics had not in his time, abandoned this oriental custom In concluding his long account of " the States and Degrees of England," from kings, princes, dukes, lords, knights, &c. he continues, " lastly, craftsmen, artizans or workmen; be they that labor for hire, and namely, such as SIT *at work, mechanicke artificers, smiths, car penters,*" &c.

ing a number of them together, as appears from the fifty-third proposition in the 'century of inventions,' "A way how to make hollow and cover a water screw, as "big and as *long* as one pleaseth, in an easy and cheap way." How, and of what materials he made this, is not known, but the fifty-fifth proposition, in the following words, has been fully and practically developed by a French engineer. "A *double* water screw, the innermost to mount the water, and the outermost for it to descend, more in number of threads, and consequently in quantity of water, though much shorter than the innermost screw by which the water ascendeth; a most extraordinary help for the turning the screw to make the water rise." In 1815, M. Pattu published an account of the following improvements, by which the ideas of Worcester are realized.

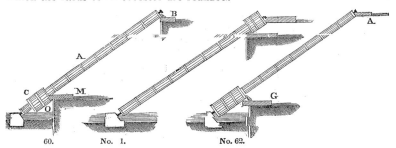

No. 60. represents two separate screws formed on the same axis, one of which, A, is long and narrow and serves for the nucleus of C, which is much wider and shorter. This is designed to propel the former. The threads of both wind round the axis in opposite directions, so that when those on one appear to be moving upwards, those on the other seem to be going downwards. The water from the stream M, is directed into the top of the large screw, and by its weight (as on an overshot wheel) puts the whole in motion, and consequently the water at O, in which the lower end of A revolves, is raised into the cistern at B. No. 61 is merely the same machine inverted. It illustrates the applications to such locations as have a short fall *above* the place to which the water is to be raised. In No. 62 the small screw drives the large one, through which the water from the lowest level is raised sufficiently high to be discharged at an intermediate one, as at G. From these figures it will be perceived that the screw has been employed like the noria and the chain of pots, to transmit power.

This machine was formerly considered as exhibiting a very singular paradox, viz. that the water "ascended by descending," and the mystery was, how both these operations could be performed at the same time, and yet produce so strange a result. It was remarked that when those formed of glass, were put in motion, the water ran *down* the under side of each turn of the tubes, and continued thus to descend until it was discharged at the top! The whole operation and the effects being visible, there seemed no room for dispute, however contrary to acknowledged principles the whole might appear. The case was apparently inexplicable, and seemed to present a parallel one to that of the asymtote; the properties of the latter being as incapable of demonstration to the *senses*, as the supposed operation of this machine could be reconciled to the *mind*. Indeed the proposition, that two geometrical lines may continue to approach each other forever, without the possibility of coming in contact, is *apparently*, quite as impossible, as that water should ascend an inclined plane, by the mere exercise of its own gravity. But the idea of water *descending* in its passage through the screw was altogether an illusion. On the contrary, it is uniformly raised by the continual elevation of that part of the tube, which is immediately behind the liquid, and which pushes it up in a manner analagous to that represented by the following diagram.

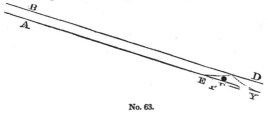

No. 63.

Suppose A Y, the edge of a wide strip of cloth or tape, secured at both ends, at an angle with the horizon, as represented, and upon which the boy's marble or ball at P, can roll. If we hold the pen with which we are writing under the tape between P Y, and raise that part into the position indicated by the dotted lines; the ball would necessarily be pushed forward to E; and if the pen were then drawn towards B on the line D B, the ball would be carried up to A, and without deviating in its path from the line Y A. If A Y were the under side of a flexible pipe or gutter, containing water at E in place of the ball, it is obvious that it would also be raised to A, in a like manner. By the same principle water is raised in the screw, and we may add, in much the same way, for the rotation of the screw is merely another mode of effecting the same thing, which we have supposed to be done more directly by the pen, i. e. by producing a continual elevation of the plane immediately behind the ball or the water. The path of the latter through a screw is the same as that of the ball, while the curves assumed by the tape, as in the dotted lines, represent sections of the helix, and the lines D B, A Y, of the cylinder within which it is formed.

All the ancient machines hitherto examined, have come down from periods so extremely remote, that not a single circumstance connected with their origin or their authors has been preserved. The screw is the first machine for raising water, whose inventor, or alleged inventor, has been named; and yet, from the imperfect and mutilated state of such ancient writings that incidentally mention it, and the loss of others which treated professedly on it, the question of its origin is far from being settled. Although it is said to have been invented by Archimedes and has long been named after him, there are circumstances which render it probable that Diodorus Siculus and Atheneus were mistaken when they attributed it to the great philosopher of Syracuse. Had the account of this machine which Archimedes himself wrote, been preserved, there would have been no occasion to reason on its origin or its author; but unfortunately this, as well as his description of pneumatic and hydrostatic engines, " concerning which he wrote some books," are among those that have perished.

There is no reason to believe that Archimedes himself ever claimed its invention; and his countryman Diodorus, who lived two hundred years after him, and upon whose authority chiefly it has been attributed to him, admits that it was invented by him in *Egypt;* thus allowing it to have been devised in that country, whence the Greeks derived all or nearly all that was valuable in their philosophy and their arts. Every person knows that Egypt was the grand school for the nations of old, in which the learned men of other countries were instructed in every branch of philosophy—for the cultivation of which the Egyptians were celebrated even in the time of Moses—hence it frequently happened, that after returning to their homes imbued with the 'wisdom of Egypt,' philosophers were considered by their countrymen as the authors of doctrines, discoveries and machines, which they had acquired a knowledge of as pupils abroad. It is not therefore impossible, that that which occurred to Thales and Pythagoras, Lycurgus and Solon, Plato and many others, may also have happened to Archimedes with respect to this machine. It has been supposed that the screw was employed in Egypt ages before he visited that country; of this, however, there is no direct proof; perhaps an examination of the immense mass of sculptures in the temples, and tombs of Thebes and Beni-Hassan, &c. may yet bring to light facts illustrative of the use of this and other machines for the same purpose in very remote times. Its ancient name of *Egyptian* screw indicates its origin.

The silence of Vitruvius respecting its origin, if Archimedes was the inventor, is singular; for through the whole of his work he appears studious to record the names of inventors. He was contemporary with Diodorus, and had therefore equal opportunities of ascertaining its history, while from his profession, and the nature of his work, a more perfect account of it would be expected from him than from the other. The Roman architect had indeed every inducement, (except such as were unworthy of him,) to record the name of the Prince of Ancient Mathematicians as its author, if such he knew him to be. The reputation of Archimedes; his splendid discoveries; his famous defence of his native city; his melancholy death; the interest which Marcellus took in his fate; the erection of his tomb by that General; and its discovery by Cicero amidst thorns and rubbish, one hundred and forty years after his death, and in the lifetime of Vitruvius—induce us to believe that, as a candid philosopher and admirer of learned men, and of Archimedes himself, (B. i, Chap. 1.) he would certainly have awarded to the latter the honor of its invention, if he believed him entitled to it, either from the testimony of ancient writers, or from traditional report.

But if this machine was not invented by him, to whom then is the world indebted for it? We reply—if it really be not more ancient than the Ptolemaic era—to a Grecian philosopher of Samos, who was contemporary with Archimedes. Some readers will recollect that when Ptolemy Evergetes, the son and successor of Philadelphus, departed on a dangerous expedition, the success of which, according to Rollin, was foretold by Daniel, (xi, 7, 9,) his wife BERENICE, influenced by a principle of superstition, that at one time was universal, vowed to sacrifice her greatest ornament, the *hair of her head*, to the Goddess Venus, if he was successful and restored to her in safety. Upon his victorious return, she cut off her locks and dedicated them in that temple which Philadelphus had founded in honor of her mother Arsinoe; the dome of which temple was intended to have been lined with loadstone, that the *iron* statue of Arsinoe might be suspended in the air; but the death both of Dinocrates the architect, and Philadelphus, prevented the completion of a building that would have rivalled the most perfect of all human productions; a work, which probably gave rise to the story of the suspension of Mahomet's coffin.[a]

---

[a] That metallic substances have been actually suspended without any tangible support appears from Poncet, to whose travels in Abyssinia we referred in the last chapter. He declares that he beheld in a monastery in that country, a golden staff about four feet long, thus suspended in the air; and to detect any deception he desired permission to examine it closely, to ascertain whether there was not some invisible prop or support. "To take away all doubt (he says) I passed my cane over it and under it, and on all sides, and found that this staff of gold did truly hang of itself in the air." Ed. Encyc. Vol. xiii, p. 46.

Sometime afterwards, this consecrated hair was missing from the temple, having been lost through the negligence of the priests, or perhaps designedly concealed. No occurrence was more likely to create alarm among a superstitious people, or to excite the ire of a despotic monarch, than such an insult to their Gods, and to his favorite queen. In this dilemma, an astronomer of Alexandria, in order to make his court to Evergetes, had the effrontery to give out publicly that JUPITER had carried off the locks of Berenice to heaven, and had formed them into a constellation! And as a proof of his assertion he pointed to an unformed cluster of stars near the tail of Leo, as Berenice's hair! And '*Coma Berenices*' is the name by which these stars are known to this day.

It was this artful courtier and astronomer who either invented or re-invented the screw. He was named CONON of *Samos*, and sometimes Conon of *Alexandria*, from his residence in Egypt. He was an intimate friend of, and greatly esteemed by Archimedes; and it would seem that they communicated their writings and discoveries to each other. When the former devised this machine, Archimedes we are told *demonstrated and fully explained its properties;* for Conon himself was not fortunate in his demonstrations. (Bayle.) From this circumstance the name of its inventor was in time forgotten, and it eventually became known as the *Archimedian* screw; but probably not till long after the death, both of its author and illustrator.

Similar instances are not uncommon in modern times; they have in fact, always occurred. Thus, the instrument known as Hadley's Quadrant was really invented by Godfrey of Philadelphia. The compass was known before Flavio Gioia, although the *Fleur de Lis*, by which he designated the north in compliment to his sovereign, is used to this day. Gunpowder was used ages before Schwartz was born—and these continents bear the name of Vespucci, not that of Columbus or Behaim.

As Conon died before Archimedes, (see Bayle) and probably in Egypt, it is very possible (supposing it originated with the former) that it was first *introduced into Europe* by the latter; a circumstance quite sufficient to connect his name permanently with it there. Atheneus mentions particularly its application by him to raise water from the hold of the ship, which was built under his directions for Hiero; and if an observation of the same author can be relied on, it is evident that he was the first to make it known to *Grecian* mariners; for he asserts, that they held his memory in great estimation, for having enabled them to carry off the water from the holds of their vessels by it.

It is greatly to be regretted that men of science should ever be found among the flatterers of despots; yet the obsequiousness of Conon has been imitated in modern as in ancient times. Custom may yet, in some degree sanction or rather screen the practice from reproach; but the period is, we believe, rapidly approaching when it will be subjected to general derision, as not only injurious to the reputation of scientific men themselves, but to science and the world at large. Our sentiments on this subject may be reprobated by some persons, and approved of by few,—still we believe they are such as conduce to the general welfare of our race, and such as will one day universally prevail, and believing this, we express them without hesitation—others may condemn them as out of place here, but in our opinion the evils they deprecate will not be removed until they are generally denounced in works devoted to the arts. Nay, we would introduce such sentiments into school books, that children may not be taught to worship a man on account of his titles, but to revere virtue and admire well cultivated talents wherever they are found. ' We might as well (says Seneca) commend a horse for his splendid trappings, as a man for his pompous ad ditions.'

Let any unsophisticated mind peruse the dedications of European works, in almost all departments of science, for the last two centuries, and he will find every *attribute of the Deity* blasphemously lavished on the *vilest* of princes, and on titled dolts, with a degree of ardor and apparent sincerity, that is as loathsome as the grossest practices of heathen idolatry. At the same time, these individuals who thus idolize, sometimes an idiot, at others an infant, and often a brute, affect pity for the ignorance and superstition of ancient pagans and modern savages.

But why this display of servile adulation? Formerly to obtain *bread:* in later times to procure *title, hereditary* title.

If there is one class of men, whose extensive knowledge of nature, and the sublimity of whose studies should lead them thoroughly to despise the tinsel and trappings of courts, and the unnatural, and to the great mass, degrading distinctions in European society, it is astronomers; men whose researches are preëminently calculated to ennoble the mind, whose labors have elicited the highest admiration of their talents, and whose discoveries have opened sources of intellectual pleasures so refined, that pure intelligences might rejoice in them. That such men should stoop to lay at the feet of ignorant and sensual despots, their fame, their learning, and in some degree the science of which they are the conservators, and accept from those, who are immeasurably their inferiors, what are preposterously named titles of *honor*, i. e. puerile and artificial distinctions, which, while they profess to *advance* those who are already in the foremost ranks of society—really lower and degrade them—titles, relics of times when men were advanced but a few steps from the savage state, and conferred by ceremonies which are the very essence of buffoonery,—is truly one of the most lamentable facts connected with the history of modern science.

Learned men by thus connecting themselves with the state, consummate an unholy, an unnatural alliance, and subject even science herself (although they may not intend it) to politicians to speculate on. They in a measure, commit suicide on their fame, by thus supporting political institutions, that can only exist by silencing the throbbings and stifling the aspirations of the general mind after knowledge; institutions, which, like the old errors in philosophy, are destined to be exploded forever. It will, we think, one day appear strangely incongruous, that some of the brightest luminaries of science should have turned to royal despots for factitious rank; as if they, in whose fair fame the world feels an interest, could descend from their radiant spheres to move as satellites around such, with an *increase* of lustre! Who can behold without sorrow, these men rendering homage by kneeling and other more disgusting mummeries, to individuals who are not only their inferiors in every attribute that adorns humanity, but often the most atrocious of criminals, and sometimes mere insensates; to beg a portion of *honor*, and a title to use it! When the world becomes free and enlightened, such examples will be adduced as illustrations of the vagaries and inconsistencies of the human mind; and patents of nobility and hereditary titles of honor, especially from such sources, will be looked upon as satires on science, on the age, and on the intellect of man.

These titles form the most conspicuous feature in that system of imposition by which the European world has too long been deluded and debased; and in a political point of view, the friends of man's inalienable rights, and of the amelioration of his condition, will always regret, that scientific men should have lent their example, to sustain distinctions that are a curse to the world. This conduct of theirs, perhaps more than any other cause, tends to uphold despotism on the earth. Of their influence in this respect, modern despots are fully aware, and which they evince by their anxiety to enlist in their train, every man eminent in any department of the arts or of science; and many of these, it is to be deplored, they too often tickle with a feather, or amuse with a trinket, while they put a bridle in their lips and yoke them to their cars.

The lust after titles and distinctions, incident to monarchical governments, is in the political and moral world, what the scrofula, or 'king's evil' is in the physical: It destroys the healthy and natural organization of society, taints its fairest features with hereditary disease, and renders the whole corrupt. The wisdom of the fathers of our republic was not more conspicuous than their virtue, when they denounced such titles and distinctions as forever incompatible with the constitution. Sweep them from the earth and man in the eastern hemisphere would become a regenerated being. Nations would no longer be kept in commotion and dread, nor their resources be consumed by political and military gladiators; nor would the abominable boast of one people in conquering and plundering another be deemed creditable; but when peace and virtue, science and the arts, would alone confer honor, and their most distinguished cultivators be deemed the most noble.

Plato was no worshipper of Dionysius, nor Solon of Crœsus; and when the talented but unprincipled Archelaus of Macedon, drew numerous philosophers around him, by his wealth and the honors he conferred on them, Socrates refused even to visit him as long, said he, as bread was cheap and water plenty at Athens.

Although the ancient world confirmed the name given to one of the constellations by Conon, the modern one refused to sanction a similar attempt to designate the remotest planet in our system, after the name of a king who was remarkable for his lack of intelligence—a bigot—and who, to preserve his prerogative, shed blood as water. Yet to that man, and to his son and successor, who, if he possessed more intelligence than the parent, was the grossest sensualist of the age, and contact with whom was pollution, did some of the votaries of science kneel as to 'THE FOUNTAINS OF HONOR!' and to receive a portion of it at their hands! while a *mechanic*, to whose glory it will ever be mentioned, could duly appreciate the offered bauble and *reject* it, if not with disdain. James Watt, the mathematical instrument maker of Glasgow, the great improver of the steam-engine, who conferred more benefits on his country than all the monarchs that ever ruled over it, and all the statesmen and warriors which it ever produced—refused a title. And who ever regretted that Milton was not a knight, or Shakespeare a marquis, or Franklin a lord; or that some of the greatest poets and philosophers, philanthropists and mechanicians, that ever lived, are known to us simply as such, without having had their names bolstered up with preposterous appendages? And who ever supposed they were less happy without them, less vigorous and successful in their researches; less respected by contemporaries, or less revered by posterity?

Long after these remarks were written, M. Arago's Memoir of Watt, reached this country, and on perusing it, we could not but smile at the disappointment expressed by the great French philosopher, that his friend was not made a *peer*. "When I inquired into the cause of this neglect, [he observes,] what think you was the response? Those dignities of which you speak, I was told, are reserved for naval and military officers;

for influential members of the House of Commons, and for members of the aristocracy. '*It is not the custom,*' and I quote the very phrase, to grant these honors to scientific and literary men, to artists and engineers." He adds, "so much for the worse for the peerage." Well be it so. In our humble opinion, it is so much the better for the memory of Watt. What had such a man to do in a house that presses like an incubus on the energies of his country, and the claims to a seat in which, are too often such as are disgraceful to our common nature? An infinitely higher honor awaits him; for both Watt and his illustrious eulogist are destined to occupy distinguished stations in that Pantheon, which is yet to be erected, whose doors will be opened only to the BENEFACTORS OF MANKIND.

There are several interesting particulars mentioned by Atheneus, respecting the magnificent ship named the 'Syracusan,' which was built under the directions of Archimedes, and to which we have alluded. From the following brief description, it will be perceived, that for richness of decoration; real conveniencies and luxuries, (for even that of a library was not overlooked,) she rivalled, if she did not excel, our justly admired packets and steam ships.

Three hundred carpenters were employed in building this vessel, which was completed in one year. The timber for the planks and ribs were obtained partly from Mount Etna, and partly from Italy; other materials from Spain, and hemp for cordage from the vicinity of the Rhone. She was every where secured with large copper nails, [bolts] each of which weighed ten pounds and upwards. At equal distances all round the exterior were statues of Atlas, nine feet in height, supporting the upper decks and triglyphs; besides which the whole outside was adorned with paintings; and environed with ramparts or guards of iron, to prevent an enemy from boarding her. She had three masts; for two of these, trees sufficiently large were obtained without much difficulty, but a suitable one for the mainmast, was not procured for some time. A swine-herd accidentally discovered one growing on the mountains of Bruttia. She was launched by a few hands, by means of a helix, or screw machine invented by Archimedes for the purpose, and it appears that she was sheathed with *sheet* lead.[a] Twelve anchors were on board, four of which were of wood, and eight of iron. Grappling irons were disposed all round, which by means of suitable engines could be thrown into enemies' ships. Upon each side of this vessel were six hundred young men fully armed, and an equal number on the masts and attending the engines for throwing stones. Soldiers, [modern marines] were also employed on board, and they were supplied with ammunition, i. e. stones and arrows, 'by little boys that were below,' [the powder monkies of a modern man of war,] who sent them up in baskets by means of pulleys. She had twenty ranges of oars. Upon a rampart was an engine invented by Archimedes, which could throw arrows and stones of three hundred pounds, to the distance of a stadium, [a furlong] besides others for defence, and suspended in chains of brass.

She seems to have been what is now called 'a three decker,' for there were 'three galleries or corridors,' from the lowest of which, the sailors went down by ladders to the hold. In the middle one, were thirty rooms, in each of which were four beds; the floors were paved with small stones

---

[a] European ships were sheathed with sheet lead in the 17th century, at which time also wooden sheathing was in vogue. See Colliers' Dict. Vol. i. Art. England.

of different colors, (mosaics) representing scenes from Homer's Iliad The doors, windows and ceilings were finished with 'wonderful art,' and embellished with every kind of ornament. The kitchen is mentioned as on this deck and next to the stern, also three large rooms for eating. In the third gallery were lodgings for the soldiers, and a gymnasium or place of exercise. There were also gardens in this vessel, in which various plants were arranged with taste; and among them walks, proportioned to the magnitude of the ship, and shaded by arbors of ivy and vines, whose roots were in large vessels filled with earth. Adjacent to these was a room, named the 'apartment of Venus,' the floor of which was paved with agate and other precious stones: the walls, roof and windows were of cypress wood, and adorned with vases, statues, paintings, and inlaid with ivory. Another room, the sides and windows of which were of box wood, contained a library; the ceilings represented the heavens, and on the top or outside was a sun dial. Another apartment was fitted up for bathing. The water was heated in three large copper cauldrons, and the bathing vessel was made of a single stone of variegated colors. It contained sixty gallons. There were also ten stables placed on both sides of the vessel, together with straw and corn for the horses, and conveniences for the horsemen and their servants. At certain distances, pieces of timber projected, upon which were piles of wood, ovens, mills, and other contrivances for the services of life.

At the ship's head was a large reservoir of fresh water, formed of plank and pitched. Near it was a conservatory for fish, lined with sheet lead, and containing salt water; although the well or hold was extremely deep, one man, Atheneus says, could pump out all the water that leaked into her, by a screw pump which Archimedes adapted to that purpose. There were probably other hydraulic machines on board, for the plants, bathing apparatus, and kitchen, &c. The upper decks were supplied with water by *pipes* of earthenware and of lead; the latter, most likely, extending from pumps or other engines that raised the liquid; for there is reason to believe that machines analogous to forcing pumps were at that time known.

The 'Syracusan' was laden with corn and sent as a present to the King of Egypt, upon which her name was changed to that of the 'Alexandria.' Magnificent as this vessel was, she appears to have been surpassed by one subsequently built by Ptolemy Philopater; a description of which is given by Montfaucon, in the fourth volume of his antiquities.

For the Spiral Pump of Wirtz, see the end of the 3d Book

CHAPTER XVII.

THE CHAIN PUMP—Not mentioned by Vitruvius—Its supposed origin—Resemblance between it and the common pump—Not used by the Hindoos, Egyptians, Greeks or Romans—Derived from China—Description of the Chinese Pump and the various modes of propelling it—Chain Pump from Agricola—Paternoster Pumps—Chain Pump of Besson—Old French Pump from Belidor—Superiority of the Chinese Pump—Carried by the Spaniards and Dutch to their Asiatic possessions—Best mode of making and using it—Wooden Chains—Chain Pump in British ships of war—Dampier—Modern improvements—Dutch Pump—Cole's Pump and experiments—Notice of Chain Pumps in the American Navy—Description of those in the United States Ship Independence—Chinese Pump introduced into America by Van Braam—Employed in South America—Recently introduced into Egypt—Used as a substitute for Water Wheels—Peculiar feature in Chinese ship building—Its advantages. .

The chain pump, although not described by Vitruvius, is introduced at this place, because it seems to be the connecting link between the chain of pots and the machine of Ctesibius. Some writers suppose it to be derived from the former; nor is the supposition improbable. Numerous local circumstances would frequently prevent the chain of pots from being used in a vertical position, and when its direction deviated considerably from the perpendicular, some mode of protecting the loaded vessels while ascending rugged banks, &c. became necessary. An open trough or wooden gutter through which they might glide, was a simple and obvious device, and one that would occur to most people; but such a contrivance could not have been long in use before the idea must have been suggested, that pieces of plank or any solid substance which would occupy the entire width of the gutter, might be substituted for the pots, since they would obviously answer the same purpose by pushing the water before them when drawn up by the chain. If this was the process by which the transition of the chain of pots into the chain pump was effected, there can be little doubt, that old engineers soon perceived the advantages of covering the top of the gutter, and converting it into a *tube;* as the machine could then be used with equal facility, in a perpendicular, as in any other position.

It may be deemed of little consequence to ascertain the circumstances which led to the invention of the chain pump; yet a knowledge of the period *when* this took place would be of more than usual interest, on account of the analogy between it and the ordinary pump, and of the relationship that appears to exist between them. The introduction of a *tube* through which water is raised by pallets or pistons, is so obvious an approach to the latter, that it becomes desirable to ascertain which of them bears the relation of parent to the other, or which of them preceded the other. But to what ancient people are we to look for its authors? Not to the Hindoos, or the Egyptians, for it is incredible that either of these people should have lost it, if it was ever in their possession. Its cheap and simple construction—its efficiency and extensive application, would certainly have induced them to retain it in preference to others of less value. Nor does it appear to have been known to the Greeks; for their navigators would never have employed the screw as a ship pump, (as Atheneus says they did,) if they had been acquainted with this machine. Of all hydraulic tube machines, the screw seems the most unsuitable for such a purpose. It requires to be inclined at an angle that is not only inconvenient but generally unattainable in ships. But if the Greeks had the chain pump, the Romans would have received it from them; whereas, from the silence of Vitruvius, it is clear that his countrymen were not acquainted with it. As an engineer, he would have been sensible of its value, and would have preferred it in many cases, in raising water from coffer-dams, docks, &c. to the tympanum and noria, which he informs us were employed in such cases.[a] Arch. Book v, Cap. 12. Moreover, if it was employed by the Romans, it would have been preserved in use, as well as other machines for the same purpose, either in Europe or in their African or Asiatic possessions; but we have no proof of its use at all in any of the latter, nor yet in the former, till comparatively modern times.

But if the origin and improvement of the chain pump is due to one nation more than another, to whom are we indebted for it? To a people as distinguished for their ingenuity and the originality of their inventions, as for their antiquity and the peculiarity of many of their customs; and who by their system of excluding all foreigners from entering the country have long concealed from the rest of the world many primitive contrivances, viz. the CHINESE. This singular people appear to have had little or no communication with the celebrated nations of antiquity, a circumstance to which their ignorance of the chain pump may be attributed. This machine has been used in China from time immemorial, and as connected with their agriculture, has undergone no change whatever. The great requisites in their husbandry "are manure and water, and to obtain these, all their energies are devoted." Of such importance is this instrument to irrigate the soil, that every laborer is in possession of one; its use being "as familiar as that of a hoe to every Chinese husbandman," "an implement to him not less useful than a spade to an European peasant." It is worthy of remark too, that they often use it, in what may be supposed to have been its original form, viz. as an open gutter; a circumstance which serves to strengthen the opinion of its origin and great antiquity among them. Like the peculiarity of their compass, which with them points to the south, it is a proof of their not having received it from other people. "The Chinese [observes Staunton] appear indeed to have strong claims to the credit of having been indebted only to themselves for the invention of the tools, necessary in the primary and necessary arts of life; these have something peculiar in their construction, some difference, often indeed slight; but always clearly indicating that, whether better or worse fitted for the same purposes as those in use in other countries, the one did not serve as a model for the other."[b]

But the general form of chain pumps in China is that of a square tube or trunk made of plank; and of various dimensions acccording to the power employed to work them. Those that are portable, with one of which every peasant is furnished, are commonly six or seven inches in diameter, and from eight to ten feet in length. Some are even longer, for Van Braam, who was several years in China, and who, as a native of Holland, was a close observer of every hydraulic device, when speaking of them, remarks, that "they use them to raise water to the height of ten or twelve feet; a single man works this machine, and even carries it wherever it is wanted, as I have had occasion to remark several times in the province of Quangtong near Vampou."[c] A small wheel or roller is attached to each end of the trunk, over which an endless chain is passed. Pallets, or

---

[a] It was preferred by the architect of Black Friars Bridge, London, to raise the water from the Caissons.
[b] Embassy to China. Lon. 1798. Vol. iii, 102.
[c] Embassy of the Dutch E. I. Company. Lon. 1798. Vol. i, 75.

square pieces of plank, fitted so as to fill (like the piston of a common pump) the bore of the tube, are secured to the chain. When the machine is to be used, one end of the trunk is placed in the water, and the other rests on the bank over which it is to be raised. The upper wheel or roller is put in motion by a crank applied to its axle, and the pallets as they ascend the trunk, push the water that enters it before them, till it is discharged above. In machines of this description one half of the chain is always outside of the tube and exposed to view, but in others the trunk is divided by a plank, so as to form two separate tubes, one above another, and hence the chain rises in the lower one and returns down the upper. These pumps are represented as exceedingly effective, delivering a volume of water equal to the bore of the trunk. Whenever a breach occurs in one of their canals, or repairs are to be made, hundreds of the neighboring peasants are summoned to the work, and in a few hours will empty a large section of it by these machines.

When a pump is intended to raise a great quantity of water at once, it is made proportionably larger, and is moved by a very simple tread wheel; or rather by a series of wooden arms projecting from various parts of a lengthened axle, which imparts motion to the chain, as represented in the figure.

No. 64. Chinese Chain Pump.

These arms are shaped like the letter T, and the upper side of each is made smooth for the foot to rest on. The axle turns upon two upright pieces of wood, kept steady by a pole stretched across them. The machine being fixed, men treading upon the projecting arms, and supporting themselves upon the beam across the uprights, communicate a rotary motion to the chain, the pallets attached to which draw up a constant and copious stream of water. Another mode of working them, which Staunton observed only at Chu-san, was by yoking a buffalo, or other animal, to a large horizontal cog wheel, working into a vertical one, fixed on the same shaft with the wheel that imparts motion to the chain, as represented in figures 49 and 54.[a] The description of this machine by Staunton is similar to that previously given by the missionaries, and they enumerate the various modes of propelling it which he has mentioned.[b] But Nieuhoff, with the characteristic sagacity of his countrymen, noticed either these, or some other machines for the same purpose, propelled by *wind*. When speaking of the populous city of Caoyeu, and its environs, he observes, "they boast likewise of store of windmills, whose sails are made of mats. The great product of the country consists of rice, which the peasant stands obliged to look after very narrowly, lest it perish upon the ground by too much moisture, or too much heat and drought. The windmills, therefore, are to *draw out the water* in a moist season, and to let it in as they think fit." That part of the country, he continues, is "full of such mills." Several of them are represented in a plate, but without showing the pumps moved by them.[c]

These were very likely to elicit the notice of a Dutchman; for draining mills, worked by horses and wind, have been used in Holland since the 14th century. They consisted however principally of the noria and chain of pots.

It is uncertain *when* the chain pump was first employed in Europe; whether it was made known by Marco Paulo, Ibn Batuta, or subsequent travelers in China, or was previously developed and introduced into use, independently of any information from abroad. An imperfect machine is described by several old authors. This was a common pump log, or wooden cylinder placed perpendicularly in a well; its upper end reaching above the level to which the water was to be raised, and having a lateral spout, as in ordinary pumps, for the discharge. A pulley was secured to one side of the log near the lower orifice, and a drum or wheel above the upper one. One end of a rope was let down the cylinder, and after being passed over the pulley was drawn up on the outside, and both ends were then spliced or united over the drum. To this rope, a number of leathern bags or stuffed globular cushions were secured at regular distances. The diameter of each was equal to the bore of the cylinder. Ribs were nailed across the periphery of the drum, and between these, the cushions were so arranged as to fall, in order to prevent the rope from slipping. When the drum was put in motion, the cushions entered in succession the lower orifice of the pump, (which was two or three feet below the surface of the water,) and pushed up the liquid before them, till it escaped through the spout.

Machines of this description were formerly employed in mines; chains of iron being substituted for the ropes, and sometimes globes of metal in place of the cushions. The latter are figured by Kircher in his Mundus Subterraneus, Tom. ii, 194. Among the earliest of modern authors who have described these pumps is Agricola. He has given five different figures of them, but they differ merely in the apparatus for working them, according to the power employed, whether of men, animals, or water. The following cut, No. 65 is from his 'De Re Metallica.' It exhibits two separate views of the lower end of the pump, showing the mode of attaching the pulley, and the passage of the rope and cushions over it. From the resemblance of the chains or ropes and cushions, to the *rosary*, or string of beads on which Roman catholics count their prayers, these machines

---

[a] Staunton, Vol. iii, 315. [b] Duhalde's China. Paris, 1735. Tom. ii, 66, 67.
[c] Ogilvy's Translation. Lon. 1673, pp. 84, 85—and Histoire Générale. Amsterdam, 1749. Tom. viii, 81, 82.

became known as '*Paternoster* pumps.' For the same reason they are named *Chapelet* by the French, in common with the chain of pots.

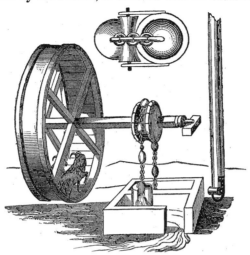

No. 65.   Chain Pump from Agricola.

The next author who describes these pumps, that has fallen in our way, is *Besson*. Plate 50, of his 'Theatre Des Instrumens,' is a representation of a double one. Two cylinders are placed parallel to each other, so that the chain passes through both. It is shown as worked by *wind*. A vertical shaft with sails is secured under the dome of an open tower; a cog wheel on the lower end of the shaft turns a trundle or pinion which is fixed on the horizontal axle of the drum, that carries the chain. Thus, when the wind turned the sails, water was raised through one of the cylinders, and when their motion was reversed by change of the wind, the liquid was elevated in the other. Instead of stuffed cushions, as in the preceding figure, pistons, resembling somewhat those of fire engines, or forcing pumps, i. e. double cupped leathers are shown, ('*Coquilles fond contre fond,*') the earliest instance of their use that we have met with. Besson, who appears to claim the addition of the second cylinder as an improvement of his own, was a French mathematician and mechanician, and spent a great part of his life in mechanical researches; in the prosecution of which he visited foreign countries. His 'Theatre' contains such devices as he collected abroad as well as those invented by himself. It was published at Lyons, with commentaries, after his decease, by Beroald, but the privilege to print was accorded to himself, ten years previous to the date of its publication, i. e. in 1568.[a]

Kircher also figures the chain pump with two cylinders. The imperfect mechanism and enormous friction of these old machines confined their application to a limited extent in Europe during the 16th and 17th centuries. Desaguliers left them unnoticed; and at the time Switzer wrote (1729) they had been discontinued in England. "I might (he observes) from *Bockler* and others, have produced almost an infinite number

---

[a] Bayle, in his dictionary, says Beroald was twenty-two years of age when he published "some commentaries on the mechanics of James Besson; but he had scarce tried his fortune that way, when he ran after the philsopher's stone."

of drafts of engines, which are placed under the terms Budromia and Hydrotechnema, &c. the first signifying the methods of raising water by buckets; and the other by globes or figures of any regular shape, fixed to a rope, which rope being fastened at each end, and passing through an elm or other pipe, which reaches from the bottom of a well to the height to which the water is to be conveyed, brings up the water with it; but these kind of engines *being out of date*, I shall pass over them."[a] Belidor has described one that was used in the ship yards and docks at Marseilles, which is represented in No. 66. The lower pulley was dispensed with; and the face of the pallets or pistons, which were hemispheres of wood, were leathered. It was worked by two galley slaves, who were relieved every hour.

No. 66. Old French Chain Pump.

Such appears to have been the general construction of the chain pump in Europe, until an increasing intercourse with the Chinese led to the introduction of the machine as made by that people. The credit of this is, we believe, due to the Dutch. From the peculiar location of Holland with regard to the sea, hydraulic engines have at all times been of too much importance to escape the examination of her intelligent travelers. But it perhaps will be said, there is no essential or very obvious distinction between the old chain pump of Europe and that of China: admitting this, still there must have been something peculiar either in the construction or mode of working the latter, to have produced the superior results ascribed to them; and to have elicited the admiration of the Jesuits and all the early travelers in China. No stronger proof of their superiority need be adduced, than the fact of their being carried in the 17th century from China to Manilla by the *Spaniards*, and to Batavia by the *Dutch*.[b] Hence they were previously unknown in those parts of Asia, as much so as in Holland and Spain. Navarrette mentions them with great praise: he thought there was not a better invention in the world to draw water from wells and tanks.[c] And Gamelli (in 1695) describes them as machines, which, in his opinion, Chinese ingenuity alone could invent.[d] Montanus mentioned them as novel. He describes one as an "engine made of four square plank, holding great store of water, which with iron chains, they

---

[a] Hydrostaticks, 313.   [b] Histoire Générale, Tom. viii, 81.   [c] Ibid.   [d] Ibid, Tom. vii, 267

hale up like buckets."[a] How such intelligent men as the Jesuits undoubtedly were, could use such language, if an effective chain pump was then known in Europe, it is difficult to conceive.

Although the Chinese pump has been mentioned by all travelers, no one has entered sufficiently into details, to enable a mechanic to realize the construction of the chain—mode of fixing the pallets—where they are attached to it, (at the centre, or on one side,)—nor how they are carried over the wheels or rollers. One cause of the superiority of these oriental machines over those of Europe, was the small degree of friction from the rubbing of the pallets, when passing through the trunk; wood sliding readily over wood, when both are wet: another was the accuracy with which the working parts were made. The experience of ages, and the immense number of workmen constantly employed in fabricating them, through every part of the empire, had brought them to great perfection: but the *position* in which they are worked, also contributed to increase the quantity of water raised by them, for except in particular locations, they are always inclined to the horizon, as shown in No. 64. Now it has been ascertained that to construct and use a chain pump to the best advantage, the distance between the pallets should be equal to their breadth, and the inclination of the trunk about $24°$, $21'$. When thus arranged, according to Belidor, it produces a maximum effect.[b] The author just named speaks of one at Strasburgh, the chain of which was made of WOOD, which being light and flexible, was very efficient, requiring much less labor to work it than those in which the chains were iron. This leads us to a remark which we do not recollect to have seen in any English work, viz. that in most if not in all the Chinese smaller pumps, the chains are of that material. One of them is thus described by the Jesuits: " Une machine hydraulique, dont le jeu est aussi simple que la composition. Elle est composée *d'une chaîne de bois*, ou d'une sorte de chapelet de petites planches quarrées de six ou sept pouces, qui sont comme enfilée parallelement á d'égales distances. Cette chaîne passe dans un tube quarré," &c.[c]

In the latter part of the 17th century, chain pumps were used in British men-of-war. In Dampier's Voyage to New Holland in the 'Roebuck,' a national vessel, he mentions one. This ship on returning home sprung a leak near the Island of Ascension, and the water poured in so fast, he relates, that " the *chain pump* could not keep her free—I set the *hand pump* to work also, and by ten o'clock, sucked her—I wore the ship and put her head to the southward, to try if that would ease her, and on that tack the *chain pump* just kept her free." English ships of war now carry four of those pumps, and three common ones, all fixed in the same well; whereas it would appear from Dampier, that they had formerly but one of each. " In the afternoon, (he observes,) my men were all employed pumping with *both* pumps." Shortly afterwards the ship foundered.[d] The vessels of Columbus were furnished with pumps; and so were those of Magalhanes; but these were probably the common instruments referred to above as ' hand pumps.'[e]

In Dampier's time chain pumps were very imperfect. The chain, and the wheel, which carried it, were inaccurately and badly made; hence when the machine was worked, the former was constantly liable to slip over the latter; and the consequent violent jerks, from the great weight of the water on the pallets, often burst the chain asunder, and under cir-

---

[a] Atlas Chinensis, translated by Ogilvy. Lon, 1671, page 675.
[b] Arch. Hydraulique, Tom. i, 363.   [c] Histoire Générale, Tom. viii, 82, and Duhalde Tom. ii, 66.   [d] Dampier's Voyages, Vol. iii, 191, 193.
[e] Irving's Columbus Vol ii, 127, and Burney's Voyages, Vol. i, 112.

cumstances which rendered it difficult and sometimes impossible to repair it. These defects, which in some cases led to the loss of vessels and of human life, at length excited the attention of European mechanics, and in the following century, numerous projects were brought forward to improve the chain pump, or to supersede it. In 1760, Mr. Abbot invented a ship pump, which was represented as of a very simple construction, and which threw "five hundred hogsheads of water in a minute; [!] the handle by which it is worked, is in the manner of a common winch, which turns with the utmost facility either to the right or the left."[a] In the following year, the States of Holland granted to M. Liniere, "an exclusive privilege for twenty-five years, for a pump, which upon trial on board a Dutch man-of-war, and in the presence of the commissioners of the admiralty, being worked by three men, raised from a depth of twenty-two feet, four tons of water in a minute, that is, 240 tons of water in an hour."[b] In 1768, Mr. Cole introduced some considerable improvements in English ship pumps. An experiment made in that year is very interesting, as it shows the imperfections of the old ones, especially the enormous amount of *friction* to which they were subject. "Lately, a chain pump on a new construction was tried on board his Majesty's ship *Seaford*, in Block House Hole, which gave great satisfaction. There were present, Admiral Sir John Moore, a number of sea officers and a great many spectators The event of the trial stands thus :

The New Pump, Mr. Cole's :	The Old Pump :
Four men pumped out one ton of water in 43½ seconds.	Seven men pumped out one ton in 76 seconds.
Two men pumped out one ton in 55 seconds."	Four men pumped out one ton in 55 seconds.
	Two men *could not move it*."[c]

The chain in Cole's pump was made like a watch chain, or those which communicate motion to the pistons of ordinary fire engines, i. e. every other link was formed of two plates of iron, whose ends lapped over those of a single one, and secured by a bolt at each end. These bolts formed a joint on which they moved; but instead of their ends being riveted, one was formed into a button head, and a slit made through the other, for the admission of a spring key, so that they could be taken out at pleasure. By this device, whenever a link or bolt was broken or worn out, another one, from a store of them kept on hand for the purpose, could be supplied in a few moments. In some experiments, the chain was purposely separated, and dropped into the well in a ship's hold, whence it was taken up, repaired, and the pump again set to work in two minutes. Chains similar to these had been previously employed by Mr. Mylne in the pumps that raised the water from the caissons at Black Friars Bridge.

The pistons were formed of two plates of brass or iron, having a disk of thick leather between them, of the same diameter as the bore of the pump. The edges of the leathers, when wet, do not bear hard against the sides of the pump; indeed it is not necessary that they should even touch; for the water that escapes past one, is received into the next compartment below; and when a rapid motion is imparted to the pistons, the inertia of the moving column prevents in a great measure any from descending. The wheel which carries the chain is generally made like the trundles in mills, viz. two thin iron disks or rings are secured about eight or nine inches apart, upon the axle, and are united by several bolts at their circumference. The

---

[a] London Magazine for 1760, p. 321   [b] Ibid. 1762, p. 283.   [c] Ibid. 1768, p. 499.

distance between these bolts is such that the pistons fall in between them, and are carried round by them. Sometimes however, the links have hooks, which take hold of the bolts. A lower wheel is now dispensed with, and the end of the pump slightly curved towards the descending chain, to facilitate the entrance of the pistons. These machines are generally worked in ships of war by means of a long crank attached to the axle, at which a number of men can work. In some vessels they are moved by a capstan.[a] The pump cylinders are of iron, and sometimes of brass, the latter being inclosed within and protected by wooden ones.

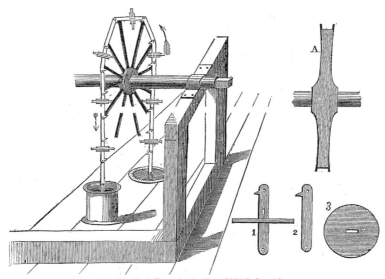

No. 67. Chain Pump in the U. S. Ship Independence.

For the following facts connected with the use of the chain pump in the United States Navy, we are indebted to Mr. Hart, Naval Constructor in the New-York Navy Yard. The first United States ship of war, which had one, was the '*Boston*,' built at Boston, in 1799. Mr. Hart's father made the pump. The chain was formed of common ox chains, and the wheel which carried it was of wood, having forked pieces of iron driven into its periphery, between which the chain was received : the cylinders were common pump logs of six inches bore. This imperfect machine was replaced the following year, by one formed after a French model, the chain and cylinders being of copper. In 1802, the Frigate *Constitution* had two similar ones placed in her; and about the same time, they were adopted in other public vessels; but in the course of a few years were discontinued generally, either from the prejudice of the seamen, or from the increased labor and expense of repairing the pistons. In ordinary pumps, a single box or piston only, has to be re-leath-

---

[a] The vessels of the exploring expedition sent out by the British admiralty, under the command of Capt. Owen, had their chain pumps fitted to work by the capstan, under the impression that it was a more economical mode than the crank, of applying personal labor. 'Narrative of Voyages to explore the shores of Africa, Arabia and Madagascar, in 1821.' Vol i, p. 14. N.York Ed. 1833.

ered; but in the chain pump, from thirty to fifty have to be renewed when worn out. The chain pumps in the British sloop of war *Cyane* were taken out when she was captured, and common pumps put in their place. Recently a change of opinion respecting these pumps has taken place, for within a few years they have to a limited extent been reintroduced into the navy. In 1837, the *Independence* was furnished with two of them; and in 1838, the same number were placed in the *Ohio:* both vessels still retain the ordinary pumps.

A description of one of those on board the Independence, which is now fitting for sea in this harbor, will give a correct idea of them all. See No. 67.

Two copper cylinders, seven inches diameter, and about twenty-two feet long, extend from the surface of the main gun-deck to the well. The one in which the chain *descends*, is continued ten or twelve inches above the deck to prevent the water that is raised, from returning through it again to the well. A horizontal wrought iron shaft is placed between the cylinders and supported by a stout frame on which its journals turn. On this shaft, a strong cast iron wheel, two feet in diameter, is secured, having twelve arms radiating like the spokes of a carriage wheel from the hub. A (No. 67) represents two of these arms with a portion of the shaft. A recess is formed at the extremities to receive the chain, and prevent it from slipping off on either side. Figs. 1 and 2, represent the links. They are of copper, seven inches long, one and a half inch wide, and one quarter of an inch thick, and are similar to those in Cole's pumps. The pallets or pistons are formed on the middle of every alternate link; that is, on those which are made of a single piece. A circular plate about a quarter of an inch thick, is cast (see fig. 1,) of a diameter rather less than the bore of the cylinders. Another loose plate of the same dimensions (fig. 3,) has an opening in its centre to allow it to pass over the link and lay upon the other. Between these, a disk of leather is introduced, i. e. a circular piece just like fig. 3, but of a diameter equal to the bore of the cylinder. This is first placed on the fixed plate in fig. 1, then fig. 3 is laid over it, and to secure the whole, a key or wedge is driven through a slit in the link, just above the surface of fig. 3, and thus compresses the leather between them. When the pumps are used, long cranks are applied to each end of the shaft, so that fifteen or twenty men can be engaged at the same time in working them. In the *Independence*, these cranks extend across the deck, and thereby interrupt the passage way. They should in all public vessels be arranged, if possible, 'fore and aft.' The arrows show the direction of the chain.

The introduction of the chain pump into ships is probably due to the Chinese, as they use it in their sea junks; and it is not likely that this application of it is of recent date among them.[a] The early missionaries thought that buckets only were employed in raising bilge water from the holds. It is not much used we believe in the French navy. In the Dictionnaire de Trevoux, it is named the *English* pump.

The Chinese chain pump was introduced into the United States by André Everard Van Braam, who was several years chief of the Dutch East India Company in China, and who settled in South Carolina at the close of the revolutionary war. In 1794, he was appointed second in the Dutch embassy to Pekin, and in 1796, he returned and settled near Philadelphia. In his account of the embassy, a translation of which was published at London in 1798, and dedicated by him to General Washington, and to which we have heretofore referred—he remarks, speaking of the

---

[a] 'The Chinese,' by J. F. Davis, vol. ii, 290

Chinese pumps, "I have introduced the use of them into the United States of America, where they are of great utility in rivers, in consequence of the little labor they require." (Vol. i, 74.) We are not aware that they are much used in this country at the present time. The chain pump is employed in the diamond districts of Brazil. M. Mawe, in his Travels, has figured and described it as used there. It has also been recently introduced into Egypt, where it is more likely to become domiciliated, than the atmospheric and forcing pump, which Belzoni endeavored in vain to establish; although St. John seems to think even it is rather too complex for the present state of the mechanic arts in the land of the Pharaohs. "Windmills for raising water, and chain pumps, have been introduced into Egypt; but as these are machines which require some regard to the principles of good workmanship, they are by no means fitted for general use." Egypt and Mohammed Ali, vol. i, 14.

The chain pump, as well as the screw, noria, chain of pots, &c. has been adopted as a first mover. Placed perpendicularly on the side of a precipice, or wherever a small stream of water can be conveyed into its upper orifice, and can escape from its lower one, the motion of the chain is reversed by the weight of the liquid column acting on the pistons. A wheel similar to the upper one is fixed below, over which the chain also passes; and from the axle of either wheel the power may be taken. A patent for this application of the chain pump was granted in England, in 1784.

There is another device of the Chinese, which is worthy of imitation; and considering the increased security it offers to floating property, and the additional safety of the lives of navigators, it is surprising that it has not been adopted by Americans and Europeans—viz. the division of the holds of ships by water-tight partitions. The Chinese divide the holds of their sea vessels into about a dozen distinct compartments with strong plank; and the seams are caulked with a cement composed of lime, oil, and the scrapings of bamboo. This composition renders them impervious to water, and is greatly preferable to pitch, tar and tallow, since it is said to be incombustible. This division of their vessels seems to have been well experienced; for the practice is universal throughout the empire. Hence it sometimes happens that one merchant has his goods safely conveyed in one division, while those of another, suffer considerable damage from a leak in the compartment in which they are placed. A ship may strike against a rock and yet not sink, for the water entering by the fracture will be confined to the division where the injury occurs. To the adoption of a similar plan in European or American merchantmen, beside the opposition of popular prejudice and the increased expense, an objection might arise from the reduction it would occasion in the quantity of freight, and the increased difficulty of stowing bulky articles. It remains to be considered how far these objections ought to prevail against the greater security of the vessel, crew and cargo. At any rate, such objections do not apply to ships of war, in which to carry very heavy burdens, is not an object of consideration. Staunton's Embassy, vol. ii, 136.

CHAPTER XVIII.

On the hydraulic works of the ancient inhabitants of America: Population of Anahuac—Ferocity of the Spanish invaders—Subject of ancient hydraulic works interesting—Aqueducts of the Toltecs—Ancient Mexican wells—Houses supplied with water by pipes—Palace of Motezuma—Perfection of Mexican works in metals—Cortez—Market in ancient Mexico—Hydraulic works—Fountains and jets d'eau—Noria and other machines—Palenque: its aqueducts, hieroglyphics, &c.—Wells in ancient and modern Yucatan—Relics of former ages, and traditions of the Indians. Hydraulic works of the Peruvians: Customs relating to water—Humanity of the early Incas—Aqueducts and reservoirs—Resemblance of Peruvian and Egyptian customs—Garcilasso—Civilization in Peru before the times of the Incas—Giants—Wells—Stupendous aqueducts, and other monuments—Atabalipa—Pulleys—Cisterns of gold and silver in the houses of the Incas—Temples and gardens supplied by pipes—Temple at Cusco: its water works and utensils—Embroidered cloth—Manco Capac.

It has been a subject of regret, that we have been unable to obtain any specific information, respecting the employment of machines to raise water on the American continents, previous to the visit of Europeans in the 15th and 16th centuries. And yet there can, we think, be scarcely a doubt, that in those countries where civilization and the arts had made considerable progress, as Peru, Chili, Guatimala, and Mexico, such machines had long been in use. Unfortunately, accounts of those countries by early European writers, contain little else than details of the successful villany of those savage adventurers, who, under the cloak of religion, and by the most revolting perfidy, robbed the natives of their independence, their property, and myriads of them of their lives.

It is impossible to reflect on the great population of ancient Anahuac—the progress which the natives had made in the arts—the separation of trades and professions—their extensive manufactures—the splendor of their buildings—their laws—the rich produce of their highly cultivated fields—the freedom and prosperity of the republics of Tlascala, and the comparative general happiness of the inhabitants; with the utter desolation brought on them and their country by the Spaniards—without feeling emotions of unmitigated indignation. No one can read even *Solis*, the advocate of Cortez and the palliator of his conduct, without being thrilled with horror at the uniform treachery, cruelty, and blasphemy of that man His watchword of 'the Holy Ghost,' while slaughtering the natives on one occasion like sheep, conveys but an imperfect idea of his ferocity and indifference to their sufferings, and of the disgusting affectation of promoting christianity, under which he pretended to act. 'Religion,' says Solis, 'was always his principal care.' The Spaniards affected to shudder at the sanguinary gods of the Mexicans, which required human sacrifices—while *they* immolated in cold blood, hecatombs of the natives to the demons they themselves worshipped—viz. *avarice* and *dominion*—until the land was filled with slaughter, and whitened with the bones of their victims. It is said, that "in seventeen years, they destroyed above six millions of them." No romance ever equalled in horror the tragedies performed by Almagro, Valdivia, Cortez and Pizarro—and yet these men have been held up as examples of heroism, and our youth have been taught to admire, and of course to emulate 'the glory of Cortez.'

It is more than probable that the people, who, in remote times, inhabited the southern continent and Mexico, remains of some of whose works, rival in magnitude those of Egypt and India, and many of them, (the roads and aqueducts particularly,) equalled in utility the noblest works of Greece and Rome—were not without hydraulic engines; and had descriptions of them been preserved, they would have furnished more interesting, and perhaps more *certain* data, respecting the peopling of America, and of the origin of the Toltec and Astec races, than any others derived from the useful arts. From the analogy there is between some of the arts, manners, and customs of the ancient people of Mexico and South America, and those of Asia, we might suppose that the swape, bucket and windlass, noria, and chain of pots, and perhaps the chain pump were known to them; but of this we have met with no direct proof. Were the fact established, that they were in possession of these machines, it would greatly tend to prove their Asiatic origin in postdiluvian times; while on the other hand, if lacking these, they had others peculiar to themselves, such a fact would be one of the most interesting circumstances connected with the early history of these continents; and might be adduced to sustain the hypothesis of those who consider this hemisphere as having been uninterruptedly occupied by man, from times anterior to Noah's flood; and consequently many of the machines, arts, and productions of the inhabitants peculiar to themselves.

The Toltecs, we are informed, introduced the cultivation of maize and cotton; they built cities, made roads, and constructed those great pyramids which are yet admired; and of which the faces are very accurately laid out; they could found metals, and cut the hardest stone—they knew the use of hieroglyphical painting, and they had a solar year, more perfect than that of the Greeks and Romans. "Few nations (says Humboldt) moved such great masses as were moved by the Mexicans," proofs of which are still found among the ruins of their temples. The calender stone, and the sacrifice stone, in the great square at Mexico, containing 282 and 353 cubic feet; a carved stone dug up, which was upwards of 22 feet in length, 19 feet in breadth, and about 10 deep—are examples; and the colossal statue of the Goddess *Teoyaomiqui*, is another.[a] And what is more to our purpose, remains of AQUEDUCTS, of surprising magnitude and workmanship, are found throughout Chili, Mexico, and Peru.

Nor had these arts been lost at the period of the Spanish invasion. At that time, agriculture, artificial irrigation, and many other of the mechanic arts, especially those which relate to the metals, appear to have been in a more advanced state, than they have ever been in Spain, during any subsequent period. When Grijalva and his companions landed in Yucatan, (in 1518) they were astonished at the cultivation of the fields, and the beauty of the edifices—as well as at the ornaments, &c. in gold, which the natives possessed, the value of the workmanship often '*exceeding that of the metal.*' Tlascala, (says Solis,) was at that period, " a very populous city;" the houses were built of stone and brick, their roofs were flat and surrounded with galleries. The Tlascalans, says Herrera, had baths, bowers, and *fountains*, and whenever a new house was finished, they had feasts and dancing, &c. like the *house warming* of old in Europe. Every house in Zempoala had a garden *with water*. Ancient *wells* are still in use in Mexico, some of which are two and three hundred feet in depth. Water is drawn from them to irrigate the soil.

The city of Cholula was located in a delightful plain; it contained 20,000 inhabitants, and the number in its suburbs was greater. The

---

[a] Clavigero says, columns of stone of one piece, 80 feet long, and 20 feet in circumference, were extant in his time, in the edifices of Mictlan. Mexico, Vol. i, 420.

Spaniards compared it to Valladolid for its beauty and magnificence. It was a great emporium of merchandise. Strangers from distant parts of the continent flocked to it. Solis says, the streets were wide and well laid out; the buildings larger and of better architecture than those of Tlascala, and the inhabitants were principally merchants and mechanics. Cortez himself, after entering this city, thus speaks of it in a letter to Charles V. "The inhabitants are better clothed than any we have hitherto seen. People in easy circumstances wear cloaks above their dress; these cloaks differ from those of Africa, for they have pockets, though the cut, cloth and fringes are the same. The environs of the city are very fertile and well cultivated. Almost *all the fields may be watered;* and the city is *much more beautiful than all those in Spain;* for it is well fortified, and built on level ground. I can assure your highness, that from the top of a mosque (temple) I reckoned more than four hundred towers, all of mosques. The number of inhabitants is so great that there is not an inch of ground uncultivated." When the Spaniards reached Tezcuco, they found it as large again as Seville. It rivalled in grandeur and extent Mexico itself, and was of a much more ancient date than that capital. Herrera says, the streets were very regular, and that *fresh water was brought in* PIPES from the mountains to *every house.* The principal front of the buildings extended on the borders of a spacious lake, where the causeway that led to Mexico began. It was from this causeway, which was built of stone and lime, that the Spaniards first beheld the distant capital, with its towers and pinnacles in the midst of the lake; and on the 8th November, 1519, Cortez and his myrmidons entered that city, which then contained a greater population than New-York does at present; for it had between three and four hundred thousand inhabitants.

When the Spaniards entered the gates, through a bulwark of stone supported by castles, they beheld a spacious street with houses uniformly built, and the windows and battlements filled with spectators. They were received into one of Motezuma's houses, which had been built by his father. This building, Solis remarks, vied in extent, with the principal palaces of emperors in Europe; and had the appearance of a fortress, with thick stone walls and towers upon the flanks. The streets of the city were straight, as if drawn by a line; and the public buildings, and houses of the nobility, which made up the greatest part of the city, were of stone and well built. The palace of Motezuma was so large a pile that it opened with thirty gates into as many different streets. The principal front took up one entire side of a spacious parade, and was of black, red and white jasper, well polished. Over the gates were the arms or symbolical figures of Motezuma or his predecessors, viz. a griffin, being half an eagle and half a lion; the wings extended and holding a tiger in its talons. The roofs of the buildings were of cypress, cedar, and other odoriferous woods, and were ornamented with carvings of "different foliages and relievos." But without referring to the splendor of this unfortunate monarch's court, his luxurious mode of living, his treasures, the chair of burnished gold in which he was carried to meet Cortez, the jewels of gold, pearls, and precious stones, that adorned his person and those of his attendants, and the "shoes of hammered gold," that were bound to his feet and legs with straps, like the Roman military sandals; it will be sufficient to notice the market of the city for the sale of merchandise, in order to realize a tolerably correct idea of the state of the arts among the Mexicans. Nothing excited the surprise of the Spaniards so much as this market—both as regarded the quantity, variety, and quality of the goods sold, and the order which prevailed.

The Mexican works of gold and silver, sent by Cortez to Charles V. says Clavigero, "filled the goldsmiths of Europe with astonishment." "Some of them were inimitable." Among others, there were *fishes* having scales alternately of gold and silver—a parrot with *moveable head, tongue, and wings*—an ape with moveable head and feet, and having a spindle in its hand, in the attitude of spinning. Vol. i, 413.

Cortez, in a letter to Charles V. dated October 1520, says, "the market place is twice as large as that of Seville and surrounded with an immense portico, under which are exposed for sale all sorts of merchandise, eatables, ornaments made of gold, silver, *lead, pewter*, precious stones, bones, shells, and feathers; delft ware, leather, and spun cotton. We find hewn stone, tiles, and timber fit for building. There are lanes for game, others for roots and garden fruits. There are houses where barbers shave the head, (with razors made of obsidian,) and there are houses resembling our apothecary shops, where prepared medicines, unguents, and plasters are sold. The market abounds with so many things, that I am unable to name them all to your highness. To avoid confusion, every species of merchandise is sold in a separate lane. Every thing is sold by the yard, (by measure) but nothing has hitherto been seen to be weighed in the market. In the midst of the great square, is a house, which I shall call *l'audencia* in which ten or twelve persons sit constantly for determining any disputes which may arise respecting the sale of goods. There are other persons who mix continually with the crowd, to see that a just price is asked. We have seen them break the false measures, which they had seized from the merchants."

Solis has recorded some facts, which are too interesting to mechanics to be omitted. "There were rows of silversmiths, who sold jewels and chains of extraordinary fashion;[a] several figures of beasts in gold and silver, wrought with so much art, as raised the wonder of our artificers; particularly some skillets with moving handles, *that were so cast;* besides other works of the same kind, with mouldings and relievos, without any signs of a hammer or graver." Herrera, speaking of these, observes, "some things were cast, and others wrought with stones, to such perfection, that many of them have surprised the ablest goldsmiths in Spain, for they could never conceive how they had been made; there being no sign of a hammer, or an engraver, or any other instrument used by them." They brought to the fair, (continues Solis) all the different sorts of cloth, made throughout this vast empire, of cotton and rabbits' fur, which the women of this country, enemies to idleness, spun extremely fine, being very dexterous in this manufacture. They had also drinking cups exquisitely made of the finest earth, different in color, and even in smell; and of this kind, they had all sorts of vessels, necessary either for the service and ornament of a house.

---

[a] These, which were worn round the neck, were doubtless similar to those known as *Panama chains;* which certainly are extraordinary specimens of workmanship. They may sometimes be met with at our jewellers, who buy them for the purity of the gold. It is said that the mode of making them has never been discovered, and that the secret is still preserved among the Indians of Panama. We have examined one which came from Carthagena, the length of which, had it been cut, was eight feet two inches; its section, which was hexagonal, did not exceed one twentieth of an inch in diameter. It was formed of one or more fine wires, which seemed to have been woven or interlaced like the platting of a whip handle. When a single thread was examined by a microscope, it was found to be composed of several smaller wires, which separate, were scarcely perceptible to our unaided vision. The weight of the chain was eleven pennyweights, and it appeared to be as flexible as a piece of twine, certainly far more so than any chain formed of links. No end of a wire could be detected, and not a particle of solder was used.

No one can doubt, that a people, thus far advanced in civilization and the useful arts, were in possession of machines of some kind or other for raising water. Indeed the location and great population of some of their cities required a familiar knowledge of hydraulic operations to supply them with water; and hence it would seem as if they had cultivated this department of the arts equally with others, for some of their aqueducts would have done honor to Greece and Rome. Nearly all the ancient cities of Mexico were supplied by them. We have already remarked that Tlascala was furnished with abundance of baths and fountains—that every house in Zempoala had water—that Tezcuco had an aqueduct, from which every dwelling was supplied *by a pipe*, as in modern cities; and we may add, Iztaclapa, which contained about ten thousand houses, had its aqueduct that conveyed water from the neighboring mountains, and led it through a great number of well cultivated gardens. In the city of Mexico, there were several aqueducts. That of *Chapultepec* was the work of Motezuma, and also the vast stone reservoir connected with it. When the Spaniards besieged the city they destroyed this aqueduct. Cortez in his first letter to Charles V. mentions the spring of Amilco, near Churubusco, of which the waters were conveyed to the city " in two large pipes, well moulded and as hard as stone, but the water never ran in more than one of them at the same time." We still perceive, says Humboldt, the remains of this great aqueduct, which was constructed with double pipes, one of which received the water, while they were employed in cleansing the other; but this aqueduct, he says, was inferior to the one at Tezcuco: of it, he observes, " we still admire the traces of a great mound, which was constructed to heighten the level of the water." The gardens of Motezuma were also adorned and nourished with streams and *fountains*, and appear to have rivalled those of Asiatic monarchs in splendor. And among the hieroglyphical ornaments of the pyramid of Xochicalco, are heads of crocodiles *spouting water*, a proof that ancient Americans were acquainted with that property of liquids by which they find their level; and applied it not merely to fountains and *jets d'eau*, but to convey water through *pipes* to their dwellings.

We cannot reflect on the progress which the ancient inhabitants of Mexico had made in the arts, and the magnitude and excellence of some of their hydraulic works, without regretting that no particular accounts of their devices for *raising* water have been preserved. Of one thing however, we may be sure, that no people ever constructed such works as they did, for the irrigation of land, and the supply of cities, who had not previously experienced the inefficiency of machines for those purposes; nor could their agriculture have been carried to the extent it was, without the aid of them in times of drought.

' The machines called *norias* (says Humboldt) are essential to Mexican agriculture.'[a] Does it not follow then that these, or others for the same purpose, were equally essential, before the conquest, when the population of the country was so much greater, and agriculture more extensively practised? There is no doubt, he observes, that all the country from the river Papaloapan " was better inhabited and better cultivated than it now is." The swape (guimbelette) is quite common in Mexico. It is there used as in this country for raising water from wells of moderate depth. A friend just returned from a tour in Texas, informs us, that among the Cooshattie tribe of Indians on the Trinity river, and in all the settlements, whether Indian, Creole, or modern Mexican; in populous villages or at

---

[a] New Spain, translated by Black, Vol. ii, 458.

solitary wigwams, the 'well pole' or swape is almost always to be seen. In wells of considerable depth, the pulley and double chain with two buckets are adopted; and the chain of pots and noria are extensively used in raising water for irrigation, being moved, as in Spain, by oxen or mules. As these are the only devices for raising water that are now in use, except the common pump to a very limited extent, and the ordinary mode of drawing it from mines, by buckets worked by animals, the question occurs, were they or any of them in use previous to the conquest? The pump excepted, we should suppose they were; but as before remarked, we have no direct evidence to establish the fact. As the Mexicans were collected in villages and accustomed to cultivate the soil, at least 400 years before the conquest, and subsequent to that event, the great mass of the farmers have been and are Indians, who adhere with "extraordinary obstinacy" to the customs of their ancestors, it seems natural to suppose that they (like the agricultural classes of all other countries) would retain some of the old modes of raising water; but as those above named are said to be the *only* ones practised, it is probable that some of them at least were known to the inhabitants of old.

Palenque is about thirty miles from Tobasco. It is surrounded by dense forests, and overgrown with the vegetation of past ages. Of its founders and inhabitants nothing is known, nor yet of the period when they flourished. The remains of this city have been traced over an extent of *twenty-four miles,* and consist of massive edifices, of a novel and very chaste style of architecture. These are accurately laid out to the four cardinal points of the compass, and are built of hewn stone. There are temples, palaces, and tombs ornamented with the richest sculptures and bas reliefs, extensive excavations, subterranean passages, bridges, dikes, AQUEDUCTS, &c. all indicative of a powerful and highly civilized people. Dupaix and his companions, who were sent out in 1805 by Charles IV. of Spain, to examine and report on these buildings, after three weeks intense labor in cutting down trees which grew over them, were enabled to examine fifteen edifices, which elicited their wonder and admiration. Mr. Waldeck, a late traveler, who has spent several years in examining and collecting evidences of early American civilization, cut down a tree, (that was growing over an ancient building at Palenque) the concentric circles in a section of which, indicated a growth of 973 years! But how many centuries had elapsed from the ruin and desertion of the city, and for the accumulation of soil over it, ere this tree took root, can only be conjectured. The sculptures on the walls are surprisingly perfect, and among them are hieroglyphics which are supposed to have phonetic power. Men and women are represented clothed in figured garments, indicating the manufacture of flowered stuffs; and various relics which have been disinterred, as toys, vessels, ornaments of dress, &c. prove considerable progress in other branches of the useful arts. But extensive as these ruins are, and pregnant with information of thrilling interest, Palenque is like *Mitla,* the partner of its glory and of its degradation, a 'city of the dead.' Not a voice is heard in it, or around it, but the hissing of serpents, the buzzing of insects, the gibbering of monkeys, and the screeching of wild birds.

There is one circumstance respecting the ancient cities and people of Yucatan which relates to our subject, that is deserving of attention. It is this—from the geographical position and physical features of the country, WELLS have *always* been of PRIMARY importance. During the greater part of the year the inhabitants have no other resource for fresh water; and this must necessarily have been the case, ever since the present organization of these continents took place. In those remote ages, then, during which the country was occupied by a numerous and civilized people, wells must have been very common; and as they are not, like structures erected on the surface, subject to decay, or obnoxious to destruction, the discovery and examination of some of them is greatly to be desired. Who can tell what stores of treasure are buried in them; what specimens of art; what means for tracing the history, and also the revolutions, through which the ancient people of America have passed; their origin, progress, and disappearance? If Palenque and its sister cities were destroyed by war, then it is almost certain that the inhabitants would have recourse to wells for the secretion of their treasures, both public and private; a practice that has been followed in all ages and by all people.

Relics of former ages, which have been found (more or less numerous) over both continents, incontestibly prove that civilized people flourished here in former ages; and that they and their progeny have disappeared, as if by some general and sudden calamity they had been swept off the stage of life, to a man. It would seem too, as if a long period of deathlike stillness had succeeded, (like that after the deluge) so that all knowledge of them had perished, when another race appeared and took possession of the soil. These were the ancestors of the present Indians, who, in their turn are rapidly becoming extinct, without our being able to tell who they were, whence they came, or *when* they first made their appearance. We see no reason to doubt their tradition respecting the great Megalonyx and Mastodon of the western prairies, having been contemporary with their forefathers, since the discovery of the bones of these animals corroborate in some degree the truth of it. Nor is it at all improbable that their accounts of the voracious and enormous *Piasâ*, ' the bird that devours men,' is fabulous; a figure of which is cut on the face of a smooth and perpendicular rock, at an elevation which no human art can now reach; near the mouth of a small stream, named the Piasâ, which enters the Mississippi between Alton and the mouth of the Illinois. See Family Mag. 1837. Vol. iv, 101.

ON THE HYDRAULIC AND OTHER WORKS OF THE ANCIENT PERUVIANS.

Molina, in his ' Natural and civil History of *Chili*,' observes that previous to the invasion of the Spaniards, the natives practised artificial irrigation, by conveying water from the higher grounds in canals to their fields. Herrera says, many of the vales were exceedingly populous and well cultivated, ' having trenches of water.' The Peruvians carried the system to a great extent. " How must we admire (says Humboldt) the industry and activity displayed by the ancient Mexicans and Peruvians in the irrigation of arid lands! In the maritime parts of Peru, I have seen the remains of walls, along which water was conducted for a space of from 5 to 6000 metres, from the foot of the Codilleras to the coast. The conquerors of the 16th century destroyed these aqueducts, and that part of Peru has become, like Persia, a desart, destitute of vegetation. Such is the civilization carried by the Europeans among a people, whom they are pleased to call barbarous."[a] These people had laws for the protection of water, very similar to those of Greece, Rome, Egypt, and all the older nations; for those who conveyed water from the canals to their own land before their turn, were liable to arbitrary punishment. Several of the ancient American customs respecting water, were identical with those of the oldest

---

[a] New Spain. Black's Trans. Vol. ii, 46, and Frezier's Voyage to the South Seas, 213.

nations. They buried vessels of water with the dead.[a] The Mexicans worshipped it.[b] The Peruvians sacrificed to rivers and fountains.[c] The Mexicans had *Tlaloc* their god of water.[d] Holy water was kept in their temples.[e] They practised divination by water.[f] The Peruvians drew their drinking water from DEEP WELLS,[g] and for irrigation in times of drought, they drew it from pools, and lakes, and rivers.

The annals of the world do not furnish brighter examples of national benevolence, than the early history of Peru. The wars of the incas were neither designed nor carried on, to gratify ambition or the lust of conquest, but to extend to the brutalized people by whom they were surrounded, the advantages of civilized life; to introduce agriculture and all its attending blessings, among hordes of savages, that were sunk in the lowest depths of bestiality. But that which sheds a peculiar glory over the ancient Peruvians, was their systematic and persevering efforts to achieve their conquests without the effusion of blood. In reading their history, the mind is not only relieved from those horrible details of carnage that constitutes so prominent a part in the historic pages of other nations, but the most agreeable emotions are excited by the benevolent and generally successful endeavors of this people, to overcome their foes by reason—by exhibiting to them the advantages of regulated society, and by invitations to embrace them. This policy was in accordance with the injunctions of their first king, whose precepts they greatly reverenced. He taught them to overcome their enemies " by love—by *the* FORCE *of benefits*," and hence we find that when they were successful, they neither robbed the inhabitants of their land, their liberty, nor their lives but used their influence and superior knowledge to ameliorate their condition. And when these efforts failed, and active warfare was the only resource, they, conscious of the wickedness of conquering men by their destruction, and that those could never be good subjects who ' obeyed from fear,' uniformly besieged them, till the latter became convinced of their own inability to resist, and of the policy of acceding to the terms of their powerful invaders.

In this manner the ' children of the sun' extended their conquests over a large part of the southern continent; and in no part of the world were provinces more loyal, or a people more attached to their institutions and to their princes; nor was there ever a people more humane. The conduct of some of the incas, when at the head of their armies, in enduring the taunts and scoffs of their ignorant and imbecile foes with philosophic forbearance, is truly admirable, and might be contrasted with that of *christian* warriors; but then their object was not to acquire fame by the destruction of their species, but to benefit them, even at the risk of their reputation. If ever offensive wars were justifiable, those of the early incas certainly were, since their object was the extension of human happiness, and which they carried on in a corresponding spirit of humanity. In neither sacred nor profane history can such examples be found.

Agriculture was the first object to which their attention was directed; hence we find engineers and other artists immediately sent into the subdued countries, or rather, among their new friends, to introduce the arts of ploughing and cultivating the soil, &c. And as large tracts of land were destitute of vegetation for *want of water*, mention is constantly made of AQUEDUCTS and RESERVOIRS among the earliest of works undertaken. In some districts, rain was, and still is, unknown. " For the space of seven

---

[a] Purchas's Pilgrim. 1080. [b] Ibid. 966. [c] Ibid. 1070. [d] Ibid. 986. [e] Ibid. 987. [f] Ibid 994. [g] Ibid. 1064.

hundred leagues along the coast (says Garcilasso) it did never rain." Contrivances to obtain and distribute water, were therefore, with the incas as with the early kings of Egypt, the most important and constant objects of their care. Nor does it appear that the Egyptians were more assiduous in this kind of labor than the people of Peru. Examples are mentioned of the latter having conveyed small streams through a space of sixty miles, to irrigate a few acres of land.

There are several points of resemblance between these two people; some of which are to be attributed to both countries being, in a great measure, destitute of rain. The first inca, like Osiris, taught the inhabitants to cultivate the land; to construct reservoirs and aqueducts; to make ploughs, harrows, and shoes for their own feet—such shoes, says Garcilasso, 'as they now wear.' The wife of Manco Capac, like Isis, taught the women to spin, to weave, and to make their own garments. Some of their fables, too, resemble those of the Egyptians respecting Isis. According to one, " the maker of all things placed in heaven a virgin, the daughter of a king, *holding a bucket of water in her hand,* for the refreshment of the earth." Both people erected stupendous structures and statues of cut and polished stone, which they wrought without iron; both shaved the head, and both embalmed the dead.

As we have no where met with any distinct notice of, or even allusion to, any Peruvian machine for *raising* water, we insert some notices of their wells and aqueducts, &c. from Garcilasso's " Royal Commentaries of Peru." The reader can then judge, whether a people who devised and constructed hydraulic works of immense magnitude for the distribution of water, were without some machines for *raising* it; and especially, when, at certain seasons, they obtained it from deep wells. The inca *Garcilasso de la Vega,* was a native of Cusco. His mother was a Peruvian princess; but his father, whose name he bore, was one of the Spanish conquerors. He was born (he informs us) eight years after the Spaniards became masters of the country, i. e. in the year 1539, and was educated by his mother and her relatives, in the Indian manner, till he was twenty years old. In 1560 he was sent to Spain, where he wrote his Commentaries. These were translated into English by Sir Paul Ricaut, and published in one volume, folio, London, 1688.

There is reason to believe that Peru, Chili, and other parts of the southern continent, were inhabited by a refined, or partially refined people, centuries before the time of Manco Capac, the first inca; and that a long period of barbarism had intervened, induced, perhaps, by revolutions similar to those which, in the old world, swept all the once celebrated nations of antiquity into oblivion. The ancient Peruvians had a tradition respecting the arrival of giants, who located themselves on the coast, and who dug WELLS of immense depth *through the solid rock;* which wells, as well as CISTERNS, still remain. When Mayta Capac, the fourth inca, reduced the province of Tiahuanacu, he found colossal pyramids and other structures, with gigantic statues, of whose authors or uses, says Garcilasso, " no man can conjecture." The ruins of these are still extant, in one of the districts of Buenos Ayres. In the same province, the writer just named mentions a monolithic temple, which, from the description, equals any of those of Egypt. These ancient buildings were supposed by the Peruvians to have furnished models for the Temple, Palace, and Fortress at Cusco, which the first incas erected. *Acosta,* in examining some of these buildings in Tiahuanacu, was at a loss to comprehend how they could have been erected; so large, well cut, and closely jointed were the stones. " I measured one myself, (he observes) which was thirty feet in length, eighteen in breadth, and six feet in thickness;" and in the Fortress of Cusco were stones, he says, *much larger*. But what adds to our surprise, many of these stones were taken from quarries at from five to fifteen leagues distance from the buildings.

There is much uncertainty respecting Manco Capac. Who he was, and from what country he came, are equally unknown. According to their *Quippus* or historical cords, and the opinion of the inca who was uncle to Garcilasso, and who communicated to the latter all the knowledge of their ancestors then extant, he made his appearance in Peru about 400 years before the invasion of the Spaniards. It is said he was whiter than the natives, and was clothed in flowing garments. Awed by his presence, they received him as a divinity, became subject to his laws, and practised the arts he introduced. He founded Cusco, and extended his influence to all the nations around. He taught them agriculture and many useful arts, especially that of irrigating land. His son succeeded him, and without violence greatly extended the limits of the kingdom; prevailing with the natives, it is said, by a peaceable and gentle manner, " to plough, and manure, and cultivate the soil." His successors pursued the same mode, and with the same success. The fifth inca, we are informed, constructed aqueducts, bridges and roads in all the countries he subdued. When the sixth inca acquired a new province, he ordered the lands to be " dressed and manured;" the fens to be drained, " for in that art [draining] they were excellent, as is apparent by their works, which remain to this day : and also they were [then] very ingenious in making *aqueducts* for carrying water into dry and scorched lands, such as the greatest part of that country is : they always made contrivances and *inventions* to bring their water. These aqueducts, though they were ruined after the Spaniards came in, yet several reliques and monuments of them remain unto this day."

The seventh inca, *Viracocha*, constructed some water works, which, in their beneficial effects, perhaps equalled any similar undertakings in any other part of the world. " He made an aqueduct 12 feet in depth, and 120 leagues in length : the source or head of it arose from certain springs on the top of a high mountain between Parcu and Picuy, which was so plentiful that at the very head of the fountains they seemed to be rivers. This current of water had its course through all the country of the Rucanas, and served to water the pasturage of those uninhabited lands, which are about 18 leagues in breadth, *watering almost the whole country of Peru*."

" There is *another* AQUEDUCT much like this, which traverses the whole province of *Cuntisuyu*, running above 150 leagues from south to north. Its head or original is from the top of high mountains, the which waters falling into the plains of the *Quechuas*, greatly refresh their pasturage, when the heats of the summer and autumn have dried up the moisture of the earth. There are many streams of like nature, which run through divers parts of the empire, which being conveyed by AQUEDUCTS, at the charge and expense of the incas, are works of grandeur and ostentation, and which recommend the magnificence of the incas to all posterity ; for these aqueducts may well be compared to the miraculous fabricks which have been the works of mighty princes, who have left their prodigious monuments of ostentation to be admired by future ages; for, indeed, we ought to consider that these waters had their source and beginning from vast, high mountains, and were carried over craggy rocks and inaccessible passages ; and to make these ways plain, they had no help of instruments forged of steel or iron, such as pickaxes or sledges, but served themselves only with one stone to break another. Nor were they acquainted with the invention of arches, to convey the water on the level from one precipice to the other, but traced round the mountain until they found ways and passages at the same height and level with the head of the springs.

"The cisterns or conservatories which they made for these waters, at the top of the mountain, were about twelve feet deep; the passage was broken through the rocks, and channels made of hewn stone, of about two yards long and about a yard high; which were cemented together, and rammed in with earth so hard, that no water would pass between, to weaken or vent itself by the holes of the channel.

"The current of water which passes through all the division of *Cuntisuyu* I have seen in the province of *Quechua*, which is part of that division, and considered it an extraordinary work, and indeed surpassing the description and report which hath been made of it. But the Spaniards who were aliens and strangers, little regarded the convenience of these works, either to serve themselves in the use of them, or keep them in repair, nor yet to take so much notice of them, as to mention them in their histories, but rather out of a scornful and disdaining humor, have suffered them to run into ruine, beyond all recovery. The same fate hath befallen the AQUEDUCTS which the Indians made for watering their corn lands, of which two thirds at least are wholly destroyed, and none kept in repair, unless some few which are so useful that without them they cannot sustain themselves with bread, nor with the necessary provisions of life. All which works are not so totally destroyed, but that there still remains some ruines and appearances of them."

The last who was independent, and by far the worst of the incas, was Atahualpa or Atabalipa, the 13th from Manco Capac. He treacherously slew his brother and murdered nearly all his relations. Garcilasso's mother and a few others escaped. He was strangled by Pizarro in May 1533, after having purchased his life of that monster, by filling the room of his prison with gold and silver vessels, and ingots, to a line chalked round the wall, at the height of about seven feet from the ground. This room was twenty-five feet by sixteen.

That the Peruvians had *wells* in the remotest times has already been noticed; and when the Spaniards invaded their country, great quantities of treasures were thrown into them. The discovery of these wells may yet bring to view numerous specimens of their works in the metals. We have not met with any intimation of their manner of *raising* water, whether by a simple cord and vessel, by means of a pulley, or a windlass, or any other machine. 'Tis true that Garcilasso, when describing the various pendants which they wore in their ears, mentions rings as large "as the frame of a pulley, for they were made in the form of those with which *we draw up pitchers from a well*, and of that compass, that in case it were beaten straight, it would be a quarter of a yard long and a finger in thickness," but in this passage we understand him to refer to the Spanish method of drawing water; and this is probable, for in another part of his work, when speaking of the large stones used in the public buildings at Cusco, he says the workmen had neither cranes nor *pulleys*. Still it is possible that he referred to the mode his countrymen employed.

There are conclusive proofs however, in some extracts that are too interesting to be omitted, that the ancient Peruvians were well acquainted with the management and distribution of water through *pipes*; and of making and laying the latter; and what is singular, both the sources of the water and the direction of the tubes under ground were kept secret, as was the custom with some people of Asia. "In many of the houses (of the incas) were great cisterns of gold, in which they bathed themselves, with COCKS *and pipes of the same metal*, for conveyance of the water." Some interesting particulars are also given by Garcilasso respecting the supply of Cusco with water. Speaking of a certain street, he says, " near thereunto are two pipes of excellent water, which pass under ground, but by whom they were laid and brought thither, is unknown, for want of writings or records to transmit the memory of them to posterity. Those pipes of water are called *silver snakes*, because the whiteness of the water resembled silver; and the windings or the meanders of the pipes were like the coils and turnings of serpents." In the fortress of Cusco was "a fountain of excellent water, which was brought at a far distance under ground, but where and from whence the Indians do not know; for such secrets as these were always reserved from common knowledge in the breasts of the inca and of his counsel." The lake *Chinchiru* near Cusco, contained good water, and "by the munificence of the inca was furnished with several pipes and aqueducts," to convey water into lower grounds, which were used till rendered useless by neglect of the Spaniards. " Afterwards, in the year 1555 and 56, they were repaired by my lord and father Garcilasso de la Vega, he being the mayor of that city, and in that condition I left them."

In describing the temple and gardens at Cusco, he observes, " there were five fountains of water, which ran from divers places through pipes of gold. The *cisterns* were some of stone, and others of gold and silver, in which they washed their sacrifices, as the solemnity of the festival appointed. In my time there was but one of these fountains remaining, which served the garden of a convent with water; the others were lost, either for want of drawing or cleansing, and this is very probable, because, to my knowledge, that which belonged to the convent was lost for six or seven months, for want of which water the whole garden was dried up and withered, to the great lamentation of the convent and the whole city; nor could any Indian understand how that water came to fail, or to what place it took its course. At length they came to find that on the west side of the convent the water took its course under ground, and fell into the brook which passes through the city; which in the times of the incas had its banks kept up with stones, and the bottom well paved, that the earth might not fall in; the which work was continued through the whole city, and for a quarter of a league without; which now by the carelessness and sloth of the Spaniards is broken, and the pavement displaced; for though the spring commonly yields not water very plentifully, yet sometimes it rises on a sudden and makes such an incredible inundation that the force of the current hath disordered the channel and the bottoms."

" In the year 1558 there happened a great eruption of water from this fountain, which broke the main pipe and the channel, so that the fury of the torrent took another course and left the garden dry; and now by that abundance of rubbish and sullage which comes from the city, the channel is filled up, and not so much as any mark or signal thereof remains. The friars, though at length they used all the diligence imaginable, yet they could not find the ancient channel, and to trace it from the fountain head by way of the pipes, it was an immense work, for they were to dig through houses and deep conveyances under ground, to come at it, for the head of the spring was high. Nor could any Indian be found that could give any direction herein, which discouraged them in their work, and in the recovery of the others which anciently belonged to the temple. Hence we may observe the ignorance and inadvertisement of those Indians, and how little the benefit of tradition availed amongst them; for though it be only forty-two years at this day since those waters forsook their course; yet neither the loss of so necessary a provision as water, which was the refreshment of their lives, nor of that stream which supplied the temple of the sun, their god, could by nature or religion conserve in them the memory of so remarkable a particular. The truth is, that it is probable that the undertakers or master-workmen of those water works did communicate or make known to the priests *only*, the secret conveyances of those waters; esteeming every thing which belonged to the honor and service of the temple to be sacred, that it was not to be revealed to common ears, and for this reason perhaps the knowledge of those waters might dye and end with the order of priests."

"At the end of six or seven months after it was lost, it happened that some Indian boys playing about the stream, discovered an eruption of water from the *broken pipe;* of which they acquainting one the other, at length it came to the knowledge of the Spaniards, who, judging it to be the water of the convent that had been lost and diverted from its former course, gave information thereof unto the friars, who joyfully received the good news, and immediately labored to bring it again into direct conveyance, and conduct it to their garden. The truth is, the pipes lying *very deep* were buried with earth, so that it cost much labor and pains to to reduce it to its right channel; and yet they were not so curious or industrious as to trace the fountain to the spring head. That garden which now supplies the convent with herbs and plants, was the garden which in the times of the incas belonged to their palace, called *the garden of gold and silver;* because, that in it were herbs and flowers of all sorts, lower plants and shrubs, and taller trees, made *all of gold and silver;* together with all sorts of wild beasts and tame, which were accounted rare and unusual. There were also strange insects, and creeping things, as snakes, serpents, lizards, camelions, butterflies, and snails; also all sorts of *strange birds*, and every thing disposed, and in its proper place with great care, and imitated with much curiosity, like the nature and original of that it represented. There was also a mayzall, which bears the Indian wheat of an extraordinary bigness, the seed whereof they call quinia. Likewise *plants* which produce lesser seeds, and trees bearing their several sorts of fruit, *all made of gold and silver*, and excellently representing them in their natural shapes. In the palace also, they had heaps or piles of *billets* and *faggots* made of gold and silver, rarely well counterfeited. And for the greater adornment and majesty of the temple of their god the sun, they had cast vast figures in the forms of *men* and *women* and *children*, which they laid up in magazines or large chambers, called pirra; and every year, at the principal feasts, the people presented great quantities of gold and silver, which were all employed in the adornment of the temple. And those goldsmiths whose art and labor was dedicated to the sun, attended to *no other* work, than daily to make *new inventions* of rare workmanship out of those metals. In short they made all sorts of *vessels* or *utensils* belonging to the temple, of gold and silver, such as *pots*, and *pans*, and *pails*, and *fire shovels*, and *tongs*, and *every thing else* of use and service, even their very *spades* and *rakes of the garden* were made of the like metal."[a]

The author of 'Italy, with sketches of Spain and Portugal,' Phil. 1834, enumerating some of the curiosities in the museum of Madrid, re-

---

[a] The mission of Messrs. James, Bowdich and Hutchinson, sent by the British government to Ashantee, found the king and all his attendants literally oppressed with embellishments of solid gold, with which their persons were nearly covered. 'Even the *most common utensils* were composed of that metal.'

marks: "what pleased me most, was a collection of Peruvian vases; a polished stone which served the incas for a mirror; and a linen mantle, which formerly adorned their copper colored shoulders, as finely woven as a shawl, and flowered in very nearly a similar manner; the colors as fresh and vivid as if new." Vol. ii, 211.

It is difficult after perusing the history of this interesting people, to reconcile the state of the arts among them at the Spanish invasion, with the opinion, that Manco Capac arrived from Asia at so late a period as the 12th century. If he was the enterprising and intelligent man that he is represented to have been, and there is every reason to believe he was, it is impossible that as an *Asiatic*, he could have been ignorant of the saw, the auger, files, of fitting wooden handles to hammers, of nails, scissors, the crane, windlass, pulley, the arch, iron, &c : or having a knowledge of these things, that he should not have introduced them, or at least some of them. But if the Peruvians were also ignorant of the swape, noria, or chain of pots, the objections to such an opinion are greatly strengthened. From what part of the eastern world could such a man have come without having a knowledge of these machines, and yet be acquainted, as he was, with all the essential features of oriental agriculture? Machines too, of the utmost importance in Peru, where rain was generally unknown, and water scarce and valuable as in Egypt itself—and machines more necessary than any other, in furthering the objects he had in view While a doubt remains respecting their employment, we should suppose that he really was, as surmised by Garcilasso, a NATIVE, who by the superiority of his understanding, and by a subtile deportment (the more effectually to carry out his measures) persuaded the people that he came from the sun. Indeed, the state of the useful arts generally among them in the 15th century, implies that they had not had any permanent connection with Asiatics for many ages; but that they were gradually recovering a knowledge of the arts, which in very remote times had been practised by nations then extinct; and hence the paucity of their tools and the *peculiarity* of some of their devices, as their *quippus* or historical cords, their modes of computation, &c. Moreover, neither the Mexicans nor Peruvians had reduced the lower animals to subjection, at any rate not for agricultural purposes; and though they had neither the horse, the ass, nor ox; yet the former had the buffalo, an animal that has been used from the remotest ages to plough the soil. This circumstance alone is sufficient to show that they did not derive their knowledge of agriculture from Asia, within the time generally supposed, if at all.

Who can reflect on the civilized people, that in remote ages inhabited these continents, without mourning over their extinction, and the loss of every record respecting them? A people, whose very existence would have been unknown, had not some relics of their labors (like the organic remains of animals whose species are extinct) yet resisted the corroding effects of time. When we examine the ruins of their temples, their cities, and other monuments of their progress in the arts, our disappointment amounts to distress, that the veil which conceals them, is not, and perhaps cannot be removed. Strange as it may appear, we are almost as ignorant of the mysterious Palenque, and hundreds of other cities, equally and some of them perhaps much more ancient—as of the builders of Babel— and we know about as little of their early inhabitants as if they had been located on another planet.

END OF THE FIRST BOOK.

BOOK II.

MACHINES FOR RAISING WATER BY THE PRESSURE OF THE ATMOSPHERE.

CHAPTER I.

ON machines that raise water by atmospheric pressure—Principle of their action formerly unknown—Suction a chimera—Ascent of water in pumps incomprehensible without a knowledge of atmospheric pressure—Phenomena in the organization, habits, and motions of animals—Rotation of the atmosphere with the earth—Air tangible—Compressible—Expansible—Elastic—Air beds—Ancient beds and bedsteads—Weight of air—Its pressure—Examples—American Indians and the air pump—Boa Constrictor—Swallowing oysters—Shooting bullets by the rarefaction of air—Boy's sucker—Suspension of flies against gravity—Lizards—Frogs—Walrus—Connection between all departments of knowledge—Sucking fish—Remora—Lampreys—Dampier—Christopher Columbus at St. Domingo—Ferdinand Columbus—Ancient fable—Sudden expansion of air bursting the bladders of fish—Pressure of the atmosphere on liquids.

WITH the last chapter we concluded our remarks on machines embraced in the first general division of the subject, (see page 8) and now proceed to those of the second; viz. such as raise water by means of the weight or pressure of the atmosphere. These form a very interesting class—they are genuine philosophical instruments, and as such may serve to exhibit and illustrate some of the most important truths of natural philosophy. The principle upon which their action depends was formerly unknown, and even now, a person, however ingenious, while ignorant of the nature and properties of the atmosphere, would be utterly unable to account for the ascent of water in them. Having no idea of the cause of this ascent, except the vague one of *suction*, he would feel greatly embarrassed if required to explain it. And when informed that there really is no such thing in nature as suction, but that it is a mere chimera, having no existence except in the imagination, the task would be attended with insuperable difficulties. Perhaps he would have recourse to a *common pump*, to *trace*, if possible, the operation in detail; if so, he would naturally begin with the first mover, or the pump handle, and would look for some *medium*, by which motion is transmitted from it, to the water in the well; but, however close the scrutiny might be made, he would be unable to detect any; and as a matter of course, while a connection between them, i. e. between the mover and the object moved, could not be discovered, it would be impossible for him satisfactorily to account for the phenomenon. If "a body cannot act where it is not present," as the sucker of a pump, on water at a distance from it, how could such a person account for the ascent of that water in obedience to the movements of the sucker? And how could he explain the process by which it was effected, while he could find no apparent communication between them? The fact is it would be difficult for him to point out any closer connection between the pump rod and the water in the well, than between a walking cane in the hands of a pedestrian, and water under the surface of the ground over which he stepped; nor could he assign a conclusive reason, why the liquid should not ascend and accompany the movements of the latter as well as of the former.

He could perceive no obvious or *adequate* cause for the elevation of water through the pipe of a pump, there being no apparent force applied to it, or in the direction of its ascent, no vessel or moveable pallet going down, as in the preceding machines, to convey or urge the liquid up—and hence he could no more comprehend how the movements of a pump box (sucker) above the surface of the ground, should induce water in a well to *rush up* towards it, than he could explain how the waving of a magician's wand should cause spirits to appear.

Long familiarity with the atmospheric pump, makes it hard for us, at the present day, to realize the difficulties formerly experienced in accounting for the ascent of water in it. Suppose the cause yet unknown and unthought of—it certainly would puzzle us to explain how a piece of leather (the sucker) moving up and down in a vertical tube, whose lower orifice is in water, some twenty-five or thirty feet below it, should conjure that water up. Such a result is opposed to all experience and observation in other departments of the arts; nor is there any thing like it, in the machines we have examined in the preceding book. The mechanism by which motion is transmitted from them to the water, is obvious to the senses—a tangible medium of communication is established between the force that works them and the water they raise; whereas in the pump, an *invisible* agent is excited, whose effects are as surprising as its mode of operation is obscure. 'Tis true, a tube (the pump pipe) is continued from the place where the sucker moves to the water, but it remains at rest, or is immoveable, and therefore cannot transmit motion from one to the other; it is merely a channel through which the water may rise—it does not raise it.

But if, in order to establish a connection between the sucker and water, the former were made to descend through the pump into the latter, still the difficulty would not be overcome. The sucker in that case would act much like one of those buckets, used in some wells, which has an opening in its bottom to admit the water, and covered by a flap to prevent its return. (The sucker is in fact, merely a small bucket of this kind, and is so named in some countries.) In both cases the water would be raised which entered through the valves—the bucket would bring up all it contained, and the sucker all that passed through it into the pump; so far the operation of both is clear, and as regards the raising of the water *above* the valves, would be the same; but it is the ascent of a column of water *behind* the sucker that requires explanation—a liquid column that follows it as closely through every turn of the tube, as if it were a rope, having its fibres at one end fastened to the sucker and *pulled up* by it. What is it that makes this water ascend against a law of its nature—against gravity? Were the cohesion of its particles such that it could be raised by a force applied only to its upper end, then indeed the difficulty would be diminished; but in that case, it would follow that a similar column would ascend after a bucket when drawn out of an open well; and further, that a traveler might then make use of a liquid walking stick, to assist him in his journeying.

Baffled thus in our attempts to find a solution here, we perhaps would begin to think, that when a liquid is raised in the pipe of a pump, it must be by some force acting *below*, or behind it, a force *a tergo*, as it is named, and of which all the preceding machines are examples. Thus when a bucket of water is raised from a well, the force is applied behind it, i. e. to the *bottom* of the bucket, through the cord, bale, and sides, to which it is attached. It is the same in the screw, the force continually elevating a portion of it immediately *behind* the water; and in the tympanum, noria, chain of pots, chain pump, &c. it is the same; the vessels or pallets go below, i. e. behind the liquid and urge it up before them. It is the same in all ordinary motions. I wish to examine a small object laying at the foot of my garden: now I cannot by moving this ruler in its direction, in the manner of a pump rod, induce it to move to me, nor can it ever be so moved, until the force of some other body in motion *behind* it impel it towards me. It is the same in the case of a stubborn boy, who not only refuses to move as directed, but opposes the natural inertia of his body to the change, and therefore can only be impelled forward by some force applied directly or indirectly behind him, by dragging or pushing him along. In this way, all the motions in the universe, according to some philosophers, are imparted or transferred; those which appear exceptions being considered modifications of it. Still however, the difficulty of establishing a connection between the movements of the sucker in the interior of the pump at one end, and this force, whatever it might be, acting on the water, *outside* of it, at the opposite end, would remain; and we should probably at last impute this ascent of water (with the ancients) to some indefinable energy of nature, both fallacious and absurd; nor would this be surprising, for in the absence of a knowledge of the atmosphere and of its properties, there really is *as great* a MYSTERY in the movements of a pump rod being followed by the ascent of the liquid, as in any thing ever attributed to the divining rod, or to the wand of Abaris.

In order to understand the operation of machines belonging to this part of the subject, and also the principle upon which their action depends, we must leave, for a few moments, the consideration of pumps and pipes, and all the contrivances of man, and turn our attention to some of the Creator's works as they are exhibited in nature. This may perhaps be deemed a departure from the subject; it is however so far from being a digression, that it is essentially necessary to ascertain the cause of water ascending in this class of machines, as well as to understand the philosophy of numerous natural as well as artificial operations, that are performed by apparatus analogous to them; as the acts of inspiration and respiration, quadrupeds drinking, the young of animals sucking their dams, children drawing nourishment from their mothers' breasts: bleeding by cupping, by leeches, or by the more delicate apparatus of a musketoe's proboscis; and if things ignoble may be named, the taking of snuff, smoking of cigars and pipes of tobacco, and also the experiments of those peripatetic philosophers, who perambulate our wharves, and imbibe nectar through straws from hogsheads of rum and molasses.

Every person is aware, that the earth on which we live is of a globular or spheroidal figure, and that it is enveloped in an invisible ocean of air or gas, which extends for a great number of miles from every part of its surface. This hollow sphere of air is named the atmosphere, and is one of the most essential parts in the economy of nature. It is the source as well as the theatre of those sublime meteorological phenomena which we constantly behold and admire. It is necessary to animal and to vegetable life. Its material is the 'breath of life' to all things living. It is moreover the peculiar element of land animals, the scene of their actions, the fluid ocean in which they only can move, and within which they are always *immersed*. It is to them, what the sea is to fish : remove them from it, and they necessarily die. In some respects nature has been more favorable to fishes than to us : most of them can ascend to the surface of the fluid in which they live, but we can only exist in the lowest depths of the atmospheric ocean that confines us : if we ascend but a little, our energies begin to fail, and we are compelled to descend to the bottom, the place she designed us to occupy.

Possibly, some people may suppose that the velocity with which the earth shoots forward in her orbit, might sometimes cause this atmosphere (which hangs as a mantle so loosely about her) to be left floating, like the tail of a comet, behind; or be entirely separated from her, like the cloud of vapor which the impetuous ball leaves at the cannon's mouth. Such however is not the fact ; on the contrary, it revolves uniformly with the earth on the axis of the latter, and accompanies her, as a part of herself, round the sun. Were it indeed separated from her, but for a moment, either by an increase or diminution of her velocity, the present organization of nature would be destroyed ; every mountain would be hurled from its base ; every house on the globe would be leveled ; and no human being could survive. Had the atmosphere not a rotatory motion also, in common with that of the earth, i. e. of the same velocity and in the same direction, a very different state of things, as regards the arts, would have subsisted than those which we behold. For example, aërial navigation would certainly have superseded nearly all traveling by land and water; and railroads, and locomotive carriages, and steamboats, would hardly have been known ; for the project of that individual who proposed to visit distant countries, by merely ascending in a balloon, till the rotation of the earth on its axis brought them under him, when he intended to descend, would have been no visionary scheme.

The air is tangible.—Although the substance of the atmosphere is not visible, it is tangible ; we *feel* it when in motion as *wind*, whether it be gently disturbed as in the evening breeze, or by the slight waving of a lady's fan; or when greatly excited, as in the hurricane, or the violent blast from a bellows' mouth. We also *see* its effects when thus in motion, in the direction of smoke, extinction of our tapers, slamming of doors, in the beautiful waving of grass, and of the full eared grain of the fields; trees yielding to its impulse, buildings unroofed, and sometimes in the prostration of large tracts of forests; in windmills, sailing of ships, and the convulsions into which it throws the otherwise placid ocean.

Air is compressible.—Indeed compressibility and expansibility are properties of all bodies; by the abstraction of heat, airs are compressed into liquids, and liquids into solids, while an increase of temperature expands solids into liquids, and these into airs. In the common air gun, four or five gallons of the dense air around us are compressed into a pint, and by further pressure they may be squeezed into a few drops of liquid, which a tea spoon might contain.

Its *expansibility* or *dilatability* is, so far as known, illimitable ; the space it occupies being always in proportion to the pressure that confines it. If a collapsed and apparently empty bladder be placed under a receiver, and the air around its exterior be removed, the small portion within will expand and swell it out to its natural shape. If it were possible to withdraw the whole of the air from this room, and a globule no larger than a pea were then admitted, it would instantly dilate and fill the room. The upper strata of the atmosphere decrease in density as they recede from the earth's surface, on account of the diminution of the pressure from super-

Air Beds.

incumbent strata, and thus at a certain height this small globule of air would occupy a space equal to the earth itself! And at the height of four or five hundred miles, it has been calculated, that less than a teacup full of the air we breathe, would fill a sphere equal in diameter to the very *orbit* of Saturn! The efficiency of the air pump in producing a vacuum depends entirely on this property.

Air at the foot of a mountain, whose elevation is between three and four miles, occupies twice the space when carried to the top. A quart of it taken from the summit would be reduced to a pint if conveyed to the bottom. From this expansive power of air arises its *elasticity*. This is familiar to most people; for when confined in flexible vessels, as air beds, pillows, life preservers, &c. as soon as any weight or pressure impinging upon them is removed, the elasticity of the confined fluid pushes up the depressed part as before. If air within a bladder were not elastic, the impressions made by the fingers in handling it would remain as in a ball of paste, and air beds would retain the form of bodies that reposed upon them, like a founder's mould of sand or plaster.[a] Those extremely light

---

[a] Air beds are not, as some persons suppose, of modern origin. They were known between three and four hundred years ago, as appears from the annexed cut, (No. 68,) copied from some figures attached to the first German translation of Vegetius, A. D. 1511. It represents soldiers reposing on them in time of war, with the mode of inflating them by bellows.

No. 68. Ancient Air Bed.

This application of air was probably known to the Romans. Heliogabalus used to amuse himself with the guests he invited to his banquets, by seating them on large bags or beds, "full of wind," which being made suddenly to collapse, threw the guests on the ground.

Dr. Arnott, the author of 'Elements of Physics,' a few years ago, proposed 'Hydrostatic beds,' especially for invalids. These are capacious bags, formed of india-rubber cloth, and filled with *water* instead of feathers, hair, &c. Upon one of these a soft and thin mattress is laid, and then the ordinary coverings. A person floats on these beds as on

balls of caoutchouc, which of late years have been introduced as parlor toys for children, rebound from the objects they strike by the spring of the air they contain. In the boy's pop gun, that is formed of a quill, the tiny pellet is sent on its harmless errand by the elastic energy of the compressed fluid. And in the air gun, it is the elasticity of the same fluid that projects balls with the force of gunpowder. If it were not elastic, people when fanning themselves would feel it thrown against their persons like water or sand. The act of inhaling it would be painful, for it would enter the chest by gluts, while its pulsations in sound would quickly destroy the membranes of the ear.

Perhaps nothing is better calculated to expand our ideas of the properties of matter, and of the wonders of creation, than the compressibility and dilatability of air. From the last named quality, it is probable that there is no such thing in nature as an *absolute* vacuum; and the best of our air pumps can scarcely be said to make even a rude approximation to one! Those, whose knowledge of nature is confined to impressions which things make on their senses, may suppose that the extremes of solidity may be found in a pig of lead and a bale of spunge; although the former is, in all probability, as full of interstices as the latter; and such persons could with difficulty be made to believe, that the entire mass of matter (air) which *fills* a space so immeasurably large as to baffle all calculation could be compressed into a lady's thimble, and even squeezed into a liquid drop, so minute, as scarcely to be perceived at the end of a needle.

Like all other matter with which we are acquainted, air has *weight*. This property is not naturally evident to our senses, but it may easily be rendered so. By accurately weighing a bladder when filled with air and afterwards when empty, it will be found heavier when full. This was an experiment of the ancients, but the moderns have ascertained its definite weight. A cubic foot of it, *near the earth's surface*, weighs about $1\frac{1}{4}$ ounces or $\frac{1}{800}$ part that of water, a cubic foot of the latter weighing 1000 ounces; hence the expression "water is 800 times heaver than air." The aggregate weight of the atmosphere has been calculated at upwards of 77 billions of tons, being equivalent to a solid globe of lead 60 miles in diameter; hence its *pressure*, for this enormous weight reposes incessantly upon the earth's surface, and upon **every** object, animate or inanimate, solid, liquid, or aëriform. The pressure it thus exerts, (in all places that are not greatly elevated above the level of the sea) is equal to to about 15 lbs. on every superficial square inch. Thus an ordinary sized person exposes so large a surface to its influence, that the aggregate

---

water alone, for the liquid in the bag adapts itself to the uneven surface of the body, and supports every part reposing upon it, with a uniform pressure. Water beds were however known to the ancients, for Plutarch (in his life of Alexander) states that the people in the province of Babylon slept during the hot months, " on skins filled with water."

The luxury of the ancients with regard to beds was carried to a surprising extent. They were of down, of the wool of Miletus, and sometimes stuffed with peacock's feathers. The Romans had linen sheets, white as snow, and quilts of needle work, and sometimes of cloth of gold. Bedsteads among the rich Greeks and Romans were sometimes of ivory, of ebony, and other rich woods, with inlaid work, and figures in relief. Some were of massive silver, and even of gold, with feet of onyx. They had them also of iron. One of that material was found in Pompeii. The earliest metallic bedstead mentioned in history is that of Og, king of Bashan. The Persians had slaves expressly for bed making, and the art became famous in Rome. *Golden* beds often formed part of the plunder which the generals exhibited at their triumphs. The Athenians put Timagoras their ambassador to Persia to death, for accepting presents from the king, among which was a "magnificent bed with servants to make it." *Plutarch in Pelopidas.*

pressure which his body sustains is not less than 14 or 15 tons. "Not less than what?" once exclaimed an elderly and corpulent lady. "Why how can that be? We could neither talk, nor walk, nor even move; and besides, sir, if that is the case, why don't we feel it?" For a very simple reason, though at the first view not a very obvious one. Air, as a fluid, presses equally in *every direction*—upwards as well as downwards—sideways and every way. Its component particles are so inconceivably minute, that they enter all substances, even liquids. Air is mixed up and circulates with the blood of all animals; it penetrates all the ramifications and innermost recesses of our porous bodies, and by the pressure of its superincumbent strata is urged through them, almost as freely as through the fleece of wool on a sheep's back, or between the fibres and threads of a ball of silk. Now, it is this circulation through the interior of our bodies that balances its pressure without. If its weight upon us were not thus neutralized, we certainly could neither talk nor walk: the lips of the loudest speaker, when once closed, could never be opened. We should be as mute and *immoveable* as if enclosed in statues of lead. And we should *feel* it, too—that is, for a moment; for it would as effectually crush us to death, as if we were placed in mortars, and *pestles*, each weighing 14 or 15 tons, were suddenly dropped upon us.

It is the air within the breast of the mother that forces milk into an infant's mouth, when the latter, by instinct, removes the external pressure from the nipple by sucking. It is the same with all mammiferous animals. The operation of cupping is another illustration of the same thing: the rarefaction of the air under the cup produces a partial vacuum within it; and as the external pressure of the atmosphere is removed from that part which is under it, the internal pressure urges the blood through the wounds. Were cupping instruments applied over the eyes, those organs would be protruded from their sockets.

As it is the *pressure* of the atmosphere upon which the action of the machines about to be described principally depends, we shall extend our remarks upon it.

Suppose a specimen of delicate fillagrane work, formed of the finest threads and plates, and of the human form and size, were sunk in water to the depth of 34 or 35 feet; it would then be exposed to the *same degree* of pressure to which our bodies are subject from the atmosphere; and when drawn up, it would be found uninjured, because the water entering into all its cavities, pressed just as much against its interior surfaces as the liquid around it against the exterior. But if it were enclosed in a skin or flexible covering, impervious to water, and then sunk as before, the pressure of the liquid around its exterior (not being balanced by any within) would crush it into a shapeless mass. Just so would it be with our bodies, and those of all terrestrial animals, if the air within them did not counteract the pressure without. And as long as this interior circulation remains, we can no more feel the pressure of the atmosphere than a fish feels that of water; nor can we be deranged or compressed by it, any more than a bundle of wool is, or a mass of entangled wire. It was ignorance of this simple fact—air in the interior of bodies exactly balancing the exterior pressure—that led the ancients astray, and induced one of the most sagacious intellects that was ever clothed in humanity (Aristotle) to ascribe this pressure to "nature's abhorrence of a vacuum."

Since the invention of the air-pump in 1654, numerous experiments are made, which demonstrate the pressure of the atmosphere. By it, this pressure may be removed from one part of the body, while it is left free to act with undiminished energy on the opposite part; as when the palm of the hand is held over the aperture of an exhausted receiver, the weight or pressure of the air on the back being no longer balanced by its action on the palm, the hand is irresistibly held to the vessel. A criminal or maniac, whose hands and feet are thus treated, would be as effectually secured as by fetters of iron. Few things are better calculated to excite wonder, and even horror, in the savage mind, than a part of the body being thus rendered helpless, as if spell-bound by some invisible agent. A few years ago, an experiment was made with the chief of a delegation of Pottawatamies to the seat of government, at which the writer was present. Although the interpreter previously endeavored to make them understand the intended operation, it will readily be supposed that such an attempt must necessarily have been fruitless. When the receiver was exhausted, he was amazed to find his hand immoveable, and that, like Jeroboam's, "he could not pull it in again to him." In his endeavors to free it, he rapidly uttered the characteristic interjections, ugh! ugh! and at last shrieked, as if in despair of being delivered from the power of the white enchanter; when his attending warriors flourished their tomahawks and rushed to his rescue, as if roused by the war-whoop.

It is not, however, necessary to have recourse to the air-pump, for proofs of atmospheric pressure. Numerous operations daily occur in common life which equally establish it. When a person washes his hands, if he lock them together so as to bring the palms close to each other, and then attempt to raise the central parts so as to form a cavity between them, at the same time keeping the extremities of the palms in close contact, he will feel the atmospheric pressure very sensibly : if the experiment be made under water, the effect will be more obvious still. Analogous to this is the attempt to open the common household bellows when the valve and nozzle are closed. The boards are then forced open with difficulty, in consequence of the pressure of the air on their exterior not being balanced by its admission within. If the materials and joints were made air-tight, and the orifices perfectly closed, the strongest man that ever lived could not force them open. This experiment, we believe, was familiar to the ancients; for we are indebted to them for both the blacksmith and domestic bellows. Another experiment of theirs, of a similar kind, was with the syringe. When the small orifice is closed with the finger, the piston is pulled up with difficulty, on account of the air pressing on its surface; and the moment we let go of the handle, it instantly drives it back, in whatever position the implement is held. The ordinary syringe seldom exceeds three-fourths or one inch in diameter, and any person can thus draw out the piston; but one of six or seven inches diameter would require a giant's strength; and one of a foot or fifteen inches would resist the efforts of two or three horses.

Numerous illustrations of atmospheric pressure may be derived from the animal kingdom. The boa constrictor when it swallows its prey affords one. As soon as this serpent has killed a goat or a deer, he covers its surface with saliva : this appears necessary to lay the long hair of these animals close, in order to prevent air from passing between the body of the victim and the interior of the devourer's throat. After taking the head into his mouth, by a wonderful muscular energy he alternately *dilates* and *contracts* the posterior portions of his body, until the pressure of the atmosphere forces into his flexible skin an animal whose bulk greatly exceeds that of his own. But if air were to pass between the body of the victim and the dilatable gullet of the boa, while the latter was making a vacuum to receive it, the pressure of the atmosphere would be neutralized as effectually as if a gash were made through his skin in front of the victim

The same process may be witnessed in ordinary snakes, for all serpents swallow their prey whole. There is no mastication to facilitate deglutition. Their upper jaws are loosely connected to the head, so that the mouth can be opened very wide, to admit larger animals than the size of the serpents would lead one to suppose.

Such examples, it must be admitted, are not very familiar ones; but there is an experiment not much unlike them, that most people have witnessed, and not a few perform it in their own persons almost daily. In every age people have been fond of oysters, and numbers of our citizens often luxuriate on a finer and larger species than those which Roman epicures formerly imported from Britain. Now, when a gentleman indulges in this food in the ordinary way, he affords a striking illustration of the pressure of the atmosphere. A large one is opened by the restaurateur, who also loosens the animal from its shell, and presents it on one half of the latter. The imitator of the boa then approaches his lips to the newly slain victim, and when they come in contact with but a portion of it, he immediately dilates his chest as in the act of inspiration, when the air, endeavoring to rush into his mouth to inflate the thorax, drives the oyster before it, and with a velocity that is somewhat alarming to an inexperienced spectator. If any one should doubt this to be effected by atmospheric pressure, let him fully inflate his lungs previous to attempting thus to draw an oyster into his mouth, and he will find as much difficulty to accomplish it as to smoke a pipe or a cigar with his mouth open.

This philosophical mode of transmitting oysters to the stomach is identical in principle with that proposed by Guericke and Papin, for shooting bullets "by the rarefaction of air."[a] A leaden ball was fitted into the breech of a gun-barrel, and the end being closed, a vacuum was produced in front of it; after which the atmosphere was allowed to act suddenly on the ball, when it was driven through the tube with the velocity of a thousand feet in a second. Just so with the oyster: it lays inertly at the orifice of the devourer's mouth—a partial vacuum is made in front of it by the act of respiration, and on dilating the chest, the atmosphere drives it in a twinkling down the natural tube in the throat—though, to be sure, with a velocity somewhat less than that of bullets through Papin's gun.

When two substances, impervious to air, are fitted so close as to exclude it from between them, they are held together by its pressure on their outsides, and with a force proportioned to the extent of the surfaces in contact. Pieces of metal have been ground together so close as to be thus united. Two pieces of common window-glass dipped in water, and pressed together, are separated with difficulty; because the water serves to expel the air, and prevent its entrance. Glass grinders are frequently inconvenienced by this circumstance. If two plates of glass were perfectly plane and smooth, so as wholly to exclude the air from between them, they would become united as one. We have heard or read of instances when they have become actually one, and were cut by a diamond as an ordinary single plate.

The boys' 'sucker,' or 'cleaver,' a circular piece of wet and thick leather, about the size of a dollar, is another illustration. This, when pressed against a smooth paving or other stone, of five or ten lbs. weight, may be used to raise it, by means of a string attached to the centre. If one of these, four inches in diameter, were applied to the cranium of a bald-headed gentleman, he might be elevated and suspended by it. Dr. Arnott recommends them to elevate depressed portions of fractured skulls,

---

[a] Phil. Trans. Abridg. vol. i, 496.

and for other surgical operations. Possibly they might be applied with some advantage to the soft and yielding skulls of infants, in order to produce those eminences upon which (according to phrenologists) the habits and character of individuals depend, as by means of them the most desirable organs of thought and passion might be developed, and the opposite ones depressed.

The principle of atmospheric pressure has been introduced by the great Parent of the universe into every department of animated and inanimate nature. Not only does it perform an important part in the vegetable kingdom, but the movements of innumerable animals, on land and in water, depend upon it; while others are enabled by it to protect themselves from enemies, and to secure their food and their prey. There is something inexpressibly pleasing in examining even the meanest specimens of the Creator's workmanship, (if such an expression may be allowed) and what is singular, the more closely we search into them, the more proofs do we meet with that the most elaborate and the most efficient of our devices are but rough copies of natural ones, which the lower animals vary and apply, according to circumstances, with inimitable dexterity. Some of these will be noticed here; others will be more appropriately introduced in subsequent chapters.

The feet of the common house-fly are constructed like the suckers above named; and hence these insects are enabled to run along, and even sleep, on the ceilings of our rooms, with their bodies hanging downwards. When in an inverted position they place a foot on an object, they spread out the sole, to make it touch at every part, so as to exclude the air from between; and when the weight of the body tends to draw it away, the pressure of the external air retains it; until the fly, wishing to move, raises the edges by appropriate mechanism, and destroys the vacuum.

There is not a more interesting subject for the contemplation of mechanics than the movements of these active little beings. To behold them running not only along the under side of a plate of glass, but also up, and more particularly *down* a vertical one, with such perfect command over their motions, is truly surprising. In the latter case, from the rapidity of their movements, and the fact that part only of their feet are in contact with the glass at the same time, one might suppose the momentum of their moving bodies would carry them over the objects they intended to reach; instead of which, they dart along with a precision and facility as if impelled by volition alone. It is strange, too, how they are enabled to produce a sufficient vacuum between their tiny feet and the asperities on an ordinary wall or ceiling! And with what celerity it is done and undone! How wonderful and how perfect must be the mechanism of these natural air-pumps; and how harmonious must that machinery work by which the energy of the insect is transmitted to them! Their movements when on the wing present another source of pleasing research. Let any ingenious person witness, without admiration if he can, a few of them in a door-way open to the sun: one or two will be found floating in the centre, as if at rest, until disturbed by the near approach of another, when they dart upon it, either in play or in anger, and drive it away; then resuming their stations, they remain as guards upon duty, till called to eject other intruders. In these combats they vary their movements into every imaginable direction; they trace in the air every angle and every curve, and change them with the velocity of thought. As they are not furnished (like most fishes and birds) with rudders in their tails, to assist in thus changing their positions, but effect it by modifying the action of their wings, how energetic must be the force that works these! And what perfect command must the insect possess over them! It would seem as if they turned their bodies in various directions, by diminishing the velocity of one wing, and increasing that of the other; and also by varying the angle at which they strike the air, and *descend* by closing them or stopping their vibrations. And with what vigor and celerity must one of these insects move its delicate wings in order to *elevate* its comparatively heavy body! Yet this movement is made quick as the others. It bounds upwards, like a balloon released from its cords; now sailing through a room, sweeping round our heads, buzzing at our ears, skimming over the floor, and anon inverting its body and resting on the ceiling! And all this within two or three seconds of time, and without any apparent exertion or fatigue. Here is a fruitful subject of inquiry to the machinist and aëronaut. All the wonders that the automatons of Maelzel and Maillardet ever wrought, are nothing compared to those that may yet be accomplished by studying the organization and motions of these living machines.

But there are larger animals than flies that suspend themselves in an inverted position. Mr. Marsden, in his "History of Sumatra," (London, 1811, p. 119) mentions *lizards* four inches long, which, he observes, are the largest reptiles that can walk in an inverted situation. One of them, of size sufficient to devour a cockroach, runs on the ceiling of a room, and in that situation seizes its prey with the utmost facility. Sometimes, however, when springing too eagerly at a fly, they lose their hold, and drop to the floor.

The *Gecko* of Java and other countries is furnished with similar apparatus in its feet, by means of which it runs up the smoothest polished walls, and even carries a load with it, equal in weight to that of its own body. Osbeck mentions lizards in China that ran up and down the walls with such agility as "they can scarce be caught." The *tree frog* of this country adheres to the leaves of trees by the tubercles on its toes: a young one has sustained itself in an inverted position against the under side of a plate of glass. From the observations of E. Jesse, author of 'Gleanings of Natural History' it appears that common frogs can occasionally do the same. His account is very interesting: "I may here mention a curious observation I made in regard to some frogs that had fallen down a small area which gave light to one of the windows of my house. The top of the area being on a level with the ground, was covered over with some iron bars through which the frogs fell. During dry and warm weather when they could not absorb much moisture, I observed them to appear almost torpid; but when it rained they became impatient of their confinement, and endeavored to make their escape, which they did in the following manner. The wall of the area was about five feet in height and plastered and white-washed as smooth as the ceiling of a room; upon this surface the frogs soon found that their claws would render them little or no assistance; they therefore contracted their large feet so as to make a hollow in the centre, and by means of the moisture which they had imbibed in consequence of the rain, they contrived to produce a vacuum, so that by the pressure of the air on their extended feet, (in the same way that we see boys take up a stone by means of a piece of wet leather fastened to a string) they ascended the wall and made their escape. This happened constantly in the course of three years."—Phil. Ed. 1833. p. 140.

Innumerable crustaceous animals adhere to rocks and stones by the same principle. But it is not the smaller inhabitants of either the land or the sea, as flies, spiders, butterflies, bees, &c. some of which scarcely weigh a grain; or lizards and frogs, &c. of five or six ounces, which thus sustain themselves against gravity; for the enormous walrus, that sometimes exceeds a ton in weight, is furnished by the Creator with analogous apparatus in his hinder feet; and thus climbs by atmospheric pressure, the glassy surfaces of ice-bergs. How forcibly do these examples illustrate the intimate connection which subsists between the various departments of Natural Philosophy. A knowledge of one *always* furnishes a key (whether it be used or not) to open some of the mysteries of another. Thus a person who understands the principle by which water is raised in a simple pump, can by it explain some of the most surprising facts in the natural history of animals; and solve problems respecting the motions and organs of motions of numerous tribes of animated beings, which two or three centuries ago, the most enlightened philosophers could not comprehend. And with a simple pump, he can moreover determine, as with a barometer, the measurement of all accessible heights, and with a degree of accuracy that, in some cases, is deemed preferable to geometrical demonstrations.

When two substances are brought together, at some distance *below* the surface of water, and so as to exclude it from between them, they are then pressed together with a force greater than when in the air, because the weight of the perpendicular column of water over them is then added to that of the atmosphere. Numerous examples of this combined pressure are also to be found in the natural world. By it, various species of fish adhere to rocks and stones in the depths of the sea, from which they cannot be separated except by tearing their bodies asunder. Some by means of it attach themselves to the bodies of others, and thereby traverse the ocean without any expense or exertion of their own, somewhat like dishonest travelers, who elude the payment of their fare. There are several species of fishes known which have a separate organ of adhesion, and there are doubtless many more which have not yet come under the observation of man. The most celebrated is the remora or *sucking* fish of Dampier and other navigators. It is, in size and shape, similar to a large whiting, except that the head is much flatter. "From the head to the middle of its back, (observes Dampier) there groweth a sort of flesh of a hard gristly substance, like that of the *limpet*. This excrescence is of a flat oval form, about 7 or 8 inches long, and 5 or 6 broad, and rising about half an inch high. It is full of small ridges with which it will fasten itself to any thing that it meets with in the sea. When it is fair weather and but little wind, they will play about a ship, but in blustering weather, or when the ship sails quick, they commonly fasten themselves to the ship's bottom, from whence neither the ship's motion, though never so swift, nor the most tempestuous sea can remove them. They will likewise fasten themselves to any bigger fish, for they never swim fast themselves, *if they meet with any thing to carry them*. I have found them sticking to a shark after it was hal'd in on deck, though a shark is so strong and boisterous a fish, and throws about him so vehemently when caught, and for half an hour together, that did not the sucking fish stick at no ordinary rate, it must needs be cast off by so much violence."[a] They are familiar to most of our seamen. Other species have a circular organ of adhesion, consisting of numerous soft papillæ, and placed on the thorax, instead of the top of the head, as in the remora. In some fish the ventral *fins* are united and are capable of adhesion. In the lamprey the *mouth* contracts and

---

[a] Dampier's Voyages, vi. edit. 1717, Vol. i, 64, and Vol. ii, part iii, p. 110. In the the plates of Vol. iii, is a figure of one. Figures of the excrescence or sucking part of the remora, and of the feet of the house-fly, may be seen in Dr. Brewster's Letters on Natural Magic.

acts as a sucker; while that curious animal the cuttle fish secures the victims that fall into its fatal embraces by the suckers on its *arms*.

The prodigious pressure that, at great depths, unites these inhabitants of the sea to their prey, led man to employ them to hunt the sea for his benefit as well as their own. Both the remora and lamprey tribe have been used for this purpose. Columbus when on the coast of St. Domingo was greatly surprised on beholding the Indians of that island fishing with them. "They had a small fish, the flat head of which was furnished with numerous suckers, by which it attached itself so firmly to any object as to be torn in pieces rather than abandon its hold. Tying a long string to the tail, the Indians permitted it to swim at large: it generally kept near the surface till it perceived its prey, when darting down swiftly it attached itself to the throat of a fish, or to the under shell of a tortoise, when both were drawn up by the fisherman." Ferdinand Columbus saw a shark caught in this manner.[a]

The same mode of fishing was followed at Zanguebar, on the eastern coast of Africa. The inhabitants of the coast when fishing for turtle, " take a living sucking fish or *remora*, and fastening a couple of strings to it, (one at the head and the other at the tail) they let the sucking fish down into the water on the turtle ground, among the half grown or young turtle; and when they find that the fish hath fastened himself to the back of a turtle, as he will soon do, they draw him and the turtle up together. This way of fishing as I have heard is also used at Madagascar."[b]

The remora was well known to the ancients. History has preserved a fabulous account of their having the power to stop a vessel under sail, by attaching themselves to her rudder. A Roman ship belonging to a fleet, it is said, was thus arrested, when she " stoode stil as if she had lien at anker, not stirring a whit out of her place." There is another illustration of the enormous pressure that fishes endure at great depths. The small volume of air that is contained in the bladder, and by the expansion and contraction of which they ascend and descend, is at the bottom of the sea compressed into a space many times smaller than when they swim near the surface. (At 33 feet from the surface it occupies but one half.) Hence, it frequently occurs that when such fish are *suddenly* drawn up, (as the cod on the banks of Newfoundland) the membrane bursts, in consequence of the diminished pressure, and the air rushing into the abdomen, forces the intestines out of the mouth. From a similar cause, blood is forced out of the ears of divers, when the bell that contains them is quickly drawn up. This pressure is also evinced in the fact that the timber of foundered vessels never rises, because the pores become completely filled with water by the pressure of the superincumbent mass, and the wood then becomes almost 'heavy as iron.'

The pressure of the atmosphere on *liquids* is equally obvious. When a bucket or other vessel is sunk in water and then raised in an inverted position, the air being excluded from acting on the surface of the liquid within, still presses on that without, so that the water is suspended in the vessel; and if the under surface of the liquid could be kept level and at rest, water might be transported in buckets thus turned upside down, as effectually as in the ordinary mode of conveying it

The experiment with a goblet or tumbler presents a very neat illustration. One of these filled with water, and having a piece of writing paper laid over it, and held close till the vessel be inverted, will retain the liquid

---

[a] Irving's Columbus, Vol. i, 273.   [b] Dampier's Voyages, Vol. ii, part ii, 108.

within it. In this experiment the paper merely preserves the liquid surface level: it remains perfectly free and loose; and so far from being close to the edge of the glass, it may, while the latter is held in a horizontal position, be withdrawn several lines from it without the water escaping; and it may be pierced full of small holes with the same effect.

If an inverted vessel be filled with any material that excludes the air, and whose specific gravity is greater than that of water, when lowered into the latter, the contents will descend and be replaced by the water. A bottle filled with sand, shot, &c. and inverted in water, will have its contents exchanged for the latter. As these substances, however, do not perfectly *fill* the vessel, and of course do not exclude all the air, the experiment succeeds better when the vessel contains heavy liquids, as mercury, sulphuric acid, &c. It is said that negroes in the West Indies often insert the long neck of a bottle filled with water, into the bung-holes of rum puncheons, when the superior gravity of the water (in this case) descends, and is gradually replaced with the lighter spirit.

In the preceding examples and those in subsequent chapters, it will be found that wherever a vacuity or partial vacuum is formed, the adjacent air, by the pressure above, rushes in and drives before it the object that intervenes, until the void is filled. If the nozzle of a pair of bellows be closed, either by the finger or by a small valve opening outwards; and a short pipe, the lower end of which is placed in water, be secured to the opening in the under board which is covered by the clapper; then if the bellows be opened, the pressure of the atmosphere will drive the water up the pipe to fill the enlarged cavity, and by then closing the boards, the liquid will be expelled through the nozzle. Bellows thus arranged become sucking or atmospheric, and forcing pumps. When the orifice of a syringe is inserted into a vessel of water and the piston drawn up, the air having no way to enter the vacuity thus formed than by the small orifice under the surface of the liquid, presses the water before it into the body of the syringe.

As every machine described in this book, and most of those in the next one, both proves and illustrates atmospheric pressure on liquids, we need not enlarge further upon it here. There are however some other particulars relating to it, which are necessary to be known: first, that its pressure is limited; and secondly, that it varies in intensity at different parts of the earth, according to their elevation above the surface of the sea. These important facts are clearly established in the accounts given of the discovery of the air's pressure, a sketch of which can scarcely be out of place here, since it was a *pump* that first drew the attention of modern philosophers to the subject, and which thereby became the proximate cause of a revolution in philosophical research, that will ever be considered an epoch in the history of science.

CHAPTER II.

Discovery of atmospheric pressure—Circumstances which led to it—Galileo—Torricelli—Beautiful experiment of the latter—Controversy respecting the results—Pascal—his demonstration of the cause of the ascent of water in pumps—Invention of the air-pump—Barometer and its various applications—Intensity of atmospheric pressure different at different parts of the earth—A knowledge of this necessary to pump-makers—The limits to which water may be raised in atmospheric pumps known to ancient pump-makers.

In the year 1641, a pump-maker of Florence made an atmospheric, or what was called a *sucking* pump, the pipe of which extended from 50 to 60 feet above the water. When put in operation, it was of course incapable of raising any over 32 or 33 feet. Supposing this to have been occasioned by some defect in the construction, the pump was carefully examined, and being found perfect, the operation was repeated, but with the same results. After numerous trials, the superintendent of the Grand Duke's water works, according to whose directions it had been made, consulted Galileo, who was a native of the city, and then resided in it. Previous to this occurrence, it was universally supposed that water was raised in pumps by an occult power in nature, which resisted with considerable force all attempts to make a void, but which, when one was made, used the same force to fill it, by urging the next adjoining substance, if a fluid, into the vacant space. Thus in pumps, when the air was withdrawn from their upper part by the '*sucker*,' nature, being thus violated, instantly forced water up the pipes. No idea was entertained by philosophers at this or any preceding period, that we know of, that this force was *limited;* that it would not as readily force water up a perpendicular tube, from which the air was withdrawn, 100 feet high as well as 20—to the top of a high building as well as to that of a low one.

When the circumstances attending the trial of the pump at Florence were placed before Galileo, (his attention having probably never before been so closely directed to the subject) he could only reply, that nature's abhorrence to a vacuum was limited, and that it " ceased to operate above the height of 32 feet." This opinion given at the moment, it is believed was not satisfactory to himself; and his attention having now been roused, there can be no doubt that he would have discovered the real cause, had he lived, especially as he was then aware that the atmosphere did exert a definite pressure on objects on the surface of the earth. But at that period this illustrious man was totally blind, nearly 80 years of age, and within a few months of his death. The discovery is however, in some measure, due to him. It has also been supposed that he communicated his ideas on the subject to Torricelli, who lived in his family and acted as his amanuensis during the last three months of his life.

It was in 1643 that Torricelli announced the great discovery that water was raised in pumps by the pressure of the air. This he established by very satisfactory experiments. The apparatus in his first one, was made in imitation of the Florentine pump. He procured a tube 60 feet long, and secured it in a perpendicular position, with its lower end in water; then having by a syringe extracted the air at its upper end, he found the water rose only 32 or 33 feet, nor could he by any effort induce it to ascend higher. He then reduced the length of the pipe to 40 feet, without any better success. It now occurred to him, that if it really was the atmosphere which supported this column of water in the pipe, then, if he employed some other liquid, the specific gravity of which, compared with that of water, was *known*, a column of such liquid would be sustained in the tube, *of a length proportioned to its gravity*. This beautiful thought he soon submitted to the test of experiment, and by a very neat and simple apparatus.

Quicksilver being 14 times heavier than water, he selected it as the most suitable, since the apparatus would be more manageable; and from the small dimensions of the requisite tube, a syringe to exhaust the air could be dispensed with. He therefore took a glass tube about four feet long, sealed at one end and open at the other. This he completely filled with quicksilver, which of course expelled the air; then placing his finger on the open end, he inverted the tube, and introduced the open end below the surface of a quantity of mercury in an open vessel; then moving the tube into a vertical position, he withdrew his finger, when part of the mercury descended into the basin, leaving a vacuum in the upper part of the tube, while the rest was supported in it at the height of about 28 inches, as he had suspected, being one-fourteenth of the height of the aqueous column. This simple and truly ingenious experiment was frequently varied and repeated, but always with the same result, and must have imparted to Torricelli the most exquisite gratification.[a]

Accounts of Torricelli's experiments were soon spread throughout Europe, and every where caused an unparalleled excitement among philosophers. This was natural, for his discovery prostrated the long cherished hypothesis of nature's abhorence of a vacuum; and at the same time, opened unexplored regions to scientific research. It met however with much opposition, particularly from the Jesuits; in many of whom it is said to have excited a degree of 'horror' similar to that experienced by them on the publication of Galileo's dialogues on the Ptolemaic and Copernican systems. They and others resisted the new doctrine with great perseverance, and even endeavored to reconcile the results of the experiments with the *fuga vacui* they so long had cherished. It was ingeniously contended that the experiment with quicksilver no more proved that the force which sustained it in the tube was the pressure of the atmosphere, than the column of water did in the first experiment; allowing this, it proved that this force, whatever it was, varied in its effects on different liquids, according to their specific gravity; a fact previously unknown, and apparently inconsistent with nature's antipathy to a void, which might be supposed to produce the same effects on all fluids—to have as great an abhorence to mercury as to water.

During the discussion great expectations were entertained by the advocates of the new doctrine from Torricelli; but unfortunately, this philosopher died suddenly in the midst of his pursuits and in the very vigor of manhood, viz. in his 39th year. This took place in 1647. The subject was however too interesting, and too important in its consequences, to be lost sight of. He had opened a new path into the fields of science, and philosophers in every part of Europe had rushed into it with too much ardor to be stopped by his decease. Among the most eminent of those

---

[a] The apparatus employed in these experiments was not original with Torricelli. The air thermometer of C. Drebble, the famous alchemist, who died in 1634, was of the same construction, except that the upper end of the inverted tube was swelled into a bulb. It is frequently figured in Fludd's works.

was Pascal, a French mathematician and divine. In 1646 he undertook to verify the experiments of Torricelli, and still further to vary them. He used tubes of glass forty feet long, having one end closed to avoid the use of a syringe. He filled one with wine and another with water, and inverted them into basins containing the same liquids, after the manner of Torricelli's mercurial experiment. As the specific gravity of these liquids was not the same, he anticipated a difference in the length of the two columns; and such was the fact. The water remained suspended at the height of thirty-one feet one inch and four lines; while the lighter wine stood at thirty-three feet three inches. Pascal was attacked with great virulence by Father Noel, a Parisian jesuit, who resisted the new doctrine with infuriate zeal, as if it also was heresy, like Galileo's doctrine of the earth's motion round the sun.

After making several experiments, one at length occured to Pascal, which he foresaw would, if successful, effectually silence all objectors. He reasoned thus: If it is really the weight or pressure of the atmosphere, that sustains water in pumps, and mercury in the tube, then, the *intensity* of this pressure will be *less* on the top of a mountain than at its foot, because there is a less portion of air over its summit than over its base; if therefore a column of mercury is sustained at 28 or any other number of inches at the base of a very high mountain, this column ought to diminish gradually as the tube is carried up to the top; whereas, if the atmosphere has no connection with the ascent of liquids, (as contended) then the mercury will remain the same at all elevations, at the base as at the summit. Being at Paris, he addressed a letter to his brother-in-law, M. Perrier, (in 1647) from which the following is an extract: "I have thought of an experiment, which, if it can be executed with accuracy, will alone be sufficient to elucidate this subject. It is to repeat the Torricellian experiment several times in the same day, with the same tube, and the same mercury; sometimes at the foot, sometimes at the summit of a mountain five or six hundred fathoms in height. By this means we shall ascertain whether the mercury in the tube will be at the same or a different height at each of these stations. You perceive without doubt that this experiment is decisive; for if the column of mercury be lower at the top of the hill than at the base, as I think it will, it clearly shows that the pressure of the air is the sole cause of the suspension of the mercury in the tube, and not the horror of a vacuum; as it is evident there is a longer column of air at the bottom of the hill than at the top; but it would be absurd to suppose that nature abhors a vacuum more at the base than at the summit of a hill. For if the suspension of the mercury in the tube is owing to the pressure of the air, it is plain it must be equal to a column of air, whose diameter is the same with that of the mercurial column, and whose height is equal to that of the atmosphere, from the surface of the mercury in the basin. Now the base remaining the same, it is evident the pressure will be in proportion to the height of the column, and that the higher the column of air is, the longer will be the column of mercury that will be sustained." This *experimentum crucis*, was made on the 19th September, 1648, the year after Torricelli's death, on the Puy de Dome, near Clermont, the highest mountain in France; and the result was just as Pascal had anticipated. The mercury fell in the tube as M. Perrier ascended with it up the mountain, and when he reached the summit it was three inches lower than when at the base. The experiment was repeated on different sides of the mountain, and continued by Perrier till 1651, but always with the same results. Pascal made others on the top of some of the steeples in Paris; and all proved the same important truth, viz. that the *pressure* of the atmosphere was that mysterious power, which under the name of nature's abhorrence to a vacuum had so long eluded the researches of philosophers. The subject was taken up in England by Boyle, who pursued it with unremitted ardor, and whose labors have immortalized his name; but it was Germany that bore off the most valuable of the prizes which the discovery offered to philosophers. The Torricellian experiment *gave rise to the* AIR PUMP; and in 1654, a Prussian philosopher, a mathematician and a magistrate, *Otto Guerricke*, of Magdeburgh, made public experiments with it at Ratisbon, before the emperor of Germany and several electors. Some authors ascribe the *invention* of the pump to Candido del Buono, one of the members of the Academie del Cimento at Florence, and intimate that the first essays with it were only made by Guerricke.

The apparatus of Torricelli, i. e. the glass tube and basin of mercury, was named a baroscope, and afterwards a barometer, because it *measured* the pressure of the atmosphere at all elevations; hence to it, engineers in all parts of the earth may have recourse, to determine the perpendicular length of the pipes of atmospheric pumps.

Another application of the barometer was the natural result of Perrier's first experiment on the Puy de Dome. As he ascended that mountain with it, the mercury kept falling in exact proportion to the elevation to which the instrument was carried; hence it is obvious, that when the tube is properly graduated, it will measure the height of mountains, and all other elevations to which it can be carried. By it, aeronauts determine the height to which they ascend in balloons. The observations of Perrier were continued daily from 1649 to 1651, during which he perceived that the height of the column slightly varied with the temperature, wind, rain, and other circumstances of the atmosphere; and hence the instrument indicated *changes of weather*, and became known and is still used as a "*weather glass.*" The extent of these variations is about three inches, generally ranging from twenty-eight to thirty-one, and are principally confined to the temperate zones. In tropical regions, the pressure is nearly uniform, the mercury standing at about thirty inches throughout the year. These facts have an important bearing on our subject; for an atmospheric pump or siphon, with a perpendicular pipe thirty-four or thirty-five feet long, might operate during certain states of the atmosphere, while in others it could not; and in some parts of the earth it would be altogether useless.

It will appear in the sequel, that the physical properties of the atmosphere which we have enumerated, must necessarily be understood, in order perfectly to comprehend the action of the machines we have to describe. As regards the aërial pressure, its limits and variation at different altitudes, we need only remark, that a sucking pump or a siphon, which raises water thirty-three feet in New-York and Buenos Ayres, London and Calcutta, St. Petersburgh and Port Jackson in New Holland, could not, in the city of Mexico, elevate it over twenty-two feet; and at Quito, and Santa Fe de Bogota in South America, and Gondar the capital of Abyssinia twenty feet, on account of the great elevation of these cities; (from the same cause, the pressure of the atmosphere on Mont Blanc is only about half that on the plains) and if Condamine and Humboldt, when on the summit of Pinchincha, had applied one to raise water there, or on the side of Antisana, at the spot where, from the great rarity or tenuity of the air, the face of the latter philosopher was streaming with blood, his attendant fainted, and the whole party exhausted, it would not have raised water over twelve or fourteen feet; (the mercury in the barometer fell to fourteen inches seven lines,) while on the highest ridge of the Himalayas, it would scarcely raise it eight or ten feet. Without a knowledge of aërial pressure, it is obvious, that engineers who visit Mexico, and the upper regions of South America, &c. might get into a quandary greatly more perplexing than that in which the Florentine was, when he applied to Galileo: but we believe the period has nearly gone by, for mechanics to remain ignorant of those principles of science, upon which their professions are based.

It perhaps may be asked, Were the limits to which water can be raised by the atmosphere not known before Galileo's time? Undoubtedly they were.. Pump makers must always have been acquainted with them; although philosophers might not have noticed the fact or paid any attention to the subject. Why then did the Italian artists make such a one as that to which we have referred? Simply because they were ordered to do so, as any mechanic would now do under similar circumstances: at the same time they declared that it would not raise the water, although they could not assign any reason for the assertion. It was indeed impossible for ancient pump makers to have remained ignorant of the extent to which their machines were applicable. A manufacturer of them would naturally extend their application, as occasions occurred, to wells of every depth, until he became familiar with the fact that the power which caused the water to ascend, was limited—and until he detected the limits. After using a pump with success, to raise water twenty-five or thirty feet, when he came to apply it to wells of forty or fifty feet in depth without lengthening the cylinder, he would necessarily learn the important, and to him mysterious fact, that the limits were then exceeded: and after probably going through similar examinations and consultations, as those which took place at Florence in the 17th century, the unvarying result would become so firmly established, that every workman would learn it traditionally, as an essential part of his profession: and if in succeeding ages, the knowledge of it became lost, the experience of every individual pump maker must have soon taught him the same truth. Attempts then similar to those of the Florentine engineer occurred frequently before, but leading to no important result, the particulars of them have not been preserved; nor is it probable that those relating to the Italian experiment would have been, had not the father of modern philosophy been consulted, and had not his pupil Torricelli taken up the subject.

CHAPTER III.

Ancient Experiments on air—Various applications of it—Siphons used in ancient Egypt—Primitive experiments with vessels inverted in water—Suspension of liquids in them—Ancient atmospheric sprinkling pot—Watering Gardens with it—Probably referred to by St. Paul and also by Shakespeare—Glass sprinkling vessel, and a wine taster from Pompeii—Religious uses of sprinkling pots among the ancient heathen—Figure of one from Montfaucon—Vestals—Miracle of Tutia carrying water in a sieve, described and explained—Modern liquor taster and dropping tubes—Trick performed with various liquids by a Chinese juggler—Various frauds of the ancients with liquids—Divining cups.

Notwithstanding the alledged ignorance of the ancients respecting the physical properties of the atmosphere, there are circumstances related in history which seem to indicate the reverse; or which, at any rate, show that *air* was a frequent subject of investigation with their philosophers, and that its influence in some natural phenomena was well understood. Thus Diogenes of Apollonia, the disciple and successor of Anaximenes, reasoned on its *condensation* and *rarefaction*. According to Aristotle, Empedocles accounted for respiration in animals by the *weight* of the air, which said he, " by its *pressure* insinuates itself with force into the lungs." Plutarch expresses in the very same terms, the sentiments of Asclepiades, representing him among other things as saying that the external air, " *by its weight*, opened its way with force into the breast." Lucretius, in explaining the property of the loadstone in drawing iron, observed that it repelled the intervening air betwixt itself and the iron, thus forming a vacuum, when the iron is " *pushed on by the air behind it*." Plutarch was of the same opinion.ᵃ Vitruvius speaks of the pressure of air, Arch. book viii, cap. 3. When Flaminius, during the celebration of the Isthmian games, proclaimed liberty to the Greeks, the shout which the people gave in the transport of their joy was so great, that some crows, which happened to be then flying over their heads, fell down into the theatre. Plutarch among other explanations of the phenomenon, suggests, that the " sound of so many voices being violently strong, the parts of the air were separated by it, and a void left which afforded the birds no support."ᵇ

But if the ancients did not detect or comprehend the direct pressure of the atmosphere, they were not ignorant of its effects, or of the means of exciting it. They in fact employed air in many of their devices as successfully as the moderns. They compressed it in air guns, and weighed it in bladders; its elasticity produced continuous jets in their fountains and force pumps, in the same manner as in ours; by its pressure, they raised water in syringes and pumps, and transferred it through siphons, precisely as we do; they excluded its pressure from the upper surface of water in their sprinkling pots, and admitted it to empty them, as in the modern liquor merchant's *taster*. That they had *condensing* air pumps is evident from the *wind guns* of Ctesibius, as well as others described by Vitruvius, b. x, cap. 13; and it is probable that they employed air in numerous other machines and for other purposes, but of which, from the loss of their writings, no account has been preserved. See Vitruvius, book x, cap. 1, where some are referred to, and Pliny Nat. Hist. book xix, cap. 4.

It would be in vain to attempt to discover the *origin* of devices for raising liquids by the pressure of the atmosphere, for it would require a knowledge of man and of the state of the arts, in those remote ages of which no record is extant. That machines for the purpose were made long before the commencement of history is certain, for recent discoveries have brought to light the highly interesting fact, that *siphons* were known in Egypt 1450 years before our era, i. e. 3290 years ago! At which period too they seem to have been in more common use in that country, than they are at this day with us. [See Book V, for an account of these instruments.]

The retention of water in *inverted* vessels while air is excluded from them, could not have escaped observation in the rudest ages. Long ere natural phenomena had awakened curiosity in the human mind, or roused the spirit of philosophical inquiry and research, it must have been noticed. When a person immerses a bucket in a reservoir and raises it in an inverted position, he soon becomes sensible that it is not the weight of the vessel merely which he has to overcome, but also that of the water within

---

ᵃ Duten's Inquiry into the Origin of the Discoveries attributed to the Moderns. Lon 1769, pp. 186, 187 203. ᵇ Life of Flaminius.

it; and not till the mouth emerges into air do the contents rush out and leave the bucket alone in his hands. This is one of those circumstances that has occurred more or less frequently to most persons in every age. It would be absurd to suppose that the groups of oriental females who, from the remotest times, have assembled twice a day to visit the fountains or rivers for water, did not often perform the experiment, both incidentally and by design. They could not in fact plunge their water pots (which were often without handles) into the gushing fount without occasionally repeating it; nor could Andromache and her maids fill buckets to water the horses of Hector, and daily charge pitchers in the stream for domestic uses, without being sometimes diverted by it. But the phenomenon thus exhibited was not confined to such occasions; on the contrary, it constantly occurred in every dwelling. An ancient domestic like a modern house-maid, could hardly wash a cup or rinse a goblet by *immersion*, without encountering it. Besides the vessels named, there were others that formed part of the ordinary kitchen furniture of the ancients, (see figures of some on page 16) the daily use of which would vary and illustrate it. These were long necked and narrow mouthed vases and bottles, that retained liquids when inverted like some of our vials. Others were still further contracted in the mouth, as the *Ampulla*, which gave out its contents only by drops. To the ordinary use of these vessels and to incidental experiments made with them, may be traced the origin of our *fountain* lamps and inkstands, bird fountains, and other similar applications of the same principle.

The suspension of a liquid in inverted vessels by the atmosphere, was therefore well known to the early inhabitants of the world, whether they understood the reason of its suspension or not; and when in subsequent times philosophers began to search into causes and effects, the phenomenon was well calculated to excite their attention, and to lead them to inquiries respecting air and a vacuum : it is probable that it did so, for the earliest experiments on these subjects, of which we have any accounts, were similar to those domestic manipulations to which we have alluded, and the principal instrument employed was simply a modification of a goblet inverted in water. This was the atmospheric '*sprinkling pot*,' or 'watering siphon,' which is so often referred to by the old philosophers, in their disputes on a *plenum* and a *vacuum*. It has long been obsolete, and not having been noticed by modern authors, few general readers are aware that such an instrument was ever in use, much less that it formed part of the philosophical apparatus of the ancient world.

The interesting associations connected with it and its modifications entitle it to a place here. Indeed were there no other reason for attempting to preserve it a little longer from oblivion, than that indicated at the close of the last paragraph, we should not feel justified in passing it by. It is moreover, for aught that is known to the contrary, the *earliest* instrument employed in hydro-pneumatical researches. Its general form and uses may be gathered from the remarks of Athenagoras respecting it. This philosopher, who flourished in the fifth century, B. C. made use of it to illustrate his views of a vacuum. " This instrument (says he) which is acuminated or pointed towards the top, and made of clay or any other material, (and used as it often has, for the watering of gardens) is, in the bottom very large and plain [flat] but full of small holes like a sieve, but at the top has only one large hole."[a] When it was plunged in water, the liquid entered through the numerous holes in its bottom; after which the single opening at top was closed by the finger to exclude the air; the

---

[a] As quoted by Switzer from Bockler, Hyd. 167.

vessel and its contents were then raised, and the latter discharged at pleasure by removing the finger.

As this was the ancient *garden* pot of the Greeks, Pliny probably refers to it when he speaks of 'sprinkling' water, oil, vinegar, &c. on plants and roots.[a] It appears to have been continued in use for such purposes in Europe, through the middle ages; and to a limited extent up to the 17th and 18th centuries.[b] Figures of it are, however, rarely to be met with, for it seems to have been nearly forgotten when the discovery of Torricelli revived the old discussions on a vacuum; and though Boyle and others then occasionally referred to it, few, we believe, gave its figure.[c] Montfaucon speaks of examining an ancient 'watering stick,' and also a 'sprinkling pot,' but unfortunately he has not described either.[d] Of

a great number of old philosophical works that we have examined for the purpose of obtaining a figure, we met with it only in Fludd's works. The annexed cut is from his De Naturæ Simia seu Technica macrocosmi historia.' Oppenheim, 1618, p. 473. The mode of using it is too obvious to require explanation. It was pushed into water in the position represented; the liquid entered through the openings in the bottom, driving the air out of the small orifice at the top; and when it was filled, the person using it placed his finger or thumb on the orifice and then moved the vessel over the plants, &c. he wished to water; discharging the contents by raising the finger.

No. 69. Ancient Watering Pot.

The application of this instrument as a 'garden pot' may sometimes be found portrayed in devices, rebuses, vignettes, &c. of old printers. In the title page of Godwin's Annals of Henry VIII, Edward VI, and Mary, (a thin latin folio published in 1616) it is represented. No. 70 is a copy. There is a similar engraving on the title page of a volume on farming, &c. entitled 'Maison Rustique,' translated from the French, and published in London by John Islip, the same year.

No. 70. Watering plants with the Atmospheric Sprinkling Pot.

Independently of the sprinkling pot, the cut is interesting as exhibiting

---

[a] Nat. Hist. xvii, 11 and 28; xix, 12; and xv, 17.   [b] Dictionnaire De Trévoux, Art. Arrosoir.   [c] Boyle's Philosophical works, by Shaw, Lon. 1725. Vol. ii, pp. 140, 144
[d] Italian Diary, Lon. 1725, 295.

Sprinkler and Wine-Taster from Pompeii.

the ancient mode of transplanting. It appears that *two* men were generally employed in the operation; one to set the trees or plants, and another to water them; a custom to which St. Paul alludes in 1 Cor. iii, 6—8. Sometimes 'he that watered' used two pots at the same time, holding one in each hand. As these vessels were not wholly disused in Shakespeare's time, it is probable, that to them he refers in Lear:

> Why, this would make a man, a man of salt, [tears]
> To use his eyes for garden water pots.   *Act* 4, *Scene* 6.

Modifications of them were adapted to various purposes by the ancients. They were used to drop water on floors in order to lay the dust, in both Greek and Roman houses. Their general form was that of a pitcher or vase, and their dimensions varied with their uses. Some of the smallest had but a single hole in the bottom. They formed part of the ordinary culinary apparatus, and were also used in religious services. Among the antiquities disinterred at Pompeii, some have been found. No. 71 represents one: it is of glass, the upper part of the tube or neck is wanting, having been broken off. Perhaps this part resembled the form indicated by the dotted lines which we have added.

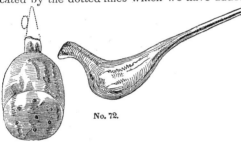

No. 72, also of glass, has been pronounced 'a wine-taster, the air having been exhausted by sucking at the small end.' It is more likely that the wide part was inserted into wine jars or amphoræ, and the cavity filled with that liquid precisely as in the sprinkling pot, and samples then withdrawn by closing the small orifice with the finger as in the modern instrument, which is shewn at No 76, and as in the dropping tube, one form of which is figured at No. 77. The general form of No. 72 assimilates it to those drinking vessels of the ancients, which they held at a distance in front, and directed the stream issuing from the small end of the vessel into the mouth; a mode still practised in some parts of the Mediterranean, and by the natives of Ceylon, Sumatra, Malabar, &c.

N 71.   Sprinkler and Wine-taster both from Pompeii.

One of the most singular facts connected with the religious institutions of the ancient heathen, was the extent to which they carried the practice of sprinkling: almost every thing was thus purified; men, animals, trees, water, houses, food, clothing, carriages, &c. In performing the ceremony various implements were used to disperse the sacred liquid. A wisp made of horse hair attached to a handle was common. A branch from certain trees, and sometimes a small broom, were used; in other cases perforated vessels were employed. Thus the Bramins in some ceremonies take a vessel of water, and after presenting it to the gods, they sprinkle the liquid with manguier leaves, on carriages, animals, &c. in others it is "sprinkled through a cullender with a hundred holes on the head of the father, mother and child."[a] The priests of the ancient Scandinavians also used a vessel "prepared like a watering pot, with which they sprinkled the altars, the pedestals of their gods, and also the men."[b] The

---

[a] Sonnerat's Voyages. Vol. i, 134 and Vol. ii, 97.   [b] Snorro's History of Scandinavia The chapter upon sacrifices is translated in Anderson's 'Bee,' vol. xvi, p. 20, Edinburgh, 1793.

Jewish priests commonly used a branch of hyssop, but occasionally a piece of wool, and sometimes the fingers. "The priest shall dip his finger in the oil and sprinkle it." The Greeks and Romans had not only founts or vases of holy water in their temples for the use of worshippers, who dipped their fingers into them, as Roman catholics and others do at this day; but on particular occasions, priests or officers attended to purify the people by sprinkling. Thus, when the Emperor Julian visited the temple at Antioch, the Neocori stood on each side of the doorway to purify with lustral water all who entered. Valentinian, who was afterwards raised to the empire, was then captain of Julian's guard, and as such walked in front. He was then a christian, and some of the water having been thrown upon him, he turned and struck the priest, saying, that the water rather polluted than purified; at which the emperor was so enraged that he immediately banished him.

Now whether the fingers or light brooms, &c. were used on such occasions we do not know; but there were others at which the former certainly were not. When the emperors dined, not only their persons and table furniture, but the *food* also was purified with lustral water. At the feast of *Daphne* near Antioch, which lasted seven days, we learn that a neochorus stood by the emperor's seat, and sprinkled the dishes and meats 'as usual.' How was this water dispersed? Certainly not by the fingers; nor is it likely that a wisp or a broom was employed, since it would be difficult to direct the small shower with sufficient precision on the smaller objects. We have made these remarks for the purpose of introducing the figure of an ancient sprinkling vessel, from the third volume of Montfaucon's Antiquities. It was supposed by him to have belonged to the table or kitchen, but its specific use he could not conjecture. It is evidently a modification of the atmospheric garden pot, and it appears admirably adapted for dispersing liquid perfumes or lustral water at the table. The ring is adapted to receive the forefinger, while the thumb could close the small orifice, and thus the contents might be retained or discharged at pleasure.

No. 73. Roman Sprinkling Vase.

Among other heathen customs that were long retained in the christian church, was this practice of sprinkling. Peter Martyr exclaims against a certain class, " who not only consecrate temples themselves, but also altars and coverings to the altars; I meane the table clothes and napkins, and also the chalices and patins, the massing garmentes, the churchyardes, the waxe candles, the frankincense, the pascal lambe, eggs, and also holie water; the boughes of their palm trees, yong springes, grass, pot-hearbes, and finally all kinds of fruites." " They doe *sprinkle* houses, deade bodies, churchyardes, eggs, flesh, pothearbes, and garmentes."[a]

Of all the transactions connected with heathen theology, few ever made a greater noise in ancient Rome than one that is connected with this part of our subject; viz. the *miracle* by which Tutia the vestal saved her life. It was a religious custom among all the nations of old, to keep *sacred fire* in the temples of their deities. In some, lamps were kept burning, in others fuel kindled on the altars. In the temples of Jupiter-Ammon, Apollo, Minerva, and some other deities were lamps constantly burning. The Israelites were to cause the lamps to " burn continually," besides which, " the fire shall ever be burning upon the altar : it shall

---

[a] Common Places, part iv, cap. 9.

never go out." Levit. vi, 13. The practice is still kept up by the Jews and also by Roman catholics. The origin of the custom is unknown; but the Jews, Persians, Greeks, &c. are generally supposed to have derived it from the Egyptians. Upon the consecration of a temple, this 'holy fire' was not obtained from ordinary sources, i. e. from other fires, but was produced by the rubbing of two sticks together; or, according to Plutarch, was drawn directly from the sun. "If it happen (he observes in his Life of Numa) by any accident to be put out, as the sacred lamp is said to have been at Athens, under the tyranny of Aristion—at Delphi, when the temple was burned by the Medes—and at Rome in the Mithridatic, as also in the civil war, when not only the fire was extinguished but the altar itself overturned—it is not to be lighted again from another fire, but *new* fire is to be gained by drawing a pure and unpolluted flame from the beams of the sun. This is done generally with concave vessels of brass."

Among the Romans a certain number of virgins were consecrated with solemn ceremonies to the Goddess VESTA. They were named vestals, and it was their peculiar duty to take charge of the sacred fire. They were greatly honored for their purity and the importance of their office. "What is there in Rome, (exclaimed Tiberius Gracchus in his address to the people) so sacred and venerable as the vestal virgins who keep the perpetual fire?" The most valuable and sacred deposites were often placed in their hands for security. The wills of rich Romans were sometimes committed to their care; hence we read of Augustus forcing from them that of Antony, while the latter was in Egypt. The vestals enjoyed many privileges; among others, when they went abroad, the fasces (emblems of authority) were carried by a lictor before them; and it was death for any one to go under the litter or chair in which they were carried; and if they met a criminal going to execution, his life was spared. The vestal daughter of Appius Claudius protected him from being arrested by the Tribunes. On the other hand, they were punished with extreme rigor if found to have broken any of their vows. To permit the perpetual or holy fire to go out was an unpardonable act, for it was believed to betoken some national calamity, and if one was found guilty of unchastity she was buried alive. "The criminal (says Plutarch) is carried to punishment through the forum in a litter, well covered without, and bound up in such a manner that her cries cannot be heard. The people silently make way for the litter and follow it with marks of extreme sorrow and dejection. There is no spectacle more dreadful than this, nor any day which the city spends in a more melancholy manner. When the litter comes to the place appointed the officers loose the cords, the high priest with hands lifted up towards heaven offers some private prayers just before the fatal minute; then takes out the prisoner, who is covered with a veil, and places her upon the steps which lead down into the cell, [grave;] after this he retires with the rest of the priests, and when she is gone down, the steps are taken away and the cell is covered with earth, so that the place is made level with the rest of the mount." [Life of Numa, Langhorne's Trans.]

Tutia, who was accused of incontinence, in order to avoid the horrid penalty, passionately called, or affected so to call, upon the goddess VESTA, to establish *by a miracle* her innocence. "Enable me (she cried) to take a SIEVE full of water from the Tyber, and to carry it full to thy temple." Upon this appeal her trial was stayed, and it was left to the deity she had invoked, to save her or not; for such a proof of the falsehood of her accusers could not, if it should take place, be resisted. The result was, she succeeded in carrying the water, and thereby not only saved her life, but greatly increased her reputation for sanctity. From the imperfect accounts of the transaction that have reached us, it may perhaps be deemed presumptuous to decide on its real character. That it actually occurred there can be no doubt. It is incorporated with both the history and the arts of the Romans. It is mentioned by Valerius Maximus, by Pliny and Livy: representations of Tutia carrying the sieve were also embodied in sculptures, in statues, and engraved on gems. The annexed figure was copied from one of the latter. It is from the first volume of Montfaucon's Antiquities, Plate 28.

No. 74. Tutia carrying Water in Sieve.

As the feat therefore was certainly performed, it must have been either by natural or by supernatural means. Some writers have admitted, and St. Augustine among them, that the *miracle* was a genuine one; but there are circumstances sufficient to show that the whole was a well conceived and neatly executed trick, on the part of Tutia and her friends; and further, that it was a much more simple one, than other deceptions to which the heathen priests sometimes had recourse. It possesses considerable interest however as furnishing another specimen of their proficiency in scientific juggling and natural magic. To say nothing of the absurdity of admitting a divine interposition, in answer to invocations addressed to a heathen goddess— and of the improbability of Tutia being condemned while innocent; there certainly was something suspicious in *her* undertaking to select the *test* for the goddess, and especially such a one as that of carrying water in a sieve. Instead of asking for a sign by water, it would have been more appropriate and more natural in her (if sincere) to have prayed for one by *fire*—by that element which was the symbol of the deity she invoked, and which it was her peculiar duty to attend at the altar and preserve pure— the element too, which, if the accusation was true, she had polluted: besides, a token by fire was always considered by the heathen as the strongest evidence of divine approbation. What prompted her then to mention the test of the sieve? Doubtless because the device by which it was to be performed was already matured; not by the assistance of Vesta, but by a very simple contrivance furnished her by the priests, from their stores of philosophical and other apparatus with which they wrought their wonders before the people.

No. 75. Supposed construction of Tutia's sieve.

The contrivance was, we presume, a modification of the ancient sprinkling pot, just described. The sieve she employed would therefore be a double one; that is, its bottom and sides were hollow, the exterior bottom only being perforated, as in the annexed cut, which represents a double metallic vessel, the inner one being capable of holding water, and the upper edges of both united and made perfectly air tight, with the exception of one or perhaps two small openings shown on the edge in the figure. Thus when such a sieve was pressed slowly under water, the liquid would enter through the perforated bottom, drive the air before it, and fill the cavity; and when the upper part was sunk below the surface, the upper or apparent sieve would also be filled. Then by *covering the small opening with the thumb*, the vessel might be raised out of the river, the water in the cavity being suspended precisely as in Nos. 69 and 70, so that Tutia might return with it to the temple, and on approaching the altar, by imperceptibly sliding her thumb to *one side*, the air would enter the opening thus exposed, and the contents of the cavity would descend in a shower, to the amazement of the spectators and to the confusion of her adversaries. With such an instrument she might go with that confidence to the trial, which she is represented to have felt, being fully convinced of success. While she was in the act of carrying the water, the spectators would be unable to detect the slightest imposition, or if, from the elevation at which she seems to have borne it, the bottom of the sieve was exposed, it would be more likely to confirm them in the belief of the miracle, as her movements would cause the suspended water to appear at the openings; but it is more probable that they were kept at too great a distance by the managers of the farce, to afford them any opportunity of exercising an undue curiosity. And when the trial was over, the sieve would be secured by those in the secret, who would have one similar in appearance ready for examination whenever required.

Few devices are better adapted to demonstrate the suspension of water by the atmosphere, than those little instruments which chemists and dealers in ardent spirit use, to examine their various liquids. Those of the former are named 'dropping tubes,' from the small quantities they are designed to take up, and the latter 'liquor tasters:' both are substantially the same, for they differ merely in form and dimensions. Some curious experiments may be made with them. For example, a series of liquids similar in appearance but differing from each other in specific gravity, and such as do not readily mix, may be placed in a glass or other vessel, so as to form separate layers, the heaviest at the bottom, and the lightest reposing on the top. An expert manipulator may then by a taster (No. 76) withdraw a portion of each, and present to the examination of his audience from the same vessel, samples of different wines, ardent spirits, water, &c. There is reason to believe that the ancient professors of legerdemain were well acquainted with such devices. It is possible that the trick performed by a Chinese juggler before the Russian embassy at Pekin, in the last century, was of the kind. It is thus described by Mr. Bell: "The roof of the room where we sat was supported by wooden pillars. The juggler took a gimblet, with which he bored one of the pillars and asked whether we chose red or white wine? The question being answered, he pulled out the gimblet and put a quill in the hole, through which ran as from a cask the wine demanded. After the same manner he extracted several sorts of liquors, all which I had the curiosity to taste, and found them good of the kinds." Bell's Travels. Lon. 1764, vol. ii, 28.

No. 76. No. 77.
Liquor Taster and Dropping Tube.

Peter Martyr speaks of old jugglers that "devoure bread, and immediately spit out meale; and when they have droonke wine, they seem presentlie to poure the same out of the midst of their forehed."

There are numerous intimations in history that hydrodynamics was one of the most fruitful sources of scientific imposture, to which ancient magicians had recourse. Besides the sieve of Tutia, the cup of Tantalus, and the Divining cup, there were "the marvellous fountain, which Pliny describes, in the island of Andros, which discharged wine for seven days and water during the rest of the year—the spring of oil which broke out in Rome to welcome the return of Augustus from the Sicilian war—the three empty urns that filled themselves with wine at the annual feast of Bacchus, in the city of Ellis—the glass tomb of Belus which was full of oil, and which when once emptied by Xerxes could not again be filled—the weeping statues, and the perpetual lamps;—all the obvious effects of the equilibrium and pressure of fluids."

The cup of Tantalus will be ound described in the Chapter on Siphons in Book V. Divining cups may be noticed here, as there is reason to believe that water was suspended in some of them by atmospheric pressure; while in others, sounds were produced by the expulsion of air through secret cavities formed within them. Divination by water has prevailed from immemorial time, and in the eastern world, has been practised in a great variety of ways. Sometimes the inquirers into futurity performed the requisite ceremonies themselves, and with ordinary instruments, as when a mirror or looking-glass was used; (see page 34) at other times professional sorcerers were employed. These men, as a matter of course, provided their own apparatus, and hence had every opportunity in its construction of concealing within some part, the device upon which their deceptions turned.

Of all the implements connected with Hydromancy, cups are the most interesting. They are among the earliest that history has mentioned, (Genesis, xliv, 5,) and they have longer retained a place in the conjurer's budget than any other. They were used by astrologers of Europe during the middle ages, and are not yet wholly abandoned in that part of the world. Like all devices of the old magicians, ingenuity seems to have been exhausted in their formation and in adapting them to different species of jugglery. They were of various materials; while some were of silver like Joseph's, others were of wood, glass, stone, &c. according to the nature of the trick to be performed by them. Sometimes presages were drawn from observing the liquid through the *sides* of the cup; for this purpose it was made of a translucent material; but then one side was left *thick* while the others were thin, so that the contents were invisible through the former, but quite plain through the latter. The indications were considered favorable when the liquid was clear and distinctly seen, and unfavorable if the inquirer could not perceive it—thus either side was presented by the conjurer as best suited his views. The same trick is still performed in some of the churches in Italy; one side of the goblet or glass is made opaque, while the other is transparent. With other cups it was the motion or *agitation* of the liquid that was looked for: if it remained at rest, the omen was bad—if violently moved, good. This kind of divination most likely depended on legerdemain or 'sleight of hand,' in dropping unperceived some substance into the vessel that produced effervescence—or by opening a secret communication with a cavity in the stem or base of the vessel, containing a liquid that had a similar effect. In Japan it is common to place a pot of water on the head; if the liquid boil over, the presage is good, "but if it stirs not, bad luck."[a] Among the prodigies mentioned by Herodotus, is one of this kind: the flesh of a victim sacrificed during the Olympic games, was placed in brazen cauldrons, and "the water boiled up and overflowed without the intervention of fire," (B. i, 59.) The emerald cup, by which the priests of Mentz deluded people in the dark ages, belongs to the same class. On certain days, two or three extremely minute fishes were secretly put in, and by their motions in the water produced such an effect that the people were persuaded "the cup was alive."[b]

---

[a] Montanus' Japan, translated by Ogilby, Lon. 1670, p. 123.
[b] Misson's Travels, vol. i, 93. See also Moreri's Dict. vol, iv. Art. Augury.

The divining cups of the Assyrians and Chaldeans appear, from imperfect accounts of them extant, to have been more artificially contrived. When one was used, it was filled with water, a piece of silver or a jewel having certain characters engraved on it was thrown in; the conjurer then muttered some words of adjuration, when the demon thus addressed, it is said, "*whistled* the answer from the bottom of the cup." These vessels were probably so contrived, that the water might compress air concealed in some cavity in the base, and force it through the orifice of a minute reed or whistle, as in the musical bottles of Peru. As Julius Cyrenius says such cups were also used by the Egyptians, it is possible that it was one of them by which Joseph divined, or affected through policy to divine. Divination by the cup is still practiced in Japan.

It is well known that the jugglers of Asia have always been unrivalled. Even in modern times, some of their tricks are beautiful applications of science, and are so neatly performed as to baffle the most sagacious of observers. A full account of them would go far to explain all the *miracles* which ancient authors have mentioned, and would afford some curious information respecting the secrets of ancient temples.

CHAPTER IV.

Suction: Impossible to raise liquids by that which is so called—Action of the muscles of the thorax and abdomen in *sucking* explained—Two kinds of suction—Why the term is continued—Sucking poison from wounds—Cupping and cupping-horns—Ingenuity of a raven—Sucking tubes original atmospheric pumps—The Sanguisuchello—Peruvian mode of taking tea, by sucking it through tubes—Reflections on it.—New application of such tubes suggested—Explanation of an ambiguous proverbial expression.

AIR is expelled from such vessels as are figured in the last chapter by thrusting them into a liquid, which entering at the bottom, drives out as it rises the lighter fluid at the top. In the apparatus now to be described, it is withdrawn in a different manner. The vessels are not lowered into water, but the latter is forced up into them. The operation by which this is accomplished was formerly named *suction*, from an erroneous idea that it was effected by some power or faculty of the mouth, independently of any other influence. A simple experiment will convince any one that the smallest particle of liquid cannot be so raised :—fill a common flask or small bottle within a quarter of an inch of the top of the neck, and place it in a perpendicular position; then let a person apply his mouth over the orifice, and he may suck forever without tasting the contents; the veriest lover of ardent spirits would die in despair ere he could thus partake of his favorite liquor; and the exhausted traveler could never moisten his parched throat, although the liquid, as in the case of Tantalus, was at his lips.

As remarked in a previous chapter, the error was not exploded till Torricelli and Pascal's experiments proved that water is not raised in pumps by suction, or any kind of attraction, but by *pulsion* from aerial pressure. Suction therefore, or that which was so called, merely removes an obstacle [air] to a liquid's ascent—it does not raise it, nor even aid in the act of raising it. In other words, it is simply that action of the muscles of the thorax and abdomen which enlarges the capacity of the lungs and chest, so that air within them becomes rarefied and consequently no longer in equilibrium with that without—hence when in this state a communication is opened between them and a liquid, the weight of the atmosphere resting upon the latter necessarily drives it into the mouth; as for example, when a person drinks water from a tumbler or tea from a cup. How singular that the rationale of taking liquids into the stomach was not understood till the 17th century—that so simple an operation and one incessantly occurring, should have remained unexplained through all previous time!

Two kinds of suction have been mentioned by some writers, but the principle of both is the same: one, the action of the chest just mentioned—the other, that of the mouth alone; viz. by lowering the under jaw while the lips are closed, and at the same time contracting and drawing the tongue back towards the throat. There is this difference between them: the former can be performed only in the intervals of respiration, while the latter may be continuous, since breathing can be kept up through the nostrils. One has been named supping, the other sucking. The term 'sucker,' commonly applied to the piston of atmospheric pumps, arose from its acting as a substitute for the mouth. With this explanation of the terms suction, sucking, &c. we shall occasionally use them, in accordance with general custom, for want of substitutes equally popular.

Infants and the young of all mammals not only practice *sucking* till they quit their mother's breasts for solid food, but most of them continue the practice through life when quenching their thirst: of this man is an example, for it is by sucking that we receive liquids into the stomach, whether we plunge our lips into a running stream, receive wine from a goblet, or soup from a spoon. As the origin of artificial devices for raising liquids by atmospheric pressure may be traced to this natural operation, some other examples may be mentioned. Of these, sucking poison from wounds is one. This has been practiced from unknown antiquity. Job, speaks of sucking the poison of asps—At the siege of Troy, Machaon 'suck'd forth the blood' from the wounds of Menelaus; and the women among the ancient Germans were celebrated for thus healing their wounded sons and husbands. The serious consequences that often attended the custom, led at an early period to the introduction of *tubes*, by means of which the operation might be performed without danger to the operator; for scrofulous and other diseases were frequently communicated to the latter, by drawing tainted blood and humors into the mouth; whereas, by the interposition of a tube, the offensive matter could be prevented from coming in contact with the lips.

Before the use of the lancet was discovered, these *cupping* tubes were applied in ordinary blood-letting. Even at the present day such is the only kind of phlebotomy practiced by the oldest of existing nations; for "the name and the use of the lancet are equally unknown among the natives of Hindostan. They scarify the part with the point of a knife and apply to it a copper cupping-dish with a long tube affixed to it, by means of which they suck the blood with the mouth."[a] It is the same with the Chinese, Malays, and other people of the east. These generally use the same kind of apparatus as the Hindoos, but sometimes natural tubes are employed, as a piece of bamboo.[b] The horns of animals, as those of oxen and goats were also much used; these on account of their conical form being better adapted for the purpose than cylindrical tubes

---

[a] Shoberl's Hindostan, v, 42. [b] Chinese Repos. iv, 44. See also Le Comte's China, and Marsden's Sumatra.

Chap. 4.] *The Sanguisuchello.* 203

Park found the negroes of Africa cupping with rams' horns; and the Shetlanders continue to use the same instrument, having derived it from their Scandinavian ancestors. Cupping was practiced by Hippocrates, and cupping-instruments were the emblems of Greek and Roman physicians.

The application of a reed or other natural tube, through which to suck liquids that cannot otherwise be reached, has always been known. The device is one which in every age, boys as well as men acquire a knowledge of intuitively, or as it were by instinct; nor does it indicate a greater degree of ingenuity than numerous contrivances of the lower animals—that of the *raven* for example, which Pliny has mentioned in the tenth book of his Natural History. This bird, during a severe drought, seeing a vase near a sepulchre, flew to it to drink, but the small quantity of water it contained was too low to be reached. In this dilemma, stimulated by want and thrown upon its own resources for invention, it soon devised an effectual mode of accomplishing its object—it picked up small pebbles and dropped them into the vessel till the water rose to the brim—an instance of sagacity fully equal to the application of a tube under similar circumstances by man.

As sucking tubes are atmospheric pumps in embryo, a notice of some applications of them will form an appropriate introduction to the latter. They constituted part of the experimental apparatus of the old Greek Plenists and Vacuists; and were used by the Egyptians as siphons. They were, and still are, employed in Peru for drinking hot liquids, and were anciently used by the laity in partaking of wine in the Eucharist. "Beatus Rhenanus upon Tertullian in the booke *De Corona Militis*, reporteth that among the riches and treasures of the church of Mense, were certain silver pypes by the which profane men, whom they call the laietie, sucked out of the challice in the holy supper."[a] The device, if not of

No. 78. Sanguisuchello.

more distant origin, was perhaps designed in the dark ages, as a check to the rude communicants, who would naturally be inclined to partake too freely of the cup. But since the laity were excluded by the Council of Constance, from sharing the wine, the use of such tubes has been retained. At the celebration of high mass at St. Denis, the deacon and sub-deacon suck wine out of the chalice by a *chalumeau* or tube of gold. [Dict. de Trévoux. Art. Chalumeau.]

'The *sanguisuchello* or blood-sucker,' says La Motraye, is a golden tube by which the *Pope* sucks up the blood [wine] at high mass; the chalice and tube being held by a deacon. The instrument, he remarks, corresponds with "the ancient *pugillaris*, or tube mentioned by Cardinal Bona in his treatise of things belonging to the liturgy, and of the leavened and unleavened bread."[b] No. 78 is a figure of the sanguisuchello. It has three pipes, but the middle or longest one is that by which the liquid is raised. The whole is of gold, highly ornamented, and enriched with a large emerald. One reason assigned for its use, is, that it is more seemly to suck the blood [wine] as through a vein, than to *sup* it.

The Peruvians make a tea or decoction of the 'herb of Paraguay,'

---

[a] Peter Martyr's Com. Places. Lon. 1583. Part 4, p. 37.   [b] La Motraye's Trav. i. 29, 31, 427, and Blainville's Trav. ii, 332.

which is common to all classes. "Instead of drinking the tincture or infusion apart, as we drink tea, they put the herb into a cup or bowl made of a calabash or gourd, tipp'd with silver, which they call *mate;* they add sugar and pour on it the hot water, which they drink immediately without giving it time to infuse, because it turns as black as ink. To avoid drinking the herb which swims at the top, they make use of a *silver pipe*, at the end whereof is a bowl full of little holes; so that the liquor *suck'd in* at the other end is clear from the herb."[a] Frezier has given an engraving of a lady thus employed, from which the annexed cut is copied.

No. 79. Peruvian female taking tea with a sucking-tube.

In Frezier's time it was the custom for every one at a party to suck out of the same tube—like Indians in council, each taking a whiff from the same calumet. With the exception of confining a company to the use of one instrument, we should think this mode of 'taking tea' deserving the consideration of the wealthy, since it possesses several advantages over the Chinese plan which we have adopted. In the first place, it is not only a more ingenious and scientific mode of raising the liquid, but also more graceful than the gross mechanical one of lifting the vessel with it. It is more economical as regards the exertion required; for in ordinary cases a person expends an amount of force in carrying a cup of tea backwards and forwards, so many times to his mouth, as would suffice to raise a bucket of water from a moderately deep well. In the use of these tubes there is no chance of verifying the old proverb—'many a slip between the cup and the lip'—And then there is no danger of breakage, since the vessel need not be removed from the table. How often has a valuable 'tea-set' been broken, and the heart of the fair owner almost with it, by some awkward visitor dropping a cup and saucer on their way to his mouth, or on their return to the table! Lastly, the introduction of these tubes, would leave the same room as at present for display in tea-table paraphernalia.

There is another application of them which some convivialists may thank us for suggesting. It has been regretted by ancient and modern epicures that nature has given them *necks* much shorter than those of some other animals; these philosophers supposing that the pleasures of eating and drinking are proportioned to the *length* of the channel through which food passes to the stomach. Now although a sucking tube will not alter the natural dimensions of a person's neck, it may be so used as to prolong the sensation of deglutition in the shortest one; for by contracting the orifice, each drop of liquid imbibed through it may be brought in contact with the organs of taste, and be detained in its passage until every particle of pleasure is extracted from it;—being the reverse of what takes place, when gentlemen swallow their wine in gulps. The most fastidious disciple of Epicurus could not object to this use of them, since nothing would touch his liquid but the tube; and as every person would

---

[a] Frezier's Voyage to the South Seas, p. 252.

provide his own, no one would ever think of borrowing his neighbor's, any more than he would ask for the loan of his tooth-pick.[a]

We are not sure that this plan of attenuating agreeable liquids, did not give rise to that mode of drinking adopted by the luxurious Greeks and Romans, to which we have before alluded. Their drinking vessels were generally horns, or were formed in imitation of them. At the small end of each a very minute opening was made, through which a stream of drops, as it were, descended into the mouth. Paintings found in Pompeii, and other ancient monuments, represent individuals in the act of thus using them—while others, whose appetite for the beverage, or whose thirst was too keen to relish so slow a mode of allaying it, are seen drinking, not out of "the little end," but out of the *large* end "of the horn." We have mentioned this circumstance because it appears to afford a solution of an old, but somewhat ambiguous saying.

CHAPTER V.

On bellows pumps: Great variety in the forms and materials of machines to raise water—Simple bellows pump—Ancient German pump—French pump—Gosset's frictionless pump: Subsequently re-invented—Martin's pump—Robison's bag pump—Disadvantages of bellows pumps—Natural pumps in men, quadrupeds, insects, birds, &c.—Reflections on them. Ancient vases figured in this chapter.

In the course of time a new feature was given to sucking tubes, by which they were converted into pumps: this was an apparatus for withdrawing the air in place of the mouth and lungs. In what age it was first devised, and by what people, are alike unknown. The circumstance that originally led to it, was probably the extension of the length of sucking tubes, until the strength of the lungs was no longer sufficient to draw water through them. In this way the bellows pump, the oldest of all pumps, we presume took its rise.

It should be borne in mind that an atmospheric pump is merely a contrivance placed at the upper end of a pipe to remove the pressure of the atmosphere there, while it is left free to act on the liquid in which the lower end is immersed; and farther, that it is immaterial what the substance of the machine is, or what figure it is made to assume. Some persons perhaps may suppose that pumps seldom vary, and then but slightly, from the ordinary one in our streets, (the ancient wooden one) but no idea could be more erroneous; for few, if any, machines have undergone a greater number of metamorphoses. The body or working part, which is named the 'barrel' and sometimes the 'chamber,' so far from being always cylindrical, has been made square, triangular, and elliptical;—it is not even always straight, for it has been bent into a portion of a circle, the centre of which formed the fulcrum of the lever and rod, both of which in this case being made of one piece: its materials have not been confined to wood and the metals, for pumps have been made of glass, stoneware, stone, leather, canvas, and caoutchouc. Some have been constructed like

---

[a] In Shakespeare's time, "every guest carried his own *knife*, which he occasionally whetted on a stone that hung behind the door. One of these whetstones may be seen in Parkinson's Museum. They were strangers at that period to the use of forks.' [Ritsons's Notes on Shakespeare's Timon of Athens. Act i, Scene 2.]

a bag, resembling the old powder-puff or the modern accordion; others in the form of the domestic and blacksmith's bellows—some in the figure of a drum, and others as a portion of one—as a simple horizontal tube suspended at the centre on a perpendicular one, and whirled round like the arms of a potter's wheel—then again as a perpendicular tube without sucker or piston, and moved like a gentleman's walking cane, from which indeed its name is derived. (See *Canne Hydraulique* in Book IV.) They have also been made of two simple tubes, one moved over the other like those of a telescope—even a kettle or cauldron has been used as a pump, and the vapor of its boiling water substituted for the sucker to expel the air it contained, after which the pressure of the atmosphere forced water into it from below. In fine, any device by which air can be removed from th interior of a vessel is, or may be used as a pump to raise water.

Nor have the 'suckers' or 'pistons' been subject to less changes than other parts of pumps. They have been made solid and hollow—in the form of cones, cylinders, pyramids, sectors, and segments of circles:—in the shape of cog-wheels, and of the arms and vanes of wind-mills, with motions analogous to such as these; and sometimes they are made in the shape of a gentleman's hat and of similar materials; while the only motion imparted to them, is the odd one of alternately pushing them inside out and outside in.

No. 80.

If a collapsed bladder or leather bag, be secured at its orifice to the upper end of a perpendicular tube whose lower end is placed in a vessel of water, (No. 80) then, if by some contrivance the bag can be distended, as shown by the dotted lines, the small quantity of air contained in it and the pipe would become rarefied, and consequently unable to balance the pressure without—hence the liquid would be forced up into the bag, until the air within became again condensed as before—that is, the bladder would be filled with water, with the exception of a quantity equal to the space previously occupied by the air within it and the pipe.

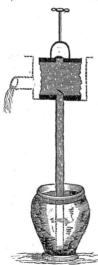

N 81. Bellows Pump.

To convert this simple apparatus into a pump, two valves or clacks only are wanting. One, opening upwards and placed in any part of the pipe or at either of its extremities. This will allow water to pass up through it, but none to descend. The other placed over an aperture made on the top of the bag, and opening outwards—through this the contents of the vessel when collapsed can be discharged; and when distended it will close, and thereby prevent the entrance of the external air. The instrument thus aranged becomes a *bellows* pump, (No. 81,) a machine, which from the obvious application of the bellows to raise and spout water as well as air, has been reinvented by machinists in almost every age.

The figure scarcely requires illustration. It repre sents a pipe attached to the under board of a circular or lantern bellows, the orifice of which is covered by a clack—the upper

Chap. 5.]   *Old German Bellows Pump.*   207

board has also an opening in its centre which is closed by a valve or clack, and also furnished with a rod and handle. The under board sometimes forms the bottom of a box, in one side of which a spout is inserted, as shown by the dotted lines.

The earliest representation of a bellows pump which we have met with in books, is among the curious cuts attached to the first German translation of Vegetius, from which No. 82 is copied. (Erffurt 1511)[a] It will suffice to show the application of this kind of pump to raise water at that time. There was of course a valve covering the interior orifice of the nozzle and opening outwards, to prevent the air from entering when the upper board was raised. This valve is not shown because the art of representing the interior of machines by section, was not then understood, or not practiced. The lower board is fastened to the ground by a post and key, and a weight is placed on the upper one to assist in expelling the water.

No. 82.   German Atmospheric Bellows Pump.   A. D. 1511.

One hundred years ago, two bellows fixed in a box and worked by a double lever, like the old fire or garden engine, was devised by M. Du Puy, Master of Requests to the king of France. The machine was recommended to raise water from the holds of ships, drain lands, &c. It appears that the widow of M. Du Puy, expected to reap great advantages from it in England; but Dr. Desaguliers, in 1744, published a description of it taken from the French account, and among other remarks he observed—" About fourteen years ago, two men here applied for a patent for this very engine, proposing thereby to drain mines;" * * * " all the difference was, that their bellows were fixed upon a little waggon; and they had a short sucking pipe under; and the force pipe went up from the two bellows. I opposed the taking out of this patent, because I thought it would be of great hurt to the undertakers, to lay out near eighty pounds for what would never bring them eighty pence; unless they made a bubble of it, and drew unwary people into a scheme to subscribe money." (Ex. Philos. ii, 501.) Bellows pumps were previously used in France. They are spoken of as common in the old Dict. de Trévoux.

[a] I am indebted to John Allan, Esq. for a copy of this scarce old work. It is the same to which Prof. Beckman refers in his article on the diving bell. Unfortunately the cuts are left without explanation.

A neat and perhaps the best modification of these machines was devised about the year 1732, in Paris, by Messrs. Gosset and Deuille. It was described by Belidor in 1739, and by Desaguliers in 1744, as "a piston without friction." It consists of a circular piece of leather pressed into the form of a deep dish, or of a low crowned hat with a wide rim. This rim is secured by bolts and screws between two flanches of a pump cylinder, forming an air tight joint—the part corresponding to the body of the hat fits loosely into the cylinder; and the crown is strengthed by a circular plate of metal of the same size and riveted to it. In the centre of this plate an opening is made and also through the leather for the passage of the water, and covered by a valve opening upwards like the ordinary sucker of a pump. The forked end of the pump-rod is secured to this plate. (See figure.) When the rod is raised, the bottom of the dish or hat is above the flanch, and when down it is pushed inside out as shown in the cut. Thus, by alternately elevating and depressing it, the water is raised as in the common pump. This piston is described in Vol. VI, of Machines approved by the French Academy for 1732, p. 85, as the invention of *M. Boulogne*.

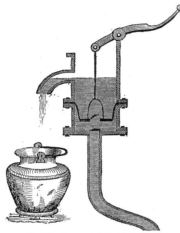

N 83. Gosset and Deuille's Pump.

The great advantage of this pump is in the sucker or piston not rubbing against or even touching the sides of the cylinder, hence there is no friction to overcome from that source, and the leather is consequently more durable; but the *length of stroke* is much less than in common pumps, it seldom exceeding six or eight inches, lest the leather should be overstrained in pressing it deeper. Large pumps of this description were worked in the mines of Brittany incessantly during three or four months without requiring any repair. India-rubber, and canvas saturated or coated with it, have been successfully used in place of leather. Some modifications of the sucker have also been introduced.

This pump was re-invented in England some years ago, and made considerable noise under a new name. See London Mechan. Magazine, and Register of Arts, 1826–29; also the Journal of the Franklin Institute for 1831, vol. vii, 193. In 1766, Mr. Benjamin Martin, the well known author of 'Philosophia Britannica' and other scientific works, proposed a good double pump of this kind for the British navy—a figure and description of it may be seen in Vol. XX. of Tilloch's Philosophical Magazine.

Dr. Robison, in the second volume of his Mechanical Philosophy, proposed what has been named an improvement on the last pump. His device is however little else than the old bellows pump. A figure of it and his description are annexed.

A, B, (No. 84) represents a wooden trunk or cylinder of metal, having a a spout at the upper part, and the lower end closed by a plate, the opening in which is covered by a clack valve E, as in No. 83. To this plate is secured the open bottom of a long cylindrical bag, the upper end being fixed to the round board F. "This bag may be made of leather or of double canvas, a fold of thin leather or of sheepskin being placed between the two folds. The upper end of the bag should be firmly tied with a cord in a groove turned out of the rim of the board at F. Into this board is fixed the fork of the piston rod, and the bag is kept distended by a number of wooden hoops or rings of wire, fixed to it at a few inches distance from one another, and kept at the same distance by three or four cords binding them together, and stretching from the top to the bottom of the bag. Now let this trunk be immersed in the water: it is evident that if the bag be stretched from the compressed form which its own weight will give it by drawing up the piston rod, its capacity will be enlarged, the valve F will be shut by its own weight, the air in the bag will be rarefied, and the atmosphere will press the water into the bag. When the rod is thrust down again, the water will come out at the valve F, and fill part of the trunk. A repetition of the operation will have a similar effect; the trunk will be filled, and the water will at last be discharged at the spout." The operation is precisely the same as in No. 81.

N 84. Bag Pump.

"Here is a pump without friction and perfectly tight; for the leather between the folds of canvas renders the bag impervious both to air and water. We know from experiment that a bag of six inches diameter made of sail cloth No. 3, with a sheepskin between, will bear a column of fifteen feet of water, and stand six hours work per day for a month, without failure; and that the pump is considerably superior in effect to a common pump of the same dimensions. We must only observe that the length of the bag must be three times the intended length of the stroke, so that when the piston rod is in its highest position, the angles or ridges of the bag may be pretty acute. If the bag be more stretched than this, the force which must be exerted by the laborer becomes much greater than the weight of the column of water which he is raising."

But after all that can be said in favor of bellows pumps, they have their disadvantages. A prominent one is this: when the leather or other material of which they are formed is worn out, a practical workman, who is not to be obtained in every place, is required to renew it. Unlike replacing the leather on an ordinary 'sucker', which a farmer or a sailor on ship-board can easily accomplish, the operation requires practice to perform it efficiently, and the expense both of time and materials is much greater than that of similar repairs to the common pump. For these and other reasons, bellows pumps have never secured a permanent place among staple machines for raising water, and the old cylindrical pump still retains the preëminence, notwithstanding the almost innumerable projects that have been brought forward to supersede it.

The preceding machines resemble in some degree the apparatus for drinking which the Creator has furnished to us and to such quadrupeds as do not lap. When an ox or a horse plunges his mouth into a stream, he dilates his chest and the atmosphere forces the liquid up into his stomach precisely as up the pipe of a pump. It is indeed in imitation of these natural pumps that water is raised in artificial ones. The thorax is the pump; the muscular energy of the animal, the power that works it; the throat is the pipe, the lower orifice of which is the mouth, and which he must necessarily insert into the liquid he thus pumps into his stomach;

and whenever the depth of water is insufficient to cover the opening between his lips, the animal instinctively draws closer those portions of them above it, and contracts the orifice below, just as we do under similar circumstances, and which we constantly practice in sipping tea or coffee from a cup, or any other beverage of which we wish to partake in small quantities. The capacious chest of the tall camel, or of the still taller cameleopard or giraffe, whose head sometimes moves twenty feet from the ground, is a large bellows pump which raises water through the long channel or pipe in his neck. The elephant by a similar pneumatic apparatus, elevates the liquid through that flexible 'suction pipe,' his proboscis; and those nimble engineers, the common house-flies, raise it through their minikin trunks in like manner.

We may here remark, that among the gigantic animals which in remote ages roamed over this planet, and which quenched their thirst as the ox does, there could have been none which stood so high as to have their stomachs thirty feet above the water they thus raised into them. And on the table lands of Mexico, and the still higher regions of Asia, Africa, and South America, animals of this kind, if such there were, must have had their stomachs placed still lower.

The mandibles of some insects are hollow, and are used as sucking pumps. They serve also sometimes as sheaths to poniards, with which nature has furnished them, as weapons of offence and defence. Those of the lion-ant are pierced, and "no doubt act as suckers." This little animal constructs a minute funnel-shaped excavation in dry sand, and covering its body at the bottom lays in wait, like an assassin, for its prey: "no sooner does an industrious ant, laden perhaps with its provision, approach the edge of the slope, than the finely poised sand gives way, and the entrapped victim rolling to the bottom, is instantly seized and *sucked to a shadow* by the lurking tyrant, who, soon after by a jerk of his head tosses out the dead body." Weasels and other animals suck the blood of their prey. The tortoise drinks by suction, for which purpose he plunges his head deep into the fluid, so as even to cover his eyes. There are several species of birds denominated '*suctorial*' on account of their obtaining food by means of atmospheric pressure, which they bring into action by apparatus analogous to the pump. The *grallatores* or *waders*, "suck up their food" out of water.

It is impossible to contemplate the structure and habits of animals, without being surprised at the *extent* to which this principle of raising liquids has been adopted by the Almighty in the formation of insects, reptiles, fishes, birds, amphibia and land animals; and also at its adaptation to their various forms, natures, and pursuits. Had we the necessary knowledge of their physiology, we would desire no greater pleasure, no other employment than to examine and describe these natural pneumatic machines, and the diversified modes of their operation.

For other natural pumps, see remarks at the end of Chapter 2, on bellows *forcing* pumps, in the next Book.

The *vessels* or vases figured in this chapter are ancient. Those in which the tubes are inserted in illustrations Nos. 80 and 81, are of glass; the one under the pump spout in No. 83, is a bronze bucket; all from Pompeii. The latter is referred to at page 67. The globular vessel in No. 84, is a figure of a brazen cauldron, also Roman, from Misson. See page 19 of this volume.

CHAPTER VI.

The atmospheric pump supposed by some persons to be of modern origin—Injustice towards the ancients—Their knowledge of hydrodynamics—Absurdity of an alledged proof of their ignorance of a simple principle of hydrostatics—Common cylindrical pump—Its antiquity—Anciently known under the name of a siphon—The *antlia* of the Greeks—Used as a ship pump by the Romans—Bilge pump—Portable pumps—Wooden pumps always used in ships—Description of some in the U. S. Navy—Ingenuity of sailors—Singular mode of making wooden pumps, from Dampier—Old draining pump—Pumps in public and private wells—In mines—Pump from Agricola, with figures of various boxes—Double pump formerly used in the mines of Germany, from Fludd's works—The wooden pump not im proved by the moderns—Its use confined chiefly to civilized states.

SOME persons are unwilling to admit that the atmospheric pump was known to the ancients, and yet they are unable to prove its origin in later times or by more recent people. The passages in ancient authors in which it is supposed to be mentioned or alluded to, are deemed inconclusive, because the terms by which it is designated were also applied to other devices.

To confine the knowledge of the ancients to such departments of the arts as are either expressly mentioned or referred to in Greek and Roman authors, and to those, specimens of which have been preserved to our times, is neither liberal nor just. Let us suppose Europe and the United States, in the course of future time, thrown back into barbarism, and all records perished, save a few fragments of the works of our dramatists, poets and historians;—and that after the lapse of some 1500 or 2500 years these should be discovered—and also some relics of our architecture, pottery, and works in the metals: Now we should think the writers of those days illiberal in the extreme, who should conclude that we were ignorant of nearly all branches of science and of the arts; and of every machine which was not *particularly* mentioned or illustrated in the former—or of which specimens were not found among the latter. And yet something like this, has been the treatment which the ancients have received at our hands.

It cannot however be denied, that remains of their works still extant, exhibit a degree of skill in architecture, sculpture, metallurgy, pottery, engraving, &c. which excels that of modern artists.[a] And as regards their knowledge of hydrodynamics—let it be remembered, that we are indebted to them for canals, aqueducts, fountains, jets d'eau, syringes, forcing pumps, siphons, valves, air vessels, cocks, pipes of stone, earthenware, wood, of lead and copper: yet notwithstanding all these, and their numerous machines for raising and transferring water, and the immense quantities of tubes for conveying it, which are found scattered over all Asia as well as Italy[b] and Greece, it has been gravely asserted, that they were ignorant of one of the elementary and most obvious principles of

---

[a] It was remarked by the late Mr. Wedgewood, who was doubtless the most skilful manufacturer of porcelain in our own times, that the famous *Barbarini Vase* afforded evidence of an art of pottery among the ancients of which we are as yet ignorant, even of the rudiments. Edin. Encyc. vol. ii, 203.

[b] The vast quantities of leaden pipes found at Pompeii induced the Neapolitan government to sell them as old metal. Pompeii, vol. i, 104.

hydrostatics: viz. that by which water in open tubes finds its own level: a fact, of which it may safely be asserted, it was *impossible* for them not to have known—a fact with which the Indians of Peru and Mexico were familiar; and one expressly mentioned by Pliny: "water, (he observes) always ascends of itself at the delivery to the height of the head from whence it gave receipt—if it be fetched a long way, the work [pipe] will rise and fall many times, but the level [of the water] is still maintained." Besides the testimony of Pliny, fountains and jets d'eau are incontrovertible proofs that a knowledge of the fact is of stupendous antiquity; they having been used in the east from immemorial ages.

But the proof adduced to establish their ignorance in this particular, is as singular as the position it is brought forward to sustain, since it equally establishes our own ignorance of the same principle! It has been said, had the ancients known that water finds its level at both extremities of a crooked tube, they would have conveyed it through pipes to supply their cities, instead of erecting those expensive aqueducts which were among the wonders of the world, and remains of which still strike the beholder with admiration:—in reply to this it need only be observed, that should any remains of the Croton aqueduct, now constructing to supply this city (New-York) with water, be found two thousand years hence, they may, *by the same argument*, be adduced as proofs that the present engineers of the United States were ignorant that water poured into an inverted siphon would stand at the same level in both its branches.

The fact is, the ancients did sometimes convey water over eminences in siphons of an easy curvature.[a] And aqueducts were in some few instances carried through valleys by *inverted* siphons. In the reign of *Claudius*, an aqueduct was formed to convey water from Fourvières to the highest part of the city of Lyons. As valleys of great depth were in the line of its course, works of an enormous expense would have been required, which might have prevented the execution of the project; consequently, instead of an elevated canal, leaden pipes were substituted, forming an inverted siphon.[b]

It is uncertain when or by whom the common atmospheric pump was invented. It is supposed to have been known to the old Egyptians, and to have been used in the ship in which Danaus and his companions sailed to Greece.[c] As the *antlia* of the Greeks, it could not have originated with Ctesibius, to whom it has sometimes been attributed, since it or some other machine or device is mentioned under that name, by Aristophanes and other writers who flourished ages before him.[d] There are other indications that it was previously known, for either it or something very like it is mentioned under the name of a *siphon*. This term it is known was a generic one, being applied to hollow vessels, as funnels, cullenders, pipes; and generally to instruments that either raised or dispersed water, as syringes, catheters, fire-engines, sprinkling-pots, &c.[e] That the machine to which we refer raised water by 'suction,' is apparent from ancient allusions to it. According to Bockler, "the Platonic philosophers asserted that the soul should partake of the joys of heaven as through a siphon;" and by it Theophrastus explained the ascent of marrow in bones; and Columella the rise of sap in trees. In these instances, it is obvious that neither the ordinary siphon nor the syringe could be intended, but the atmospheric pump; a machine that Agricola described as a

---

[a] Fosbroke's Encyc. Antiq. i, 41. [b] Hydraulia, Lon. 1835, p. 254. [c] See Edin. Encyc. Art. Chronology, vol. vi, 263. [d] Robinson's Antiquities of Greece, cap. 4 On Military Affairs. [e] See Ainsworth's Dict.

siphon; and one to which the remark of Switzer only can apply—"the siphon was undoubtedly the chief instrument known in the first ages of the world, (besides the draw-well) for the raising of water."[a]

Nor is there any thing in the account given by Vitruvius of 'the Machine of Ctesibius,' which indicates that the atmospheric pump was not in previous use. His description is obviously that of a *forcing* pump, (and appears to have been so understood by all his translators,) one whose working parts were placed not above but in the water it was employed to elevate; whose piston was solid, and which by means of pipes forced the water *above itself;* that raised the water "very high;"—attributes which do not belong to the common pump. It is true he has not mentioned the latter, perhaps because it was not then employed as now in civil engineering, and therefore not within the scope of his design in writing his work. The manner in which Pliny speaks of it, shows that it was an old device in his time, since it was one with which even country-people or farmers, (the last to adopt new and foreign inventions) were familiar. In his 19th Book, 'On Gardens,' cap. 4, he observes: when a stream of water is not at hand, the plants should be watered from tanks or wells, the water of which may be drawn up by plain poles, hooks and buckets, by swapes or cranes, [windlass] "or by *pumps* and such like." And that these were no other than the old wooden pump of our streets and such as our farmers use, is obvious from a passage in his 16th Book, cap. 42, where speaking of the qualities and uses of different kinds of wood, he remarks, "pines, pitch trees and allars, are very good to make PUMPS and conduit pipes to convey water; and for these purposes their wood is bored hollow."

Although sufficient time may be supposed to have elapsed from the age of Ctesibius to that of Pliny for the introduction of the atmospheric pump to the countrymen of the latter, (supposing it to have been invented by the former) we can hardly believe, if it were not of more remote origin, that it could even in that time have found its way into Roman farm-yards and gardens; much less that it should have superseded, (as it appears to have done) every other device on board of their ships. New and foreign inventions were neither circulated so easily nor adopted so readily in ancient as in modern days; and even now a long time would elapse before inventions of this kind would find their way through the world and longer before they became generally adopted. But had the pumps of which Pliny speaks been of recent introduction, he would certainly have said so; and had they been the 'water forcers' of Ctesibius, to which he alludes in his 7th Book, he could scarcely have avoided recording the fact.

That the antlia was the atmospheric pump would also appear from its employment in ships. There is no reason to suppose that more than *three* kinds of marine pumps were ever in use—the chain pump, the screw, and the common pump. In the chapter on the former we have shown that it was not known or used by the Greeks and Romans. The screw was first adopted as a ship pump by Archimedes, (see page 133) and hence it would seem that the last only could be intended by more ancient as well as subsequent authors when speaking of the antlia: that it was so, antiquarians generally admit. "The well, (says Fosbroke in his article on the vessels of the classical ancients) was emptied by the winding screw of Archimedes now in use; but in other ships by the antlia or pump." It is of the latter that Pollux speaks, and to it Tacitus refers when mentioning

---

[a] Hydrostatics, 294.

the wreck of some vessels in which Germanicus and his legions sailed down the Amisia into the German ocean: "the billows broke over them with such violence, that all the pumps at work could not discharge the water." [B. ii, 23. Murphy's Translation.]

Martial, the Roman poet, speaks of the antlia as a machine 'to draw up water;' according to Ainsworth, 'a pump.' Kircher figures and describes the old wooden pump as the antlia. [Mundus Subterraneus, tom. ii, 196.]

The Romans appear to have employed it exclusively or nearly so in their navy; and even in that of the Greeks it is not probable that the screw was extensively adopted, on account of its not being so well adapted for ships as the other. Of this the former people seem to have been convinced; they preferred the pump and all modern nations have confirmed their judgment. Had they used the screw to any extent it would have been continued in European vessels after the fall of the Empire, when most of their arts and customs were naturally and necessarily continued—their ship pumps as well as their ships. But as the atmospheric pump only has so come down, we infer that the machine now commonly used to discharge water from the holds of our vessels is identical, or nearly so, with that employed by Roman sailors of old.

The oldest modification of the ship pump appears to have been that formerly known as the 'bilge' or 'burr' pump; and it was the simplest, for it had but one distinct valve, viz. 'the lower box,' as the one which retains the water in a pump is sometimes named. This pump kept its place in ships till the last century, and may yet occasionally be met with in those of Europe. It was often worked without a lever, but its peculiarity consisted principally in the construction of the piston or sucker.[a] It differed from the ordinary pump "in that it hath a staff, six, seven or eight foot long, with a bur of wood whereunto the leather is nailed, and this serves instead of a box; so two men standing over the pump, thrust down this staff, to the middle whereof is fastened a rope for six, eight or ten to hale by, and so they pull it up and down." This account published nearly 200 years ago, might be sufficiently descriptive then, when the pump was in common use, but few persons could now realize from it a correct idea of the substitute for the ordinary sucker. It is however rather more explanatory than the accounts given in later works. In some it has been described as "a long staff with a burr at the end to pump up the bilge water." Here the burr only is mentioned, not the leather, and the idea imparted is that of a solid piston, such as are used in forcing pumps.

The sucker of the bilge pump consists of a hollow cone or truncated cone of strong leather, the base being equal in diameter to that of the pump chamber or cylinder. It is inverted and nailed to the lower end of the rod. The lower edge of the leather resting against the burr. When thrust down it collapses and permits the water to pass between it and the sides of the chamber, and when its motion is reversed, the weight of the liquid column above it, presses it out again. To prevent the cone from sagging, three strips of leather are sewed to its upper part at equal distances from each other, and their other ends nailed to the rod. (See No. 85.) The action of this sucker is something like moving a parasol up and down in water; the sides close as the rod descends and open when it rises. It is the simplest modification of the sucker known and probably the most ancient. It is figured by Agricola, (*vide* C in No. 88) but

---

[a] This part of an atmospheric pump is sometimes named the sucker, the bucket, the upper box, the piston;—we shall generally use the first when speaking of the atmospheric pump; and the last when referring to forcing pumps.

is not mentioned by Belidor, Switzer, Desaguliers or Hachette; nor has it been noticed by more recent writers, with the exception of Mr. Millington[a] and perhaps one or two others. It has long been known in some parts of the United States. We noticed it twenty years ago at New Rochelle, Westchester county, in this state, (New-York) and were informed by a pump maker there, that they "always had it." It is not however universally known, for in 1831 a patent was taken out for it.[b]

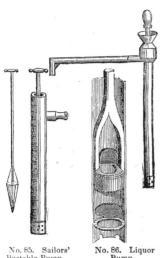

No. 85. Sailors' Portable Pump. No. 86. Liquor Pump.

There is another application of the burr pump in ships that is probably due to old navigators. We allude to the use of those portable instruments which, says an old author, "are made of reed, cane, or laten, [brass] that seamen put down into their casks to pump up the drink, for they use no spickets." No. 85 represents one, with a separate view of the sucker, from an illustrated edition of Virgil, of the 16th century. They appear to be of considerable antiquity and were perhaps used for the same purpose by the ancient sailors of Tyre and Carthage, Greece and Rome. No. 86 is a figure of the common liquor pump, derived from the former. It is from *L'Art du Distillateur*, in '*Descriptions des Arts et Mètiers*,' folio, Paris, 1761. The separate section of the lower part shows the 'boxes' to have been similar to those now often used. Another sucker is figured with a spherical valve; a boy's marble, or a small ball of metal being placed loosely over the orifice, instead of a clack. It was at that time made both of tin plate and of copper as at present. One of these pumps is mentioned by Conrad Gesner, as constituting part of a portable Italian distillery, in the former part of the 16th century, at which period it seems to have been common. See a reference to it, page 218.

Ship pumps seem to have been made of bored wooden logs since the days of the elder Pliny, and probably were so by both Greeks and Romans long before his time. We learn that they were made by *shipwrights*, i. e. by a certain class of them.[c] At the present day, every person knows that wooden pumps are oftener to be found in ships than any other: this has always been the case. It is to them only that reference is made in the relations of early voyages. The vessels of Columbus,[d] Vasco de Gama and Magalhanes, were furnished with them; indeed no other kind appears to have been used by old European navigators. From the importance of efficient machines to raise water from ships, it may reasonably be supposed that if any nation had possession of a superior one, it would soon have been adopted by the rest; but there is not the slightest intimation of any difference between them. The pump in Spanish, Portuguese, English and French vessels, is spoken of as common; as much so as the anchor or rudder: thus—when the *Vitoria* one

---

[a] Epitome of Philosophy, Lon. 1823, p. 199.   [b] Journal of the Franklin Institute, vol. ix, 235.   [c] Lardner's Arts of the Greeks and Romans, vol. i, 350.   [d] Irving's Columbus, vol. ii, 127.

of Magalhanes' fleet, put into St. Jago on her return in 1522, a boat was sent ashore for provisions, and "some negroes to assist in *working the pumps*, many of the ship's company being sick, and the leaks having increased."[a] In the account of Frobisher's third voyage in search of a north-west passage in 1578, the *Anne Francis* having run on a sunken rock, "they had above two thousand strokes together at *the pumpe* before they could make their shippe free of the water, so sore she was brused."[b] In the voyage of some English vessels to the north the following year, one was nearly lost; "by mischance the shippe was bilged on the grapnell of the Pavos, [another vessel] whereby the company [owners] had sustained great losses, if the chiefest part of their goods had not been layde into the Pavos; for notwithstanding their *pumping* with *three pumpes*, heaving out water with buckets and all the best shifts they could make, the shippe was halfe full of water ere the leake could be found and stopt."[c] In November 1599, two large Portuguese ships arrived at Terceira in distress, having been separated in a storm, during which they "were forced to use *all their pumpes*" to keep afloat.[d] Tavernier sailed in 1652, from the Persian Gulf to Maslipatan in a large ship belonging to the king of Golconda;—a storm arose and became so violent that the water "rowl'd in from stem to stern, and the mischief was that *our pumpes* were nought." Fortunately several bales of leather were on board, of which they made bags or buckets, "which being let down from the masts with pulleys through certain great holes which were cut in the deck, drew up a vast quantity of water."[e]

Wooden pumps, with and without metallic cylinders and boxes, are still common in European and American ships of war. The latter with few exceptions have no other. A description of those on board the North Carolina, a ship of the line, may possibly interest some readers. This vessel has six. They are large trees bored out and lined with lead. They reach from the surface of the main gun deck to the well, a distance of twenty-three feet. A brass cylinder, 2 feet 9 inches long and 9 inches bore, in which the piston works, is let into the upper part of each; The piston rods (of iron) pass through the centre of a guide piece, secured over every pump, and are thus kept from deviating from a perpendicular position. They are connected to the levers by slings as in the common brass lifting pump and some others. The levers are double, and shaped like those of fire-engines, staves of wood being slipped through the rings whenever the pumps are worked. Each lever works two pumps; and the length of stroke, or the distance through which the pistons move in the cylinders, is 14 inches. The pistons or upper boxes are of brass with butterfly valves; the band of leather round each is secured by screws, (in place of nails in the wooden box.)

'Necessity is the mother of invention:' the truth of this proverb is often illustrated by seamen, especially as regards the raising of water. Numerous are the instances in which they have relieved themselves from situations so alarming as to paralyze the inventive faculties of most other men; either by devices to work the ordinary pumps when their strength was exhausted, or in producing substitutes for them when worn out. A singular example of the latter is mentioned by Dampier, which may be of service to sailors. It is attributed to a people who are not remarkable for their contributions to the useful arts, and on that account

---

[a] Burney's Voyages, vol. i, 112. [b] Hackluyt's Collection of Voyages, &c. Lon. 1598, black letter, vol. iii, 88. [c] Ibid, vol. i, 421. [d] Astley's Collection of Voyages Lon. 1746, vol. i, 227. [e] Travels in India, Lon. 1678, p. 90.

Chap. 6.] *Various modes of working Ship Pumps.* 217

it would hardly be just to omit it. In the course of Dampier's voyage round the world, while sailing (in 1687) along the west side of Mindanao, one of the Philippine Islands, he concluded to send the carpenters ashore to cut down some trees for a bowsprit and topmast. "And our pumps being faulty and not serviceable, they did cut a tree to make a pump. They first squared it, then sawed it in the middle, and then hollowed each side exactly. The two hollow sides were made big enough to contain a pump-box in the midst of them both, when they were joined together; and it required their utmost skill to close them exactly to the making a tight cylinder for the pump-box, being unaccustomed to such work. We learned this way of pump-making from the *Spaniards*, who make their pumps that they use in their ships in the South Seas after this manner; and I am confident that there are no better hand-pumps in the world than they have." (Dampier's Voyages, vol. i, 443.) In the absence of tools to bore logs the device is an excellent one, and in some particulars such a pump would be superior to the common one. It is not so readily made as one of planks, but it is more durable.

Various ingenious modes of working their pumps have been devised by seamen and others; the power of the men has been applied as in the act of rowing—this plan by far the most efficient is adopted in the French navy. A rope crossed over a pulley and continued in opposite directions on a ship's deck, so that any number of men may be employed at the same time, has been extensively used in pumps with double suckers, as shown at No. 92. Ropes passed through blocks and connected to the brake of the common pump have also been worked in a similar way. Captain Leslie, in a voyage from Stockholm to this country, adopted the following plan, which in a heavy gale, may be very efficient: 'He fixed a spar aloft, one end of which was ten or twelve feet above the top of his pumps, and the other projected over the stern: to each end he fixed a block or pulley. He then fastened a rope to the pump rods, and after passing it through both pulleys along the spar, dropped it into the sea astern. To the rope he fastened a cask of 110 gallons measurement and containing about 60 gallons of water. This cask answered as a balance weight, and every motion of the ship from the roll of the sea made the machinery work. When the stern descended, or when a sea or any agitation of the water raised the cask, the pump rods descended; and the contrary motion of the ship raised the rods, when the water flowed out The ship was cleared out in four hours, and the exhausted crew were of course greatly relieved.'

A ship pump made of such boards or plank, as are commonly found on board of large vessels, was devised by Mr. Perkins, for which he received a gold medal from the London Society of Arts. It is figured and described in the 38th volume of the Society's Transactions.

The facility with which wooden pumps are made and repaired, the cheapness of their material, the little amount of friction from pistons working in them, and their general durability, have always rendered them more popular than others. Like many of our ordinary machines, they seem to have been silently borne down the stream of past ages to the 15th and 16th centuries, when, by means of the printing press, they first emerge into notice in modern times. The earliest representation of one we have met with in print is in the German translation of Vegetius, on the same page with No. 82, the bellows pump: No. 87, on next page, is a copy. It is square, made of plank and apparently designed to drain a pond or marsh. The piston or sucker, which is separately represented, is cylindrical and was perhaps intended to show a variation in the construction

28

No. 87. A. D. 1511.

of that instrument. It has no valve or clack, but appears to be a modification of the one used in the old bilge pump, which was sometimes compared to a 'gunner's sponge.'

There are numerous proofs in old authors, that pumps were common in *wells* in the 15th century, since they are mentioned in the early part of the following one, as things in ordinary. In 1546, they were used to some extent in those of London. In the 'Practice of the New and old Phisicke,' by Conrad Gesner, (who died in 1565) translated by George Baker, 'one of the Queene's maiesties chiefe chirurgians in ordinary,' and dedicated to Elizabeth, (Lon. black letter, 1599,) is a description of a Florentine distilling apparatus, to which a portable pump was attached; the latter is described as "an instrument which is so formed that the water by sucking is forced to rise up and run forth, as the like practice is *often used in pits of water* or *welles.*" Folio 215. The celebrated mathematician, necromancer, and alchymist, Dr. John Dee, who was frequently consulted by queen Elizabeth, had a pump in the well belonging to his house. In Beroald's commentary on the 44th proposition of Besson, (the chain of pots) he observes that it "operè sans intermission en tirant l'eau de tout puits facilement *sans pompes.*"[a] Sarpi, who first discovered the valves of the veins, compared them to those of a pump, 'opening to let the blood pass, but shutting to prevent its return.'

But pumps had not wholly, in the 16th century, superseded the old mode of raising water with buckets in European cities. At that time a great portion of the wells were open—of this, numberless intimations might be found. Thus in Italy, the poet Aurelli, who was made governor of a city by Leo X. was murdered by the inhabitants on account of his tyranny, and his body with that of his mule thrown into a well. In London, it was not till the latter part of the following century that the chain and pulley disappeared. This is evident from the following enactment of the common council of that city the year after the great fire. (1667) "And for the effectual supplying the engines and squirts with water, pumps are to be placed in all wells:"[b]—a proof that many were open and the water raised in buckets.

Pumps are also described in old works on husbandry, gardening, &c. from which it appears that they were often used to raise water for irrigation. In the 'Systema Agriculturæ, being the mystery of Husbandry discovered and laid open,' Lon. 1675, directions are given respecting various modes of making and working them; and it is particularly directed that the rods be made of such a length as to permit the suckers or 'upper boxes' to descend at every stroke below the surface of the water in the well; this it is observed, 'saves much trouble.' The same remark accompanies an account of windmills for watering land [pumps driven by them] in the old 'Dictionarum Rusticum.'

In the mines of Hungary pumps were early introduced, but at what period is uncertain. It is not improbable that those described by Agri-

---

[a] Theatre des Instrumens, 1579. [b] Maitland's History of London, p. 297.

Chap. 6.] *Pumps in German Mines.* 219

cola, were similar to such as were used in some of the same mines by the ancients, and have always formed part of the machinery for discharging water from them since the fall of the Roman empire. All that are figured in the De Re Metallica, are extremely simple, and with one exception are atmospheric or sucking pumps. They are all of bored logs. Some are single pumps, and are worked by men with levers, cranks, and also by a kind of pendulum. Others are double, triple, &c., and worked by water wheels. Of the last some are arranged in rows, and the piston rods raised by cams as in a stamping mill; the weight of the rods carrying them down. Others are placed in tiers one above another; the lowest one raises the water from the bottom of the shaft or well, and discharges it into a reservoir at its upper end: into this reservoir the next pump is placed, which raises it into a higher one, and so on to the top. A pump of this kind from Agricola, has been often republished. It was copied by Bockler and others. A figure of it is inserted in Gregory's Mechanics, Jamieson's Dictionary, &c.

N 88. Pump and Pistons from Agricola.

We have selected No. 88, as a specimen of a single pump, and of upper and lower boxes. A, A, represent two of the latter; the upper part of one is tapered to fit it into the lower end of the pump log as is yet sometimes done. D, B, an upper box, of a kind occasionally used at the present time. The valve or clack is a disk laid loosely over the apertures, and is kept in its place by the rod, which passes through its centre and admits it to rise and fall. C, the conical sucker referred to, p. 214.

The annexed figure of a double pump is from Fludd's works. It appears to have been sketched by him while in Germany, from one in actual use. It is represented as worked by a water wheel, that, by means of cog wheels transmitted motion to the horizontal shaft; the cams on which alternately depressed one end of the levers to which the pump rods were attached, and thus raised the latter. They descended by their own weight, as will appear from an inspection of the figure. The separate view of a rod is intended to show the application of cranks on the horizontal shaft, in place of cams and le-

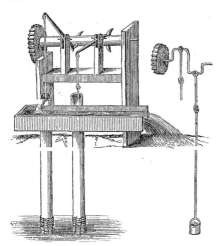

No. 89. Double Pump in German mines. A. D. 1618.

vers. The lower ends of the pumps are inserted in baskets which act as strainers. A double series of pumps, (one over the other) as employed in a mine at Markirch in Germany, is also figured by Fludd. It is interesting on account of the mode of communicating motion to the rods. A crank on the axle of a water wheel imparts motion to a walking beam, as in a steam engine; (in the latter the operation is reversed) and the pump rods are attached to both ends of the beam.[a]

The idea may probably occur to the general reader, that the mechanical talent and enterprise of the preceding and present century, which have produced so many original machines and scarcely left an ancient device unimproved, must have imparted to the old atmospheric pump new features, and made it capable of increased results. It is true that few devices have occupied a greater share of attention, and on none have more efforts to improve them been bestowed; but how far these have been successful may be inferred from the fact—that notwithstanding the endless variety of forms into which its working parts have been changed, and the great number of alledged improvements in suckers, pistons, valves, &c. the machine as made by the ancients, still *generally* prevails; so that were some of their pump makers to reäppear, and visit their fellow craftsmen throughout the world, they would find little difficulty in resuming their occupation.

The pump, although a simple instrument, is confined chiefly to civilized states, while the *extent* to which it is employed, indicates pretty correctly the degree of refinement attained by the people who possess it. Whether it was known to the Egyptians under the Pharaohs or not, may be a question; but when Egypt under the Greeks realized a partial revival of her former glory, the forcing pump we know made its appearance there; and under the second Ptolemy, when that country was a school for the rest of the world, its most valuable modifications were known. In succeeding ages, the atmospheric pump has been a regular attendant on the revival of learning and of the arts. Wherever these have made the most progress, *there* the pump is mostly used. In Germany, France, Holland, Great Britain, and the United States, it is most extensively employed. In Spain, Portugal, Mexico and South America, but partially so. In Turkey, Egypt, Greece, &c. still less; while in Asia and Africa, generally, it is unknown.[b] Egypt, even under the auspices of Mohammed Ali, is not yet prepare to receive it again. Its history in any country is that of the people. Take Russia for an example: of the devices for raising water there, we are informed the inhabitants use the swape, a rope passing over a pulley, (Nos. 13 and 14) a drum on which a rope is wound, (No. 23) horizontal and vertical wheels, and lastly pumps; these last it is said, were formerly *very rare*, but *are now become common*.[c] Just so of the people, they were formerly very rude and ignorant, but are now becoming enlightened.

---

[a] De Naturæ Simia seu Technica macrocosmi historia, pp. 453, 455.
[b] As regards a knowledge of the pump in China, see remarks on Chinese bellows, in the next Book.
[c] Lyell's Character of the Russians, and a detailed history of Moscow. Lon. 1823, p. 63.

CHAPTER VII.

Metallic pumps—Of more extended application than those of wood- Description of one—Devices to prevent water in them from freezing—Wells being closed, no obstacle in raising water from them—Application of the atmospheric pump to draw water from great distances as well as depth—Singular circumstance attending the trial of a Spanish pump in Seville—Excitement produced by it—Water raised to great elevations by atmospheric pressure when mixed with air—Deceptions practised on this principle—Device to raise water fifty feet by atmospheric pressure—Modifications of the pump innumerable—Pumps with two pistons—French marine pump—Curved pump—Muschenbroeck's pump—Centrifugal pump—West's pump—Jorge's improvement—Original centrifugal pump—Ancient buckets figured in this chapter.

THAT the public hydraulic machinery of the Romans was of the most durable materials sufficiently appears from Vitruvius. The chain of pots described by him was, contrary to the practice in Asia and Egypt, wholly of *metal*—the chain was of iron and the buckets of brass. The pumps of Ctesibius that were employed in raising water to supply some of the public fountains, he informs us, were also of brass and the pipes of copper or lead. Some of the oldest pumps extant in Europe are formed altogether of the latter. Leaden pumps were very common in the 16th century. They are mentioned by old physicians among the causes of certain diseases in families that drank water out of them. The pump of the celebrated alchymist, Dee, alluded to in the last chapter, was a leaden one; and which he expected to be able to transmute into gold, by means of the elixir or the philosopher's stone, which he spent his life and fortune in seeking. In the vicinity of some English lead mines such pumps have for many centuries been in use. The Italian pump that led to the discovery of atmospheric pressure was also a metallic one.

The introduction of metals in the construction of pumps greatly extended their application and usefulness, for they were then no longer required to be placed directly *over* the liquids they raised. Those of wood were necessarily placed within the wells out of which they pumped water; but when the working cylinder and pipes were of copper or lead, the former might be in the interior of a building, while the reservoir or well from whence it drew water, was at a distance outside; the pipes forming an air-tight communication between them under the surface of the ground.

The following figure, (No. 90) represents a common metallic sucking pump; the cylinder of cast-iron or copper, and the pipes of lead. It will serve to explain the operation of such machines in detail, and to show the extent of their application. When this pump is first used, water is poured into the cylinder to moisten the leather round the sucker, and the pieces which form the clacks or valves; it also prevents air from passing down between the sucker and the sides of the cylinder when the former is raised. Now the atmosphere rests equally on both orifices of the pipe, the open one in the well, and the other covered by a valve at the bottom of the cylinder: in other words, it presses equally on the water in the cylinder and in the well which covers both;[a] but when by

---

[a] Not absolutely so, or in a strict philosophical sense, but the difference is so slight in an altitude of 25 or 28 feet, (the ordinary limits) as to be inappreciable in a practical point of view.

the depression of the handle or lever, the sucker is raised, this equality is destroyed, for the atmospheric column over the cylinder, and consequently over the valve O is lifted up, and sustained by the sucker alone; it therefore no longer presses on the upper orifice, while its action on the lower one remains undiminished. Then as the external air cannot enter the pipe to restore the equilibrium except through its orifice immersed in the well; in its efforts to do so, (if the expression is allowable) it necessarily drives the water before it on every ascent of the sucker, until the air previously contained in the pipe is expelled, and both pipe and cylinder become filled with water.

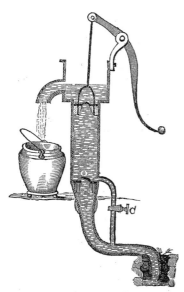

No. 90. Common Metallic Pump.

The subsequent operation is obvious. When the sucker descends, the clack on its upper surface is raised by the resistance of the water through which it passes; and when at the bottom of the cylinder, this clack closes by its own weight: so that when the sucker is again elevated, besides overcoming the resistance of the atmosphere, it carries up all the water above it, and which it discharges at the spout—at the same time the atmosphere resting undisturbed on the water in the well, pushes up a fresh portion into the vacuity formed in the cylinder, and the valve O prevents its return.

The *horizontal* distance between the cylinder or working part of the pump and the well is, in theory unlimited, but in practice it seldom exceeds one or two hundred feet. In all cases where long pipes are used, their bore should be enlarged in proportion to their length, or the velocity with which the sucker is raised, should be diminished; and for this reason—*time* is required to overcome the inertia and friction of long columns of water in pipes; hence a sucker should never be raised faster than the pipe can furnish water to fill the vacuity formed by its ascent. In pumps whose pipes have too small a bore, it frequently happens that the sucker is forcibly driven back when quickly raised, because the water had not time to rush through the pipe and fill the vacuity in the cylinder *as rapidly* as it was formed. The bore of wooden pumps being equal throughout, the water is not pinched or wire-drawn while passing through them, as in most of those of metal. This is one reason why they generally work easier than the latter. It is immaterial in what part of the pipe the valve O is: it is usually placed at the upper end for the convenience of getting to it when requiring repairs. When it fits close to its seat, the water always remains suspended in the pipe, (unless the latter should be defective) as mercury is sustained in a barometer tube.

In cold climates it is a matter of some importance to prevent water in pumps from freezing. Metallic pumps are, from the superior conducting property of their material, more subject to this evil than those of wood. Of various devices a few may be mentioned. The old mode of enclosing

Chap. 7.] *Limits of the perpendicular length of Suction Pipes.* 223

the pump in a case containing tanners' bark, charcoal, the dung of horses, &c. is continued. Others are to prevent the valve O from sitting close to its seat, or to open it, by pressing the sucker upon a pin attached to it, so that the contents of the cylinder and pipe may descend into the well; hence every time the pump is used a fresh portion is required to 'prime it.' A more common method is to connect the lower part of the cylinder with the suction pipe by a stop cock and short tube, as at C. By opening the cock the water in the pump descends through it into the pipe. But the usual practice in this country, is to make the cylinder of such a length that two or three feet of it may be below the surface of the ground, and out of the reach of the frost; about a foot above the valve O or lower box, a plain cock is inserted: in winter this cock is left partially open, and the water above escapes slowly through it into the ground; while that below, into which the sucker is made to extend at its lowest position, serves instead of fresh 'priming.'

A similar device is attached to the lateral pipes that convey the water of the Schuylkill into the houses of Philadelphia.

Some persons can scarcely conceive how the atmosphere can have access to a well, while the latter is covered with slabs of stone or timber, and a thick bed of clay or mould over all. They forget that it is the rarity of air, the extreme minuteness of its particles, which enables it to circulate through the finest soils, as freely as people pass through the various chambers and passages of their dwellings. Were the sides of a well coated, and its mouth covered with the best hydraulic cement—no sooner could the sucker or piston of a pump produce a partial vacuum within it, than the air would stream through the cement as water through a colander or shower bath. And if the top and sides were rendered perfectly air-tight, it would then enter the bottom and ascend through the water without any perceptible obstruction. If it were possible to make a well impervious to air, no water could be raised from it by one of these pumps: no movement of the sucker could then bring it up. We might examine the apparatus with solicitude—remove its defects with care—consult the learned with the Florentines, or get enraged like the Spanish pump maker of Seville;—still, the water, like Glendower's spirits of the deep, would in spite of all our efforts refuse to rise.

When the atmospheric pump is required to raise water from a perpendicular depth, not exceeding 26 or 28 feet, (i. e. in those parts of the earth where the mercury in the barometer generally stands at 30 inches) the length of the cylinder need not exceed that which is required for the stroke of the sucker. In ALL cases, the *perpendicular* distance between the sucker, when at the highest point of its stroke and the level of the water, should never exceed the same number of feet as the tube of a barometer, at the place where the pump is to be used, contains *inches* of mercury. But in the temperate zones where pumps are chiefly used, the pressure of the air varies sometimes to the extent of two inches of mercury, or between two and three feet of water; hence the distance should be something less. And as the level of water in wells is subject to changes, it is the laudable practice of pump makers to construct the cylinder and rod of the sucker, of such a length, that the latter may always work within 26 or 28 feet of the water.

By keeping the above rule in view, water may be raised by these pumps from wells of all depths; for after it has once entered the cylinder, it is raised thence by the sucker independently of the atmosphere, and to any height to which the cylinder is extended. This seems to have been well understood by old engineers. The remark of those who made the Florentine pump is a proof; and others might be adduced from much older authorities. Plate 48, in Besson's Theatre, represents an atmospheric pump raising water from a river to the top of a high tower. The cylinder is square, formed of plank and bound with iron clamps. It is shown as nearly *four times* the length of the suction pipe, which is round. When pump rods are required of great length, they should be made of pine. This wood does not warp, and as it is rather lighter than water, its weight has not to be overcome (like iron rods) when raising the sucker.

A circumstance to which we have slightly alluded, was announced in the public papers of Europe, in the year 1766, which roused the attention of philosophers; for it seemed to threaten a renewal of the disputes about a vacuum, and the ascent of water in pumps and siphons, &c. A tinman of Seville, in Spain, undertook to raise water from a well 60 feet deep, by the common pump. Instead of making the sucker play within 30 feet of the water, he made the rod so short, that it did not reach within 50 feet of it. As a necessary consequence, he could not raise any. Being greatly disappointed, he descended the well to examine the pipe, while a person above was employed in working the pump; and at last in a fit of despair, at his want of success, he dashed the hatchet or hammer in his hand, violently against the pipe. By this act a small opening was made in the pipe about ten feet above the water—when, what must have been his surprise! the water instantly ascended and was discharged at the spout!

The fact being published, it was by some adduced as a proof that the pressure of the atmosphere could sustain a perpendicular column of water much longer than 32 or 34 feet, and consequently that the experiments of Torricelli and Pascal were inconclusive. M. Lecat, a surgeon at Rouen in Normandy, repeated the experiment with a pump in his garden: he bored a small hole in the suction pipe ten feet above the water, to which he adapted a cock. When it was open, the water could be discharged at the height of 55 feet, instead of 30 when it was shut.

As might be supposed, these experiments when investigated, instead of overthrowing the received doctrine of atmospheric pressure, more fully confirmed it. It was ascertained that the air on entering the pipe became mixed with the water; and which therefore, instead of being carried up in an unbroken column, was raised in disjointed portions, or in the form of *thick rain*. This mixture being much lighter than water alone, a longer column of it could be supported by the atmosphere: and by proportioning the quantity of air admitted, a column of the compound fluid may be elevated one or two hundred feet by the atmospheric pump; but there is no advantage in raising water in this manner *by the pump*, and we believe it is seldom or never practiced. In a paper, on the duty performed by the Cornwall Steam Engines in raising water, in the Journal of the Franklin Institute for May, 1837, it is stated that a little air is sometimes admitted in the pump pipes, which it is alledged, "made the pump work more lively, in consequence of the spring it gave to the column of water, and caused less strain to the machinery." In the same paper Mr. Perkins states that forty years before, an attempt was made to impose upon him in this country, a pump which raised water by atmospheric pressure 100 feet: but he detected "a small pin hole" in the pipe through which the air was admitted.

The same deception it seems gave rise to the humorous poetical satire, ' *Terrible Tractoration.*' The ingenious author states in his preface, that he was employed in 1801, as agent for a company in Vermont, and of which he was a member, to proceed to London, and secure a patent for 'a new invented hydraulic machine.' "I was urged to hurry my departure in consequence of a report in circulation, that certain persons by stealth had made themselves masters of the invention, and were determined to anticipate us in our object of securing a patent in London. In consequence of this report, the experiments made with this machine were performed in a hasty manner. By it, water was raised through *leaky* tin pipes in a hasty experiment, 42 feet from the surface of the fountain to the bottom of the cylinders, in which the pistons were worked. I embarked from New-York the 5th of May, and arrived in London after a tedious passage the 4th of July. I waited on Mr. King, then ambassador from the United States, to whom I had letters, and was by him favored with a letter to Mr. Nicholson, an eminent philosopher and chemist. With this gentleman I had several interviews on the subject of my hydraulic machine, and from him received an opinion in writing unfavorable to its merits. I likewise made a number of experiments in London, with a different result from what I had seen in Vermont. In this desperate situation of the adventure, I received a letter from one of the Vermont Company, informing me there was a deception in the patent—that from experiments made subsequent to my departure, it appeared that no water could be raised by *Langdon's invention* higher than by the common pump, unless by a perforation in the pipe, which made what the inventor called an air hole, and which by him *had been kept a secret*. Mr. Nicholson informed me that a similar deception had been practised on the Academicians of Paris, but that the trick was discovered by the hissing noise made by the air rushing into the aperture." From the disappointment Mr. Fessenden turned to his pen, and wrote ' *The modern Philosopher or Terrible Tractoration.*' See preface to 2nd American ed. Phil. 1806.

It is possible however to raise water by a short cylinder, fifty or even a hundred feet high, but for all practical purposes the device is useless. The first thing of the kind that we know of, was accomplished nearly forty years ago by a boy. He fixed a small pump (the cylinder was 12 inches in length) in the garret of a high dwelling, and a tub of water in the cellar, the perpendicular distance being nearly 50 feet. About half way up the stairs, he placed a close vessel, (a three gallon tin boiler) from the bottom of which a small leaden tube was continued to the pump cylinder; and another tube being soldered to the top, descended into the tub of water. A third tube was soldered to the top of the vessel, and terminated near the pump, having a cock soldered to the end. This cock being shut and the pump worked, the air in the pipes and the vessel was withdrawn, and the latter consequently filled with water by the atmosphere; he then opened the cock which admitted the atmosphere to act on the surface of the water in the vessel, and by again working the pump the contents of the vessel were raised and discharged in the garret. By a series of close vessels placed at distances not exceeding 30 feet above each other, water may be raised in this manner to any elevation.

It is impossible to notice here a moiety of the projects for improving the atmospheric pump and the various parts of which it is composed; their name is *legion*, and this volume is far too limited to comprise an account of them all. Those that we are about to describe are of modern date, but it does not therefore follow that they were unknown to the ancients. Men in every age, when striving to accomplish a specific object, naturally fall into similar trains of thought, and hit upon the same or nearly the same devices. Could the ancient history of this machine be procured, it would we have no doubt prove, that (like the instruments invented by a celebrated French surgeon, fac-similes of which of exquisite finish, were subsequently found in Pompeii) not a few of its diversified modifications were anticipated by Greek and Roman machinists. Why then were they not preserved or continued in use? For the same reason that the old pump is still generally preferred: and were it not for the art of printing it is probable that not one of the modern improvements of this machine would be known 2000 years hence, any more than those devised by the ancients are now known to us. Those persons who are familiar with it, well know that a large majority of its supposed improvers, have returned from long and laborious mental pilgrimages in its behalf, laden, like old devotees, with little else than stores of worthless relics.

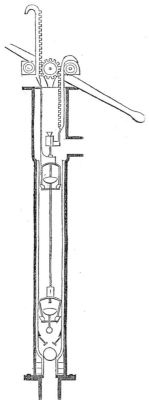

No. 91. Double Piston Pump.

Of innumerable variations in its construction, the greater part consists of different modes of communicating motion to the rod, by wheels, cranks, racks and pinions, cams, plain and jointed levers, pendulums, balance poles, vibrating platforms, &c. Of these it would be useless to speak. Others consist in two or more suckers in the same cylinder; in altering the form of the latter; and some in imparting motion to the cylinder, and dispensing with the sucker. We shall notice some of these here, and others in the next Book.

The introduction of two suckers or pistons into one cylinder has long been a favorite project. Dr. Conyers in 1673 proposed a pump of this kind. He made it of plank, square and *tapered*, (in the form of an inverted and truncated pyramid,) $8\frac{1}{2}$ feet long, 20 inches square at the upper end, and 8 at the bottom where the valve or lower box was placed. He fixed *two* suckers on the *same* rod, one at its lower end and the other so as to play half way down the trunk. This pump he said, raised "at least twice as much water as the ordinary one of the same size." If such was the fact, it was by the expenditure of twice as much force. Had the bore of the trunk, where the upper sucker played, been uniform throughout, and the lower sucker laid aside, and with it the force expended in moving it, the result would clearly have equalled that of both. Phil. Trans. Abridg. Vol. i, 545.

About the year 1780, Mr. Taylor of Southampton, Eng. introduced two suckers or pistons into one cylinder, each united to a separate rod, that one might ascend as the other descended, and thus discharge double the quantity of water: No. 91 is a figure of it. The rod of the lower sucker slides through the centre of the upper one; and also through its valve, which is a spherical or hemispherical piece of brass, placed loosely over its seat and to which the rod acts as a guide. The upper parts of the rods terminate in racks, between which a cog wheel is placed, having an alternate movement imparted to it, by a lever attached to its axis, as in the common air pump.

Another mode of working this pump, is by means of a drum fixed to one end of the shaft of the cog wheel; over this a rope is passed and crossed below, to which any number of men, on each side, may apply their strength. Both parties pull the rope towards them by turns, and thereby impart the requisite movement to the cog wheel, and consequently to the pump rods and suckers, as shown in No. 92. Mr. Adams, in his Lectures on Natural Philosophy, published in 1794, observed that these kind of pumps had been "in general use in the royal navy for five or six years." Vol. iii, 392.

No. 92. Working Ship Pumps by Ropes.

In 1813, the London Society of Arts awarded a medal and twenty guineas to Mr. P. Hedderwick, for various modes of imparting motion to two pistons in the same cylinder, by a series of *levers*, instead of cog wheels and racks. Trans. vol. xxxii, 98.

Atmospheric pumps with two pistons are used in the French marine, and are arranged so as to be worked by the men as in the act of rowing. Neither racks nor pinions are used in communicating motion to the rods. The upper ends of these are continued outside the cylinders and bent a little outwards, and then connected by a bolt to each end of a short vibrating beam which is moved by the men. The rods do not descend in the centre of the cylinder, as in the preceding figure, but are attached to one side of the suckers. The lower rod passes through an opening in the upper sucker, which is closed by a collar of leather. Hachette's Traité Elémentaire des Machines. Paris, 1819, p. 153.

Pumps with double pistons are not of modern date: there is one figured in Besson's Théatre des Instrumens.

The alledged superiority of these pumps is more specious than real. It is true the inertia of the water in ascending the pipes has not to be overcome at every stroke, as in the common pump, since its motion through them is continuous; nor is its direction changed, as when two separate cylinders are used, being then diverted into them from the pipes at angles more or less acute. These are real advantages; but if we mistake not, they are the only ones, unless taking up less room on ship board be another. But from the cylinders being twice the ordinary length, these machines are really double pumps; having not only two suckers and two rods, but also two cylinders, and requiring twice the power to work them. The principal difference between them and the usual double pump, is that the cylinders are united together on the same axis, while in the latter, they are placed parallel to each other. In point of economy, we think pumps with two distinct cylinders are preferable; they are less complex, and of course less liable to derangement: a longer stroke can be obtained in them, and, what is of more importance, when one is disordered, the other can be continued in use. On these considerations we believe double piston pumps were abandoned in the British navy.

A singular modification of the common pump was devised in England in 1819, for which the Society of Arts awarded a premium of twenty guineas. The chamber was curved, and the centre of the circle, of which it formed a part, served as a fulcrum on which the rod and handle (both of one piece) moved. The rod was curved so as to move in the centre of the chamber.

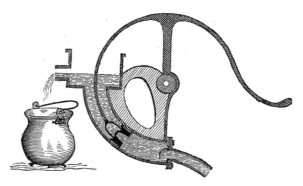

No. 93. Curved Pump.

The objects supposed to have been attained by this arrangement, were "greater simplicity of workmanship," and "greater steadiness and precision of action" (of the sucker.) The device is ingenious, but can never be generally adopted. The spring of the rod with the wear of the bolt on which it turns, must soon render the play of the sucker and wear of the chamber unequal: the difficulty and expense of making the latter curvilinear, and of repairing it when bruised or otherwise injured, are fatal objections. The pipe must be separated from the chamber to get at the lower box or valve; and the application of the pump is limited to depths within 30 feet. We have noticed it, lest the same idea occurring to some of our mechanics, should lead them to a useless expenditure of time and money. In the same year a patent was issued in England for making the cylinder in the form of a ring, or nearly so, the centre of which was the fulcrum on which the piston turned, and an alternating motion was imparted to the latter. Repertory of Arts, vol. xxxv. 1819.

An interesting modification of the atmospheric pump was described by Muschenbroeck in his Natural Philosophy. Instead of a piston or sucker working inside of the cylinder, the latter itself is moved, being made to slide over the pipe somewhat in the manner of telescope tubes. No. 94 represents this pump. The upper end of the suction pipe, being made of copper or brass, and its exterior smooth and straight, is passed

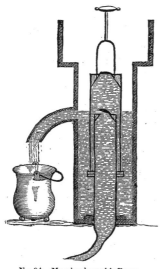

No. 94. Muschenbroeck's Pump.

through the bottom of a small cistern. Its orifice is closed by a valve opening upwards. A short cylinder whose diameter exceeds that of the suction pipe is slipped over the latter; and to its lower end a stuffing box is adapted to prevent air or water from passing between them. Its upper

Chap. 7.] Centrifugal Pump. 229

end is covered by a valve also opening upward. The pump rod is attached to the same end by a fork, as represented in the figure. By moving the cylinder up and down, the air within it and the pipe is soon expelled, and its place occupied by a portion of the water in which the lower end of the suction pipe is immersed. When the cylinder is then raised the atmosphere forces up water into it, and when it is depressed, the water being prevented by the valve on the end of the pipe from descending into the well, escapes out of that on the top of the cylinder, precisely as in the bellows pump. (p. 206.) By keeping water in the cistern, air is effectually prevented from entering between the pipes at the stuffing box, even if it be not perfectly tight. A cup or dish formed on the upper end of the cylinder to contain a little water over the valve, would be an advantage in this description of pumps, for any defects in it by which air is admitted would be fatal, as a vacuum could not then be formed within the cylinder, and of course no water raised by it. Our common pumps would be almost useless if water was not kept over the valves; it is that which renders them air tight, and consequently efficient.

In the early part of the 18th century, a new method of exciting the pressure of the atmosphere for the purpose of raising water was adopted. Its discoverer burst the fetters with which long established modes of accomplishing this object had embarrassed common minds. He left the old track entirely, and the result of his researches was a philosophical machine that bears no resemblance to those by which it was preceded.

Most people are practically acquainted with the principle of the *Centrifugal pump*, viz. that by which a body revolving round a centre tends to recede from it, and with a force proportioned to its velocity: thus mud is thrown from the rims of carriage wheels, when they move rapidly over wet roads; a stone in a sling darts off the moment it is released; a bucket of water may be whirled like a stone in a sling and the contents retained even when the bottom is upwards. A sailor on ship board, or a housemaid, dries a wet mop by whirling it till the force communicated to the watery particles overcomes their adhesion to the woolen fibres. Boys sometimes stick pellets of tough clay to the end of a switch or flexible rod, and then drawing it quickly through the air, the force imparted to the balls sends them to their destination. If a tube be substituted for the rod, and the end that is held in the hand closed, by a similar movement, balls dropped or water poured into it, would be thrown forward in like manner; and if by some arrangement the movement of the tube was made continuous, projected streams of either balls or water might be rendered constant: the centrifugal gun is a contrivance to accomplish the one—the centrifugal pump the other.

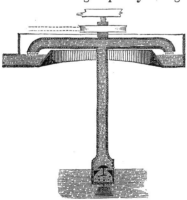

No. 95. Centrifugal Pump.

This pump generally consists of tubes, united in the form of a cross or letter T, placed perpendicularly in the water to be raised. (No. 95.) The lower end is supported on a pivot; perforations are made to admit the water, and just above them a valve to retain it when the pump is not in motion. The ends of the transverse part are bent downwards to discharge the water into a circular trough, over which they turn. To charge it, the orifices may be closed by loosely inserting a cork into each, and then filling the pump through an opening at the top which is then closed by a screw cap. A rapid rotary motion is imparted to the machine by a pulley fixed on the axis and driven by a band, from a drum, &c. The centrifugal force thus communicated to the water in the arms or transverse tube, throws it out; and the atmosphere pushes up the perpendicular one fresh portions to supply the place of those ejected. These pumps are sometimes made with a single arm like the letter L inverted; at others quite a number radiate from the upright one. It has also been made of a series of tubes arranged round a vertical shaft in the form of an inverted cone. A valuable improvement was submitted by M. Jorge to the French Academy in 1816. It consists in imparting motion to the arms only, thus saving the power consumed in moving the upright tube, and by which the latter can be *inclined* as circumstances or locations may require.

A combination of the centrifugal pump with Parent's or Barker's mill, was proposed by Dr. West, which in some locations may be adopted with advantage. It is simply a vertical shaft round which two tubes are wound: (No. 96) the upper one is the pump; the lower one the mill. The area of the lower one should be to that of the upper in the inverse ratio of the perpendicular height, and as much more as is necessary to overcome the friction. The cup or basin into which the stream (part of which is to be raised) is directed, may be attached to the shaft and turn with it, or the latter may pass through it. Tilloch's Phil. Mag. vol. xi.

N 96. West's Pump.

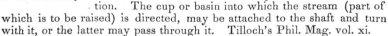

The first centrifugal pump appears to have been invented by M. Le Demour, who sent a description of it to the French Academy in 1732. (Machines approuvé. Tom. vi, p. 9.) It was merely a straight tube attached in an inclined position to a vertical axis, and whirled round by the handle—the tube was fastened by ligatures to three strips of wood projecting from the axis, as shown at No. 97.

With this pump we close our remarks on devices for raising water by atmospheric pressure; more might have been added, but as nearly all the machines yet to be described illustrate the same principle, the reader is referred to the following Books, and particularly to the atmospheric and forcing pumps described in the next one.

No. 97. Le Demour's Pump.

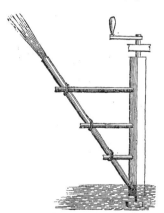

[The vessels under the pump spouts in Nos. 90, 93 and 94, are Roman bronze buckets from Pompeii.]

END OF THE SECOND BOOK.

BOOK III.

MACHINES FOR RAISING WATER BY COMPRESSURE INDEPENDENTLY OF ATMOSPHERIC INFLUENCE.

CHAPTER I.

DEFINITION of machines described in this Book—Forcing Pumps—Analogy between them and bellows—History of the bellows that of the pump—Forcing pumps are water bellows—The Bellows of antediluvian origin—Tubal Cain—Anacharsis—Vulcan in his forge—Egyptian, Hindoo, and Peruvian blowing tubes—Primitive bellows of goldsmiths in Barbary—Similar instruments employed to eject liquids—Devices to obtain a continuous blast—Double bellows of the Foulah blacksmiths, without valves—Simple Asiatic bellows—Domestic bellows of modern Egypt—Double bellows of the ancient Egyptians—Bellows blowers in the middle ages—Lantern bellows common over all the East—Specimens from Agricola—Used by negroes in the interior of Africa—Modern Egyptian blacksmiths' bellows—Vulcan's bellows—Various kinds of Roman bellows—Bellows of Grecian blacksmiths referred to in a prediction of the Delphic oracle—Application of lantern bellows as forcing pumps—Sucking and forcing bellows pumps—Modern domestic bellows of ancient origin—Used to raise water—Common blacksmiths' bellows employed as forcing pumps—Ventilation of mines.

MACHINES of the third class described in this Book, are such as act by compressure: the water is first admitted into close vessels and then forcibly expelled through apertures made for that purpose. This is effected in some by compressing the vessels themselves, as in bellows pumps—in others by a solid body impinging on the surface of the liquid, as in fire engines—sometimes a column of water is used for the same purpose, at others the expansive force of compressed air. Of the last two, Heron's fountain, air engines, and soda fountains, are examples. Strictly considered, these machines have nothing to do with the pressure of the atmosphere, (the active principle of those of the second class,) but in practice it is very generally employed. When the working cylinder of a forcing pump is immersed in the water it is intended to raise, or when the latter flows into it by gravity, it is a forcing pump simply; but when the cylinder is elevated *above* the water that supplies it, and consequently is then charged by atmospheric pressure, the machine is a compound one, embracing the peculiar properties of both sucking and forcing pumps. The latter therefore differ from the former in raising water above their cylinders; and to elevations that are only limited by the strength of their materials and the power employed to work them. They have been considered by some writers as the oldest of all pumps. We shall consider their varieties in the order in which we suppose they were developed.

An intimate connection has ever subsisted between the forcing pump and the bellows; they are not only identical in principle, but every form adopted in one has been applied to the other. The bellows, from the simple sack or skin employed by the negroes of Africa to the complex and efficient instrument of China, and the enormous blowing machines of our foundries, has been used to raise water: and every modification of the pump, not even excepting the screw, has been applied as a bellows.[a] A singular proof of the analogy between them and of their connection in ancient times, is, that in one of the earliest accounts we have of the *cylindrical* pump, (viz. by Vitruvius) it was used as a bellows "to supply wind to hydraulic organs." And that rotary pumps are as numerous as rotary bellows, is known to every mechanic. Thus, while pumps have been used as bellows, bellows have been employed as pumps; and every device to obtain a continuous current of air in the one, has been adopted to induce an unbroken stream of water in the other.

The history of the bellows is also that of the pump; and if we mistake not it affords the only legitimate source now open in which the origin of the latter can be sought for with any prospect of success. Under this impression we shall examine the bellows of various people, and in doing so the reader will find an auxiliary, but very important branch of the subject, illustrated at the same time, viz. that which relates to VALVES, for the bellows was probably the first instrument of which they formed a part. No other machine equally ancient can be pointed out in which they were required. In fine, the forcing pump is obviously derived from the bellows, or rather it is an application of that instrument to blow water instead of air—an application probably coeval with its invention.

The origin of the arts is generally considered as a subject of mere conjecture. Antiquarians and historians despair of discovering any thing of importance relating to the early history of any of the simple machines. In the present case, however, there can be no doubt that the *first* bellows was the *mouth*; and it was the first pump too, both atmospheric and forcing. The representation of it when employed as a bellows was a favorite subject with ancient statuaries and painters. Pliny gives several examples, and among others, Stipax the Cyprian, who cast an elegant figure of a boy "roasting and frying meat at the fire, puffing and blowing thereat with his mouth full of wind, to make it burn." Aristoclides, was also celebrated for a painting of a boy, "blowing hard at the coals; the whole interior of the room appeared to be illuminated with the fire thus urged by the boy's breath, and also what a mouth the boy makes." Holland's Translation.

That the bellows is of antediluvian origin, there can be little doubt, for neither Tubal Cain nor any of his pupils could have reduced and wrought *iron* without it. The tongs, anvil and hammer of Vulcan, (or Tubal Cain) have come down to our times, and although the particular form of his bellows be not ascertained, that instrument is, we believe, as certainly continued in use at the present day, as the tools just named. Nor is there any thing incredible in such belief, for if even the common opinion, that the whole globe was enveloped in the deluge, be true, Noah and his sons, aware that the destinies of their posterity, so far as regarded the arts of civilization, must in a great measure depend upon them, would naturally secure the means of transmitting to them the knowledge of those machines that related to metallurgy, as among the most essential of all. Of these, the bellows was quite as important as any other; without it, other tools would have been of little avail. Now if we refer to oriental machinery, (among which the bellows of the son of Lamech is to be found if at all,) we shall find, in accordance with its characteristic unchangeableness, that the instrument now used over all Hindostan and Asia in general, and by the modern blacksmiths of Cairo and Rosetta, is identical with

---

[a] Hachette's Traité élémentaire des Machines, p. 142.

that with which the smiths of Memphis, and Thebes, and Heliopolis, urged their fires, between three and four thousand years ago, and is similar to those found figured in the forges of Vulcan on ancient medals and sculptures. Numerous were the forms in which the bellows was anciently made, but the general features of the one to which we allude, (the lantern bellows) have remained as unchangeable as those of blacksmiths themselves.

Strabo attributed the bellows to Anacharsis who lived about 600 years B. C. but it is probable that some particular form of it only was intended, for it is not credible that the Greeks in Solon's time could have been ignorant of an instrument that is coeval with the knowledge of metals; and without which the *iron money* of Lycurgus, two centuries before, could never have been made. Pliny (B. vii, 56) attributes it with greater propriety to the Cyclops, who are supposed to have flourished before the deluge. The prophet Jeremiah, who lived long before Anacharsis, speaks of it in connection with metallurgical operations. "The bellows are burned, the lead is consumed of the fire, the founder melteth in vain." Isaiah, who lived still earlier, viz. in the 8th century B. C. alludes to the blacksmith's bellows—"the smith that bloweth the coals in the fire." And Job, nine or ten centuries before the Scythian philosopher flourished, speaks of "a fire not blown." The prophet Ezekiel also speaks of the blast furnace as common—"they gather silver, and brass, and iron, and lead, and tin, into the midst of the furnace, to *blow the fire* upon it to melt it." xxii, 20. Homer, as might be supposed, could not fully describe the labors of Vulcan, without referring to this instrument. His account of the great mechanic at work, is equally descriptive of a smith and his forge of the present day.

> Obscure in smoke, his forges flaming round,
> While bathed in sweat from fire to fire he flew;
> And puffing loud, the roaring bellows blew.
> * * * * *
> Just as the god directs, now loud, now low,
> They raise a tempest, or they gently blow.
> *Iliad*, xviii, 435, 545. Pope.

The first approach made to artificial bellows was the application of a reed or other natural tube, through which to direct a stream of air from the mouth—a device that has never passed into desuetude. Such was the origin of the modern blow-pipe, an instrument originally designed to increase the intensity of ordinary fires, but which subsequently became (as the arts were developed) indispensible to primitive workers in metal. How long blowing tubes preceded the invention of other devices for the same purpose is uncertain; but from the fact that oriental jewelers and goldsmiths still fuse metal in pots by them, it may be inferred they were the only instruments in use for ages, before the bellows proper was known: a circumstance to which their universal employment over all Asia at the present time may be attributed, and the skilful management of them by mechanics there. As the only contrivance for urging fires in primitive times, men would naturally become expert in using them, and, as in all the arts of the East, their dexterity in this respect would be inherited by their children, and be retained in connection with their use, with that tenacity that has scarcely ever been known to give up an ancient tool or the ancient mode of using it: hence the paucity of their implements; a file, a hammer, a pair of tongs, and a blowing tube, being in general all that the budget of an African or Asiatic jeweler contains.

As we have given figures of sucking tubes to illustrate the origin of the atmospheric pump, we here insert some of blowing tubes, as showing the incipient state of the forcing pump.

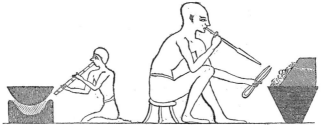

No. 98. Egyptian using a reed. 1600 B. C.   No. 99. Ancient Egyptian Goldsmith.

No. 98, represents an Egyptian blowing a fire with a reed. It is from the paintings at Beni Hassan, and extends back through a period of 3,500 years. According to Mr. Wilkinson, the figure is that of a goldsmith, "blowing the fire for melting the gold," but from the comparative large size of the vessel, it would seem rather to be a cauldron in which the articles were *pickled*. No. 99, is the figure of a goldsmith either soldering or fusing metal with the blow-pipe, from the sculptures at Thebes. The portable furnace has raised cheeks to confine and reflect the heat. The pipe is of metal with the end enlarged and pointed.[a]

No. 100.   Goldsmith of Hindostan

Sonnerat, has given (in the volume of illustrations to his voyages,) a plate representing modern goldsmiths of Hindostan, from which the annexed figure (No. 100) is copied. It will serve to show, when compared with the preceding cuts, what little changes have taken place in some mechanical manipulations in the East, from very remote times. A similar figure is in Shoberl's Hindostan. The same mode of fusing their metals was practiced by the ancient gold and silver smiths of Mexico and Peru. Instead of bellows, says Garcilasso, the latter had blow-pipes "made of copper, about a yard long, the ends of which were narrow, that the breath might pass more forcibly by means of the contraction, and as the fire was to be more or less; so accordingly they used eight, ten, or twelve of these pipes at once, as the quantity of metal did require." (Commentaries on Peru, p. 52.)

The next step was to apply a leathern bag or sack, formed of the skin of some animal, to one end of the tube (shown in No. 80) as a substitute for the mouth and lungs. The bag was inflated by the act of opening it, or by blowing into it, and its contents expelled by pressure. To such Homer seems to allude in his account of Eolus assisting Ulysses:

> The adverse winds in leathern bags he braced,
> Compressed their force, and locked each struggling blast.   *Odys.* 10.

---

[a] Ancient bronze tongs or forceps, similar to those in the cut, have been found in Egypt, which retain their spring perfectly. Crucibles similar to those used at the present day have also been discovered. Wilkinson's Manners and Customs of the Ancient Egyptians. vol. iii, 224.

Chap. 1.]   *Origin of the Valve.—African Bellows.*   235

And Ovid
A largess to Ulysses he consigned,
And in a steer's tough hide enclosed a wind. **Met.** xiv.

The goldsmiths' bellows of Barbary consists of a goat's skin, having a reed inserted into it: 'he holds the reed with one hand and presses the bag with the other.' (Ed. Encyc. vol. iii, 258.) The Damaras, a tribe of negroes in Southern Africa mentioned by Barrow, manufacture copper rings, &c. from the ore. The bellows they use, he observes, "is made of the skin of a gemsbok, (a species of deer) converted into a sack, with the horn of the same animal fixed to one end for a pipe."

Simple instruments of this description have always been applied to eject liquids. Small ones were commonly used by ancient physicians in administering enemas; a purpose for which they are still used. Large ones were recommended by Apollodorus the architect, a contemporary of Pliny and Trajan, as a substitute for fire engines, when the latter were not at hand. When the upper part of a house was on fire, and no machine for throwing water to be procured, hollow reeds, he observed, might be fastened to leathern bags filled with water, and the liquid projected on the flames by compressing them.

As the current of wind from a single sack or bag, necessarily ceased as soon as it was collapsed, some mode of rendering the blast *continuous* was desirable; and in the working of *iron* indispensible. The most obvious plan to accomplish this was to make use of *two* bags, and to work them so that one might be inhaling the air, while the other was expelling it—that is, as one was distended, the other might be compressed. This device we shall find was very early adopted, and by all the nations of antiquity.

But by far the most important improvement on the primitive bellows or bag, was the admission of air by a *separate opening*—a contrivance that led to the invention of the VALVE, one of the most essential elements of hydraulic as well as pneumatic machinery. The first approach to the ordinary valve, was a device that is still common in the bellows of some African tribes. A bag formed of the skin of a goat, has a reed attached to it to convey the blast to the fire; and the part which covered the neck of the animal is left open for the admission of air. This part is *gathered up* in the hand when the bag is compressed, and opened when it is distended.

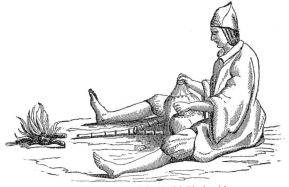

No. 101.   Bellows of the Foulah Blacksmiths.

An improvement upon this primeval device is exhibited in the bellows of the Foulah blacksmiths, on the western coasts of Africa. It consists of two calabashes connected together by two hollow bamboos or reeds, inserted into their sides, and united at an angle to another which leads to the fire, as represented in the figure. A large opening is made on the top of each, and a cylindrical bag or tube made of soft goats' skins stitched or otherwise secured round the edges. The workman seats himself on the ground, and placing the machine between his legs, he grasps the ends of the bags, and by alternately raising each with the mouth *open*, and pushing it into the calabash when *closed*, the air in the latter is forced into the fire, and a uniform blast maintained. The action is very similar to that of gathering in the hands the lower edges of two hat linings, and constantly drawing one out and thrusting the other in.

The negroes of the Gold Coast are represented to have other kinds of bellows. The principal tools of their smiths, are "a hard stone for an anvil, a pair of tongs and a small pair of bellows, with three or more pipes, which blow very strong—an invention of their own." We have not been able to find any description of these. See Grand Gazetteer, Art. Guinea; and Histoire Générale, tom. v, 214.

Another species equally simple but more efficient, is common in Asia, Africa, and also in Wallachia, Greece and other parts of Europe. The contrivance for admitting the air is an improvement upon the last, but the orifice is still opened and closed by the fingers of the blower. Instead of the mouth of each bag being drawn up in the hand, it is stretched out in the form of a long slit; to the lips of which two strips of wood are sewed. The inner side of each strip is made straight and smooth, so that when brought together, they form a close joint. They are grasped in the middle by the workman, who alternately opens them when he raises the mouth to admit the air, and closes them when he expels it.

No. 102. Primitive Bellows of Asia.

No. 102 represents the assistant of a Hindoo blacksmith, urging his fire with a pair of these instruments, (copied from the volume of plates to Sonnerat's Voyages.) From an inspection of the figure, it will be perceived that the strips facilitate the act of compressing each bag, by their extending quite over it, as well as by their stiffness: in these respects they may be considered as the nucleus of the boards in the common bellows. In this device, the *valve* becomes further developed.

To similar instruments, Mr. Emerson refers in his 'Letters from the Egean.' The crew of a Hydriot vessel having taken her ashore at Paros to repair the iron clasp of her rudder, an opportunity occurred of examining their bellows. Mr. E. describes them as "a very antique device," consisting of "two sheepskins, united by an iron pipe introduced into the fire, which were alternately dilated with air and compressed, by an Arab slave who knelt above them." With the exception of their not being made of bull's hide but of sheepskin, he observes they would completely suit the description of the bellows given by Virgil in the Fourth Georgic. Blacksmiths in Ceylon use the same kind, but made of bullocks' hides, and furnished with nozzles of bamboo. The blower seats himself on the ground between the two bags, and works them with his hands, pulling up one and pushing down the other. (See a figure in Davis' history of that island, and also in the Register of Arts, vol. i, 300.) The *domestic* bellows of Egypt is made in the same way, and probably has always been so: to it, Job most likely alluded, (chap. xx, 26.) "The ordinary hand bellows now used for small fires in Egypt, (says Mr. Wilkinson) are a sort of bag made of the skin of a kid, with an opening at one end like the mouth of a common carpet bag, where the skin is sewed upon two pieces of wood; and these being pulled apart by the hands and closed again, the bag is pressed down and the air thus forced through the pipe at the other end."

The next improvement seems to have been that by which the slit was superseded by a flap or clack, so as to be *self-acting*, as in the ordinary European or American bellows—in other words a *valve*, that opened by the pressure of the atmosphere when the bag was raised, and which was closed by its own weight or by the elasticity of the confined air. Among the interesting discoveries which recent examinations of Egyptian monuments have brought to light, figures of such bellows have been found sculptured in a tomb at Thebes, which bears the name of Thothmes III, one of the Pharaohs who was contemporary with Moses. No. 103 represents four employed at one fire, each pair being worked by the hands and feet of a laborer, and in a manner singularly ingenious and effective; proving that the Egyptians of those times well knew how to combine muscular energy with the weight of the body to produce a maximum effect.

No. 103. Egyptian Bellows and Bellows Blowers. 1500 B. C.

The bags were secured to frames or to the ground, and appear to have had rings of cane within them to keep the leather extended in a horizontal direction. A separate pipe proceeded from each to the fire. The valves or clacks are not shown because being placed underneath they were out of sight. In working them a laborer stood upon two, one under each foot, and taking two cords in his hands, the lower ends of which were secured to the top of the bags; he alternately rested his weight upon each to expel the air, and inflated them when exhausted by pulling the cords; thus the whole weight of his body was uninterruptedly employed in closing one bellows, while the muscular force of his arms was incessantly engaged in opening another. We question if a more simple and efficient application of human effort can be produced.

Such bellows were used in Egyptian kitchens, and were indeed necessary when the massive cauldrons and huge joints of meat boiled in them, are considered.[a] The same practice continued through the middle ages, in Europe, when 'bellows blowers' formed part of the establishment of

---

[a] Wilkinson's Manners and Customs of the Ancient Egyptians. vol. ii, 384. Vol. iii, 339

royal kitchens, and whose duty it was "to see that soup when on the fire was neither burnt nor smoked."[a] Among the relics that formerly belonged to Guy, the famous Earl of Warwick, is a cauldron or kitchen boiler, made of bell-metal, which contains 120 gallons, but whose capacity does not equal that of more ancient ones.[b] To the old custom of employing persons exclusively at the bellows, as in the preceding cut, Virgil alludes in the following line:

> One stirs the fire and one the bellows blows. *Æn.* viii.

Every modern bellows maker would be convinced from an inspection of the last figures, that valves were employed, since the instruments could not possibly have acted without them; but all doubts respecting an acquaintance with the valve in those remote ages when the sculptures were executed, is removed by two other bellows portrayed in the same tomb, and shown in the next cut. These differ from the preceding and were

No. 104. Egyptian Bellows in use before the Exodus.

perhaps intended to show another variety of the instrument as made in those times. Their upper surfaces seem to have been of wood, in the centre of which, the orifices of the valves are distinctly shown; the valves or clacks were therefore inverted, as in our ordinary bellows turned upside down. To persons not familiar with the subject, this circumstance might excite surprise, but the class to which these belong have almost *always* had the valve in the *moveable* board; and in whatever position they were used—whether horizontally as in these figures, or vertically as in the next. In Ceylon and other parts of the East they are used as shown in No. 104.

But are not both bellows in the last cut double-acting, that is, impelling air from them both when moved up as well as when pushed down? From the figures it would seem that such were intended; for *two pipes* are represented as proceeding from *each*, while one only is connected to those in No. 103; and one instrument was deemed sufficient to occupy one laborer—to this there possibly may be an allusion in the knots on the ends of the cords, which, in the hieroglyphical language of Egypt may signify the greater liability of slipping through the hands, in consequence of the superior force required to work them. Indeed *four* different bellows are represented. In No. 103, two are made of single bags, and two of double ones, as appears by the bands around them: and in No. 104, one is round like the lantern bellows, and the other oblong, both kinds of which are common at this day in the East; and both, as already remarked, seem to be double acting like those of our smiths.

This variety was probably designedly introduced into the sculptures to aid in conveying to posterity a knowledge of the state of the arts at that time in Egypt. The circumstance is an interesting one, and should lead to a more thorough examination of those wonderful, those eternal records of the arts and sciences of past ages, than has ever been given them; not only every group but every figure among the millions imprinted on these

---

[a] Fosbroke's Encyc. Antiq.   [b] Moule's English Counties. Lon. 1831.

imperishable pages, deserves not merely to be scrutinized, but accurately copied. Many of them are fraught with information of the highest interest to the arts; and whether the mass of hieroglyphical records be ever understood or not, there is no difficulty in comprehending the most interesting of these.

One of the figures in the last illustration is obviously a modification of the old *lantern* bellows (so named from its resemblance to the paper lantern, still common in Egypt:) they consist of two circular boards united to the ends of a cylindrical bag of flexible leather. In the centre of one board is an opening covered by a flap opening inwards, and to the other the tuyere is attached. In working them, the board, through which the air is admitted, is moved, and the other kept stationary. They are quite common in Asia, Egypt, and generally throughout the oriental world; and appear to have undergone no change whatever, either in their materials, form, or modes of working them, since the remotest times: even working them by the feet, as practiced by Egyptians under the Pharaohs, is still common at the native iron-forges of Ceylon. Dr. Davy in his account of that island has given a figure of them, a copy of which is inserted in the Register of Arts for 1828, page 267: the cords for raising them are attached to an elastic stick, instead of being held in the hands as in the two last cuts.

They are used by modern blacksmiths of Egypt in a horizontal position, (as in the next figure) and worked by an upright lever, which the assistant pushes from and draws towards him. M. P. S. Girard has given a figure and description of them in the *Grande Description*, tom. ii, E. M. p. 618, planche 21. He observes that the coppersmiths of Cairo and Alexandria use the same; and further that they are common in the interior of Africa: "leur forme est probablement tres anciene. Il résulte en effet de quelques reseignemens que m'ont donnés des marchands venus avec les caravanes de Dârfour, que des soufflets de la même forme sont employé par les peuples de *l'interieur de l'Afrique*."

Lantern bellows were formerly common in Europe. They were employed in old organs. See L'Art du Facteur d'Orgues, Arts et Métieres, p. 667, plates 132 and 135. Sometimes the blowers had their feet fixed upon the upper boards, and holding by a horizontal bar they inflated one bellows by raising one foot, and compressed the other by pushing down the other foot. (Encyc. Antiq.) The scabilla of the Romans were small bellows of the same kind, one of which was attached to one foot for the purpose of beating time, and with castanets were used to animate dancers. Several are figured by Montfaucon. The ancients varied the form of the bellows almost infinitely in adapting them to various purposes. Some were attached to altars to aid in the combustion of victims: one for this purpose is represented on one of the Hamilton vases. Lantern bellows were also common in European *blast* furnaces. No. 105 shows their application to this purpose, copied from the De Re Metallica of Agricola. Similar bellows, except the boards being of an oblong form like the one in 103, are common in Hindostan, and worked by hand as in the next figure, but without any frame to support them; the blower kneels and works them in nearly a vertical position. See a figure in Shoberl's Hindostan, vol. v, p. 9.

The bellows of Vulcan were probably of the same kind. Those represented in tom. i, p. 24, of Montfaucon's Antiquities, appear, from that portion of them which is seen projecting from the back of the forge, to be identical with those in No. 105, and worked in precisely the same way. In plate xx, on Painting, of D'Agincourt's History of the Fine

Arts, which contains some illustrations of the Eneid, executed in the 4th and 5th centuries, Vulcan's forge is represented and the bellows blower behind it, apparently with the same kind of instrument as here shown.

No. 105. Lantern Bellows from Agricola.

That Vulcan's bellows were not permanent fixtures as those of our smiths are, but were similar to those figured above, appears from their having been laid aside when not in use, in common with other implements of the forge; a practice usual at the present time in various parts of the east: and we may add that like modern blacksmiths of Asia, he *sat* at work. Thus, when his wife Charis informed him of the arrival of Thetis at their dwelling, he replied:—

> Haste, then, and hospitably spread the board
> For her regale, while with my best despatch
> I *lay my bellows* and my tools *aside*,
> He spake, and vast in bulk and hot with toil,
> *Rose* limping from his anvil-stock,
> Upborne with pain on legs tortuous and weak:
> First, from the forge dislodg'd he *thrust apart*
> *His bellows*, and his tools collecting all,
> Bestowed them careful in a silver chest.

And when he subsequently returned to make the armor which Thetis required for her son, he

> . . . . . to his bellows quick repaired,
> Which *turning to the fire*, he bade them move.—*Il.* xviii.—*Cowper.*

A singular circumstance is related by Herodotus, which shows that the same mode of obtaining a continuous blast, viz. by two bellows, (and in all probability by the same kind as those above figured) was employed by blacksmiths in ancient Greece. The Lacedemonians having been repeatedly defeated by the Tegeans, sent an embassy to the Delphic oracle, to ascertain the means by which they could overcome them. The Pythian assured them of success if they recovered the body of Orestes, the son of Agamemnon, which had been buried several centuries somewhere in Arcadia, the land of their enemies. Being unable to discover the tomb they sent a second time to inquire concerning the place of his interment, when they received the following answer·

> A plain within th' Arcadian land I know,
> Where *double winds* with *forced* exertion blow,
> Where form to form with mutual strength replies,
> And ill by other ills supported lies;
> That earth contains the great Atrides' son;
> Take him and conquer: Tegea then is won.

On the receipt of this, search was again made for the body without intermission, and at last it was discovered in a singular manner. At the time a commercial intercourse existed between the two countries, a Spartan cavalry officer, named LICHAS, being in Tegea, happened to visit a smith at his forge, and observing with particular curiosity the process of working the iron, the smith desisted from his labor and addressed him thus: "Stranger of Sparta, you seem to admire the art which you contemplate; but how much more would your wonder be excited, if you knew all that I am able to communicate! Near this place, as I was sinking a *well*, I found a coffin seven cubits long. I never believed that men were formerly of larger dimensions than at present, but when I opened it, I discovered a body equal in length to the coffin—I correctly measured it, and placed it where I found it." Lichas, after hearing this relation, was induced to believe that this might be the body of Orestes, concerning which the oracle had spoken. He was further persuaded, when he recollected that the *bellows* of the smith might intimate the *two winds;* the anvil and the hammer might express one form opposing another; the iron also, which was beaten, might signify ill succeeding ill, rightly conceiving that the use of iron operated to the injury of mankind. The result proved the sagacity of the Spartan: the body was recovered, and finally the Tegeans, says Herodotus, were conquered. Clio, 67, 68.

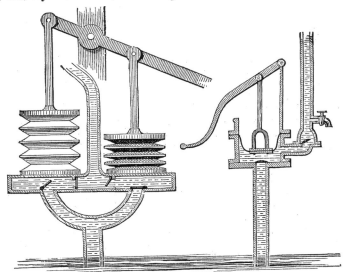

No. 106. Double Lantern Bellows Pump.   No. 107. Single Forcing Pump.

The application of lantern bellows as forcing pumps is, without doubt, of great antiquity: their adaptation to raise water was too obvious not to have been early perceived, and hence we infer that they were at least occasionally employed for that purpose by most of the nations of old. Such pumps are mentioned in old works on hydraulics; but as they have never come into general use, even in modern times, a particular account of them previous to the art of printing, is not to be expected. A writer in the *Grande Description of Egypt*, describing the smith's bellows of that country, observes :—" Ces sortes de soufflets ètoient employés verticalement dans le seizieme siècle tant pour animer le feu des forges que *pour èlever l' eau*, soit en rarefiant l'air soit en le comprimant; ils sont décrits dans l'ouvrage de Ramelli, imprimé en 1558."

No. 106 represents a double lantern bellows-pump, as used in the 16th century. The mode of its operation is too obvious to require detailed description. As one bellows is distended by working the lever, the atmosphere drives water up the suction-pipe into its cavity; and the other at the same time being compressed, expels its contents through the ascending or forcing pipe: the valves at the lower part of the latter, and those over the orifices of the two branches of the suction-pipe opening and closing, as shown in the figure. There is a pump similar to this, but geared in a different manner, in Hachette's Traite elèmentaire des machines. Papin, in a way to raise water, which he proposed enigmatically in the Philosophical Transactions in 1685, used the lantern bellows as a forcing-pump. In a solution by another writer, it is said:—" A vessel made like the body of a pair of bellows, or those puffs heretofore used by barbers being filled with water, a piece of clockwork put under it, may produce the jets." Phil. Trans. Abridg., vol. i. 539. A similar application of the bellows was described in Besson's Theatre, in 1579, the moveable board being impelled by a spring.

No. 107 is another example of bellows forcing-pumps. It consists of the frictionless piston of Gosset and Deville, (No. 83,) but without a valve; a forcing or ascending pipe, having its lower orifice covered by a valve, is attached to the cylinder below the piston. Pumps of this kind have also been made double acting, by passing the piston rod through a stuffing box on the top of the cylinder, and by a double set of valves arranged as in the pump of La Hire.

Of late years machines like those figured in the two last cuts, have been reintroduced into Europe and this country.

Although we have not heard of any one having run out of his wits for joy at their discovery, like the blacksmith mentioned by Cardan, we have heard of some who were nearly in that predicament from disappointment in having found themselves anticipated. A few years ago they were announced in this city as a new and very important discovery; and several gentlemen allowed their names to go abroad as vouchers of their originality and superiority over the common pump.

The proofs of the antiquity of many of our ordinary utensils are derived from representations of them on vases, candelabra, and other works of art that have come down. Of this, the domestic bellows is an example; the only evidence of its having been known to the Greeks or Romans, is furnished by a *lamp;* but for the preservation of which, it might have been deemed a modern invention. Of no other article of ancient household furniture are more specimens extant than of lamps, and not a little of the public and private economy of the ancients has been illustrated by them. Among those in private collections and public museums, are some that were once suspended in temples, others that illuminated theatres and baths—that decorated the banqueting-rooms of wealthy patricians, as well as such as glimmered in the dwellings of plebeians; the former are of bronze, elaborately wrought and enriched, the latter mostly of earthenware. The fertility of conception displayed in these utensils is wonderful. All nature seems to have been ransacked for devices, and in modifying them, the imaginations of the designers ran perfectly wild; while many are in their forms and decorations exquisitely chaste, others are bizarre and some are obscene. There is one of bronze on which an individual is represented blowing the flame with his mouth, as in the act of kindling a fire; and in another the artist has introduced, as an appropriate embellishment, a person performing the same operation with a pair of bellows, of precisely the same form as those in our kitchens. No. 108 is a figure of this lamp, from the 5th volume of Montfaucon's Antiquities.

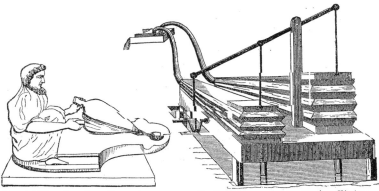

No. 108. Bellows figured on Roman Lamp.   No. 109. Bellows Forcing Pumps from Kircher.

An example of the application of such bellows as atmospheric pumps has already been given, page 207. The adjoining figure (No. 109) is copied from Kircher's Mundus Subterraneus, tom. i., p. 230, Amsterdam, 1665: it represents two large bellows employed as sucking and forcing pumps, being worked by a water wheel, to the axis of which the crank represented was attached.

Bellows like the last and worked in a similar manner, were among ancient devices for ventilating mines: the various modes of adapting them to the purpose may be adduced as another example of their analogy to pumps. Sometimes they were used to force down fresh air in sufficient quantities to render the impure and stagnant atmosphere below respirable; at others they drew the foul air up. In the first case, they were placed near the mouth of the shaft, a pipe was attached to the nozzle and continued down to the place where the miners worked, and when the bellows were put in motion, currents of fresh air were supplied. In the latter case, the pipe was connected to the opening in the under board, i. e. to the aspirating valve, through which the impure air was drawn, and then expelled out of the nozzle; but in this case an expiring valve was required in the nozzle, opening outwards to prevent air from entering through it when the bellows were again distended. The same result was sometimes obtained in the following manner: An opening was made and covered by a valve in the upper board instead of the lower one, and when the bellows were distended, the impure air rushed up the pipe which was attached to the nozzle, and was expelled through the opening covered by the flap when the bellows were closed. Several figures representing these and other applications of bellows are given by Agricola.

Goguet observes that draft furnaces were probably invented early, *but bellows were not.* We should suppose the reverse was the fact; for the advantages of an artificial blast must have been obvious from the first use of fire, and *naturally* led to the use of the mouth to blow it, then the reed, sack, and subsequently a slit or valve in the latter, would follow as an almost necessary sequence; and long before the idea of increasing the intensity of heat by flues or chimneys could have been thought of. No natural occurrence could have led to the invention of these before the other, nor has there, as yet, been found any account or representation of draft furnaces of equal antiquity with those of bellows.

CHAPTER II.

PISTON Bellows: Used in water organs—Engraved on a medal of Valentinian—Used in Asia and Africa. Bellows of Madagascar. Chinese bellows: Account of two in the Philadelphia Museum—Remarks on a knowledge of the pump among the ancient Chinese—Chinese bellows similar in their construction to the water-forcer of Ctesibius, the double acting pump of La Hire, the cylindrical steam-engine, and condensing and exhausting air-pumps. Double acting bellows of Madagascar—Alledged ignorance of the old Peruvian and Mexican smiths of bellows: Their constant use of blowing tubes no proof of this—Examples from Asiatic gold and silver smiths—Balsas—Sarbacans—Mexican Vulcan. Natural bellows-pumps: Blowing apparatus of the whale—Elephant—Rise and descent of marine animals—Jaculator fish—Llama—Spurting Snake—Lamprey—Bees—The heart of man and animals—Every human being a living pump: Wonders of its mechanism, and of the duration of its motions and materials—Advantages of studying the mechanism of animals.

THE bellows described in the last chapter are all formed of leather or skins, and are obvious modifications of the primitive bag or sack; the wooden ends of some of them being adopted merely to facilitate their distension and collapse. From the simplicity of their construction and general efficiency they still retain a place in our workshops and dwellings, and are in no danger of being replaced by modern substitutes: but the ingenuity of ancient bellows makers was not exhausted on these, for they had others, differing both in form, materials and mode of action; viz: *piston bellows;* machines identical with cylindrical forcing-pumps. At what time these were first devised we have no account; but as they are described by Vitruvius, in his account of hydraulic organs, without the slightest intimation of their being then of recent date, they may safely be classed among those inventions, the origin of which is too remote to be discovered.

No. 110. Roman Piston Bellows.

No. 110 represents a person working two of them to supply wind for a water organ, from Barbaro's Vitruvius, Venice, 1567. They are substantially the same as those figured by Perrault and Newton in their translations, and by Kircher in his Musurgia Universalis, (tom. ii, 332.)

The blower, by alternately raising one piston and depressing the other, pumped air into a large reservoir: this was an open vessel inverted into another containing water, and as the air accumulated in the former, the liquid was gradually displaced and rose in the latter, as in a gas holder. It was the constant pressure exerted by this displaced water that urged the air through the pipes of the organ, whenever the valves for its admission were opened. The question, perhaps may be asked, Why did the ancients prefer these bellows in their organs to those formed of leather and boards, such as are figured at Nos. 105, 108, 109 ? Probably because the pressure required to be overcome in forcing air into the reservoirs was greater than the form and materials of the latter could safely bear. It is very obvious from the brief description of the piston bellows of the Romans, that they were calculated to produce much stronger blasts than could be obtained from those made of leather. Vitruvius informs us that the cylinders and valves were made of *brass*, and the pistons were *accurately turned* and covered (or packed) with strips of unshorn sheepskins. They seem to have been perfect condensing air-pumps.

A figure of an ancient hydraulic organ is preserved on a medal of Valentinian: two men, one on each side, are represented as pumping and listening to its music. This medal is engraved in the third volume of Montfaucon's Antiquities, (plate 26,) but the piston rods only are in sight; the top of the cylinders being level with the base on which the blowers stand.

As piston bellows were known in the old world, it might be supposed they would still be employed in those parts of the East where the arts and customs of former ages have been more or less religiously retained. Such is the fact; for like other devices of ancient common life, they are used by several of the half civilized tribes of Asia and Africa—people, among whom we are sure to meet with numerous primitive contrivances, embodied in the same rude forms and materials as they were before Grecian taste or Roman skill improved them. It is chiefly to the incidental observations of a few travelers that we are indebted for a knowledge of these implements in modern days; but when the times arrive for voyages of discovery to be undertaken for the purpose of describing the machines, manufactures and domestic utensils of the various nations of the earth; (undertakings of equal importance with any other,) these bellows and their numerous modifications will furnish materials for a chapter in the history of the useful arts that will be replete with interesting information. As they are clearly identified with the forcing-pump, an account of some of them will not be out of place.

Dampier thus describes the bellows used by the blacksmiths of *Mindanao*. " They are made of a wooden cylinder, the trunk of a tree, about three feet long, bored hollow like a pump, and set upright on the ground; on which the fire itself is made. Near the lower end there is a small hole in the side of the trunk next the fire made to receive a pipe; through which the wind is driven to the fire by a great bunch of fine feathers, fastened to one end of a stick, which closing up the inside of the cylinder, drives the air out of the cylinder through the pipe. Two of these trunks, or cylinders, are placed so nigh together, that a man standing between them may work them both at once, one with each hand."[a] Here we have both the single and double chambered forcing-pump; and although Dampier has not noticed the valves, the instruments were certainly furnished with them, or with some contrivance analogous to them, but being out of

---

[a] Dampier's Voyages, i. 332.

sight, were left unnoticed by that intelligent sailor. The bellows of *Madagascar*, says Sonnerat, "is composed of the hollow trunks of two trees tied together. In the bottom there are two iron funnels, and in the inside of each trunk a sucker furnished with raffia, which supplies the place of tow. The apprentice, whose business it is to use this machine, alternately sinks one of the suckers while he raises the other."[b] Similar implements are also used in smelting iron as well as in forging it. In the first volume of Ellis's "History of Madagascar," Lon. 1838, there is a representation of two men reducing iron ore by means of four piston bellows. No. 111 is a copy.

No. 111. Piston Bellows of Madagascar.

The furnace is described as a mere hole dug in the ground, lined with rude stonework and plastered with clay. It was filled with alternate layers of charcoal and ore, and covered by a conical roof of clay, a small opening being left at the apex. The bellows were formed of the trunks of trees, and stood five feet above the ground, in which they were firmly imbedded. The lower ends were closed "air tight," and a short bamboo tube conveyed the wind from each to the fire, as represented. "A rude sort of piston is fitted to each of the cylinders, and the apparatus for raising the wind is complete." As no mention is made of valves nor of the openings through which air entered the cylinders, it is probable that the pistons were perforated for that purpose, and the passages covered by flaps or valves opening downwards, a device which the artificers of Madagascar are acquainted with. See No. 114. These bellows are of various sizes, though generally from 4 to 6 inches in diameter. Sometimes only one is used, but it is then made of larger dimensions, and the blower *stands* and works it with both his hands. To do it conveniently, he raises himself on a bank of earth. The bellows are not always perpendicular, but are inclined as figured in the back ground of the cut.

---

[b] Sonnerat's Voyages, iii. 36.

The blacksmiths of *Java* use the same kind. Raffles, in his History of the Island, (2 vol. 193,) after quoting Dampier's description of the bellows of Mindanao, observes his account "exactly corresponds" with that of Java. "The blacksmiths' bellows of *Sumatra*," says Mr. Marsden, "are thus constructed: two bamboos of about four inches diameter and five feet in length, stand perpendicularly near the fire, open at the upper end and stopped below. About an inch or two from the bottom a small joint of bamboo is inserted into each, which serve as nozzles, pointing to and meeting at the fire. To produce a stream of air, bunches of feathers, or other soft substance, are worked up and down in the upright tubes like the piston of a pump. These, when pushed downwards, force the air through the small horizontal tubes; and by raising and sinking each alternately, a continual current of air is kept up."[c] The *Bashee* Islanders use the same kind of bellows.[d] The smiths of *Bali* have them also: "their instruments are few and simple, their forge small, and worked by a pair of upright bellows, such as we find described in Raffles' Java."[e] They are not confined to southern Asia and the Ethiopian Archipelago, but are used in continental Africa. "The bellows of the negro artificers on the *Gambia*, are a thick reed or a hollow piece of wood, in which is put a stick wound about with feathers, [a piston,] which by moving of the stick, makes the wind."[f]

Without entering into the controversy respecting the origin of wooden bellows, it may be inferred from the preceding extracts, that such have been in use from remote times; and that the cylindrical forcing-pump, so far as regards the principle of its construction, is equally ancient: of this, the instrument now to be described, affords another indication. It is the bellows of the most numerous and most singular of all existing people—a people, the wisdom of whose government has preserved them as a nation, through periods of time unexampled in the history of the world, and which still preserves them amidst the prostration by European cupidity of nearly all the nations around them; a people, too, who notwithstanding all that our vanity may suggest to depreciate, have furnished evidence of an excellence in some of the arts that never has been surpassed. The *Chinese*, like the ancient Egyptians, whom they greatly resemble, have been the instructors of Europeans in several of the useful arts; but the pupils, like the Greeks of old, have often refused to acknowledge the source whence many inventions possessed by them were derived, but have claimed them as their own: of the truth of this remark, we need only mention *printing*, the *mariner's compass*, and *gunpowder*.

In the bellows of the Chinese, we perceive the characteristic ingenuity and originality of that people's inventions. A description and figure of their bellows were published in London, 1757, by Mr. Chambers, in a work entitled "Designs of Chinese Buildings, Furniture, Dresses, Machines and Utensils, from drawings made in China." The following account from the fourth volume of the "Chinese Repository," a very interesting work published at Canton in China, is substantially the same. "The bellows used by them is very aptly called '*Fung Seăng*,' 'wind box,' and is contained in an oblong box about two feet long, ten inches high, and six inches wide. These dimensions, however, vary according to the whim of the maker, and they occur from *eight inches*, to four feet or more in length, and so of the width and height. The annexed profile view will give some idea of the principle upon which it is constructed."

---

[c] History of Sumatra, p. 181.   [d] Dampier's Voyages, i. 429.   [e] Chinese Repository, iv. 455.   [f] Ogilvy's Africa, Lon. 1670, p. 356.

"A, B, C, D, is a box divided into two chambers at the line O H. In the upper one is the piston E, which is moved backwards and forwards by means of the handles attached to it; and is made to fit closely by

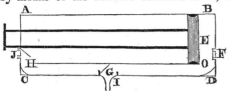

No. 112. Section of a Chinese Bellows.

means of leather or paper. The lid of the box slides upon the top, and is sufficiently thick to allow the workman to labor upon it. At F J are two small holes each covered with a valve; and just below them, at O H in the division of the two chambers, are larger holes, for the entrance of the wind into the lower chamber. This part of the bellows is made of a thick plank, hollowed into an ovoid form, and is about an inch thick. The clapper G is fastened to the back side of the box, and plays horizontally against the two stops placed near the mouth I. It is made as high as the chamber, and when forced against the stop, it entirely closes the passage of air beyond. When the piston is forced inwards, as represented in the cut, the valve at F is closed, and that at J is opened; and thus the upper chamber is constantly filled with air. The wind driven into the lower chamber by the piston urges the clapper G against the stop, and is consequently forced out at the mouth. The stream of air is uninterrupted, but not equable, though in the large ones the inequality is hardly perceived. An iron tube is sometimes attached to the mouth which leads to the furnace, and in other cases the mouth itself is made of iron." The Chinese generally use them in an inclined or horizontal position, frequently making use of the upper side as a work bench. In the figure (and the one given by Chambers) two rods are connected to the piston to prevent it from springing when used : this appears to be the practice with regard to those of large dimensions. In small ones a single rod is sometimes used, and the chamber is cylindrical. In the collection of M. Bertin, (a French minister and secretary of state in the former part of the last century,) which contained " about 400 original drawings, made at Pekin, of the arts and manufactures of China" a portable and *single-acting* bellows is represented as in the next figure.[a]

No. 113. Chinese Single Bellows, and Tinker.

" This instrument is made like a box in which is a piston, so constructed that when it is drawn out behind, the vacuum which it occasions in the box makes the air rush in with great impetuosity through a lateral opening, to which a sucker [a valve] is affixed: and when the piston returns in an inverse direction, the sucker [valve] closes itself, and the air is forced out by the opposite extremity." Navarette preferred the

---

[a] China, its Customs, Arts, &c., translated from the French. Lon. 1824 ; vol. i. 17.

Chinese bellows to the European one, he said it was more commodious and efficient.[b] It is employed to some extent in Java, having been introduced from China.[c]

Since the preceding remarks were written, we have examined two bellows from China, in the splendid "Chinese Collection" in Philadelphia. One of them belonged to a traveling blacksmith. It is formed of a cylindrical joint of bamboo, $2\frac{1}{2}$ feet in length, and between five and six inches diameter. The piston rod is a wire $\frac{1}{4}$ or $\frac{3}{8}$ of an inch thick, with a small gimlet handle. Air is admitted through a cluster of five or six small holes in each end, which are covered in the inside by paper flaps: these are the induction valves, marked J F in No. 112. Along one side of the bellows a strip of wood $2\frac{1}{2}$ inches wide and $1\frac{1}{4}$ thick, is secured by what appears to be eight small thumb screws, and the junction made tight by cement or wax; this projecting piece resembles those on the sides of high pressure steam-engine cylinders, and is intended for a somewhat similar purpose, its interior being hollowed into a passage for the wind when expelled by the piston from either end of the cylinder. A short metallic tube conveys the wind from the middle of this piece to the furnace as in No. 112. The ends of the bellows are secured from splitting by two thin and narrow iron hoops, and at one place a small clamp is driven across a crack, as is sometimes practiced in mending wooden bowls. The instrument resembles the one in the last figure, but is double acting: the figure of the artist accompanying it is seated on the ground and works it with one hand while he attends the fire with the other.

The other bellows consists of a long box like the one figured at No. 112. From the circumstance of its not being confined in a glass case, and permission to examine its interior having been politely accorded, we had an opportunity of ascertaining some particulars that are not mentioned in any published account of these instruments that has fallen in our way. It is twenty-two inches long, seven deep, and five wide, made of thin boards of a species of fir and extremely light: the sides and ends are dovetailed together; and the bottom appeared to be intended to slide over the sides, having strips projecting from it and no pins or nails visible; this arrangement enables a person to examine the interior, and to replace or repair the valves, &c. with great facility. The boards of which the machine is made are of a uniform thickness (about $\frac{3}{8}$ of an inch) except the top, which is $1\frac{1}{4}$ inches. The reason for this extra thickness was perceived as soon as it was removed, (it was secured to the sides and ends by long wooden pins,) for a deep and wide groove is made through its whole length with the exception of $\frac{1}{4}$ of an inch at each end, and at the middle of the groove a passage is cut at right angles to it through one side for the air to pass into the tuyere. Upon the removal of this thick cover, the inside of the box was not exposed, for another thin one was found inserted within the sides, and flush with their edges. This was a board slipped between the sides and resting upon the upper edge of the piston, having two openings, one at each end, which coincided with the groove in the outer cover, (the inner cover is represented by the line H O in No. 112;) hence the wind is driven by the piston alternately through each opening into the groove, and, by the action of the valve in the middle of the latter, is compelled to pass into the tuyere. This valve is represented at G in No. 112, and from an inspection of that cut, it will be apparent that some contrivance of the kind is absolutely necessary, in order to prevent the wind when forced from one end of the bellows, from passing along the groove

[b] Histoire Gènèrale, Tom. viii. 106.   [c] Raffles' Java, ii. 193.

into the other end: it consists of a narrow piece of hard wood of the same depth as the groove, and of a length that rather exceeds the width of the groove. A hole is drilled through one end and a pin driven through it into the solid part of the cover, so that it turns freely on this pin, and closes and opens a passage for the escape of the wind into the tuyere. It is driven by the wind at every stroke of the piston against the opposite cheek of the groove, and thus prevents the wind from passing into the other end of the cylinder, as shown at G in No. 112. It is surprising how easily this valve plays although its upper and lower edges rub against the surfaces of the two covers—a trifling movement of the piston drives it against the cheek, and occasions a snapping sound somewhat like that from the contact of metal.

When the inner cover was raised out of its place, the piston and induction valves were exposed to view, and the simplicity and efficiency of these parts were in keeping with the rest: the two valves are mere flaps of paper, glued at their *lower* edges to the under side of the openings, and hence they stand nearly perpendicular, instead of being suspended from above; the slightest impulse of air closed them. The piston is half an inch thick, but is reduced at the edges to a quarter of one; it appears to be formed of two thin pieces which, united, are equal in thickness to that mentioned; and between them are inserted two small sheets of moderately stiff paper, which project an inch over every side. The part that projects is folded at the corners and turned over the edges of the piston; one sheet being turned one way, and the other the contrary, so that when the piston is moved, the air presses the paper against the sides of the bellows and renders the piston perfectly tight, on the same principle as the double cupped leathers of fire-engines and other forcing-pumps; and at the same time without any perceptible increase of friction. The two piston rods are half inch square, and work through holes in one end of the box without any stuffing-box. The whole machine is of wood, except the paper for the piston and valves. Although the instrument appears to be a rectangular box, it is not exactly so, the bottom being a little wider than the top.

It would be superfluous to point out the application of piston bellows to raise water, since they are perfect models of our atmospheric and forcing-pumps. Why, then, it may be asked are not the Chinese found in the possession of the latter? In reply to this question, it may be observed: 1. That from our imperfect knowledge of the people, it is not certain that such machines have not been, and are not used to a limited extent in the interior of that great empire. 2. That custom, and probably experience, have induced them, in common with other nations of the Oriental world, to give the preference to more simple devices—to their chain pump, bamboo wheel, &c., a preference which we know is in some instances based on solid grounds: for example, the chain pump as used by them, raises more water with the same amount of labor, than any atmospheric or forcing-pump, if placed under the same circumstances. And as for the noria or bamboo wheel, which driven by a current, raises water night and day, and from 20 to 50 feet, we are told that it answers the purpose "as completely as the most complicated European machine could do; and I will answer for it [says Van Braam] that it does not occasion an expense of ten dollars." 3. A circumstance connected with one of their ancient as well as modern scenic representations, shows that when the *forcing* or spouting of water is required, their artists are at no loss for devices to effect it; and that, too, under very unusual circumstances. One of the pantomimes performed at Pekin is the "*Marriage of the Sea with the Land.*" The latter divinity made a display of his wealth and productions, such as dragons, elephants, tigers, eagles, ostriches, chestnut and pine trees, &c. The *Ocean*, on the other hand, collected whales, dolphins, porpoises and other sea monsters, together with ships, rocks, shells, &c., " all these objects were represented by performers concealed under cloths, and who played their parts admirably. The two assemblages of productions, terrestrial and marine, made the tour of the stage, and then opened right and left to leave room for an immense whale, which placed itself directly before the emperor, and *spouted out several hogsheads of water*, which inundated the spectators who were in the pit."[a] As both the water and forcing apparatus were contained within the moving figure, we can only imagine the jets to have been produced by means of piston or bellows forcing-pumps, or something analogous to them—or by air condensed in one or more vessels containing water, like soda fountains. 4. If Chinese lads never discovered a source of amusement in the application of their bellows (some of which are only *eight inches* long) as squirts or pumps, they must differ essentially from lads of other nations—a position that few judges of human nature would admit. Boys are the same in all ages, and the mischievous youngsters of the Celestial Empire have doubtless often derived as much pleasure from annoying one another with water ejected from these implements, as those of Europe and this country do with similar devices. Such an application of them was sure to be found out by boys, if by no one else. Whether the bellows-pump originated in this manner or not, may be uncertain, but several useful discoveries have been brought to light in much the same way : it was a youth who changed the whole character of the steam-engine, by giving it that feature upon which its general utility depends—his ingenuity, stimulated by a love of play, rendered it self-acting.

The antiquity of the Chinese bellows is a subject of much interest. It may have been the instrument which Anacharsis introduced into Greece, it having, perhaps, been employed by his countrymen, the ancient Scythians, as well as by their descendants, the modern Tartars. If it has been in use, as supposed, from times anterior to Grecian and Roman eras, the origin of the pump in the second century B. C. can hardly be sustained; for when the induction valves of one of these bellows are placed in water, (as we suppose has occasionally been done ever since its invention,) it is then the "water forcer" of Ctesibius; and if pipes be connected to F and J, (No. 112,) and their orifices placed in a liquid, the apparatus becomes the double acting pump of La Hire. But what may be surprising to some persons, its construction is identical with that of the steam-engine; for let it be furnished with a crank and fly wheel to regulate the movements of its piston, and with apparatus to open and close its valves, then admit steam through its nozzle, and it becomes the double acting engine of Boulton and Watt. Again, connect its induction orifices to a receiver, and it becomes an exhausting air-pump; apply its nozzle to the same vessel, and it is a condensing one. The most perfect blowing machine, and the *chef d'œuvre* of modern modifications of the pump, are also its fac-similes.

It would seem that the Chinese have other kinds of bellows, or different modes of working these. Bell, in his account of the Russian embassy in 1720, says that he was lodged in a village twelve miles from Pekin in a cook's house, which gave him an opportunity of observing the customs of the people even on trifling occasions : " My landlord," he observes, " being in his shop, I paid him a visit, where I found six kettles placed in a

---

[a] China, its Costumes, &c., iii. 34.

row on furnaces, having a separate opening under each of them for receiving the fuel, which consisted of a few small sticks and straw. On *his pulling a thong*, he blew a pair of bellows which made all his kettles boil in a very short time."[c] Like other Asiatics, the Chinese have probably a variety of these instruments. The van, or winnowing machine, which we have received from them, is a *rotary* bellows. See page 70 of this volume.

Various rotary bellows are described by Agricola, as employed in the ventilation of mines, and worked by men with cranks, and in one instance by a horse treading on the periphery of a wheel.[d] Rotary blowing machines have been represented as of more recent origin, but they are in all probability of great antiquity. The Spaniards introduced them into Peru as early as 1545, to reduce the silver ores, but they were soon abandoned.[e] For rotary pumps, see a subsequent chapter of this book.

We are indebted for some interesting information respecting the arts of various islanders of the Indian ocean to Mr. William Clark of Philadelphia, who, besides spending several years in whaling voyages, resided two years in Southern Africa. The vessel to which he was attached having on one occasion touched on the coast of Madagascar, some native smiths were found using bellows that excited particular attention; some were cylindrical, being formed of bored logs, others were square trunks, five or six inches in diameter, and about five feet long; but the internal construction of both was the same. The ship's carpenter was permitted to open one. It was composed of four planks that had been split from trees, the insides shaved smooth and straight, and the whole put together with wooden pins instead of nails or screws. It was divided into two parts by a partition or disk, which was permanently secured in its place, (shown at A in the annexed cut,) where, like a piston, it occupied the entire space across. On one side of the trunk, and opposite the edge of A, an opening was made for the insertion of the tube C that conveyed the wind to the fire, the edge of A at this place being feathered, and a small projecting piece added to it, in order to direct the current of air from either side of the partition into C. An opening was made in the centre of A, through which a smooth piston rod B, played; two pistons or boards, P P, accurately fitted to work in the trunk, were attached on opposite sides of the partition to B; these pistons were perforated, and the openings covered by flaps or valves like those of a common pump box, but the upper one was secured to the *under* side of the piston as shown in the figure. The trunk rested on four short pieces of wood pegged to it. In some, holes were made at the lower part for the admission of air. These bellows were therefore double acting, and consequently one of them was equal in its effects to two of those represented at No. 111, which drive the air out only on the descent of the piston, whereas these forced it into the fire both on ascending and descending. Thus, when the blower raised the rod B, the flap on the lower piston closed, and the air in that division of

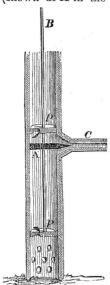

No. 114. Double Acting Bellows of Madagascar.

[c] Bell's Travels, i. 312. [d] De Re Metallica, pp. 162, 163, 164, 169. [e] Garcilasso's Commentaries, p. 347.

the trunk was expelled through C ; at the same time the flap of the upper piston was opened by its own weight and the air passing through it, and on the descent of B all the air in the upper part of the trunk was forced into the fire in like manner; hence an uninterrupted, though not an equable blast of wind was kept up. The whole apparatus was of wood except the flaps, which were pieces of green hide rendered pliable by working them in the hands; and they were prevented from opening too far by narrow slips of the same material pegged over them. There was no packing to the pistons, but they were moved with great rapidity.

These bellows are different from those described by Dampier, Sonnerat and Ellis, as used in the same island; but they serve to corroborate a remark that has been made by several travelers, viz: that the negroes of Africa are in possession of a great variety of those instruments. The one above described is a fine specimen of their ingenuity, for there can be little doubt that it is original with them—it evidently is not derived from the double acting bellows of China, nor can it have been procured from Europe, since nothing of the kind has, we believe, ever been used there. It is the only bellows that we have met with having valves in the pistons.

It need hardly be observed that double *pumps* have been made on the same principle. There is one figured by Belidor in the second volume of his Architecture Hydraulique, which differs from the above figure only in having two short piston rods connected together by a frame on the outside of the cylinder, instead of one long one working through the disk.

No stronger proofs could possibly be adduced of the analogy between pumps and bellows, than what the figures in this and the preceding chapter afford.

While engaged on this part of the subject we were induced to refer again to the accounts of the old Mexicans and Peruvians, in hopes of finding some indications of the pump in the instruments employed to urge air into their furnaces; but, strange as it will appear to modern mechanics, they are said to have been wholly ignorant of the bellows. This, if true, is a very singular fact; and, considering the extent to which they practiced the arts of metallurgy, one that is unexampled in the history of the world. It appears, moreover, irreconcilable with the opinion of their oriental origin; for it is difficult to conceive how emigrants or descendants of emigrants from Asia, could have been ignorant of this simple instrument which has been used in one form or another on that continent from the earliest times, and which is still employed by the rudest tribes there, and also in all those parts whence the early Peruvians are supposed to have come. The bellows is common almost as the hammer, from the peninsular of Malacca to that of Kampschatka, and from the Phillippine islands to those of Japan. In Africa, too, it is used in great variety and by people whose progress in the arts is far behind that of the ancient smiths of America.

How little is known respecting the mechanical implements of Mexican and Peruvian workmen and of their processes, and yet but three centuries have elapsed since the latter were in full operation ! We are not aware that a single tool has been preserved, much less their modes of manufacture; nor is this much to be wondered at when the spirit that animated the conquerors is considered—it was the acquisition of gold, not the tools for or manner of working it, that they had in view; and had it not been for the prodigious amount of bullion which they found worked into various figures and utensils, we should scarcely have ever heard of the latter; and yet the workmanship on some of them, exceeded the value of the metal. That there are errors in the accounts of early writers on the arts and apparatus of old American mechanics is unquestionable, and among them may be mentioned that which confined the materials of their cutting instruments to obsidian and other stones; whereas it is now certain that they had chisels, &c. of bronze or alloys of copper and tin; and probably of a similar composition to those of Egyptian workmen. As for bellows, it was no easy task (supposing it had been undertaken by the old historians of Mexico and Peru) to determine positively that they were *unknown* throughout those extensive countries. To ascertain what tools were and were not used, required something more than a superficial knowledge of the people. Before a stranger could speak decidedly on the subject of bellows, it was necessary that he should become familiar with their modes of working the metals, by frequently visiting them in their workshops and dwellings; and, from an intimate knowledge of their language, making inquiries respecting the tools and details of the processes adopted by artisans of distant tribes; for bellows might be used to a limited extent in one country, and (from variety in the ores, articles manufactured or customs of workmen) not at all in another. But there does not appear to have been any efforts made to collect information of this kind by the conquerors—its value was not appreciated by them or by their immediate successors, and hence the opportunity was neglected and could never be recalled; for other historians agree with Clavigero, that the wonderful arts of the Mexican and Peruvian founders were soon *lost*, "by the debasement of the Indians and the indolent neglect of the Spaniards." Even Garcilasso, although a native Indian, by his mother's side, does not seem to have possessed any particular knowledge on the subject of working the metals: he derived his information from Acosta, to whose work he refers his readers.

But what are the proofs that bellows were unknown to the subjects of Manco Capec and Motezuma? The principal one is derived from their fusing metals without them: they kept their furnaces in blast, it is alledged, by the breath of a number of men who blew on the fires through tubes of bamboo. That this was often practiced there is no doubt, and that it was the general custom is admitted; but it does not therefore follow that they had no contrivances for producing artificial blasts: this will appear from the practice of oriental gold and silver smiths, both of ancient and modern times. The fusion of gold and silver with blowing tubes is a device of remote antiquity, and like all ancient customs relating to the useful arts, it is still practiced by the Hindoos, Malays, Ceylonese, Persians and other Asiatics; and also by Egyptians and numerous African tribes. The goldsmiths of Sumatra, Mr. Marsden observes, "in general *use no bellows*, but blow the fire with their mouths through a joint of bamboo; and if the quantity of metal to be melted is considerable, three or four persons sit round the furnace, which is an old broken *kwali*, or iron pot, and blow together: at Padang alone, where the manufacture is more considerable, they have adopted the Chinese bellows."[a] We have already described the single and also a double acting bellows of these people; besides which they have that of China, and yet it seems that all the working goldsmiths of the country, except those of a single town, still melt their metal as the Mexicans and the Peruvians did: hence the mere fact of the old smiths of these continents using blowpipes to fuse metal, is no more a proof of their ignorance of bellows, than the like practice is of modern Asiatics being also ignorant of them.

Nothing is easier than to err respecting a knowledge of bellows in former times, by inferences drawn from the use of blowpipes. In the oldest

---

[a] History of Sumatra, 179.

monuments of Egypt (those of Beni Hassan) the latter are represented in the remote age of Osirtasen, 1700 B. C. which to a superficial observer might lead to the supposition that the former were then unknown; but a close examination of the sculptures shows the fallacy of such a conclusion, since blowing tubes are also figured long after the reign of Thothmes in whose time bellows were certainly common.[a] Again, on the last day of the feast of Tabernacles, the Jews were allowed by rabbinical precepts to light one fire from another, but not to strike new fire from stone or metal, nor to quench it, although to save their goods, " nor to blow it with bellowes, but with a *reede*."[b] Now a stranger, having an imperfect knowledge of Jewish customs, upon witnessing fires thus blown would, in some parts of the world, be very apt to conclude that they had no bellows. And again, if we had not a proof that our domestic bellows was known to the Romans, we might have inferred from Pliny's account of statuaries and painters representing individuals blowing fires with their *mouths*, that artificial instruments for the purpose were then unknown.

Enough may be gathered from early writers on America to account for bellows not being employed in those operations in which they would seem to have been most required, viz: in smelting of metals. According to Acosta, some ores could not be reduced by bellows, but only by air furnaces. Garcilasso, in the last chapter of the eighth book of his Commentaries, makes the same remark. In smelting the silver ore of Potosi, he says the Indians used neither bellows nor blowing tubes, but a natural wind, which, in their opinion, was the best; they therefore fused the ore in small furnaces placed on the hills in the night time, whenever the wind was sufficient for the purpose; and it was a pleasant sight, he observes, " to behold eight, ten, or twelve thousand of those fires at the same time, ranged in order upon the sides of the mountains." The Spaniards suspecting that the metal, when thus diffused among a great number of hands, might be more readily purloined, and that much of it was wasted in so many fires, introduced blast furnaces, the fires in which were urged "by large bellows," but these not succeeding, (the blast being too strong,) they had recourse to rotary bellows, (" engines with wheels, carried about with sails like a windmill which fanned and blowed the fire,") but these also failed to accomplish the purpose, " so that the Spaniards despairing of the success of *their* inventions, *made use of those which the Indians had framed and contrived*." No stronger reason could be adduced why the bellows was not previously used in the reduction of ores.

At a subsequent fusion of the metal in their dwellings, the workmen (says Garcilasso) instead of bellows, continued to use blowing tubes, "though our [Spanish] invention of bellows is much more easie and forcible to raise the fire." Supposing they were ignorant of bellows before the arrival of the Spaniards, here is a proof that after they became acquainted with these instruments, they still preferred their tubes, as the gold and silver smiths of Asia generally do at this day; and hence the use of such tubes does not show, as has been stated, " that they were unacquainted with the use of bellows."

If there was nothing else to adduce in favor of the old Peruvians being acquainted with bellows, or with the principle of their construction and application, than the balsas or blown floats which their fishermen and those of Chili used instead of boats, we should deem them sufficient. These were large bags made of skins of the sea wolf and filled with air. They

---

[a] Wilkinson's Manners and Customs of the Ancient Egyptians, iii. 339. [b] Purchas' Pilgrimage, 223.

were "so well sewed, that a considerable weight could not force any of it out." They carried from twelve to fourteen hundred pounds, and if any air escaped, there were two leathern pipes through which the fishermen "blow into the bags when there is occasion." Frezier's Voyage to the South Seas, page 121. These were real bellows, only applied to another purpose. Had they not been found less efficient or less economical than blowing tubes, they would doubtless have been used as substitutes for the latter in the fusion and reduction of ores. It may here be noticed as a singular fact, and one which may possibly have reference to bellows, that *Quetzalcoatl*, the Mexican God of the air or wind, was also the Vulcan of all the nations of Anahuac.

Both Mexicans and Peruvians were accustomed from their youth to use blowing tubes, for the primitive air gun, through which to shoot arrows and other missiles by the breath, was universally used, and the practice is still kept up by their descendants. Motezuma, in his first interview with Cortez, shrewdly compared the Spanish guns, as tubes of unknown metal, to the *sarbacans* of his countrymen. From the expertness acquired by the constant employment of these instruments in killing game, it was natural enough to use them instead of bellows to increase the heat of their furnaces, and custom rendered them very efficient.

We have prolonged our remarks on this subject because it has been concluded that remains of furnaces, found far below the surface in various parts of this continent and in regions abounding with iron, could never have been employed in reducing that metal; for in those remote ages in which such furnaces were in action, the bellows, it is said, was *unknown;* a position that we think untenable, and quite irreconcilable with the advanced state of metallurgy in those times.

Before leaving the subject of bellows and bellows pumps, we may remark that numerous illustrations of the latter may be found in the natural world. To an industrious investigator, the animal kingdom would furnish an *endless variety,* for every organized being is composed of tubes and of liquids urged through them. The contrivances by which the latter is accomplished may be considered among the prominent features in the mechanism of animals; and although modified to infinitude, one general principle pervades the whole; this is the *distension and contraction* of flexible vessels or reservoirs in which fluids are accumulated and driven through the system. On the regular function of these organs the necessary motions of life chiefly depend; by them urine is expelled from the bladder, blood from the heart, breath from the lungs, &c.; they are natural bellows pumps, while other devices of the Divine Mechanician resemble syringes or piston pumps.

The whale spouts water with a bellows pump, and in streams compared with which the jet from one of our fire-engines is child's play. His blowing apparatus consists of two large membranous sacs; elastic and capable of being collapsed with great force. They are connected with two bony canals or tubes whose orifices are closed by a valve in the form of two semicircles, similar to those known to pump makers as butterfly valves. When the animal spouts, he forcibly compresses the bags, already filled with water, and sends forth volumes of it to the height of 40 or 50 feet. The roaring noise that accompanies this ejection of the liquid is heard at a considerable distance, and is one of the means by which whalers, in foggy weather, are directed to their prey. The proboscis of the elephant is sometimes used as a hose pipe, through which he plays a stream in every direction by the pump in his chest. Numerous insects that live in water move their bodies by the reaction of that liquid on streams they eject from their bodies: oysters and some other shell fish move in this manner. Myriads of marine animals also ascend and descend in their native element by means of forcing pumps: when about to dive, they admit water into certain receptacles, and in such quantities as to render their bodies specifically heavier than the fluid they float in; and when they wish to ascend, they pump out the water which carried them down.

That expert gunner, the jaculator fish, shoots his prey with pellets or globules of water as from a piston pump. When an insect hovers near or rests on some aquatic plant within five or six feet of him, he shoots from his tubular snout a drop of water, and with so "sure an aim as generally to lay it dead." The habit of ejecting saliva, which some persons acquire, is by making a pump of the mouth and a piston of the tongue. Other animals practice the same; thus the llama of Chili and Peru, when irritated, "ejects its saliva to a considerable distance"—Frezier says ten paces, or thirty feet. The spurting snake of Southern Africa, it is said, ejects its poison into the eyes of those who attack it with unerring aim. The tongue of the lamprey moves backwards and forwards like a piston, and produces that *suction* which distinguishes this animal and others of the same family. The sting of some insects, that of the bee, for example, is a very complex apparatus, consisting of a lancet with its sheath, to penetrate the bodies of their enemies; first acting as a trocar and canular, and then as a pump to force poison into the wound—" an awl to bore a hole, [says Paley,] and a syringe to inject the fluid."

It perhaps may be supposed from the *form* of common pumps, that there is little resemblance between them and these natural machines, but it should be remembered that this form is purely arbitrary, (they are, as we have already seen, sometimes made of flexible materials, and alternately dilated and collapsed like the chests of animals.) The general custom of making them of hollow cylinders and of inflexible materials, arose from experience having proved that when thus made, they are more durable and less liable to derangement than any others that have yet been devised.

The circulation of the blood in man and other animals is effected by apparatus strikingly analogous to sucking and forcing bellows pumps. The heart is one of these—the arteries are its forcing, the veins its suction pipes, and both pump and pipes are furnished with the most perfect valves. By contraction, this wonderful machine forces the blood through the former to the uttermost parts of the system; and by distension, draws it back through the latter.[a] They vary in dimensions as in construction. Some are adapted to the bodies of animals so minute as to be imperceptible to unaided vision, and from these to others of every size up to the huge leviathan of the deep. The aorta of the whale, says Paley, " is larger in the bore than the main pipe of the water-works at London bridge; and the water roaring in its passage through that pipe, is inferior in impetus and velocity to the blood rushing from the whale's heart."

Every human being may be considered as, nay is, a living pump. His body is wholly made up of it, of the tubes belonging to it, and the liquid moved by it—with such additions as are required to communicate the necessary motion and protect it from injury. Health, life itself, every thing, depends upon keeping it in order. If one of its forcing pipes, (an artery,) be severed, we bleed to death; are any of its sucking tubes (the veins)

---

[a] In the 6th vol. of Machines approved by the French Academy, is the description of a bellows pump, made in imitation of the heart, by M. Bedaut, who named the working part of it " *La Cœur*," the heart—of which it was a rude resemblance.

choked, the parts around them become diseased, like sterile land for want of nourishment; does the pump itself stop working, we instantly die. The regularity and irregularity of its motions are indicated by the pulse, which has always been adopted as the unerring criterion of health and disease, or as an engineer would say, the number of its strokes per minute, is the proof of its state whether in good or bad working order. The pulse not only indicates incidental disorders in this hydraulic machine, but is a criterion of its age, as well as of its constant condition: the movements are strong and uniform in youth, feeble and uncertain in sickness and age, and as the machine wears out and the period of its labor approaches, its strokes at last cease and its vibrations are then silent for ever.

What mechanic can contemplate this surprising machine without being electrified with astonishment that it should last so long as it does in some people! Formed of materials so easily injured, and connected with tubes of the most delicate texture, whose ramifications are too complex to be traced, their numbers too great to be counted, and many of them too minute to be perceived, and the orifices of all furnished with elaborate valves; that such complicated machinery should continue incessantly in motion, sixty, eighty, and a hundred years, not only without our aid, but in spite of obstructions that are daily thrown in its way, is as inexplicable and mysterious as the power that impels it.

Few classes of men are more interested in studying natural history, and particularly the structure, habits, and movements of animals, than mechanics; and none can reap a richer reward for the time and labor expended upon it. It presents to the studious inquirer sources of mechanical combinations and movements so varied, so perfect, so novel, and such as are adapted to every possible contingency, as to excite emotions of surprise that they should have been so long neglected. There is no doubt that several modern discoveries in pneumatics, hydraulics, hydrostatics, optics, mechanics, and even of chemistry, might have been anticipated by the study of this department of science. Of this truth examples might be adduced from every art, and from every branch of engineering: the flexible water-mains (composed of iron tubes united by a species of ball and socket joint) by which Watt conveyed fresh water under the river Clyde were suggested by the mechanism of a *lobster's tail*:—the process of tunneling by which Brunel has formed a passage under the Thames occurred to him by witnessing the operations of the *Teredo*, a testaceous *worm* covered with a cylindrical shell, which eats its way through the hardest wood—and Smeaton, in seeking the form best adapted to impart stability to the light-house on the Eddystone rocks, imitated the contour of the bole of a *tree*. The fishermen's boats of Europe, adapted to endure the roughest weather, are the very model of those formed for her progeny by the female *gnat;* " elevated and narrow at each end, and broad and depressed at the middle"—the beaver when building a dam—but it is vain to quote examples with which volumes might be filled.

CHAPTER III.

Forcing Pumps with solid pistons: The Syringe: Its uses, materials and antiquity—Employed by the Hindoos in religious festivals—Figured on an old coat of arms—Simple Garden Pump—Single valve Forcing-pump—Common Forcing-pump—Stomach pump—Forcing-pump with air-vessel—Machine of Ctesibius: Its description by Vitruvius—Remarks on its origin—Errors of the ancients respecting the authors of several inventions—Claims of Ctesibius to the pump limited—Air vessel probably invented by him—Compressed air a prominent feature in all his inventions—Air vessels—In Heron's fountain—Apparently referred to by Pliny—Air gun of Ctesibius—The Hookah.

THE earliest machine consisting of a cylinder and piston that was expressly designed to force liquids was probably the *syringe*, an instrument of very high antiquity: see its figure in the foreground of the next illustration. To the closed end a short conical pipe is attached whose dimensions are adapted to the particular purpose for which the instrument is to be used. The piston is solid and covered with a piece of soft leather, hemp, woollen listing, or any similar substance that readily imbibes moisture, in order to prevent air or water from passing between it and the sides of the cylinder. When the end of the pipe is placed in a liquid and the piston drawn back, the atmosphere drives the liquid into the cylinder; whence it is expelled through the same orifice by pushing the piston down: in the former case the syringe acts as a sucking pump; in the latter as a forcing one. They are chiefly employed in surgical operations, for which they are made of various dimensions—from the size of a quart bottle to that of a quill. They are formed of silver, brass, pewter, glass, and sometimes of wood. For some purposes the small pipe is dispensed with, the end of the cylinder being closed by a perforated plate, as in those instruments with which gardeners syringe their plants.

It has been said that the syringe was invented by Ctesibius, being the result of his first essays in devising or improving the pump; but such could not have been its origin, since it is mentioned by philosophers who flourished centuries before him. It was known to Theophrastus, Anaxagoras, Democritus, Leucippus, Aristotle, and their pupils: to the rushing of water into it when the piston was drawn up, these philosophers appealed to illustrate their opposite views respecting the cause of the liquid's ascent, some contending that it proved the existence of a vacuum, others that it did not. To this ancient application of the syringe, most of the early writers on atmospheric pressure allude.[a] "It is pretty strange [observes Desaguliers] that the ancients, who were no strangers to the nature of winds, and knew a great deal of their force, were yet entirely ignorant of the weight and perpendicular pressure of the air. This is evident, because they attribute the cause of water rising up in pumps, or any liquors being drawn up into syringes (commonly called *syphons* on that account, while pumps were call'd sucking-pumps) to nature's abhorrence of a vacuum; saying, that it fill'd up with water the pipes of pumps under the moving bucket or piston, rather than suffer any empty space. The syringe was in use, and this notion concerning its suction obtain'd *long before* Ctesibius, the son of a barber at Alexandria, invented the pump."[b]

---

[a] See Rohault's Philosophy with Clarke's Notes. Lon. 1723; vol. i. 172. Switzer's Hydrostatics, Preface and 172. Chambers' Dict. Articles Syringe, Embolus, Vacuum.
[b] Ex. Philos. vol. ii, 249.

There is reason to believe that the syringe was employed by the Egyptians in the process of embalming. In various translations of the account given by Herodotus (Euterpe, 87) it is expressly named: "They fill a syringe with germe of cedar wood and inject it."[a] Dr. Rees, in his edition of Chambers' Dictionary, (Art. Embalming,) uses the terms "infusing by a syringe," and "syringing a liquid," &c. The least expensive mode of embalming was "infusing by a syringe a certain liquid extracted from the cedar."[b] Beloe, in his translation, does not indicate the instrument used—they "inject an unguent made from the cedar." As clysters originated in Egypt, and were used monthly by the inhabitants as a preservative of health, (Herod. ii, 77,) we are most probably indebted to the people of that country for the syringe. Had it been a Grecian or Roman invention, the name of its author would have been known, for from its utility and application to various useful purposes, an account of the circumstances connected with its origin was as worthy of preservation, as those relating to the pump or any other machine. Suetonius uses the term "clyster" to denote the instrument by which it was administered; and Celsus by it, refers to "a little pipe or squirt." (Ainsworth.) Hippocrates and the elder Pliny frequently mention clysters, but without describing distinctly the instrument employed: the latter in his 30th book, cap. 7, seems to refer to the common pewter syringe, "*an instrument or pipe of tin :*" this is at least probable, for pewter, according to Whittaker, was borrowed from the Romans. It is well ascertained that pewterers were among the earliest workers of metal in England. A company of them was incorporated in 1474; but at what time the syringe became a staple article of their manufacture is uncertain.

No. 115. Syringes used by Hindoos in celebrating some religious festivals.

Had the syringe not been mentioned by ancient authors, its antiquity might be inferred from a particular employment of it by the Hindoos. The arts, manners and customs of these people have remained unchanged from very remote times; and such is their predilection for the religious institutions of their ancestors, that nothing has, and apparently nothing can induce them to admit of the slightest change in the ceremonies that pertain to the worship of their deities: hence the same rites are still performed,

---

[a] Quoted in Ogilby's Africa. Lon. 1670, p. 81. [b] Historical Description of Egypt, Newcastle: vol. i. 602.

Chap. 3.] *And its Applications.* 261

and by means of the same kind of instruments as when Alexander or even Bacchus invaded India. In some of their religious festivals the syringe is made to perform a prominent part; for a red powder is mixed with water, with which the worshipers " drench one another by means of a species of squirt; to represent *Parasou Rama*, or some other hero returning from battle covered with blood." Some writers suppose the ceremony is designed to celebrate " the orgies of *Krishna* with his mistresses and companions." No. 115 represents a rajah and some of his wives engaged in this singular species of religious worship and connubial exercise, in honor of Krishna. The instruments are clearly garden syringes, and probably of the same kind as are mentioned by Heron of Alexandria, as used in his time for sprinkling and dispersing water.

The *Hohlee* is another Hindoo festival which resembles in some measure the Saturnalia of the Romans. It is observed through all Hindostan, and in celebrating it, the syringe is put in requisition. Mr. Broughton, who, with some other Europeans, visited a Mahratta rajah to witness the ceremony, observes—" A few minutes after we had taken our seats, large brass trays filled with *abeer*, and the little balls already described were brought in and placed before the company, together with a yellow-coloured water, and *a large silver squirt* for each individual. The Muha Raj himself began the amusements of the day, by sprinkling a little red and yellow water upon us from the *goolabdans*, small silver vessels kept for the purpose of sprinkling rose-water at visits of ceremony. Every one then began to throw about the *abeer*, and to *squirt* at his neighbour as he pleased." (Shoberl's Hind. vol. ii, 241, and vol. vi, 14.) A somewhat similar custom prevails in Pegu. At the *feast of waters*, the king, nobles, and all the people sport themselves by throwing water upon one another; and " it is impossible to pass the streets without being soundly wet." (Ovington's Voyage to Surat in the year 1689. Lon. 1696: page 597.)

The syringe in front of No. 115, is copied from Rivius' German Translation of Vitruvius, A. D. 1548. It is from a view of the barber's shop belonging to the father of Ctesibius. (See pp. 121 and 122 of this volume.) Across the shop is a partition, behind which the young philosopher is seen intently perusing a book, and on the floor around him are a flute, a syringe, a pair of bellows, bagpipes, &c.; while in front, the old gentleman in the European costume of the 16th century, and with a sword at his side! is actively engaged in purifying the head and face of a customer.

In the third volume of a Collection of "Emblems, Human and Divine" in Latin: Prague, 1601, page 76, a pair of bellows, a *syringe*, and a flying eolipile are represented as forming the device of some old Italian family, with the singular motto, " *Todo est viento.*"

Few ancient devices could be pointed out that have given rise to more important improvements in the arts than the primitive syringe. Its modifications exert an extensive and beneficial influence in society. As a piston bellows it is still extensively used in oriental smitheries—and as the same, it contributed to one of the most refined pleasures of the ancients, by supplying wind to their organs. It may be considered as the immediate parent of the forcing if not of the atmospheric pump—in both of which it has greatly increased the comforts and conveniencies of civilized life; in the fire-engine it protects both our lives and our property from the most destructive of the elements; and in the hands of the surgeon and physician it extends the duration of life by removing disease. The modern philosophical apparatus for exhausting air, and the ancient one for condensing it; the mammoth blowing machines in our founderies, and the steam engine itself, are all modifications of the syringe.

A forcing pump differs but little from a syringe: the latter receives and expels a liquid through the same passage, but the former has a separate pipe for its discharge, and both the receiving and discharging orifices are covered with valves. By this arrangement it is not necessary to remove a pump from the liquid to transfer the contents of its cylinder, as is done with the syringe, but the operation of forcing up water may be continuous while the instrument is immoveable. A forcing pump, therefore, is merely a syringe furnished with an induction and eduction valve—one through which water enters the cylinder, the other by which it escapes from it. Of the process or reasoning which led to the application of valves to the syringe, history is silent; but as has been remarked in a previous chapter, their employment in bellows or air forcing machines, probably opened the way to their introduction into water forcing ones. The ordinary bellows has but one valve, and the simplest and most ancient forcing pumps have no more. One of these is shown at No. 116. It represents a syringe having the orifice at the bottom of the cylinder covered by a valve or clack, opening upwards; and a discharging pipe connected to the cylinder a little above it: when placed in water the orifice of this pipe is closed with the finger, and the piston being then drawn up, the cylinder becomes charged, and when the piston is pushed down the valve closes and the liquid is forced through the pipe. In this machine the finger performs the part of a valve by preventing air from entering the cylinder when the piston is being raised. Such pumps made of tin plate were formerly common, and were used to wash windows, syringe plants and garden trees, &c. The figure is from plate 57 of "L'Exploiter des Mines," in ARTS ET METIERS, and is described (page 1584) as a *Dutch* pump, " pour envoyer commodément de l'eau dans les différents quartiers de l'attelier."

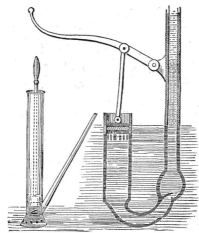

No. 116.   No. 117.
Single-Valve Forcing Pumps.

No. 117 is another single-valve forcing pump from the second volume of a Latin treatise on Natural Philosophy, by P. P. Steinmeyer, Friburgh, 1767. It is secured in a cistern, the surface of the water in which is always kept above the small openings made through the upper part; so that when the piston is drawn up, as in the figure, the liquid flows in and fills it; and on the descent of the piston the water is forced up the ascending pipe, the valve preventing its return. This is a very simple and efficient forcing pump; and having no induction valve and the piston being always under water, it is not very liable to derangement. It has, however, its defects; for in elevating the piston the whole weight of the atmosphere above it has to be overcome, a disadvantage that in large machines would not be compensated by the saving of a valve. As the piston has to pass the holes in the upper part of the cylinder, its packing would be injured if their inner edges were not rounded off. This pump has been erroneously attributed to a modern European engineer: see the London Register of Arts, v, 154, and Journal of the Franklin Institute, viii, 379.

Common Forcing Pump.

The ordinary forcing pump has two valves, as in the annexed figure, which represents it as generally made. The cylinder is placed above the surface of the water to be raised, and consequently is charged by the pressure of the atmosphere; the machine, therefore, is a compound one, differing from that last described, which is purely a forcing pump, the water entering its cylinder by gravity alone. The action of the machine now under consideration is similar to that of the syringe: when the piston is raised the air in the pipe below the cylinder rushes through the valve and is expelled on the descent of the piston through the other valve in the ascending or discharging pipe; and on a repetition of the strokes of the piston, water rises in the suction pipe, enters the cylinder, and is expelled in the like manner. Pumps of this kind are sometimes placed in the yards of dwelling houses, the suction pipe extending into a well, and the ascending one to a cistern in the upper parts of the building. In these cases a cock is generally inserted a little above the valve in the ascending pipe to supply water if required in the vicinity of the pump.

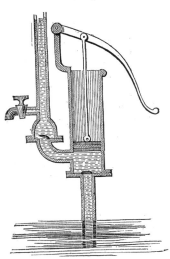

No. 118. Common Forcing Pump.

The beautiful instrument used of late years to transfer liquids into and from the human stomach is a modification of the above machine. It cannot with propriety be named a syringe, for as it is furnished with valves, it is, in every respect, a pump. Having been employed with much success in withdrawing poison from the stomach, it is now justly classed among the essential apparatus of the surgeon. Its origin and history are detailed in a pamphlet published by its inventor, Mr. John Read, of England, who devised it in 1819, and in the following year obtained a patent for it under the name of a "Stomach and Enema Pump." After visiting London twice in vain for the purpose of procuring suitable tubes, he tried to get some made in the country, but failed. On a third visit to the metropolis he obtained an indifferent one which he thought might answer, and after adapting it to a pump, "I then [he observes] presented it to Sir Astley Cooper, who asked me for what purpose it was intended; I told him it was intended for the removal of fluid poisons from the human stomach; after a few minutes inspection of the instrument, Sir A. made the following reply :—' about three weeks ago I was called to attend a young lady about 10 o'clock in the morning who had taken opium; I gave her sulphate of copper, sulphate of zinc and other things: I sat by her until eight in the evening, when she died! If I had been in possession of this instrument at the time, I could have relieved her in five minutes, and have saved her life.' After many questions how I came to think of such a thing, which I satisfactorily explained, he said ' what can I do for you ?' my answer was—the publicity of your opinion is all I wish: he replied, ' that you shall soon have ;' and he ordered me to meet him the next day at Guy's Hospital at one o'clock, when he proposed to try an experiment on a dog; but as no dog could be procured, [that day,] Sir Astley proposed Friday at the same hour; when I attended as before, and a dog was then ready for the experiment in the operating theatre, which was crowded to excess. The dog was brought to Sir A. who gave him four drachms of opium dissolved in water. The dog's pulse was first at 120; in seven minutes it fell to 110, and from that to 90. The poison was suffered to remain in the dog's stomach 33 minutes, till he appeared to be dead, and I was doubtful it would be the case before Sir A. would let me use the pump. I must confess I was very impatient to be at work on the dog with my instrument in hand ready for action. Sir A. kept his finger on the dog's pulse, then at 90, and said very deliberately, 'I think it will do now, as it is 33 minutes since I gave him the dose.' A basin of warm water being then brought, Sir A. passed the tube I had provided into the dog's stomach: I immediately pumped the whole contents of the basin [the warm water] *into* the stomach, and as quickly repumped the whole from the stomach, containing the laudanum, back again into the basin. Sir A. observed, while I was emptying the dog's stomach, the laudanum swimming on the surface, and said '*It will do;*' a second basin of water was then injected and withdrawn by the pump as before : I asked for a *third*, but Sir Astley said it was unnecessary, as the laudanum had all been returned in the first basin." In half an hour the animal was completely revived and running about the theatre.

It may be of use to state, that the quickest and easiest mode of employing a stomach pump (according to the inventor) is to use it only as a *forcing* pump—that is to inject warm water or other dilutents into the stomach until that organ becoming surcharged, the fluid regurgitates by the mouth; in other words to fill the stomach to overflowing—the liquid passing down the tube and rising through the œsophagus by the side of it; the operation being continued till the fluid returns unchanged. In the absence of a pump, a tunnel or other vessel attached to a flexible tube might answer.

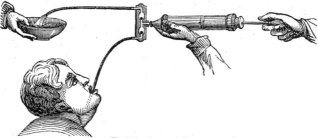

No. 119. Stomach Pump.

There are numerous varieties in stomach pumps, arising from the different modes of constructing and arranging the valves, so as either to inject or withdraw liquids through the same tube without shifting the apparatus. No. 119 represents one that is described in the Journal of the Franklin Institute, (vol. xiii, 223.) It consists of an ordinary syringe screwed to a cylindrical valve box which contains two egg-shaped cavities. In each cavity is a small and loose spherical valve that fits either of the orifices. Two flexible tubes are attached to each cavity as represented. Suppose the upper tube inserted into a person's stomach and the lower one into a basin of warm water; if the syringe were then worked, the liquid would be forced into the stomach and the poison diluted : then by turning the instrument in the hand so as to bring the upper tube down, (without withdrawing the one in the stomach,) the valves would drop upon the other orifices in each cavity, and the syringe would raise the contents of the stomach into the basin, as represented in the figure.

We have no idea that the inventor of the stomach pump was indebted to *Hudibras* for the hint, yet that old warrior seems not only to have been a proper subject for its occasional application, but he appears to have had some notions that might eventually have led to it. Those readers who are familiar with Butler's account of him will remember that when he was insulted by Talgol the butcher, the knight, as he justly might,

> . . . . . . "grew high in wroth,
> And lifting hands and eyes up both,
> Three times he smote on stomach stout
> From whence at last these words broke out:
> * * * * * * *
> Nor all that farce that makes thee proud,
> Because by bullocks ne'er withstood,
> Shall save, or help thee to evade
> The hand of justice, or this blade.
> Nor shall those words of venom base,
> Which thou hast from their native place
> *Thy stomach, pump'd* to fling on me
> Go unrevenged : * * * *
> Thou down the same throat shall devour 'em,
> Like tainted beef, and pay dear for 'em."—*Canto* II, *Part* I.

It was a common practice with the ancient Roman epicures to empty the stomach by an emetic before dinner. Had the application of the pump for such a purpose been then known, it would of course have been preferred as the more agreeable and certain device of the two. But if the ancients had no apparatus for withdrawing the contents of the stomach, they were not destitute of means for conveying nauseous or corroding liquids *into* it. Pliny, in his Nat. Hist. xxx, 6. says such medicines were swallowed "through a pipe or tunnel" inserted into the mouth for that purpose.

The pump figured at No. 118 ejects water as a syringe and only when the piston is forced down; but by the addition of what is called an *air-vessel*, the stream from the discharging pipe may be made continuous: this vessel is closed at its upper part, and open at bottom, where it is connected by screws to the forcing pipe directly over the valve, as represented in the annexed illustration. A discharging pipe may then be connected to the lower part of the vessel, or it may be, as it often is, inserted through the top, in which case its lower end should extend nearly to the bottom. When by the descent of the piston water is forced out of the cylinder, part of it enters the pipe, and part rushes past it and *compresses the air* confined in the upper part of the vessel; and when the piston is raised to draw a fresh portion into the cylinder, this air expands and drives out the water that compressed it and thus renders the stream constant. It will be perceived that the quantity of water raised is not increased by this arrange-

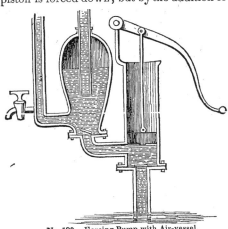

No. 120. Forcing Pump with Air-vessel.

ment; its flow from the discharging orifice being merely rendered uniform, or nearly so.

In the ordinary use of forcing pumps a constant instead of an interrupted flow of water from the discharging orifice, may be a matter of no importance; but when those of large dimensions are required to raise it to great elevations, air vessels are not only valuable but indispensable adjuncts; for the elastic fluid within them forms a medium for gradually overcoming the inertia of the ascending liquid columns, and thereby prevents those jars and shocks which are incident to all non-elastic substances in rapid motion, when brought suddenly to a state of rest. A column of water moving with great velocity through a pump, produces, when instantly stopt, a concussion like that of a solid rod of the same length, when its end is driven against an unyielding object; but with an air-vessel, the effect is like that of the same rod when brought in contact with a bale of cotton or caoutchouc. Less force is required also to work pumps that have air-vessels, because in them the column of water in the discharging pipe is continued in motion during the ascent of the piston, hence it has not to be moved from a state of rest on the piston's return. When two or more cylinders are connected to one discharging pipe, one air-vessel only is required, as in fire-engines, water-works, &c.

It is this kind of forcing pump that is generally adopted in water-works for the supply of towns and cities; the piston rods being moved by cranks or levers attached to water wheels: sometimes they are driven by windmills, steam-engines, and by animals. The cylinders are commonly used perpendicularly as in the figure, but they are sometimes worked in an inclined and also in a horizontal position.

The celebrated pump of Ctesibius was constructed like that represented in the last figure, except that it had two cylinders. It seems to have been almost identical in its construction with our fire-engines. "It remains now [says Vitruvius] to describe the machine of Ctesibius which raises water very high. This is made of brass; at the bottom a pair of buckets [cylinders] are placed at a little distance, having pipes like the shape of a fork annexed, meeting in a basin in the middle. At the upper holes of the pipes within the basin, are made valves, hinged with very exact joints; which, stopping the holes, prevent the efflux of the water that will be pressed into the basin by the air. Upon the basin a cover like an inverted funnel is fitted, which is adjoined and fastened to the basin by a collar, riveted through, that the pressure of the water may not force it off: and on the top of it, a pipe called the *tuba*, is affixed perpendicularly. The buckets [cylinders] have valves placed below the lower mouths of the pipes, and fixed over holes that are in their bottoms: then pistons turned very smooth and anointed with oil, being inclosed in the buckets [cylinders] are worked with bars and levers from above; the repeated motion of these, up and down, pressing the air that is therein contained with the water, the holes being shut by the valves, forces and extrudes the water through the mouths of the pipes into the basin; from whence rising to the cover, the air presses it upwards through the pipe; and thus from the low situation of the reservoir, raises it to supply the public fountains." Book x, cap. 12. Newton's Trans.

The machine as thus described is a proof of the progress which the ancients had made in hydraulics: the whole appears to have been of the most durable materials, and of the best workmanship. Although the figures of this and other machines which Vitruvius inserted in his work are lost, there is little difficulty in realizing its construction from the text. Transla-

Chap. 3] *Machine of Ctesibius.* 267

tors and commentators have generally agreed in their views of it as represented below, viz: two ordinary forcing pumps connected to an airvessel and one discharging pipe.

No. 121. Machine of Ctesibius.

The cylinders are secured in a frame of timber, and the piston rods are attached by joints to levers, one end of which are depressed by cams on the axis of the wheel, as shown above and also at No. 89. Barbaro has figured a crank at the axis which gives a reciprocating motion to a horizontal shaft placed over the pumps, and projecting pieces from which impart motion to the piston rods. Vitruvius informs us that when machines were employed to raise water from rivers, they were worked by undershot wheels impelled by the stream, and hence the pumps of Ctesibius were believed to have been moved by the same means.

But for Vitruvius we should not have known that forcing pumps constituted part of the water works of antiquity; and had he not remarked that they were employed to supply "public fountains," it might have been supposed that water never rose higher in the dwellings of ancient cities than that which was drawn directly from the aqueducts.

It would be almost unpardonable to pass over this celebrated machine without further remark, since it is, in several respects, one of the most interesting of all antiquity. An account of its origin and early history would form a commentary on *most* of the arts and sciences of the ancients, and would, we believe, furnish evidence of their progress in some of them that few are willing to believe. Although it was attributed to Ctesibius, there is some uncertainty respecting the extent of his claims. It may appear invidious to attempt to rob this illustrious man of inventions ascribed to him, but our object is to ascertain, not to depreciate them or diminish their number. It has frequently been remarked that little dependence can be placed on ancient writers as regards the authors of the useful machines. Generally those who introduced them from abroad, who improved them, increased their effects, or extended their application, were reputed their inventors. This has been the case more or less in every part of the world, and is so at the present day. The Greeks found authors among themselves for almost every machine, although most of them were certainly derived from Egypt. Thus, the sails and masts of ships, the wedge, auger, axe and level, were known before Dædalus. The saw, drill, compasses, glue and dovetailing, before Talus. Cast iron was employed, and moulding practiced, and the lathe invented, long before Theodorus of Samos lived; and the screw and the crane before Archytas. The last individual was celebrated for various inventions, and among others, Aristotle mentions the child's rattle, from which it may be inferred that he was an amiable man and fond of children—but Egyptian children were amused with various species of toys, centuries before he flourished; and they then had dolls whose limbs were moved by the pulling of strings or wires, as ours have at this day. Wilkinson's Manners and Customs of the Ancient Eygptians. Vol. ii, 426–7.

As regards machines for raising water, we have already seen, that some have been ascribed to others than their authors. Even the siphon has been attributed to Ctesibius, (Adams's Lectures, vol. iii, 372,) because it was found in the construction of his clepsydra, and no earlier application of it was then known; but it is now ascertained to have been in common use among his countrymen in the remote age of Rameses—in the Augustan era of Egypt, when the arts, we are informed, " attained a degree of perfection, which no after age succeeded in imitating." Had the "Commentaries of Ctesibius" to which Vitruvius referred his readers for further information, been preserved, we should have had no occasion to attempt a definition of his claims to the forcing pump; unfortunately, however, these and Archimedes' Treatise on Pneumatic and Hydrostatic Engines have perished, and have left us in comparative ignorance of the history of such machines among the ancients.

We have already seen that the syringe was in common use ages before Ctesibius, and that it was employed by philosophers to illustrate their hypothesis of water rushing into a vacuum. Now a forcing pump is merely a syringe with an additional orifice for the liquid's discharge, and having both its receiving and discharging orifices covered by valves or clacks. Ctesibius therefore did not invent the piston and cylinder, nor was he the first to discover the application of these to force water, for they were in previous use and for that purpose. Was he the inventor of valves? No, for they were used in the Egyptian bellows thirteen or fourteen hundred years before he lived, and appear always to have been an essential part of those instruments. They were employed in clepsydra; and were most likely used in the hydraulic organ of Archimedes, which Tertullian has described. Is the *arrangement* of the valves, by which water is admitted through one and expelled by the other, to be ascribed to him? We believe not, for the *same arrangement* was previously adopted in the bellows, so far as regards the application of one of them, and the principle of both: and if it could be shown that the Chinese bellows was then in use, as we suppose it was, and possibly known in Egypt, (for that some intercourse did take place in ancient times between Egypt and China, even if one people be not a colony of the other, is proved by Chinese bottles and inscriptions found in the tombs at Thebes,) then the merit of Ctesibius would seem to be confined principally to the construction of *metallic* bellows as "*water* forcers," or, *to the application of valves to the ordinary syringe*, by which it was converted into a forcing pump, either for air or water. But it is not certain that the last was not done before, for neither Vitruvius nor Pliny asserts that "water forcers" were not in previous use. The former says he applied the principle of "compressed air" to them, in common with "hydraulic organs," "automatons," "lever and turning machines," and "water dials," (Book ix, cap. 9;) hence it may as well be concluded from this passage, that he invented these as the pump. It is, indeed, almost impossible to believe that the Egyptians, of whose sagacity and ingenuity, unrivalled monuments have come down, did not detect the application both of the bellows and syringe to raise water long before Ctesibius lived; hence we are inclined to place the forcing pump in its simplest form, with the syringe and atmospheric pump, among the works—

> "Of names once famed, now dubious or forgot
> And buried 'midst the wreck of things that were."

That the forcing pump was greatly improved by Ctesibius, there can be no question; but that which gave celebrity to his machine was probably the air-vessel, an addition, which though not very clearly described by Vitruvius, appears to have originated with him. By it the pump instead of acting as before like a squirt or syringe produced a continuous stream as in a *jet d'eau*, a result well adapted to excite admiration, and to give eclat to his name. The whole account of his machine shows its connection with and dependence upon air; whereas had it been simply a forcing pump it would have had nothing to do with it: it would have raised water independently of it; and without an air-vessel Vitruvius never could have asserted that it forced water up the discharging tube by means of "air pressing it upwards." Compressed air acted a prominent part in all his machines. In his wind guns, water clocks, and numerous automata; some of the latter in the shape of birds, &c. appeared to sing, others "sounded trumpets," and these results are said to have been produced with "fluids compressed by the force of air." We may add that he compressed air in his hydraulic organs and precisely in the same manner as in the pump, viz: by water, and by either air or water forcing pumps. The commencement of his discoveries was the experiment on air with the weight and speculum in his father's shop, (see page 122) in which the descending weight "compressed the inclosed air" and forced it through the several apertures into the open air, and thereby produced distinct sounds. "When therefore Ctesibius observed that sounds were produced from the compression and concussion of air, he first *made use of that principle* in contriving hydraulic organs, also water forcers, automatons," &c. What principle was this which Vitruvius says he applied to water forcers in common with organs, &c.? That of compressed air, as we understand it; and the employment of which is so evident, in the description of his machine already given.

Does any one doubt that the air-vessel was known to, and used by Ctesibius? Let him recollect that Heron, his disciple and intimate friend, has also described it; for the celebrated fountain of this philosopher, which still bears his name, and remains just as he left it, is simply an air-chamber, in which the fluid is compressed by a column of water instead of a pump; and one of his machines for raising water by steam, was another, in which the elasticity of that fluid was used in a similar manner. Besides these, there are others represented in the *Spiritalia;* indeed, a great portion of the figures in that work are modifications of air chambers. At pages 42 and 118, of Commandine's Translation, are shown spherical vessels containing water, into which perpendicular discharging tubes descend: to expel the liquid, syringes or minute pumps are adapted to the vessels, for the purpose of injecting air or water, and by that means to produce *jets d'eau*. The common syringe is also figured at large and in section, p. 120.[a] Pliny also seems to refer to air-vessels in his xix book, cap. 4, where he speaks of water forced up " by pumps and such like, going with *the strength of wind enclosed.*" Holland's Trans.

As the ancients have not particularized the claims of Ctesibius to the pump, it is impossible to define them with precision at this distance of time. Perhaps the instrument had been laid aside, or the knowledge of it almost lost when he revived and improved it, as some of his own inventions have been in modern times—his gun, for example, of which Philo of Byzantium has given a description, and which " was constructed in such a manner as to carry stones with great rapidity to the greatest distance."[b] Its invention has been claimed by the Germans, the French, Dutch, and from the following remark of Blainville, by the Swiss also: speaking of Basil, he observes, " They make a great noise here about a hellish invention of a gunsmith, who invented wind guns and pistols. This invention may be truly called diabolical, and the use of it ought to be forbid on pain of death."[c] Now if the modern inventor of the air gun, an instrument which, two centuries ago, was spoken of as " a *late* invention,"[d] cannot with certainty be ascertained, it can hardly be expected that the specific claims of Ctesibius to the pump can be pointed out after a lapse of 2000 years. If he was the first to combine two or more cylinders to one discharging pipe— to form them of metal, as well as the valves and pistons—and the first to invent and apply air-vessels, his claims are great indeed, and for aught that is known to the contrary he is entitled to them all. His merits as respects the latter will be apparent, if we call to mind the fact that their application to pumps has not been known in Europe for two centuries; and that their introduction was in all probability derived from him, for it was not till a hundred years after Vitruvius's description of his machine had been translated, printed and circulated, that we first hear of air-vessels in modern times.

We may here remark that at whatever period *tobacco* was first smoked in the *Hookah*, (and according to some authors, this weed was used in Asia before the discovery of America,) the air-vessel was known; for that instrument is a perfect one, as any person may prove by the following experiment: let a smoker, instead of sucking at the end of the tube which he inserts in his mouth, *blow* through it, and the liquid contents of the hookah will be forced out through the perpendicular tube on which the weed is placed as in a miniature fire-engine, carrying up with it the pellet of tobacco, somewhat in the manner of those light-balls which are sometimes placed on *jets d'eau*, or the boy's pea playing on a pipe stem. An operation, in the opinion of some physicians, *more beneficial* to the performer than the ordinary one, and disposing of the scented material in a manner more suited to its value.

---

[a] Heronis Alexandrini Spiritalium liber. A Federico Commandino urbinate, ex Græco nuper in Latinum conversus. 1583.
[b] Duten's Inquiry into the Origin of the Arts attributed to the Moderns, p. 186. Travels, i, 388. [d] Wilkins' Mat. Magic.

CHAPTER IV.

FORCING pumps continued: La Hire's double acting pump—Plunger pump: Invented by Moreland; the most valuable of modern improvements on the pump—Application of it to other purposes than raising water—Frictionless plunger pump—Quicksilver pumps—Application of the principle of Bramah's press by bees in forcing honey into their cells. Forcing pumps with hollow pistons: Employed in French water-works—Specimen from the works at Notre Dame—Lifting pump from Agricola—Modern lifting pumps—Extract from an old pump-maker's circular—Lifting pumps with two pistons—Combination of hollow and solid pistons—Trevethick's pump—Perkins' pump.

OF the various modifications which the forcing pump has undergone in recent times we can notice but a few, and of these the greater part were most likely known to ancient engineers. The most prominent one is that by which the machine is made double acting. Now the device by which this is effected has not only frequently occurred to quite a number of ingenious men in their endeavours to improve the pump who were ignorant of its having been accomplished; but it is an exact copy of one that has been applied to the wind pump of China from time immemorial, (see No. 112;) it probably therefore did not escape such men as Ctesibius, and Heron, and others who appear to have exercised their ingenuity and sagacity to the utmost in order to improve this machine, and who were enthusiastically attached to such researches. The remarks on modern improvements of the atmospheric pump, pages 225–6, are equally applicable to those of the forcing one; and it is worthy of remark, that notwithstanding the present improved state of mechanical science, the ancient forms of both now prevail—for the forcing pump as made by Ctesibius in Egypt, and as described by Vitruvius as used by the Romans, is still more common than any other.

The double acting pump represented in the figure, was devised by M. La Hire in the early part of the last century. His description of it was published in the Memoirs of the French Academy in 1716; and from one of his expressions we perceive (what was indeed very natural) that if he was not indebted for the improvement to the contemplation of bellows, these instruments were at least closely associated with it in his mind. The pump I propose [he observes] furnishes water continually, "just as the double bellows makes a continual wind." The piston rod passes through a stuffing box or collar of leathers on the top of the cylinder. The latter has four openings covered by valves or clacks; two for the admission of water and the same number for its discharge. A B is the suction pipe, and C D the ascending or discharging one. Suppose the lower end of the suction pipe in water; then if the piston be thrust down, the valve near B will close, and the air in the lower part of the cylinder will be forced through the valve at D and up the pipe D C, and in consequence of the rarefaction of the air above the piston, the valve at C will be

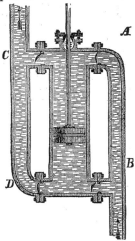

No. 122. Double Acting Pump.

closed, and water will ascend through B A and enter the cylinder at A; then if the piston be raised it will force all the water above it through the valve at C, the only passage for it, while at the same time a fresh portion will enter the cylinder through the valve at B. Thus at every stroke of the piston, whether up or down, the contents of the cylinder are forced out at one end, and it is replenished at the same time through the other; this pump therefore discharges double the quantity of water that an ordinary one of the same dimensions does. The piston rod may be inserted through either end of the cylinder, as circumstances may require. These pumps are frequently used in a horizontal position.

Another variation of the forcing pump consists in making the piston of the same length as the cylinder but rather less in diameter, so that it may be moved freely in the former without touching the sides. These pistons are made wholly of metal and turned smooth and cylindrical, so as to work through a stuffing box or cupped leathers. The quantity of water raised at each stroke has therefore no reference to the capacity of the cylinder, however large that part of one of these pumps may be, for the liquid displaced by the piston can only be equal to that part of the latter that enters the cylinder. Switzer has given a figure and description of an old engine composed of three of these pumps " that has been some years erected in the county of Surrey." Newton has figured the piston bellows described by Vitruvius as furnishing wind to hydraulic organs in a similar way. In Commandine's translation of Heron's Spiritalia, page 159, the same kind of plunger is figured in a pump belonging to a water organ; and at p. 71, a fire-engine, with two working cylinders, has pistons of the same kind. These pistons were formerly named plungers, and the pumps plunger-pumps. Their construction and action will be understood by the figure, which represents one of a number that were employed in the water-works, York Buildings, London, in the last century. The piston was of brass, cast hollow and filled with lead, the outside being "turned true and smooth." A short rod attached to the upper end of the piston was connected by a chain to the arched end of a vibrating beam, that was moved by one of Newcomen's engines. The piston was therefore merely raised by the engine, while its own weight carried it down: to render it sufficiently heavy for this purpose, a number of leaden disks (or cheeses, as they were named from their form) having holes in their centres, were slipped over the rod and rested upon the piston, as in the figure. These were increased until they were found sufficient to press down the piston and force the water up the ascending pipe. The cupped leathers through which the piston worked, were similar to those now used in the hydrostatic press. A small cistern was sometimes formed on the top of the pump, that the water it contained might prevent air from entering through the stuffing box or between the cupped leathers: it served also to charge the pump through a small pipe or cock. A valve opening upwards was sometimes placed just above the plug of the cock, and the latter left open when the machine was started, that the air within the cylinder might escape; and as soon as the water rose and filled the pump, the cock was shut. It is immaterial at what part of the cylinder the forcing or ascend-

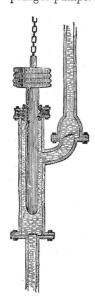

No. 123. Plunger Pump.

ing pipe is attached, whether at the bottom, near the top, or at any intermediate place. Small pumps of this kind are now commonly employed to feed steam boilers and for other purposes, and are worked by levers like the ordinary lifting and forcing pumps, the pistons being preserved in a perpendicular position by slings, &c.

These pumps are believed to be of English origin, having been invented by Sir Samuel Moreland, "master of mechanics" to Charles 2d. Like some old philosophers, he exercised his ingenuity in improving hydraulic and other engines, for raising water. Besides the plunger pump, for which he obtained a patent in 1675, he invented a "cyclo-elliptic movement" for transmitting motion to piston rods, a figure of which is inserted by Belidor in the second volume of his Arch. Hydraulique. He is also the reputed inventor of the speaking trumpet,[a] of a capstan, and a steam-engine. In 1681 he made experiments with an engine consisting of two or more of his pumps at Windsor, in presence of the king and court, during which he forced water from the Thames in a continual stream to the top of the castle; and according to Dr. Hutton, "sixty feet higher." Moreland visited France the same or the following year, by order of the king, to examine the famous water-works at Marli, and while in Paris he exhibited models of his pump before the French court, and also constructed several for his friends. In 1683 he presented an account of various machines for raising water to Louis 14th, in a manuscript volume written and ornamented with much elegance; and in 1685, an account of his improvements was published in Paris in a work entitled, " Elévation des eaux par toute sorte de machines, réduite à la mesure, au poids, à la balance, par le moyen d'un nouveau piston et corps de pompe ; et d'un nouveau mouvement cyclo-elliptique, et rejetant l'usage de toute sorte de manivelle ordinaires, par le Chevalier Moreland." It does not appear that he ever published this work in England, for Switzer had recourse to *Ozanam*, a French writer, for a description of Moreland's pump; as he could procure no English account of it, "having taken great pains to find out what Sir Samuel had left on that head to no purpose." Ozanam states that Moreland spent "twelve years study and a great deal of money" to bring this pump to perfection; " and without this *new* invention it would have been impossible to have reduced the raising of water to weight and measure, as he has done." The latter observation refers to the leaden weights placed on the piston rod, and the quantity of water raised by them: the water and the elevation to which it was raised being compared with the sum of the weights employed to force it up.[b]

If we mistake not this is the most valuable and original modification of the forcing pump that modern times have produced. The friction of the piston is not only greatly reduced, but the boring of the cylinder is dispensed with; an operation of considerable expense and difficulty, particularly so, before efficient apparatus for that purpose was devised. Another advantage is the facility of tightening the packing without taking out the piston or even stopping the pump. The value of Moreland's invention in

---

[a] There is an instrument very like a speaking trumpet in the hands of a figure in one of the illustrations of the Eneid, executed in the fourth or fifth century, in the 25th plate of "Painting" in D'Agincourt's History of the Fine Arts. It is a conical tube, the length being equal to that of the individual using it; and by which he appears to direct, from the top of a tower, the combatants below. Kircher has given a figure of a trumpet through which he supposed Alexander spoke to his army.

[b] See Switzer's Hydrostatics, plate 25, pp. 302, 357. La Motraye's Travels, vol. iii, Lon. 1732. Desaguliers' Philos. vol. ii, 266. Belidor's Architecture Hydraulique, tom. ii, 61, and L'Art D'Exploiter Les Mines, in Arts et Metiers, page 1058, and planche 47.

the estimation of engineers appears from the increasing employment of it. It is, moreover, for aught that is known to the contrary, the parent of the common lifting pump; and to its inventor the double acting steam-engine of Watt is in some measure due, the efficiency of that noble machine depending entirely upon closing the top of the cylinder and passing the piston rod through a stuffing box—both of which had already been done in this pump. Steam-engines have also been constructed on the same plan as these pumps; one long piston playing in two horizontal cylinders, and the power transmitted from it by means of a cross-head attached to the middle of its length, and on that part which moves between the stuffing boxes. Another celebrated machine is also copied from them —Bramah's hydrostatic press is one of Moreland's pumps.

There is another species of plunger pumps in which the stuffing box is dispensed with, and consequently the piston works without friction. A square wooden tube, or a common pump log of sufficient length, and with a valve at its lower end is fixed in the well as shown in the figure. The depth of the water must be equal to the distance from its surface to the place of delivery; and a discharging pipe having a valve opening upwards is united to the pump tree at the surface of the water in the well. The piston (a solid piece of wood) is suspended by a chain from a working beam, and loaded sufficiently with weights to make it sink. As the liquid enters the pump through the lower valve, and stands at the same level within as without, whenever the piston descends, it necessarily displaces the water, which has no other passage to escape but through the discharging pipe, in consequence of the lower valve closing. And when the piston is again raised as in the figure, a fresh portion of water enters the pump and is driven up in like manner.

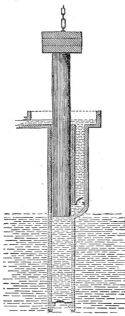

No. 124. Frictionless Plunger Pump.

Dr. Robison observes that he has seen a machine consisting of two of these pumps, made by an untaught laboring man. The plungers were suspended from the ends of a long beam, on the upper surface of which the man walked, as on the picotah of India. He stood on one end till one plunger descended to the bottom of its tube, and he then walked to the other end, the declivity at first being about 25°, but gradually growing less as he advanced. In this way he caused the other plunger to descend, and so on alternately.

By this machine a feeble old man whose weight was 110lbs. raised 7 cubic feet of water 11½ feet high in a minute, and wrought eight or ten hours every day. A stout young man weighing 134lbs. raised 8⅓ cubic feet to the same height in the same time. The application of this pump is extremely limited, and there is a waste of power in the water that is uselessly raised around the piston at every stroke.

The pistons of preceding machines are made of solid materials; but the pump now to be described has a *liquid* one. It was invented about the year 1720, by Mr. Joshua Haskins, who made the first experiment with it in the house and presence of the celebrated Desaguliers. His design was to avoid the friction and consequent loss of power in common pumps, he therefore "contrived a new way of raising water without any friction of solids; making use of quicksilver instead of leather, to keep the air or water from slipping by the sides of the pistons." Various modifications of it were soon devised by the inventor, by Dr. Desaguliers, and by Mr. William Vreem, the assistant of the latter, "who was an excellent mechanic." One form of it is represented by the figure. A is the suction pipe, the lower end of which is inserted in the water to be raised. Its upper end terminates in the chamber C, and is covered by a valve. The forcing pipe B, with a valve at its lower end, is also connected to the chamber.

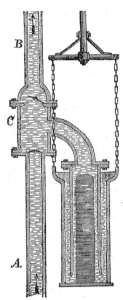

No. 125. Mercurial Pump.

Between these valves a pipe, open at both ends, is inserted and bent down, as in the figure. The straight part attached to it is the working cylinder of the pump and should be made of iron. Another iron pipe, a little larger in the bore than the last, and of the same length, is made to slide easily over it. This pipe is closed at the bottom and suspended by chains or cords, by which it is moved up and down. Suppose this pipe in the position represented, and filled with mercury—if it were then lowered, the air in the cylinder and between the valves would become rarified, and the atmosphere pressing on the surface of the water in which the end of A is placed, would force the liquid up A till the density of the contained air was the same as before; then by raising the pipe containing the mercury, the air, unable to escape through the lower valve, would be forced through the upper one; and by repeating the operation, water would at last rise and be expelled in the same way; *provided* the elevation to which it is to be raised does not exceed thirteen times the depth of the mercurial column around the cylinder; the specific gravity of quicksilver being so many times greater than that of water. When the depth of the former is 30 inches, the latter may be raised as many feet in the suction pipe and forced up an equal distance through the forcing one, making together an elevation of sixty feet; but if water be required higher, the depth of the mercurial column in the moveable pipe must be proportionably increased. To make a small quantity of mercury answer the purpose, a solid piece of wood or iron that is a little less than the cylinder, is secured to the bottom of the moveable vessel as shown in the centre: this answers the same object as an equal bulk of mercury.

These pumps have their disadvantages: they are expensive; and however well made, the quantity of quicksilver required is considerable—the agitation consequent on the necessary movement soon converts it into an oxide and renders it useless—great care is also required in working these machines; if the movements are not slow and regular, the mercury is very apt to be thrown out; to prevent which the upper end of the vessel containing it is dished or enlarged. For experimental researches modifications of such pumps may be useful, but for the reasons above stated, they have never been extensively employed in the arts. A simple form of one is described in a late volume of the London Mechanics' Magazine, and also in the 22d vol. of the Journal of the Franklin Institute, p. 327. See also vol. xxxii, Phil. Transactions, and Abridg. vol. vi, 352. Desaguliers' Phil. vol. ii, 491. In Jamieson's Dictionary, p. 852, a mercurial pump in the form of a wheel is described.

The *hydrostatic press* is simply a cylindrical forcing pump, whose piston is moved by the water, instead of the latter by it. A platen, on which are placed the articles to be pressed, is connected to the upper end of the piston rod; water is then injected into the cylinder by a much smaller pump, and as this liquid is, to all practical purposes, *incompressible*, the piston is necessarily raised, and the articles brought against an immoveable plate, between which and the platen they are compressed. The degree of pressure thus excited depends upon the difference between the area of the pistons of the pump and of the press. The apparatus exhibits in another form, the celebrated hydrostatic paradox by which the pressure of a liquid column however small, is made to counterbalance that of another however large. Hydrostatic presses have been applied with advantage in numerous operations, as, expressing oil from seeds, pressing paper, books, hay and cotton; tearing up trees by the roots, proving the strength of steam boilers, metallic water-pipes, and even cannon. In this city (New-York) ships of a thousand tons are raised out of the water to repair, by one of these machines erected at the head of one of the docks. The cylinder is secured in a horizontal position, and the pumps are worked by a steam-engine. The frame on which the vessel floats, and by which it is raised, is suspended by a number of chains on each side that pass over pulleys and terminate at the end of the piston.

There is a very interesting and beautiful illustration of the principle of Bramah's hydrostatic press in the contrivance by which bees store their honey. The cells, open at one end and closed at the other, are arranged horizontally over each other, and in that position are *filled* with the liquid treasure. Now suppose a series of glass tumblers or tubes laid on their sides and piled upon one another in like manner were required to be then filled with water, it certainly would require some reflection to devise a plan by which the operation could be performed; but whatever mode were hit upon, it could not be more ingenious and effective than that adopted by these diminutive engineers. At the further or closed extremity of each cell, they fabricate a *moveable piston* of wax which is fitted air tight to the sides, and when a bee arrives laden with honey, (which is contained in a liquid form, in a sack or stomach,) she penetrates the piston with her proboscis and through it injects the honey between the closed end of the cell and the piston, and then stops the aperture with her feet. The piston is therefore pushed forward as the honey accumulates behind it, till at last it reaches the open end of the cell, where it remains, hermetically sealing the vessel and excluding the air.[a] As soon as one cell is thus charged, the industrious owners commence with another. It will be perceived that these pistons are propelled precisely as in the hydrostatic press, the liquid honey being incompressible, (with any force to which it is there subjected,) every additional particle forced in necessarily moves the piston forward to afford the required room. Without such a contrivance the cells could no more be filled, and kept so, than a bucket could be, with water, while laying on one side. Were the organization

---

[a] To keep the honey pure, and preserve it from evaporation, in the high temperature of a hive, the air must be kept from it. Could human ingenuity have devised a more perfect mode of accomplishing the object? The fact is, bees in this matter, might long ago have taught man the practice which is now pursued of preserving both liquid and solid aliment *fresh* for years—in tin cases impervious to the air, and from which it has been excluded.

of bees closely examined, it would doubtless be found that the relative diameters of their proboscis and of the cells, and the area of the (bellows) pumps in their bodies, are such as are best adapted to the muscular energy which they employ in working the latter. Were it otherwise, a greater force might be required to inject the honey and drive forward the piston, than they possess. In the case of a hydrostatic press, when the resistance is too great to be overcome by an injection pump of large diameter, one of smaller bore is employed.

We shall now produce a few specimens of forcing pumps with hollow pistons, or such as admit water to pass through them. If a common atmospheric pump be inverted, its cylinder immersed in water, and the valves of the upper and lower boxes reversed as in the figure, it becomes a forcing, or, as it is sometimes named, a *lifting* pump; because the contents of the cylinder are lifted up when the piston is raised, instead of being driven out from below by its descent, as in Nos. 116, 117. In a lifting pump the liquid is expelled from the top of the cylinder—in a forcing one from the bottom—it is the water above the piston that is raised by the former; and that which enters below it, by the latter. The piston rod in the figure is attached to an iron frame that is suspended to the end of a beam or lever as in Nos. 123, 124. The valve on the top of the piston, like that at the end of the cylinder,

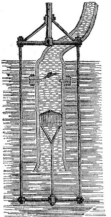

No. 126. Lifting Pump.

opens upwards. When the piston descends (which it does by its own weight and that of the frame) its valve opens and the water enters the upper part of the cylinder, then as soon as it begins to rise its valve closes, and the liquid above it is forced up the ascending pipe. Upon the return of the piston the upper valve is shut by the weight of the column above it, the cylinder is again charged and its contents forced up by a repetition of the movements. Machines of this description are of old date. They were formerly employed in raising water from mines. They were adopted by Rannequin in the celebrated water-works at Marli; and by Lintlaer in the engines he erected during the reign of Henry 4th, at Pont Neuf, to supply the Louvre from the river Seine.

As they cannot in all locations be inserted conveniently in the reservoir containing the water to be raised, they have sometimes been placed in cisterns erected above the original source, and supplied by atmospheric pumps extending to it, as in No. 127. The cylinder of the atmospheric pump terminates in the bottom of the cistern, and is placed directly under that of the lifting one; the pistons of both being attached to the same rod and worked by the same frame. Such was the construction of the old Parisian water-works at the bridge of Notre Dame. These consisted of a series of pumps arranged as in the figure, and worked by an undershot wheel.

No. 127. Pumps from water-works at Notre Dame, Paris.

If the head of a common pump (No. 90) be closed, except an opening through which the rod works, or may be worked, it is then converted into a lifting pump, and will raise water to any elevation through a pipe attached to the spout. The earliest specimen that we have met with is represented by the 128th figure, from Agricola. Although a rude device, it is interesting as illustrative of the resources of old mining engineers, in modifying and applying the common wooden pump under a variety of circumstances. The upper parts of two atmospheric pumps terminate in a close chamber or strong box, (two sides of which are removed in the figure to show its interior,) their lower ends extending into water collected at a lower depth in the mine. From the top of the box a forcing pipe is continued to the surface of the ground, or to another level in the mine, from which the water raised through it can be discharged. The piston rods are worked by a double crank, one end of which turns in a socket formed in the inside of the chamber, and the other is continued through the opposite side and bent into a handle by which the laborer works the machine. Two collars are formed on the crank axle, one close to the outside, and the other to the inside of that part of the chamber through which it passes, and some kind of packing seems to have been used to prevent the water from leaking through. Four iron arms with heavy balls at their ends are secured to the axle to equalize the movement. These were the old substitutes for the modern fly-wheel: they were quite common in all kinds of revolving machinery in the 15th and 16th centuries.

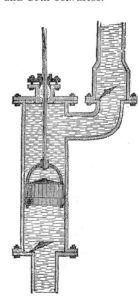

No. 128. Lifting Pump from Agricola.   No. 129. Modern Lifting Pump.

The modern form of the lifting pump is represented in figure No. 129. The working cylinder being generally brass or copper, and having a strong flanch at each end : the upper one is covered by a plate with a stuffing box in the centre, through which the polished piston rod moves ; and the under one by another to which the suction pipe is attached, and whose orifice is covered by a valve. To the forcing or discharging pipe a cock is commonly soldered as in No. 118, to supply water when required at the pump. This is one of the most useful forms of the pump for household purposes: it may be placed in the kitchen, cellar or yard, and will not only draw water from a well, but will force it up to every floor of a dwelling, and still answer every object of the ordinary atmospheric pump; and if an air-vessel be connected to the pipe, as in No. 120, it will then become a domestic fire-engine; and when a sufficient length of hose pipe is kept at hand, water may, in case of fire, be conveyed in a few moments to any part of the building. Desaguliers, a century ago, recommended this application of it, and it is surprising that it has not become more general. The following extract from a pump-maker's circular, 120 years since, refers to it. " Pumps which may be worked by one man, for raising water out of any well, upwards of 120 feet deep, sufficient for the service of any private house or family, and so contrived that by turning a cock, may supply a cistern at the top of the house, or a bathing vessel in any room; and by screwing a leather pipe the water may be conveyed either up stairs, or in at a window, in case of any fire." Switzer's Hydrostatics, 352.

Although the valve in the ascending pipe is not an essential part of these pumps, it is a valuable addition, since it removes the pressure of the liquid column above it from the stuffing box, when the pump is not in use. The inventor of these pumps (and of the stuffing box) is unknown. They are described by Desaguliers, Belidor, and other writers of the last century as then common, and they are figured in the 6th volume of machines and inventions approved by the French Academy, p. 19.

Sometimes the cylinder itself has been made to answer the purpose of an air-vessel. With this view it is made longer than usual, and the discharging pipe is connected to the middle of its length, *below* which the piston works. The air is therefore compressed in the upper part of the cylinder, but as it is liable to escape at the joints and through the stuffing box, a separate vessel is far preferable. Mr. Martin, in the 2d vol. of his Philosophy, has figured and described a pump of this kind, which he says was the invention of Sir James Creed.

In 1815, the London Society of Arts awarded a silver medal and fifteen guineas for a lifting pump with two pistons. The cylinder was made twice the usual length, and each end furnished with a stuffing box through which two separate rods worked. The suction pipe being attached, like the forcing one, to the side of the cylinder; the lower piston was inverted having its valve on the top as in No. 126. The outer ends of the rods were connected to the centre of two small wheels or friction rollers which moved between two guide pieces, and thus prevented the rods from deviating from the centre of the cylinders; the upper wheel was connected by a short rod to the pump lever as in the common pumps, and the other one by a longer rod (bent at its lower part) to the same lever, but on the opposite side of the fulcrum; so that as one was raised the other was lowered; hence the two pistons alternately approached to and receded from each other, and consequently one of them was always forcing up water whenever the machine was at work. Transactions Soc. Arts, vol. xxxiii. 115. We believe these pumps have never been much used, nor do we think they possess any advantages over two separate ones; for they are to all intents and purposes double pumps. The cylinders are twice the length of single ones—they have two pistons, two rods, two stuffing boxes, and double the amount of friction of single ones. Two distinct pumps are more economical. After one of the above has been a little while in use, air will unavoidably insinuate itself through the lower stuffing box and diminish or destroy the vacuum upon which the efficiency of the machine depends. The same remarks apply to these that were made on atmospheric pumps with two pistons, at page 227.

There is a pump with two pistons in Besson's Theatre des Instrumens, which shows that such devices were known in the 16th century. It consists of a square trunk four or five feet in length, and the bore five or six inches across, immersed perpendicularly in water at the bottom of a well; its lower end being open and the upper one closed, except at the centre, where an opening is left and covered by a valve. A square piston, with its valve opening upwards, is fitted to work in the trunk from below by a rod connected to its under side, as in No. 126. A lever passes through the lower part of the trunk, (through slits made for it in two opposite sides,) one end of which is secured to a piece of timber walled in the well, by a pin, on which it moves; and the other end extends to the opposite side of the trunk, where it is hooked to a chain that reaches from the pump brake at the top of the well. The lower end of the piston rod is connected by a bolt to that part of the lever that is within the trunk. This apparatus forms the lifting or forcing part of the machine. A common pump tree or bored log extends from the place to which the water is to be raised, to the top of the trunk, and the junction with the latter made perfectly tight: an upper box or piston with its rod is fitted to work in the tree like an ordinary wooden pump, while the valve on the trunk answers the purpose of a lower box. This rod is attached to the brake on one side of the fulcrum and the chain that is connected to the lever and lower rod to the opposite side, so that as one piston rises the other descends and a constant stream of water is discharged above.

This is the oldest pump with two pistons that we know of, and it has one advantage over others, viz: in raising water without changing its direction. We at first intended to insert a figure of it, but the apparatus for working it is too complicated for popular illustration. Although motion is imparted to the piston as noticed above, it is not done directly, but by means of such an enormous amount of complex and useless machinery as would excite amazement in a modern mechanician. There is an assemblage of rods and levers, tongs and lazy tongs, chains, right and left handed screws, a heavy counterpoise and a massive pendulum, &c., all of which are required to be put in motion before the pistons can be moved. A figure of such a pump would possibly interest some readers as a matter of curiosity, for certainly a rarer example of the waste of power could not well be imagined: it presents as clumsy and "round-about" a mode of accomplishing a very simple purpose, as that of the genius who, in tapping a cask of wine, never thought of inserting the spigot into the barrel, but attempted to drive the barrel on the spigot.

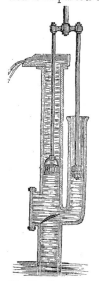

No. 130. Trevethick's Pump.

Sometimes pumps with solid and hollow pistons are combined as in No. 130, a contrivance of Mr. Trevethick. The cylinder of a forcing pump communicates with that of an atmospheric one; both piston rods are connected to a cross-bar and rise and fall together. When the pistons are raised the water above that in the long (or atmospheric) cylinder is discharged at the spout, and the space below them is filled by the atmosphere forcing up fresh portions through the suction pipe. When the pistons descend, the valve on the suction pipe closes, and the solid piston drives the water in its cylinder through the hollow one in the other. so that whether rising or falling the liquid continues to flow. As both cylinders are filled at the same time, the bore of the suction pipe should be proportionably enlarged. The plate bolted over the opening at the lower part of one cylinder is to give access, in large pumps, to the lower valve.

In some pumps both a solid and a hollow piston are made to work in the same cylinder. Such were those that constituted the "single-chamber fire-engine" of Mr. Perkins. A plunger worked through a stuffing box as in No. 123, and its capacity was about half that of the cylinder; consequently on descending it displaced only that proportion of the contents of the latter. The apertures of discharge were at the upper part of the cylinder, and a single receiving one at the bottom. From the lower end of the plunger a short rod projected, to which a hollow piston or sucker was attached, fitted to work close to the cylinder, so that when the plunger was raised, this piston forced all the water above it through the discharging apertures. To convert one of these pumps into a fire-engine, the cylinder of the pump was surrounded by a shorter one of sheet copper, the lower end of which was left open, and its upper one secured air-tight to the flanch of the pump; the space left between the two forming a passage for the water expelled out of the inner one. A larger and close cylinder encompassed the last, and the space between them was the air chamber, to the lower part of which a hose pipe was attached by a coupling screw in the usual way.

Such pumps are more compact than those with two cylinders, but they are more complex, less efficient, and more difficult to keep in order and to repair. The friction of the plunger and sucker is much greater than that of the piston of an ordinary double acting pump of the same dimensions; and the latter discharges double the quantity of water; for although double acting, the effect of these pumps is only equal to single acting ones. For the above reasons they have, we believe, become obsolete or nearly so.

CHAPTER V.

ROTARY or rotatory pumps: Uniformity in efforts made to improve machines—Prevailing custom convert rectilinear and reciprocating movements into circular ones—Epigram of Antipater—Ancie opinion respecting circular motions—Advantages of rotary motions exemplified in various machines Operations of spinning and weaving; historical anecdotes respecting them—Rotary pump from Sa viere—Interesting inventions of his—Classification of rotary pumps—Eve's steam-engine and pump Another class of rotary pumps—Rotary pump of the 16th century—Pump with sliding butment—Trot ter's engine and pump—French rotary pump—Bramah and Dickenson's pump—Rotary pumps with pis tons in the form of vanes—Centrifugal pump—Defects of rotary pumps—Reciprocating rotary pumps A French one—An English one—Defects of these pumps.

No one can study the past and present history of numerous machines and devices without perceiving a striking uniformity in the efforts made to improve them in distant times and countries; the same general defects and sources of defects seem always to have been detected, and similar methods hit upon to remedy or remove them: the same ideas, moreover, led inventors to modify and apply machines to other purposes than those for which they were originally designed, and also to increase their effect by changing the nature and direction of their motions. So uniform have been the speculations of ingenious men in these respects, that one might be almost led to suppose they had reasoned, like the lower animals from a common instinct; and that the adage of Solomon, "there is no new thing under the sun," was as applicable to the inventions of man, as the works of nature. It would indeed be no very hard task to show that the preacher was correct, to an extent not generally believed, when he penned the following interrogatory and reply—"Is there any thing whereof it may be said,—See, this is new?—*it hath been of old time* which was before us." Did a modern savan invent some peculiar surgical instruments of great merit?—similar ones were subsequently discovered in the ruins of Pompeii. Have patents been issued in late years for economizing fuel in the heating of water, by making the liquid circulate through hollow grate bars?—the same device has been found applied to ancient Roman boilers.— And the recent practice of urging fires with currents of steam (also patented) was quite common in the middle ages. (See remarks on the Eolipile in the next book.) Numbers of such examples might be adduced from almost every department of the useful arts.

From the earliest times it has been an object to convert, whenever practicable, the rectilinear and reciprocating movements of machines into circular and continuous ones. Old machinists seem to have been led to this result by that tact or natural sagacity that is more or less common to all times and people: thus the dragging of heavy loads on the ground led to the adoption of wheels and rollers—hence our carts and carriages:—the rotary movements of the drill and the wimble superseded the alternating one of the punch and gouge, in making perforations:—the horizontal wheel of the potter rendered modeling of clay vessels by hand no longer necessary:—the whetstone gave way to the revolving grindstone:—the turning lathe produced round forms infinitely more accurate, and expeditiously than the uncertain and irregular carving or cutting away with the knife. The *quern*, or original hand mill, was more efficacious than the alternate action of the primitive pestle and mortar for bruising grain; and the various forces by which corn mills have subsequently been worked, have always been applied through revolving mechanism. The short handles, on the moveable stone, by which females and slaves moved it round, became in time lengthened into levers, and being attached to the peripheries of larger stones, slaves were sometimes yoked to them, who ground the grain by walking round a circular path. Subsequently slaves were replaced by animals, and these, in certain locations, by inanimate agents—wind and water. The period is unknown when man first derived rotary motion from the straight currents of fluids, for there is no sufficient reason to believe that the water mill located near the residence of Mithridates was the first one ever used in grinding corn: *that* may have been the one first known to the Romans; but it is very probable that such machines as well as wind mills were in use in Egypt, Syria, China, and other parts of Asia, in times that extend far beyond the confines of authentic history. An epigram of Antipater, a contemporary of Cicero, implies that water mills were not then very common in Europe. "Cease your work, ye maids, ye who laboured in the mill: sleep now, and let the birds sing to the ruddy morning, for Ceres has commanded the water nymphs to perform your task; these, obedient to her call, throw themselves on the wheel, force round the axle-tree, and by these means the heavy mill."

Rotary motions were favorite ones with ancient philosophers: they considered a circle as the most perfect of all figures, and erroneously concluded that a body in motion would naturally revolve in one.

To the substitution of circular for straight motions, and of continuous for alternating ones may be attributed nearly all the conveniences and elegancies of civilized life. It is not too much to assert that the present advanced state of science and the arts is due to revolving mechanism; we may speak of the wonders that steam and other motive agents have wrought, but what could they have done without this means of employing them? The application of rotary, in place of other movements, is conspicuous in modern machinery; from that which propels the stately steam ship through the water, and those flying chariots named "locomotives" over the land, to that which is employed in the manufacture of pins and pointing of needles. It is by this that the irregular motion of the ancient flail and primeval sieve, have become uniform in thrashing, bolting and winnowing machines; —hence our circular saws, shears and slitting mills;—the abolition of the old mode of spreading out metal into sheets with the hammer, by the more expeditious one of passing it through rollers or flatting mills:— and hence revolving oars or paddle wheels for the propulsion of vessels— the process of inking type with rollers in place of hand balls—rotary and power printing presses—and revolving machines for planing iron and other metals instead of the ancient practice of chipping off superfluous portions with chisels, and the tedious operation of smoothing the surfaces with files.

But in few things is the effect of this change of motion more conspicuous than in the modern apparatus for preparing, spinning and weaving vegetable and other fibres, into fabrics for clothing. The simple application of rotary motions to these operations has in a great degree revolutionized the domestic economy of the world, and has increased the general comforts of our race a hundred fold. From the beginning of time females have spun thread with the distaff and spindle : Naamah the antediluvian, and Lachesis and Omphale the mythological spinsters, have been imitated in the use of these implements by the industrious of their sex in all ages and countries to quite modern days, and even at present they are employed by a great part of the human family. In India, China, Japan, and generally through all the East, as well as by the Indians of this hemisphere, this mode of making thread is continued : the filaments are drawn from the distaff and twisted by the finger and thumb, the thread being kept at a proper tension by a metallic or other spindle, suspended to it like the plummet of a builder's level, and the momentum of which, while turning round, keeps twisting the yarn or thread in the interim of repeating the operation with the fingers.[a] The thread as it is

---

[a] There are numerous allusions in history to this primitive mode of spinning, that are highly illustrative of ancient manners. At the battle of Salamis, Queen Artemisia commanded a ship in the Persian fleet, and Xerxes, as a compliment to her bravery, sent her a complete suit of armor, while to his general (who was defeated) he presented a distaff and spindle. When Pheretine applied to Euelthon of Cyprus for an army to recover her former dignity and country, he intimated the impropriety of her conduct by sending her a distaff of wool and a golden spindle. Herod. iv, 162. Hercules attempted to spin in the presence of Omphale, and she bantered him on his uncouth manner *of holding* the distaff. Among the Greeks and Romans, the rites of marriage directed the attention of women to spinning ; a distaff and fleece were the emblems and objects of the housewife's labors; and so they were among the Jews—" A virtuous woman [says Solomon] layeth her hands to the spindle and her hands hold the distaff." When Lucretia was surprised by the visit of Collatinus and his companions of the camp, although the night was far advanced, they found her with her maids engaged in spinning. A painting found in Pompeii represents Ulysses seated at his own gate, and concealed under the garb of a beggar ; Penelope, who is inquiring of the supposed mendicant for tidings of her husband, holds in one hand a spindle, as if just called from spinning. On the monuments of Beni Hassan both men and women are represented spinning and weaving. Several Egyptian spindles are preserved in the museums of Europe. See an account of a female in Sardis spinning while going for water, at page 22.

formed is wound round the spindle. In 1530 Jurgen of Brunswick devised a machine which dispensed with this intermitting action of the fingers, *i. e.* he invented the spinning *wheel*, which rendered the operation of twisting the filaments uniform. The wheel, however, like the primitive apparatus it was designed to supersede, produced only one thread at a time; but in the last century Hargreaves produced the " spinning jenny," by which a single person on *turning a wheel*, could spin eighty-four threads at once; then followed the "rollers" of Arkwright, the "mule" of Crompton, to which may be added the " gin" of Whitney, and also "carding engines," in place of the old hand cards, all composed of and put in motion by revolving machinery :—these have indefinitely extended the spinning of thread, and relieved females from a species of labor that, more than any other, occupied their attention from the beginning of the world; and lastly " power looms" impelled by water, wind, steam, or animals (through the agency of circular movements) are rapidly superseding the irregular and alternating motion of human hands in throwing the shuttle to and fro.

The conversion of intermitting into continuous circular movements is also obvious in ancient devices for raising water. The alternate action of the swape, the jantu and vibrating gutter thus became uninterrupted in the noria and tympanum—the irregular movement of the cord and bucket became uniform in the chain and pots; and so did the motion of the pitcher or pail as used by hand, when suspended to the rim of a Persian wheel. And when the construction of machines did not allow of a suitable change or form, they were often worked by cranks or other similar movements; (of this several examples are given in preceding pages;) but in no branch of the arts has this preference for circular movements over straight ones been so signally exhibited as in the numerous rotary pumps and steam engines that have been and still are brought forward; and in no department of machinery has less success attended the change. Most of these machines that have hitherto been made may be considered as failures; this result is consequent on the practical difficulties attending their construction, and the rendering of them durable. These difficulties (which will appear in the sequel) are to a certain extent unavoidable, so that the prospect of superseding cylindrical pumps and steam-engines is probably as remote as ever.

At an early stage in the progress of the machines last named, it became a desideratum with engineers to obtain a continuous rotary movement of the piston rod in place of the ordinary rectilinear and reciprocating one, that the huge " walking beam," crank and connecting shaft might be dispensed with, the massive fly-wheel either greatly reduced or abandoned, and the power saved that was consumed in overcoming their inertia and friction at every stroke of the piston. Reasoning analogous to this had long before led some old mechanicians to convert the motion of the common pump rod into a circular one; in other words, to invent rotary or rotatory pumps. By these the power expended in constantly bringing all the water in the cylinder and suction pipe, alternately to a state of rest and motion was saved, because the liquid is kept in constant motion in passing through them. The steam engine was not only originally designed as a substitute for pumps to raise water, but in all the variety of its forms and modifications it has retained the same analogy to pumps as these have to bellows. One of the oldest of modern rotary steam engines, that of Murdoch, was a copy of an old pump, a figure of which, No. 131, is taken from Sérviere's collection.

Two cog wheels, the teeth of which are fitted to work accurately into each other, are enclosed in an elliptical case. The sides of these wheels turn close to those of the case, so that water cannot enter between them. The axle of one of the wheels is continued through one side of the case, (which is removed in the figure to show the interior,) and the opening made tight by a stuffing box or collar of leather. A crank is applied to the end to turn it, and as one wheel revolves, it necessarily turns the other; the direction of their motions being indicated by the arrows. The water that enters the lower part of the case is swept up the ends by each cog in rotation, and as it cannot return between the wheels in consequence of the cogs being there always in contact, it must necessarily rise in the ascending or forcing pipe. The machine is therefore both a sucking and forcing one. Of rotary pumps this is not only one of the oldest, but one of the best. Fire engines made on the same plan were patented about twenty-five years ago in England, and more recently pumps of the same kind, in this country. We have seen one with two *elliptical* wheels, which were so geared that the longer axis of one wheel might coincide (in one position) with the short one of the other. Sometimes a groove is made along the face of each cog, and a strip of leather or other packing secured in it.

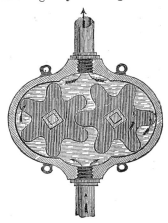

No. 131. Rotary Pump from Serviere.

The pump figured above is believed to have been known long before Sérviere's time, a model of it, as of other interesting machines, having been placed in his museum without regard to its origin. Ramelli is said to have described some similar ones in the middle of the 16th century, but we have not been able to procure a copy of his book. The suction pipe in the preceding figure has been added. In the original the water entered directly through the bottom of the case: the *model* was probably so made that part of the wheels might be visible, and the construction and operation of the machine more easily comprehended.

Sérviere was a French gentleman, born at Lyons in 1593. Independent in his circumstances, and inclined to mechanical researches, he was led to establish a cabinet of models of rare and curious machines—of these some were invented by himself and displayed uncommon ingenuity. It does not appear that any account of the whole was ever published; a part only being included in the small volume edited by his grandson, the title of which we have given at the foot of page 63. That work is divided into three parts: the first relates to figures formed in the lathe, as spheres, cubes, ellipses, &c.; some, being hollow and containing others within them, like Chinese balls, are extraordinary specimens of workmanship. There are also vases, urns, &c. not only round and elliptical, but angular, so that not only were the oval and eccentric chucks known to Sérviere, but the lathe for turning irregular surfaces appears to have been used by him. The second part contains an account of clocks all made by himself, the mechanism of which is exceedingly ingenious. Some are moved by springs, others by weights, water, sand, &c. They are fully equal to any thing of the kind at the present day: indeed that beautiful device by which a small brass ball is made to traverse backwards and forwards across an inclined plane, which still retains a place among mantel clocks, is one of Sérviere's, besides several modifications of it equally interesting. One half of the third part is occupied with descriptions of machines for raising water: these consist of gutters, swapes, chain of pots, gaining and losing buckets, norias, tympanums and other wheels; and lastly, pumps, among which is the rotary one figured above. Breval, in his "Remarks on Europe," part ii, page 89, mentions several machines in Sérviere's "famous cabinet of mechanicks" that are not noticed in the volume published by his grandson; while others are inserted that were not invented till after his death, as Du Fay's improvement on the tympanum.

Rotary pumps may be divided into classes according to the forms of and methods of working the pistons, or those parts that act as such : and according to the various modes by which the *butment* is obtained. It is this last that receives the force of the water when impelled forward by the piston ; it also prevents the liquid from being swept by the latter entirely round the cylinder or exterior case, and compels it to enter the discharging pipe. In these particulars consist all the essential differences in rotary pumps. In some the butments are moveable pieces that are made to draw back to allow the piston to pass, when they are again protruded till its return ; in others, they are fixed and the pistons themselves give way. It is the same with the latter ; they are sometimes permanently connected to the axles by which they are turned, and sometimes they are loose and drawn into recesses till the butments pass by. In another class the pistons are rectangular, or other shaped, pieces that turn on centres, something like the vanes of a horizontal wind mill, sweeping the water with their broad faces round the cylindrical case, till they approach that part which constitutes the butment, when they move edgeways and pass through a narrow space which they entirely fill, and thereby prevent any water passing with them. In other pumps the butment is obtained by the contact of the peripheries of two wheels or cylinders, that roll on or rub against each other. No. 131 is of this kind—while the teeth in contact with the ends of the case act as pistons in driving the water before them, the others are fitted to work so closely on each other as to prevent its return. The next figure exhibits another modification of the same principle.

In 1825 Mr. J. Eve, a citizen of the United States, obtained a patent in England for a rotary steam-engine and pump. No. 132 will serve to explain its application to raise water. Within a cylindrical case a solid or hollow drum A is made to revolve, the sides of which are fitted to move close to those of the case. Three projecting pieces or pistons, of the same width as the drum, are secured to or cast on its periphery ; they are at equal distances from each other, and their extremities sweep close round the inner edge of the case, as shown in the figure. The periphery of the drum revolves in contact with that of a smaller cylinder B from which a portion is cut off to form a groove or recess sufficiently deep to receive within it each piston as it moves past. The diameter of the small cylinder is just one third that of the drum. The axles of both are continued through one or both sides of the case, and the openings made tight with stuffing boxes. On one end of each axle is fixed a toothed wheel of the same diameter as its respective cylinder ; and these are so geared into one another, that when the crank attached to the drum axle is turned, (in the direction of the arrow,) the groove in the small cylinder receives successively each piston ; thus affording room for its passage, and at the same time by the contact of the edge of the piston with its curved part, preventing water from passing. As the machine is worked the water that enters the lower part of the pump through the suction pipe, is forced round

Chap. 5.] Rotary Pumps. 287

and compelled to rise in the discharging one, as indicated by the arrows. Other pumps of the same class have such a portion of the small cylinder cut off, that the concave surface of the remainder forms a continuation of the case in front of the recess while the pistons are passing; and then by a similar movement as that used in the figure described, the convex part is brought in contact with the periphery of the drum till the piston's return.

All rotary pumps are both sucking and forcing machines, and are generally furnished with valves in both pipes, as in the ordinary forcing pumps. The butments are always placed between the apertures of the sucking and forcing pipes.

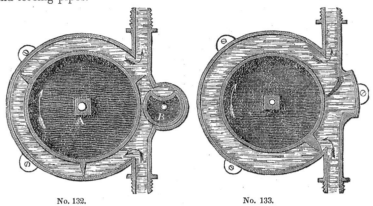

No. 132.   No. 133.

There is another class of pumps that bears some relationship to the preceding—the *eldest* branch, we believe, of the same family. One of these is figured in the 133rd illustration: the butment consists of a curved flap that turns on a hinge; it is so arranged as to be received into a recess formed on the rim or periphery of the case, and into which it is forced by the piston. The concave side of the flap is of the same curve as the rim of the case, and when pushed back forms a part of it. Its width is, of course, equal to that of the drum, against the rim of which its lower edge is pressed; this is effected in some pumps by springs, in others by cams cog wheels, &c., fixed on the axles, as in the last one. The force by which the flap is urged against the drum must exceed the pressure of the liquid column in the discharging pipe. The semicircular pieces on the outer edge of the case represent ears for securing the pump to planks or frames, &c., when in use. The arrows in the figures show the direction in which the piston and water is moved.

Such machines have often been patented, both as pumps and steamengines. In 1782 Mr. Watt thus secured a "rotative engine" of this kind, and in 1797 Mr. Cartwright inserted in the specification of his metallic piston a description of another similar to Watt's, except that the case had two flaps, and three pistons were formed on the drum. In 1818 Mr. Routledge patented another with a single flap and piston, (Rep. of Arts, vol. xxxiii, 2d series;) but the principle or prominent feature in all these had been applied long before by French mechanicians. Nearly a hundred years before the date of Watt's patent, *Amontons* communicated to the French Academy a description of a rotary pump substantially the same as represented in the last figure. It is figured and described in the first volume of *Machines Approuv.* p. 103: the body of the pump or case is a short cylinder, but the piston is elliptical, its transverse diameter being equal to that of the cylinder, hence it performed the part of two pistons. There are also two flaps on opposite sides of the cylinder. A pump not unlike this of Amontons, with an elliptical case, is described in vol. iv. of Nicholson's Phil. Journal 466. Several similar ones have since been proposed.

In other pumps the flaps, instead of acting as butments, are made to perform the part of pistons; this is done by hinging them on the rim of the drum, of which, when closed, they also form a part: they are closed by passing under a permanent projecting piece or butment that extends from the case to the drum.

In No. 134 the butment is movable A solid wheel, formed into three spiral wings that act as pistons, is turned round within a cylindrical case. The butment B is a piece of metal whose width is equal to the thickness of the wings, or the interior breadth of the cylinder: it is made to slide through a stuffing box on the top of the case, and by its *weight* to descend and rest upon the wings. Its upper part terminates in a rod, which, passing between two rollers, preserves it in a perpendicular position. As the wheel is turned, the point of each wing, (like the cogs of the wheels in No. 131,) pushes before it the water that enters the lower part of the cylinder, and drives it through the valve into the ascending pipe A: at the same time the butment is gradually raised by the curved surface of the wing, and as soon as the end of the latter passes under it, the load on the rod causes it instantly to descend upon the next one, which in its turn produces the same effect. This pump is as old as the 16th century, and probably was known much earlier. Besides the defects common to most of its species, it has one peculiar to itself:—as the butment must be loaded with weights sufficient to overcome the pressure of the liquid column over the valve, (otherwise it would itself be raised and the water would escape beneath it;) the power to work this pump is therefore more than double the amount which the water forced up requires. The instrument is interesting, however, as affording an illustration of the early use of the sliding valve and stuffing box; and as containing some of the elements of recent rotary pumps and steam-engines.

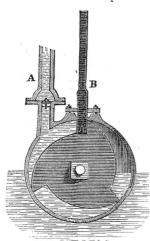

No. 134. Rotary Pump of the 16th century.

The pump represented by No. 135 consists also of an exterior case or short cylinder within which a small and solid one A is made to revolve. To the last an arm or piston is attached or cast in one piece with it, the sides and end of which are fitted to bear slightly against the sides and rim in the case. A butment B B slides backwards and forwards through a stuffing box, and is so arranged (by means of a cam or other contrivance connected to the axle of the small cylinder on the outside of the case) that it can be pushed into the interior as in the figure, and at the proper time be drawn back to afford a passage for the piston. Two openings near each other are made through the case on opposite sides of B B, and to these the suction and forcing pipes are united. Thus when the piston is moved in the direction of the arrow on the small cylinder, it pushes the water before it, and the vacuity formed behind is instantly filled with fresh portions driven up the suction pipe by the atmosphere; and when the piston in its course descends past B B it sweeps this water up the same way. Bramah and Dickenson adopted a modification of this machine in 1790, as a steam-engine and also as a pump. Rep. of Arts, vol. ii, 73.

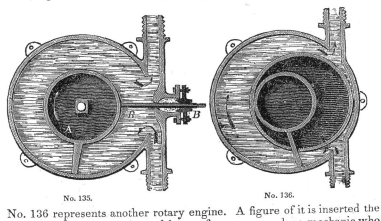

No. 135.   No. 136.

No. 136 represents another rotary engine. A figure of it is inserted the rather because it was reïnvented here a few years ago by a mechanic who was greatly distressed on finding that he had been anticipated. A notice of it may therefore prevent others from experiencing a similar disappointment. Like most others it consists of two concentric cylinders or drums, the annular space between them forming the pump chamber; but the inner one, instead of revolving as in the preceding figures, is immovable, being fixed to the sides of the outer one or case. The piston is a rectangular and loose piece of brass or other metal accurately fitted to occupy and move in the space between the two cylinders. To drive the piston, and at the same time to form a butment between the orifices of the induction and eduction pipes, a third cylinder is employed to which a revolving motion is imparted by a crank and axle in the usual way. This cylinder is eccentric to the others, and is of such a diameter and thickness that its interior and exterior surfaces touch the inner and outer cylinders as represented in the cut, the places of contact preventing water from passing: a slit or groove equal in width to the thickness of the piston is made through its periphery, into which slit the piston is placed. When turned in the direction of the large arrow, the water in the lower part of the pump is swept round and forced up the rising pipe, and the void behind the piston is again filled by water from the reservoir into which the lower pipe is inserted. This machine was originally designed, like most rotary pumps, for a steam engine. It was patented in England by Mr. John Trotter, of London, in 1805, and is described in the Repertory of Arts, vol. ix, 2d series. As a matter of course, he contemplated its application to raise water:—"The said engine [he observes] may be used to raise or give motion to fluids in any direction whatever."

In others the pistons slide within a revolving cylinder or drum that is concentric with the exterior one. No. 137 is a specimen of a French pump of this kind. The butment in the orm of a segment is secured to the inner circumference of the case, and the drum turns against it at the centre of the chord line: on both sides of the place of contact it is curved to the extremities of the arc, and the sucking and forcing pipes communicate with the pump through it, as represented in the figure. To the centre of one or both ends of the case is screwed fast a thick piece of brass whose outline resembles that of the letter D: the flattened side is placed towards the butment and is so formed that the same distance is preserved between it and the opposite parts of the butment, as between its convex surface and the rim of the case. The pistons, as in the last figure, are rectangular pieces of stout metal, and are dropped into slits made through the rim of the drum, their length being equal to that of the case, and their width to the distance between its rim and the D piece. They are moved by a crank attached to the drum axle. To lessen the friction and compensate for the wear of the butment, that part of the latter against which the drum turns is sometimes made hollow; a piece of brass is let into it and pressed against the periphery of the drum by a spring.

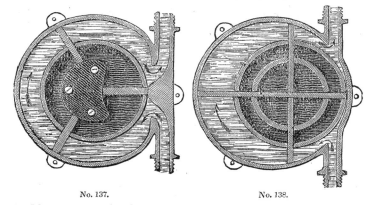

No. 137.　　　　　　　　No. 138.

In No. 138 the axis of the drum or smaller cylinder is so placed as to cause its periphery to rub against the inner circumference of the case. Two rectangular pistons, whose length are equal to the internal diameter of the case, cross each other at right angles, being notched so as to allow them to slide backwards and forwards to an extent equal to the widest space between the two cylinders. The case of this pump is not perfectly cylindrical, but of such a form that the four ends of the pistons are always in contact with it. An axle on the drum is moved by a crank. This pump, and another similar to it, were described in Bramah and Dickenson's patent for three rotative *steam-engines* in 1790. Rep. of Arts, vol. ii, 85. Fire engines have been made on the same principle.

Another class of rotary pumps have their pistons made somewhat like the vanes of wind mills. They were originally designed as steam-engines, and were, if we mistake not, first introduced by Hornblower, in the latter part of the last century. He employed four revolving vanes which were so arranged that, while one passed edgeways through a narrow cavity which it filled, the opposite one presented its face to the action of the steam. These machines have been variously modified as pumps, but generally speaking they are more complex and of course more liable to derangement than others: we have known two of them, fifteen inches diameter and apparently well made, (at a cost of 150 dollars,) which a friend used to force water to an elevation of twenty feet, become deranged, and thrown aside as useless in the course of three or four weeks.

A centrifugal forcing pump may be made by enclosing the arms of an atmospheric one, (such as represented at No. 95, page 229,) in a close drum or case, to which an ascending or forcing pipe is attached: the water would rise through the pipe, provided the velocity of the arms was increased according to the elevation of its discharging orifice. In place of tubular arms, two or more vanes radiating from a vertical axis and turned rapidly in the case would produce the same effect; the suction pipe being connected to the bottom at the centre and the forcing pipe to the rim or the top. Such pumps are in their construction simpler than other rotary ones, besides which no particular accuracy is required in fitting their working parts; nevertheless, they are as liable to derangement as others, for the velocity required to be given to the arms is so great, that the teeth of the wheels and pinions by which motion is transmitted to them are soon worn out.

Centrifugal pumps like those just described have been tried as substitutes for paddle wheels of steam-vessels: *i. e.* the wheels were converted into such pumps by inclosing them in cases made air-tight, except at the bottom through which the ends of the paddles slightly projected; a large suction pipe proceeded from one side of each case (near its centre) through the bows of the vessel and terminated below the water line: by the revolution of the wheels water was drawn through these tubes into the cases and forcibly ejected below in the direction of the stern, and by the reaction moved the vessel forward.

It must not be supposed that the preceding observations include an account of *all* rotary pumps. We have only particularized a few out of a great multitude, such as may serve as types of the various classes to which they belong. Were a detailed description given of the numerous forms of these machines, modes of operation, devices for opening and closing the valves, moving the pistons, diminishing friction, compensating for the wear of certain parts, for packing the pistons, &c. &c., those readers who are not familiar with their history would be surprised at the ingenuity displayed, and would be apt to conclude that all the sources of mechanical combinations had been exhausted on them. We would advise every mechanic who thinks he has discovered an improvement in rotary pumps, carefully to examine the Repertory of Arts, the Transactions of the Society of Arts, the London Mechanics' Magazine, and particularly the Journal of the Franklin Institute of Pennsylvania, before incurring the expenses of a patent, or those incident to the making of models and experiments.

Rotary pumps have never retained a permanent place among machines for raising water: they are, as yet, too complex and too easily deranged to be adapted for common use. Theoretically considered they are perfect machines, but the practical difficulties attending their construction have hitherto rendered them (like rotary steam engines) inferior to others. To make them efficient, their working parts require to be adjusted to each other with unusual accuracy and care, and even when this is accomplished, their efficiency is, by the unavoidable wear of those parts, speedily diminished or destroyed: their first cost is greater than that of common pumps, and the expense of keeping them in order exceeds that of others; they cannot, moreover, be repaired by ordinary workmen, since peculiar tools are required for the purpose—a farmer might almost as well attempt to repair a watch as one of these machines. Hitherto, a rotary pump has been like the Psalmist's emblem of human life:—" Its days are as grass, as a flower of the field it flourisheth, the wind [of experience] passeth over it, and it is *gone*." Were we inclined to prophecy, we should predict that in the next century, as in the present one, the cylindrical pump will retain its preëminence over all others; and that makers of the ordi-

292   *Reciprocating Rotary Pumps.*   [Book III

nary wooden ones will then, as now, defy all attempts to supersede the object of their manufacture.

RECIPROCATING ROTARY PUMPS :—One of the obstacles to be overcome in making a rotary pump, is the passage of the piston over the butment, or over the space it occupies. The apparatus for moving the butment as the piston approaches to or recedes from it, adds to the complexity of the machine; nor is this avoided when that part is fixed, for an equivalent movement is then required to be given to the piston itself in addition to its ordinary one. In reciprocating rotary pumps these difficulties are avoided by stopping the piston when it arrives at one side of the butment and then reversing its motion towards the other; hence these are less complex than the former: they are, however, liable to some of the same objections, being more expensive than common pumps, more difficult to repair, and upon the whole less durable. Their varieties may be included in two classes according to the construction of the pistons; those that are furnished with valves forming one, and such as have none the other. The range of the pistons in these pumps varies greatly; in some the arc described by them does not exceed 90°, while in others they make nearly a complete revolution. They are of old date, various modifications of them having been proposed in the 16th century. No. 139 consists of a close case of the form of a sector of a circle, having an opening at the bottom for the admission of water, and another to which a forcing pipe with its valve is attached. A movable radius or piston is turned on a centre by a lever as represented; thus, when the latter is pulled down towards the left, the former drives the contents of the case through the valve in the ascending pipe.

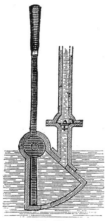
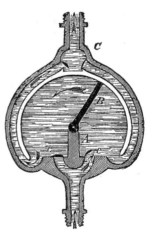

No. 139.   No. 140.

Belidor has described a similar pump in the first volume of his Arch. Hydraul. 379. The case is a larger portion of a circle than that of No. 139, and the piston is furnished with a valve. A pump on the same principle was adopted by Bramah as a fire-engine in 1793: His was a short cylinder, to the movable axle of which two pistons were attached that extended quite across, and had an opening covered by a clack in each.

No. 140 consists of a short horizontal cylinder: a portion of the lower part is separated from the rest by a plate where the suction pipe terminates in two openings that are covered by clacks *c c*. The partition *A* extends through the entire length of the cylinder and is made air and water tight to both ends, and also to the plate upon which its lower edge rests. The upper edge extends to the under side of the axle to which the piston B is united. One end of the axle is passed through the cylinder and the opening made tight by a stuffing box; it is moved by a crank or lever. Near the clacks *c c* two other openings are made through the plate, to which two forcing pipes are secured. These tubes are bent round the outside of the cylinder and meet in the chamber C where their orifices are covered by clacks. Thus when the piston is turned in either direction, it drives the water before it through one or other of these tubes; at the same time the void left behind it is kept filled by the pressure of the atmosphere on the surface of the liquid in which the lower orifice of the suction pipe is placed. The edges of the pistons are made to work close to the ends and rim of the cylinder by means of strips of leather screwed to them. Modifications of these pumps have also been used in England as fire-engines.

Reciprocating rotary pumps have sprung up at different times both here and in Europe, and have occasionally obtained " a local habitation and a name," but have never become perfectly domesticated, we believe, in any country. We have seen some designed for ordinary use that were elegantly finished, and decorated with *gilding* and japan—they resembled those exotic plants which require peculiar care, and are rather for ornament than for use.

Reciprocating rotary pumps have also been proposed as steam-engines. Watt patented one in 1782.

CHAPTER VI.

APPLICATION of pumps in modern water-works: First used by the Germans—Water-works at Augsburgh and Bremen—Singular android in the latter city—Old water-works at Toledo—At London bridge —Other London works moved by horses, water, wind and steam—Water engine at Exeter—Waterworks erected on Pont Neuf and Pont Notre Dame at Paris—Celebrated works at Marli—Error of Rannequin in making them unnecessarily complex. American water-works: A history of them desirable—Introduction of pumps into wells in New-York city—Extracts from the minutes of the Common Council previous to the war of Independence—Public water-works proposed and commenced in 1774—Treasury notes issued to meet the expense—Copy of one—Manhattan Company—Water-works at Fair Mount, Philadelphia.

BEFORE noticing another and a different class of machines, we propose to occupy this and the two next chapters with observations on the employment of pumps in "water-works," and as engines to extinguish fires —both in this country and in Europe.

The hydraulic machinery for supplying modern cities with water generally consists of a series of forcing pumps very similar to the machine of Ctesibius, (No. 120;) and when employed to raise water from rapid streams, or where from tides or dams a sufficient current can be obtained, are worked like it by under or by overshot wheels. An account of old European water-works is an important desideratum, for it would throw light on the history of pumps in the middle ages, during which little or nothing respecting them is known. The older cities of Germany were the first in modern days that adopted them to raise water for public purposes; but of their construction, materials, and application under various circumstances, we have no information in detail. Rivius, in his Commentary on the machine of Ctesibius, speaks of pumps worked by water wheels as then common, (A. D. 1548.) The hydraulic engines at *Augsburgh* were at one time greatly celebrated. They are mentioned, but not described, by Misson and other travelers of the 17th century. They raised the water 130 feet. Blainville, in 1705, speaks of them as among the curiosities of the city. He observes—" The towers which furnish water to this city are also curious. They are near the gate called the *Red Port*, upon a branch of the *Leck* which runs through the city. Mills which go day and night, by means of this torrent, work a great many pumps, which raise water in large leaden pipes to the highest story in these towers. In the middle of a chamber on each of them, which is very neatly and handsomely ceiling'd, is a reservoir of a hexagonal figure, into which the water is carried by a large pipe, the extremity of which is made like a dolphin, and through an urn or vase held by a statue sitting in the middle of the reservoir. One of these towers sends water to all the public fountains by smaller pipes, and the three others supply with water a thousand houses in the city; each of which pays about eight crowns yearly, and receives a hundred and twenty pretty large measures of water every hour." Travels, vol. i, 250. Misson's Travels, 5 ed. vol. i, 137.

Contemporary with the engines at Augsburgh was one at *Bremen* that is mentioned by several writers of the 17th century. It was erected on one of the bridges and moved by a water wheel: it raised water into a reservoir at a considerable elevation, whence the liquid was distributed to all parts of the city. An old author when speaking of it, mentions an android in Bremen, a species of mechanism for which the Germans were at one time famous. At the entrance of the arsenal, he observes, " stands the figure of a warrior arm'd cap-a-pe, who, by mechanism under the steps, as soon as you tread on them, lifts up the bever of his helmet with his truncheon to salute you."

There was also a celebrated water-engine at Toledo, the former capital of Spain. It raised the water of the Tagus to the top of the Alcazar, a magnificent palace erected on the summit of the declivity on which the city built; the elevation being " five hundred cubits from the surface of the river." What the particular construction of this machine was we have not been able to ascertain, nor whether it was originally erected by the Moors who built the palace. It is mentioned by Moreri as a "wonderful hydraulic engine which draws up the water from the river Tagus to so great a height, that it is thence conveyed in pipes to the whole city;" but in the middle of the last century (1751) the author of the Grand Gazetteer, or Topographic Dictionary, remarks (page 1289) that this " admirable engine" was then " entirely ruined."

The introduction of pump engines into the public water-works of England and France is sufficiently ascertained. This did not take place till long after they had been employed in Germany; and both London and Paris were indebted to engineers of that country for the first machines to raise water from the Thames and the Seine. Previous to their introduction, cities were commonly supplied from springs by means of pipes. As early as A. D. 1236, the corporation of London commenced to lay a six inch leaden pipe from some springs at Tyburn, a village at that time some miles distant from the city. This is supposed to have been the first attempt to convey water to that city through *pipes*, and fifty years elapsed before the whole was completed. These pipes were formed of sheet lead and the seams were soldered: part of them was accidentally discovered in 1745 while making some excavations, and another portion in 1765. (London Mag. for 1765, p. 377.) In 1439 the abbot of Westminster, in whom the right of the soil was vested, granted " to Robert Large the mayor and citizens of London, and their successors, one head [reservoir] of water, containing twenty-six perches in length and one in breadth, together with all its springs in the manor of Paddington : in consideration of which grant, the city is for ever to pay to the said abbot or his successors, at the feast of St. Peter, *two pepper corns*." This grant was confirmed by Henry VI, who at the same time authorized the mayor and citizens, by a writ of the privy seal, to purchase two hundred fothers of lead " for the intended works of pipes and conduits, and to *impress* plumbers and labourers." Maitland's Hist. of London, pp. 48, 107.

In the 33d year of Henry VIII, the mayor of the city of *Glocester*, with the dean of the church there, were authorized to " convey water in pipes of lead, gutters and trenches" from a neighbouring hill, " satisfying the owners of the ground there for the digging thereof."[a] In the following year, the mayor and burgesses of *Poole* were authorized to erect a wind mill on the king's waste ground, and a conduit head sixteen feet square, " and to dig and draw [water] in, by, through and upon all places meet and convenient, into and from the same, &c.—yielding yearly to the king and his heirs one pepper corn."[b] It would appear that the reservoir was in too low a situation for its contents to flow through pipes to the town, and hence the wind mill to raise it sufficiently for that purpose. The machine used was probably the chain of pots, which, as remarked page 125, was at that time often employed in such cases. In the 35th of Henry VIII, the corporation of London was authorized to draw water through pipes from various villages and other places within five miles of the city, and for this purpose to enter any grounds not enclosed with " stone, brick or mud walls, and there to dig pits, trenches and ditches ; to erect heads, lay pipes, and make vaults and *suspirals*," &c. Two years afterwards, (A. D. 1546,) a law was passed by which those who destroyed conduit heads and pipes, were put to death.[c] In 1547, William Lamb conveyed water in a leaden pipe from a conduit or spring, which still bears his name.[d]

In 1582, the first pump machines were used in London. In that year Peter Maurice, a German engineer, proposed to erect a machine on the Thames for the more effectual supply of the city, " which being approved of, he erected the same in the river near London bridge, which by suction and pressure, through pumps and valves, raised water to such a height as to supply the uppermost rooms of the loftiest buildings, in the highest part of the city therewith, to the great admiration of all. This curious machine, *the first of the kind that ever was seen in England*, was so highly approved of, that the lord mayor and common council, as an encouragement for the ingenious engineer to proceed in so useful an undertaking, granted him the use of one of the arches of London bridge to place his engine in, for the better working thereof."[e] Maurice's engine consisted of a series of forcing pumps (similar to Nos. 118 and 121) seven inches in diameter, and the pistons had a stroke of thirty inches; they were worked by an undershot wheel that was placed under one of the arches and turned by the current, during the rise and fall of the tide ; the water was raised to an elevation of 120 feet. The number of pumps and wheels was subsequently increased; but in 1822, when the old bridge was taken down, the whole were removed.[f]

Two years before Maurice undertook to raise water from the Thames, Stow says—" One Russel proposed to bring water from Isleworth, viz :

---

[a] Statutes at large. Lon. 1681. [b] Ibid. [c] Ibid. [d] Maitland, 158. [e] Ibid. 160.
[f] A description of the London Bridge Water-works, by Beighton, may be seen in the Philos. Trans. vol. vi, 358, and in Desaguliers' Philos. ii, 436.

the river Uxbridge to the said north of London; and that by a *geometrical instrument:* he propounded the invention to Lord Burleigh." In 1594, Bevis Bulmer, an English engineer, undertook to supply a small district of the city with Thames water, which he raised by four pumps that were worked by horses. They were continued in use till the time that Maitland commenced his history, viz: to 1725. The pumps of other London works were moved by horses, by wind mills, and others by the current of the common sewer.[a] About the year 1767, one of Newcomen's steam-engines was erected at the London bridge works to raise water at neap tides, and also as a security against fire during the turning of the tide, when the wheels were consequently at rest. A company was incorporated in 1691 to supply the neighbourhood of York Buildings with Thames water: Newcomen's engines were employed; and the pumps had solid plungers, one of which we have figured and described at page 272—Maitland enumerates them among other causes of the company's embarrassments: " the directors, by purchasing estates, erecting new water-works [new machines for raising water] and other pernicious projects, have almost ruined the corporation; however, their chargeable engines for raising water by fire, *being laid aside,* they continue to work that of horses, which, together with their estates in England and Scotland, may in time restore the company's affairs."[b] A figure of this *chargeable* engine is inserted in the second volume of La Motraye's Travels.

The author of the Grand Gazetteer, a folio of nearly 1500 pages, published in 1751, was a native of Exeter, on which account he excuses himself for describing that city at large; after mentioning some ancient conduits he observes :—" this city is otherwise well watered, and not only by most houses of note having wells and pumps of their own, but by the river water being forced by a *curious water-engine,* through pipes of bored trees laid under ground, even up to the very steep hill at Northgate Street; and then by pipes of lead into the houses of such inhabitants as pay a very moderate price for such benefit. The said water house and engine were *begun about Anno.* 1694." This extract shows that at the close of the 17th century, such works were not very common in English cities: of this there are numerous indications: thus at Norwich " the water-works at the new mills were undertaken in 1697, and completed in about two years."[c]

During the reign of Henry IV of France, John Lintlaer, a Fleming, erected an engine consisting of lifting pumps (such as No. 125) at the Pont Neuf which were worked by the current of the Seine. The water was raised above the bridge and conveyed in pipes to the Louvre and Tuilleries. This engine received the appellation of *The Samaritan,* from bronze figures of Christ and the woman of Samaria, which decorated the front of the building in which it was enclosed. The success that attended this experiment, led to the erection of similar engines at Pont Notre Dame, a figure of one of the pumps of which is inserted at page 277.

The most elaborate machine ever constructed for raising water was probably the famous one at *Marli,* near Paris, for supplying the public gardens at Versailles from the Seine. It was designed by Rannequin, a Dutch engineer, and set to work in 1682, at a cost of eight millions of livres—about a million and half of dollars.[d] We are not aware that any description of it in detail was ever published till Belidor inserted a short account in the second volume of his Architecture Hydraulique in 1739; and such was its magnitude and the multiplicity of its parts, that he was

---

[a] Maitland, pp. 622, 628.   [b] Ibid 634.   [c] Norfolk Tour, Norwich, 1795.

[d] Desaguliers says "*eighty* millions, about four millions of pounds sterling," but Belidor has only eight.

for a long time unwilling to undertake its elucidation, on account of the difficulty of describing it with sufficient precision. Its general features may be sketched in a few words, but a volume of letter-press and another of plates, would be required to explain and delineate the whole minutely.

The reservoir or head of the aqueduct, into which water from the Seine was raised by this machine, was constructed on the top of a hill, 614 toises, or three quarters of a mile, from the river, and at an elevation of 533 feet (English) above it. To obtain a sufficient motive power, the river was barred up by a dam, and its whole width divided, by piles, into fourteen distinct water courses, into each of which a large undershot wheel was erected. The wheels, by means of cranks attached to both ends of their axles, imparted motion to a number of vibrating levers, and through these to the piston rods of between 200 and 300 sucking and forcing pumps! The pumps were divided into three separate sets. The first contained 64, which were placed near the river, and were worked by six of the wheels: they drew the water, by short suction pipes, out of the river and forced it through iron pipes, up the hill; but instead of these pipes being continued directly to the reservoir, (which might have been done by making them and the machinery of sufficient strength,) Rannequin made them terminate in a large cistern, built for the purpose, at the distance from the river of 100 fathoms only, and at an elevation of about 160 feet. In this cistern he then placed 79 other pumps (the second set) to force the water thence to another cistern 224 fathoms further up the hill, and at an elevation of 185 feet above the other. In this last cistern 82 pumps more (the third set) were fixed, which forced the contents to the reservoir.

In thus dividing the work, Rannequin made a mistake for which no ingenuity could compensate: as the second and third sets of pumps containing no less than *one hundred and sixty-one*, with all the apparatus for working them, merely transferred through a part of the distance, the water which the first set drew directly from the river, they were in reality unnecessary, because the first set might have been made to force it through the whole distance; hence they not only uselessly consumed (at least) four fifths of the power employed, but they rendered the whole mass of machinery cumbersome and complicated in the highest degree; and consequently extremely inefficient, and subject to continual repairs. The first set of pumps, as already observed, were worked by the wheels near which they were placed, and the remaining wheels imparted motion to the piston rods of the second and third sets, in the two cisterns on the hill: of these, therefore, eighty-two pumps were stationed at an elevation of upwards of three hundred feet *above* the power that worked them; and nearly *half a mile from it!* and seventy-nine were one hundred fathoms from the wheel, and 160 feet above them! To work these pumps, a number of *chains*, or *jointed iron rods*, were extended on frames above the ground, all the way from the cranks on the water wheels in the river to both cisterns, where they were connected to the vibrating beams to which the piston rods were attached. It was the transmission of power to such elevations and extraordinary distances by these chains, that acquired for the machine the title of "a monument of ignorance."

A writer in the Penny Magazine (vol. iv, page 240) who examined the machine in 1815, says the sound of these rods working was like that of a number of wagons loaded with bars of iron running down a hill with axles never greased. The creaking and clanking (he observes) must have convinced the most ignorant person that the expenditure of power was enormously beyond what was required for the purpose effected. It has been estimated that 95 per cent of the power was expended in communicating motion to the apparatus!

The evil of working the pumps with shafts and chains at such great distances from the power, was seen a few years after the machine was completed. In 1738 an attempt was made by M. Camus to raise the water to the reservoir by a single lift. The attempt succeeded but partially, and the machine was much strained by the extraordinary effort, chiefly because only a small portion of the power was used; viz: those wheels that raised the water into the first cistern; the others which moved the shafts and chains abovementioned, not being applicable for the purpose. But even this comparatively small power forced the water to the reservoir, and thus demonstrated the practicability of completing the work at one throw, if the whole apparatus had been adapted accordingly. Nothing more was done for nearly forty years, and the machine proceeded as before till 1775, when another trial was made to raise the water only to the second cistern: this succeeded, and it was then hoped that the first cistern would be dispensed with; but many of the old pipes burst from the undue strain upon them, financial difficulties impeded their renewal, and the old plan was once more resorted to. The water wheels at last fell into decay and were replaced by a steam engine, of sixty-four horse power, by order of Napoleon; but the old shafts, chains, pipes and cisterns, &c. still remain.

We have mentioned only 225 pumps, but there were in all upwards of 250; some being feeders to others, and to keep water always over the pistons of those near the river. As each pump had two valves, an immense quantity of water must have escaped at every stroke on the opening and closing of 500 of these; to which may be added that which leaked past the leathers or packing of the pistons, and through the innumerable joints. The 64 pumps near the river were placed in a perpendicular position and had solid pistons. They resembled No. 118, except that the sucking as well as forcing pipes were united to the sides of the cylinders: those in the cisterns had hollow pistons, and the cylinders were inverted and immersed in the water: one of them is represented at No. 126.

AMERICAN WATER-WORKS.—A history of these is desirable and is certainly due to posterity. There are circumstances connected with their origin, plans, progress and execution, especially in the older cities of the Union, of Mexico and the Canadas, that ought to be preserved. An account of them would be useful to future engineers, and, as a record of historical and statistical facts, would include matter of general interest in coming times. The circumstances attending the first use of pumps and fire-engines, &c. may *now* be deemed too trifling to deserve particular notice, but they will increase in interest as time grows older. When the destiny that awaits the republic is accomplished—when the continent becomes studded with cities from one ocean to the other, and civilization, science and self government pervade the whole, *then* every incident relating to the early cultivation of the useful arts and improvements of machinery will be sought for with avidity and be dwelt upon with delight. Why should not the introduction of the most useful materials, manufactures and implements into this mighty continent form episodes in its history, as well as the fleece, the auger, saw, or bellows in that of classic Greece? And why should not the names of those persons be preserved from oblivion who *here* made the first pump and fire-engine, the first cog wheel and steam-engine—who built the first ship, forged the first anchor, erected the first saw, slitting, or grist mill—who made the first plough, grew the first wheat, raised the first silk, wove the first web, cast the first type, made the first paper, printed the first book, &c. &c.? It is such men as these and their successors, that found, strengthen and enrich a nation—who, without ostentation or parade, promote its real independence—men, whose labors should be mentioned in the national archives with honor, and whose statues and portraits should occupy the niches and panels of the capitol.

The precise time when pumps were first introduced into New-York is uncertain. This city, as is well known, was founded by the Dutch in 1614, who gave it the appellation of New-Amsterdam, and to the colony that of New-Netherlands; names that were continued till the British, in 1664, took possession of both and imposed the present ones. In examining the manuscript Dutch records in the office of the clerk of the Common Council: (a volume of which including the period that extends from May 29, 1647, to 1661, has been translated,) we have not met with any reference to pumps, either in wells or as fire-engines. In the first volume of "*Minutes of the Common Council*" (in manuscript) which embraces the transactions from October 1675 to October 1691, are several ordinances relating to *wells*, but no mention is made of pumps or other devices by which the water was raised. In the second volume under the date of August 31, 1694, a resolution directed that " the public wells within the city be repaired as formerly." From the following extract it appears that the water was raised by a cord and bucket, a windlass, or a swape: September 24, 1700, " Ordered that the neighbourhood that live adjacent to the king's farm and have benefit of the public well there built, do contribute to the charge thereof in proportion, or else be debarred from *drawing* water there."

In the third volume, containing minutes from February 1702 to March 1722, are notices respecting wells to be dug and others to be filled up, but nothing is said respecting pumps being placed in any. The same remark applies to the fourth volume, including a period of eighteen years, viz: from April 1722 to September 1740; and yet it would seem that pumps were at this last date used in some of the public wells, for in the fifth volume under the date of October 25, 1741, they are referred to in a "draft of a bill for mending and keeping in repair the public wells AND PUMPS in this city;" and again November 8, 1752, a bill was before the corporation "for keeping in repair the public wells AND PUMPS; and January 10, 1769, two hundred pounds [were] ordered to be raised " for mending and keeping in repair the public wells AND PUMPS." The precise period when pumps were first introduced is therefore uncertain; but from the language of the minute of October 1741, it would appear that they had then been some time in use in public wells; and from another minute in the same volume, in *private* wells also, for it was ordered that "the pump" of an individual should be deemed a public one and kept in repair at the public expense, on an application to that effect being made by the owner.

From the rapid growth of the city[a] the number of wells was increased, as now, every year, and in 1774 measures were taken to insure a more abundant supply from a large well in the Collect, the water to be raised by machinery and distributed through the city in wooden pipes. On the 22d April of that year, *Christopher Coles* proposed to the corporation " to erect a reservoir and to convey water through the several streets of this city." The proposition was subsequently approved of, and Mr. Coles directed " to enlarge the well and proceed." A committee was appointed

In 1696 the population was 4,302	1790 . . 33,131	1825 . . 167,059
1731 6,628	1800 . . 60,489	1830 . . 203,007
1756 10,381	1810 . . 96,373	1835 . . 270,089
1773 21,876	1820 . . 123,706	1840 . . 312,932
1786 23,614		

to assist him and to superintend the works, and several contracts were made for materials. To meet the expense £2500 in treasury notes were ordered to be issued, and subsequently further amounts were printed and issued. One of the small notes is now in the possession of John Lozier, Esq., superintendent of the Manhattan water-works, and is in these words:

NEW-YORK WATER-WORKS.
No. 3842.

This note shall entitle the bearer to the sum of TWO SHILLINGS, *current money of the colony of* NEW-YORK, *payable on* DEMAND, *by the* MAYOR, ALDERMEN *and* COMMONALTY *of the city of New-York, at the office of chamberlain of the said city, pursuant to a vote of the said Mayor, Aldermen and Commonalty of this date. Dated the second day of August, the year of our Lord one thousand seven hundred and seventy-five.*

By order of the Corporation,

ii s.   WM. WADDELL,
J. H. CRUGER.

It appears that the well (near White street) was enlarged, and a reservoir built, but no pipes were laid nor machinery to raise water erected before the war broke out and put a stop to the work. The project was not again revived till 1797, when the Manhattan Company was incorporated: the present wells were then made and the water raised by three or four common forcing pumps, worked by horses. These pumps raised the water by atmospheric pressure twenty-five feet, and forced it forty feet higher, into a reservoir in the Park where the post office is now (1840) located. In 1804 the pumps were replaced by two double acting ones (No. 122) fifteen inches in diameter and with a stroke of four feet. They were and still are worked by one of Watt's steam-engines. The water is raised to the same elevation as before. These works will probably be discontinued as soon as the Croton aqueduct, now being constructed, is finished.

The first water-works of Philadelphia were commenced in 1799, and consisted of forcing pumps, worked by steam-engines which raised water from the Schuylkill into a reservoir constructed, at an elevation of 50 feet, on the banks of that river; and from which it was conveyed to the city in pipes of bored logs. In 1811 the "city councils" appointed a committee to devise means for procuring a more perfect supply than these works afforded: and shortly after it was determined to erect two steam-engines and pumps on another location, viz: at *Fair Mount*, two miles and a half from the city, and near the upper bridge that crosses the Schuylkill. A reservoir 318 feet in length, 167 in width, and 10 in depth, was made at an elevation of 98 feet, into which the pumps forced water from the river.

The great expense attending the employment of steam-engines led to the adoption (in 1819) of water as the moving power. A dam was erected, and in 1822 three water wheels were put in operation; these, by cranks on their axles imparted motion through a connecting rod to the pistons of the pumps. In addition to the water consumed in turning these wheels, a surplus remained to work five additional ones, whenever the wants of the city might require them. An additional reservoir was also made, which contains four millions of gallons. The water in both is 102 feet above low tide, and 56 above the highest ground in the city. Iron pipes were also substituted for the old wooden ones. The whole was executed under the directions of F. Graff, Esq.

We took the opportunity while at Philadelphia in October of the present year (1840) to visit Fair Mount. Six breast wheels (15 feet long and 16 feet in diameter) were in operation; each, by a crank on one end of its axle, communicating motion to the piston rod of a single pump.[a] The pumps are double acting, the same as figured and described at page 271. They are placed a little below the axles of the wheels and in nearly a horizontal position. The cylinders are 16 inches diameter; and, that the water may not be pinched in its passage into and escape from them, the induction and eduction pipes are of the same bore; and all angles or abrupt changes in their direction and those of the mains are avoided. The stroke of two or three of the pumps was four feet, and their wheels made fourteen revolutions per minute: the others had a stroke of five feet ten inches, and the wheels performed eleven revolutions in a minute, consequently the contents of the cylinders of the latter were emptied into the reservoirs twenty-two times in the same period, and those of the former twenty-eight times. The cylinders are fed under a head of water from the forebays and they force it to an elevation of 96 feet, through a distance of 290. An air chamber is adapted to each.

It is impossible to examine these works without paying homage to the science and skill displayed in their design and execution; in these respects no hydraulic works in the Union can compete, nor do we believe they are excelled by any in the world. Not the smallest leak in any of the joints was discovered; and, with the exception of the water rushing on the wheels, the whole operation of forcing up daily millions of gallons into the reservoirs on the mount, and thus furnishing in abundance one of the first necessaries of life to an immense population—was performed with less noise than is ordinarily made in working a smith's bellows! The picturesque location, the neatness that reigns in the buildings, the walks around the reservoirs and the grounds at large, with the beauty of the surrounding scenery, render the name of this place singularly appropriate.

Dr. T. P. Jones, the talented editor of the Journal of the Franklin Institute, promised his readers " A history of the origin, progress and present state of the Water-works at Fair Mount," some years ago, but which has not yet been published. His familiarity with the subject in general, and with those works in particular, would make the history highly interesting to the present generation, and a source of valuable information to future ones. See Journal of the Franklin Institute, vol. iii, first series; which contains a plan and section of one of the wheels and one of the pumps.

---

[a] What a contrast with the old works at London bridge, where one wheel worked *sixteen* small pumps; the friction of the numerous pistons and the apparatus for moving them consuming a great portion of the power employed.

CHAPTER VII.

FIRE-ENGINES: Probably used in Babylon and Tyre—Employed by ancient warriors—Other devices of theirs—Fire-engines referred to by Apollodorus—These probably equal in effect to ours: Spiritalia of Heron: Fire-engine described in it—Pumps used to promote conflagrations—Greek fire, a liquid projected by pumps—Fires and wars commonly united—Generals, the greatest incendiaries—Saying of Crates respecting them—Fire pumps the forerunners of guns—Use of engines in Rome—Mentioned in a letter of Pliny to Trajan, and by Seneca, Hesychius and Isidore. Roman firemen—Frequency of fires noticed by Juvenal—Detestable practice of Crassus—Portable engines in Roman houses—Modern engines derived from the Spiritalia—Forgotten in the middle ages—Superstitions with regard to fires—Fires attributed to demons—Consecrated bells employed as substitutes for water and fire-engines—Extracts from the Paris Ritual, Wynken de Worde, Barnaby Googe and Peter Martyr respecting them—Emblematic device of an old duke of Milan—Firemen's apparatus from Agricola—Syringes used in London to quench fires in the 17th century—Still employed in Constantinople—Anecdote of the Capudan Pacha—Syringe engine from Besson—German engines of the 16th century—Pump engine from Decaus—Pump engines in London—Extracts from the minutes of the London Common Council respecting engines and squirts in 1667—Experiment of Maurice mentioned by Stow the historian—Extract from 'a history of the first inventers.'

OF the machines described in the 1st and 2d books some are employed in raising water for the irrigation of land, and for numerous purposes of rural and domestic economy; others in various operations of engineering and the arts, but with the exception of the centrifugal pumps, (Nos. 95, 6, and 7,) the liquid falls inertly from them all—*i. e.* it is not forcibly ejected as from a forcing pump or syringe: whether it be poured from a bucket, drawn from a gutter, escape from a noria, or from the orifice of a screw, or the spout of an atmospheric pump, it flows from each by the influence of gravity and consequently *descends* as it flows—such machines are therefore inapplicable for projecting water on fires, because for this purpose the liquid is required to *ascend* after leaving the apertures of discharge and with a velocity sufficient to carry it high into the air; and also when conveyed to a distance through flexible or other tubes, to be delivered from them at elevations far above the machine itself. As these effects are produced by the pumps described in the present division of the subject, most of them have at different times been adopted as fire-engines; some account of these important machines may therefore be inserted here.

Water is the grand agent that nature has provided for the extinguishment of flames, and contrivances for applying it with effect have, in every civilized country, been assiduously sought for. In the absence of more suitable implements, buckets and other portable vessels of capacity, at hand, have always been seized to convey and throw water on fires; and when used with celerity and presence of mind at the commencement of one have often been sufficient; but when a conflagration extends beyond their reach, the fate of the burning pile too often resembles that of the ships of Eneas:

> Nor buckets poured, nor strength of human hand
> Can the victorious element withstand.   *Eneid*, v.

The necessity of some device by which a *stream* of water might be forced from a distance on flames must have been early perceived, and if we were to judge from the frequency and extent of ancient conflagrations, the prodigious amount of property destroyed, and of human misery induced by them, we should conclude that ingenious men of former times were stimulated in an unusual degree to invent machines for the purpose. That this was the case cannot well be questioned, although no account of their labors has reached our times. It seems exceedingly probable that some kind of fire-engines were used in the celebrated cities of remote antiquity —in Nineveh, Tyre, Babylon and others. It is scarcely possible that the Tyrian and Babylonian mechanicians, whose inventive talents and skill were proverbial, should have left their splendid cities destitute of such means for preserving them from the ravages of fire. If the great extent of Babylon, for example, be considered, its location, (on an extensive plain,) the length of its streets, (fifteen miles,) the height of its buildings, (three and four stories,) and its unrivaled wealth, together with the heat and dryness of the climate; the necessity of such machines will be apparent, and what appears necessary to us, we may rest assured, appeared equally so to its mechanicians, and that they were quite as capable of providing by their ingenuity for the emergency. Nor are we left wholly to conjecture respecting their knowledge of hydraulic or pneumatic machinery, since the most memorable machine for raising water in the ancient world was made and used at Babylon, and one which, as has been elsewhere observed, greatly exceeded in the elevation to which it raised it, all, or nearly all the water-works of modern days. Had they engines like ours then? We dare not say they had, although we see nothing improbable in the opinion: the antiquity of the syringe is unquestionable; and its application to project water on flames must have been as obvious in remote as in present times; and people would as naturally be led then as now, to construct large ones for that purpose.

There are other reasons for believing that syringes or pumps for squirting water on fires were in use previous to the time they are first mentioned in history. *Fire* was one of the most common and most destructive agents employed in ancient wars. When a city was besieged or assaulted, it was the first object with the assailants to protect the moving towers, in which their battering engines, &c. approached the walls, from being consumed by fire, oil and pitch, &c. thrown upon them from the ramparts. Every source was examined that ingenuity could unfold, for materials and devices to protect them; and as not only the lives and property of the inhabitants, but often the destinies of armies and even of nations were on such occasions at stake, it is reasonable to conclude that the most perfect apparatus which could then be procured, were employed both for destroying buildings by fire, and also for preserving them from it. We know that men were specially trained to fire buildings, and that they were expert in their profession, especially in shooting lighted arrows and darts into and upon structures that could not be approached; hence the necessity of devices for throwing water upon these missiles and the places inflamed by them. There is an allusion to both practices in the Epistle to the Ephesians, vi, 16. Such a system of warfare could never have been carried to the extent that it was, and for so many ages too, among the celebrated nations of old, without forcing pumps or something like them being used to squirt water on such parts as could not be reached by it when thrown from the hand. We cannot conceive how the constant repetition of one army applying its energies to the destruction of another by means of fire, and the latter equally intent on devising and applying means to extinguish it, without the application of the syringe and of machines on the principle of the bellows occurring to them—an application so obvious (even then) that the slightest mental effort to produce a contrivance for the purpose could not have overlooked it, even if the occasions were of little moment, much less, when the inventive powers of armies, and of military engineers in particular, were engaged in the research, and the fate of nations depended upon the result. From a remark in one of Pliny's letters, to which we shall presently refer, it appears that among the Romans individuals were brought up to the *profession* of extinguishing fires.

The *Helepoles*, or 'town takers' of Demetrius, although proofs of his mechanical genius, would have availed him little at the siege of Rhodes, nor the movable towers of Hannibal at Saguntum, if these warriors had not been in possession of means to prevent them from being consumed by the fire of the besieged—of materials to resist its effects, and apparatus to extinguish it. That the resources of the ancients in these respects were not inferior to ours, may be inferred from several historical facts respecting their modes of securing these towers. They were generally covered with raw hides, leather soaked in water, or cloth made of hair, and sometimes, although seldom, they were plated with metal. Such were some of those employed by Titus at the siege of Jerusalem. They were seventy-five feet high and were covered all over with sheets of iron; perhaps nothing else could have resisted the incessant torrents of fire which the infuriated Jews showered upon them. But a singular proof of the sagacity and researches of the ancients is, that the modern application of *alum* to render wood incombustible was also known; for Archelaus, one of the generals of Mithridates in a war with the Romans, washed over a wooden tower with a solution of it and thereby defeated all the attempts of Sylla to set the structure on fire. Thus we see that when mechanical means failed them, or were not at hand, they had recourse to chemical ones. But that water and machines for dispersing it, were extensively employed on such occasions appears from a remark of Vitruvius. He observes that the lower stories of the towers contained large quantities of water for the purpose of extinguishing fire thrown upon them. Of course they had means of projecting it wherever required, but of these unfortunately he is silent. Montfaucon has engraved a figure of a species of wheel for the purpose, but its representation is too imperfect to indicate the nature of the machine of which it seems to have formed a part.

That machines of the *pump* kind were used on these occasions is evident from the temporary contrivance of Apollodorus, mentioned in the remains of a work of his *On War Machines*, and quoted by Professor Beckman. We have noticed, at page 235, one of his plans for extinguishing fire in the upper parts of a building, and that to which we now refer is from the same passage. Water, he observes, may be conveyed to *elevated places* when exposed to *fiery darts*, by means of the entrails of an ox: these natural tubes being connected to a bag filled with water; by compressing the bag the liquid will be forced through them to its place of destination. This device, he says, may be adopted when the *machine* called SIPHO *is not at hand*. Now if we had not known that the term sipho was anciently used to designate syringes and other tubular instruments, the substitute which Apollodorus here proposes sufficiently proves that it was a forcing pump to which he refers, and one too that, like our fire-engines, was furnished with leathern hose through which the water was conveyed to the "elevated places" he mentions. The importance of flexible pipes accompanying the pump or sipho, when employed in war, is obvious; for one of the objects of those who threw "fiery darts" on the towers and other structures, was to fire them, if possible, at places inaccessible to water for the most difficult to be reached—hence the necessity not only of engines, to project streams of that liquid, but also of such tubes to *direct* it to the places inflamed: and hence the suggestion of the tubes mentioned by Apollodorus when artificial ones were not to be procured: an ox was always within the reach of an army.

As these engines would of course be similar to such as were used to

Chap. 7.]   *Fire-Engine described by Heron.*   305

extinguish fires in cities in times of peace, it is to be regretted that neither Apollodorus nor Vitruvius has described them: perhaps they were too common to have been thought worthy of particular notice. In the design and execution of their essential parts, they were probably equal to our best engines. Some persons may doubt this, but it should be remembered that the nature of ancient wars naturally led to the best construction of all military machinery; and of defensive apparatus, engines to extinguish *fire* could not have been the least important, when that element was universally employed. The contests of the ancients were often those of mechanical skill rather than of fighting—conflicts of talent in engineering than in generalship; hence the ingenuity displayed in their machinery and the wonders wrought by it. Archimedes, by superior machines, protected Syracuse for eight months against all the efforts of the legions of Marcellus and the Roman engineers. The successes of Demetrius and Hannibal were often due to the novelty of their engines: the Carthagenian machinists were indeed proverbially skilful, so much so, that in Rome itself any curious piece of mechanism was, by way of eminence, named *punic*. Ancient armies were also often employed in obtaining, raising and cutting off water; the hydraulic engines of Ganymede nearly ruined Cæsar and his army in Alexandria, Cyrus took Babylon by diverting the course of the Euphrates, &c. The frequent use of hydraulic engines in war either to extinguish fires or for other purposes, would naturally lead to skill in making as well as in using them.

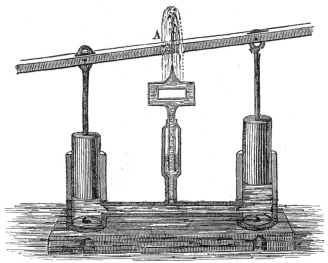

No. 141. Egyptian Fire-Engine of the 2d century before Christ, from Heron's Spiritalia.

That the idea of employing forcing pumps as fire-engines was not new in the time of Apollodorus or Vitruvius, we have conclusive evidence. Among the small number of ancient writings that escaped destruction in those dark and turbulent ages that intervened between the decline of the Roman power and the introduction of printing into Europe, was a Greek manuscript, containing an account of various devices for the application of water, and among them an engine for extinguishing fires. This small work was illustrated with figures, like the original work of Vitruvius. Several Latin translations were made and published in the 16th and 17th centuries, and most of them were ornamented with copies of the original illustrations. This was the *Spiritalia* of Heron, to which we have already referred, (page 270.) As the engine may interest some readers, a figure of it is annexed. See No. 141 on the preceding page.

To persons not familiar with hydraulic machinery this figure will appear a rude and imperfect affair; but notwithstanding its antiquity and the mutilations which it has unquestionably sustained in passing through the hands of copyists, it exhibits nearly all the essential elements of a modern engine. Like the machine of Ctesibius, Heron's engine consists of two brass forcing pumps connected to one discharging pipe. The cylinders are secured to a base of wood and are partly immersed in water; they are described in the text as turned or bored very smooth, with pistons accurately fitted to work in them. The piston rods are attached by bolts to a double lever at equal distances from the centre or fulcrum at A. The carriage not being necessary to elucidate the principle of the machine was omitted by Heron. The rectangular figure into which the upper part of the discharging pipe is formed, has certainly been added by some transcriber of the manuscript. Neither Heron nor his contemporaries could have made such an obstacle to the issuing fluid, and nothing of the kind is mentioned in the text. There is, moreover, conclusive evidence that the figure has been altered; for example, there is no provision represented by which the direction of the perpendicular jet can be changed, and hence an engine made according to it, would, on this account alone, be useless; now Heron not only describes a movable tube, fitted by a joint (goose neck) to the perpendicular one, by turning of which the water could be discharged on any given place, but he refers his readers to the figure of it in the illustration.

Had Heron's machine an *air chamber?* This is an interesting question, since if it were determined in the affirmative, there would be little left for the moderns to claim in fire-engines except details in the construction of the carriages and other matters of minor importance, that have been left unnoticed in the Spiritalia. The accounts of machines by ancient authors are generally very concise; they did not think it necessary to enter into that minutiæ of narration that characterizes the specifications of modern patents, nor would it have been of much use to us if they had, but the contrary, for the multiplicity of mere technical terms would rather have increased than removed our embarrassments. This is evident from the variety of explanations given of a few such terms that Vitruvius employs in describing some of the inventions of Ctesibius and other mechanicians: hence in all the accounts of ancient *machinery*, it was of more importance to preserve the figures or illustrations than the text from corruption.

The description of Heron's engine which the text and the figure afford, is, to persons conversant with such machines, sufficiently explicit, with the exception of that part of both which relates to the discharging pipe and apparatus connected to it—or in other words, to the air vessel, for that there was one, we think every intelligent reader will presently admit. Had the figure been always exactly copied by the multipliers of manuscripts, of course no obscurity would here have been felt, but even in the state in which it has reached us, an air vessel *is* certainly portrayed. It may be asked, If this be so, why was it not discovered before? Possibly because no one sought *particularly* for it: its diminutive size and general resemblance to a plain tube would prevent any one else from recognizing it. It will be seen in the figure that one part of the discharging pipe descends into an enlarged portion of that below it, and that a space is left between them; thus constituting an air chamber, and precisely of the same plan as those generally used in engines at this day. This part of the figure (and this alone) in Commandine's translation of the Spiritalia is not shown in section, but the arrangement of the pipes is precisely as shown in the cut. Now this addition to the discharging pipe could not have been made in the 16th century, when the work fell into the hands of printers and engravers, for at that time the use of it was not known, while from the small dimensions figured it could have been of no service. That it originated with Heron and formed a prominent feature in the original figure, is evident from the text: when speaking of the escape of the water from this part of the machine, he expressly states that it was forced out, in the same manner as out of a vase or fountain, which he had previously described, *by means of compressed air*—'*per aerem in ipso compressum.*'[a] Nothing can be plainer than this; for every manufacturer of pumps knows that in the absence of an air vessel there could have been no air to compress.

It is an interesting circumstance in the history of this ancient engine that the air vessel should have been preserved through so many ages while its use was not known. While its size was diminished its form was retained. It is no wonder that the old copyists considered it an unsightly and unnecessary enlargement of the discharging pipe, and hence they reduced it accordingly—certainly the fancy that could add the rectangular twist to the upper part, would not hesitate to remove the supposed deformity from the lower one. Some persons, deceived by the imperfect representation, have supposed that such engines were not used in the time of Heron, and that the figure and description were inserted in his work as mere hints for future mechanicians to improve on; but the description sufficiently indicates that similar machines were in actual use.[b] The materials and workmanship of the pumps—metallic pistons and spindle valves, with guards to prevent the latter from opening too far; the mode of forming the goose-neck by a kind of swivel joint, somewhat like the union or coupling screw; the application of an air vessel; two pumps forcing water through one pipe, and both worked by a double lever, are proofs that the machine described by Heron was neither an ideal one, nor of recent origin or use. There are features in it that were very slowly developed by manufacturers in modern times. It is not at all improbable that ancient engines were equal in effect to the best of ours; but, whether they were or not, one thing is certain, that to the ancients belongs the merit of discovering the principles employed in these machines and of applying them to practice. It is remarkable too, that fire engines made their first appearance in *Egypt*, thus adding another to the numerous obligations under which that wonderful country has placed civilized nations in all times to come.

Having noticed the use of pumps to *extinguish* fires in ancient warfare, we may remark that they were also employed in the middle ages, if not before, to *promote* conflagrations, viz: to lanch streams of *Greek fire*. This mysterious substance is represented as a liquid: Beckman says it certainly was one; and so far from being quenched, its violence was augmented by contact with water. It was principally employed in naval combats, being enclosed in jars that were thrown into the hostile vessels. It was also blown through iron and copper tubes planted on the prows of galleys and fancifully shaped like the mouths of animals, which seemed to vomit streams of liquid fire. There is among the figures of war machines in the old German translation of Vegetius already mentioned, one that

---

[a] Spiritalia, p. 70.
[b] Siphones autem quibus utuntur ad incendia hoc modo construuntur.—Ibid.

(judging from the flames issuing from monstrous animals' mouths) seems to have been designed for projecting Greek fire, though it is difficult to perceive how it was done. Another mode of using this terrible material, was by forcing it in jets " by means of large fire-engines," and sometimes " the soldiers squirted it from hand engines." Its effects upon those on whom it was thrown, seem to have been somewhat similar to those produced by the composition of alcohol and spirits of turpentine recently adopted as a substitute for oil in lamps, and which has occasioned so many fatal disasters, by the explosion of vessels containing it and its consequent dispersion over the persons of the sufferers. It was easy (says Beckman) to conceive the idea of discharging Greek fire by means of forcing pumps, because the application of them to extinguish fires was known *long before* its invention. It is supposed to have originated with Callinicus, a Syrian engineer of Balbec, in the 7th century. It may however have been known to the old Greeks and Romans, for they made use of similar devices for projecting fire: Montfaucon, in describing their marine combats, observes " another mode of annoying enemies' ships was by throwing fire therein, which they did after different ways, some using for that purpose *siphones*, and fire buckets, others threw in pots filled with fire." From an expression of Dr. G. A. Agricola, a physician of Ratisbon of the last century, in a work on Gardening, (see page 127 in Bradley's translation) it would appear as if something like the Greek fire was then in use. Enumerating several pernicious inventions, he notices " That infernal one of gunpowder. How many cities and fortresses has it ruined? How many thousands of men has it destroyed? And what is most deplorable is, that this art grows more and more complete every day, and is brought to that perfection, that in *Holland* and some other parts they have FIRE PUMPS filled with burning compositions, wherewith they eject fiery torrents to a great distance, which may occasion dreadful and irreparable damages to mankind."

Fires and wars have ever been deemed the most awful of earthly calamities, and, unfortunately for our race, they have too often been united, for warriors have generally had recourse to the former to multiply the miseries of the latter; and in almost every age cities have, like Jericho and Ai, Hebron and Ziglag, Troy and Thebes, Carthage and Athens, Saguntum and Bagdat, been burnt with fire; and in some cases " all the souls therein destroyed"—" cities burned without inhabitants." It was, we believe, from the horrible, the inconceivable sufferings endured on such occasions, that much of the thrilling imagery of the Bible was derived. To the offending Jews, God was represented as "a consuming fire," and they were urged to repentance " lest his fury come forth like fire, and burn, that none can quench it—lest he break out like fire in the house of Joseph and there be none to quench it in Bethel;" and some of the sublimest effusions of the prophets have reference to " firebrands, arrows and death" —to " blood and fire and pillars of smoke." In modern times, too, warriors have been the greatest incendiaries: hamlets, towns and cities have been wantonly consumed, and the " gallant" actors have made the air shiver with their shouts of acclamation on witnessing the spreading conflagration. Well did the ancients represent Mars fierce in aspect, brandishing a spear, and driving in his chariot o'er mangled corses, amid the clangor of arms and the shrieks of the dying—Fear, Terror and Discord in his train, while before went Bellona, with her hair loose and clotted with gore, and a firebrand in her hand. And these are the demons that men professing christianity worship with all the fervor of deluded heathen; and, what will in future times appear incredible, they demand reverence for the act, and they—receive it! Strange, that notwithstanding the boasted superiority of the age and the benign spirit and precepts of religion—the *profession* of war—the most prolific source of human misery and crime, is still deemed honorable; and men under whose tyranny nations and provinces groan, and by whom human life is extinguished not only without remorse but with indifference, are permitted to take precedence in *moral* society. Crates was certainly correct when he intimated that wars would never cease till men became convinced of the folly and wickedness of allowing themselves to be driven as soldiers like sheep to the slaughter, or like wolves to devour each other—but as he expressed it, not till men become sensible that generals are only *ass drivers*.

As Greek fire preceded gunpowder in Europe, so pumps or the 'spouting engines' for projecting it may be considered the forerunners of guns: it is even possible that the first idea of the latter (supposing they were not introduced from the east) might have been derived from accidental explosions of the liquid in the pump cylinders, when the pistons would of course be driven out of them like balls out of cannon. But be this as it may, enough has been adduced to show that the forcing pump and its modifications have exerted no small degree of influence in ancient wars and consequently in the affairs of the old world.

Although the police and other arrangements for the actual suppression of fires in ancient *Rome* are not well ascertained, some interesting particulars are known. A body of *firemen*, named matricularii, was established whose duty it was to extinguish the flames. Similar companies were also organized in provincial cities. This appears from Trajan's reply to Pliny respecting the formation of one in Nicomedia, and from which we learn that these ancient firemen frequently created disturbances by their dissentions and tumults. Pliny (the younger) was governor of Bithynia; after giving the emperor an account of a fire in Nicomedia, a town in his province, he continues, "You will consider, sir, whether it may not be advisable to form a company of firemen, consisting only of one hundred and fifty members. I will take care none but those of *that business* shall be admitted into it; and that the privileges granted them shall not be extended to any other purpose. As this corporate body will be restricted to so small a number of members, it will be easy to keep them under proper regulations." In answer the emperor sent the following letter: "TRAJAN TO PLINY.—You are of opinion it would be proper to establish a company of firemen in Nicomedia, agreeably to what has been practiced in several other cities. But it is to be remembered that societies of this sort have greatly disturbed the peace of the province in general, and of those cities in particular. Whatever name we give them, and for whatever purpose they may be instituted, they will not fail to form themselves into factious assemblies, however short their meetings may be. It will therefore be safer to provide such machines as are of service in extinguishing fires, enjoining the owners of houses to assist in preventing the mischief from spreading, and, if it should be necessary, to call in the aid of the populace." Pliny's Letters, B. x. Ep. 42 and 43. Melmoth's Translation.

The direction to procure "machines as are of service in extinguishing fires" was in consequence of Nicomedia being destitute of them—an unfortunate circumstance for the inhabitants, but one that is hardly now regretted by those who are in search of information respecting fire-engines among the ancients; since it led Pliny to mention them, and thereby afford us a proof of their employment by the Romans. "While I was making a progress [he writes to Trajan] in a different part of the province, a most destructive fire broke out in Nicomedia, which not only consumed several private houses, but also two public buildings, the town house and the temple of Isis, though they stood on contrary sides of the street. The occasion of its spreading thus wide was partly owing to the violence of the wind, and partly to the indolence of the people, who, it appears, stood fixed and idle spectators of this terrible calamity. The truth is, the city was not furnished with either *engines*, buckets, or any single instrument proper to extinguish fires; which I have now however given directions to be provided." It has been generally imagined [observes Melmoth] that the ancients had not the art of raising water by engines, but this passage seems to favor the contrary opinion. The word in the original [for *engine*] is *sipho*, which Hesychius explains *instrumentum ad jaculandus aquas adversus incendia*—an instrument to throw up water against fires. But there is a passage in Seneca which seems to put the matter beyond conjecture, though none of the critics upon this place have taken notice of it. Solemus (says he) duabus manibus inter se junctus aquam concipere et compressa utrinque palma in modum siphonis exprimere. Q. N. ii, 16, where we plainly see the use of this sipho was to throw up water. In the French translation of De Sacy, (Paris 1809,) the word is rendered *pumps* :—" D'ailleurs, il n'y a dans la ville, ni *pompes* ni seaux publics, enfin nul autre des instrumens nécessaires pour éteindre les embrasemens." And Professor Beckman quotes both Hesychius and Isidore to prove that " a fire-engine, properly so called, was understood in the 4th and in the 7th centuries by the term *sipho*," and we may add that Agricola in the 16th century designated syringes for extinguishing fires by the same term. Heron's engine is also named a siphon. See note p. 307.

From an expression in the letter of Pliny just quoted, we learn that men were regularly brought up to the art of extinguishing fires, the same as to any other profession: Of the company that he proposed to establish, he remarks, " I will take care that none but those of *that business* shall be admitted into it." The buildings in ancient Rome were very high, the upper stories were mostly of wood, and the streets and lanes were extremely narrow, hence the suppression of conflagrations there must have been an arduous business, and one that required extraordinary intrepidity and skill; qualifications that could only be obtained by experience. Besides engines for throwing water, the firemen used sponges or mops fixed to the end of long poles, and they had grapples and other instruments by means of which they could go from one wall to another, (Encyc. Antiq.) Of the great elevation of the houses several Roman writers speak. Seneca attributed the difficulty of extinguishing fires to this cause. Juvenal mentions

> Roofs that make one giddy to look down. *Sat.* vi.

When the city was rebuilt after the great conflagration, (supposed to have been induced by Nero,) the height of the houses was fixed at about seventy feet. These were raised to a certain height without wood, being arched with stone, and party walls were not allowed. That fires were constantly occurring in old Rome is well known. Juvenal repeatedly mentions the fact: Thus in his third satire:—

> Rome, where one hears the *everlasting* sound
> Of beams and rafters thundering to the ground,
> Amid alarms by day and fears by night.

And again

> But lo! the flames bring yonder mansion down!
> The dire disaster echoes through the town;
> Men look as if for solemn funeral clad,
> Now, now indeed these *nightly* fires are sad.

Their frequency induced Augustus to institute a body of watchmen to guard against them, and, from the following lines of Juvenal, it appears that wealthy patricians had servants to watch their houses during the night:

> With buckets ranged the ready servants stand,
> Alert at midnight by their lords' command. *Sat.* xiv.

As every calamity that befalls mankind is converted by some men to their own advantage, so the numerous fires in Rome led to the detestable practice of speculating on the distresses they occasioned. Thus Crassus, the consul, who, from his opulence was surnamed the *Rich*, gleaned his immense wealth, according to Plutarch "from war and *from fires;* he made it a part of his business to buy houses that were on fire, and others that joined upon them, which he commonly got at a low price on account of the fear and distress of the owners about the event." But the avarice of Crassus, as is the case with thousands of other men, led to his ruin. With the hope of enlarging his possessions, he selected the province of Syria for his government, or rather for his extortion, because it seemed to promise him an inexhaustible source of wealth: but by a retributive Providence his army was overthrown by the Parthians, whom he attempted to subdue, and who cut off his head, and in reference to his passion for gold fused a quantity of that metal and poured it down his throat.

Among other precautions for preventing fires from spreading that were adopted in Rome on rebuilding the city, was one requiring every citizen to keep in his house "a machine for extinguishing fire." What these machines were is not quite certain, whether buckets, mops, hooks, syringes or portable pumps. That they were the last is supposed to be proved by a passage in the writings of Ulpian, a celebrated lawyer and secretary to the Emperor Alexander Severus, wherein he enumerates the things that belonged to a house when it was sold, such as we name fixtures, and among them he mentions SIPHONES *employed in extinguishing fires.* Beckman thinks the leaden pipes which conveyed water into the houses for domestic purposes might be intended; but they would hardly have been designated as above, merely because the water conveyed through them was occasionally used to put out fires. This was not their chief use, but an incidental one. That they were pumps or real fire-engines was the opinion of Alexander ab Alexandro, a learned lawyer of the 15th century; an opinion not only rendered probable by the terms used and the necessity of such implements for the security of the upper stories, which neither public engines nor streams from the aqueducts could reach, but also from the apparent fact, that syringes or portable pumps have always been kept (to a greater or less extent) in dwellings from Roman times. And a sufficient reason why they should generally be sold with the houses, might be found in their dimensions being regulated according to those of the buildings for which they were designed.

The population of Rome was so great that the area of the city could not furnish sites sufficient for the houses; and hence (as Vitruvius has observed, B. ii, cap. 8) the height of the walls was increased in order to multiply the number of stories—'for want of room on the earth the buildings were extended towards the heavens.' Portable fire-engines were therefore particularly requisite, in order promptly to extinguish fires on their first appearance, whether in the upper or lower floors. In the latter case, when this was not done, the people in the higher stories would be cut off from relief and the means of escape. Were some of our six and seven story buildings in the narrow streets, densely filled with human beings, and a raging fire suddenly to burst out on the ground floors, the probability is that many lives would be lost, notwithstanding the great number of our public engines, and hose and ladder companies. Juvenal intimates the distressed situation of those dwelling above under such circumstances.

> Hark! where Ucalegon for water cries
> Casts out his chattels, from the peril flies,
> Dense smoke is bursting from the *floor below*.   Sat. iii.

However perfect or imperfect hydraulic and hydro-pneumatic engines in ancient Alexandria and Rome may have been, it is certain that these machines and the arts related to them experienced the withering influence of that moral and mental desolation which raged throughout Europe during the *dark* ages. The decline of learning was necessarily accompanied with a corresponding decay in all the useful and ornamental arts: some of these have disappeared altogether, and have never been recovered, so that the attainments of the ancients in them have perished. But the connection between literature and the arts was as apparent in their restoration as in their declension—if they departed together they also returned in company. The revival of learning not only led to the introduction of printing and the invention of the press, but it furnished, in the multiplication of ancient manuscripts, then extant, immediate employment for both; and although it may be supposed that there can be little or no relation between Greek or Latin manuscripts and modern fire-engines, yet there really is an intimate one, for it is all but certain that the first idea of these machines as now made, was derived from Heron's Spiritalia; just as the application of double and treble forcing pumps in modern water-works, was from Vitruvius' treatise on architecture. The printing press, therefore, not only opened the literary treasures of the ancients to the world at large, which had previously been confined to a few, but at the same time it made us acquainted with some of their machinery and their arts, that had long been forgotten or lost sight of.

Fire-engines were nearly or altogether forgotten in the middle ages: portable syringes seem to have been the only contrivances, except buckets, for throwing water on fires, and from their inefficiency and other causes, their employment was very limited. The general ignorance which then pervaded Europe not only prevented the establishment of manufactories of better instruments; but the superstitions of the times actually discouraged their use. There is not a more singular fact (and it is an incontrovertible one) in the history of the human mind, than that the religious doctrines and opinions of a large portion of mankind should have in every age produced the most deplorable results with regard to conflagrations. The Parsees, Ghebres, &c. of Asia, and other religious sects, which have subsisted from the remotest ages, never willingly throw water upon fires—they consider it criminal to quench it, no matter how disastrous the result may be: they had rather perish in it than thus extinguish the emblem of the Deity they worship. " They would sooner be persuaded to pour on oyl to increase, than water to assuage the flame."[a] Among such people fire-engines of course were never used. Another and a larger part of the human race though they entertain no such reverence for fire, are so far influenced by the pernicious doctrine of *Fatalism*, as to make little or no efforts to suppress it. They look upon fires as the act of God! determined by him! and therefore conclude it useless to contend with him, in attempting to extinguish those which He has kindled! Hence the proverbial indifference of Mahommedans in the midst of conflagrations. What

---

[a] Ovington's Voyages to Surat in 1689, page 372.

Toreen has said of Surat in particular, is applicable to every city of Asia and of the East. " Many fine buildings have been destroyed by fire, which, according to the Mahommedan doctrine of predestination, it is in vain to withstand." Of the Chinese, by far the shrewdest of Asiatics, Mr. Davis remarks, " The foolish notion of fatalism which prevails among the people, makes them singularly careless as regards fire ; and the frequent occurrence of accidents, has no effect upon them, although the fearful conflagration of 1822, went far to destroy the whole city," (Canton.)

The miserable delusions which ecclesiastics established in Europe during the middle ages were quite as preposterous, and equally effective in paralizing the energies of the people. It is difficult to reflect on them without feeling emotions of wonder as well as pity, at the wretched condition of our race when void of knowledge ; and of gratitude, that in our times the shackles of ignorance and superstition are rapidly rusting away. It was a common belief that fires (and various other calamities) were induced by wicked spirits, and that the best mode of removing the evils was by driving the authors of them away! These intangible workers of mischief, according to the demonologists of the times, consisted of numerous classes, and the labors of each were confined to certain elements. It was those who roamed in the *air* that were the greatest incendiaries. " *Aeriall spirits, or divells, are such as keep quarter most part in the aire* [they] *cause many tempests, thunder and lightnings, teare oakes, fire steeples, houses,*" &c. (See Burton's Anatomy of Melancholy.) When a house, therefore, was on fire, the priests, instead of stimulating by their example the bystanders to exert themselves in obtaining water, &c. had recourse to the images and pretended relics of saints, which they brought out of the churches, in order to exert their influence in stopping the progress of the flames, and expelling the invisible authors of them. The pall, or sacred covering of the altar, was also frequently carried in procession, to contribute to the overthrow of the fiends. But when a church itself took fire, (such was the ignorance of the times,) the people then heartily blasphemed the saint to whom it was dedicated, for not preventing the mischief; (Encyc. Antiq.) like Sylla abusing the image of Apollo when he was defeated in battle.

Other curious but popular substitutes for water and fire-engines, were church *Bells*: these were consecrated with imposing ceremonies. They were washed inside and out with holy water—perfumed with censers—anointed with sacred oil—named and signed with the cross, that devils (says the ritual) " hearing this bell may tremble and flee from the banner of the cross designed upon it." Besides striking demons with horror and driving them from the vicinity, these bells had the wonderful power of allaying storms, tempests, thunder and lightning, and extinguishing fires; and some of them had the rare gift of ringing on important occasions of their own accord.[a] M. Arago, in a paper on Thunder and Lightning, inquires (among other alledged means of dissipating thunder clouds) into this old superstition of " Ringing of Bells ;" and he cites specimens of prayers, *still* offered up, on their consecration, according to the Paris Ritual, " O eternal God ! grant that the sound of this Bell may put to flight the fire strokes of the enemy of man, the thunder bolt, the rapid fall of stones, as well as all disasters and tempests." In the "Golden Legend" of Wynken de Worde, the old English printer, it is said " the evil spirytes that ben in the region of th' ayre, doubte moche when they here the Belles ringen:

---

[a] See a particular account of the ceremonies of consecrating bells as witnessed by the author of " Observations on a Journey to Naples." Lon. 1691.

and this is the cause why the Belles ringen whan it thondreth, and whan grete tempeste and rages of wether happen, to the end that the feinds and wycked spirytes should ben abashed and flee, and cease of the movynge of tempeste." The following lines to the same effect, are from Barnaby Googe, an old British poet:

> If that the thunder chaunce to rore,
> And stormie tempestes shake,
> * * * * *
> The clarke doth all the belles forthwith
> At once in steeple ring :
> With wondrous sound and deeper farre
> Than he was wont before,
> Till in the loftie heavens darke,
> The thunder bray no more,
> For in these christned belles they thinke
> Doth lie such powre and might
> As able is the tempeste great,
> And storme to vanquish quight.

The application of bells to the purposes of fire-engines is also mentioned by Peter Martyr, in his " Common Places," a work dedicated to Queen Elizabeth. Black letter, 1583. Speaking of things consecrated by papists in common with the ancient heathen, he says of *bells*—" they be washed, they be annointed, they be conjured, they are named and handled with far greater pomp and ambition, than men are when they are baptized, and more is attributed to them than to the prayers of godly men. For they say, that by the ringing of them—the wicked spirits, the host of adversaries, the laying await of enemies, tempestes, hayle, stormes, whirlwindes, violent blastes and hurtfull thunderclaps, are driven away, FLAMES and FIRES *are extinguished*, and finally whatever else soever!" Part iv, cap. 9, p. 125.

There is no small ringing of bells in this city (New-York) during fires; but their unaided effects on the devouring element, ere other means have arrived, has, we believe, been but small. Few have, however, been consecrated ; but as from one to two hundred Spanish bells have recently been sold here, (having been taken from the convents in consequence of the civil war which has so long raged in that country,) *this* virtue of sacred bells may soon be tested. Certainly, if they can do a moiety of the good things mentioned above, they were worth much more than forty cents per lb. the average price at which they were sold.

We have had recourse in a few instances to heraldry, or rather to the emblems or personal devices of ancient families, for information respecting machines, some of which are no longer in use ; as the eolipile, and the atmospheric sprinkling pot : see pages 261 and 396. Besides these the syringe and the bellows have also been adopted on such occasions; and it may be here observed that the device of Galeaz, duke of Milan, the second of the name, was a brand burning and two fire buckets.[a] This, although no proof that machines of the pump kind were not in use to extinguish fires in Italy during the 15th century, is an indication that none were employed at the time when the device was adopted.

The oldest sketch of a complete set of apparatus for extinguishing fire that we have seen, is in a cut representing the interior of a laboratory or smelting furnace, in the De Re Metallica of Agricola, page 308. The implements are, a syringe, a sledge hammer, two fire hooks and three leathern buckets ; conveniently arranged against a wall. See the annexed illustration. These figures seem to have escaped the notice of Beckman

---

[a] Devices Heroïques. A Lyon. 1577, page 50.

and subsequent authors, nor is this surprising since they form a very small and obscure part of the original engraving. We noticed the latter several times before observing them. The syringe was made of brass; it is designated *siphunculus orichalceus, cujus usus est in incendiis*.

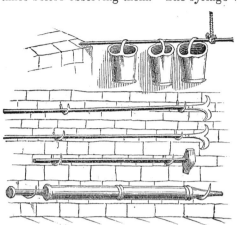

No. 42. Firemen's Apparatus from Agricola.

In these figures we behold all that was preserved through the middle ages of ancient firemen's machinery: the engine of Heron seems to have been quite forgotten. Indeed the syringe itself was not generally used in Europe till late, for it was not till the close of the 16th century that "hand squirts," as they were named, were introduced into London. Previous to that time watchmen, buckets, hooks and ladders, only were in use. Cutting away with axes and throwing water from buckets are mentioned (observes Fosbroke) by Petronius and Gervase of Canterbury. The owners of houses or chimneys that took fire were fined; and men were appointed to watch for fires and give the alarm. In 1472 a night bellman was employed in Exeter to alarm the inhabitants in case of fire, and in 1558, leathern buckets, ladders and crooks, were ordered to be provided for the same city; no application of the pump seems to have been then thought of.

Syringes continued to be used in London till the latter part of the 17th century, when they were superseded by more perfect machines. An account of them and the mode of working them would make a modern fireman smile. They were usually made of brass and held from two to four quarts. The smaller ones were about two feet and a half long, and an inch and a half in diameter; the bore of the nozzles being half an inch. *Three men* were required to work each, which they achieved in this manner: two, one on each side, grasped the cylinder with one hand and the nozzle with the other; while the third one worked the piston! Those who held the instrument plunged the nozzle into a vessel of water, the operator then drew back the piston and thus charged the cylinder, and when it was raised by the bearers and in the required position, he pushed in the piston and forced, or rather endeavoured to force, the contents on the fire. We are told that some of these syringes are preserved in one or two of the parish churches. It can excite no surprise that London should have been almost wholly destroyed in the great fire of 1666, when such were the machines upon which the inhabitants chiefly depended for protecting their property and dwellings. If the diminutive size of these instruments be considered, the number of hands required to work each, beside others to carry water and vessels for them, the difficulty and often impossibility of approaching sufficiently near so as to reach the flames with the jet, the loss of part of the stream at the beginning and end of each stroke of the piston, and the trifling effect produced—the whole act of using them, appears rather as a farce, or the gambols of overgrown boys at play, than the well directed energies of men to subdue the raging element.

In Asia syringes have probably been always in limited use. They are the only instruments of the pump kind now known there, if China be excepted. Very effective engines on the European plan are made by the Chinese. (Chinese Repos. vol. iv.)

The fire-engine of the Turks is an improvement on the syringe, but not much more effective. The author of "Sketches of Turkey" observes, when speaking of fires in Constantinople, " Indeed, when we afterwards saw the machines used by the Turks to extinguish fires, we were not surprised at the feeble resistance which they could oppose to the progress of the devouring element. The engines, in fact, are not larger than those employed with us to water gardens: they have but a single chamber, which is about eight inches long by three or four in diameter; they are readily carried about by hand." Commodore Porter, in his interesting account of "Constantinople and its Environs," says their fire-engines "are like those we use in our gardens, for watering the beds and walks, and deliver about as much water as a good large syringe. When an alarm of fire is given, a man seizes on one of these and runs to the spot indicated, with the engine on his shoulder, another brings a skin of water, pours it into the reservoir and they pump away." A characteristic anecdote is thus facetiously related by Commodore Porter. " They had heard of the fire-engines and fire companies of the United States—how half a shingle could be burnt, and the engines save the other half from the flames. They could not understand it. Mr. Eckford fortunately arrived with his beautiful ship, having one of our engines on board, requiring some twenty men to work it. The Capudan Pacha heard of it—' Mash Allah! let us see it,' exclaimed the old man. The engine was brought on shore and placed in the Navy Yard; a short suction was fixed to it and put into the Bosphorus; men were set to work it—the Navy Yard was soon inundated, and the Bosphorus began to run dry! ' Mash Allah!' said he, ' very good —but it will require a sea to supply it with water. It won't do for us, for there is no sea in the middle of the city.' They therefore have thought best to stick to their squirts, and to let the fire spread until the wind changes, or it is tired of burning."

Sandys, in the beginning of the 17th century, visited Constantinople, and speaks of the frequency of fires in that city : he observes, " It is not to be marvelled at, for the citizens dare not quench the fire that burneth their own houses, because officers are appointed for that purpose." He is silent respecting the instruments then used.

When the useful arts began to excite attention, the defects of portable syringes were too apparent to be neglected, hence in the early part of the 16th century several attempts were made to remedy them, by those noble spirits who burst through the prejudice that had so long consigned the subjects of practical mechanics to the mere makers of machines, as one unworthy of a philosopher's pursuit; and from the cultivation of which no distinction, save such as was allied to that of a skilful artisan, could be derived—a species of fame from which professors of philosophy shrunk, like Plato, with feelings of horror. To render the syringe an efficient fire-engine, would seem to be impossible, except by converting it into a forcing pump, and in that case it would be no longer a syringe. As long, therefore, as such an idea did not occur to engineers, they had no resource but to improve the " squirt" as well as they could ; and however hopeless the task may now appear, it was not only attempted, but to a certain extent accomplished, and with considerable ingenuity too, as will appear from the following figure, No. 143. It is described in Besson's "Theatre," and must therefore have been invented previous to 1568, the date of the permission to print his work.

No. 143. Syringe Engine from Besson. A. D. 1568.

"*Proposition De L'Autheur :*—Artifice autant singulier (comme je " pense) que non point commun, pour jecter l'eau contre un grand feu, " mesmement lors que pour la grandeur de la flamme, nul ne peut entrer " ny approcher de la maison qui brusle. *Declaration de la mesme figure :* " C'est instrument, qui est faict en forme de Cone, se soustient sur deux " Rouës : ayant sa bouche tournée vers le septentrion : et aupres de sa " base il y a des demi cercles, qui servent à l'hausser, au baisser, d'avan- " tage vers sa dicte bouche septentrionale est un Entonnoir, pour y verser " l'eau dedans : et en sa base, ou bien partie meridionale, est une vis, dont " est poussé dedans et reculé un Baston auquel sont des Estouppes, ainsi " qu'aux siringues. Le reste appert."

In reading the above, it should be remembered that *letters of reference* to designate the different parts of machines were not then in general use, but the sides and angles of the pages were marked with various points of the compass; and particular parts pointed out by their position with regard to these, and by the intersection of lines drawn between them. In this engine several defects of the "hand squirts" are avoided; as the necessity of inverting the instrument to refill it by plunging the nozzle into the vessel of water, the small quantity contained in the former, and the consequently incessant repetition of the operation and interruption of the jet, and the difficulty of directing it on the flames with certainty or precision. Besson, (if he was the inventor,) therefore, greatly enlarged the capacity of the cylinder, making it sufficient to contain a barrel, or more; and as a matter of necessity, placed it on a carriage. To eject the water uniformly, he moved the piston by a screw; and when the cylinder was emptied, it was refilled through the funnel by an attendant, as the piston was drawn back by reversing the motion of the crank. When recharged, the stop cock in the pipe of the funnel was closed and the liquid forced out as before. As flexible pipes of leather, the "ball and socket" and " goose-neck" joints had not been introduced, some mode of *changing the direction of the jet* of this enormous syringe was necessary. To effect this, it is represented as suspended on pivots, which rest in two upright posts: to these are secured (see figure) two semicircular straps of iron, whose centres coincide with the axis, or pivots, on which the syringe turns. A number of holes are made in each, and are so arranged as to be opposite each other. A bolt is passed through two of these, and also through a similar hole, in a piece of metal, that is firmly secured to the upper part of the open end of the cylinder; and thus holds the latter in any position required. The iron frame to which the box or female part of the screw is attached, is made fast to the cylinder; and it is through a projecting piece on the end of this frame that the bolt is passed. By these means, any elevation could be given to the nozzle, and the syringe could be secured by passing the bolt through the piece just mentioned, and through the corresponding holes in the straps. When a *lateral* change in the jet was required, the whole machine was moved by a man at the end of the pole, as in the figure. To the frame, jointed feet were attached, which were let down when the engine was at work. The women represented (one only is given in our figure) reminds us of a remark by Fosbroke: " In the middle ages during fires women used to fetch water in brazen pails to assist." Considering the age when this engine was devised and the objects intended to be accomplished by it, it certainly has the merit of ingenuity as well as originality. Beroald says of it: " Ceste noble invention est si souvent requise, pour esteindre les grand feux desquels on ne peut approcher; que sans faute elle merite d'estre plus au long, et plus ouvertement expliqueé, afin qu'elle soit mieux entendue." It will be obvious to every practical mechanic that engines of this kind, of large dimensions, must have been at best but poor affairs. To make the piston work sufficiently accurate and tight, and to keep it so, must have been a work of no small difficulty.

A correspondent, in a late number of the Lon. Mechanics' Magazine, vol. xxx, has communicated a very imperfect figure of this engine to that work, extracted from an English book, published in 1590, entitled " A Treatise named LUCARSOLACE, divided into four books, which in part are collected out of diverse authors, diverse languages, and in part devised by CYPRIAN LUCAR, Gentleman." London: 1590. It is very obvious that Lucar copied the engine in question from Besson's work, which was published in 1579, but was authorized to be printed in 1568; and which Besson's death then prevented. The following extract from Lucar's book is not without interest. "And here at the end of this chapter I will set before your eyes a type of a 'squirt' which hath been devised to cast much water upon a burning house, wishing a like squirt and plenty of water to be alwaies in a readinesse where fire may do harme; for this kind of squirt may be made to holde an hoggeshed of water, or if you will, a greater quantity thereof, and may be so placed on his frame, that with ease and a smal strength, it sahl be mounted, imbased or turned to any one side, right against any fired marke, and made to squirt out the water upon the fire that is to be quenched."

The Germans were proverbially in advance of the rest of Europe in the 15th, 16th and 17th centuries, in almost every department of the arts. " The excellency of these people [observes Heylin in his Cosmography] lieth in the mechanic part of learning, as being eminent for many mathematical experiments, *strange water-works*, medicinal extractions, chemistry, the art of printing, and inventions of like noble nature, to the no less benefit than admiration of the world." As early as A. D. 1518, some kind of fire-engines were used in Augsburg, being mentioned in the building accounts of that city. They were named "instruments for fires," and "water *syringes* useful at fires." Their particular construction is unknown; but from a remark in the accounts respecting wheels and poles, they are supposed to have been placed on carriages: they were probably large syringes and mounted like the one represented in the last figure.

The oldest *pump* engines of modern times were certainly made in Germany, and about the close of the 16th or beginning of the next century. The first one noticed by Beckman is that of Hautsch, which the Jesuit

Schottus saw tried at Nuremberg in 1656. In giving an account of it, Schottus remarks that the invention was not then new, it being known in other cities, and he himself remembered having seen a small one in his native city (Konigshofen) forty years before, consequently about 1617. We are not informed by either the professor or jesuit of the particular construction of this small engine, but there is a book extant that was published in 1615, which contains a figure and description of a German engine of that time, and which furnishes the information desired. This book is the "Forcible Movements" of Decaus, a work which, like the Theatre des Instrumens of Besson, escaped the notice of Beckman.[a]

No. 144. German Pump Engine from Decaus. A. D. 1615.

This machine is named "A rare and necessary Engin, by which you may give great reliefe to houses that are on fire:" we give the whole of the explanation: "This engin is *much practiced in Germany*, and it hath been seen what great and ready help it may bring; for although the fire be 40 foot high, the said engin shall there cast its water by help of four or five men lifting up and putting down a long handle, in form of a lever, where the handle of the pump is fastned: the said pump is easily understood: there are two suckers [valves] within it, one below to open when the handle is lifted up, and to shut when it is put down; and another to open to let out the water: and at the end of the said engin there is a man which holds the copper pipe, turning it to and again to the place where the fire shall be." In other words, this was a single forcing pump, such as figured at No. 118, and secured in a tub. For the convenience of

---

[a] Of Decaus' history scarcely any thing is known—even his name is left in doubt, for he is sometimes named *Isaak*, at others *Solomon* de Caus. An account of his book may be seen in Stuart's Anecdotes of the Steam-Engine, vol. i, p. 27. But there seems to be an error in the note given of the English translation by Leak, which is stated to have been made in 1707, whereas the copy in our possession is dated nearly fifty years earlier. It is entitled "New and rare inventions of Water-works, shewing the easiest waies to raise water higher then the spring; by which invention the perpetual motion is proposed, many hard labours performed and varieties of motions and sounds produced. A work both usefull, profitable and delightfull for all sorts of people: first written in French by *Isaak de Caus*, a late famous Engenier, and now translated into English by John Leak." London: printed by and for Joseph Moxon. 1659.

transportation the whole was placed on a sled, and dragged to a fire by ropes. The bore of the forcing pipe seems to have been small compared with that of the pump cylinder, a circumstance combined with the long lever and the number of men employed in working the latter, that contributed to increase the elevation of the jet. This machine exhibits a decided improvement on the primitive syringe, and constitutes a great step towards the modern engine. In the short angular tube to which the jet pipe is attached, we behold the germ of the more valuable goose-neck.

Notwithstanding the superiority of pump engines over the syringe, many years elapsed before they were generally adopted. " The English [observes a British writer] appear to have been unacquainted with the progress made by the German engineers; or to have been very slow in availing themselves of their discoveries, for at the close of the 16th century "*hand squirts*" were first introduced in London for extinguishing fires; and it was not till the beginning of the next, that they began to place them in portable and larger reservoirs—when placed in the latter and worked by a lever, the engines thus obtained were considered a great mechanical achievement; for when in 1633, three of them were taken to extinguish a large fire on London bridge, they were considered " such excellent things, that nothing that was ever devised could do so much good, yet none of them did prosper, for they were all broken." The observation that "hand squirts" or syringes were placed in reservoirs and then worked by a lever is not strictly correct: they were small forcing pumps that were employed. A syringe could not act at all if permanently fixed in a vessel, because it discharges the water through the same orifice by which it receives it. Some improvements were made on fire-engines by Greatorix in 1656, as mentioned by Evelyn: what they were is not known. The probability is, that they related to the carriage or sled. If his engines were the same that were advertised in 1658, this was the case, for they were recommended as " more traversable in less room, and more portable than formerly used." Fosbroke's Encyc. Antiq.

But the fire-engine as thus improved had still many imperfections: the water was projected in spurts as from a syringe; and the jet not only ceased with the stroke of the piston, but a portion of the water was in consequence lost by falling between the fire and engine at the termination of each stroke. An obvious mode of rendering the jet constant was by connecting two pumps to one discharging pipe, (as in the figure of Heron's,) and working the pistons alternately either by a double lever or two single ones. This was first adopted by the old German engineers, and thus another step was taken towards perfecting these useful instruments. Instead of a circular tub, a square box or cistern was adopted and mounted on four solid wheels in place of a sled; and a strainer, or false bottom, perforated with numerous small holes, was placed within the cistern to prevent gravel or dirt, thrown in with the water, from entering the pump. Such appear to have been the best fire-engines in England when the great fire in London occurred in 1666. They are referred to in the official account of the fire, dated Whitehall, September 8th, of the same year— " this lamentable fire in a short time became too big to be managed by *any engines.*" But nothing can show their general inefficiency in a stronger light than the measures adopted by the city government the following year to guard against a similar calamity. Instead of relying upon engines, they seem to have retained their confidence in the old syringe.

1. By an act of the Common Council, the city was divided into four districts, and " each thereof was to be provided with eight hundred leathern buckets—fifty ladders, of different sizes, from twelve to forty-two

feet in length—*two brazen hand squirts* to each parish—four-and-twenty pickax sledges—and forty shod shovels.

2. That each of the twelve companies provide themselves with *an engine*—thirty buckets—three ladders—six pickax sledges—and *two hand squirts;* to be ready upon all occasions. And the inferior companies such a number of *small engines* and buckets, as should be allotted them by the Lord Mayor and court of Aldermen.

3. That the Aldermen passed the office of Shrievalty, do provide their several houses with four-and-twenty buckets, and *one hand squirt* each and those who have not served that office, twelve buckets and *one hand squirt* each.

4. And for the effectual supplying the *engines* and *squirts* with water, pumps were to be placed in all wells; and fire plugs in the several main pipes belonging to the New River and Thames Water-works." Maitland.

The oldest account of English fire-engines that we have seen is in a small old quarto in our possession, the title page of which is wanting. From two poetical addresses to the author, it appears that the initial letters of his name were I. B., and that the work was entitled "A Treatise on Art and Nature." Two thirds of it are occupied with "water-works," and the rest with "fier-works," except four or five pages "on voyces, cals, cryes and sounds;" *i. e.* on making of whistles, &c. for sportsmen to imitate the voices of certain birds and other game. The date of publication was about 1634: this, we infer from page 51, where, speaking of " The engin near the north end of London bridge, [he observes] which engin I circumspectly vieued as I accidentally passed by, immediately after the late fier that was upon the bridge. *Anno* 1633." Shops and dwelling houses were built on both sides of the bridge at that time.

After describing several modes of raising water by sucking, forcing and chain pumps, he continues:—" Having sufficiently spoken concerning mils and engins for mounting water for meer conveyance, thence we may derive divers squirts and petty engins to be drawn upon wheeles from place to place, for to quench fier among buildings; the use whereof hath been found very commodious and profitable in cities and great townes." Hence engines were at this time not uncommon in England. No less than seven are figured by the author, and all are placed in cisterns or tubs mounted on wheels: neither air vessels nor hose pipes are described or mentioned. Five of the engines consist of single cylinders; of these some are in a perpendicular position, others are laid horizontally, and one is inverted, and fed by a branch pipe covered by a valve. The last one figured has two horizontal cylinders, a suggestion of the author's, and the piston rods are shown as worked alternately by pallets or arms on a vertical shaft, to which a reciprocating rotary movement was imparted by pushing a horizontal lever to and fro. One of these old fire-engines is a species of bellows pump, the construction of which we will endeavour to explain: Two brass vessels were connected at their open ends to a bag of leather: they resemble, both in shape and size, two men's hats, the linings of which being pulled out and sewed together form a cylindrical bag between them. A circular opening, six or seven inches in diameter, was made through a horizontal piece of plank fixed in the cistern of the engine, and over this opening one of the vessels, with its crown upwards, was placed, and made fast by screws through the rim: the other vessel being suspended from it by the bag and hanging loosely in the water. Within the lower vessel (in the centre of its bottom) a valve opening upwards admitted the water, and on the top or crown of the upper vessel, another valve, also opening upwards, was placed. Over the last valve the base of the jet pipe was secured. To work this machine, the rim of the lower vessel was connected at opposite points, by two iron rods or slings and a cross head, to the end of a lever, by which the lower vessel was moved up and down—compressing the bag when raised, and stretching it to its natural length when lowered; like the lantern bellows No. 105, or the bellows pump No. 106. To make the vessel rise and fall perpendicularly, the two rods were passed through holes in the plank. Water was kept in the cistern as high as the plank; so that when the movable vessel was raised the contents of the bag would be forced into the upper vessel and expelled through the jet pipe, and when it was again lowered, the water would enter through its valve and fill both as before. These engines, he observes, had sometimes two levers and were worked by two men, "the lower brasse [vessel] being poysed with two sweeps."

The goose-neck was used in England at this time. It is not represented in the figures, which are very indifferently executed, but is sufficiently well defined in the description of one of the engines. The author directs a hollow ball to be placed on the orifice of the forcing pipe, "having a [jet] pipe at the top of it, and made to screw another pipe [elbow] upon it, *to direct the water to any place.*"

Small or *hand* engines continued to be employed in London in the 18th century. This appears from a law passed in the 6th year of Queen Anne's reign, by which it was enacted that "each parish shall keep a large engine, and an *hand engine,* and a leather pipe, and socket of the same size as the plug or fire cock, [of the water mains,] that the socket may be put into the pipe to convey the water clear to the engine," under a penalty of ten pounds. In case of a fire, the first person who arrived with a parish engine to extinguish it was entitled to thirty shillings—the second twenty, and the third ten, provided the engines were in good order, "with a socket or hose, or leather pipe." The following year, the owners or keepers of "other large engines," (not parish engines,) were entitled to the same reward upon arriving with them and assisting in extinguishing a fire.

It is a singular proof of the general ignorance of hydraulic machinery, or want of enterprise in London pump makers of the 16th and 17th centuries, that they so long continued the use of "squirts" and engines with single cylinders, when they had daily before their eyes in the Thames Water-works examples of the advantages of combining two or more to one pipe. The application also of such machines as fire-engines was obviously enough shown to them; for when Maurice had finished his labors in 1582, the mayor and aldermen went to witness an experiment with his pumps at London bridge: "and they saw him throw the water over Saint Magnus's steeple, before which time [says Stow] no such thing was known in England as this raising of water." Immediately subsequent to the above date, the "squirt" manufacturers might surely have imitated Maurice's machine, but they did not for nearly a hundred years afterwards; that is, not until such engines had been introduced a second time from Germany, and designed expressly to put out fires.

Before the improvements of Newsham and his contemporaries of the 18th century, some important additions would seem to have been made in England, since, previous to 1686 "the engine for extinguishing fire" was claimed as an English invention. This is stated in a small volume published that year in London by John Harris, and apparently edited by him. It is entitled "A pleasant and compendious history of the first inventers and instituters of the most famous arts, misteries, laws, customs and manners in the whole world, together with many other rarities and remarkable things rarely made known, and never before made public : to which is added several curious inventions, peculiarly attributed to England and English men." We shall offer no apology for closing this chapter with the following abstract, although the concluding part only refers to our subject. " Fine Spanish needles were first made in England by a Negro in Cheapside, who refused to communicate his art; but in the eighth year of Queen Elizabeth's reign, Elias Corous, a German, made it known to the English. About the fifth year of Queen Elizabeth, the way of making *pins* was found out by the English, which before were brought in by strangers to the value of 60,000 pound a year. Watches were the invention of a German, and the invention brought into England Anno 1580. The famous inventers and improvers were Cornelius Van Dreble and Janus Torrianellus. The first clocks were brought into England much about the same time. Chaines for watches are said to be the invention of Mr. Tomackee. The engine for clock wheels is an English invention of about one hundred years standing, as likewise that for the speedy cutting down wheels for watches. Other late inventions there are, to whom as their inventers the English lay claime, as an engine for raising glass, an engine for spinning glass, an engine for cutting tobacco, the rouling press, the art of damasking linnen, and watering of silks, the way of separating gold from silver and brass, boulting mills, making caine chairs, the curious art of colouring and marbling books, making of horn ware, and the *engine to extinguish fire*, and the like."

CHAPTER VIII.

FIRE-ENGINES continued: Engines by Hautsch—Nuremberg—Fire-engines at Strasbourg and Ypres —Coupling screws—Old engine with air chamber—Canvas and leather hose and Dutch engines—Engines of Perier and Leopold—Old English engines—Newsham's engines—Modern French engine—Air chambers—Table of the height of jets—Modes of working fire-engines—Engines worked by steam. FIRE ENGINES IN AMERICA: Regulations respecting fires in New Amsterdam—Proclamations of Governor Stuyvesant—Extracts from old minutes of the Common Council—First fire-engines—Philadelphia and New-York engines—Riveted hose—Steam fire-engines now being constructed. Devices to extinguish fire without engines—Water bombs—Protecting buildings from fire—Fire escapes—Couvre feu—curfew bells—Measuring time with candles—Ancient laws respecting fires and incendiaries—The dress in which Roman incendiaries were burnt retained in the auto da fe.

THE fire-engine mentioned in the previous chapter, which Schottus witnessed in operation at Nuremberg in 1656, appears to have been equal to any modern one in the effects ascribed to it, since it forced a column of water, an inch in diameter, to an elevation of eighty feet. One German author says a *hundred feet*. It was made by John Hautsch, who, like most of the old inventors, endeavored to keep the construction of his machine a secret. He refused to allow Schottus to examine its interior; though the latter it is said readily conceived the arrangement, and from his account it has been supposed the cylinders were placed in a horizontal position. The cistern that contained the pumps was eight feet long, two in breadth, and four deep; it stood on a sled ten feet in length and four in width, and the whole was drawn by two horses. The levers were so arranged that twenty-eight men could be employed in working them. The manufacture of these engines was continued by George Hautsch, the son, who is supposed to have made improvements in them, as some writers ascribe the invention of fire-engines to him.

In the 16th century no place could have furnished equal facilities with Nuremberg for the fabrication of, and making experiments with, hydraulic machines. It was at that time the Birmingham of Europe. "Nuremberg brass" was celebrated for ages. Its mechanics were so numerous that, for fear of tumults, they were not allowed to assemble in public "except at worship, weddings and funerals." No other place, observes an old writer, had "so great a number of curious workmen in all metals." The Hautschs seem to have been favorites with the genius of invention that presided over the city; an aptitude for and an inclination to pursue mechanical researches were inherited by the family. From a remark of Dr. Agricola of Ratisbon, in his curious work on Gardening, we learn that one of them did not confine himself to devices for throwing streams of water into the air; for he contrived a machine by means of which he intended to raise himself into the upper regions. "What can be more ridiculous [exclaims the author just named] than the art of flying, sailing or swimming in the air? Yet we find there have been some who have practiced it, particularly one *Hautsch* of Nuremberg, who is much spoken of for his *flying engine*. In the mean time it is well for the world that these attempts have not succeeded; for how should we seize malefactors? They would fly over the walls of towns like Apelles Vocales, who they tell us saved himself by flying over the walls of Nuremberg, and the print of whose feet is there shown to strangers to this day." The art of flying was a standard subject with Nuremberg mechanics for centuries, and several curious results are recorded, but perhaps nothing more so than the above objection to it.

No. 145. Fire-engine belonging to Strasbourg, A. D. 1739.

For nearly a hundred years after the date of Hautsch's engine those used throughout Europe, with the exception perhaps of a few cities in Germany, were very similar to those described by Belidor, as employed in France in his time. They consisted simply of two pumps placed in a chest or cistern that was moved on wheels or sleds, and sometimes carried by men like the old sedan chair. These engines differed from each other only in their dimensions and the modes of working them. Nos. 145 and 146 will convey a pretty correct idea of them during the early part of the 18th century. The former belonged to Strasbourg, the latter to Ypres.

Chap. 8.] Fire-Engine at Ypres. 325

The front part of the cistern in which the pumps are fixed, is separated by a perforated board from the hinder part, into which the water was poured from buckets. The cylinders were four inches in diameter, and the pistons had a stroke of ten inches. Each pump was worked by a separate lever, A A; an injudicious plan, since a very few hands could be employed on each; and as the engine had no air vessel it was necessary, in order to keep up the jet, that the piston should be raised and depressed alternately—a condition not easily performed by individuals unused to the operation, and acting under the excitement of a spreading conflagration. The contrivance for changing the direction of the jet was very defective, and considering the date of this engine it is surprising that such a one was then in use. A short leathern pipe would have been much better. It will be perceived that the jet pipe is connected to the perpendicular or fixed one by a *single* elbow, instead of a double one, like the ordinary goose-neck. The joints were also made differently. The short elbow piece had a collar or ring round each end, and the jet and perpendicular pipes, where they were united to the elbow, the same. The faces of these collars were made smooth, so as to fit close to and at the same time turn on each other : loose flanches on the pipes were bolted to others on the elbow, and thus drew the collars together so as to prevent water from leaking through. Now it will be seen that although the joint which unites the elbow to the perpendicular pipe would allow the jet pipe to be turned in a lateral or horizontal direction, there appears no provision to raise or to lower it, and no apparent use at all for the other joint. We were at first at a loss to divine how the stream could be directed up and down as occasions might require, for Belidor has not explained it; but on examining more closely the figure in his work, we found that the jet pipe itself was not straight, but bent near its junction with the elbow: this dissolved the mystery, for it was then obvious that by twisting this pipe round in its joint, its smaller orifice could be inclined up or down at pleasure. This very imperfect device is also shown in the next figure, the jet pipe being curved through its whole length, instead of a single bend as in the last one.

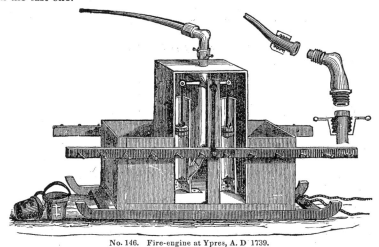

No. 146. Fire-engine at Ypres, A. D 1739.

The pumps of this engine are substantially the same as those of the last, but the piston rods are moved by a short vibrating beam placed directly over the cylinders. The axle of the beam is continued through both sides of the wooden case, and to its squared ends two iron rods are fitted, like crank handles on the axles of grindstones. To the lower ends of these rods are attached, by bolts, two horizontal bars of wood, on the outside of which a number of long pins are inserted, as shown in the cut. When the engine was in use men laid hold on these pins, one man to each, and pushed and pulled the bars to and fro, somewhat as in the act of rowing, and thus imparted the requisite movement to the pistons: a mode of working fire-engines that might, we think, be adopted with advantage in modern ones; for the vigorous working of these is so exhausting, that the strongest man can hardly endure it over a minute at a time. The jet pipe of this engine is connected to the other by coupling screws or "union joints," the most useful and ingenious device for joining tubes that ever was invented; and one which, from its extensive application in practical hydraulics, in gas and steam works, and also in philosophical apparatus, has become indispensable. We notice it here on account of its having been erroneously attributed to a modern engineer; whereas it was not *new* when introduced into Ypres fire-engines above a hundred years ago.

Two of the greatest improvements ever made in these machines were introduced about the same time, viz: the air chamber and flexible pipes of leather and canvas; upon these principally the efficiency of modern engines depends. By the former the stream ejected from a single pump is rendered continuous; and by the latter, it is no longer necessary to take the engine itself into, or close to, a building on fire; where in most cases it is impossible, from the heat of the flames and from smoke, to use it with effect. The modern author, or rather introducer, of the beautiful device for rendering the broken or interrupted jets of old engines uniform, is not known. In accordance with the customs of the age, he probably kept it secret as long as he could. We suspect that Hautsch's engine was furnished with an air chamber, and that it was on that account chiefly that he was so anxious to prevent its construction from becoming known. Beckman states that Hautsch used a flexible pipe to enable him readily to change the direction of the jet, "but not an air chamber, which Schottus certainly would have described." How Schottus could have done this, when according to Prof. B. himself, Hautsch refused to let him see the interior of the engine, it is difficult to imagine; and unless he had been acquainted with the properties of an air vessel, had the engine even been thrown open to his inspection, he could hardly have comprehended its action, unless explained to him by the manufacturer; at any rate, the secret, if it was in Hautsch's possession, was not long after divulged; for in 1675 an anonymous writer in the *Journal des Scavans* figured and described an engine with this appendage. The account was the same year translated and published in volume xi of the Philosophical Transactions, p. 679. As this is the earliest notice of the application of an air vessel to pumps in modern times that we have met with, it is entitled to a place here.

"This engine [No. 147] is a chest of copper, pierced with many holes above, and holds within it the body of a pump whose sucker is raised and abased by two levers. These levers having each of them two arms, and each arm being fitted to be laid hold on by both hands of a man. Each lever is pierced in the middle by a mortaise, in which an iron nail [bolt] which passes through the handle [rod] of the sucker, turns when the sucker is raised or lowered. Near the body of the pump there is a *copper pot*, I, [air vessel] joined to it by the tube G, and having another tube K N L, which in N may be turned every way. To make this engine play, water is poured upon the chest to enter in at the holes that are in the cover thereof. The water is drawn in to the body of the pump at the hole F, at the time when the sucker is raised; and when the same is let down, the valve of the same hole shuts, and forces the water to pass through the hole into the tube G of which the valve being lifted up, the water enters into the pot, and filling the bottom it enters through the hole into the tube K N L in such a manner, that when the water is higher than the [orifice of the] tube K, and the hole of the tube G is shut by the valve, the air inclosed in the pot hath no issue, and it comes to pass, that when you continue to make the water enter into the pot by the tube G, which is much thicker [larger] than the aperture of the end L, at which it must issue, it must needs be, that the surplus of the water that enters into the pot, and exceeds that which at the same times issues through the small end of the jet, *compresses the air* to find place in the pot; which makes that, whilst the sucker is raised again to make new water to enter into the body of the pump, the air which has been compressed in the pot drives the surplus of the water by the force of its spring, meantime that a new compression of the sucker, makes new water to enter and causes also a new compression of the air. And thus the course of the water, which issues by the jet, is always entertained in the same state." The box or chest had two projecting pieces on each side, through which two staves were passed for the convenience of carrying it. This small engine appears to have been in every respect an effective one; the whole of the parts, both of the pump and apparatus for working it, were well adapted to produce the best effect. The goose-neck seems to have been formed of a species of ball and socket joint.

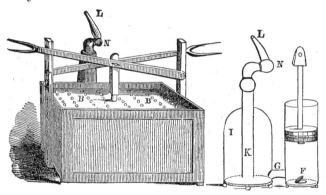

No. 147. View and Section of a Fire engine with Air Vessel. A. D. 1675.

One might suppose that when this account of the construction and effects of air chambers was published to the world, and in the standard journals of France and England, that they would speedily have been adopted in fire-engines throughout Europe. Such, however, was not the fact; on the contrary, they appear to have remained comparatively unknown for nearly fifty years longer; for it was not till the expiration of the first quarter of the 18th century that they began to be much used, and some years more elapsed before they were generally employed. We can only account for this by the limited circulation of the scientific journals named, and their being confined principally to learned men; who then as formerly felt indifferent towards mechanical researches: mechanics in those days were no great readers, and the few who possessed a taste for books were commonly without the means to gratify it. It is however, singular that this account of the air vessel should have escaped the researches of Beckman, and especially so as it was republished in 1704 by Harris in his Lexicon Technicum, and in 1705 by Lowthorp in the abridgment of the Philosophical Transactions. He observes, "I can find *no older* engine with an air chamber than that described by Perrault, and of which he has given a figure. He says it was preserved in the king's library at Paris; that it was employed for throwing water to a great height during fires; and that it had only one cylinder, and yet threw out a continued jet of water. He neither mentions the period of the invention nor the name of the inventor, and I can only add that his book was printed in 1684." Beckman, in a note, states that he had not seen the first edition of Perrault's work, and therefore knew not whether the French engine was described in it. We may here make the same remark, since the only copy in our possession is of the edition of 1684, having endeavored, but without success, to procure an impression of the previous one.

In 1672 *hose* or leathern tubes were first publicly used, in modern times, to convey water *from* engines to fires by John and Nicholas Van der Heide, in Amsterdam, of which city they were inspectors or superintendents of fire apparatus. They made the tubes in fifty feet lengths, with brass screws fitted to the ends, so that any number could quickly be connected together, as occasions might require. The introduction of hose pipes forms an epoch in the history of fire-engines, for they wonderfully increased the effect and extended the application of these machines. Previous to their adoption large engines could not be used to extinguish fires in the interior of dwellings—it was only when the flames burst through the windows or roof, that they came into play; and even then, it was often with difficulty and danger that they could be brought sufficiently near to discharge the water with effect, while in most cases the jet was so much diffused by the resistance of the air or wind as to descend rather in a shower of spray than in a compact stream. For want of hose the engines themselves were also frequently burnt; this was indeed a common occurrence, and is often mentioned in the notices of conflagrations. In the great fire of London the rapid spread of the flames drove the firemen from their engines, and many were consumed. In 1731 a great part of the town of Blandford, England, was destroyed, and in an account published by one of the sufferers, it is said "the engines were play'd, but were soon burnt." This loss of engines was invariably caused by the want of hose; for when plenty of the latter is at hand, the former can be placed and worked at any convenient distance from the fire, and the liquid discharged upon almost any part of it.

Another advantage resulting from the introduction of leathern pipes was in making the engines supply themselves. Before the use of hose, water was poured from buckets into the cistern in which the pumps were placed; hence when a fire broke out, one of the first objects was to form a lane of men, extending from the engine to the nearest rivulet, pond, well, or other source of water; those on one side passed along the full buckets to the engine, while those on the other returned the empty ones. To dispense with this number of men, the Van der Heides screwed one end of a hose pipe to the lower part of the cistern and extended the other to the edge of a pond or well, where its orifice was widened into a bag that was kept open by a frame. Into this bag the labourers poured the contents of their buckets, and sometimes portable pumps were used to raise water into it, for it was necessary that it should be sufficiently elevated above the cistern of the engine that its contents might readily flow into the latter. This was the first step towards using *suction* hose, and consequently towards making an engine supply itself. Perhaps it may be thought strange that they did not adopt this plan at first instead of the device just described, but in point in fact they could not, for before suction pipes could be used, a radical change was required in the construction of the lower parts of the pumps, and one that could not without much difficulty and expense be made in the old engines. Hence the Van der Heides very properly preferred making new machines altogether; to which they adapted suction pipes. These great improvements were made about 1675. In 1677, one or both of the Van der Heides obtained an exclusive privilege to construct such engines for twenty-five years. In 1695 there were in Amsterdam upwards of sixty of their engines, and when a fire broke out, the six that were located nearest were taken to extinguish it. The use of leather and canvas hose became general in the next century. In 1720 the latter was woven without seams in Leipsic and other places in Germany.

Whether the engines of the Van der Heides had *air vessels* is not ascertained; Professor Beckman says their internal construction is no where represented. There is strong evidence that they had none; for so late as the first quarter of the 18th century, the Dutch engines were not generally furnished with them, and this would certainly not have been the case had they ever been " common in all the towns of the Netherlands," as Van der Heide's engines were. Mr. Chambers, in his Cyclopedia, A. D. 1728, observes, article *Hydrocanisterium,* " The Dutch and others use a long flexible tube of leather, sail-cloth, or the like, which they carry or conduct in the hand, from one room to another, as occasion requires; so that the engine may be applied where the fire is only withinside, and does not burst out to expose it to its external action. To improve on this original fire-engine, they *have since* contrived to make it yield a continual stream." At the time Belidor wrote, air vessels were not common in Holland, and in 1744, Desaguliers speaking of their advantages, remarks, " In the use of engines to put out fires which have no air vessels, like the *Dutch engines,* or old parish engines, a great deal of water is lost at the beginning and end of the jet or spouting of the water." Philos. ii, 164. Beckman says, it is certain that air vessels were not common in Germany till after they were used by Leopold.

Perier in France, Leopold in Germany, and Newsham in England, contemporaneous engine makers in the early part of the 18th century, were greatly celebrated in their respective countries. They were sometimes considered inventors of the fire engine, though very erroneously, for so far as the principle of its construction, application of the air vessel, goose-neck, flexible pipes of leather and canvas, the connection of these by screws, &c. were concerned, the engine was perfected before their time; indeed not one of them contributed any thing essential to it. Their merit consisted in improving these machines in various *minor details;* in the arrangement of the different parts, construction of the carriages, mode of communicating motion to the pistons, and in rendering the whole more durable and efficient by superior workmanship and materials. In these respects the English engineer, we believe surpassed his competitors, but then he was the last of the three that entered the field, for Perier started before Leopold, and both were some years in advance of Newsham.

No account of Perier's engines is to be found in modern books; even Belidor has taken no notice of them. To supply this deficiency, we intended to insert a figure of one, taken from the 2d ed. of Poliniere's "Experiences de Physique," Paris, 1718, (the only work with which we are acquainted that contains a representation of them,) but on account of the unusual number of illustrations required in this chapter, it is omitted. A short description will suffice. After describing an atmospheric pump belonging to the arsenal of Paris, and another attached to a hotel in the faubourg St. Antoine, which had two spouts and two valves in the suction pipe, the author observes, J'ay vû à Paris des pompes dont on se sert pour tâcher d' éteindre le feu quand il arrive des incendies; and he then enters into a minute description of one of these Parisian engines. In its general appearance it resembled the Dutch one No. 148, consisting of two working cylinders with an air vessel between them, the piston rods moved by a double lever, through the ends of which staves four feet in length were inserted. The pump cylinders were sixteen inches long and four in diameter, but instead of being placed in a square wooden box or cistern, they were secured in an open copper pan, of an oval shape, and the same depth as the cylinders, and fastened by bolts to a base of wood or piece of plank, to the four corners of which short ropes were fastened. At one end of the pan, the leather hose which conveyed the water to the fire was connected by a screw to a copper pipe that communicated with the lower part of the air chamber. The leather tubes, Poliniere observes, were lubricated with a composition of tallow and wax to render them pliable; and, to prevent mice and other vermin from destroying them, soaked in an infusion of colycinth or bitter apple. In furnishing the pumps with water, Perier adopted the first device of the Van der Heides, and hence we infer that he was ignorant of the better mode of making them supply themselves through suction pipes. As they could only draw water out of the vessel in which they were placed, and it being too small and inconvenient for numbers of people to pour the contents of their buckets into it when the engine was in use, a canvas or sail cloth bag, coated with pitch or tar, was connected by a flexible pipe of the same material, to the lower part of the pan. This bag was of a conical form, the wide end being uppermost, and supported with the mouth open on a folding frame, something like a high camp stool. Into this bag the water for the supply of the pumps was poured. It might of course be placed at any convenient distance from the engine, by means of additional lengths of pipes that were always kept ready and which were connected together by screws. These engines, Poliniere says, forced the water through the orifice of the jet pipe to a surprising distance. He observes also that smaller ones were in use; which consisted of a single cylinder and air chamber, and were worked by a single lever.

The following extract relating to Perier's engine is from the Dictionnaire Œconomique, 3d. edit. Paris, 1732, from which it appears that at that date they were small affairs, and differed but little from our garden engines; in other words, they were then nothing more than *pompes portative*, the name by which they were designated at the first. " La pompe que le Sieur du Perier a inventée ou perfectionnée est très commode dans les incendies. Deux hommes la peuvent aisément transporter avec tout son attirail, et la placer dans tel lieu que l'on voudra. Il n'est pas nécessaire qu'elle soit dans l'endroit où se trouve l'eau, il y a un canal de coutil ciré en dedans, qui sert à conduire l'eau jusqu'à la pompe. Ce canal peut être augmenté en y adaptant d'autres canaux faits de la même façon. La pompe étant placée dans le lieu le plus commode, ou peut encore porter l'eau dans le plus fort de l'incendie par le moien d'un canal, qui est fait de cuir, et qu'on augmente, autant qu'on veut, en y ajoûtant d'autres canaux par le moien de quelques vis. La matiere dont est compose, ce canal donne la facilité de passer d'un appartement dans l'autre pour appliquer l'eau dans l'endroit le plus nécessaire. Les circonvolutions du canal n'empêchent point l'eau d'agir avec violence, et la force avec laquelle elle agit est d'autant plus grande, que les hommes qui font aller la pompe, emploient eux-mêmes plus de force, la quantité d'eau dépend encore du nombre de pistons."

In 1699 Perier obtained from the king an exclusive privilege to construct fire-engines, which Professor Beckman thinks were the first public ones employed in Paris. In 1716 an ordinance of the king directed a larger number than those already in use, to be distributed in different parts of the city, and public notice to be given where they could be found in case of fire.[a] In 1722 there were thirty in use, besides others belonging to public buildings. As these machines had air vessels, it is strange that Belidor neither mentions the fact nor refers to Paris engines at all. After describing a Dutch one, No. 148, he quotes (as if he knew of no others with air vessels) Perrault's description of the one that was in the king's library fifty years before, and an account of another that Du Fay saw at Strasbourg in 1725.

Leopold's engines do not appear to have possessed any peculiar feature to which he could lay claim as inventor. They seem to have been identical or nearly so with the one described in the Journal des Savans forty years before, (No. 147.) Each consisted of a single pump with an air vessel enclosed in a copper chest. One man raised a jet by it to the height of from twenty to thirty feet. Leopold kept the construction for some time a secret, and with this view the pump was entirely enclosed in the chest; a cover being soldered on the latter. Beckman says he made and sold a great number of them. In 1720 he published a description of them in a pamphlet; and in 1724 he inserted an account of them in his *Theatrum Machinarum Hydraulicarum*, a work published that year at Leipsic in three volumes folio.

The annexed figure, No. 148, exhibits an improvement on Leopold's engine, having two cylinders and working by a double lever. Small engines seem to have been preferred to those of large dimensions, such as were made by Hautsch, or those of modern times. Before the introduction of hose pipes, small ones were certainly more useful, since they could be carried into any part of a house when on fire, but when flexible pipes of leather and canvas became common, their efficiency was not to be compared with that of the large sizes.

No. 148. Dutch Fire-Engine. A. D. 1739.

English fire-engines were much the same dimensions as those used on the continent till Newsham and contemporary engineers introduced others that approached in size those in present use; but for several years after the smaller ones retained the preference. The London manufacturers made six different sizes, the larger one only being placed on wheels. Even in the middle of the 18th century such as are represented by the figure on the next page were common in that city. A similar figure was published by Mr. Clare in 1735 in his work on the motion of fluids, and so late as 1765 it was described (in the London Magazine for that year) as the engine in common use. As an indication that

---

[a] Supplement to Dict. Œconomique. Amsterdam, 1740. Tom. ii, 163.

air vessels were not used in England before the 18th century, it may be observed that in the year last named, those engines which had them were named "constant stream'd engines," to distinguish them from those that had none—such being called squirting engines.

No. 149. English Fire-Engine of the middle of the 18th century.

In 1729 Switzer published his System of Hydrostatics, in which he in serted the circulars of two rival engine makers—Fowke and Newsham. As these documents contain some interesting particulars respecting the state of practical hydraulics at the time, as well as of fire-engines, we insert some extracts from each, previous to introducing Newsham's engine.

"MR. FOWKE, *Nightingale Lane, Wapping*: makes

"1. *Constant stream'd* engines for extinguishing fires, the large sizes play two streams at once, being the first and only of their kind, and does the office of two engines, and so contrived as to be drawn through, (and if occasion requires,) worked in a passage three feet wide, which no other can, and will feed themselves with a sucking pipe. Their movements are easy and natural, having a perpendicular stroke, and are without either rack, wheel, chain or crank, whereby the friction is lessened more than any others, and consequently requires less strength, are more useful, and less liable to disorder and decay, and much cheaper than any other; and therefore are by judicious persons esteemed preferable to all others. By screwing a pipe they water gardens, dispersing the particles of water for about fourteen yards square, like small rain. The four larger sizes run on wheels, and the other two carried by two men like a chair.

"2. ENGINES which will work either by water, wind, horses or men, and so contrived that either may work at a time, or be assistant to each other, whereby large quantities of water may be raised, so that if the height, distance and quantity required be known, the expense and strength may be calculated so as to serve cities, towns, noblemen and gentlemen's seats and fountains, brewers, distillers, dyers; and for draining of lands, ponds, and mines of lead, coal, &c.

"3. PUMPS which may be worked by one man, for raising water out of any well upwards of one hundred and twenty feet deep, sufficient for the service of any private house or family; and so contrived that by turning a cock, may supply a cistern at the top of the house, or a bathing vessel in any room; and by screwing on a leather pipe, the water may be conveyed either up stairs or in at a window, in case of any fire.

"4. All manner of fancies in fountains."

After referring to a number of machines erected by him in London and its vicinity, Mr. Fowke concludes with a table of prices of fire-engines, the smallest being £14 and the largest £60. Newsham's circular is obviously designed to counteract the effect of Fowke's.

"RICHARD NEWSHAM, of Cloth Fair, London, engineer, makes the most useful, substantial, and convenient engines for quenching fires, which *carries continual streams with* great force. He hath play'd several of them before his majesty, and the nobility, at St. James's, with so general an approbation, that the largest was at the same time ordered for the use of that royal palace. And as a further encouragement (to prevent others from making the same sort, or any imitation thereof) his majesty has since been graciously pleas'd to grant him his *second* letters patent, for the better securing his property in this, and several other inventions for raising water from any depth, to any height required.

"The largest engine will go through a passage about three foot wide, in complete working order, without taking off or putting on any thing : and may be worked with ten men in the said passage. One man can quickly and with ease, move the largest size about, in the compass it stands in : and is to be play'd without rocking, upon any uneven ground, with hands and feet, or hands only, which cannot be parallel'd by any other sort whatsoever. There is conveniency for above twenty men to apply their full strength, and yet reserve both ends of the cistern clear from incumbrance, that others at the same time may be pouring in water, which drains through large copper strainers. The staves that are fixed through the leavers, along the sides of the engine, for the men to work by, though very light, as alternate motions with quick returns require ; yet will not spring and lose time the least : but the staves of such engines as are wrought at the ends of the cistern, will spring or break, if they be of such a length as is necessary for a large engine, when a considerable power is apply'd : and cannot be fix'd fast, because they must at all times be taken out before that engine can go through a passage. The playing two streams at once, do neither issue a greater quantity of water, nor is it new, or so useful, there having been of the like sort at the steel-yard, and other places, thirty or forty years ; and the water being divided, the distance and force are accordingly lessen'd thereby.

"Those who pretend to make the forcers work in the barrels, with a perpendicular stroke, without rack, wheels, chains, crank, pully, or the like, by any kind of contrived leavers, or circular motion whatsoever, with less friction, than if guided and work'd by wheel and chains, (which of all methods is the best,) do only discover their ignorance ; they may as reasonably argue, that a great weight can be dragg'd upon a sledge, with as little strength, as if drawn upon wheels.

"As to the treddles, on which the men work with their feet, there is no method so powerful, with the like velocity or quickness, and more natural and safe for the men. Great attempts have been made to exceed, but none yet could equal this sort; the fifth size of which hath play'd above the grasshopper upon the Royal Exchange ; which is upwards of fifty-five yards high, and this in the presence of many thousand spectators.

"Those with suction feed themselves with water from a canal, pond, well, &c. or out of their own cisterns, by the turn of a cock, without interrupting the stream. They are far less liable to disorder, much more durable in all their parts, than any extant, and play off large quantities of water to a great distance, either from the engine, or a leather pipe, or pipes of any length requir'd ; (the screws all fitting each other.) This the cumbersome squirting engines, which take up four times more room, cannot perform; neither do they throw one fourth part of their water on the fire, at the like distances, but lose it by the way; nor can they use leather pipe with them to much advantage, whatever necessity there may be for it. The five large sizes go upon wheels, well box'd with brass, fitted to strong iron axles, and the other is to be carried like a chair."

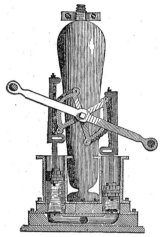

No. 150. Section of Newsham's Engine.

No. 150 is a vertical section of the pumps in Newsham's engine, with the air vessel between them, and showing also the sectors and chains by which motion is transmitted from the levers to the piston rods, and the latter preserved in a perpendicular position. The chains are similar to watch chains in their construction, and the length of each is equal to the arc of one of the sectors. Four are used, two to each sector. Their mode of operation is in this manner: One end of a chain is fastened to the top of a piston rod, by a bolt and nut as represented, and the other end riveted to the lower extremity of the sector; so that when the latter is turned down by depressing the lever, it necessarily draws, by this chain, the piston down with it. Another chain is fastened in the same manner to the lower part of the piston rod, (that is above the cylinder,) and the upper extremity of the sector, and hence when the lever is elevated, this chain raises the piston with it. He probably derived the idea of thus working them from Newcomen's mode of working pumps by the atmospheric steam-engine. The round opening below the valves in the above figure, is where the suction pipe is continued to the hose, shown at one end of the cistern in the next figure

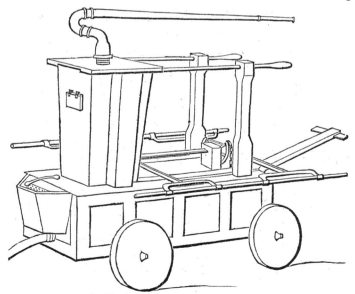

No. 151. Newsham's Fire-Engine, A. D. 1740.

No. 151 is an external view of one of Newsham's engines at the time of his death, as drawn by Mr. Labelye, the engineer of Westminster bridge, and inserted by Desaguliers in the second volume of his Philosophy, in 1744. Its general appearance is far inferior to modern ones, but the essential parts—the pumps—were equal to those now used. The strong iron shaft by which the pistons were raised and depressed was continued along the top of the cistern, and to it the levers were secured as at present; but in addition to the levers, sectors, like those that moved the pistons, were also fastened to it—portions of two of these are shown in the cut, and there were two others near the upright case : to their upper parts, two long strips of plank, or *treddles*, were suspended by short chains, and on these planks, six men, who stood upon the cistern and held by the hand rails, alternately threw their weight; first on the treddle on one side of the carriage, and then on the other, and thus aided the firemen at the levers in working the engine. The box or trough, with a grate within it, at the end of the cistern, was for the purpose of emptying buckets of water to supply the pumps, when the suction pipe (figured below it) was not used. The small flap on the end of the upright case covered printed directions how to use and keep the engine in order.

If the section, No. 150, be compared with English engines in previous use, one of which is figured at No. 149, it will be seen at a glance how great were the improvements that Newsham introduced. Independently of the three most original of his contributions—the sectors and chains—treddles—and working the pumps with long staves at the sides of the carriage instead of short ones at the ends—the whole machine was improved more or less in every part. To keep the cistern and levers as low as possible, the carriage was placed on bent axles. He introduced and improved the three-way cock, and the goose-neck was perfected in his hands; the elbows being jointed to each other by very fine screws. Desaguliers thought that no part of the engine could be altered for the better. A writer in the London Magazine for 1752, (page 395,) says that Newsham in these machines gave " a nobler present to his country than if he had added provinces to Great Britain." Their merits were generally acknowledged: he received orders for them from various parts of Europe, and it will be seen in a subsequent part of this chapter that those first used in this city were made by him.

The celebrity his engines acquired had a blighting effect on other manufacturers—like Aaron's rod swallowing up those of his competitors. His engines were purchased for the use of the parishes throughout the country generally, and also by the various insurance companies, which, unlike ours, are at the sole expense of extinguishing fires, and of providing the means to effect it. Every insurance company in English cities keeps in its pay a number of firemen to take charge of and work its own engines. Two horses are attached to each engine to draw it to and from fires. The height of the jet from Newsham's engines was about fifty feet. He mentions in his circular having thrown it to an elevation of *fifty-five yards*, but he was certainly mistaken.

Several improvements have been made in English fire-engines since Newsham's time, but they are chiefly confined to the carriage, and to details and arrangements of the various parts. Treddles are dispensed with, and the carriages are made longer, so that a greater number of men can be employed in working them. They resemble American engines so closely, that a separate figure of a modern English engine is unnecessary. The reader is therefore referred to No. 154. Others on the principle of the semi-rotary pump, (No. 140,) are also used to a limited extent in London.

The following figure of a modern French fire-engine is from the *Manuel du Fondeur*; Paris, 1829. It consists of two cylinders and an air vessel arranged in the usual way. One of the pumps, and half of the air chamber, is shown in section. The cistern is more elevated than in English or American engines, and from the consequent height of the levers would seem more inconvenient to be worked. The suction pipe is of copper with folding joints, and a perforated hollow ball at the extremity to prevent dirt or gravel from entering with the water. A short leathern tube connects this pipe with the suction cock. This engine is worked at the ends of the carriage, and the piston rods are connected to the lever by slings, and made to rise and fall in a perpendicular position by radius bars jointed to the upper ends of the latter, and to permanent pieces that project from the frame that supports the fulcrum.

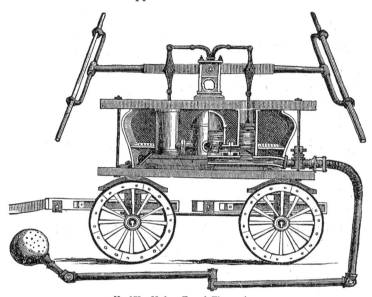

No. 152. Modern French Fire-engine.

The elevation of the jet depends upon the pressure to which the air in the air chamber is subjected; the elasticity or spring of that fluid being inversely as the space it is made to occupy. Before an engine is set to work the interior of the chamber, like that of all *empty* vessels, (to use a vulgar solecism,) is filled with common air, of that degree of density in which it appears near the earth's surface; but when the pumps are set to work, the water forced by them into the chamber crowds the air into the dome or upper part of that vessel, whence there is no passage for its escape; and, as the liquid accumulates, the air is condensed more and more, until, by its reaction on the surface of the water, it drives the latter through the jet or hose pipe, and with a force exactly proportioned to the degree of its compressure. Thus if the volume of air in the chamber be compressed into half its bulk, the jet would rise to about 32 or 33 feet, (if not retarded by friction, angles or other imperfections in the pipe;) and if it were made to occupy one third of its former space, its spring would be three times greater than common air, and would force the jet to an elevation of about 64 or 66 feet; and so on. A tabular statement, similar to the following, exhibiting the relation between the height of a jet and the air's compressure, has long been published. It is, however, of little use to practical men. We doubt if a column of water of the size of those thrown by ordinary engines could be raised by any means, two hundred feet above the orifice of the pipe whence it issued: the resistance of the atmosphere would disperse it before it could reach that elevation.

Volume of air contained in the air chamber compressed to	Ratio of the air's elasticity.	Height to which it is said the water will spout.
$\frac{1}{2}$	2	33 feet
$\frac{1}{3}$	3	66 "
$\frac{1}{4}$	4	99 "
$\frac{1}{5}$	5	132 "
$\frac{1}{6}$	6	165 "
$\frac{1}{7}$	7	198 "
$\frac{1}{8}$	8	231 "
$\frac{1}{9}$	9	264 "
$\frac{1}{10}$	10	297 "

Great as are the advantages derived from air chambers, some attention to them is required in order to secure at all times the benefit they are designed to impart. When neglected (and we believe few parts of an engine exercise the attention of firemen less) they often become actually injurious, for when no advantage is derived from the elasticity of the confined air, the water is impeded in its progress by passing through them. Upon the trial of engines it sometimes occurs that the water is thrown higher at their first working than after they have been a few minutes in use, and this notwithstanding all the efforts of the firemen to make the jet reach the first elevation. This result has sometimes been attributed to fatigue in the men—to obstacles in the pipes—to grit or sand under the valves, &c. whereas in fact it was often due to the air vessel alone ; *i. e.* to the escape of air from it. This escape may be occasioned by minute leaks in the chamber, but when no such imperfections exist the air frequently makes its exit, and its place becomes occupied by the liquid. Whenever air is subjected to great pressures in contact with water, it is quickly absorbed by the latter, and in this way it is that it often disappears from the air chambers of fire-engines, and also from those of pressure-engines, Heron's fountain, water rams, &c. When a long suction hose is attached to an engine and the latter worked at a moderate velocity, a sufficient supply of air to replace that taken up by the water, commonly enters, unknown to the firemen, through the seams and joints; but when one engine is fed by another pouring water into its cistern, there is little chance for the requisite supply of air, unless a minute opening were left in the cap that screws over the orifice of the suction pipe, at one end of the engine.

The suction cocks of some engines diminish their useful effect in consequence of the holes through the plugs being smaller than other passages for the water.

The great desideratum in modern fire-engines is an improved mode of working them. At page 72 we remarked that experimental researches have shown the useful effect of a man working a pump, in the ordinary way with a lever, to be fifty per cent less than when he turns a crank; and that when his strength is applied as in the act of rowing, the effect is nearly one hundred and fifty per cent more than in moving a pump lever. This is sufficient to induce efforts to supersede the present mode of working the pumps of fire-engines, and particularly so, as the labor is so severe that few can continue it above a minute or two at a time, when if relays of men are not ready, buildings on fire are left to fate. The jars or concussions produced by the violent contact of the levers with the sides of the carriage at every stroke, is a source of waste of firemen's energy, and want of uniformity in their movements when at work, is another. In the 29th vol. of the London Mechanics' Magazine, a contrivance is described for diminishing the shocks consequent on the contact of the levers with the carriage. It consists of three spiral springs enclosed in cylindrical cases secured on each side of the carriage; pads rest on the springs and project above each case, and upon them the levers strike when pulled down. Blocks of caoutchouc were previously tried, but the violence of the blows soon rendered that material useless. The velocity with which engines are sometimes worked also occasions a useless expenditure of their strength; we have seen some drawing water through long suction pipes, and the pumps worked so quickly that the water certainly had not time to pass through the hose and *fill* the cylinders, ere the pistons began to descend.

If some mode of making the carriage immovable, and the pumps were worked by long cranks on each side, the firemen could not only perform fifty per cent more labor, but they could do it with less exertion, and consequently endure it longer. A modification of the plan adopted in the Ypres engine, page 325, would be still more effective; in addition to which ropes might be attached to the bars, and any number of spectators could then assist.

If we review the progress of fire-engines in modern times, from the simple syringe to the splendid machines of the present day, we shall find that every important improvement in the apparatus for raising the water, was a nearer approach to the engine described by Heron. Previous to the 16th century, syringes or squirts only were in use, and not till the Spiritalia had been translated and printed do we meet with the application of pumps. At first a single working cylinder was employed, and the piston moved by a single lever as in No. 144; then two cylinders, each worked by a separate lever, were united to one discharging pipe—next the double lever, as figured by Heron, by which an alternating movement of the pistons, and a more efficient application of the force employed was secured; then the goose-neck, also mentioned by Heron—and lastly, the air vessel made its appearance. If the beautiful and philosophical device last mentioned, be, as some persons have supposed, a modern invention, why is it that no one has ever rose up to claim it? Is not this a tacit admission that it was derived directly from the Spiritalia, or from Vitruvius's description of the machine of Ctesibius? To the ancients, then, we are indebted for the most valuable features in our fire-engines, and it is not unreasonable to conclude that those used in ancient Egypt and old Rome were as effective as ours. If they were not, it is very strange that Heron should have hit upon that construction of them and that arrangement of their parts, which we have only acquired after a century spent in experiments.

Of late years "steam fire-engines" have been introduced with success in some parts of Europe: a small horizontal steam-engine with its boiler, being arranged on the carriage of the fire-engine. One large pump cylinder only is used, and its piston and that of the steam cylinder are attached to the same rod. Mr. Braithwaite, a London engineer, was, we believe, the first who made one of these machines. The steam cylinder was seven and a half inches diameter, and the pump six and a half; the water was forced through an ajutage of seven-eighths of an inch, to an elevation of ninety feet. The time of getting the apparatus into play from the moment of igniting the fuel, was eighteen minutes. When an alarm of fire was given, the fuel was kindled and bellows attached to the engine were worked by hand. When the horses were harnessed to drag the machine to the fire, the bellows were worked by the motion of the wheels. (See London Mechanics' Magazine for 1830, and in volume xviii, for 1832, there is a figure and description of one made by Mr. B. for the Prussian government, being designed to protect the public buildings of Berlin.)

One or two of these machines on an improved plan by Mr. Ericsson, are now being constructed in this city.

FIRE-ENGINES IN AMERICA.—The first use of fire-engines is an important event in any country, and may be considered as constituting an epoch in the history of its useful mechanism: moreover, wherever they are *made*, they indicate a certain degree of refinement in civilization and an advanced state of the mechanic arts. To their introduction into this continent, future historians may, and probably will, have recourse for data respecting the state of society in the early days of the republic, and the still earlier times during which the country was subject to Europe; for the circumstances which precede, and eventually lead to the adoption of fire-engines, invariably reflect light on the manners and customs, the police and other municipal regulations of the times, as well as on many of the arts, particularly those connected with *building*. The following extracts from official records in the clerk's office, respecting their introduction into the city of New-York, will be found to illustrate some of the above remarks.

It does not appear that either squirts or engines were used during the time the city remained in possession of its founders; viz: from A. D. 1614 to 1664. The volume of Dutch records preserved in the clerk's office, to which we referred, page 299, contains several enactments relating to fires and fire wardens, but no mention is made of instruments for extinguishing fires until 1648, when ladders, hooks and buckets were ordered from Holland. As these records have never been printed, a few extracts from the "Ordinances of the Director-General and the Council of the New Netherlands," will be acceptable to most readers. The first one is dated May 29, 1647: it cannot, perhaps, be strictly considered as related to our subject, although it was designed to remove a fruitful source of fires, viz: *inebriety*. On the above date the Director-General, *Petrus Stuyvesant*, issued a proclamation, addressed to certain of the inhabitants " who are in the habit of getting drunk, of quarrelling, fighting, and of smiting each other on the Lord's day of rest, of which on the last Sunday, we ourselves witnessed the painful scenes." It appears from this and other edicts to the same effect, that the governor had considerable difficulty in keeping a portion of his people sober; and from following a practice which he denounces as the " dangerous, injurious, and damnable selling, giving out, and dealing out, wines, beers, and ardent spirits to the Indians or natives of this land."

Another proclamation is more to our purpose. " *Whereas* it has come to the knowledge of his excellency, the Director-General of New Netherlands, Curacoa, &c. and of the Islands of the same, and their Excellencies the Councillors, that certain careless persons are in the habit of neglecting to clean their chimnies by sweeping, and paying no attention to their fires; whereby lately fires have occurred in two houses; and whereas the danger of fire is greater as the number of houses increases here in New-Amsterdam; and whereas the greater number of them are built of wood and are covered with reeds, together with the fact that some of the houses have wooden chimnies, which are very dangerous: Therefore, by the prompt and excellent Director-General and their honours the Councillors, it has been deemed advisable and highly necessary to look into this matter, and they do hereby ordain, enact, and interdict, that from this time forth no wooden or platted chimnies shall be permitted, . . . Those already standing shall be permitted to remain during the good pleasure of the *fire wardens*. . . . . As often as any chimnies shall be discovered to be foul, the fire wardens aforesaid shall condemn them as foul, and the owner shall immediately, and without any gainsaying, pay the fine of three guilders, for each chimney thus condemned as foul; to be appropriated to the maintenance of *fire ladders, hooks*, and *buckets;* which shall be provided and procured [from Holland] the first opportunity. And in case the house of any person shall be burned, or be on fire, either through his own negligence, or his own fire, he shall be mulcted in the penalty of *twenty-five guilders*, to be appropriated as aforesaid. Thus done, passed and published at Fort Amsterdam, this 23d day of January, 1648."

This ordinance does not appear to have produced the desired effect, since a similar one was published in September of the same year. In February 1656 another was issued, by which the fire wardens were directed to establish such penalties for chimneys or houses taken fire "*as shall be found among the customs of our Fatherland.*" At the close of the following year the use of squirts or engines does not appear to have occurred to the inhabitants, a circumstance from which it may be inferred that such machines were at that time little used in Holland, and this also appears from an allusion to the practice of quenching fires there, in a proclamation prohibiting wooden chimneys, flag roofs, &c. " In all well regulated cities and corporations, it is customary that fire buckets, ladders and hooks, are in readiness at the corners of the streets, and in public houses, for the time of need. [Here is no mention of engines, although the instruments used in Holland are obviously alluded to.] The Director-General and the councillors do ordain and authorize in these premises, the burgomasters of this city, either personally or by their treasurer, promptly to demand for every house, whether small or large, *one beaver*, or eight guilders in seawant, according to the established price; for the purpose of ordering from the revenue of the same, by the first opportunity, *from Fatherland, two hundred and fifty leather fire buckets;* and out of the surplus, to have made some *fire ladders* and *fire hooks*: and in addition to this, once a year, to demand for every chimney, one guilder for the support and maintenance of the same. Thus done in the session of the director-general and councillors, held in the fort of Amsterdam, in New Netherlands, this 15th day of December, A. D. 1657."

After New Netherlands became a British province, similar ordinances continued to be enacted till the year 1731, when two of Newsham's engines were ordered from London. These were probably the first fire-engines used on this continent. The following extracts are from the minutes of the common council.

" At a common council held the 16th day of February 1676–7, in the 28*th year of Charles* II. Ordered that all and every person and persons that have any of the city's ladders, buckets or hooks in their hands or custody, forthwith bring the same unto the mayor, as they will answer the contrary at their peril." The same date sundry wells were ordered to be made "for the public good of the city," among which was " one over against Youleff Johnson's the butcher; and another in Broadway against Mr. Vandike's." " At a common council held the 15th day of March 1683, *in the 36th of the reign of Charles* II. Ordered that provision be made for hooks, ladders and buckets, to be kept in convenient places within this city, for avoyding the peril of fire." No mention is here made of engines, nor in the next extract, wherein the want of instruments to quench fire is especially referred to. "Feb. 28, 1686: Whereas great damages have been done by fire in this city, by reason there were not instruments to quench the same. It is ordered that every inhabitant within the city whose dwelling house has two chimnies shall provide one bucket for its use: and every house having more than two hearths, shall have two buckets." Every brewer was to provide six, and every baker three buckets, under a penalty of six shillings for every bucket ordered. "January, 1689: Ordered that there be appointed five *Brent masters* for the city of New-York, as follows: Peter Adolf, Derek Vanderbrink, Derek Ten Eyk, Jacob Borlen, Tobias Stoutenburgh; and that five ladders be made to serve upon occasion of fire, with sufficient hooks thereto."

November 16, 1695: Every dwelling in the city was to be provided with one or more buckets by New-Year's day. The tenants were to provide them for the houses they occupied, and the cost to be deducted from the rent. Every brewer was again ordered to procure for his premises six, and every baker three. Several buckets were lost, and the public crier was directed to give notice. These "orders" do not appear to have been implicitly obeyed, for they were frequently repeated, and in November 1703, a penalty was attached for noncompliance. "October 1, 1706: Ordered that Alderman Vanderburgh do provide for the public use of this city, eight ladders and two fire hooks, and poles of such length and dimensions as he shall judge to be convenient, to be used in case of fire." November 20, 1716, a committee was appointed "to provide a sufficient number of ladders and hooks for the public use of this city in case of fire." In November 1730, FIRE-ENGINES are first mentioned. On the 18th of that month among other provisions enacted for the prevention and extinguishment of fires, one is in the following words: "And be it ordained by the authority aforesaid, that forthwith provision be made for hooks, ladders and buckets, and FIRE-ENGINES, to be kept in convenient places within the city for avoiding the peril of fire." At the same time the inhabitants were again directed to provide and keep buckets in their houses. It does not appear that any active measures to procure the engines were taken till the next year, for under the date of May 6, 1731, the common council "Resolved that this corporation do with all convenient speed procure *two complete fire-engines, with suction and all materials thereunto belonging, for the public service: that the sizes thereof be the fourth and sixth sizes of Mr. Newsham's fire-engines*: and that Mr. Mayor, Alderman Cruger, Alderman Rutgers and Alderman Roosevelt, or any three of them, be a committee to agree with some proper merchant or merchants to send to London for the same by the first conveyance, and report upon what terms the said fire-engines, &c. will be delivered to this corporation."

On the 12*th of June* the committee reported that the engines could be imported at an advance of 120 per cent on the invoice; and they were ordered accordingly. They seem to have arrived about the 1*st of December*, for on that day, a room in the City Hall was ordered to be fitted up "for securing the fire-engines." On the 14*th of December* a committee of two was appointed "to have the fire-engines cleaned and the leathers oiled and put into boxes, that the same may be fit for immediate use." January 2*d*, 1732. The mayor and four members of the court were authorized to employ persons to put the fire-engines in good order, and also to agree with proper persons to look after and take care of the same. It appears that Anthony Lamb was the first superintendent of fire-engines, for on the 24*th of January* 1735, the mayor was ordered " to issue his warrant to the treasurer to pay Mr. Anthony Lamb, overseer of the fire-engines, or order, the sum of three pounds, current money of this colony, in full of one quarter of a year's salary, due and ending the first instant." On the same date a committee was appointed to employ workmen " to put them in good repair, and that they have full power to agree with any person or persons by the year, to keep the same in such good plight, repair and condition, and to play the same as often as there shall be occasion upon any emergency."

April 15, 1736. " A convenient house [was ordered] to be made contiguous to the watch-house in the Broad street for securing and well keeping the fire-engines of the city." This seems to have been the first engine house. May 1, 1736. Jacobus Turk, a gunsmith, was appointed to take charge of the fire-engines and to keep them in repair at his own cost, for a salary of ten pounds current money. Mr. Turk undertook during the next year *to make an engine*, for May 15, 1737, the common council ordered the sum of ten pounds to be advanced " to the said Jacobus Turk, to enable him to go on with finishing a small fire-engine he is making for an experiment :" probably the first made in America.

November 4, 1737. The common council drew up a petition to the legislature to enable the corporation " to appoint four-and-twenty able bodied men, inhabitants within this city, who shall be called the *firemen* of this city, to work and play the fire-engines within the same, upon all occasions and emergencies, when they shall be thereunto required by the overseer of the said engines, or the magistrates of the said city : and that the said firemen as a recompense and reward for that service, may by the same law be excused and exempted from being elected and serving in the office of a constable, or being enlisted, or doing any duty in the militia regiment, troop, or companies, in the said city, or doing any duty in any of the said offices during their continuance as firemen aforesaid." This law was passed by the assembly in September following, and the duty of firemen defined. The next notice of engines occurs ten years afterwards, in March 1748, when the corporation " ordered that one of the fire-engines of this city, of the *second size*, be removed to Montgomery's Ward of this city, near Mr. Hardenbrooks; and that a shed be built thereabouts at the charge of this corporation for the securing and keeping the same." By this it appears that several engines besides the two original ones were then in use. The one just named was a different size (much smaller) than those first ordered. It is uncertain whether the additional ones were made by Mr. Turk, but probably not, since both large and small ones were ordered from London for several years after this date. From the following extract we find that several of the large fire-engines (the sixth size of Newsham) belonged to the city. *February* 28, 1749, " Ordered that Major Vanhousand and Mr. Provost do take care to get a sufficient house built for *one of the large fire-engines*, to be kept in some part of Hanover square at the expense of this corporation, and that there be a convenience made therein for hanging fifty buckets : and also ordered that there be one hundred new fire buckets made for the use of this corporation with all convenient speed."

May 8, 1752. " Ordered that Jacob Turk have liberty to purchase six small *speaking trumpets* for the use of this corporation," *i. e.* for the purpose of giving directions to firemen during conflagrations. *June* 20, 1758. " One large fire-engine, one small do. and two hand do." were ordered to be procured from London. *July* 24, 1761. Mr. Turk, after superintending the engines for twenty-five years, was superseded by Jacobus

Stoutenburgh, who was directed to take charge of them at a salary of thirty pounds; and "the late overseer, Mr. Jacobus Turk, [was ordered to] deliver up to the said Jacobus Stoutenburgh, the said several fire-engines." *November* 19, 1762. The firemen were directed to wear *leather caps* when on duty. *May* 7, 1772. An engine was ordered to be provided for the Out ward. *July* 10, 1772. "Alderman Gautier laid before this board an account of the cost of two fire-engines belonging to Thomas Tillier: and Alderman Gautier is requested to purchase the same." *September* 9, 1772. A committee was authorized "to purchase one other fire-engine of David Hunt." The three engines last named were probably from England, for at the time these machines were in the list of ordinary imported manufactures.

It is not impossible that some engines were made in Massachusetts about the time of the Revolutionary war. In October 1767, the people of Boston, irritated by the exactions and disgusted with the parasites of monarchy, determined in a town meeting to cease importing from the 31st of December following, numerous articles of British manufacture, among which were enumerated anchors, nails, pewter-ware, clothing, hats, carriages, cordage, furniture, and *fire-engines*. And in March 1768, the Assembly resolved, "that this house will, by all prudent means, endeavour to discountenance the use of foreign superfluities, and to encourage the manufactures of this province;" hence it is reasonable to suppose that engines either had been, or then could be made in the province; otherwise it is not likely that their importation would have been denounced. As an article of trade they were, from the limited number required, insignificant—they had no connection with luxury; and so far from being "superfluities," they were necessary to protect the property of the people from destruction—they would therefore be among the last things that a people would cease to import while unable to make them.

It was not till several years after the close of the struggle for independence that fire-engines were made in this and some other cities. They have, however, long been made here and in Philadelphia, Boston, &c. Small engines were formerly used, but they have gradually disappeared, the manufacturers confining themselves principally to the largest. The use of buckets has also been discontinued on account of the extensive application of hose. Village engines are sometimes constructed with single cylinders and double acting, but being more liable to derangement, they are not extensively used. Rotary engines are also made in some parts of New-England, on the principle of Bramah and Dickenson's pumps, (No. 138.) As ordinary fire-engines are merely forcing pumps, arranged in carriages and furnished with flexible pipes; it is not to be supposed that any radical improvement upon them can be effected. The pump itself is, perhaps, not capable of any material change for the better; and it is at present essentially the same as when used by Ctesibius and Heron in Egypt, twenty centuries ago: hence fire-engines, since hose pipes and air chambers were introduced, have differed from each other chiefly in the carriages and in the arrangement and dimensions of the pumps—as the position of the cylinders, modes of working the pistons, bore and direction of the passages for the water, &c. In these respects there is not much difference between European and American engines; nor in the varieties of the latter. Those made in Philadelphia rather resemble French and German engines, in working the pumps at the ends of the carriages, and without the sectors and chains; while New-York engines are precisely the same as Newsham's, both in the arrangement of the pumps and mode of working them, with the exception of treddles, which are not used.

No. 153 represents an external view of a Philadelphia engine : the pumps and air vessel are arranged as in No. 150, but the piston rods are connected directly to the bent lever, which is moved by a double set of handles or staves. A number of men stand upon each end of the cistern and work the engine by the staves nearest to them, while others on the ground apply their strength to the staves at the extremities of the lever. The staves turn upon studs at the centre of the cross bars, and when put in operation, fall into clasps that retain them in their places. Provision is made to convey the stream either from the lower or from the upper part of the air chamber. Hose companies supply the engines with water. The firemen, as in all American cities, are volunteers, and generally consist of young tradesmen and merchants' clerks, &c. They are exempt from militia and jury duty. Each member pays a certain sum on his admission, and a small annual subscription. A fine is also imposed upon any one appearing on duty without his appropriate dress (see figure in the cut) as well for being absent. A generous spirit of rivalry exists among the different companies, which induces them to keep their engines in a high state of working order.

No. 153.   External View of a Philadelphia Fire-engine.

No. 154 exhibits a New-York engine. The pump cylinders are arranged and worked precisely as shown in the section No. 150. They are six and a half inches diameter, and the pistons have a stroke of nine inches. Previous to the formation of hose companies, each engine was provided with a reel of hose; this, when not in use, was covered by a case of varnished cloth or leather. Most of the engines still have reels, which are carried as shown in the cut. The stream of water is invariably taken from the top of the air chambers, which resemble the one at No. 150. This practice is bad, because in most cases that part of the hose between the goose-neck and the fire descends to the ground, and hence the water in the pipe is unnecessarily diverted from its course and a corresponding diminution of effect is the result. In all cases the hose had better be connected to the bottom of the air chamber, or to its side near the bottom, as in Nos. 148, 152, 155. Very *long* chambers (as the one in No. 150) retard the issue of the liquid more than others which discharge it from the top, because the water has to descend in them nearly perpendicularly to enter the orifice of the pipe, and its direction is then precisely *reversed*, for it has to rise perpendicularly in order to escape.

No. 154. New-York City Fire-engine.

In exterior decoration American engines are probably unrivaled: the firemen take pride in ornamenting their respective machines, and hence most of them are finished in the most superb and expensive manner. The whole of the iron work, except the tire of the wheels, is frequently plated with silver; every part formed of brass is brought to the highest polish; and while all the wood work, including the wheels, is elegantly painted and gilded, the backs, fronts, and panels of the case that encloses the air chamber and pumps, are enriched with historical and emblematical paintings and carved work, by the first artists.

A new organization of the fire department of New-York has long been in contemplation, and the project of a law to that effect, is at this time under the consideration of the legislature of the state, and of the corporation of the city.

The most valuable contribution of American mechanicians to the means of extinguishing fires is the *riveted* hose, invented by Sellers and Pennock of Philadelphia. It is too well known both here and in Europe to require particular description. No modern apparatus is complete without it.

The Mechanics' Institute of New-York offered a gold medal (in January 1840) for the best plan of a *Steam* fire-engine. The publication of the notice was very limited, and but two or three plans were sent in. Of these, one by Mr. Ericsson, a European engineer now in this country, received the prize. No. 155 represents a view of the engine. No. 156 a longitudinal section of the boiler, steam engine, pump, air vessel and blowing apparatus. No. 157, plan. No. 158, a transverse section of the boiler through the furnace and steam chamber. No. 159, the lever or handle for working the blowing apparatus by hand. The following is the inventor's description, in which the same letters of reference denote the same parts in all the figures.

"A the double acting force pump, cast of gun metal, firmly secured to the carriage frame by four strong brackets cast on its sides. $a, a,$ Suction valves. $a', a',$ Suction passages leading to the cylinder. $a'',$ Chamber containing the suction valves, and to which chamber are connected suction pipes $a''', a''',$ to which the hose is attached by screws in the usual manner, and closed by the ordinary screw cap. The delivering valves and passages at the top of the cylinder are similar to those just mentioned.

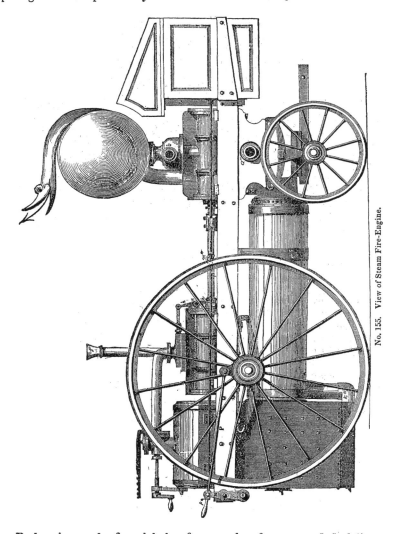

No. 155. View of Steam Fire-Engine.

B the air vessel, of a globular form, made of copper. $b\ b$ delivery pipes, to which the pressure hose is attached: when only one jet is required, the opposite pipe may be closed by a screw cap, as usual. The piston or bucket of the force pump to be provided with double leather packing: [cupped leathers] the piston rod to be made of copper.

Chap. 8.] *Steam Fire-Engine.* 347

"C the boiler, constructed on the principle of the ordinary locomotive boiler, and containing 27 tubes of $1\frac{1}{2}$ inch diameter. The top of the steam-chamber and the horizontal part of the boiler should be covered with wood, prevent the radiation of heat. c the fire door. c' the ash pan.

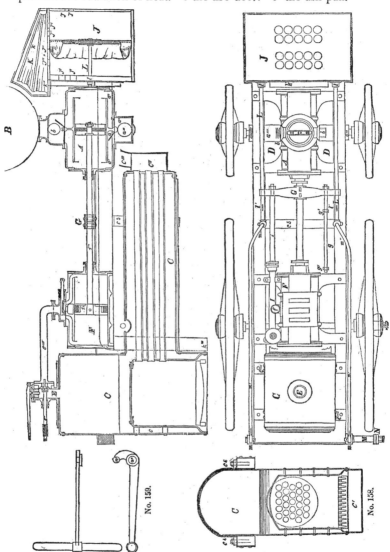

c'' a box attached to end of boiler, inclosing the exit of the tubes. The hot air from the tubes received by this box is passed off through smoke pipe c''', which is carried through D D, making a half spiral turn round the air vessel in the form of a serpent. c^4, iron brackets riveted to the boiler, and bolted to the carriage frame. c^5, a wrought iron stay, also bolted to the carriage frame, for supporting the horizontal part of the boiler.

"E, a cylindrical box attached to the top of the steam chamber, containing a conical steam valve e, and also safety valve e'. e'' screw with handle connected to the steam valve, for admitting or shutting off the steam. e''' induction pipe, for conveying the steam to

F, the steam cylinder, provided with steam passages and slide valve of the usual construction, and secured to the carriage frame in the same manner as the force pump. f Eduction pipe, for carrying off the steam into the atmosphere. f' Piston, provided with metallic packing, on Barton's plan. f'', Piston rod of steel, attached to the piston rod of the force pump by means of

G. a crosshead of wrought iron, into which both piston rods are inserted and secured by keys. g, Tappet rod attached to the crosshead, for moving the slide valve of the steam cylinder by means of nuts g', g', which may be placed at any position on the tappet rod.

H. Spindle of wrought iron, working in two bearings attached to the cover of the steam cylinder, the one end thereof having fixed to it, h a lever, moved or struck ultimately by the nuts g', g'. h' a lever, fixed to the middle part of the spindle H, for moving the steam valve rod.

I. Force pump for supplying the boiler, constructed with spindle valves on the ordinary plan; the suction pipe thereof to communicate with the valve chamber of the water cylinder, and the delivering pipe to be connected to the horizontal part of the boiler. i, Plunger of force pump, to be made of gun metal or copper, and attached to the crosshead G.

J. Blowing apparatus, consisting of a square wooden box, with paneled sides, in which is made to work a square piston j, made of wood, joined to the sides of said box by leather. j', Circular holes or openings through the sides, for admitting atmospheric air into the box; these holes being covered on the inside by pieces of leather or India rubber cloth to act as valves. j'', are similar holes through the top of the box, for passing off the air at each stroke of the piston, into

K. Receiver or regulator, which has a movable top k, made of wood, joined by leather to the upper part of the box; a thin sheet of lead to be attached thereto, for keeping up a certain compression of air in the regulator. k', Box or passage made of sheet iron, attached to the blowing apparatus, and having an open communication with the regulator at k'': to this passage is connected a conducting pipe, as marked by dotted lines in No. 156, for conveying the air from the receiver into the ash pan, under the furnace of the boiler at k'''; this conducting pipe passes along the inside of the carriage frame on either side.

L, L. Two parallel iron rods, to which the piston of the blowing apparatus is attached: these rods work through guide brasses l, l, and they may be attached to the crosshead G, by keys at l', l'. The holes at the ends of the crosshead for admitting these rods are sufficiently large to allow a free movement whenever it is desirable to work the blowing apparatus independently of the engine.

M. Spindle of wrought iron, placed transversely, and working in two bearings fixed under the carriage frame: to this spindle are fixed two crank levers m, m, which by means of two connecting rods m' m', give motion to the piston rods L, L, by inserting the hooks m'', m'', into the eyes at the ends of the said piston rods.

N. Crank lever, fixed at the end of spindle M, which by means of

O. Crank pin, fixed in the carriage wheel, and also

P. Connecting rod, will communicate motion to the blowing apparatus, whenever the carriage is in motion, and the above parts duly connected.

"A pin n is fixed in lever N, placed at such distance from the centre of spindle M, that it will fit the hole n' of the lever shown in No. 159, whilst n'' receives the end of the spindle M. Whenever the blowing apparatus is to be worked by the engine or by manual force, the connecting rod P should be detached by means of the lock at p. The carriage frame should be made of oak, and plated with iron all over the outside and top; the top plate to have small recesses, to meet the brackets of the cylinders, as shown in the drawing. The lock of the carriage, axles, and springs to be made as usual, only differing by having the large springs suspended *below* the axle. The carriage wheels to be constructed on the suspension principle; spokes and rim to be made of wrought iron, and very light.

The principal object of a steam fire-engine being that of not depending on the power or diligence of a large number of men, one or two horses should always be kept in an adjoining stable for its transportation. The fire grate and flues should be kept very clean, with dry shavings, wood and coke, carefully laid in the furnace ready for ignition; and a torch should always be at hand to ignite the fuel at a moment's notice. To this fire-engine establishment the word of fire should be given, without intermediate orders: the horses being put to, the rod attached connecting the carriage wheel to the bellows, and the fuel ignited, the engine may on all ordinary occasions be at its destination, and in full operation in ten minutes."

Attempts to supersede fire-engines were formerly common. Zachary Greyl is said to be the first who, in modern times, devised a substitute. This consisted of a close wooden vessel or barrel, containing a considerable quantity of water, and in the centre a small iron or tin case full of gunpowder: from this case a tube was continued through the side or head of the barrel, and was filled with a composition that readily ignited. When a room was on fire, one of these machines was thrown or conveyed into it, and the powder exploding dispersed the water in the outer case in every direction, and instantly extinguished the flames although raging with violence a moment before. In 1723, Godfrey, an English chemist, copied this device, and impregnated the water with an "antiphlogistic" substance. He named his machines "water bombs." In the year 1734, the States of Sweden offered a premium of twenty thousand crowns for the best invention of stopping the progress of fires; upon which M. Fuches, a German chemist, introduced an apparatus of the same kind. Similar devices have been brought forward in more recent days; but after making a noise for a time, they have passed into oblivion. (See London Magazine for 1760 and 1761.)

Among the devices of modern times for securing buildings from fire, may be mentioned the plan of Dr. Hales, of covering the floors with a layer of earth; and that adopted by Harley in 1775, of nailing over joists, floors, stairs, partitions, &c. sheet iron or tin plate. To increase the effect of fire-engines, the author of this work devised in 1817, and put in practice at Paterson, New Jersey, in 1820, the plan of fixing perforated copper pipes over or along the ceilings of each floor of a factory or other buildings, and connecting them with others on the outside, or at a short distance from the walls, so that the hose of a fire-engine could be readily united by screws; but the plan had been previously developed by Sir W. Congreve. It has recently been brought before the public as a new invention.

Of the numerous *Fire Escapes* that have been brought forward in modern times, the greater part are such as were employed by the ancients to scale walls and to enter fortresses, &c. in times of war. It is indeed obvious that the same devices by which persons entered buildings, would also answer the purpose of escaping from them : and as the utmost ingenuity of the ancients was exercised in devising means to accomplish the one, it was exceedingly natural that modern inventors should hit upon similar contrivances to effect the other. In the cuts to the old German translation of Vegetius, to which we have so often referred, there are ladders of rope and leather, in great variety, with hooks at the ends which when thrown by hand or an engine, were designed to catch hold of the corners and tops of the walls or windows—folding ladders of wood and metal, some consisting of numerous pieces screwed into each other by the person ascending, till he reached the required elevation; others with rollers at their upper ends to facilitate their elevation by rearing them against the front of the walls—baskets or chests containing several persons raised perpendicularly on a movable frame by means of a screw below, that pushed out several hollow frames or tubes contained within each other, like those of a telescope, whose united length reached to the top of the place attacked—sometimes the men were elevated in a basket suspended at the long end of a lever or swape. Several combinations of the lazy tongs, or jointed parallel bars are also figured—one of these moved on a carriage raised a large box containing soldiers, and is identical with a fire escape described in volume xxxi of the Transactions of the London Society of Arts.

Anciently the authors of accidental fires were punished in proportion to the degree of negligence that occasioned them; and they were compelled to repair to the extent of their means, the damage done to their neighbors. A law of this kind was instituted by Moses, probably in imitation of a similar one in force among the Egyptians. Other preventive measures consisted in the establishment of watchmen, whose duty it was to arrest thieves and incendiaries, and to give alarm in case of fire. From the earliest days, those who designedly fired buildings were put to death. A very ancient custom which related to the prevention of fires, is still partially kept up in Europe, although the design of its institution is almost forgotten, viz : the ringing of town bells about eight o'clock in the evening, as a signal for the inhabitants to put out their lights, rake together the fire on their hearths, and cover it with an instrument named a *curfew;* a corruption of *couvre feu,* and hence the evening peal became known as "the curfew bell." It has been supposed that the custom originated with William the Conqueror, but it prevailed over Europe long before his time, and was a very beneficial one, not only in constantly reminding the people to guard against fire, but indicating to them the usual time of retiring to rest; for neither clocks nor watches were then known, and in the absence of the sun they had no device for measuring time. Alfred the Great, who measured time by candles,[a] ordered the inhabitants of Oxford to *cover* their fires on the ringing of the bell at carfax every night. The instrument was made of iron or copper. Its general form may be understood by supposing a common cauldron turned upside down and divided perpendicularly through the centre; one half being furnished with a handle riveted to it would be a couvre feu. When used it was placed over the ashes with the open side close to the back of the hearth. (See Dict. Trevoux: Hone's Every Day Book, vol. i. 243, and Shakespeare's Romeo and Juliet, Act iv, scene 4.)

In the thirteenth year of Edward I. (A. D. 1285,) an act was passed

---

[a] In Shakespeare's play of Richard III. act v. scene 3, there is a reference under the name of a *watch* to these candles. They were marked in sections, each of which was a certain time in burning, and thus measured the hours during the night or cloudy weather.

against incendiaries, and night watchmen were ordered to be appointed in every town and city. In 1429 another act declared, "If any threaten by casting bills to burn a house, *if money be not laid in a certain place*, and after do burn the house, such burning shall be adjudged high treason." Beckman says that regulations respecting fires were instituted in Frankfort in 1460. In 1468 straw thatch was forbidden, and in 1474 shingle roofs were prohibited. The first general order respecting fires in Saxony are dated 1521, those for Dresden in 1529, and there is one respecting buildings in Augsburg, dated 1447. The following preamble to an act passed in the 37th year of Henry VIII. by which those found guilty of the crimes enumerated, were to suffer "the pains of death," is interesting in more respects than one. "Where divers and sundry malicious and envious persons, being men of evil perverse dispositions, and seduced by the instigation of the devil, and minding the hurt, undoing and impoverishment of divers of the kings true and faithful subjects, as enemies to the commonwealth of this realm, and as no true and obedient subjects unto the kings majesty, of their malicious and wicked minds, have of late invented and practised a new damnable kind of vice, displeasure and damnifying of the kings true subjects and the commonwealth of this realm, as in secret burning of frames of timber, prepared and made by the owners thereof, ready to be set up and edified for houses : cutting out of heads and dams of pools, motes, stews and several waters : cutting of conduit-heads, or conduit-pipes : burning of wains and carts loaden with coals or other goods : burning of heaps of wood, cut, felled and prepared for making coals : cutting out of beasts tongues : cutting off the ears of the kings subjects : barking of apple trees, pear trees, and other fruit trees; and divers other like kinds of miserable offences, to the great displeasure of Almighty God and of the kings majesty," &c. (Statutes at large.)

The crime of arson was rife in old Rome, and it is singular that the mode of punishing those found guilty of it, is among the numerous ancient customs that have been retained by Roman Catholics in their religious institutions. The *tunica molesta* of the Romans was a garment made of paper, flax, or tow, and smeared with pitch, bitumen or wax, in which incendiaries were burnt; and hence arose the peculiar dress worn by the victims in those horrible, those demoniacal "Acts of faith!" the *Auto da Fe*, of Italian, Spanish, and Portuguese inquisitions, (to which the scenes in Smithfield and other parts of England may be added,) acts, in which the order of justice was completely reversed—the sufferers being the innocents, and the court and judges the real criminals.

CHAPTER IX.

PRESSURE ENGINES: Of limited application—Are modifications of gaining and losing buckets and pumps—Two kinds of pressure engines—Piston pressure engine described by Fludd—Pressure engine from Belidor—Another by Westgarth—Motive pressure engines—These exhibit a novel mode of employing water as a motive agent—Variety of applications of a piston and cylinder—Causes of the ancients being ignorant of the steam engine—Secret of making improvements in the arts—Fulton, Eli Whitney and Arkwright—Pressure engines might have been anticipated, and valuable lessons in Science may be derived from a disordered pump—Archimedes—Heron's Fountain—Portable ones recommended in Flower Gardens and Drawing-rooms, in hot weather—Their invention gave rise to a new class of hydraulic engines—Pressure engine at Chemnitz—Another modification of Heron's fountain—Spiral pump of Wirtz.

PRESSURE ENGINES, named by the French *Machines à colonne d'eau*, form an interesting variety of hydraulic devices belonging to the present division of the subject. They consist of working cylinders with valves and pistons, and resemble forcing pumps in their construction, but differ from them in their operation; the pistons not being moved by any external force applied to them through cranks, levers, &c. but by the weight or pressure of a column of water acting directly upon or against them. Pressure engines are not very common, because they are only applicable to particular locations—such as afford a suitable supply of water for the motive column; but wherever refuse, impure, salt or other water can be obtained from a sufficient elevation, such water may be used to raise a quantity of fresh by these machines.

In some forms pressure engines appear rather complicated, but when analyzed, the principle of their action and mode of operation will be found extremely simple :—If two buckets, partly filled with water, be suspended and balanced at the ends of a scale beam, and a stream be directed into one of them, that one will preponderate, and consequently the other with its contents will be raised, and to a height equal to the descent of the former; but when it is required to raise water in this manner to an elevation that exceeds the distance through which the descending vessel can fall; then the capacity of the latter is enlarged, and it is suspended nearer to the fulcrum or centre on which the beam turns, as in the gaining and losing bucket, page 66 :—It is virtually the same principle that is employed in pressure engines; the difference is principally in the manner of performing the operation. Instead of vessels suspended as above, two solid pistons, moving in cylinders, are attached by rods or chains to the ends of a beam, or to the ends of a cord passed over a pulley, so that the pressure of a longer or heavier column of water resting upon one piston forces it down, and thereby raises the other and with it the lighter or shorter column reposing upon it.

By referring to the 16th illustration on page 64, it will be apparent that if a cylinder extended from B to the top of the cistern Z, and a hollow piston like the upper box of an atmospheric pump fitted to work in it were substituted for the bucket B, the effect produced would be much the same as with the two buckets, for the same quantity of water could be raised through the cylinder into the cistern Z, if allowance were made for an increase of friction in the passage of the piston.ᵃ And if another cylinder extended from F to the bottom of the pit at O, and a solid piston fitted to it were used instead of the bucket A, with a contrivance at the bottom to allow the water to escape ; the apparatus would then be a pressure engine, although the principle of the motive part of it would not have been essentially changed. The cylinders in this example would perform the part of the buckets A B ; they might be considered as permanent or fixed, and very long buckets with movable bottoms, *i. e.* the *pistons*, which by ascending and descending in them received and discharged their contents. And as with the buckets A and B, the quantity of water expended from the motive or descending column would be proportionate to the dimensions of the other and the elevation to which it was raised. Pressure engines may therefore be considered as a peculiar modification of the gaining and losing bucket machines, and as a combination of these with atmospheric and forcing pumps. They admit of various forms according to the location in which they are used and the objects to be accomplished by them. As liquids press equally in all directions, the cylinders may be placed in any position—horizontal, inclined, or vertical. Sometimes a pressure engine consists of a single cylinder with its appropriate pipes and valves like a double acting pump. The water to be raised enters at one side of the piston, and the motive column at the other ; but more commonly a distinct cylinder and piston receives the impulse of the motive column, and in order to transmit it to the other, the two pistons are some times connected to the same rod as in No. 161—at other times to opposite ends of a vibrating beam as in No. 162—so that while one cylinder and its apparatus act as a pump to raise water, the other is exclusively employed to work it. In this respect pressure engines may be considered rather as devices for communicating motion to machines proper for raising water, than as the latter, and they are sometimes used as propellers of other machines, but in whatever light they may be viewed, they are too interesting to be omitted.

There are two kinds of pressure engines, but they differ from each other only in that part which receives the impulse of the motive column and transmits it to the other. In one, a solid body (a piston) is used for that purpose—in the other, a volume of air; but while a slight variation is thus caused in the two machines, the essential features, as well as the moving principle of both remains the same. Piston pressure engines are said to have been invented in the 18th century by M. *Hoell*, a German engineer.ᵇ It is more probable that he improved them only, for they certainly were known much earlier : still it may be that he was ignorant of their previous use, and was led to their reïnvention by his efforts to raise water from the Hungarian mines, in which he erected several pressure machines on the principle of Heron's fountain : the transition from these to the other was easy and natural, and may have resulted from his endeavours to avoid a defect to which the former are subject, viz : the absorption of the air by the water. About A. D. 1739, an improved form of pressure engines was devised, and introduced into some mines in France by Belidor, which he has described in his Architecture Hydraulique. Some writers have considered him the author of piston pressure engines, but

---

ᵃ Pumps in certain locations are sometimes worked in that manner : the pistons or rods being loaded with weights sufficient to depress them, are raised by a bucket of water suspended at the opposite end of the beam, which when it reaches the bottom its contents are discharged, like the bucket A in No. 16, D in No. 17, or G in No. 162.
ᵇ " Machine à colonne d'eau. German : Wasser Saulene. Machine ; inventée par M. Hoell, premier Machiniste d'L'Impératrice." Arts et Metieres. L'Art d'exploiter Les Mines. Folio ed., Paris 1779, page 1449. Quarto ed. tom. xviii, p. 131.

the honor of first inventing them is not, in fact, due to either Belidor or Hoell, as the following figure, No. 160, from a work published a century before either of those engineers flourished, will show. It is from Fludd's *Naturæ simia seu technica macrocosmi*, 467. The character of this author as an astrologer and alchymist, and that of his works, which abound with absurdities, have probably caused the figure to be overlooked by modern writers on hydraulic and hydrostatic apparatus. Chemistry, however, is not the only science that is indebted to the shrewd but mistaken seekers of universal panaceas and of the philosopher's stone.

The lower end of the pipe B D having a valve opening upwards, is inserted into the water to be raised. The pipe A receives the descending or motive liquid column, which in this case was refuse or stagnant water, flowing from a source of sufficient altitude. This pipe may be at a considerable distance from B D, and is so represented in the original figure. It terminates below in a pit, drain, or low ground, whence the water discharged from it may escape. The end of it should be lower than that of D, and should be sealed or covered by water as represented, to prevent the entrance of air. A communication is made between both these pipes by the horizontal one C. This last is connected to A at one of the apertures of a three-way cock, the upper and lower part of A being united to the other two. The other end of C terminates at the bottom of a working cylinder, which is closed at the top, by a short tube communicating with B D, immediately below a valve placed in the latter. In the cylinder, a piston (indicated by the dotted lines) is fitted. It is described as a wooden plug covered with leather and loaded with lead, so as to make it descend in the cylinder by its weight.

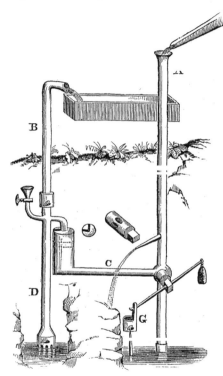

No. 160. Pressure Engine from Fludd. A. D. 1618.

To put this machine in operation, the cylinder and pipe D are first filled with water through the funnel and small cock, after which the latter is closed. The plug of the three-way cock is then so arranged, that its two orifices coincide only with the *upper part* of A, and with C, when the pressure of the column in A will force up the piston, and with it all the water previously in the cylinder, which is thus compelled to ascend through the upper valve into the discharging pipe B. When this has taken place, the vessel G, suspended at one end of the rod that passes through the shank of the plug, has become filled with water, from the small jet issuing from A, and descending, *turns the plug* of the cock, so as to close the communication of the upper part of A with C, and open it be-

Chap. 9.]   *Improved Pressure Engine from Belidor.*   355

tween C and the lower part of A, upon which the piston descends in the cylinder and the foul water in C escapes through the lower end of A and runs to waste. By the time the piston reaches the bottom of the cylinder, the latter is refilled with water by the pressure of the atmosphere, as in a common pump; and the contents of G have escaped through an orifice in its bottom, which is closed by a valve—this valve being opened by a projecting pin upon which the vessel descended, as shown in the figure. As soon as G is emptied, the weight on the opposite end of the rod preponderates, and turns the plug of the cock into its former position; and thus the play of the machine is continued without intermission. The operation of filling the cylinder through the funnel, is required only at the first, like the priming of a new pump.

The origin of this machine is uncertain. It does not appear to have been invented by Fludd himself, but is inserted among others, which he copied from older authors; and such as he examined abroad. As he traveled in Germany and has described some of the hydraulic machines used in the mines there, (see one figured on page 219,) it is probable that he derived a knowledge of it in that country. It possesses considerable interest—it is *self acting*, and that by a very simple device—it shows an old application of the three-way cock—it exhibits the application of refuse or putrid water, to raise fresh, and in a way somewhat similar to one recently proposed—and it is the oldest piston pressure engine known.

The next figure from Belidor, shows a great improvement on the last, so much so, that in some respects it may be considered a new machine.

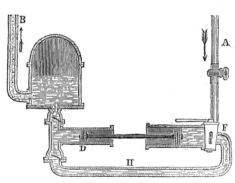

No. 161. Pressure Engine from Belidor. A. D. 1739.

A, conveys the descending column from its source to the three-way cock F; to one of the openings of which it is united. This cock is connected, at another opening, to the horizontal cylinder C, whose axis coincides with that of a smaller one D. Both cylinders are of the same length; and their pistons are attached to a common rod, as represented. Two valves are placed in the ascending pipe B—one below, the other above its junction with the cylinder D. The horizontal pipe H connects B and D with the third opening of the cock. By turning the plug of this cock, a communication is opened alternately between each cylinder and the water in A. Thus when the water rushes into C, it drives the piston before it to the extremity of the cylinder, and consequently the water that was previously in D is forced up the ascending pipe B; then the communication between A and C is cut off, (by turning the cock) and that between A and D is opened, when the pistons are moved back towards F by the pressure of the column against the smaller piston—the water previously in C escaping through an opening shown in front of the cock and runs to waste, while that which enters D is necessarily forced up B at the next stroke of the pistons. The cock was opened and closed by levers, connected to the middle of the piston rod, and was thus worked by the machine itself. By the air chamber, the discharge from B is rendered continuous.

Suppose the water in A has a perpendicular fall of thirty-four or thirty-five feet, and it were required to raise a portion of it to an elevation of seventy feet above F; it will be apparent that if both pistons were of the same diameter, such an object could not be accomplished by this machine—for both cylinders would virtually be but one—and so would the pistons; and the pressure of the column on both sides of the latter would be equal. A column of water thirty-five feet high presses on the base that sustains it with a force of 15lbs. on every superficial inch; and one of seventy feet high, with a force of 30lbs. on every inch—hence without regarding the friction to be overcome, which arises from the rubbing of the pistons; from the passage of the water through the pipes: and from the necessary apparatus to render the machine self-acting—it is obvious in the case supposed that the area of the piston in C must be more than double that in D, or no water could be discharged through B. Thus in all cases, the relative proportion between the area of the pistons, or diameter of the cylinders, must be determined by the difference between the perpendicular height of the two columns. When the descending one passes through a perpendicular space, greatly exceeding that of the ascending one, then the cylinder of the latter may be larger than that of the former: a smaller quantity of water in this case raising a larger one: it, however, descends like a small weight at the long end of a lever, through a greater space.

In 1769 the London Society of Arts, awarded to Mr. Westgarth a premium of fifty guineas, for his invention of a pressure engine. It is described by the celebrated Smeaton, in vol. v, of their Transactions, as "one of the greatest strokes of art in the hydraulic way, that has appeared since the invention of the fire [steam] engine." Several were erected by Mr. W. in 1765, to raise water from lead mines in the north of England. They were simple in their construction, and somewhat resembled the engines of Newcomen. They differed from those of Belidor in the position of the cylinder; the introduction of a beam; the substitution of cylindrical valves in the place of cocks; and using the motive column to move the piston in one direction only. The cylinder of Westgarth's engine was placed in a *vertical* position, the piston rod of which was suspended by a chain from the arched end of a "walking" or vibrating beam; while the other end of the latter, projected over the mouth of the mine or pit, and was connected (by a chain) to the rod of an *atmospheric pump* placed in the pit. This rod was loaded as in Newcomen's engine, so as to descend by its weight and thereby raise the piston of the pressure engine when the column of water was not acting on the latter. Thus, when the motive column of water was admitted into the cylinder, the piston was depressed, and the end of the beam also, to which it was connected; consequently the pump rod and its sucker were raised, and with them water from the mine. Then as soon as the piston reached the bottom of the cylinder, the motive column was cut off, by closing a valve; and a passage made for the escape of that within the cylinder, by opening another—upon which the loaded pump rod again preponderated—the valve to admit the column on the piston of the pressure engine was again opened, and the operation repeated as before.

In another form these machines have been adopted, in favorable locations, as first movers of machinery, and when thus used, they exhibit a very striking resemblance to high pressure steam-engines. Indeed, the elemental features of steam and pressure engines are the same, and the modes of employing the motive agents in both are identical—it is the different properties of the agents that induces a slight variation in the machines—one being an elastic fluid, the other a non-elastic liquid. In steam-

Chap. 9] *Motive Pressure Engine.* 357

engines a piston is alternately pushed up and down in its cylinder by steam; and by means of the rod to which the piston is secured, motion is communicated to a crank and fly-wheel, and through these to the machinery to be propelled: it is the same with pressure engines when used to move other machines, except that instead of the elastic vapor of water, a column of that liquid drives the pistons to and fro, as will be perceived by an examination of the following figure.

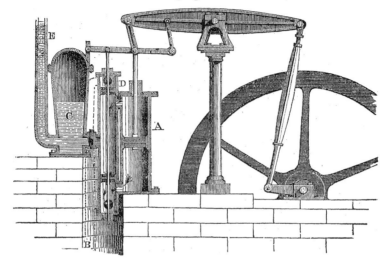

No. 162. Motive Pressure Engine.

E represents the lower part of the pipe which conveys the water down from its source into the air chamber C, from the lower part of which it passes through a short tube and stop cock into the valve case, or "side pipe" D. This pipe is parallel with the working cylinder of the engine A, and rather longer: it communicates with A through two passages for the admission of the water to act upon both sides of the piston. The ends of D are closed by stuffing boxes, through which a rod in the direction of its axis is made to slide, and upon this rod are secured two plugs, shown in the cut, that fill the interior of the pipe like pistons, and as the rod is raised or lowered, alternately open and close the passages into the cylinder. Suppose the position of the various parts of the engine as indicated in the figure, and the stop-cock in the short tube that connects the "side pipe" to the air vessel be opened, the water would then rush into the upper part of the cylinder A, as shown by the arrows, and by its statical pressure force down the piston; while any water previously below the piston would escape through the lower passage into the side pipe (beneath the plug) and run off to waste through the tube B, marked by dotted lines, and the circular orifices of which are also figured. When the piston has reached the bottom of the cylinder, the rod to which the plugs are attached is drawn down, so as to close the upper passage and open the lower one, upon which the water enters through the latter and drives up the piston as before, the previous contents of the cylinder being forced through the circular orifice in the upper part of the side pipe into B. In this manner the operation is continued and motion imparted to the beam, crank and fly-wheel. The apparatus for moving the rod that opens and closes the passages into the cylinder is analogous to that of steam engines, being effected by an eccentric on the crank shaft. It is omitted in the cut, that the essential features of the machine might appear more conspicuous.

It is obvious that engines of this kind may be employed to impart motion to pumps or any other machinery. The intensity of the force transmitted by them depends upon the perpendicular height of the motive column and the area of the piston. The use of the air vessel is, as in the hydraulic ram and other machines, to break the force of the blow or concussion consequent on the sudden stoppage of the descending column by closing the passages. Wherever the waste pipe B can descend thirty-five or thirty-six feet, the engine may derive an additional power from the *vacuum* thus kept up behind the piston, as in low-pressure steam-engines. The application of this feature to pressure engines was included in an English patent granted to John Luddock in February 1799. (Repertory of Arts, vol. xi, page 73.)

The invention of pressure engines brought to light a *new* mode of employing water as a motive agent; and also the means of applying it in locations where it could not otherwise be used. When water moves an under or overshot wheel, the machinery to be propelled must be placed in the immediate vicinity—hence saw, grist, and fulling mills, &c. are erected where the falling liquid flows; and when steam is the moving force, the engines are located where the fluid is generated; but with pressure engines it is different, for the motive agent may be taken to the machine itself. In valleys or low lands, having no natural fall of water, but where that liquid can be conveyed in tubes from a sufficient elevation, (no matter how distant the source may be,) such water, by these machines, may be made to propel others. And by means of them the small lakes often found on mountains, and water drawn from the heads of falls and rapids, may furnish power for numerous operations in neighboring plains. When cities are supplied from elevated sources, an additional revenue might be derived from the force with which the liquid issues from the tubes: the occupant of a house into which a lateral pipe from the mains is conveyed, might connect the pipe to a pressure engine, and thereby impart motion to lathes, or printing presses; raise and lower goods on different stories; press cotton, paper, books, &c. as by a steam-engine. But unlike the machine just named, a pressure engine is inexpensive, and simple in its construction—it requires neither chimneys, furnaces, nor loads of fuel; neither firemen nor engineers, nor is there any danger of explosions. It may be placed in the corner of a room, or be concealed under a counter or a table. It may be set in operation in a moment, by opening a cock, and the instant the work is done, it may be stopped by shutting the same, and thus prevent the least waste of power—and when the work is accomplished, the water can be used for all ordinary purposes as if just drawn from the mains, for the engine might be considered as merely a continuation of the lateral tube.

Pressure engines afford another illustration of the variety of purposes to which a *piston and cylinder* may be applied. These were probably first used in piston bellows; next in the syringe; subsequently in pumps of every variety; and then in pressure and steam-engines. The moving piston is the nucleus or elemental part that gives efficiency to them all; and the apparatus that surround it in some of them, are but its appendages. To what extent it is destined to be employed when steam becomes superseded by other fluids, time only can reveal; but if we may judge of the future by the past, this simple device will perform greater wonders in the world than it has yet accomplished. It is by it only that the energy of elastic fluids can be economically employed.

Those ingenious men who first constructed a bellows, a syringe, or a pump, little thought that similar implements should become self-acting, and even be motive engines to drive others. What weary laborer at the pump in ancient Greece or Rome, ever dreamt, while indulging in those reveries that the mind conjures up to divert attention from toil or pain, that a machine similar to the one upon which his strength was expended, should be devised to work without human aid:—and that a modification of it, excited by the vapor of a boiling cauldron, should exert a force compared with which the power of the Titans was impotence—a force that should drive fleets of gallies through a storm—hurl missiles like the balistæ—propel chariots "without horses"—polish a mirror—forge a hatchet, a tripod or a vase—and spin thread and weave it into veils, fine as those worn by the vestal virgins—and yet should *never tire!* Could the imaginations of the depressed plebeians and slaves of antiquity have had a glimpse of such a machine, and had they been informed that it would in some future time, which the oracles had not revealed, be generally employed—how vehemently would they have importuned the gods to send it in their days! And why did they not have it? Because the useful arts were neglected and their professors despised—while those professions the most destructive of human felicity were cultivated. War was accounted honorable, and hence nations were incessantly engaged in conflicts with each other—a military spirit pervaded the minds of the people, and it rewarded them by soaking every land with their blood.

The history of machines composed of pistons and cylinders also illustrates the process by which some simple inventions have become applied to purposes, foreign to those for which they were originally designed— each application opening the way for a different one. In this manner devices apparently insignificant have eventually become of the utmost value, and it is probable that there is no mechanical combination or device, however useless it may now appear, but which will be thus brought into play. These machines also teach us how new discoveries are made in the arts, viz: by *observing common results*, and applying the principles or processes by which they are induced, to other objects or designs. Every mechanical movement and manufacture—an unsuccessful experiment—defects or derangements of ordinary machines, &c. are all practical demonstrations that indicate the means to produce analogous effects, or to avoid them. Fulton employed steam-engines to turn paddle wheels—Eli Whitney adopted circular saws as cotton gins; and both became benefactors of their country—a poor barber in England, after exercising his ingenuity on the perpetual motion, applied some of his devices to cotton spinning, and not only became one of the most opulent of manufacturers, but secured a place in the biography of eminently useful men.

Nearly all modern improvements and inventions have been brought about in a somewhat similar manner, and there are few but what might have been anticipated by attention to every-day facts. Suppose pressure engines had not yet been known: they might be developed by reflecting on a very common circumstance connected with ordinary pumps. When one of these no longer retains water in the cylinder and trunk, it is necessary to *prime* it, by pouring in a quantity sufficient to fill the space in which the sucker moves: this water resting upon the latter presses it down, and consequently raises the lever or pump handle, which again descends as soon as the water escapes below; thus illustrating the principle by which pressure engines act—the lever being moved by the water instead of the water by it. How many ages have elapsed, and how many millions of people have witnessed this operation, without a useful idea having been derived from it? And without any one thinking that valuable lessons in science might be learnt from a disordered pump, or from the irregular movements of a pump handle? Those observing minds, however, that are constantly on the alert for facts—like bees incessantly on the wing for honey—would not now suffer even such an occurrence to pass unnoticed; nor would they hesitate to consider those unpleasant knocks which hundreds of people (and the writer among them) have occasionally experienced from the unexpected descent of a heavy pump handle on their persons, and in some instances more unpleasant ones from its sudden ascent—as admonitions to turn the experiment to advantage. The simple rise of water which his body displaced in a bathing tub, was seized in a twinkling by the mathematician of Syracuse to solve a new and difficult problem; yet the same thing had been previously witnessed for thousands of years, but no one ever thought of applying the result to any such purpose.

It perhaps may be a question whether the machines already described in this chapter were known to the engineers of antiquity, but there is no room to doubt their acquaintance with another variety of pressure engines, since we have obtained a knowledge of them from the Spiritalia of Heron, whose name they still bear. It is obvious that a liquid may be forced out of a vessel by pressing into the latter any other substance, no matter what the nature of it may be, whether solid or fluid, liquid or aëriform: thus, the solid plunger or piston of a pump does not more effectually expel the contents of the cylinder in which it moves, than the elastic fluid in a soda fountain drives out the aerated water; hence, if air be urged by the pressure of a liquid column, or by any other force, to occupy the interior of a vessel containing water, the liquid may be raised through a tube to an elevation equal to the force that moves it; the air in this case performing the part of pistons in the pressure engines already described; and its effects are greater than can be produced by solid pistons, for the friction of these consumes a considerable portion of the motive force, so that a column of water raised by them can never equal the one that raises it; whereas air, from its extreme mobility, receives and transmits the momentum of the motive column undiminished to the other.

The fountain of Heron is the oldest pressure engine known, and in it a volume of air is used as a substitute for a piston. It is not certain that it was invented by him, for it may have been an old device in his time, and one which he thought worthy of preservation, or of being made more extensively known, and therefore inserted an account of it in his book. See No. 163. The two vessels A B, of any shape, are made air tight. The top of the upper one is formed into a dish or basin; in the centre of which the jet pipe is inserted, its lower end extending to near the bottom of A: a pipe C, whose upper orifice is soldered to the basin extends down to near the bottom of the lower vessel, either passing through the top of B, as in the figure, or inserted at the side. Another pipe D is connected to the top of B, and continued to the upper part of A. This pipe conducts the *air* from B to A. Now suppose the vessel A filled with water, through an aperture made for the purpose, and which is then closed; the object is to make this water ascend through the jet, and it is accomplished thus:—water is poured into the basin, and of course it runs down the pipe C into B; and as it rises in the latter, the air within is necessarily compressed, and having no way to escape but up the pipe D, it ascends into the upper part of A, where being pressed on the surface of the water, the latter is compelled to ascend through the jet pipe, as shown in the cut. The water thus forced out, falls back into the basin, and running down C into B continues the play of the machine, until all the water in A is expended. The elevation to which water in A can be thus raised through a tube, will be equal to the perpendicular distance between the two orifices of C. To persons who are ignorant of the construction of these fountains, the water in the basin appears to descend, and to *rise again* through the jet. Such is not the fact; were it so, this machine would be a perpetual motion, or something very like one. Some persons beguiled by the apparent possibility of inducing it to ascend, have attempted the solution of that problem by a similar apparatus. We may as well confess that in our youth we were of the number. The younger Pliny seems to have fallen into the same mistake respecting a fountain belonging to his country seat.

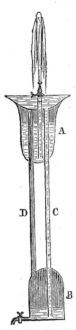

No. 163. Heron's Fountain.

Portable fountains of this kind might be adopted as appropriate appendages to flower gardens, and even drawing rooms. The pipes might be concealed within, or modeled into a handsome column, whose pedestal formed the lower vessel, while the upper end assumed the figure of a vase. Such an addition to the furniture of an apartment would be a useful acquisition at those seasons when the atmosphere, glowing like the air of an oven, scorches our bodies during the day, and in the evening we gasp in vain for the cooling breeze: at such times a minute stream of water spouting and sparkling in a room would soon allay the heat and invigorate our drooping spirits—imparting the refreshing coolness of autumn amid the burning heats of summer; and if the liquid were perfumed with attar of roses, or oil of lavender, we might realize the most innocent and delicious of oriental luxuries. The play of such a fountain might be continued for two or three hours at a time, for the size of the stream need hardly exceed that of a thread, and by a slight modification, the jet could be renewed as often as the upper vessel was emptied, by simply inverting the machine: or, the whole might be arranged without, except the ajutage and the vase in which the jet played. (See remarks on fountains in the fifth book.)

This fountain has been named *a toy*, but it is by such toys that important discoveries have been made in every age. It is clearly no rude or imperfect device: not a first thought; on the contrary, it bears the evidence of a matured machine, and of being the result of a familiar acquaintance with the principles upon which its action depends. Unlike older hydraulic machines, it requires no distinct vessel within which to raise a a liquid; nor does it resemble pumps, since neither cylinders, suckers, valves or levers are required, nor any external force to keep it in motion.

Its invention may be considered as having opened a new era in the history of machines for raising water, for it is susceptible of an almost endless variety of modifications, and of being applied to a great number of purposes. To understand this it is only necessary to bear in mind that the relative position of the two columns is immaterial: they may be a mile distant from each other, or they may be nearly together. The one that raises the other may be above, below, or on a level with the latter; both may be conveyed in pipes along or under the surface of the ground, and in any direction: the only condition required is, that the perpendicular distance between the upper and lower orifices of the pipe in which the motive column flows, shall be equal to the force required to raise the other to the proposed elevation.

A pressure engine on the principle of Heron's fountain, erected by M. Hoell in 1755, to raise water from one of the mines in Hungary, has long been celebrated. In the vicinity of one of the shafts at Chemnitz, there is a hill upon which is a spring of water, one hundred and forty feet above the mouth of the shaft. This spring furnishes more water than that which rises at the bottom of the mine, which is one hundred and four feet below the mouth of the shaft. The water in the mine is raised by means of that on the hill by an apparatus similar to the one figured in the annexed cut.

A represents a strong copper vessel eight feet and a half high, five feet diameter, and two inches thick. A large cock marked 3 is inserted near the bottom, and a smaller one 2 near the top. From this vessel a pipe D, two inches in diameter, reaches down and is connected to the top of the vessel B at the bottom of the shaft. This vessel is smaller than the upper one, being six feet and a half high, four feet diameter, and two inches thick, and of the same material as the other. A pipe E, four inches diameter, rises from near the bottom of B to the surface of the ground, where it discharges the water. The pipe C conveys the water from the spring on the hill; it is also four inches diameter, and descends to near the bottom of A. It is furnished with a cock 1. Water is admitted into B through a cock 4, or a valve opening inwards, which closes when B is filled. The vessel A is supposed to be empty, or rather filled with air, and its two cocks shut. The cock 1 is then opened, when the water rushing into A condenses the air within it and the pipe D, and this air pressing on the water in B, forces it up the pipe E. As soon as it ceases to flow through E, the cock 1 is shut and 2 and 3 are opened, when the water in A is discharged at 3. The cock or valve at the bottom of B is opened, and the water entering drives the air up D into A where it escapes at 2. The operation is then repeated as before.

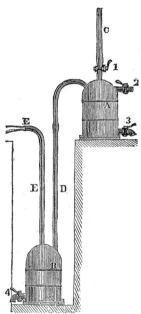

No. 164. Pressure Engine at Chemnitz.

If, when water ceases to run at E, the cock 2 be opened, both water and air rush out of it together, and with such violence that the liquid is, by the generation of cold consequent on the sudden expansion of the condensed air, converted into hail or pellets of ice. This fact is generally shown to strangers, who are usually invited to hold their *hats* in front of the cock so as to receive the blast; when the hail issues with such violence, as frequently to pierce the hats, like pistol bullets. This mode of producing ice was known to the marquis of Worcester, who refers to it in the eighteenth proposition of his Century of Inventions, relating to an " artificial fountain, holding great quantity of water, *and of force sufficient* to make snow, ice, and thunder." Some additions to the machine at Chemnitz, by which it might be rendered self-acting, were proposed in 1796. They consisted of small vessels suspended from levers that were secured to the shanks of the cocks, which they opened and shut in the same manner as shown in No. 160. A similar contrivance may be seen in several old authors—it is in the Spiritalia: Decaus, Fludd, Moxon and Switzer have all given figures of it. The quantity of water raised from the shaft compared with that expended from the spring was as 42 to 100.

By arranging a series of vessels above each other and connecting them by pipes as in No. 163, water may be raised to almost any height, in locations that have the advantage of a small fall. The distance between the vessels not exceeding the perpendicular descent of the motive column, which last is made to transmit its force to each vessel in succession—forcing the contents of one into the next above, and so on. Such a machine is interesting as showing the extent to which the principle of Heron's fountain may be applied, but for practical purposes it is of little value. It is too complex (if made self-acting) and too expensive for common use; and it is far inferior to the water ram. It was described by Dr. Darwin, in his *Phytologia*, to which modern writers generally refer, but it is an old affair. It is figured by Moxon in his "Mechanick Powers," Lon. 1696, and is mentioned by older authors. It is substantially the same as the double fountain of Heron, as found in the Spiritalia and the works of most writers on hydraulics.

By far the most novel and interesting modification of Heron's fountain was devised in the year 1746 by *H. A. Wirtz*, a Swiss pewterer or tin-plate worker of Zurich. It is sometimes named a spiral pump, and was made to raise water for a dye house in the vicinity of that city. What the circumstances were that led Wirtz to its invention we are not informed—whether it was suggested by some incident, or was the result of reasoning alone. It is represented in the illustrations Nos. 165 and 166, the first being a section and the latter an external view.

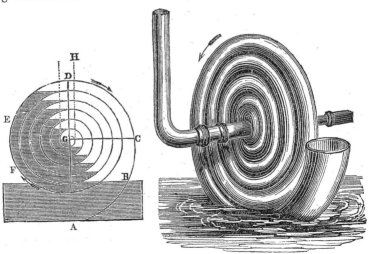

No. 165. Section of Wirtz's Pump.   No. 166. View of Wirtz's Pump.

Wirtz's machine consists either of a helical or a spiral pipe. As the former it is coiled round in one plane as A B C D E F in No. 165. As a spiral it is arranged round the circumference of a cone or cylinder, and then resembles the worm of a still. The interior end at G is united by a water tight joint to the ascending pipe H. See No. 166. The open end of the coil is enlarged so as to form a scoop. When the machine, immersed in water as represented, is turned in the direction of the arrow, the water in the scoop, as the latter emerges, passes along the pipe driving the air before it into G H, where it escapes. At the next revolution both air and water enter the scoop; the water is driven along the tube as before, but is separated from the first portion by a column of air of nearly equal length. By continuing the motion of the machine another portion of water and another of air will be introduced. The body of water in each coil will have both its ends horizontal, and the included air will be of about its natural density; but as the diameters of the coils diminish towards the centre, the column of water which occupied a semicircle in the outer coil, will occupy more and more of the inner ones as they approach the centre G, till there will be a certain coil, of which it will occupy a complete turn. Hence it will occupy more than the entire space within this coil, and consequently the water will run back over the top of the succeeding coil, into the right hand side of the next one and push the water within it backwards and raise the other end. As soon as the water rises in the pipe G H, the escape of air is prevented when the scoop takes in its next quantity of water. Here, then, are two columns of water acting against each other by hydrostatic pressure, and the intervening column of air. They must compress the air between them, and the water and air columns will now be unequal. This will have a general tendency to keep the whole water back and cause it to be higher on the left or rising side of each coil, than on the other. The excess of height will be just such as produces the compression of the air between that and the preceding column of water. This will go on increasing as the water mounts in H. Now at whatever height the water in H may be, it is evident that the air in the small column next to it will always be compressed with the weight of the water in H—an equal force must therefore be exerted by the water in the coils to support the column in H. This force is the sum of all the differences between the elevation of the inner ends of the water in each coil above the outer ends; and the height to which the water will rise in H will be just equal to this sum. Dr. Gregory observes that the principles on which the theory of this machine depends are confessedly intricate; but when judiciously constructed, it is very powerful and effective in its operation. It has not been ascertained whether the helical or spiral form is best. Some of these machines were erected in Florence in 1778. In 1784, one was made at Archangelsky, that raised a hogshead of water in a minute to an elevation of seventy-four feet, and through a pipe seven hundred and sixty feet long. See Gregory's Mechanics, vol. ii.

It perhaps may facilitate an understanding of this curious machine, by remarking that the pressure exerted by the column of water in one side of each coil is proportioned to its length, and that this pressure is transmitted, through the column of air between them, to that of the next: the combined force of both is then made to act, by the revolution of the tubes, upon the third column, and so on, till the accumulated force of them all is communicated to the water in H; and hence the elevation to which water can be thus raised, can never exceed the sum of the altitudes of the liquid columns in the coils.

END OF THE THIRD BOOK.

BOOK IV.

MACHINES FOR RAISING WATER, (CHIEFLY OF MODERN ORIGIN) INCLUDING EARLY APPLICATIONS OF STEAM FOR THAT PURPOSE.

CHAPTER I.

DEVICES of the lower animals—Some animals aware that force is increased by the space through which a body moves—Birds drop shell fish from great elevations to break the shells—Death of Æschylus—Combats between the males of sheep and goats—Military ram of the ancients—Water rams—Waves—Momentum acquired by running water—Examples—Whitehurst's machine—Hydraulic ram of Montgolfier—" Canne hydraulique" and its modifications.

Of the machines appropriated to the fourth division of this work, (see page 8,) centrifugal pumps and a few others have already been described. There remain to be noticed, the water ram, canne hydraulique, and devices for raising water by means of steam and other elastic fluids.

If the various operations of the lower animals were investigated, a thousand devices that are practised by man would be met with, and probably a thousand more of which we yet know nothing. Even the means by which they defend themselves and secure their food or their prey, are calculated to impart useful information. Some live by stratagem, laying concealed till their unsuspecting victims approach within reach—others dig pitfalls to entrap them; and others again fabricate nets to entangle them, and coat the threads with a glutinous substance resembling the birdlime of the fowler. Some species distill poison and slay their victims by infusing it into their blood; while others, relying on their muscular energy, suffocate their prey in their embraces and crush both body and bones into a pulpy mass. The tortoise draws himself into his shell as into a fortress and bids defiance to his foes; and the porcupine erects around his body an array of bayonets from which his enemies retire with dread. The strength of the ox, the buffalo and rhinoceros is in their necks, and which they apply with resistless force to gore and toss their enemies. The elephant by his weight treads his foes to death; and the horse by a kick inflicts a wound that is often as fatal as the bullet of a rifle; the space through which his foot passes, adding force to the blow.

There are numerous proofs of some of the lower animals being aware that the momentum of a moving body is increased by the space through which it falls. Of several species of birds which feed on shell fish, some, when unable to crush the shells with their bills, carry them up in the air, and let them drop that they may be broken by the fall. (The Athenian poet Æschylus, it is said, was killed by a tortoise that an eagle dropped upon his bald head, which the bird, it is supposed, mistook for a stone.)

When the males of sheep or goats prepare to *butt*, they always recede backwards to some distance; and then rushing impetuously forward, (accumulating force as they go,) bring their foreheads in contact with a shock that sometimes proves fatal to both. The ancients, perhaps, from witnessing the battles of these animals, constructed military engines to act on the same principle. A ponderous beam was suspended at the middle by chains, and one end impelled, by the united efforts of a number of men at the opposite end, against walls which it demolished with slow but sure effect. The battering end was generally, and with the Greeks and Romans uniformly, protected by an iron or bronze cap in the form of a ram's head; and the entire instrument was named after that animal. It was the most destructive of all their war machinery—no building, however solid, could long withstand its attacks. Plutarch, in his life of Anthony, mentions one *eighty feet* in length.

The action of the ram is familiar to most people, but it may not be known to all that similar results might be produced by a liquid as by a solid—that a long column of water moving with great velocity might be made equally destructive as a beam of wood or iron—yet so it is. Waves of the sea act as water-rams against rocks or other barriers that impede their progress, and when their force is increased by storms of wind, the most solid structures give way before them. The old lighthouse on the Eddystone rocks was thus battered down during a storm in 1703, when the engineer, Mr. Winstanley, and all his people, perished.

The increased force that water acquires when its motion is accelerated, might be shown by a thousand examples: a bank or trough that easily retains it when at rest, or when slightly moved, is often insufficient when its velocity is greatly increased. When the deep lock of a canal is opened to transfer a boat or a ship to a lower level, the water is permitted to descend by slow degrees: were the gates opened at once, the rushing mass would sweep the gates below before it, or the greater portion would be carried in the surge quite over them—and perhaps the vessel also. A sluggish stream drops almost perpendicularly over a precipice, but the momentum of a rapid one shoots it over, and leaves, as at Niagara, a wide space between. It is the same with a stream issuing from a horizontal tube—if the liquid pass slowly through, it falls inertly at the orifice, but if its velocity be considerable, the jet is carried to a distance ere it touches the ground. The level of a great part of Holland is below the surface of the sea, and the dykes are in some parts thirty feet high: whenever a leak occurs, the greatest efforts are made to repair it immediately, and for the obvious reason that the aperture keeps enlarging and the liquid mass behind is put in motion towards it; thus the pressure is increased and, if the leak be not stopped, keeps increasing till it bears with irresistable force all obstructions away. A fatal example is recorded in the ancient history of Holland:—an ignorant burgher, near Dort, to be revenged on a neighbour, dug a hole through the dyke opposite the house of the latter, intending to close it after his neighbor's property had been destroyed; but the water rushed through with an accelerating force, till all resistance was vain, and the whole country became deluged. The ancients were well aware of this accumulation of force in running waters. Allusions to it are very common among the oldest writers, and various maxims of life were drawn from it. The beginning of strife, says Solomon, "is as when one letteth out water"—the "breach of waters"—"breaking forth of waters"—"rushing of mighty waters," &c. are frequently mentioned, to indicate the irresistable influence of desolating evils when once admitted.

That the force which a running stream thus acquires may be made to drive a portion of the liquid far above the source whence it flows, is obvious from several operations in nature. During a storm of wind, long swelling waves in the open sea alternately rise and fall, without the crests or tops of any being elevated much above those of the rest; but when they meet from opposite directions, or when their progress is suddenly arrested by the bow of a ship, by rocks, or other obstacles, part of the water is driven to great elevations. There is a fine example of this at the Eddystone rocks—the heavy swells from the Bay of Biscay and from the Atlantic, roll in and break with inconceivable fury upon them, so that volumes of water are thrown up with terrific violence, and the celebrated light-house sometimes appears from this cause like the pipe of a fountain enclosed in a stupendous *jet d'eau*. The light room in the old light-house was sixty feet above the sea, and it was often buried in the waves, so immense were the volumes of water thrown over it.

The hydraulic ram raises water on precisely the same principle: a quantity of the liquid is set in motion through an inclined tube, and its escape from the lower orifice is made suddenly to cease, when the momentum of the moving mass drives up, like the waves, a portion of its own volume to an elevation much higher than that from which it descended. This may be illustrated by an experiment familiar to most people. Suppose the lower orifice of a tube (whose upper one is connected to a reservoir of water) be closed with the finger, and a very minute stream be allowed to escape from it in an upward direction—the tiny jet would rise nearly to the surface of the reservoir; it could not, of course, ascend higher—but if the finger were then moved to one side so as to allow a free escape till the whole contents of the tube were rapidly moving to the exit, and the orifice then at once contracted or closed as before, the jet would dart far *above* the reservoir; for in addition to the hydrostatic pressure which drove it up in the first instance, there would be a new force acting upon it, derived from the *motion* of the water. As in the case of a hammer of a few pounds weight, when it rests on the anvil it exerts a pressure on the latter with a force due to its weight only, but when put in motion by the hand of the smith, it descends with a force that is equivalent to the pressure of perhaps a ton.

Every person accustomed to draw water from pipes that are supplied from very elevated sources, must have observed, when the cocks or discharging orifices are suddenly closed, a jar or tremor communicated to the pipes, and a snapping sound like that from smart blows of a hammer. These effects are produced by blows which the ends of the pipes receive from the water; the liquid particles in contact with the plug of a cock, when it is turned to stop the discharge, being forcibly driven up against it by those constituting the moving mass behind. The philosophical instrument named a *water hammer* illustrates this fact. The effect is much the same as if a solid rod moved with the same velocity as the water through the tube until its progress was stopped in the same manner, except that its momentum would be concentrated on that point of the pipe against which it struck, whereas with the liquid rod the momentum would be communicated equally to, and might be transmitted from *any* part of, the lower end of the tube; hence it often occurs that the ends of such pipes, when made of lead, are swelled greatly beyond their original dimensions. We have seen some $\frac{3}{4}$ of an inch bore, become enlarged to $1\frac{1}{4}$ inches before they were ruptured. At a hospital in Bristol, England, a plumber was employed to convey water through a leaden pipe from a cistern in one of the upper stories to the kitchen below, and it happened that the lower end of the tube was burst nearly every time the cock was used. After several attempts to remedy the evil, it was determined to solder one end of a smaller pipe immediately behind the cock, and to carry the other end to as high a level as the water in the cistern; and now it was found that on shutting the cock the pipe did not burst as before, but a jet of considerable height was forced from the upper end of this new pipe: it therefore became necessary to increase its height to prevent water escaping from it—upon which it was continued to the top of the hospital, being twice the height of the supplying cistern, but where to the great surprise of those who constructed the work, some water still issued: a cistern was therefore placed to receive this water, which was found very convenient, since it was thus raised to the highest floors of the building without any extra labor. Here circumstances led the workmen to the construction of a water-ram without knowing that such a machine had been previously devised.

The first person who is known to have raised water by a ram, designed for the purpose was, Mr. Whitehurst, a watchmaker of Derby, in England. He erected a machine similar to the one represented by the next figure, in 1772. A description of it was forwarded by him to the Royal Society, and published in vol. lv, of their Transactions.

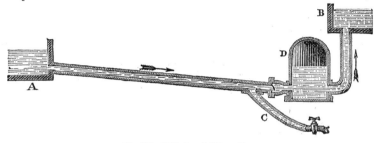

No. 167. Whitehurst's Water-Ram.

A, represents the spring or reservoir, the surface of the water in which was of about the same level as the bottom of the cistern B. The main pipe from A to the cock at the end of C, was nearly six hundred feet in length, and one and a half inches bore. The cock was sixteen feet below A, and furnished water for the kitchen offices, &c. When it was opened the liquid column in A C was put in motion, and acquired a velocity due to a fall of sixteen feet; and as soon as the cock was shut, the momentum of this long column opened the valve, upon which part of the water rushed into the air-vessel and up the vertical pipe into B. This effect took place every time the cock was used, and as water was drawn from it at short intervals for household purposes, "from morning till night—all the days in the year," an abundance was raised into B, without any exertion or expense.

Such was the first water-ram. As an original device, it is highly honorable to the sagacity and ingenuity of its author; and the introduction of an air vessel, without which all apparatus of the kind could never be made durable, strengthens his claims upon our regard. In this machine he has shown that the mere act of drawing water from long tubes for ordinary purposes, may serve to raise a portion of their contents to a higher level; an object that does not appear to have been previously attempted, or even thought of. The device also exhibits another mode, besides that by pressure engines, of deriving motive force from liquids thus drawn, and consequently opens another way by which the immense power expended in raising water for the supply of cities, may again be given out with the liquid from the lateral pipes. Notwithstanding the advantages derived from such an apparatus, under circumstances similar to those indicated by the figure, it does not appear to have elicited the attention of engineers, nor does Whitehurst himself seem to have been aware of its adaptation as a substitute for forcing pumps, in locations where the water drawn from the cock was not required, or could not be used. Had he pursued the subject, it is probable the idea of opening and closing the cock (by means of the water that escaped) with some such apparatus as figured in No. 160, would have occurred to him, and then his machine being made self-acting, would have been applicable in a thousand locations. But these additions were not made, and the consequence was, that the invention was neglected, and but for the one next to be described, it would most likely have passed into oblivion, like the steam machines of Branca, Kircher, and Decaus, till called forth by the application of the same principle in more recent devices.

Whenever we peruse accounts of the labors of ingenious men, in search of new discoveries in science or the arts, sympathy leads us to rejoice at their success and to grieve at their failure : like the readers of a well written novel who enter into the views, feelings and hopes of the hero ; realize his disappointments, partake of his pleasures, and become interested in his fate ; hence something like regret comes over us, when an industrious experimenter, led by his researches to the verge of an important discovery, is, by some circumstance diverted (perhaps temporarily) from it ; and a more fortunate or more sagacious rival steps in and bears off the prize from his grasp—a prize, which a few steps more would have put him in possession of. Thus Whitehurst with the water-ram, like Papin with the steam-engine, discontinued his researches at the most interesting point—at the very turning of the tide that would have carried him to the goal ; and hence the fruit of both their labors has contributed but to enhance the glory of their successors.

The *Bèlier hydraulique* of Montgolfier was invented in 1796. (Its author was a French paper maker, and the same gentleman who, in conjunction with his brother, invented balloons in 1782.) Although it is on the principle of Whitehurst's machine, its invention is believed to have been entirely independent of the latter. But if it were even admitted that Montgolfier was acquainted with what Whitehurst had done, still he has, by his improvements, made the ram entirely his own. He found it a comparatively useless device, and he rendered it one of the most efficient —it was neglected or forgotten, and he not only revived it, but gave it a permanent place among hydraulic machines, and actually made it the most interesting of them all. It was, previous to his time, but an embryo; when, like another Prometheus, he not only wrought it into shape and beauty, but imparted to it, as it were, a principle of life, that rendered its movements *self-acting ;* for it requires neither the attendance of man, nor any thing else, to keep it in play, but the momentum of the water it is employed to elevate. Like the organization of animal life, and the mechanism by which the blood circulates, the pulsations of this admirable machine incessantly continue day and night, for months and years; while nothing but a deficiency of the liquid, or defects in the apparatus can induce it to stop. It is, compared to Whitehurst's, what the steam-engine of Watt is to that of Savary or Newcomen.

Montgolfier positively denied having borrowed the idea from any one— he claimed the invention as wholly his own, and there is no reason whatever to question his veracity. The same discoveries have often been, and still are, made in the same and in distant countries, independently of each other. It is a common occurrence, and from the constitution of the human mind will always be one. A patent was taken out in England for self-acting rams in 1797 by Mr. Boulton, the partner of Watt, and as no reference was made in the specification to Montgolfier, many persons imagined them to be of English origin, a circumstance that elicited some remarks from their author. " Cette invention (says Montgolfier) n'est point d'origine Anglaise, elle appartient toute entière à la France ; je déclare que j'en suis le seul inventeur, et que l'idée ne m'en a été fournie par personne ; il est vrai qu'un de mes amis a fait passer, avec mon agrément, a MM. Watt et Boulton, copie de plusieurs dessins que j'avais faits de cette machine, avec un mémoire détaillé sur ses applications. Ce sont ces mêmes dessins qui ont été fidèlement copiés dans la patente prise par M. Boulton à Londres, en date du 13 Décembre 1797 ; ce qui est une vérité dont il est bien éloigné de disconvenir, ainsi que le respectable M. Watt." We have inserted this extract from Hachette, because we really supposed on reading the specification of Boulton's patent in the Repertory of Arts, (for 1798, vol. ix,) that the various modifications of the ram there described were the invention of that gentleman. The patent was granted to "Matthew Boulton, for *his* invention of improved apparatus and methods for raising water and other fluids."

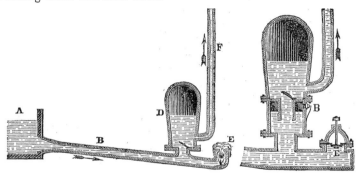

No. 168. Montgolfier's Ram.   No. 169. The same.

No. 168 represents a simple form of Montgolfier's ram. The motive column descends from a spring or brook A through the pipe B, near the end of which an air chamber D, and rising main F, are attached to it as shown in the cut. At the extreme end of B, the orifice is opened and closed by a valve E, instead of the cock in No. 167. This valve opens downwards and may either be a spherical one as in No. 168, or a common spindle one as in No. 169. It is the play of this valve that renders the machine self-acting. To accomplish this, the valve is made of, or loaded with, such a weight as just to open when the water in B is at rest; *i. e.* it must be so heavy as to overcome the pressure against its under side when closed, as represented at No. 169. Now suppose this valve open as in No. 168, the water flowing through B soon acquires an additional force that carries up the valve against its seat; then, as in shutting the cock of Whitehurst's machine, a portion of the water will enter and rise in F, the valve of the air chamber preventing its return. When this has taken place the water in B has been brought to rest, and as in that state its pressure is insufficient to sustain the weight of the valve, E opens; (descends) the water in B is again put in motion, and again it closes E as before, when another portion is driven into the air vessel and pipe F ; and thus the operation is continued, as long as the spring affords a sufficient supply and the apparatus remains in order.

The surface of the water in the spring or source should always be kept at the same elevation, so that its pressure against the valve E may always be uniform—otherwise the weight of E would have to be altered as the surface of the spring rose and fell.

This beautiful machine may be adapted to numerous locations in every country. When the perpendicular fall from the spring to the valve E is but a few feet, and the water is required to be raised to a considerable height through F, then, the *length* of the ram or pipe B, must be increased, and to such an extent that the water in it is not forced back into the spring when E closes, which will always be the case if B is not of sufficient length. Mr. Millington, who erected several in England, justly observes that a very insignificant pressing column is capable of raising a very high ascending one, so that a sufficient fall of water may be obtained in almost every running brook, by damming the upper end to produce the reservoir, and carrying the pipe down the natural channel of the stream until a sufficient fall is obtained. In this way a ram has been made to raise one hundred hogsheads of water in twenty-four hours to a perpendicular height of one hundred and thirty-four feet, by a fall of only four feet and a half. M. Fischer of Schaffhausen, constructed a water-ram in the form of a beautiful antique altar, nearly in the style of that of Esculapius, as represented in various engravings. A basin about six inches in depth, and from eighteen to twenty inches in diameter, received the water that formed the motive column. This water flowed through pipes three inches in diameter that descended in a spiral form into the base of the altar; on the valve opening a third of the water escaped, and the rest was forced up to a castle several hundred feet above the level of the Rhine.

A long tube laid along the edge of a rapid river, as the Niagara above the falls, or the Mississippi, might thus be used instead of pumps, water wheels, steam-engines and horses, to raise the water over the highest banks and supply inland towns, however elevated their location might be; and there is scarcely a farmer in the land but who might, in the absence of other sources, furnish his dwelling and barns with water in the same way, from a brook, creek, rivulet or pond.

If a ram of large dimensions, and made like No. 168, be used to raise water to a great elevation, it would be subject to an inconvenience that would soon destroy the beneficial effect of the air chamber. When speaking of the air vessels of fire-engines, in the third book, we observed that if air be subjected to great pressure in contact with water, it in time becomes incorporated with or absorbed by the latter. As might be supposed, the same thing occurs in water-rams; as these when used are incessantly at work both day and night. To remedy this, Montgolfier ingeniously adapted a very small valve (opening inwards) to the pipe beneath the air chamber, and which was opened and shut by the ordinary action of the machine. Thus, when the flow of the water through B is suddenly stopped by the valve E, a partial vacuum is produced immediately below the air chamber by the recoil of the water, at which instant the small valve opens and a portion of air enters and supplies that which the water absorbs. Sometimes this *snifting* valve, as it has been named, is adapted to another chamber immediately below that which forms the reservoir of air, as at B in No. 169. In small rams a sufficient supply is found to enter at the valve E.

Although air chambers or vessels are not, strictly speaking, constituent elements of water-rams, they are indispensable to the permanent operation of these machines. Without them, the pipes would soon be ruptured by the violent concussion consequent on the sudden stoppage of the efflux of the motive column. They perform a similar part to that of the bags of wool, &c. which the ancients, when besieged, interposed between their walls and the battering rams of the besiegers, in order to break the force of the blows.

The ram has also been used in a few cases to raise water by atmospheric pressure from a lower level, so as to discharge it at the same level with the motive column or even higher. See *Siphon Ram,* in next book.

The device by which Montgolfier made the ram self-acting, is one of the neatest imaginable. It is unique : there never was any thing like it in practical hydraulics, or in the whole range of the arts ; and its simplicity is equal to its novelty, and useful effects. Perhaps it may be said that he only added a valve to Whitehurst's machine : be it so—but that simple valve instantly changed, as by magic, the whole character of the apparatus—like the mere change of the cap, which transformed the Leech Hakim into Saladin.[a] And the emotions of Cœur de Lion, upon finding his great adversary had been his physician in disguise, were not more exquisite than those, which an admirer of this department of philosophy experiences, when he contemplates for the first time the metamorphosis of the English machine by the French Savan. The name of Montgolfier will justly be associated with this admirable machine in future ages. When all political and ecclesiastical crusaders are forgotten, and the memories of all who have hewed a passage to notoriety merely by the sword, will be detested—the name of its inventor will be embalmed in the recollections of an admiring posterity.

The water cane, or *canne hydraulique,* raises water in a different manner from any apparatus yet described. A modification of it in miniature has long been employed in the lecture room, but it is seldom met with in descriptions of hydraulic machines. It is represented at No. 170 ; and consists of a vertical tube, in outward appearance like a walking cane, having a valve opening upwards at the bottom, and placed in the liquid to be raised. Suppose the lower end twelve or fifteen inches below the surface, the water of course would enter through the valve and stand at the same height within as without : now if the tube were raised quickly, but not entirely out of the water, the valve would close and the liquid within would be carried up with it ; and if, when the tube was at the highest point of the stroke, its motion was *suddenly reversed* (by jerking it back) the liquid column within would still continue to ascend until the momentum imparted to it at the first was expended ; hence a vacuity would be left in the lower part of the instrument into which a fresh portion of water would enter, and by repeating the operation the

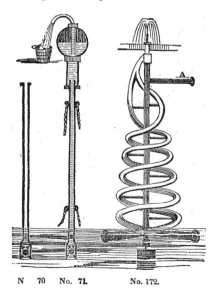

N 70   No. 71.   No. 172.

---

[a] Walter Scott's Tales of the Crusaders.

Chap. 1.] And its Modifications. 373

tube would become filled, and a jet of water would then be thrown from the upper orifice at every stroke. This effect obviously depends upon the rapidity with which the instrument is worked, i. e. a sufficient velocity must be given to the water by the upward stroke to prevent it descending, till the tube again reaches the lowest point, and consequently receives another supply of water. The instrument should be straight and the bore smooth and uniform, that the liquid may glide through with the least possible obstruction. As its length must be equal to the elevation to which the water is to be raised, it is necessarily of limited application, and especially so since the whole (both water and apparatus) has to be lifted at every stroke—not merely the liquid that is discharged, but the whole contents of the machine.

By making the upper part of the tube slide within another that is fixed, a short part only of the apparatus might then be moved, and by connecting an air chamber as in No. 171, a continual stream from the discharging orifice might be produced. A stuffing box should be adapted to the end of the fixed tube. Hachette suggested the application of a spring pole (like those used in old lathes) to communicate the quick reciprocating motion which these machines require.

No. 172 represents another form of the instrument. Two spiral tubes coiled round in opposite directions are secured to and moved by a vertical shaft. Their upper ends are united and terminate in one discharging orifice; the lower ones are enlarged, and each has a valve or clack opening inwards to retain the water that enters. By means of the handle A, which is mortised to the shaft, an alternating circular motion is imparted to the whole, and the water thereby raised through these coiled tubes on precisely the same principle as through the perpendicular ones just described. Thus, when the handle is moved either to the right hand or to the left, one valve closes, and the water within receives an impulse that continues its motion along the tube after the movement of the latter is reversed; and by the time its momentum is expended a fresh portion of water has entered that prevents its return. In this manner all the coils become filled, and then every additional supply that enters below drives before it an equal portion from the orifice above. This machine, therefore differs from Nos. 170 and 171 only, in being adapted to a horizontal instead of a perpendicular movement. Each tube in the figure forms a distinct machine, and should be considered without reference to the other Their discharging orifices are united to show how a constant jet may be produced. By making the upper part turn in a stuffing box in the bottom of a fixed tube, as in No. 171, water might then be raised higher than the movable part of the apparatus.

That property by which all bodies tend to continue either in a state of rest or motion, viz: *inertia*, increases the effect of these machines, for when the momentum imparted to the liquid in the tubes is exhausted, inertia alone prevents it from instantaneously flowing back, and hence there is time for an additional portion of water to enter at the valve. The action of the canne hydraulique is similar to that by which persons throw water to a distance from a bucket, or a wash-basin. The momentum given to these vessels and their contents carries the latter to a distance, while the former are held back or retained in the hands. Coals are thus thrown from a scuttle, earth from a shovel, and it is the same when a traveler on a galloping horse, or when drawn furiously in an open carriage, continues on his journey after the animal suddenly stops—his adhesion to his seat, not being sufficient to resist the motal inertia of his body.

CHAPTER II.

Machines for raising water by fire: Air machines—Ancient weather-glasses—Dilatation of air by heat and condensation by cold—Ancient Egyptian air-machines—Statue of Memnon—Statues of Serapis and the Bird of Memnon—Decaus' and Kircher's machinery to account for the sounds of the Theban Idol—Remarks on the Statue of Memnon—Machine for raising water by the sun's heat, from Heron—Similar machines in the sixteenth century—Air-machines by Porta and Decaus—Distilling by the sun's heat—Musical air machines by Drebble and Decaus—Air machines acted on by ordinary fire—Modifications of them employed in ancient altars—Bronze altars—Tricks performed by the heathen priests with fire—Others by heated air and vapor—Bellows employed in ancient altars—Tricks performed at altars mentioned by Heron—Altar that feeds itself with flame, from Heron—Ingenuity displayed by ancient priests—Secrets of the temples—The Spiritalia—Sketch of its contents—Curious Lustral Vase.

A separate book might with propriety have been devoted to machines which raise water through tubes by means of the weight, pressure, momentum, or other natural properties of liquids, without the necessary intervention of wheels, cranks, levers, &c. With such, those now to be described might also have been classed, since they too require neither external machinery nor force. They differ however from pressure-engines and water-rams, and every other device yet noticed, in bringing into action a new element, viz. *heat* or *fire*. It is by this that the force upon which their movements depend is generated, viz. in the expansion of elastic fluids. There are two kinds of these machines which differ according to the fluid medium upon which the fire is made to act. In some this is common *air*, in others *steam* or vapor of water, and sometimes both steam and air have been employed. The present chapter is appropriated to air machines. These might be divided into two classes, according to the nature of the heat employed; in some this is derived from the sun; in others, from ordinary fire. Those in each class might also be arranged according to that property of the air upon which their action depends, viz. 1. the force developed by its expansion; 2. the vacuum formed by its condensation; 3. those in which both are combined. The first might be compared to forcing pumps, the second to sucking or atmospheric ones, and the third to those which both suck and force up the water.

It was observed in the second book (page 176) that all gases or airs are expanded by heat and contracted by cold. A proof of this is afforded by the usual mode of employing cupping-glasses; a minute piece of cotton or sponge dipped in alcohol is inflamed and placed in a glass; upon which the air becomes dilated or increased in bulk, so that a great part is driven out to make room for the rest; the mouth of the instrument is then applied to the place from which blood is to be withdrawn; the flame of the cotton is thereby extinguished and as the remaining air becomes cool it cannot resume its previous state of density, and consequently a vacuity or void is left in the glass. Plumbers sometimes make small square boxes of sheet lead; and on soldering in the covers the temperature of the contained air is so greatly increased, that before the soldering is completed a large portion is expelled, and when the boxes become cool every side is found slightly collapsed. This result is the required proof of the vessels being tight. Now it is clear that if a communication was opened by a tube between the interior of one of these boxes and a vessel of water placed a few feet below, that the liquid would be forced into it (by the atmosphere) until the contained air occupied no greater space than it did before any part was driven out by the heat. This mode of raising liquids may be illustrated at the tea-table : Let a saucer be half filled with cold water, hold an inverted cup just over it and apply for a moment a small slip of lighted paper to the interior of the cup, drop the paper on the water and cover it with the cup, when the liquid contents of the saucer will be instantly forced up into the inverted vessel.

If an inverted glass siphon be partly filled with water and the orifice of one leg be then closed and that leg be held to the fire, the air expanding will drive out the liquid and cause it to ascend in the other leg. Several philosophical instruments illustrate the same thing. Previous to the discovery of atmospheric pressure and the invention of the barometer, the expansion of air by heat was the principle upon which the ancient weather glasses were constructed. They were made in great variety. The simplest consisted of a glass tube having a bulb blown on the closed end. It was held over a fire to dilate the air, and the open end was then plunged into a vessel of water. Its construction was the same as the modern barometer. Variations in the temperature and density of the atmosphere caused the water to rise and fall in the tube, as the contained air was dilated and contracted, and thus changes in the weather were indicated. From these instruments the barometer received its former name of " the weather glass."[a]

The degree of elevation to which water can be thus raised depends upon the temperature to which the contained air is subjected ; its dilatation or increase of bulk being, according to some authors, in common with

---

[a] The following extract from a book published ten years before the discovery of atmospheric pressure, may interest some readers. Although the instruments to which it refers are no longer in use, they ought not to be entirely forgotten.

"A weather-glasse is a structure of at the least two glasses [a tube and the vessel containing the water] sometimes of three, foure, or more as occasion serveth, inclosing a quantity of water, and a porcion of ayer proporcionable ; by whose condensacion or rarifaction the included water is subject unto a continual mocion, either upward or downward ; by which mocion of the water is commonly foreshown, the state, change, and alteracion of the weather;—for I speak no more than what my own experience hath made me bold to affirm ; you may (the time of the year, and the following observacions understandingly considered) bee able certainly to foretell the alteracion or uncertainty of the weather a good many hours before it come to pass.

There are divers severall fashions of weather-glasses, but principally two. 1. The circular glasse. 2. The perpendicular glasse. The perpendiculars are either single, double or treble. The single perpendiculars are of two sorts, either fixt or moveable : The fixt are of contrary qualities ; either such whose included water doth move upward with cold, and downward with heat, or else upward with heat and downward with cold. In the double and treble perpendiculars, as the water ascendeth in one, it descendeth as much or more in the other. In the moveable perpendiculars, the glasse being artificially hanged, it moveth up and down with the water."

The author then describes the various kinds mentioned and tells his reader " if you doe well observe the form of the figures you cannot go amisse." He also gives directions for making coloured water for the tubes, such as " may be both an ornament to the work and delectable to the eye." Treatise on Art and Nature, A. D. 1633 or 4. See account of this book page 321. A modification of an air-glass may be found in the Forcible Movements of Decaus, (plate viii,) which he names *an Engine that shall move of itself*. Lord Bacon, in whose time these air glasses were common, presented what appears to have been an improved one and of his own invention, to the Earl of Essex, who it is said, was so captivated with it that he presented the donor with Twickenham Park and its garden, as a place for his studies. The instrument was named ' *A secret curiosity of nature, whereby to know the season of every hour of the year, by a Philosophical Glass, placed (with a small proportion of water) in a chamber.* An account of Lord Bacon's Works, London, 1679.

other permanently elastic fluids, in a geometrical progression to equal increments of heat. A volume of air at ordinary temperatures is increased over one third, if raised to 212° Fahrenheit. At the fusing point of lead (about 600°) it is more than doubled, and at the heat of 1100°, it would be tripled. Let a small glass tube be attached to the neck of a Florence flask, and heat both in boiling water; if the end of the tube be then placed in mercury, the latter will as the air becomes cooled rise in the tube, to the height of ten inches—equal to about eleven feet of water. If they were heated to 600° it would rise to fifteen inches; and if to 1100° to twenty-one or two inches. We have connected a tube to the mouth of a common quart bottle, and after heating the latter over a fire, placed the end of the tube in mercury, and on removing the whole to the open air, then at 50°, the mercury in a few minutes rose to sixteen inches; hence rather more than one half of the air had been expelled by the heat. These effects take place when the enclosed air is *dry;* but if it be moist, or if a drop or two of water be in the vessel, the results are greater, because the vapor of the liquid would alone fill or nearly fill the vessel and would drive out a corresponding quantity of air.

This mode of creating a vacuum and raising water by the dilatation and condensation of air is now seldom used, because superior results, as just intimated, are obtained from steam and with less expense. Air machines are however interesting in several respects. They are among the earliest examples of elastic fluids being employed as a moving power, induced by alternate changes of temperature. They constitute the first link in that chain of devices that has now terminated in the steam engine, but which will probably be prolonged through future ages by the addition of even more efficient mechanism. In this view of the subject, air machines will connect the researches and inventions of antiquity, in the development and applications of the most valuable because most pliable of all motive forces, with every improvement future engineers may make to the end of time.

The oldest air-machines known were made in Egypt, and the oldest account extant of such devices is also derived from that country, viz. from the *Spiritalia*. It is also worthy of remark that they are associated by Heron with other devices of the priesthood, for exciting wonder and performing prodigies before the people; thus affording a collateral proof that occupants of the ancient temples at Thebes, Memphis and Heliopolis were intimately acquainted with the principles of natural philosophy; and fully capable of teaching those who flocked to them for information, from Greece and neighboring countries. There is a circumstance too that indicates a more thorough and *practical* acquaintance with the mechanical properties of elastic fluids, and the means of exciting those properties than might at first be supposed, viz. in the substitution of the sun's heat for that of ordinary fires. This seems to have been adopted in cases where the miracle to be wrought could not be accomplished by the latter without danger of detection, or when it could not be so secretly effected, or could not be performed with such imposing effect. Of this, the vocal statue on the plain of Thebes is an example. This gigantic idol saluted the rising sun and continued to utter sounds as long as the solar beams were shed over it, and while surrounded by the myriads that worshipped at its shrine. Now these sounds were produced, according to Heron, by the dilatation of air, or by vapor evolved by the sun's heat from water contained in close vessels, that were concealed in or connected to the base of the statue, and exposed to the solar rays. The expanded fluid, it is supposed, was conveyed through tubes whose orifices were fashioned to produce the required sounds.

Cambyses desirous of ascertaining the concealed mechanism, it is said, broke the statue from the head to the middle. According to some writers he discovered nothing; while others mention an opinion prevalent among the Egyptians that the image previously uttered the seven mysterious vowels, but never afterwards. Strabo has recorded a tradition that the injury was caused by an earthquake. He visited Egypt in the first century, and remarks, that early one morning as he and Gallus the prefect, with many other friends, and a large number of soldiers were standing by the statue, they heard a certain sound, but could not determine whether it came from the trunk or the base; there was however a prevailing belief that it proceeded from the latter or its vicinity. The sounds finally ceased in the fourth century, when christianity became established in the country. Some authors have supposed two different devices were employed; one previous to, and the other subsequent to the mutilation of the statue; and that one or both consisted of springs, &c. on the principle of some of the speaking heads of the middle ages. Heron, however, who must have been familiar with the image and the sounds uttered by it, attributed the latter to air or vapor, evolved and expanded by solar heat; so that, however we may speculate on the subject, in his opinion the Pharaonic priesthood were well acquainted with the dilatation and contraction of airs by heat and cold, and with various modes of employing them. Moreover, the movements of the famous statue of Serapis and also those of the Bird of Memnon (an image which we have previously mentioned) were also produced by air or vapor dilated by the sun's heat; and we shall presently see that tricks on the same principle were frequently performed at ancient altars.

Modern expositions of the mechanism or supposed mechanism of the Theban Idol are derived from the Spiritalia. That of Decaus consists of a close vessel, of the form of a pedestal, having a partition across it by which two air-tight compartments are formed. One of these is half filled with water and exposed to the solar rays—the other contains air and to its upper part are connected two organ pipes or reeds that communicate with the statue. A communication is formed between the two compartments by a siphon, the legs of which are inserted at the top and descend nearly to the bottom of each compartment. Thus when the sun warmed the vessel containing water, the air and vapor within, became expanded and pressing on the surface of the liquid forced part of the latter through the siphon into the other compartment, by which a corresponding portion of air was forced through the organ pipes. A figure and details of this apparatus form the 23rd plate of Decaus' Forcible Movements. Pausanias and some other ancient authors compared the sounds to those produced by the vibration of harp strings; and Juvenal, who was exiled to Egypt by Domitian, seems to have been of the same opinion.

- - - - when the radiant beam of morning rings
On shatter'd Memnon's still harmonious strings.—xv. *Sat.*

Hence Kircher in his explanation, instead of conveying the rarefied air, or vapor through a wind instrument, made it act against the vanes of a wheel, as in Branca's steam machine, and as the wheel was thus blown round, a number of pins attached to its periphery struck a series of wires so arranged as to receive the blows. A figure of this device is inserted in his *Œdipus Ægyptiacus*. Rome, 1652, Tom. iii, page 326.

Whatever the device was, it seems to have been as admirable in its execution and the disposition of the mechanism, as in its conception. We are not certain that it was ever fully understood except by the priests of the adjoining temple. The Romans do not appear to have had sufficient curiosity to give it a critical examination. What a contrast it forms with some modern wonders! These have puzzled people only while examination was prohibited—while access to them was denied, as the chess-player of Kempelen; but the colossal android of Thebes defied the scrutiny of the world through unknown periods of time. In it, the old priests of Egypt have sent down a surprising specimen of their skill. We know from the Bible that they had a profound knowledge of Natural Magic; i. e. of the applications of science to purposes of deception, and this statue confirms the scriptural account:—it shows us what an amount of labor and ingenuity was expended in the fabrication of idols, and to what a prodigious extent the ancient systems of delusion were carried—how the very magnitude and even *sublimity* of the impostures were calculated to bear down the intellect and establish an unshaken belief in the communion of the priests with the gods.

If a close metallic vessel containing water be exposed to the sun, the air in the upper part will become dilated by the heat and may be employed to raise the water: for if a tube be inserted at the top, and the lower end reach nearly to the bottom, the elasticity of the air will be expended in forcing the liquid up the tube and to an elevation according to the increase of its temperature. A device of this kind is described by Heron which is represented in the annexed cut.

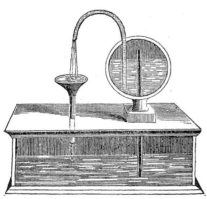

No. 173. Raising Water by the Sun. From Heron.

On the lid of a box or cistern containing water is placed a globe, also partly filled with the same fluid. A pipe rises from the cistern to about the centre of the globe. Another pipe through which the water is to be raised proceeds from near the bottom of the globe and terminates over a vase or cup, which communicates with the cistern as represented. When the sun beams fall on the globe, the air within is rarefied and by its expansion forces the water through the pipe into the vase, through which it descends again into the cistern. When the sun beams are withdrawn and the surface of the globe becomes cool, a partial vacuum is formed in the globe, and the pressure of the atmosphere then drives a fresh portion into it from the cistern below; when it is again ready to be acted on by the sun as before. In addition to the air, at first contained in the globe, a quantity of vapor or low steam would be evolved by the heat and contribute greatly to the result. The cistern represents an open reservoir which may be at a distance from the globe, and the vase merely exhibits the place of discharge —having no necessary connection with the reservoir. The apparatus as a model, is figured just as philosophical instruments still are. Thus in modern books, a pump (for example) is often shown as discharging water into the reservoir from which it raises it. We make these remarks because some persons are too apt to consider these ancient figures as literal representations of working machines, whereas, they were designed merely to illustrate the principles upon which the movements depended; and as specimens from which others for practical purposes might be deduced. It is quite a common remark in old authors, after describing a device, to observe that various machines for other purposes may be derived from it, and excuse themselves for not pointing out particular modes of doing this, because they considered them too obvious to require it.

Whether such modes of raising water were practised in Europe previous to the sixteenth century, we have no means of ascertaining; but in the middle of that and the beginning of the following one they are frequently to be met with in old authors.

Baptist Porta, in his *Natural Magic*, after describing a method of raising water from the bottom to the top of a tower, by means of a vacuum formed by water flowing from a close vessel;—next proposes a mode of accomplishing the same object "*by heat alone.*" A close vessel of brass was to be placed upon the tower, having a pipe connected to its upper part and extending down to the water to be raised; the orifice being a short distance below the surface. The vessel was then "to be made hot by the sun, *or fire,*" to rarefy the contained air and expel a portion of it through the pipe. As the vessel grows cold, he observes, the remaining air is condensed, and because it cannot then fill the vacuity, "the water is called in and ascends thither."[a] (Book xix. chap. 3.) He does not mention the height of the tower because the philosophers of that age had no idea that the elevation to which water would ascend into a vacuum had any limits—and hence in another part of the same work Porta uses the following language—" A vacuum is so abhorred by nature that the world would sooner be pulled asunder than any vacuity can be admitted." (Book xviii. chap. 1.) There is another passage in the 5th chapter of the 19th Book, from which it seems that he employed the elastice force of air or steam, or a mixture of both as in No. 174—and generated either by the heat of the sun, or by that of lamps or candles, as shown at No. 189.

After describing a fountain of compression, which he exhibited to some of the great Lords of Venice, and the operation of which he says caused great surprise as there was no visible cause for the water flying so high— he continues, " I also made another place near this fountain that *let in light,* and when the air was extenuated, *so long* as *any light lasted* the fountain threw out water, which was a thing of much admiration, and yet but little labor." This passage is probably imperfectly translated.

No. 174, on the next page, forms the 9th plate attached to the " Forcible Movements" of Decaus. (Translated by Leak, London, 1659.) It exhibits an extension of Heron's machine already noticed, (No. 173.) Decaus says " this engine hath a great effect in hot places as in Spain and Italy.'

Four air tight copper vessels, a foot square and 8 or 9 inches deep, are so arranged that the sun may shine strongly upon them. A pipe, having a valve *o*, opening upwards, communicates with the lower part of each, to supply them with water from the spring below. Another pipe passes over the upper surfaces, having branches which descend nearly to the bottom of each vessel. A valve is also placed in this pipe, from the upper end of which the jet of the fountain issues. At the commencement, each vessel is about one third filled with water, through openings on the top, which are then plugged up. " Then the sun shining upon the said engine shall make an expression by rarefying the enclosed air and force the

---

[a] NATURAL MAGICK *in twenty books by John Baptist Porta, wherein are set forth all the riches and delights of the Natural Sciences.*"—London, 1658.
It contains, beside a multitude of absurdities, many ingenious devices. The *trombe,* camera obscura, air gun, repeating guns, air tubes, ear trumpets, &c. &c. are described. The work was first published in 1560.

water to flow out as in the figure. And after the heat of the day is passed and the night shall come, the vessels shall draw the water of the cistern [spring] by the pipe and sucker [lower valve] and shall fill the vessels as before. - - - - - - - And you must observe that the two suckers [valves] must be made very light and likewise very just, so as the water may not descend by them after it is raised."

No. 174. Air Machine, from Decaus.

An improvement upon the preceding machine is next given by Decaus. The form of the vessels is altered, and double convex lenses or "burning glasses" are so arranged on their covers as to collect " the raies of the sun within the said vessels, the which will cause a great heat to the water, and by that means make it spring forth with great abundance, and also higher if it be required." (See the figure below.) It is, we think, in the range of probability that the heat of the solar rays may yet be applied in some situations to raise water with effect

No. 175. Air Engine, from Decaus.

Whether this application of lenses " to encrease the force of the sun" in air engines, was a device of Decaus, we know not. The idea however

Distilling by the Sun's Heat.

would naturally occur to any engineer of the time, engaged on the improvement of such machines, because distilling by the sun both with lenses and without them was a common practice with chemists in that age and some centuries before. Baptist Porta described the process in the tenth book of his *Natural Magic*, and observes that "the waters extracted by the sun are the best." See also *Maison Rustique*, Paris, 1574, page 211. Kircher's *Mundus Subterraneus*, Tom. ii, 392. Other authors also describe the application both of convex and concave lenses to concentrate the solar rays on distilling vessels; a practice probably as old as the time of Archimedes, or even older. We give an extract from an English translation of one of Gesner's works, who died in 1545.

"Further, although that the Chimisticke authours doe teach and shew diverse fashions of distilling by ascension, yet may all these waies be brought into three orders. - - - - - - - - The first manner is, when we distill anie liquide substance or flowers in the sunne by force of his heate. - - - - - - - - The singular man *Adam Loucier*, in his treatise on the arte of distilling, setteth forth an easie maner of distilling by the heate of the sunne beames - - - - - and the same is to be wrought on this wise: take, saith Loucier, a hollowe burning glasse, which directlie place towarde the hote beames of the sunne, after (betweene the beams of the sunne and the burning-glasse,) set the glass bodie [Retort] - - - - in such manner, that the beames of the hote sunne falling into the hollow glasse, maie so beate backe and extende to the glasse bodie with the proper matter (as to the object standing righte against) - - - - as more livelie appeareth by this figure here described."[a]

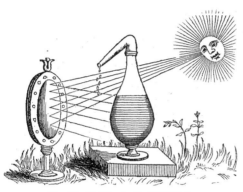

No. 76. Distilling by the Sun's Heat.

Air dilated and vapor evolved by the sun's heat were also used to produce music in the middle ages, a device which often caused that celestial melody which, like the harp of Dunstan, acquired for its authors a reputation, sometimes of superior sanctity, and at others of dealing with the wicked one. The musical machine of the famous *Drebble*, according to Bishop Wilkins, was of this kind: i. e. a modification of the supposed one in the Statue of Memnon. Drebble's machine, says the Bishop, "would of itself render a soft and pleasant harmony when exposed to the sun's rays, but being removed into the shade would presently become silent. The reason of it was this; the warmth of the sun working upon some

---

[a] *The practice of the new and old phisicke, wherein is contained the most excellent secrets of phisicke and philosophie, devided into foure Books—in the which are the best approved remedies for the diseases, as well inward as outward, of al the parts of mans body: treating very amplie of al distillations of waters, of oyles, balmes, quintessences, with the extraction of artificial saltes, the vse and preparation of antimony, and potable gold, gathered out of the best and most approved authours, by that excellent Doctor Gesnerus. Also the pictures and maner to make the vessels, furnaces and other instruments thereunto belonging. Newly corrected and published in Englishe, by George Baker, one of the Queenes Majesties chiefe Chirurgians in ordinary. London, Black Letter, 1599.*

moisture within it, and rarefying the inward air unto so great an extension that it must needs seek for vent or issue, did thereby give several motions unto the instrument." (Math. Magic, Book ii, chap. 1.)

Decaus, besides his explanation of the vocal statue of Egypt, has given a description of *a musical summer*, a device apparently similar to Drebble's; and in the twenty-second plate of his work he has figured another which Switzer has copied into his system of hydrostatics.

The heat of the sun is too uncertain to be relied upon in those projects that require immediate and certain results. During the evening, night, and early dawn, nothing could be effected; and even in mid-day, clouds and showers often intercept or divert the rays: moreover a machine when placed so as to be heated directly by the sun, soon experiences a diminution of its influence by the motion of the earth. Those rays which fall directly upon it becoming, in consequence of this motion, oblique. These and other unfavorable circumstances are common to most countries where the solar heat is sufficiently intense, while in others it is too feeble to be used with effect; hence, in the temperate zones, within which the arts have at all times been chiefly cultivated, the application of ordinary fire has superseded, for nearly all practical purposes, that derived from the sun. In some parts of the earth saline waters are concentrated, and salt produced by the heat of the glowing orb of day, but for every thing like the devices belonging to our subject it is now seldom employed, if at all.

The oldest applications of fire to raise liquids are, singularly enough, also to be found among the philosophical tricks of ancient priests, and among the prodigies which they performed at the altar itself. The selection of altars for such displays was natural, because it was at them the will of the gods was more particularly expected to be made known. It must not be supposed that ancient altars were all simple structures of wood, stone, brick, or marble; on the contrary, many of them were elaborately designed, and constructed entirely of *metal*. Every one knows that bronze or brazen altars are of frequent occurrence in the Old Testament, and the descriptions of some prove them to have been splendid specimens of workmanship and design. The altar for "burnt offerings," being upwards of eight feet square and five deep, was covered with plates of brass. The grate, fire place, vessels, &c. were also of the same material. One of the numerous brazen altars built by Solomon was an extraordinary affair, being twenty cubits, or *thirty-three feet* square, and *sixteen* feet high. The large number of victims consumed on it and the necessary fires account for these dimensions.

As some of the most effectual frauds were consummated at and by means of altars, the civil governors of the heathen, and some of the worst princes of the Jews, made use of them for the performance of state tricks, to intimidate the people and subdue them to their will. In such matters a collusion between the priests and statesmen of antiquity is very obvious. (By a similar combination of church and state it is that the people of Europe are still oppressed.) When Themistocles could not otherwise carry out his measures, he did not fail to make the oracles interfere. There are some interesting particulars in 2d of Kings, chap. xvi. respecting a brazen altar which Ahaz examined at Damascus, and an exact copy of which he had made and erected in Jerusalem. It evidently was of a novel construction and was probably designed for working pretended miracles for state purposes, for it was among those destroyed by his son Hezekiah. Montfaucon in the supplement to his antiquities describes some singular altars, and among others, one on which an eagle was made suddenly to rise as in the act of flying away. This he observes was effected by machinery moved by a person appointed for the purpose.

There are numerous intimations in history of frauds practiced at altars by fire, and by *water* and other *liquids*. We shall notice a few here, and others in the next chapter. A very ancient tradition taught that those were the greatest gods that answered their worshippers by fire. This was a prevailing belief among the ancient heathen, and hence the ingenuity of the priests was particularly exercised in devising means to produce a spontaneous or "divine fire," to consume the sacrifices. Servius, a Roman writer, affirms that in ancient times fire was never kindled on the altars, but was drawn down from heaven by prayers. Solinus another Roman author, who wrote in the first century, speaks of one in Sicily upon which the fuel, though ever so green, would kindle of itself if the sacrifice was acceptable to the gods. Pausanius relates an example of which he was a witness. Some of the devices are known. When the victim was laid on the altar and the fuel ready to be kindled, a libation of wine or oil was poured upon it; streams of the liquid trickled through fissures or secret channels into a pan of coals concealed below, and instantly the sacrifice was enveloped in flames, and the desired proof of its acceptance given. At other times naptha, a mineral oil that takes fire on being exposed to the air, was adroitly dropt on the fuel by the priests as they officiated. This is the substance by which Medea is supposed to have destroyed Creusa, by impregnating with it the enchanted gown which she presented to her. When Creusa had put it on and was approaching the altar, it burst into flames, and she expired in excruciating torments. The Druids had the art of kindling without fire a sulphurous substance by which they struck terror into their enemies. There are presumptive proofs that both they and the priests of Delphos had gunpowder, with which they imitated thunder and lightning; and this accords with a remark of Pliny, in the second book of his Natural History, (chap. 53.) "It appeareth upon record in chronicles that by certain sacrifices and prayers, *lightnings* may either be compelled or easily intreated to fall upon the earth." And he observes that there was an old tradition in Etruria, that lightning was procured "by exorcisms and conjurations." The ancient priests of Ethiopia worshiped the sun, and at the close of harvest they separated a portion of the fruits from the rest as a sacrifice to the Deity: if the offering was acceptable it instantly took fire. The Vestal Æmylia rekindled the sacred fire on the altar of Vesta, by putting her veil over it, that is, by some device which the act of adjusting her veil concealed: in fact enough is known to convince us that old temples were perfect laboratories. (See an expose of the pretended descent of celestial fire on Good Fridays, into the holy sepulchre of Jerusalem (which is, we believe, still kept up) in Motraye's Travels, vol. i, page 79.)

But it was not by the sudden appearance of flames only that fire was employed as an agent of deception. Equally surprising effects, and as secretly produced, were derived from the *heat* which the fuel and burning sacrifice gave out. It will readily be imagined that this heat must have been intense when a bullock, a sheep, or a goat, was consumed; and sometimes several animals were offered at once upon the same altar. In burnt-offerings, every part was to be reduced to ashes, and hence particular care was required that the fuel should be dry as well as in abundance; otherwise the mass of flesh and juices might extinguish the fire—a circumstance that was deemed very inauspicious. It was also customary to pour wine and oil upon the sacrifice, and spices and perfumes to correct the odor. These, of course, increased the heat, and in addition to which, it appears from one of the Hamilton Vases, that large bellows were some times used to promote the combustion by a blast.

A modern mechanician will at once perceive that the radiation of heat from such fires into the *interior* of altars offered an effective and unsuspicious source of fraud—one from which a distinct series of prodigies might be derived. Let us see how they could be realized. Suppose a bronze altar made air-tight, with a cylindrical or other opening through its centre, in which to place the fire and to afford a draft, (as in those wooden boilers in which water is heated by a fire in the centre; the liquid being in contact with the heated sides of the furnace, and the ashes from the grate falling through the draft opening, which is continued through the bottom of the boilers) or the passage for the draft might be made at right angles to the furnace or fire-place, and terminate at one side of the altar; the upper part of the furnace would then be level with the top of the altar upon which the victim was laid. Suppose the air-tight cavity round the furnace filled to a certain height with wine, oil, or other inflammable liquid, a vapor would then be evolved by the heat, and mixing with the contained air would press upon the surface of the liquid, which, by concealed tubes might be conveyed to the fire and thus sustain it without any additional fuel. The vapor might also be made to produce sounds as in Drebble's machine—images of birds might by it be made to sing—dragons and serpents to hiss. The current, like the blast of a bellows, might be made to excite the flames; and by appropriate mechanism impart motion to various automata—cause the doors of the temples mysteriously to fly open and to close, &c. &c. Now it so happens that these very things were done and by means of air and vapor.

The annexed figure, from Problem XI, of the Spiritalia, will serve as a specimen of the ingenuity of the ancients in these respects. It is merely one of a number that Heron has given. The altar was of metal, hollow and air-tight, and placed on a hollow base or pedestal (also air-tight) which contained a quantity of oil or wine. Upon the base stood two statues, each holding a vase in one hand as represented. Pipes, as shown by the dotted lines, communicated through the statues with the liquid. As the air within the altar became dilated by the heat, it necessarily forced the liquid up the pipes and drove it out of the mouths of the vases in which the pipes terminated. It is not easy to see why the bottom of the altar did not open directly into the base or reservoir of wine, instead of the pipe that connects them, since it would have promoted the evolution of vapor; but the figure represents only one of the numerous modifications employed.

No. 177. Liquids raised by heat in ancient altars—from the Spiritalia.

It is obvious from this and some other devices described by Heron, (as No. 173) that vapor from the contained liquids contributed *chiefly* to the result, although he has not in all cases mentioned it. Indeed it is not certain that he did not confound steam with air, as the philosophers of the sixteenth century did, of which some examples are given in the next two chapters. Had *air* alone been used in the above altar, the effect could only have have been momentary; for part of it would be soon absorbed by the liquid and carried out with it, and there appears no provision for a fresh supply. Besides, as the liquid was expelled, the higher would the remainder have to be raised, and consequently unless the air received a corresponding increase of temperature the discharge from the vase might cease.

Had it not been for the Spiritalia, we should never have suspected that *air* was made to perform so important a part in ancient frauds, nor that its expansion and contraction had been employed to raise liquids. Notwithstanding the high opinion which history gives us of the philosophical knowledge of old priests, we should hardly have surmised that they had the art of applying this subtil fluid so ingeniously. They seem, however, to have ransacked all nature for devices; and to have become familiar with the principles upon which the most valuable of our arts and machinery are based. Astronomy, acoustics, chemistry, optics, hydrostatics, pneumatics, and hydraulics, were all pressed into their service. Even the application of *steam*, as a source of motive force, did not escape them; so that had their energies been devoted to the development of useful mechanism, the world would probably have been indebted to them for the steam engine itself.

What wonders would an insight into the old temples have revealed! To have had an opportunity of inspecting the machinery, new and old—to have been present at the consultations of the priests—witnessed their private experiments—heard them expatiate on the defects of this device and the perfect working of that—suggesting a wheel here and a spring there—to have been present at their consultations respecting the suspension of water in Tutia's sieve, and witnessed the congratulations exchanged at the eclat with which that and many other trials came off, &c. &c.—would have made us acquainted with discoveries both in science and mechanical combinations that would throw some modern inventions into shade:—But the tremendous evils which their impostures induced rendered concealment on the part of the priests indispensable. Exposure would not only have endangered their wealth and influence, but might have led to their extermination by an outraged and plundered people—hence the veil of religion was interposed to screen the operators and their apparatus, and inevitable *death* was the consequence of undue curiosity: witness that of *Alcithœ*, a female of Thebes, who ridiculed the orgies of Bacchus, and was represented by the priests as having been changed into a *bat;* a fiction of theirs, most likely, to conceal their having taken her off. *Æpytus* might be adduced as another example—he forcibly entered the temple of Neptune and was *struck blind* by a sudden eruption of *salt-water* from the altar; probably sulphuric or other acid secretly ejected by the priests. In this chapter we have seen they had the means of doing this by the dilatation of air within the cavities of altars.

We shall conclude this chapter with some remarks on the *Spiritalia*, a work that had more influence in reviving the study of hydrodynamics in modern times, than any other. This little book, like a rivulet, sent its streams of knowledge over all Europe in the sixteenth century. It stimulated, if it did not create that spirit of investigation and experimental research which then commenced and has continued unimpaired to the present time. It seems to have caused an unusual degree of excitement. Philosophers, chemists, and physicians, as well as engineers, illustrated their writings by its problems and figures. Porta, Decaus, Fludd, and others, avowedly transferred its pages to their works, while many writers with less candor and less ingenuity made use of it without acknowledgment. Of all the old mechanicians, Besson seems to have been less indebted to it than any other.

The Spiritalia formed but a small part of the writings of Heron: had all of them reached our times, we should have possessed an almost perfect system of ancient mechanical philosophy. He wrote books on clepsydra, automata, dioptrics, war machinery, engines for raising weights; and an introduction to mechanics, which is said to have been the most complete work on the subject which the ancients possessed. Taken as a whole, the Spiritalia seems more like the manual of an ancient magician than any thing else—a collection of deceptions with the processes by which they were matured. In it Heron, instead of appearing in the character of a philosopher, rather assumed (perhaps for amusement or to expose the frauds of the Egyptian hierarchy) that of a minister of Isis, initiating an acolyte into the mysteries of his profession. And numerous as are the devices described, they doubtless formed but a small part of those which constituted the active and efficient capital of the Egyptian priesthood. With the exception of an hydraulic and another organ, a syringe, fire engine, fountain of compression, three lamps and two eolipiles, (and most of which were also used for unworthy purposes) the whole may be considered as a text book for conjurers. Of the seventy-six problems contained in the book, twelve relate to the working of prodigies at the altars, by air dilated by the heat of the sacred fires, &c. as already noticed; upwards of forty relate to sacrificial vases, Tantalus' cups, magic pitchers, &c. In some of these were concealed cavities, in which the liquid was retained or discharged, by closing with the thumb a minute opening in the handle. Water was poured into some and they gave out wine, and *vice versa.* In these we have a solution of the trick by which water was changed into wine in the temple of Bacchus, on the 7th of January at the annual feast of the god, as mentioned by Pliny. In others were disguised partitions forming various compartments in which different liquids were retained, and all discharged at one orifice (by a species of three or four-way cock) so that those in the secret could draw wine, oil, or water, at pleasure; besides many other *merry conceits,* as the old authors name them. There is we think among them abundant evidence that our solution of Tutia's miracle of carrying water in a sieve was the true one. It is probable that in some of these vases, specimens of the old divining cups may be found.

The ingenious reader will not repine at our inserting a specimen of a lustral vase. We have selected this because it shows that mechanical as well as hydrodynamical devices were adopted as occasions required. It shows also that the mode of increasing or diminishing the pressure of a valve to its seat, by a loaded lever, as in the safety valve of a steam engine, was known—a circumstance that may be deemed quite insignificant by some persons; but attention to such little things often enables us to arrive at correct estimates of an ancient device, and of the ingenuity and fertility of conception of ancient devisers.

Most readers are aware that *holy water* was derived from that of the heathen. When a worshiper was about to enter the temple, he sprinkled himself from a vase of it placed near the entrance. On some particular occasions the people were sprinkled by priests. (See an example at page 196.) Those who celebrated the Eleusinian mysteries were particularly required to wash their hands in holy water. In the middle ages the liquid was a source of considerable profit to monks, and it was even a custom for clerks and scholars to hawk it for sale. From Heron's description of the following figure, (No. 178,) we learn that heathen priests also made it a source of revenue; the vessels containing lustral water not being always open for public use, free of charge, but closed, and like a child's money box provided with a slit at the top, through which a certain sum was to be put before the donor could receive any of the purifying contents. In the vase before us *five drachmæ*, or about seventy-five cents, were required, and it will be perceived from the construction of the apparatus that no less sum could procure a drop, although as much more might be put in as the donor thought proper.[a]

No. 178. Ancient Vase of Lustral Water.

The device is a very neat specimen of religious ingenuity, and the more so since it required no attending minister to keep it in play. We may judge of other apparatus belonging to the old temples by the talent displayed in this. A portion of the vase is removed in the figure to show the interior. Near one side is seen a cylindrical vessel at A. It is this only that contained water. A small tube attached to the bottom is continued through the side of the vase at *o*, where the liquid was discharged. The inner orifice of the tube was formed into the seat of a valve, the plug of which was fixed on the lower end of the perpendicular rod, whose upper end was connected by a bolt to the horizontal lever or vibrating beam R. One end of R is spread out into a flat dish and so arranged as to receive on its surface every thing dropped through the slit. The lever turns on a pin or fulcrum very much like a pump handle, as represented. The operation will now be understood. As the weight of the rod kept the valve closed while nothing rested upon the broad end of the lever, so no liquid could escape; but if a number of coins of sufficient weight were dropped through the slit upon the end of R, the valve would then be opened and a portion of liquid escape at *o*;—the quantity flowing out would however be very small, not only from the contracted bore of the tube, but from the fact that the valve would be open only a moment; for as the lever became inclined from its horizontal position the pieces of money would slide off into the mass accumulated at H, and the efflux would as quickly be stopped: the apparatus would then be ready to supply the next customer on the same terms. This certainly was as simple and ingenious a mode of dealing out liquids as it was a profitable one, and after all was not half so demoralizing as the retailing of ardent spirits in modern times.

One would suppose the publication of such a work as Heron's Spiritalia must have been as distasteful to the occupants of ancient temples, as some of Luther's writings were to Leo X and his associates of the Vatican.

---

[a] In spondea, hoc est in vasa sacrificii injecto quinque drachmarum numismate aqua ad inspergendum effluit. Spiritalia, xxi.

CHAPTER III.

ON STEAM: Miserable condition of the great portion of the human race in past times—Brighter prospects for posterity—Inorganic motive forces—Wonders of steam—Its beneficial influence on man's future destiny —Will supersede nearly all human drudgery—Progress of the arts—Cause why steam was not formerly employed—Pots boiling over and primitive experiments by females—Steam an agent in working prodigies—Priests familiar with steam—Sacrifices boiled—Seething bones—Earthquakes—Anthemius and Zeno—Hot baths at Rome—Ball supported on a jet of steam, from the Spiritalia—Heron's whirling eolipile—Steam engines on the same principle—Eolipiles described by Vitruvius—Their various uses—Heraldic device—Eolipiles from Rivius—Cupelo furnace and eolipile, from Erckers—Similar applications of steam revived and patented—Eolipiles of the human form—Ancient tenures—Jack of Hilton—Puster, a steam deity of the ancient Germans—Ingenuity of the priests in constructing and working it—Supposed allusions to eolipilic idols in the bible—Employed in ancient wars to project streams of liquid fire—Draft of chimneys improved, perfumes dispersed, and music produced by eolipiles—Eolipiles the germ of modern steam engines.

If we contemplate the past history of man, we shall find that, with a few insignificant exceptions, the entire race has been, as it were, doomed to support an existence surcharged with misery. From the earliest periods of recorded time, we behold the great mass slaves to an organized despotism which a few crafty spirits entailed upon the species—a despotism both mental and physical—to subdue the body and enthrall the mind —political and ecclesiastical despotism. To the neglect of mental cultivation alone, these evils are to be attributed; for in every age men have had the *same* elements of prosperity and of happiness. The earth and its treasures have always been at their disposal, and the natural capacities of the human intellect, have probably always been the same. It is the improvement of these capacities by culture, and their degeneracy by neglect, that make all the differences in men's condition. The horrible sufferings of the myriads of human beings who have passed through a life of unceasing and unrequited toil, were owing to their ignorance, and hence the tyrants of the earth have always labored, and still labor, to keep those uninformed that are subject to their sway. Ignorance was the grand engine by which the most atrocious systems of tyranny, superstition and magic were established in ancient times; and whose influences are not yet done away.

But within the last two centuries a new era has opened with brighter prospects for the human family at large, than has ever yet dawned upon it. An era that has been ushered in by the discovery, or rather application, of a new motive agent, viz. STEAM. The wonderful effects which this fluid has been made to produce, are so creditable to the human intellect, and so fraught with consequences of the highest import to our race in all times to come, as to excite even in the most torpid minds emotions of stirring interest. Steam is changing every thing, and every thing for the better. It has opened new sources of social and individual happiness: nor is its influence confined to the physical condition of man, for by its connection with the manufacture of paper and with the printing press, it has done more to rouse and exercise the moral and intellectual energies of our nature than any thing else; and has imparted a vigorous impulse to them, as well as to the useful arts. As all the advantages derived in modern times from steam originated in attempts to *raise water* by it, we need offer no apology for indulging in some preliminary remarks.

What a proof is steam of the stores of *motive forces* that are to be found in the inorganic world! Forces that can render us incalculable service, if we would but open our eyes to detect, and exercise our energies to employ them. Who could have supposed two centuries ago, that the simple vapor of water would ever be used as a substitute for human exertions, and should relieve man from a great portion of the physical toil under which he has groaned from the beginning of the world? That it would arm him with a power which is irresistible, and at the same time the most pliant—one that can uproot a mountain, and yet be controlled by a child! Who could have *then* imagined that a vessel of boiling water should impart motion to machinery in every department of the arts, and be equally adapted to all—should spin and weave threads fine as those of the gossamer; and forge tons of iron into single bars with almost equal rapidity and ease—raise water from mines, in streams equal to rivers; and extract mountains of mineral from the bowels of the earth—should propel carriages, such as no horses could move, with the velocity of wind; and urge ships of every class through the ocean, in spite both of winds and waves—should be the means of circulating knowledge at the price of waste paper, and of awakening and stimulating the mental capacities of men! In a word, that a little aqueous vapor should revolutionize the whole social and political condition of man: and that after having done all this, that it should probably give place to other agents, still more powerful and beneficial, which science and observation should discover.

What a proof is steam of the high destiny that awaits our species! The most fervid imagination cannot realize the importance of those discoveries in science and the arts, of which it is merely the forerunner; the first in that new catalogue of motive agents that are ordained to change the condition of men, and to regenerate the earth; for all that is yet done is but as the twilight that ushers in the orb of day. Hitherto man has been comparatively asleep, or in a state resembling it—insensible of the rich inheritance which the Creator has placed at his disposal in the elastic fluids, and of their adaptation to impart motion to every species of mechanism. How few persons are aware that the grand invention of imparting motion to a piston by steam and other elastic fluids, is the pivot on which the chief affairs of the world is destined hereafter to turn? And the time is not distant when, by means of it, the latent energy of the gases, or other properties of inert matter, will supersede, in a great degree, the drudgery of man—will perform nearly all the labor which the bones and sinews of our species have hitherto been doomed to accomplish. There are persons, however, whose minds biased by the eternal bondage in which the mass of our race has always been held, who will startle at the idea of the whole becoming an intelligent and highly intellectual body. They cannot conceive how the affairs of life are to be continued —the execution of innumerable works which the constitution of society requires should be performed, if these helots become free. But can they, can any one, seriously believe that the all-wise and benevolent Creator could possibly have intended that the highest class of beings which he has placed on this planet—the only one capable of appreciating his works and realizing correct ideas of his attributes—that the great portion of these, should pass through life in incessantly toiling for mere food;—and undergoing privations and sufferings to obtain it, from which the lowest animals are exempt? Assuredly not. Had such been his design, he would not have created them with faculties expressly adapted for nobler pursuits.

It is the glory of modern science, that it calls into legitimate use both the physical and mental powers of man. It rewards him with numerous forces derived from inanimate nature, and instructs him in the application of them, to all, or nearly all, the purposes of life; and eventually it will require from him no greater amount of physical toil, than what conduces to the full development of all the energies of his compound nature. It is destined to awaken that mass of intellect which has hitherto lain dormant, and been all but buried in the laboring classes; and to bring it into active exercise for the benefit of the whole. And for aught we know, the "*new earth*," spoken of in the scriptures, may refer to that state of society, when science has thus relieved man from all injurious labor—when he will walk erect upon the earth and subdue it, rather by his intellect than by the sweat of his brow—when the curse of ignorance will be removed, and with it the tremendous punishment that has ever attended it. Then men will no longer enter in shoals into a new state of existence in another world, as utterly ignorant of the wonders of creative wisdom in this, as if they had never been in it, and had not possessed faculties expressly adapted to study and enjoy them.

There is no truth in the observation of some people, that all discoveries of importance are already made; on the contrary, the era of scientific research and the application of science to the arts may be considered as but commenced. The works of creation will forever furnish materials for the exercise of the most refined intellects, and will reward their labors with a perpetual succession of new discoveries. The progress which has been made in investigating the laws that govern the aqueous, atmospherical, mineral and vegetable parts of creation, is but a prelude to what is yet to be done—it is but the clearing of the threshold preparatory to the portals of the temple of science being thrown open to the world at large. There is no profession however matured, no art however advanced, that is not capable of further improvement; or that, so far as we can tell, will not always be capable of it. If an art be carried to the utmost perfection it is capable of in one age, discoveries in others will in time be made, by means of which it will be still further advanced; for every improvement in one has an effect, more or less direct, on every other.

The benefits already derived from steam, then, are but as a drop to the ocean when compared with those that posterity will realize; for if such great things have been accomplished by it in one century, what may not be expected in another? and another? It has been calculated that two hundred men, with machinery moved by steam, now manufacture as much cotton as would require twenty millions of persons without machines; that is, one man by the application of inorganic motive agents can now produce the same amount of work that formerly required one hundred thousand men. The annual product of machinery in Great Britain, a mere spot on the earth, would require the physical energies of one half the inhabitants of the globe, or four hundred millions of men: and the various applications of steam in different parts of the world now produce an amount of useful labor, which if performed by manual strength would require the incessant exertions of every human being. Hence this great amount of labor is so much gained, since it is the result of inorganized forces, and consequently contributes so much to the sum of human happiness. Now if such results have been brought about so quickly and by steam alone, what may not be expected from it, and other aëriform fluids, in ages to come, when the progressive improvement of every art and every science shall have brought to light not only other agents of the kind, but more efficient means of employing them? There is no end to the beneficial applications of the gases as motive agents, and no limits to the power to be derived from them. As long as rain falls or rivers flow—while trees (for fuel) grow, or mineral coal is found, man can thus wield a power that renders him almost omnipotent.

The question may be asked, why was not the elastic force of steam earlier used as a source of motive power? Because, as we observed before, men neglected to employ those powers of reflection and invention which God had given them. It certainly formed no part of the Creator's plan of governing the world that they should have so long remained ignorant of its application. He has placed man at the head of creation and furnished him with powers appropriate to his position. Every object in nature he can use for good or for evil. They are the materials from which he may, as an expert machinist, fabricate at will all that his wants require: he may prostitute them to the miseries of himself and his fellows; or he may neglect them to the injury of all. It is the order of nature that her latent resources shall be discovered and applied by *diligent research*. Hence some of the finest specimens of the Creator's wisdom can only be appreciated after careful study, a fact which is itself a proof of his wisdom and beneficence, since their realization is thus held out as an inducement to investigate them.

Steam has of course been noticed ever since the heating of water and boiling of victuals were practiced. The daily occurrence implied by the expression "the pot boils over" was as common in antediluvian as in modern times; and hot water thus raised was one of the earliest observed facts connected with the evolution of vapor. From allusions in the most ancient writings, we may gather that the phenomena exhibited by steam were closely observed of old. Thus Job in describing Leviathan alludes to the puffs or volumes that issue from under the covers of boiling vessels. "By his neesings a light doth shine, and his eyes are like the eyelids of the morning; out of his nostrils goeth smoke [steam] as out of a seething pot or cauldron." In the early use of the vessels last named, and before experience had rendered the management of them easy and safe, females would naturally endeavour to prevent the savory contents of their pots from flying off in vapor; hence attempts to confine it by covers; and when these did not fit sufficiently close, a cloth or some similar substance interposed between it and the edge of the vessel, would readily occur; and a stone or other weight placed upon the top *to keep all tight* would also be very natural. Then as the fluid began again to escape, further efforts would be made to retain it by additional weights. In this manner, doubtless many a contest was kept up between a pot and its owner, till one gained the victory; and we need not the testimony of historians to determine which this was. In those times it was not generally known that a boiling cauldron contained a spirit, impatient of control—that the vessel was the generator of an irresistible power, and the cover a safety-valve; and that the preservation of the contents and the security of the operator depended upon letting the cover alone, or not overloading it:—hence it no doubt often happened that the confined vapor threw out the contents with violence, and then it was that primitive cooks began to perceive that there was death as well as life in a boiling pot. In this manner, we suppose females were the first experimenters on steam, and the earliest witnesses of steam boiler explosions.

The domestic exhibitions of the force of steam must have excited the attention of mechanicians in every age, nor could its capabilities of overcoming resistances opposed to it, have escaped them. Thus we find that experimenters are almost always said to have derived the first hint from a culinary vessel: hence the Marquis of Worcester, according to a tradition, had his attention drawn to the use of this fluid to raise water, by witnessing, while a prisoner in the Tower of London, the lid of a boiler thrown off by the vapor—but the anecdote is of much older date, and was applied to many others before his time as well as since. Vitruvius illustrates his views respecting the appearance of springs on mountains, by a cauldron which, he says, when two thirds filled with water and heated by the fire, " communicates the heat to the water; and this on account of its natural porosity, receiving a strong inflation from the heat, not only fills the vessel, but swelling with the steam and *raising the cover*, overflows," &c. (Book viii, chap. 3.) Such occurrences are nature's hints, by attention to which important discoveries have always been made. Even when people in former times were injured by the explosion of a cauldron, the misfortune should have been considered as an *indication of nature* to employ the force thus developed—and also as a punishment for having neglected to do so. Nay, we don't see why such occurrences may not, in this view of them, be considered *providential*, as well as similar ones, which theological writers avail themselves of, to establish a similar doctrine.

There are intimations that the elastic force of steam was employed by several people of antiquity, but the details of its application are unfortunately not known. Some relics of its use, as well as that of heated air, are to be found in the deceptions practiced by the heathen priesthood. Its application for similar purposes was continued till comparatively modern times, for it was the animating principle in the eolipilic idols of the middle ages; and, from an incidental notice of some experiments of a Greek architect, it is probable that the trembling of the earth, and other horrors experienced by those who were initiated into the greater mysteries of ancient worship, were also effected by steam. Artificial thunder, lightning from the vapor of inflammable liquids, and unearthly music, were produced by its means. Some of the tricks performed by the Pythoness and her coadjutors at Delphos seemed to have been matured by it. The famous tripod against which she leaned is represented as a brazen vessel from which a miraculous vapor arose. Steam was one of the agents of deception in trials of ordeal. Those persons condemned to undergo that of boiling water, were protected by the priests (when it was their interest or inclination to do so) by admitting a concealed stream of steam into the lower part of the cauldron containing tepid water—the consequent agitation of the liquid and the ascent of the vapor that escaped condensation presented to the ignorant and unsuspecting beholders every appearance of genuine ebullition. On similar occasions air was forced through the liquid in the dark ages.

Ancient priests, both among the Jews and heathen, were from their ordinary duties necessarily conversant with the generation of steam. Its elastic force could not therefore escape the shrewd observers among them. Sacrifices were frequently *boiled* in huge cauldrons, several of which were permanently fixed in the vicinity of temples—in " the boiling places" as their locations are named by Ezekiel, " where the ministers of the house shall boil the sacrifice of the people." (See an example from Herodotus at page 200.) It would seem moreover as if some of the boilers were made on the principle of Papin's Digester, in which bones were softened by 'high steam'—at any rate a distinction is made between seething pots and cauldrons, and from the manner in which both are mentioned they seem to have been designed for different purposes; the former to seethe or soften bones, the latter to boil the flesh in only. " They roasted the passover with fire, but the other offerings sod they in pots and in cauldrons." (2 Chr. chap. xxxv, 13.) " Set on a pot, set it on, and also pour water into it. Gather the pieces thereof into it, even every good piece, the thigh and the shoulder; fill it with the choice bones. Take the choice of the flock and burn [or heap] also the bones under it, and make it *boil well*, and let them *seethe the bones* of it therein." (Ezek. xxiv. 3, 5.) The opinion of the Jews having close vessels in which steam was raised higher than in common cauldrons is also rendered probable from the fact that the Chinese, a contemporary people, employ similar ones, and which from their tenacity to ancient devices have probably been used by them from times anterior to those of the prophet. (Davis' Chinese, ii, 271. John Bell's Travels, i, 296 and ii, 13.)

Similar processes have been common with chemists in all ages, in the making of extracts, and sometimes in preparing food. There is an example in Porta's *Natural Magic*. He tells us (in the xiii chap. on distillation) that he has restored persons at the point of death to health by " an essence extracted out of flesh." He directs three capons to be dressed and boiled " *a whole day* in a glass vessel *close stopt*, until the bones and flesh and all the substance be dissolved into liquor."

Some of the ancient philosophers, who were close observers of nature, compared the earth to a cauldron, in which water is heated by internal fires; and they explained the phemonena of earthquakes by the accumulation of steam in subterraneous caverns, until its elastic energy rends the superincumbent strata for a vent. Vitruvius explains by it the existence of boiling springs. In the reign of Justinian, Anthemius, an architect and mathematician illustrated several natural phenomena by it; but of this we should probably never have heard, had it not been for a quarrel between him and his next door neighbor, Zeno, the rhetorician. This orator appears to have inherited a considerable share of credulity and superstition, which gave his antagonist the advantage. Anthemius, we are informed, had several steam boilers in the lower part of his house, from each of which a pipe conveyed the vapor above, and by some mechanism, of which no account has been preserved, he shook the house of his enemy as by a real earthquake, upon which the frightened Zeno rushed to the senate " and declared in a tragic style that a mere mortal must yield to the power of an antagonist who shook the earth with the trident of Neptune."

There are reasons for believing that the expansive force of the steam which was evolved in heating the immense volumes of water for the hot baths at Rome, was employed to elevate and discharge the contents of the boilers. Sir W. Gell has given, in his Pompeiana, a representation of a set of cauldrons belonging to the Thermæ, at Pompeii, derived from impressions left in the mortar or cement in which they were embedded. It would seem that several series or sets were used, each consisting of three *close* boilers (in shape not unlike modern stills,) placed directly upon, and connected by pipes to each other. The manner in which they were connected is not known; Gell says by a species of siphon. The lowest boiler was the largest and was placed directly over the furnace; and the arrangement was such, that when any part of the boiling liquid was withdrawn, an equal quantity, already warmed, entered from the next boiler above, which at the same time derived a supply from the uppermost one;

this last being always kept filled by a pipe from the aqueduct or castellum. Remains of the pipes, cocks, copper flues, &c. have been found in abundance, but the details of the heating apparatus and those connected with the elevation and distribution of the liquid have not been ascertained: this is to be regretted because, from the number and magnitude of the hot baths at Rome, the operations of boiling and dispersing the water must have been conducted on a scale far more extensive than any thing in modern times—the most extensive breweries and distilleries not excepted. Some idea of the operations may be derived from the fact that a single establishment could accommodate two thousand persons with warm, or rather *hot*, baths at the same time. Seneca, in a letter to Lucilius, says " there is no difference between the heat of the baths and a *boiling furnace;*" and it would, he observes, appear to a reasonable man as a sufficient punishment to wash a condemned criminal in them. The persons who had the charge of heating in close vessels and distributing daily such large quantities of water, must necessarily have been conversant with the mechanical properties of steam, and with economical modes of generating it. In some cases the water was heated by passing through a coiled copper tube, like a distiller's worm, which was embedded in fire. We have previously remarked that the Romans also heated water by making it pass through the hollow grates of a furnace. (See Pompeii. vol. i, 196, and Gell's Pompeiana.)

Besides the various applications of heated air and of vapor already noticed, there is in problem XLV of Heron's Spiritalia, a description of a close boiler, from the upper part of which a current issues that supports at some distance above the boiler a light ball like those that are made to play on jets of water. (See the annexed figure, No. 179.) The whirling eolipile, No. 180, is the subject of problem L—and is the earliest representation of a machine moved by steam that is extant. It consists of a small hollow sphere, from which two short tubes proceed in the line of its axis, and whose ends are bent in opposite directions. The sphere is suspended between two columns, their upper ends being pointed and bent towards each other. One of these columns was hollow and conveyed steam from the boiler into the sphere, and the escape of the vapor from the small tubes by its reaction imparted a revolving motion to the sphere. These two applications of steam have been considered the result of a fortunate random thought, which Heron, or some other old mechanic, stumbled on by a species of chance medley, whereas they certainly indicate an intimate though it may be a limited acquaintance with the mechanical properties of that fluid. We should never suppose that this elegant application of the jet to sustain a ball in the air was the fruit of a first attempt to use steam, much less that the complex movement of the whirling eolipile was another thought of the moment. Did any modern experimenter in hydraulics ever hit upon the suspension of a ball by a jet of water in his first essays, or devise Barker's mill at a sitting,

Eolipiles, from Heron.

No. 179.   No. 180.

without having ever heard of either? No more than any old mechanician ever invented the above before experimental researches on steam had became familiar to him, if not to his contemporaries. Besides, there have been within the last half century not less than half a dozen *patents* taken out for rotary steam engines identical in principle with the whirling eolipile. The fact seems to be that Heron selected the two devices above, on the same principle as the rest of the illustrations, i. e. such as in his judgment would be the most interesting.[a]

From a remark of Vitruvius in the first book of his Architecture, chap. 6, we learn that those portable steam machines named *Eolipiles*, (from Eolus the god of wind, and their application to create artificial winds) were in common use in his time. Speaking of the town of Mytilene, he observes that the inhabitants were subject to colds, in those seasons when certain winds blew; and which might have been in some degree avoided by a more proper disposition of the streets. " Wind, [he remarks] is only a current of air, flowing with uncertain motion;—it arises from the action of heat upon moisture—the violence of the heat forcing out the blasts of air. That this is the fact, *the brass eolipiles* make evident;—for the operations of the heavens and nature may be discovered by the action of artificial machines. *These brass eolipiles are hollow and have a very narrow aperture, by which they are filled with water, and then placed on the fire:—before they become hot, they emit no effluvia, but as soon as the water begins to boil, they send forth a vehement blast.*" As these instruments have been adapted to a great variety of purposes, as well as being intimately connected with this part of our subject, we shall notice them with some detail. From the times of Vitruvius to those of Des Cartes, and up to the present century, they have been used as philosophical instruments to illustrate the nature of winds and meteors, as well as for other scientific pursuits. They were used as substitutes for bellows in blast furnaces and ordinary fires. The draft of chimneys was increased by means of them. They were made to produce music and disperse perfumes. They constituted the distilling vessels of the alchymists, and in another form were employed as weapons of war, and were even deified in the *steam idols* of old. They were the first instruments employed to raise water by steam, and the first to produce motion by it; and hence they constitute the germ of modern steam engines, to which we may add that they led to the invention of steam guns. (See Martin's Philosophy, vol. ii, 90.) They are commonly made of iron, brass, or strong copper, having a short neck in which a very minute opening is made. In order to charge one with water (or other liquid) it is placed on a fire until nearly red hot; it is then taken off, and the neck placed in water or the whole plunged in it, which, as the vessel cools, takes the place of the air driven out by the heat. It is then placed on a brazier of charcoal or other fire until steam is rapidly evolved and discharged with violence at the orifice.

Vitruvius has not mentioned the particular purposes for which eolipiles were used by the Romans. It is however known that they were employed as bellows for exciting fires; and as this was not for want of the latter instruments, they must have had properties which rendered them

---

[a] Balls dancing on jets of water and *air*, were a favorite accompaniment of the old garden water works, and hydraulic organs, &c. of Italy, where the device has probably been in use since the times of the Republic: the amusement of children with peas on the ends of tobacco pipes or reeds is in imitation of it: the current of air blown through the perpendicular tube keeps the tiny globe some inches above the orifice, where its motions, varying with the force of the current, produce a very agreeable effect.

preferable, on some occasions, to bellows. One perhaps was their occupying little room on the hearth; and another, their requiring no attendant to keep up the blast. It has already been observed (page 237-8,) that human bellows-blowers formed part of the large domestic establishments in ancient Egypt, and Nos. 103 and 104 of our illustrations represent some at work in one of the kitchens of the Pharaohs. The practice was probably common among all the celebrated nations of old, and we know that it was continued in Europe till the sixteenth century if not later. To supersede these workmen might therefore have been one reason for the employment of eolipiles.

In a Latin collection of " *Emblems human and divine*," (Prague, 1601,) there is a device of one of the old Counts of Hapsburg, which consists of a blowing eolipile with a stream of vapor issuing from it, and the motto *Læsus Juvo*. (Vol. ii, 372.) The same device is also given in a treatise on *Heroic Symbols*, Antwerp, 1634. Hence this ancient domestic instrument was adopted on such occasions as well as the bellows, syringe, watering pot, &c.

Rivius in commenting on the eolipiles mentioned by Vitruvius describes those in use in his own time, (A. D. 1548,) and gives several figures, from which we have selected the first three of the following ones.

Eolipiles, from Rivius and Cardan.

No. 181.   No. 182.   No. 183.   No. 184.

Rivius names them " wind holders" and " fire blowers." He says they were made in various shapes and of different materials, and were used "to blow the fire like a pair of bellows." Some, designed for other purposes, that will presently be mentioned, were made of gold or silver and richly ornamented, as represented above. At a subsequent period of the sixteenth century, Cardan gave a figure of one. (See No. 184.) Fludd, Porta and other old writers also describe them. The latter, in book xix, chap. 3, of his Natural Magic, speaks of them as used in houses to blow fires. Sir Hugh Platte, in 1594, published a figure and description of " a rounde ball of copper, or latton [brass] that blows the fyre verie stronglie by the attenuation of the water into ayre."

Bishop Wilkins, in his Mathematical Magic, (published in 1648) speaks of eolipiles as then common. They are made, he observes, " of some such material as may endure the fire, having a small hole, at which they are filled with water, and out of which (when the vessels are heated) the air doth issue forth with a strong and lasting violence. These *are frequently used* for the exciting and contracting of heat *in the melting of glasses or metals*. They may also be contrived to be serviceable for sundry other pleasant uses, as for the moving of sails in a chimney corner, the motion of which sails may be applied to the turning of a spit, or the like." (Book ii, chap. 1.) Kircher has given a figure of an eolipile turning a joint of meat, (as indicated by the Bishop) in the first volume

Chap. 3.]   *Smelting Ore with the blast of an Eolipile.*   397

of his *Mundus Subterraneus*, page 203.) We do not remember to have met with a figure of an eolipile applied to the fusing of glass or metal, except in the *Aula Subterranea* of Lazarus Erckers (or Erckern) on Metallurgy, published in German, in 1672, and which, like that of Agricola, is illustrated with numerous cuts. The author was superintendent of the mines of Hungary, Germany, and the Tyrol, under three Emperors, and his work is said to contain every thing necessary to be known in the assaying of metals. The annexed figure is copied from the fifth edition, (with notes) published at Frankfort on the Mayn, in 1736. It is named *Eine tupfferne kugel darinn wasser ist, wird ubers feuer gesekt, und an statt Eines blas-balgs gebraacht,* and is represented as smelting copper ore in a cupelo furnace. Erckers has figured it twice—at pages 1 and 136.

No. 185. Smelting ore with the blast of an Eolipile.

It is not a little singular that this mode of increasing the intensity of fires by a jet of steam directed into the burning fuel has recently been patented both in this country and Europe. It does not however appear to have answered the expectations formed of it, since it has never come into general use, nor are we aware that it is at present employed at all. Two obvious discrepancies between ancient and modern applications of steam for such purposes may here be noticed, since they will, we think, account for the failure of the latter: one is in the nature of the fuel—the other in the temperature of the blast. In the old eolipiles, the steam, having but a very minute passage through which to escape, was raised to a temperature which far exceeded that which is generated in ordinary steam engine boilers—the vapor was perfectly invisible, and its escape only known by the sound of the blast, and its effect on the fire. But in late experiments the current consisted of steam loaded with moisture—a mass of aqueous globules poured into the fire, instead of the rarefied and glowing *aura* that rushed with such impetuous velocity from eolipiles. The powerful effect of the latter on fires of wood and charcoal is unquestionable, but the results of similar blasts on other kinds of fuel (as stone coal) has not yet we believe been sufficiently ascertained. Another difference consisted in the dimensions of the volumes of the blasts :—the one from the eolipile was small and compact—that of the other large and diffuse, a circumstance that may account still further for the different results; for it should be remembered that in using an eolipile it is not the jet of steam alone that is impelled against the burning fuel, but a volume of atmospheric air is set in motion by the blast and carried into the fire along with it: the same thing takes place in using a common bellows, more air being forced against the fire than what issues from the nozzle ; and hence as the velocity of the jet from an eolipile was much greater and the jet itself smaller than those of modern applications of steam for the same purpose, a much larger proportion of air was also borne along with it. It is probable that on particular occasions the ancients filled them with oil or spirituous liquors instead of water, in order to promote a more rapid combustion.

The idea of increasing the heat of fires by water is very old. Pliny says that charcoal which has been wetted gives out more heat than that which is always kept dry. (Nat. Hist. B. xxxiii, cap. 5.) Dr. Fryer speaks of "water cast on sea coal" rendering the heat more intense. (Travels, Lon. 1698, p. 290.)

If there was no evidence that eolipiles had been moulded into various shapes of men, animals, &c. we might have concluded that such was the fact from what is known of the practice of the ancients. Whenever the design, action or movement of a machine or implement corresponded at all with those of men, it was sure to resemble them in form, if its use could possibly admit of it. The taste for such things was universal in former times, and is to a certain extent indulged in all times. It seems inherent in savage people; hence their grotesque and monstrous statues or idols, speaking heads and other androidii of the old mechanics. There has in fact always been a predominating disposition to imitate the human form; and in accordance with it, eolipiles were made to assume the figures of men, boys, &c. the blast escaping from the eyes, mouth, or other parts of the figure. Even so late as the seventeenth century we are told that "to render eolipiles *more agreeable,* they commonly make them in the form of a head, with a hole at the mouth." (Ozanam's *Mathematical and Physical Recreations,* English translation, London, 1708, 419.) It was indeed natural that these machines should be made to resemble figures of the god from whom they were named. An old one is described in the Encyclopedia of Antiquities as "made in the shape of a short fat man with very slender arms, in a curious wig, cheeks extremely swollen, a hole behind for filling it, and a small one at the mouth for the blast."

Most readers are aware that *tenures* by which lands were held in the middle ages were often based on the most trifling and ridiculous considerations. Camden has noticed a great number in his Brittannia, and in the description of the county of Suffolk, there is one which seems to have had reference to the employment of an eolipile; but whether it had or not, there is in Dr. Plott's History of Shropshire an account of one of these "merry tenures" in which blowing the fire with an eolipile formed part of the duty required. The instrument was of the human form and designated, like many other domestic utensils, by the soubriquet "*Jack.*" "Jack of Hilton, a little hollow image of brass, about twelve inches high, with his right hand on his head and his left on pego," *blows the fire* in Hilton-hall every new year's day, while the Lord of Essington drives a goose three times round it, before it is to be roasted and eaten by the Lord of Hilton or his deputy. In some accounts it is stated that the image blew the fire while the goose was roasting, which is more probable than the other. The custom is supposed to have been continued at Hilton-hall from the tenth or eleventh to the seventeenth century. This image is considered by some writers as an ancient idol.

From the above use of eolipiles it will be perceived that there is a similar analogy between them and machines to raise water by steam, as between pumps and bellows; every device for blowing a fire having been used to raise liquids.

It will readily be imagined that these blowing images offered too many advantages to escape being pressed into the secret services of the temples, even supposing they did not originate in them. By charging the interior with different fluids the results could be varied according to circumstances, and if an inflammable liquid was employed, as oil, spirits of wine, turpentine, &c. &c. streams and flashes of fire could be made to shoot from any or every part of the figure. Enough is known to convince us that such things were often done. Notwithstanding all the care of the old priests to conceal, and when concealment was impracticable to destroy their apparatus, some specimens of their machinery have come down. In the fifteenth or early part of the sixteenth century, an eolipilic idol of the ancient Germans was found in making some excavations, and is we believe still extant. A figure of it is inserted in the second volume of Montfaucon's Antiquities. It is made of a peculiar species of bronze and is between three and four feet in height, and the body two and a half in circumference. Its appearance is very uncouth. It is without drapery, with one knee on the ground, the right hand on the head, and the left, which is broken off, rested upon the thigh. The cavity for the liquid holds about seven gallons, and there are two openings for the escape of the vapor, one at the mouth and the other in the forehead. These openings were stopped with plugs of wood, and the priests had secret means of applying the fire. It appears from Weber and other German writers on the subject that this idol was made to express various passions of the deity it represented, with a view to extort offerings and sacrifices from the deluded worshippers; and that the liquid was inflammable. When the demands of the priests were not complied with, the ire of the god was expressed by sweat (steam) oozing from all parts of his body; and if the people still remained obdurate, his fury became terrible: murmurs, bellowings, and even thunderbolts (the wooden plugs) burst from him; flashes or streams of fire rushed from his mouth and head, and presently he was enveloped in clouds of smoke; when the people, horror stricken, consented to comply with the requisitions. It is very evident from the accounts that the priests had the means of rapidly increasing or diminishing the intensity of the fire, as the disposition of the worshippers required the idol to express approbation or displeasure. It further appears that the monks in the middle ages made use of this idol, and found it not the least effectual of their wonder-working machines. It was in fact in this manner chiefly that the great body of ecclesiastics then maintained their influence over the multitude. The very same devices which their predecessors had found effectual in the temples of Osiris, Ceres, and Bacchus, were repeated; and such images of the heathen gods and goddesses as had escaped destruction were converted into those of Christian saints, and being repaired were made to perform the same miracles which they had done before in pagan Greece and Rome. Monks, as we have before observed, were then the most expert mechanicians, and some of their most elaborate productions were imitations of ancient androidii—and the speaking heads of Bacon, Robert of Lincoln, Gerbert and Albertus, were considered proofs of an intercourse subsisting between their owners and spirits, as much so as in the cases of Orpheus and Odin, and other magicians of old.

The name of the German idol is written differently: Puster, Pluster, Plusterich, Buestard, Busterich, are all names given to it and the deity it represents. The name is said to be derived from the Saxon verb *pusten*, to blow—or *puster*, a bellows: this shows its connection with the eolipile as a "fire blower"; and it is probable that from these eolipilic idols the term *Æolist*, "a pretender to inspiration," is derived. (See Dictionary Trevoux. Art. Puster.) This ancient steam idol was, A. D. 1546, placed for safe keeping in the fortress of Sunderhausen, where it remained during the last century.

How singular that steam should have been among the motive agents of the most ancient idols of Egypt (as the Statue of Memnon and others) and in some of the deified images of Europe! That it should formerly have been employed with tremendous effect to delude men, to lock them in ignorance; while it now contributes so largely to enlighten and benefit mankind. These instances of early applications of steam make us regret that detailed descriptions of the various apparatus have not been preserved. Many ingenious devices were evidently employed, and although we condemn the contrivers of such as were used for purposes of delusion, we cannot but admire the ingenuity which even these men displayed, in exhibiting before a barbarous people their gods in the most imposing manner and with such terrific effect—in making idols express by means of steam approbation and anger with the voice of thunder or the hissing of dragons, and causing them to appear and disappear in clouds of smoke and sheets of flame.

It is probable from the antiquity of these idols and of eolipiles that allusions to both might be found in the Bible. May not such expressions as "the blast of his mouth," "the blast of the terrible ones," "the blast of his nostrils," &c. have reference to eolipiles or steam idols of old? "Their molten images [says Isaiah] are wind and confusion." Hospitably receiving a traveler into the house during a storm, and protecting him from the inconvenient heat of the fire when urged by an eolipile, *may* be alluded to by the same prophet in the following passage: "Thou hast been a strength to the poor, a strength to the needy in his distress, a refuge from the storm, a shadow from the heat when the blast of the terrible ones is as a storm against the walls." The expression 'terrible ones,' probably referring to the *hideous forms* into which we have already seen those blowing instruments were moulded. Eolus the god of winds was represented "with swoln cheeks, like one who with main force blows a blast, with wings on his shoulders and a fiery countenance." Idols were always made of a terrific form, and are so made by barbarous people at the present day. When God is personified as *blowing on the fire*, is there not an allusion to these instruments?

Eusebius, in the third book of his life of Constantine, says that when images were subverted, among other things found in some of them were "small faggots of sticks"—perhaps the remains of fuel employed to raise steam in them.[a]

From the observation of one of the early travelers into the East, it seems that eolipiles were employed even in war and with great effect. Carpini, in the account of his travels, A. D. 1286, describes a species of eolipile of the human form, and apparently charged with an inflammable liquid, as having been used in a battle between the Mongals and the troops of Prester John. The latter, he says, caused a number of hollow figures to be made of copper, which resembled men, and being charged with some combustible substance, "were set upon horses, each having a man behind on the horse with a pair of bellows, to stir up the fire. When approaching to give battle, these mounted images were first sent forward against the enemy, and the men who rode behind set fire by some means to the combustibles, and blew strongly with their bellows; and the Mongal men and horses were burnt with wild fire and the air was darkened with smoke."[b] Supposing these eolipiles to have been charged with alcohol or spirit of wine, they must have been (as we see they were) of terrible effect, since, as modern experiments show, a jet of flame from each might have extended to a distance of twenty-five or thirty feet.

Besides blowing directly upon or against a fire, eolipiles were employed to *increase the draft* of chimneys, for which purpose the jet rose perpendicularly from the centre of the dome, as in No. 181. One or two stand-

---

[a] Peter Martyr's Common Places, Part ii, 336. [b] Kerr's Collection of Voyages, vol. 1, 135.

ing on the hearth and heated by the fire, close to which they were placed, the vapor rushed through the orifice and drove the smoke before it; and at the same time induced a current of atmospheric air to follow in the same direction. Sometimes those designed for this purpose had a handle or bail to suspend them over the fire, as No. 183. As several ancient domestic customs still prevail in Italy, and numerous culinary and other implements found in Herculaneum and Pompeii are similar to those now used, it might be supposed that some relics of eolipiles and their uses would be still met with in that country. The supposition has been verified; for we are informed that these instruments are, or were in the seventeenth century, " commonly made use of in Italy to cure smoaky chimneys, for being hung over the fire, the blast arising from them carries up the loitering smoke along with it"—and again, " an eolipile has been sometimes placed in a chimney where it can be heated, the vapor of which serves to drive the smoke up the chimney." This application of steam, it will be perceived, is similar to that lately adopted to increase the draft of chimneys of locomotive carriages.

Rivius mentions another use of eolipiles. He says some were made of gold, silver and other costly metals, and were filled with scented water, " to cause a pleasant temperature, to refresh the spirit and rejoice the heart, not only of the healthy but also of the sick." He observes that they were used for these purposes in the halls and chambers of the wealthy. Rhenanus, an old German writer, who died in 1547, enumerating the treasures belonging to the ancient church at Mentz, mentions eolipiles in the form of " silver cranes, in the belly of which was put fire" and which gave out " a sweete savour of perfumes by the open beake." Seneca has observed that perfumes were sometimes disseminated in the amphitheatres, by being mixed with boiling water, so that the odor rose and was diffused by the steam. We learn from Shakespeare that perfuming rooms was common in his time, the neglect of cleanliness rendering such operations necessary. It is probable that he refers to the same process as that mentioned by Rivius. "Being entertained for a perfumer, as I was smoking a room." " *Much ado about Nothing*," Act 1, Scene 3.

Eolipiles were also employed to produce music. By adapting trumpets, flutes, clarionets, and other wind instruments to the neck or orifice of one, they were sounded as by currents of air. This application of eolipiles is probably coëval with their invention. It is indeed only a variation of the supposed musical apparatus of the Memnonian Statue, and of the devices described by Heron. All the old writers on eolipiles mention it. Fludd figures a variety of instruments sounded by currents of steam; and Rivius, after noticing the use of eolipiles for blowing fires and fumigating rooms, observes "they are also made to produce music, the steam passing through reeds or organ pipes, so as to cause astonishment in those who have no idea of such wonderful operations." Gerbert applied eolipiles in place of bellows to sound an organ at Rheims in the tenth century; and the instrument according to William of Malmsbury was extant two hundred years afterwards. (During the middle ages, the churchmen were the only organ makers; and even so late as the sixteenth century, they retained the manufacture chiefly in their own hands: in the household book of Henry VIII. mention is made of two payments of ten pounds each to John, or " Sir John, the organ maker," of whom the editor says, ' it is almost certain that he was a priest.')

The preceding notice of eolipiles is due to them as the true germ of modern steam engines, for such they were, whether the latter be considered as devices for raising water only, or as machines to move others. We have seen that the oldest apparatus moved by steam, of which there is any account, was an eolipile suspended on its axis, at once both boiler and engine, (No. 180) and we shall find that the first attempts to raise water by the same fluid were made with the same instruments. Indeed, all the early experiments on steam were made with eolipiles, and all the first steam machines were nothing else.

CHAPTER IV.

Employment of steam in former times—Claims of various people to the steam engine—Application of steam as a motive agent, perceived by Roger Bacon—Other modern inventions and discoveries known to him—Spanish steam-ship in 1543—Official documents relating to it—Remarks on these—Antiquity of paddle-wheels as propellers—Project of the author for propelling vessels—Experiments on steam in the sixteenth century—Jerome Cardan—Vacuum formed by the condensation of steam, known to the Alchymists—Experiments from Fludd—Others from Porta—Expansive force of steam illustrated by old authors—Interesting example of raising water by steam from Porta—Mathesius, Canini and Besson—Device for raising hot water from Decaus—Invention of the steam engine claimed by Arago for France—Nothing new in the apparatus of Decaus, nor in the principle of its operation—Hot springs—Geysers—Boilers with tubular spouts—Eolipiles—Observations on Decaus—Writings of Porta—Claims of Arago in behalf of Decaus untenable—Instances of hot water raised by steam in the arts—Manufacture of soap—Discovery of iodine—Ancient soap makers—Soap vats in Pompeii—Manipulations of ancient mechanics—Loss of ancient writings—Large sums anciently expended on soap—Logic of Omar.

It will have been perceived from the preceding chapter that eolipiles for blowing fires and for other purposes were formerly common, and consequently that people were familiar with the generation of steam, and of *high* steam too, long before modern steam engines were known. Of the applications of this fluid to produce motion or raise liquids, during the long period that intervened between the time of Heron and the introduction of printing into Europe, scarcely any thing is known; yet there can be no doubt that it was occasionally used to a limited extent for one purpose or the other, and perhaps for both.

As the origin and early progress of the steam engine are necessarily connected with this part of our subject, the inquisitive reader will not object to dwell a little upon it, although some parts of the detail do not relate directly to the elevation of liquids.

From the important and increasing influence of the steam engine on human affairs, a controversy has arisen between writers of different nations respecting the claims of their countrymen to its invention; and some acrimonious feelings have been displayed. This is to be regretted as fostering prejudices and passions which it is the province of philosophers to eradicate—not to cherish. National vauntings may form articles in the creed, as they are made to contribute to the capital of politicians; but should find no place in that of a savan. Philosophy, like Christianity, contemplates mankind as one family, and recognizes no sectional boasting. Neither science nor the arts are confined by degrees of longitude, nor are the scintillations of genius to be measured by degrees from the equator. As in the republic of letters, so in that of science and the arts, geographical distinctions respecting the abode of its citizens should be unknown.

A few scattered relics of ingenious men who flourished in the dark ages are still extant, which serve to convince us that experimental researches of some of the monks and other ardent inquirers after knowledge in those times were more extensive, and evinced a more thorough acquaintance with the principles of natural philosophy, than is generally surmised. The following remarks of *Roger Bacon* are an instance. From them we may safely infer that he was aware of the elastic force of steam and its applicability to propel vessels on water and carriages on land. That he was acquainted with gunpowder is generally admitted, and it would seem that neither diving bells nor suspension bridges escaped him: " Men may construct for the wants of navigation such machines that the greatest vessels, directed by a single man, shall cut through the rivers and seas with more rapidity than if they were propelled by rowers; chariots may be constructed which, without horses, shall run with immeasurable speed. Men may conceive machines which could bear the diver, without danger, to the depth of the waters. Men could invent a multitude of other engines and useful instruments, such as bridges that shall span the broadest rivers without any intermediate support. Art has its thunders more terrible than those of heaven. A small quantity of matter produces a horrible explosion, accompanied by a bright light; and this may be repeated so as to destroy a city or entire battalions."

Bacon was not a man to speak or write in this manner at random. His *experiments* led him to the conclusions he has thus recorded, for he was by far the most talented and indefatigable experimental philosopher of his age. His discoveries however were not understood, or their importance not appreciated, for he was imprisoned ten years as a practiser of magic, &c. There is a remark in his treatise " on the secret works of art and nature," that is too valuable to be omitted: he says a person who is perfectly acquainted with the manner that *nature observes in her operations*, can not only rival but surpass her. " That he was acquainted with the *rarefaction of air*, and the structure of the *air pump*, is past contradiction." He was (says Dr. Friend) the miracle of the times he lived in, and the greatest genius perhaps for mechanical knowledge which ever appeared in the world since Archimedes. The camera obscura and telescope were known to him, and he has described the mode of making reading glasses. Most of the operations now used in chemistry are said to be described or mentioned by him. A description of his laboratory and of the experiments he made, with a sketch of the various apparatus employed, would have been infinitely more valuable than all the volumes on scholastic divinity that were ever written.

In 1543, a naval officer under Charles V. is said to have propelled a ship of two hundred tons, by steam, in the harbor of Barcelona. No account of his machinery is extant, except that he had a large copper boiler, and that paddle wheels were suspended over the sides of the vessel. Like all old inventors he refused to explain the mechanism. The following account was furnished for publication by the superintendent of the Spanish royal archives. " *Blasco de Garay,* a captain in the navy, proposed in 1543, to the Emperor and King, Charles the Fifth, a machine to propel large boats and ships, even in calm weather, without oars or sails. In spite of the impediments and the opposition which this project met with, the Emperor ordered a trial to be made of it in the port of Barcelona, which in fact took place on the 17th of the month of June, of the said year 1543. Garay would not explain the particulars of his discovery: it was evident however during the experiment that it consisted in a large copper of boiling water, and in moving wheels attached to either side of the ship. The experiment was tried on a ship of two hundred tons, called the Trinity, which came from Colibre to discharge a cargo of corn at Barcelona, of which Peter de Scarza was captain. By order of Charles V, Don Henry de Toledo the governor, Don Pedro de Cordova, the treasurer Ravago, and the vice chancellor, and intendant of Catalonia witnessed the experiment. In the reports made to the emperor and to the prince, this ingenious invention was generally approved, particularly on account of the promptness and facility with which the ship was made to go about. The treasurer Ravago, an enemy to the project, said that the vessel could be propelled two leagues in three hours—that the machine was complicated and expensive—and that there would be an exposure to danger in case the boiler should burst. The other commissioners affirmed that the vessel tacked with the same rapidity as a galley manœuvred in the ordinary way, and went at least a league an hour. As soon as the experiment was made Garay took the whole machine with which he had furnished the vessel, leaving only the wooden part in the arsenal at Barcelona, and keeping all the rest for himself. In spite of Ravago's opposition, the invention was approved, and if the expedition in which Charles the Vth was then engaged had not prevented, he would no doubt have encouraged it. Nevertheless, the emperor promoted the inventor one grade, made him a present of two hundred thousand maravedis, and ordered the expense to be paid out of the treasury, and granted him besides many other favors."

"This account is derived from the documents and original registers kept in the Royal Archives of Simuncas, among the commercial papers of Catalonia, and from those of the military and naval departments for the said year, 1543.   THOMAS GONZALEZ.

"Simuncas, August 27, 1825."

From this account it has been inferred that *steam vessels* were invented in Spain, being only revived in modern times; and that Blasco de Garay should be regarded as the inventor of the first *steam engine*. As long as the authenticity of the document is admitted and no earlier experiment adduced, it is difficult to perceive how such a conclusion can be avoided; at least so far as *steam vessels* are concerned. It may appear singular that this specimen of mechanical skill should have been matured in that country; but at the time referred to, Spain was probably the most promising scene for the display of such operations. Every one knows that half a century before, Columbus could find a patron no where else. The great loss which Charles sustained in his fleet before Algiers the previous year, must have convinced him of the value of an invention by which ships could be propelled without oars or sails; and there is nothing improbable in supposing the loss on that occasion (fifteen ships of war and one hundred and forty transports, in which eight thousand men perished and Charles himself narrowly escaped) was one principal reason for Captain Garay to bring forward his project. M. Arago, who advocates with peculiar eloquence and zeal the claims of Decaus and Papin, as inventors of the steam engine, thinks the document should be set aside for the following reasons: 1st. Because it was not *printed* in 1543. 2d. It does not sufficiently prove that steam was the motive agent. 3d. If Captain Garay really did employ a steam engine, it was "according to all appearance" the reäcting eolipile of Heron, and therefore nothing new. To us there does not appear much force in these reasons. M. Arago observes, " manuscript documents cannot have any value with the public, because, generally, it has no means whatever of verifying the date assigned to them." To a limited extent this may be admitted. Respecting private MSS. it may be true; but surely *official* and national records like those referred to by the Spanish secretary should be excepted. We have in the eighth chapter of our Third Book quoted largely from official MS. documents belonging to this city, (New-York:) now these are preserved in a public office and may be examined to verify our extracts as well as their own authenticity: and the Spanish records we presume are equally accessible, and their authenticity may be equally established. The mere printing of both could add nothing to their credibility, although it would afford to the public greater facilities of judging of their claims to it. So far from rejecting such sources of information respecting the arts of former times, we should have supposed they were unexceptionable.

But it is said—although a boiler is mentioned, that is not sufficient proof that *steam* was the impelling agent, since there are various machines in which fire is used under a boiler, without that fluid having any thing to do with the operations: Well, but the account states that which really appears conclusive on this point, viz. that this vessel contained "*boiling water*" and that Ravago the treasurer, opposed the scheme on the ground that there would be an exposure to danger "in case the boiler should burst." As this danger could not arise from the liquid contents merely, but from the accumulation of steam, (the irresistible force of which was, as has been observed, well known from the employment of eolipiles) it is obvious enough that this fluid performed an essential part in the operation—in other words was the source of the motive power. Had it not been necessary, Garay would never have furnished in it such a plausible pretext for opposition to his project. It has been also said "if we were to admit that the machine of Garay was set in motion by steam, it would not necessarily follow that the invention [steam engine] was new, and that it bore any resemblance to those of our day." True, but it would at least follow that Garay should be considered the father of *steam navigation*, until some earlier and actual experiment is produced. Arago further thinks, that if Garay used steam at all, his engine was the whirling eolipile (No. 180)—" every thing" he observes would lead us to believe that he employed this. We regret to say there are strong objections to such an opinion. That an engine acting on the same principle of recoil as Heron's eolipile *might* have been made to propel a vessel of two hundred tons is admitted; but from modern experiments with small engines of this description, we know; 1st, that in order to produce the reported result, the elasticity of the steam employed *must* have been equivalent to a pressure of several atmospheres; and 2d, that the enormous consumption of the fluid when used in one of these engines must have required either a number of boilers or one of extraordinary dimensions. Had Garay employed several boilers, the principal difficulty would be removed, as he might then have made them sufficiently strong to resist the pressure of the confined vapor; he however used but one, and every person who has witnessed the operation of reäcting engines will admit that a single boiler could hardly have been made to furnish the *quantity of steam* required, *at the requisite degree of tension.*

As the nature of this Spanish engine is not mentioned, every person is left to form his own opinion of it. We see no difficulty in admitting that he employed the elastic force of steam to push a piston to and fro—or that he formed a vacuum under one by condensing the vapor. Such applications of steam were as likely to occur to a person deeply engaged in devising modes of employing it, in the sixteenth as well as in the seventeenth century, notwithstanding the objection so often reiterated, that the arts were not sufficiently matured for the fabrication of a metallic cylinder and piston, and apparatus for transmitting the movements of a piston to revolving mechanism. The casting and boring of pieces of ordnance show that the construction of a steam cylinder was not beyond the arts of the sixteenth century, or even of the two preceding ones; while the water-works, consisting of forcing pumps worked by wheels, and also numerous other machines put in motion by cranks, (and the irregularity of their movements being also regulated by fly wheels) described in the works of Besson, Agricola, &c. show that engineers at that time well understood the means of converting rotary into rectilinear motions, and rectilinear into rotary ones.

Had Garay used a steam apparatus on the principle of Savery's, Papin's, or Leopold's, to raise water upon an overshot wheel fixed on the same axle as the paddles, we should probably have heard of it, since such a wheel would have been a more prominent object than the paddles or the boiler itself.

It need not excite surprise that Garay adopted paddle wheels as propellers, since they were well known before his time, being of very ancient date. Roman galleys were occasionally moved by them, and they have probably never been wholly laid aside in Europe since the fall of the empire. Stuart, in his Anecdotes of the Steam Engine, observes that the substitution of them for oars is mentioned in several old military treatises. In some ancient MSS. in the King of France's library it is said that boats, in which a Roman army under Claudius Candex were transported into Sicily, were propelled by wheels moved by oxen. An ancient bas-relief has also been found representing a galley with *three* wheels on each side; the whole being moved by three pair of oxen. Robertus Valturius, in his *De Re Militari*, Verona, 1472, gives a figure of a galley with *five* paddle wheels on each side. Another is portrayed with one on each side. To these we add another from the Nuremburg Chronicle, published in 1497; at folio XCVIII a vessel is figured with *two* wheels on the side that is represented. An old English writer mentions them in 1578; and in 1682, a horse tow-boat with paddle wheels was used at Chatham, England.

Of various substitutes for revolving oars or paddle wheels, there is one which, among other things, we have long purposed to try. It consists in protruding into the water, in a horizontal direction from close receptacles formed in the stern and below the water line, a series of two or more solid, or tight hollow bodies, of such dimensions that the water displaced might afford a resistance sufficient to drive forward the boat. Some idea of this resistance may be obtained by attempting to sink an empty barrel or hogshead, or by pushing a bucket or washing tub into a liquid, bottom downwards. The moveable bodies or propellers might be square boxes of wood, closed tight and made to slide in and out at the stern like the drawers of a bureau; their outer ends being flush with the stern when drawn in, and the joint (at the stern) made tight by some contrivance analogous to a stuffing box; their velocity and length of stroke being proportioned to the size of the vessel and its required speed. The water itself would drive or help to drive back each propeller at the termination of its stroke, just as a hollow vessel is pushed up when thrust under water. The receptacles might be open at the top so as to allow any water which leaked in at the joint to be readily discharged. We are not aware that such a plan has ever been proposed.

Chap. 4.] *Raising Water by the condensation of Steam.* 407

There are several indications that mechanicians in different parts of Europe, were alive to the power developed by steam at the time Garay was making his experiments; and we have little doubt that interesting information respecting it will yet be obtained from the obsolete tomes of the XV and XVI centuries. Those old authors, whose works are generally quoted on the subject, obviously derived their information principally from those of their predecessors as well as from the laboratories of the alchymists.

Jerome Cardan, an Italian, born in 1501 and died in 1575, one of the most eccentric geniuses that ever lived, in whom was united "the most transcendent attainments with the most consummate quackery, profound sagacity with the weakest superstition; who on one page is seen drawing the horoscope of Christ, and in another imploring his forgiveness for the sin of having eaten a partridge on Friday; unfolding the most beautiful relations in algebraic analysis, and foretelling from the appearance of specks on his nails his approach to some discovery; above all, eloquently enforcing the obligations of a pure religion and expressing the finest sentiments in morals, while his long life was one continued exertion, grossly outraging both. Here, this philosopher, juggler and madman, is entitled to brief mention from displaying in his writings a knowledge of what has been called the capabilities of steam, and more particularly with the fact of a vacuum being speedily procured by its condensation."

N 186. Fludd.

That the alchymists were familiar with the formation of a vacuum by the condensation of steam, and with raising water into it by atmospheric pressure is certain. Their ordinary manipulations necessarily made them acquainted with both. In Fludd's *Integrum Morborum Mysterium*, page 462, he illustrates his notions respecting fever and dropsy, by what he calls a *common* experiment, and with the apparatus figured in the cut. An empty retort or one containing a little water was suspended over a fire with the neck turned down into a vessel of water: when the retort was heated the air or vapor became expanded and part of it driven out through the liquid. Upon removing the fire, the water was forced by the atmosphere through the neck to supply the place of the air or vapor expelled by the heat. This although nothing more than the old process of filling eolipiles, most of which could be charged in no other way, shows that the principle was well understood and adopted in various operations. We add another and earlier example from Porta's *Natural Magic*, a work first published in 1560, where he distinctly shows the formation of a vacuum by the condensation of steam, and raising of water into it by the atmosphere. "Make a vessel with a very long neck; the longer it is, the greater wonder it will seem to be. Let it be of transparent glass that you may see the water running up: fill this with *boiling water*, and when it is very hot, or setting the bottom of it to the fire that it may not presentlie wax cold; the mouth being turned downwards that it may touch the water, it will suck it all in." Discharging the hot water is not mentioned, but that is of course implied, and before the vessel was placed on the fire—while full of hot water, it could not suck up any of the cold. (Book 19, cap. 3.)

That the same laborious experimenters were acquainted with the property of steam to displace liquids from close vessels is equally clear. Many of their operations made them familiar with the fact that in this respect its effects were similar to those of compressed air. Portions of their apparatus were admirably adapted to produce jets of water by means of steam—the mere opening of a cock to draw off the liquid contents of a heated alembic would often illustrate the operation, just as the overturning of an eolipile, or inclining one till the orifice was covered with water, would do.

So far as relates to the principles of raising liquids into a vacuum formed by the condensation of steam, and forcibly ejecting them by its elasticity, nothing new was discovered by Decaus, Worcester, Savery, or Papin: both operations had long been performed with eolipiles, and were of common occurrence in laboratories. It was in the extension of these operations to hydraulic purposes that the merits of those last named consisted. 'Draining machines' were wholly out of the track of the transmuters of metals—the design of such contrivances was one which few if any of them would have stooped to pursue. Had they made the raising of water by steam a subject of particular study, hardly one of them could have failed to produce a machine similar to Savery's, for every element of it was in their possession and in constant use. 'Tis true we have as yet referred only to the expulsion of *hot* water from close vessels,

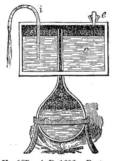

No. 187. A. D. 1606. Porta.

but the application of steam to drive cold liquids from a separate vessel was not unknown. Of this there is an incidental but very conclusive proof in a book of Porta's, entitled *Spiritali*, (named after Heron's work) originally published in Latin in 1601, and five years after in Italian and Spanish. In the translation of 1606, is the annexed figure No. 187, designed to show "into how many parts a simple portion of water may be transformed" i. e. by measuring the quantity expelled from a close vessel, by vapor evolved from a certain quantity heated in a retort. " Make a box of glass or tin, (c) the bottom of which should be pierced with a hole, through which shall pass the neck of a bottle (a) used for distilling, containing one or two

ounces of water. The neck shall be soldered to the bottom of the box so that nothing can escape there. From the same bottom shall proceed a pipe, (*i*) the opening of which shall almost touch it, leaving just room enough between them for the water to run. This pipe shall pass through an opening in the lid of the box, and extend itself on the outside to a small distance from its surface. The box must be filled with water by a funnel (*e*) which is afterwards to be well closed, so as not to allow the air [steam] to escape:—finally, the bottle must be placed upon the fire and heated a little; then the water, changed into steam, will act violently upon the water in the box, and will make it pass through the pipe (*i*) and flow off on the outside," &c.[a] This apparatus although designed merely to illustrate the relative bulk of a volume of water and that of the steam into which it might be converted, yet exhibits in the clearest light the principle afterwards adopted for raising liquids by the elasticity of steam.

---

[a] Arago's History of the Steam Engine, translated by Lieut. Harewood, U. S. N., Journal of the Franklin Institute, Vol. XXV. This device of Porta's was, we believe, first brought forward by Mr. Ainger, an English writer, whose work we have not seen.

The diagram and description, observes Stuart, are so complete, that the application to such a purpose of a similar apparatus could not be considered even as a variation of Porta's idea.

In the first histories of the modern steam engine, its origin was traced to a device for raising water, proposed by the Marquis of Worcester, in his Century of Inventions, a tract written in 1655 and published in 1663. Subsequent researches have brought to light facts (some of which have just been noticed) which prove that steam was applied to that and other purposes long before; and future inquiries will probably produce still earlier examples. Previous to describing other old applications of this fluid, we shall notice some experiments which historians of the steam engine have introduced. Thus Rivault, a French courtier, is said to have "discovered" in 1605 that a tight bomb shell containing water and thrown into a fire would be exploded by the confined vapor—and by Decaus in 1615, that a close copper ball partly filled with water and heated, would be rent asunder with a noise resembling that of a petard—and by the Marquis of Worcester in 1663, that a piece of ordnance would also be exploded, if treated in the same way with its mouth and touch-hole plugged up. Now, the fact which these experiments established (if they were all made) was one with which every person who ever used an eolipile was familiar; and which was no more a new discovery in the beginning of the seventeenth century, than experiments to prove that the cover of a common cauldron might be blown off by the same agent, could have been in the middle of it. It was a knowledge of the same fact that led ancient philosophers to account for the phenomenon of earthquakes—which induced the ministers of the steam deities, mentioned in the last chapter, to regulate the resistance of the plugs which closed the mouth and eyes of the idols, so as to give way before the tension of the steam exceeded the strength of the metal, and blew both them and their gods to atoms. When the Spanish treasurer objected to the project of Garay that the boiler might "burst," he did not dream of having made a discovery of the danger arising from imprisoned steam : had such been the case his objection would have had no force till the fact upon which it was based, had been tested and become generally known—but the ground of his opposition every person of that age could appreciate as well as we can; and it is not improbable that on that ground only was the project abandoned. The same objection still prevents thousands from traveling either in steam boats or steam carriages.

Examples to show that old chemists were as familiar with the same fact almost as with "the cracklings of thorns under a pot," might be quoted in abundance—they are not necessary, but we shall adduce one or two. In Porta's Natural Magic, (Book X, chap. 1, on *Distillation*,) he speaks of regulating the capacity of stills to the various substances treated in them. Such as were of "a flat and vapourous nature" require, he observes, large vessels, "for when the heat shall have raised up the flatulent matter, and it finds itself straighten'd . . . . it will seek some other vent, *and so tear the vessels in pieces*, which will fly about with a great bounce and crack, and not without endamaging the standers by." Again, in the ninth chapter of the same book, he directs that particular care should be taken to make the joints tight "lest the force of the vapours arising may *burst* it [the still] open and scald the faces of the by standers." That such occurrences were not uncommon may be inferred from another remark; (in the 21st chapter of the 10th book,) speaking of "the separation of the elements" and of the various substances distilled, he observes, "we account those *airy* which fill the vessels and receivers and easily *burst* them, and so flie out." These examples are sufficient to prove that the *irresistible* force of steam when confined, was known in the middle of the sixteenth century—in fact it always has been known since distillation was practised or an eolipile used. Particular care was always required to keep the orifice of the latter instrument open when on the fire.

Besides the Natural Magic of Porta and the writings of Cardan, there were other works published in the sixteenth century in which steam is either incidentally mentioned, or expressly treated on. About the year 1560, Mathesius, a German preacher, in order to illustrate the enormous force of a little imprisoned vapor, introduced into a sermon a description of an apparatus "answering to a steam engine"—an instance of ingenuity equal to that of Cardan, who contrived to swell the contents of a treatise on arithmetic, (which he wrote for the booksellers by the page) by expatiating on the motions of the planets, on the creation, and the tower of Babel. Canini, a Venetian, made experiments on steam in 1566. In 1569, an anonymous tract, printed at Orleans, and ascribed to Besson, contains an account of the expansion of water into steam, and the relative volumes of each. About 1597, a German writer proposed the whirling eolipile of Heron, as a substitute for dogs in turning the spit, and recommended it in a passage, an extract from which may be seen at page 76 of this volume.

The "Forcible Movements" of Decaus, or de Caus, is the next authority for early notices of steam. This work was first published at Frankfort in 1615, and in Paris in 1624. It is entitled *Les Raisons des Forces Mouvantes*, avec diverses machines tant utiles que plaisantes, &c:—Reasons of moving forces, with various machines both useful and interesting. The title seems to have been slightly changed in different editions; and, as noticed at page 319, the name of the author also; a circumstance that has led Mr. Farey to suppose there were two books, written by different authors of the same name. In the English translation of 1659, which consists of two parts: "The theorie of the conduct of water" and the "Forcible movements," the theorems on steam are omitted. By these theorems Decaus intended to show that heat carries off water by evaporation—that steam when condensed returns to its original bulk—and that a hollow ball or eolipile may be exploded by it. The only device for employing this fluid which he has given, is in illustration of the fifth theorem, viz: *Water may be raised above its level by means of fire:* "The third method of raising water is by the aid of fire, whereby diverse machines may be made. I shall here give the description of one. Take a ball of copper marked *A*, well soldered at every part. It must have a vent hole marked *D* by which water may be introduced; and also a tube marked *C*, soldered into the top of the ball, and the end *C* reaching nearly to the bottom, but not touching it. After filling this ball with water through the vent hole, stop it close and

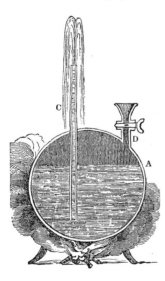

No. 188. Decaus, A.D. 1615.

put the ball on the fire, then the heat acting against the said ball, will cause all the water to rise through the tube   C."

On the supposition that this apparatus was originally designed by Decaus, M. Arago has claimed for France the invention of the steam engine. The English, he observes in his Memoir of Watt, have ascribed the honor to the Marquis of Worcester; but on this side the channel, " we maintain that it belongs to a humble engineer, almost forgotten by our biographers, viz. Solomon de Caus."   And in his 'History of the Steam Engine,' he asserts that " the idea of raising water by the elastic force of steam" belongs to the same individual.   With the disposition and even an anxiety to give to every inventor his full meed of praise, we confess that we cannot perceive in the figure and description before us, sufficient ground from which such inferences could fairly be drawn. The fact is, to no one age or people can the origin of the steam engine be attributed—nor yet its various applications. That some have contributed greatly more than others to develope, mature and apply it, no person doubts.

Were it even admitted that no apparatus precisely like that represented in the figure was previously known, it would be difficult to establish the claims put forward in behalf of Decaus. But there was nothing novel either in its construction or in the principle of its operation; while for nearly all practical purposes it was valueless.

So far as respects the apparatus simply, no part of it was invented by him. It is figured in the Spiritalia as an illustration of Problem IX, viz. a hollow sphere partly filled with water, and resting upon a tripod, with a jet pipe extending down into the liquid. Instead of fire under it to raise steam, a syringe is connected to the upper part, by which to inject air or water. This figure is copied in Plate VII of the " Forcible Movements," (Leak's Trans.) and of it Decaus observes, " as concerning the figure of the globe, it may serve for pleasure to cast the water very high by the pipe, after that you have forced it in with violence with the syringe."   Had not this device of raising water by air compressed with a syringe been found in the Spiritalia, it might also have been deemed the invention of Decaus, for he does not mention the source whence he derived it; and as it is, we think he may with as much reason be considered the author of *air*-engines, as the first inventor of steam engines. The apparatus is also a modification of that by which Heron raised water by the heat of the sun, but the author of the Spiritalia was too well versed in the subject, to introduce in that work such a device as that of Decaus.

The elevation of water by the elastic force of steam was also well known before the time of Decaus. Nature herself has always presented striking proofs of it in boiling springs, and in the magnificent fountains and jets that are thrown up in various parts of the earth from subterraneous cauldrons by imprisoned steam; as in the Geysers of Iceland, where the hot liquid is thus violently forced through natural tubes, of nine or ten feet in diameter, to heights varying from twenty to ninety feet, and accompanied with intermitting volumes of the vapor; phenomena the philosophy of which was well understood by the ancients. But if such examples are deemed too indirect, and as known only to a few, there are others with which people generally have always been conversant: Vessels for heating water, with tubular spouts, whose upper orifices stand higher than the top of the vessels or the liquid within them, are of extreme antiquity; some that resemble our tea-kettles and coffee-pots are found portrayed on the paintings and sculptures of Egypt. Now every one knows that when the covers of these fit so close as to prevent the steam from escaping as fast as it is generated, the confined vapor forces up the hot liquid through the spouts; and in a manner precisely the same as described by Decaus, for the effect is the same whether the discharging tube be connected to the lower side of a boiler like a tea-kettle spout, or inserted through the top and continued within to the liquid. From such domestic exhibitions of the effects of steam, the devices of Heron and other ancient experimenters were probably derived: a person whose thoughts were turned to the subject of raising water by it could not fail to profit by them, or to hit upon so slight a modification of the apparatus as shown in the last figure.

The same application of steam was often exhibited by alchymists as already observed in their manipulations, and in drawing off the contents of their stills and retorts; but it was still more clearly illustrated in common life in the employment of eolipiles, and the copper ball of Decaus was merely one of these with a jet pipe prolonged into the liquid. The very terms "ball of copper," "ball of brass," were those by which eolipiles were designated. (See page 396.) Now no one was ignorant that an opening on the top of one of these instruments let out steam, and that through one near the bottom hot water would be violently expelled through a vertical tube, if attached to the opening. Suppose the one figured at No. 185 either accidentally or designedly placed on the fire with the tube inclined upwards, and heated in that position while two thirds filled with water; the vapor would then accumulate in the dome, and would necessarily drive out the boiling liquid until the lower orifice of the tube was no longer covered with water: or imagine No. 184 inclined till water rushed out instead of steam. That such experiments were not only frequent but common, no person can reasonably doubt, although no notice of them may be found in books. Such a mode of raising water was of little value and not thought worth recording, and but for its introduction into some histories of the steam-engine, we should not have deemed it of sufficient importance to notice. Moreover, the ordinary mode of charging eolipiles which had but one minute orifice, viz. by heating and then plunging them in water, must have frequently caused them to produce liquid jets, in consequence of their imbibing too much, and there being no other way of expelling the surplus than by placing the instrument on the fire. Probably an eolipile was never used that was not occasionally overcharged with the liquid, and thus made to raise a portion of it by the elastic force of steam. At any rate, no one who was familiar with these instruments, from Heron to Decaus, could have been ignorant of the fact that they might be applied to produce jets of hot water as well as of vapor; and few ever used them who did not occasionally make them produce both.

It would be an unjust reflection on Decaus to suppose he could not have given a better plan than No. 188 for raising water *by steam*, if the project had been seriously entertained by him; but there is not the slightest ground to believe he ever dreamt of applying that fluid to hydraulic purposes, or as a substitute for pumps, chains of pots, &c. He certainly would have laughed at any one proposing a device by which water could not be raised until the whole of it was boiled, whether the quantity was a pint, a hogshead, or a million of gallons; and in some cases not until its temperature far exceeded that at which ebullition in open vessels takes place. Why then, it may be asked, did he mention the device at all? Simply to show that "water *may be* raised above its level by means of fire." Well, but he says that "*diverse* machines" may be deduced from it. True, and he has given a description of one, from which we may judge of the rest: these were most likely mere trifles—whims that suited the taste of the age. No. 189 is probably one of them, which a contemporaneous English author adduces under "Experiments of mocions by rarefying water with fier," and of which he also observes, *other devices* may be derived from it.

Decaus appears to have read and traveled much, and to have collected knowledge from every source within his reach. He describes saw-mills that were used in Switzerland, fire-engines of Germany, canal-locks which he noticed between Venice and Padua: he cites Tacitus, Pausanias and Pliny; quotes largely from Heron, and refers his more learned readers to Archimedes, a commentary upon whose writings he promises to undertake. Of course he was acquainted with the works of Porta, for this Neapolitan philosopher and his writings were greatly celebrated throughout Europe. Now had Decaus turned his thoughts at all to the elevation of water by steam, he would at once have perceived the advantages of a device like No. 187, by which the liquid could be raised in unlimited quantities without being heated at all, as well as under all possible circumstances: and having perceived this, would he not (if the project of thus raising water had ever entered his head) have given it, or a modification of it, instead of No. 188? It is clear that he wanted an illustration of a proposition merely, and the one he has given he considered as good as any other.

As long as the *Natural Magic* and the *Spiritali* of Porta are admitted to have been published, the former about fifty and the latter at least ten years before the work of Decaus, there is little if any thing whereon to found a claim for the latter. If we were to concede, what certainly is *not* "established beyond dispute, that the first idea of raising a weight by means of the elastic power of steam belongs to the French author," the *fact* would still remain that the Neapolitan had long before shown *how this could be* DONE; and M. Arago has himself observed, that "in the arts, as in the sciences, the last comer is supposed to be acquainted with the labors of those who preceded him—all denial in this respect is without value." The object of Porta in introducing the device referred to was not to show its application to raise water, and it is not fair to conclude that he was ignorant of its adaptation for that purpose because he has not gone out of his way to point it out. It has also been objected, that his apparatus raised the liquid to a very limited height. We do not know that Decaus's did more, for we are only told that the contents of the ball would be driven out, without the slightest intimation of an elevated discharge. Well, (an advocate of the latter will say) but his apparatus is capable of raising water to all heights. And so is Porta's. But had Porta "the least idea of the great power which steam is susceptible of acquiring?" The extracts which we have given from his Natural Magic, on the rupture of vessels by steam, prove that he was well aware of it; and the book from which these extracts are taken was his earliest production, being published in 1560, at which time (he observes in the preface) he was only about fifteen years old. To conclude, we are constrained to embrace the opinion, notwithstanding the arguments and eloquence of M. Arago, that the device described by Decaus brought to light no new fact, and gave rise to no new or useful result.

Although instances rarely occur in the arts in which the elevation of hot water by the steam evolved from it could be of service, there are some, as in chemical manipulations, in a few breweries and distilleries, and also in soap manufactories. The operation in the latter is worth noticing:—
In the ordinary process of manufacturing common hard soap, three or four tons are often made at once in a deep iron vat or boiler. Into this several

hundred gallons of ley, with the other ingredients, tallow, rosin, lime, &c. are put. After the whole has been several times boiled, the semi-fluid mass is suffered to remain some time at rest, when the ley collecting at the bottom leaves a thick stratum of soap formed above. As no openings are made in the sides or bottom of the boiler, the hot ley is drawn off at the top, and is usually done by a common pump. Long after the fire is withdrawn steam continues to rise from the liquid below; for, from the vast mass of heated materials, their non-conducting property, and that of the furnace, the heat is slowly dissipated. The soap in the mean time acquires a firmer consistence, and prevents the vapor from escaping above almost as effectually as the bottom of the boiler does below; so that not until the steam attains considerable elastic force can it open a passage through, and when it does the opening is instantly closed as before. When therefore the pump is pushed through the whole to the bottom of the vat, and started to work, the liquid continues of itself to pass up, being urged by the steam. (It is necessary to work the pump at first because the openings in its end become stopped with soap in passing it down. The end of a plain tube would be choked in the same way; but by a pump attached to it, the pressure of the atmosphere is added to that of the steam to force the passage open.) The large body of soap keeps settling down as the ley is discharged, and thus preserves the steam at the same degree of tension until all the ley is ejected, when the steam itself escapes also through the pump. The soap, it will be perceived, acts as a flexible piston, its adhesion to the sides of the boiler and its spissitude and weight effectually confining the vapor below.

Of the origin of this mode of raising ley, and the extent to which it is practiced, we are not informed. It affords however an example of the truth of the remark, that important results may be deduced from attention to simple facts, as well as from the observance of common products. An examination of the residuum of a soap-boiler's kettle, it is well known, led to the discovery of a new chemical element, (iodine,) and of its virtues as a specific in the cure of the *goitre;* and from the preceding remarks it may be inferred, that an observing Greek or Roman soap-boiler *might* have discovered the applicability of steam to raise water, since he possessed all the requisite machinery in his ordinary apparatus, and might have performed the operation as often as he made soap. His ingenuity would also have been rewarded by a diminution of his labor. And who can prove that such a plan was *not* in use in some of the old soap-factories of former times? In that, for example, which has been discovered in Pompeii, in one apartment of which are the vats, placed on a level with the ground, and in another were found heaps of lime of so superior a quality as to have excited the admiration of modern manufacturers.[a]

Of the manipulations of ancient mechanics and manufacturers we know little or nothing. Of the thousands of their devices, many valuable ones have certainly been lost. Some of these have been revived or rediscovered in modern times, among which we think may be mentioned various applications of steam. There were indeed so many occasions for the employment of this fluid by the ancients, and particularly in raising of water, that, taken in connection with the information respecting it in the Spiritalia, the part it was made to perform in the temples, the traces of it in the hot

---

[a] Soap must have been an expensive article among the Greeks, at least such as was used in the toilette, if we were to judge from the amount that Demetrius extorted from the Athenians, viz. 250 talents, which, says Plutarch, "he gave to Lamia and his other mistresses *to buy soap.*"

baths at Rome, and the apparatus of Anthemius, by which last it was adapted to a very novel purpose as a motive agent, thus exhibiting resources in its apppplication that could only be derived from experience—we cannot divest ourselves of the idea that the ancients were better acquainted with the mechanical properties of steam and its application to the arts than is commonly supposed.

But for the destruction of the numerous libraries of the ancients, some of which contained volumes that treated on every subject, we should have been intimately acquainted with their arts and machinery; and but for the logic of Omar,[a] we might have been in possession of those treatises on mechanics that Ctesibius studied, and which supplied Heron with materials for the Spiritalia; for the latter refers to inventions and writings of his predecessors, and admits having incorporated some of their productions with his own. Possibly the very books out of which he selected the applications of steam No. 179 and No. 180 might now have been extant. The destruction of such works as these was a severe loss to the world. Had they been saved, the state of society would not, in the following ages, have been so greatly degenerated, nor would the arts have sunk to so low an ebb. Mechanics have therefore as much reason, if not more, to deplore the loss of those volumes that treated on the subjects of their pursuits, as learned men have to regret the destruction of those that related to literature only. It was also easier to replace the latter than the former—to revive the literature than the arts of the ancients; for reflections on history, politics, morals, literature, romance, &c. are more or less common to our race in all times, and in every age men will be found to clothe them, or selections of them, in glowing language; whereas mechanical inventions, though often brought about by the observance of common facts, are frequently the results of fortuitous thoughts which local occurrences or singular circumstances induce, and if once lost can hardly be revived, except by congenial minds under similar circumstances. Besides, it is not by the mere arresting an idea as it floats through the mind that discoveries or improvements in mechanism are effected: on the contrary, it requires to be cultivated and matured by reflection; the accuracy of the device suggested by it has to be tested by models, and these by experiment, before the incipient thought becomes embodied in a working machine.

---

[a] Amrou, his general, having taken Alexandria, wrote for directions respecting the disposition of the famous library which it had been the pride of the Ptolemies to collect. The reply was—if the writings agreed with the doctrines of the Koran, they were *useless*; and if they did not, they ought to be *destroyed*. The argument was irresistible, and the whole were burnt.

CHAPTER V.

Few Inventions formerly recorded—Lord Bacon—His project for draining mines—Thomas Bushell—Ice produced by hydraulic machines—Eolipiles—Branca's application of the blast of one to produce motion—Its inutility—Curious extract from Wilkins—Ramseye's patent for raising water by fire—Manufacture of nitre—Figure illustrating the application of steam, from an old English work—Kircher's device for raising water by steam—John Bate—Antiquity of boys' kites in England—Discovery of atmospheric pressure—Engine of motion—Anecdotes of Oliver Evans and John Fitch—Elasticity and condensation of steam—Steam-engines modifications of guns—A moving piston the essential feature in both—Classification of modern steam engines—Guerricke's apparatus—The same adopted in steam-engines—Guerricke one of the authors of the steam-engine.

How few, how exceedingly few of the conceptions and experimental researches of mechanics have ever been recorded! How many millions of men of genius have passed through life without making their discoveries known! Even since printing was introduced, not a moiety of those who possessed in an unusual degree the faculty of invention have preserved any of their ideas on paper. Of some men celebrated for the novelty of their devices, nothing is known but their names; they have gone, and not a trace of their labors is left. Of others, the title by which they designated their inventions is nearly all that has come down—no particulars by which we might judge of their merits. This is the case with many of the old experimenters on steam, especially those who raised or attempted to raise water by it. Among these we have sometimes thought *Lord Bacon* should have a place, under the impression that he employed, or designed to employ, that fluid to raise water from the deluged mines which he undertook to recover. He obviously had some *new* modes and machines for the purpose. An account of these he laid before the King, (James I) who approved of the project, and consented that the aid of parliament should be invoked. In the "Speech touching the recovery of drown'd mineral works," which Bacon prepared to be delivered before parliament, is the following passage : "And I may assure your Lordships that all my proposals, in order to this great architype, seemed so rational and feasable to my Royal Sovereign, our Christian Salomon, [!] that I thereby prevailed with his Majesty to call this honorable Parliament, to confirm and impower me, in *my own way of mining*, by an act of the same."[a] This great man was therefore in possession of a novel plan of accomplishing one of the most arduous undertakings in practical hydraulics; and so impressed with a belief in its efficiency that the king was induced by him to call, or agree to call, a parliament, chiefly it would seem to give sanction to it. What the plan was, we are not informed, nor is any account of it believed to be extant. Dr. Tenison, (Archbishop of Canterbury) the author of "Baconiana," alluding in 1679 to Bacon's "Mechanical Inventions," observes, "His *instruments* and ways in recovering deserted mines, I can give no account of at all; though certainly, without *new tools*, and *peculiar inventions*, he would never have undertaken that new and hazardous work."[b] That the project consisted chiefly in some peculiar mode of raising the water is certain; and it is worthy of remark that a member of

---

[a] Baconiana. Lond. 1679, p. 133.   [b] "An Account of all the Lord Bacon's Works," subjoined to Baconiana, p. 17.

his household was a mining engineer, and celebrated for the invention or construction of hydraulic engines, viz. "Mr. *Thomas Bushell,* one of his lordship's menial servants; a man skilful in discovering and opening of mines, and famous for his curious *water-works* in Oxfordshire, by which he imitated rain, hail, the rainbow, thunder and lightning."[a] This was probably the same individual who is mentioned in some biographies as "Master of the Royal Mines in Wales," under Charles I.

That the application of steam to drain mines and impart motion to machinery had begun to excite attention in England before the death of Bacon, (in 1626) is very obvious. Of this there are several indications; and within four years of his demise, a *patent* was granted for a method of discharging water "from low pitts *by fire."* Then he was acquainted with the writings of Porta, and consequently with the apparatus No. 187. No experiment or fact of the kind illustrated by this could have escaped him, even if he had not been engaged in the project of recovering flooded mines; and he was, to say the least, as likely as any other man of his age to perceive the adaptation of such an apparatus as No. 187 for raising water, and also to apply it. We hear of no such uses of steam in England before his time, but soon after his death they make their appearance without any one very distinctly to claim them. It may however be said, if Bacon raised water by steam, Bushell, his engineer, would most likely have done the same after the death of the chancellor, and proofs of this fact might be obtained from an examination of the water-works of the latter. Had we any account of these, the question most likely could be settled; but almost the only information we have respecting the machines and labors of Bushell is contained in the extract above, and there is but one particular from which any thing respecting their construction can be inferred, viz.—*hail* is said to have been produced by them. How this was done we know not; possibly by admitting high steam into a close vessel, from which water mixed with air[b] was expelled with a velocity sufficient to produce ice, somewhat in the same manner as the operation is performed by compressed air in the pressure engine described at page 362. The same thing was done by others who we know did experiment on steam, and who performed the operation without the aid of a great fall of water. The Marquis of Worcester makes it the subject of the 18th proposition of his "Century of Inventions," in a fountain which he says a child could invert. And a century before, Cornelius Drebble "made certain machines which produced rain, *hail* and lightning, as naturally as if these effects proceeded from the sky."

But whether Lord Bacon used steam or not—and it must be admitted that there is no direct evidence that he did—it is interesting to know that his great mind was bent to the subject of raising water on the most extensive scale, and this too at the time when steam first began to be proposed for that purpose in England. On this account, if on no other, are his labors entitled to notice here.[c]

---

[a] Account of Lord Bacon's Works, p. 19.

[b] Dr. (afterwards Bishop) Burnett, in his Letters from Italy, noticing the water-works at Frescati, observes, "the mixture of wind with the water and the thunder and storms that this maketh, is noble." 3d edition, Rotterdam, 1687, p. 245.

[c] Lord Bacon seems to have been greatly interested in mining and in the reduction, compounding and working of metals. In his treatise on the Advancement of Learning he divides natural philosophy into the mine and furnace, and philosophers into pioneers and smiths, or diggers and hammerers; the former being engaged in the inquisition of causes, and the latter in the production of effects. In his "Physiological Remains," we find the saving of fuel thus noticed under the head of "Experiments for Profit:" "Building of chimneys, furnaces and ovens, to give out heat with *less* wood."

Three years after Bacon's death, the first printed account was published of any modern attempt (yet discovered) to communicate motion to solids by steam, and as usual an eolipile was employed. Occupying a place on the domestic hearth, as this instrument did, the shrill current proceeding from it must have often excited attention, and led ingenious men to extend the application of the blast to other purposes. The first idea that would occur to a novice when attempting to obtain a rotary movement from a current of vapor, would be that of a light wheel, having its wings or vanes placed so as to receive the impulse, in a similar manner as little paper wheels are made to revolve, which children support on a pin or wire and blow round with the mouth—or those which resemble ventilators and revolve when held against the wind at the end of a stick. These toys are vertical and horizontal windmills in miniature, and windmills and smoke-jacks were the only instruments in the 16th century that revolved by currents of air. Hence it was natural to imitate the movements of these in the first applications of steam; and the more so since steam at that time was generally considered to be nothing but air.[a] Such was the device of *Giovanni Branca*, as described in a work entitled *The Machine*, written in Italian and Latin, and published at Rome in 1629. The volume contains sixty-three engravings. The twenty-fifth represents an eolipile, in the form of a negro's head, and heated on a brazier: the blast proceeds from the mouth, and is directed against pallets or vanes on the periphery of a large wheel, which he thus expected to turn round; and by means of a series of toothed wheels and pinions, to communicate motion to stampers for pounding drugs. He proposed also to raise water by it with a chain of buckets, to saw timber, drive piles, &c.

It is hardly necessary to observe that the apparatus figured by Branca in all probability never existed except in his imagination, and that his stampers, buckets, saws, piles, &c. could no more have been moved by the blast of his eolipile, than those venerable trees were which Wilkins and older writers have represented being torn up from the earth by a man's *breath*—the blast being directed against the vanes of a wheel, and the force multiplied by a series of toothed wheels and pinions, until its energy could no longer be resisted by the roots.[b] Branca seems to have had these childish dreams in his mind when he proposed a continuous stream of steam from an eolipile, in lieu of intermitting puffs of air from a person's mouth. Italian writers have however claimed for him the invention of the steam-engine, a claim quite as untenable as that put forth in behalf of Decaus; for, in the first place, his mode of producing a rotary motion by a current of vapor was not new: all that can be accorded to him in this respect is, that he perhaps was the first to *publish* a figure and description of it. Then it indicates neither ingenuity nor research. There probably never was a boy that made and played with "paper windmills" who would not have at once suggested it, had he been consulted; and when eolipiles were common, many a lad doubtless amused himself by making his "mills" revolve in the current of vapor that issued from them. Moreover, the device is of no practical value. How infinitely does it fall short when compared with that of Heron, (No. 180.) The philosophical principle of

---

[a] A horizontal and a vertical windmill are figured at folio 49 of Rivius' translation of Vitruvius, A. D. 1548.

[b] By the multiplication of wheels and pinions it were easy to have made, says Wilkins, "one of Sampson's hairs that was shaved off, to have been of more strength than all of them when they were on: by the help of these arts it is possible, as I shall demonstrate, for any man to lift up the greatest oak by the roots with *a straw*, to pull it up with *a hair*, or to blow it up with his *breath*." *Math. Magic*, book i, chap. 14.

recoil by which the Alexandrian engineer imparted motion by steam, has often been adopted, and engines resembling his are made even at this day; but one on the plan of Branca never was, and, without presumption it may be said, never will be. The principle being bad, no modification or extension of it could be made useful. No boiler could by it be made to work even a pump to inject the necessary supply of water.

Mr. Farey has well observed that steam has so little density, that the the utmost effect it can produce by *percussion* is very trifling, notwithstanding the great velocity with which it moves. The blast issuing from an eolipile, or from the spout of a boiling tea-kettle, appears to rush out with so much force that at first sight it might be supposed its power, on a larger scale, might be applied in lieu of a natural current of wind to give motion to machinery; but on examination it will be found, that the steam being less than half the specific gravity of common air, its motion is impeded and resisted by the atmosphere. As steam contains so little *matter* or *weight*, it cannot communicate any considerable force by its impetus or concussion when it strikes a solid body. The force of a current of steam also soon ceases. This may be observed in a tea-kettle: the vapor which issues with great velocity at the spout, becomes a mere mist at a few inches distance, and without any remaining motion or energy; and if the issuing current were directed to strike upon any kind of vanes, with a view of obtaining motion from it, the condensation of the steam would be still more sudden, because the substance of such vanes would absorb the heat of the steam more rapidly than air.

Branca's apparatus has been made to figure in the history of the steam-engine, but with equal propriety might the child's windmill be introduced into that of air-engines, for the analogy is precisely the same in both. His device had no influence in developing modern engines. Instead of leading to the employment of the fluid in close vessels, and to the use of a piston and cylinder, its tendency was the reverse: hence so far from indicating the right path, it diverted attention from it.

At the time Branca was preparing his book for the press, some experiments on steam were being made in England—or so it would seem from Sanderson's edition of Rymer's Fœdera. In vol. xix is a copy of a patent or special privilege granted by Charles I to David Ramseye, one of the grooms of the privy chamber, for the following inventions; and dated January 21, 1630:

"1. To multiply and make saltpeter in any open field, in fower acres of ground, sufficient to serve all our dominions. 2. *To raise water from low pitts by fire.* 3. *To make any sort of mills to goe on standing waters, by continual motion, without the help of wind, waite* [weight] *or horse.* 4. To make all sorts of tapistrie without any weaving loom, or waie ever yet in use in this kingdome. 5. *To make boats, shippes and barges to goe against strong wind and tide.* 6. To make the earth more fertile than usual. 7. *To raise water from low places, and mynes, and coal pitts, by a new waie never yet in use.* 8. To make hard iron soft, and likewise copper to be tuffe and soft, which is not in use within this kingdome. 9. To make yellow wax white verie speedilie." The privilege was for fourteen years, and the patentee was to pay a yearly rent of 3*l*. 6*s*. 8*d*. to the king. Mr. Farey says that Ramseye had patents for other inventions from Charles I, but does not enumerate them. As it was not then customary to file specifications, there is no record of the details of his plan.

It is singular that English writers have passed over this patent almost without comment, and yet it contains the first direct proposal to raise water in that country by steam of which any account has yet been produced.

It may perhaps be said, that steam is not mentioned; still it is clearly implied in the second device, and was probably used in the third, fifth and seventh. The very expression "to raise water by fire," is the same that Porta, Decaus, and other old authors, used when referring to such applications of steam. Worcester, Papin, Savery and Newcomen, all described their machines as inventions for "raising water by fire;" and hence they were named "fire water-works," "fire machines," and "fire engines." It should moreover be remembered that the word *steam* was not then in vogue. It is not once used by the translators of the Bible. The fluid was generally referred to as air, or wind, or smoke, according to the appearances it presented. "Rarefying water into ayer by fier," and similar expressions, were common. The idea of air in motion, or wind, was also applied to currents of steam: thus we read of "heating water to make wind," and eolipiles were designated "vessels to produce wind." From the form of clouds which steam assumes when discharged into the atmosphere, it was also named *smoke:* thus Job calls it, in a passage already quoted; and Porta, in describing the apparatus No. 187, speaks of it both as smoke and air. "The water [in the bottle] must be kept heated in this way until no more of it remains; and as long as the water shall *smoke*, (sfumera) the *air* will press the water in the box," &c.—and again, "from that you can conclude how much water has run out, and into how much *air* it has been changed." Had Ramseye therefore called his device a *steam* machine, its nature would not have been so well understood as by the title he gave it, if indeed it could have been comprehended at all by the former term. The expression "raising water *by fire*" appears to have as distinctly indicated, in the 17th century, a steam-machine, as the term *steam-engine* does now; and there is no account extant of any device either proposed or used, in that century, for raising water from wells and mines *by fire*, except it was by *means of steam*.

The date of this patent being so near that of the publication of Branca's book, it may perhaps be thought that Ramseye derived some crude notions from it of applying a blast of steam to drive mills and raise water, as suggested by the Italian; but we should rather suppose some modification of, or device similar to, Porta's (see page 408) was intended in No. 2, and that Nos. 3, 5 and 7 were deduced from it. When once an efficient mode of raising water by steam (like No. 187) was realized, some application of it to propel machinery would readily occur. We know that both Savery and Papin and others proposed to work mills, by discharging the water they raised upon overshot wheels; and this idea was so obvious and natural, that hundreds of persons have proposed it in later times without knowing that it had previously been done.

From the order in which the first three devices are noticed in the privilege, it is possible that they were all modifications of the same thing; that the second and third were deduced from the first, and consequently invented independently of any previous steam machines. The operation of making saltpetre or nitre consists principally in boiling, in huge vats or cauldrons, the lixivium containing the nitrous earth; and from the large quantities of water and fuel required, was formerly carried on in such places only as afforded these in abundance. At such works, the idea of employing the vast volumes of vapor (which escaped uselessly into the air) to raise the hot, and subsequently cold, liquids, would naturally occur to an observing mind, and especially when the subject of raising water by steam was exciting attention. Certainly the idea was as likely to occur to practical men while engaged in the manufacture of nitre in the beginning of the 17th century, as it was to Worcester and others in the middle of it and to Papin and Savery at the close. Perhaps it will be said, nitre was not made in England at that time, and therefore Ramseye could not have taken the hint from such works; and that the suggestion could only have been derived from a long practical experience in them, which he probably never had. This may be true, and it is not improbable that he was merely an agent in the business, having by his influence at court obtained the patent for his own as well as the inventor's benefit. The clause attached to the 8th device, "not in use within *this kingdome*," implies that they were not all of English origin. But whatever were the origin and details of those for raising water, it is clear that the subject of steam was then abroad in the world, and ingenious men in various parts of Europe were exercising their wits to employ it.

It appears to us from the caption of Ramseye's patent, that No. 2 (raising water by fire) was not the first thing of the kind proposed in England, since if it were he would have said so, as well as of No. 8, (softening iron and copper)—and this further appears from what he remarks of No. 7, "raising water from low places, mynes and coal pitts," probably an improvement upon No. 2, and differing from all previous applications of steam for the purpose; hence we are told that it was a "*new* waie," one "*never yet in use.*" Had not steam therefore been previously applied to raise water, it is exceedingly probable that he would have attached a similar remark to No. 2.

The Treatise on Art and Nature, mentioned page 321, is the oldest English book we have met with that illustrates the raising of water by steam with a cut. The annexed figure is from page 30. It possibly may have been deduced from the one given by Decaus, (No. 185) but we should think not; since, although the volume is a compilation, and two thirds of it taken up with "water-works," there is nothing except this from which to infer even the slightest acquaintance with Decaus's book. It seems to have been copied without alteration from some other author. It is named "*A conceited*[a] *Lamp, having the image of a cock sitting on the top, out of whose mouth by the heat of the lamp either water or ayer may be sent.*" The device consists of an eolipile containing water and heated by a lamp of several wicks. The image of the bird is hollow, and communicates by a species of three-way cock with the steam, and also with a pipe that descends into the liquid; so that when the bird is turned round till an opening in the moveable disk to which its lower part is attached coincides with another which communicates with the steam in the upper part of the vessel, vapor issues from the mouth; and when it is turned till the upper orifice of the pipe corresponds with the opening in the disk, then hot water is driven out; and when the opening in the disk does not coincide with either, nothing can escape. After observing that an opening with a proper stopper should be made in the vessel, to charge it with water, the writer con-

No. 89.   A. D. 1633—4.

---

[a] No. 45 of Worcester's Century of Inventions, is named "A most *conceited* Tinder-Box;" No. 71 "A Square Key *more conceited* than any other;" and No. 74 "A *conceited* Door."

tinues—"The larger you make this vessel, the more strange it will appear in its effects, so the lights [wicks] be proporcionable. Fill the vessell halfe full of water, and set the lights on fire underneath it, and after a short time, if you turn the holes that are on the sides of the pipes, that they may answer one another, the water being by little and little converted into ayer [steam] by the heat of the lights that are underneath, will breath forth at the mouth of the cock : but if you turn the mouth of the cock the other way, that the holes at the bottom of the pipes may answer each to other, then there being no vent for the ayer to breath out, it will presse the water and force it to ascend the pipe, and issue out where the ayer breathed out before. This is a thing may move great admiracion in the unskilfull, and such as understand it not. *Other devices, and those much more strange in their effects, may be contrived from hence.*"[a]

No. 190. Kircher. 1641.

Kircher, in 1641, described in his *Ars Magnetica*[b] the device for raising water figured in the margin, a model of which was found in his museum after his death. A close vessel containing the water to be elevated is connected by a pipe that proceeds from its upper part to the top of the boiler, which is supported on a trevet. When the boiler was heated, steam ascended through the pipe, and accumulating in the upper vessel, forced the water up the jet-pipe as represented.

This, it will be seen, is Porta's machine (No. 187) adapted to the operation of raising liquids, which it exhibits in a very neat and satisfactory manner. It is not however equally clear that Kircher had any idea of adapting the plan to the draining of mines, or other hydraulic purposes in the arts. Had such been the case, he would most likely have mentioned it in his *Mundus Subterraneus*, a work published some years afterwards, and in the second volume of which he figures and describes the ordinary machines then in use, viz. the bucket and windlass, chain of pots, chain pump, and atmospheric pumps. The form of the model (an imitation of a vase supported on a column) rendered it an appropriate addition to his philosophical apparatus.

In 1643, the great discovery of atmospheric pressure was made; a discovery whose influence, like that of the atmosphere itself, is felt more or less in every art and every science. It led in a very short time to a series of inventions of the highest value, among which the reciprocating steam-engine should probably be placed. We mention it here in chronological order, that its influence in developing and improving the machine just named may be more readily appreciated when we come to notice subsequent attempts to impart motion by steam.

---

[a] *John Bate*, who published a treatise on Fire-works in 1635, was perhaps the compiler of this curious volume. Strutt, in his Sports and Pastimes of the People of England, quotes Bate's book, but it would seem that the same cuts were not in both, for when speaking of *boys' kites*, Strutt observes that the earliest notice of them that he could find in books was in an English and French Dictionary of 1690; whereas there is a figure of a man flying one, with crackers and other fire-works attached to the tail, in the second part of "Art and Nature."

[b] This work was published in quarto, at Rome and Colonne, in 1641; and in folio, at Rome, in 1654. Catalogue of Kircher's Works at the end of the first volume of *Mundus Subterraneus*. Amsterdam, 1665.

Some remarkably ingenious experimentalists flourished about the middle of the 17th century, whose names have perished; and of their labors nothing is known, except an enumeration of the uses to which some of their inventions could be applied. An example of this is furnished by an anonymous pamphlet,[a] published in 1651, from which the following extract is taken. The device referred to seems to have possessed every attribute of a modern high-pressure engine, and the various applications of the latter appear to have been anticipated. " Whereas, by the blessing of God, who only is the giver of every good and perfect gift, while I was searching after that which many, far before me in all humane learning, have sought but not yet found, viz. a perpetual motion, or a lessening the distance between strength and time; though I say, not that I have fully obtained the thing itself, yet I have advanced so near it, that already I can, with the strength or helpe of four men, do any work which is done in England, whether by winde, water or horses, as the grinding of wheate, rape, or raising of water; not by any power or wisdome of mine own, but by God's assistance and (I humbly hope, after a sorte,) immediate direction, I have been guided in that search to treade in another pathe than ever any other man, that I can hear or reade of, did treade before me; yet, with so good success, that *I have already erected one little engine, or great model, at Lambeth*, able to give sufficient demonstration to either artist or other person, that my invention is useful and beneficial, (let others say upon proof how much more,) as any other way of working hitherto known or used." And he proceeds to give " a list of the uses or applications for which these engines are fit, for it is very difficult, if not impossible, to name them all at the same time. To grind malt, or hard corne; to grind seed for the making of oyle; to grind colours for potters, painters, or glasse-houses; to grind barke for tanners; to grind woods for dyers; to grind spices, or snuffe, tobacco; to grind brick, tile, earth, or stones for plaster; to *grind sugar-canes;* to draw up coales, stones, ure, or the like, or materials for great and high buildings; to *draw wyre;* to draw water from mines, meers, or fens; to draw water to serve cities, townes, castles; and to draw water to flood dry grounds, or to water grounds; *to draw or hale ships, boates, &c. up rivers against the stream; to draw carts, wagons, &c. as fast without cattel: to draw the plough without cattel to the same despatch if need be;* to brake hempe, flax, &c.; *to weigh anchors with less trouble and sooner;* to spin cordage or cables; to bolt meale faster and fine.; to saw stone and timber; to polish any stones or mettals; *to turne any great works in wood, stone, mettals, &c.* that could hardly be done before; *to file much cheaper in all great works; to bore wood, stone, mettals;* to thrashe corne, if need be; to winnow corne at all times, better, cheaper, &c. For paper mills, *thread mills,* iron mills, *plate mills; cum multis aliis.*" If this extraordinary engine of motion, observes Mr. Stuart, to whom we are indebted for the extract, was not some kind of a steam-engine, the knowledge of an equally plastic and powerful motive agent has been utterly lost.

Steam is not here indicated, but it is difficult to conceive any other agent, unless some explosive compound be supposed, by which the pressure of the atmosphere was excited. That the engine consisted of a working cylinder and piston, and the latter moved by steam, must we

---

[a] *Invention of Engines of Motion lately brought to perfection;* " whereby may be despatched any work now done in England, or elsewhere, (especially works that require strength and swiftness,) either by water, wind, cattel, or men, and that with better accommodation and more profit than by any thing hitherto known and used." London, 1651.

think be admitted; for although most of the operations mentioned might have been performed by forcing up water on an overshot wheel, by an apparatus similar to Papin's or Savery's steam-engines, there are others to which such a mode was quite inapplicable, as raising of anchors, or propelling carts, wagons and ploughs. The inventor, whoever he was, has given proofs of an extraordinary sagacity, for every operation named by him is now effected by the steam-engine, except raising the anchors of steam-vessels and ploughing. The latter is at present the subject of experiment, and the former will in all probability be soon adopted. The author's labors were most likely not appreciated by his contemporaries, and as the world is always too apt to think the worst in such cases, the whole will probably *now* be set down by some persons as the dream of a sanguine projector—the judgment commonly passed upon those who are in advance of the age they live in. Of this lamentable truth several examples will be found in this volume, and in the history of every important invention. We shall notice two here, as they relate to two of the most valuable applications of steam. *Oliver Evans*, in 1786, urged upon a committee of the legislature of Pennsylvania, the advantages to be derived from steam-boats and "steam-wagons," and predicted their universal adoption in a short time. The opinion which the committee formed of him was expressed a few years afterwards, by one of its members, in the following words: "To tell you the truth, Mr. Evans, we thought you were deranged when you spoke of making *steam-wagons*." The other relates to *John Fitch*, a clock and watch maker, than whom a more ingenious, persevering and unfortunate man never lived. In spite of difficulties that few could withstand, he succeeded in raising the means to construct a steam-boat, which he ran several times from Philadelphia to Burlington and Trenton in 1788. As a first attempt, and from the want of proper manufactories of machinery at the time, it was of necessity imperfect: then public opinion was unfavorable, and the shareholders finally abandoned the scheme. His feelings may be imagined, but not described; for he saw and predicted the glory that awaited the man who should succeed in introducing such vessels in more favorable times. "The day will come [he observes] when some more powerful man will get fame and riches by my invention, but nobody will believe that poor John Fitch can do any thing worthy of attention." He declared that within a century the western rivers would swarm with steam-vessels, and he expressed a wish to be buried on the margin of the Ohio, that the music of marine engines in passing by his grave might echo over the sods that covered him. In a letter to Mr. Rittenhouse, in 1792, he shows the applicability of steam to propel ships of war, and asserts that the same agent would be adopted to navigate the Atlantic, both for packets and armed vessels. Descanting on one occasion upon his favorite topic, a person present observed as Fitch retired, "poor fellow! what a pity he is crazy!" He ended his life in a fit of insanity by plunging into the Allegany.[a]

In tracing the progress of discovery which resulted in the steam-engine, we have seen that the two grand properties of aqueous vapor—its elastic energy, and the instant annihilation of this energy by condensation—were well known in the 16th century. On these properties of steam were based all the efforts of experimenters to accomplish the two great objects they had in view; i. e. to impart motion by it to general mechanism, and to employ it as a substitute for pumps to raise water. Before either the

---

[a] Supplement to Art. "Steam-Boat," Ed. Encyclopedia, by Dr. Mease; and Watson's "Early Settlement and Progress of Philadelphia," &c. Phil. 1833.

elastic force or the condensation of steam could be beneficially applied to give motion *directly* to solids, some plan very different from that of Branca was required—one by which the fluid could be used in *close* vessels. Now there is in the whole range of mechanical combinations but one device of the kind yet known, and it has but few modifications, viz. *a piston and cylinder*. Experience has proved, that of all contrivances for transmitting the force of highly elastic fluids to solid bodies, this is the best. Thus guns are cylinders, and bullets are pistons, fitted to fill the bore and at the same time to move through the straight barrels. It is the same, whatever the impelling agent may be; whether gun-powder, steam, or compressed air. The air-guns of Ctesibius are the oldest machines of the kind on record, and from them we see that the ancients had detected this mode of employing aëriform fluids.

Steam-engines simply considered are but modifications of guns. In the latter, the bullet or piston is driven entirely out of the cylinder, and in one direction only, because the intention is to impart the momentum to a distant object at a blow: but by the former the design is to derive from the moving bullet a *continuous* force; hence it is not allowed to leave the cylinder, but is made to traverse incessantly backwards and forwards within. In order to transmit its impetus to the outside of the cylinder and to the objects to be acted upon, a straight rod is attached to it, and made to slide through an opening in one end of the cylinder. It is by means of this rod that motion is imparted to the machinery intended to be moved. All the mechanism, the wheels, cranks, shafts, drums, &c. of steam-engines are but appendages to the cylinder and piston; they may be removed and the energy of the machine still remains; but take away either cylinder or piston and the whole becomes inert as the limbs of an animal whose heart has ceased to beat. Therefore it is the working cylinder and piston alone that give efficiency to modern steam-engines; and it is to those persons who contributed to introduce them, that the glory attending the invention of these great prime movers is chiefly due.

Whatever may be said respecting more ancient applications of steam as a moving power, modern engines are one of the results of the discovery of atmospheric pressure. All the early ones of which descriptions are extant were rather *air* than steam machines, not being moved by the latter fluid at all. Their inventors had no idea of employing the elastic force of steam, but confined themselves to the atmosphere as a source of motive force: hence they merely applied steam in lieu of a syringe to displace air from a cylinder, that, when the vapor became condensed by cold into a liquid, the atmosphere might force down the piston. That this was the way in which modern engines took their rise appears, further, from the same feature being retained in a great portion of them to this day. They are now ranged in three classes—1st *atmospheric*, 2d *low pressure*, and 3d *high pressure* engines; and this we know is the order in which they were developed. In the first, the power is derived exclusively from the atmosphere, the vapor employed being used only as a substitute for an air-pump in making a vacuum under the piston. In process of time the second was devised, in which the elastic force of steam is made to act against one side of the piston, while a vacuum is formed on the opposite side. The next step was to move the piston by the steam alone, and such are named high-pressure engines. The term *steam*-engine is therefore not so definite as some persons might suppose, since it is not confined to those in which steam is the prime mover. Had it not been for Torricelli's discovery, it is possible that we should never have known any other species of steam-engine than those of the third class; and hence we repeat, that whatever may be thought of engines made previous to the 17th century, those of modern days were obviously derived from atmospheric ones of the first class, while these in their turn were very likely deduced from the apparatus described in the next paragraph.

Otto Guerricke, of whom we spoke at page 190, one of the earliest, and as far as mechanical ingenuity went perhaps the most gifted, of the early elucidators of atmospheric pressure, exhibited in his public experiments at Ratisbon, in 1654, the following application of that pressure as a motive force. A large cylinder, A, was firmly secured to a post or frame. It was open at the top and closed at the bottom, and had a piston accurately fitted to work in it. A rope was fastened to the piston-rod and passed over two pulleys, B C, as represented, by which was suspended a scale, D, containing several weights. When the air was withdrawn from the lower part of the cylinder, the pressure of the atmosphere depressed the piston and raised the scale and weights. To vary the experiment, the weights were removed and twenty men were employed to pull at the rope with all their strength; but as soon as a vacuum was made by the small air-pump attached to the bottom of the cylinder, the piston descended, notwithstanding all their efforts to prevent it.

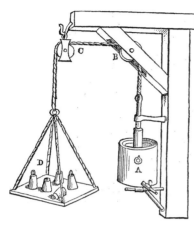

No. 191. Guerricke. A. D. 1654.

This is the oldest apparatus on record for transmitting motion to solids by a piston. We can however hardly believe that it was the first devised for the purpose. It would be strange if it were; for whatever may have been the nature of Anthemius's, Garay's, and other old machines in which steam was the active principle, pumps and syringes had been too common, and experiments with them too frequent, for such a device to have been unknown. Such men as Aristotle and Archimedes, Ctesibius, Heron, Roger Bacon and their successors, were all aware that a syringe presented the same phenomenon as Guerricke's apparatus, when the piston was drawn up while the discharging orifice was closed: the same thing was also observed with common pumps when the suction-pipes were either closed or choked. Experiments therefore to illustrate the force thus excited were in all probability made, and with apparatus similar to that of the Prussian philosopher, long before his time, although no account of them is extant. But if even such had been made, they would not lessen in any degree the merit of Guerricke, since his experiment undoubtedly originated with himself, and all knowledge of similar ones had been lost.

In this device we behold the same moving force, and the same mechanism for applying it, as were subsequently adopted in steam-engines, which at first were little more than copies of this: for example, had a loaded pump-rod been suspended to the rope instead of the scale and weights, the apparatus would have differed from Newcomen's engine only in the mode of exciting the atmospheric pressure. To Guerricke, therefore, is due the credit of having not only pointed out the power which alone gave efficiency to the first steam-machines, but also of devising the most effectual means of employing it. No one could, we think, claim an equal degree of merit for simply applying (not inventing) another mode of producing a vacuum under the piston; but without insisting on this, it may be observed that even at present, in all low-pressure engines, the vacuum is made just as Guerricke made it, viz. by an *air-pump;* so that the impress of his genius on the steam-engine is no more obliterated in this respect than it is in others. Every unbiassed mind will therefore admit, that an honorable place in its history should be assigned to the philosopher of Magdeburg.

CHAPTER VI.

Reasons of old Inventors for concealing their discoveries—Century of Inventions—Marquis of Worcester—His Inventions matured before the Civil Wars—Several revived since his death—Problems in the "Century" in older authors—Bird roasting itself—Imprisoning Chair—Portable Fortifications—Flying—Diving—Drebble's Submarine Ship—The 68th Problem—This remarkably explicit—The device consisted of one boiler and two receivers—The receivers charged by atmospheric pressure—Three and four-way cocks—An hydraulic machine of Worcester mentioned by Cosmo de Medicis—Worcester's machine superior to preceding ones, and similar to Savery's—Piston Steam-Engine also made by him—Copy of the last three Problems in the Century—Ingenious mode of stating them—Forcing-Pumps worked by Steam-Engines intended—Ancient Riddle—Steam-Boat invented by Worcester—Projectors despised in his time—Patentees caricatured in a public procession—Neglect of Worcester—His death—Persecution of his widow—Worcester one of the greatest Mechanicians of any age or nation—Glauber.

As yet we have not met with any definite description of a steam-engine in actual use. This can only be accounted for from the fact that old inventors were all jealous of the printing-press. They believed their interest required concealment on their part, that pirates might not rob them of their labors. They have been blamed for this, and so have some modern mechanics, but we think without reason; for, to obtain satisfaction at *law* in such cases, was formerly as difficult as it is now in most cases. To have to purchase *justice*, as in a lottery, with money, is bad in itself, and worse because those without money cannot obtain it; but to have to give more for it than it is worth, if perchance it be awarded, is a disgrace to enlightened nations—an evil that savages would not for a moment endure. It is thus that law, though ordained to promote justice, is so prostituted as not only to defeat the object for which it was designed, but to cherish the grossest injustice. It has always been a bar to the progress of the arts. The difficulty and expense of obtaining and preserving an exclusive right to their inventions—that is, to their own property—have induced inventors more or less, in every age, to conceal their discoveries till death, and even then to destroy all records respecting them.

When old inventors were solicitous of *public* patronage, instead of establishing their claims to it by explaining the principles and operations of their machines, they contented themselves with enumerating their uses and good qualities merely. They proclaimed the great things that could be done, but studiously concealed the modes and means of doing them: hence new inventions were sometimes announced enigmatically, the moving or constituent principles being so obscurely hinted at that few readers could apprehend them. Of this mode of exciting public attention, the account of the engine of motion in the last chapter is an example; and several more may be seen in the pamphlet published by the Marquis of Worcester, in 1663, entitled " A Century of the Names and Scantlings [outlines or hints] of such Inventions as at present I can call to mind to have tried and perfected; which, my former notes being lost, I have, at the instance of a powerful friend, endeavored, now in the year 1655, to to set down in such a way as may sufficiently instruct me to put any of them in practice." This book is made up of one hundred inventions, numbered from one upwards. It contains a distinct reference to a working steam-machine for raising water, and also hints by which its nature and construction are pretty clearly ascertained. There is some reason to believe that the modern high-pressure engine is also referred to. From the circumstance of the author having figured largely in the civil wars, he having been an enthusiastic adherent of Charles I. and of monarchy, his character and that of his book have been represented in the best and worst of lights. By his enemies he was held up as false and unprincipled in the highest degree; by his friends, as chivalrous and of unspotted honor. The " Century" has been denounced as a scheme to impose on the credulity of mankind—the dream of a visionary—and Hume, in his History, goes so far as to name it " a ridiculous compound of lies, chimeras, and impossibilities." On the other hand, it has been received by many (and generally by practical men) in the light in which the author represents it, viz. as a memorial of inventions actually put in practice by him—such as he had really " tried and perfected."

With the political conduct of Worcester we have nothing to do. He naturally enough supported that system by which he and the rest of the Lords acquired and entailed their exclusive privileges; among which the abominable one of being legislators by *birth* was perhaps the most odious and unnatural. On the fall of the king he retired to the continent, but, at the request of Charles II, ventured to visit London in disguise in 1656. Being discovered, he was arrested, and confined in the Tower until the reëstablishment of monarchy in 1660. He died in 1667.

We have no positive information respecting the time when he commenced his mechanical researches. There is however reason to believe that most, if not all, the inventions enumerated in the " Century" were matured before the civil wars broke out, and consequently that the account of them was drawn up, as he declares, in 1655.[a] No. 56 he observes was tried before Charles I, Sir William Balfour, and the Dukes of Richmond and Hamilton; and this could not have been later than 1641, for Balfour was dismissed that year. In addressing the Century to Parliament, he mentions having had " the unparalleled workman, Caspar Kaltoff," in his employment " these five and thirty years," and who was at that time (1663) engaged in his service. This carries back his experiments to 1628. Some of his " water-works" were in operation in his father's castle (at Ragland, in Wales) at the commencement of the Long Parliament, (1640) for by their sudden movements he is said to have frightened certain adherents of the Parliament, who went to search the castle for arms. The nature of these works is not indicated, except that they consisted of "several engines and wheels," and that large quantities of water were contained in reservoirs *on the top of a high tower.* Whether steam was the agent employed to raise this water is unknown. It could not have been if the tradition, credited by some writers, was true, viz. that his attention was first

---

[a] The Century is copied in vol. xiii of Tilloch's Phil. Mag.; and the editor remarks, " this little tract was first published in 1655."

drawn to the employment of steam by observing, *while a prisoner in the Tower*, a pot-lid raised or thrown off by it. If this was the case, then no dependence can be placed on Worcester's assertion, that the whole Century was written in 1655; but there is no reason to question his veracity in this respect. On the contrary, the tradition is obviously a fable; one that has been applied to others as well as to him.

Although many of the devices in the Century appear at first sight extremely absurd, and others impossible, yet every year is producing a solution of one or more of them. One half, at least, of the number have been realized; among which are telegraphs, floating baths, short-hand, combination locks, keys, escutcheons and seals, rasping mills, candle-moulds, engines for deepening harbors and docks, contrivances for releasing unruly horses from carriages, torpedoes, diving apparatus, floating gardens, bucket engines, (see page 64 of this volume) universal lever, repeating guns and pistols, double water screws (p. 140 of this vol.) abacus, portable bridges, floating batteries &c. besides his applications of steam, which will be noticed more at large farther on.

It must not be supposed that Worcester was the first projector of every problem in the Century, although his solutions may have been peculiar to himself. The greater part may be found in the works of Porta, Fludd, Wilkins, and others of his predecessors and contemporaries; so that the charges of absurdity brought against many of them are not attributable to him alone. Indeed, the Century is in a great measure free from those puerile conceits that abound in old authors.[a] No. 3 he names "a one-line cypher," that is, a character composed of a single *line*, which by its position was made to represent each and every letter of the alphabet. (Now used in short-hand.) No. 4 is an improvement, and consists in substituting *points* or *dots* in place of lines. No. 5, " a way by circular motion, either along a rule or ringwise, to vary any alphabet, even this of points," &c. Now these three systems were explained and illustrated by diagrams in detail, twenty-two years before the publication of Worcester's book, by Bishop Wilkins, in his " Mercury, or secret and swift Messenger," a tract printed in 1641. The eleventh chapter treats " of writing by invented characters "—" how to express any sense either by *lines, points* or *figures*." The last was by arranging the points or dots in the forms of circles, squares, triangles, &c. Wilkins speaks of the whole as an old device. Another problem in the Century is " a universal character." This had been often attempted, and Wilkins wrote also upon it. Another, "a waterball," to show the hour of the day. There were some singular specimens of these clocks in Serviere's museum, which was celebrated for its collection of mechanical devices, and which doubtless Worcester had often visited. (See page 285, and note foot of page 63.) The universal lever, No. 26 of the Century, he admits having seen at Venice, and the bucket engine (No. 21) at Rome. It is probable he derived his " imprisoning chair" from the same place; for there was in his time, as well as since, a

---

[a] There is a singular one in book xiv of Porta's Magic, " Of a bird which roasts itself;" which, had Worcester mentioned, few would have credited without the explanation. " Take a wren and spit it on a hazel stick, and lay it down before the fire, the two ends of the hazel spit being supported by something that is firm; and you'll see with admiration the spit and the bird turn by little and little, without discontinuing, till 'tis quite roasted." This, says Ozanum, was first found out by Cardinal Paloti, at Rome. The motion may be accounted for on a similar principle as the rotation of glass tubes when supported at each end before a fire, and even when inclined against the fire-place with one end on the hearth, viz. : the heat, being applied to one side only, causes the tubes to *bend*, and consequently to preponderate and thus turn round. See Phil. Trans. vol. xxx—Abridg. vol. x 551.

famous machine of the kind exhibited in the Borghese villa, which could not have escaped his notice. It is described by Blainville as "very artfully contrived; and strangers, who are not acquainted with the trick, are infallibly caught as in a trap when they are prevailed upon to sit in it." Travels, vol. ii, p. 35. We shall notice a few more: "A little engine portable in one's pocket, which placed to any door, without any noise but one crack, openeth any door or gate." A similar device is quoted by Wilkins from Ramelli, thus: "A little pocket engine wherewith a man may break or wrench open any door." (Math. Magic, book i, chap. 13, first published in 1648.) Again—"An instrument whereby an ignorant person may take any thing in perspective as justly and more than the skilfullest painter can do by the eye." Probably the camera obscura, which Baptist Porta had described, about a century before, in his Natural Magic. See page 364 of the English translation of 1658, and also Fludd's *Natural Simia seu Technica*, 1618, page 308, for another mode. No. 29 of the Century relates to "A moveable fortification—as complete as a regular one, with half-moons and counterscarps." Such a one is figured in Fludd's Simia. It is of a triangular form, with breast works and cannon ranged along two sides. The whole is made of thick timber clamped together, and moved by horses, which are yoked to a long pole or mast, also supported on wheels and attached to the rear or base of the triangle, so as to be out of the reach of shot from the enemy. The horses have their faces to the fortification, just as if yoked to the pole of a common carriage and fronting it—or, according to the old saying, "the cart is put before the horse."

His modes of discoursing by knotted strings, gloves, sieves, lanterns, &c. are similar to others mentioned in the Natural Magic of Porta, and in other works. Wilkins's Secret and Swift Messenger also contains much curious information on such subjects. Several numbers of the Century relate to repeating guns. These, as is well known, exercised the wits of inventors long before his time. Porta, in his Magic, book xii, speaks of "great and hand guns, discharged ten times" although loaded but once. They are even of much older date. Sometimes several barrels were joined together. The "arithmetical instrument, whereby persons ignorant of arithmetic may perfectly observe numerations and subtractions of all sums and fractions," was in all probability the abacus, or Chinese *swan-pan*, now used in schools.

Flying and diving, also mentioned by him, have occupied the ingenuity of inventors in every age. Cornelius Drebble constructed a diving-vessel which was propelled by oars worked through openings in the sides. Short conical tubes of leather, through which the oars were passed, were connected to the openings so as to exclude the water; hence the joints somewhat resembled those of the feet of a tortoise when protruded from the shell. The vessel was lowered by admitting water, and raised by pumping it out. (The distance of diving-vessels below the surface is easily and accurately ascertained by a curved tube containing a little mercury, one end being within the vessel and the other without.) Charles, Landgrave of Hesse Cassel, hearing of Drebble's diving-ship, requested Papin to contrive one. Papin's machine is figured and described in the Gentleman's Magazine for 1747, page 581. Drebble's vessel did not require a constant supply of fresh air, for he had, or pretended to have, an elixir in a small vial, a few drops of which restored the vitiated air so as to make it again fit for respiration. Something of this kind was known even before Drebble's time, if we may judge from one of several illustrations of diving in the old German translation of Vegetius, A. D. 1511. A man clothed in a dress of thin skin or oiled silk fitted close to his body, and covering every part except his head and hands, is represented walking on the bottom of a river. In his left hand he holds a leathern flask, through the contracted neck of which he is drawing a portion of the contents with his mouth. Wilkins devoted a chapter of his *Math. Magic* to diving. He notices Drebble's machine, and many other curious devices; so that on this subject Worcester had an abundance of materials and hints to work upon.

No. 50 of the Century relates to portable ladders. A variety of these are figured in the old translation of Vegetius just referred to. There are several other things named in the Century which might be traced to older sources, but it is not necessary; for Worcester has not, that we are aware of, ever claimed all the devices he has named. He mentions two whose authors he recollected, but as the account was drawn up from memory, he could hardly recall to mind the sources whence all were derived. He says they were such as he could call to mind to have *tried* and *perfected*: he does not say invented. While many originated with himself, others were such as he improved only. That he had sources of information which have not been discovered, there can be little doubt. Of the thousands of old treatises on the "Mysteries of Nature and Art," a staple subject, and title too, from Roger Bacon to Moxon, how few are extant! But some will perhaps yet be met with on the shelves of antiquaries and the lovers of old books in Europe.

Those numbers of the Century which relate to steam are 68, 98, 99 and 100; but it is in 68 only that steam is clearly indicated. The device is named "a fire water work," and is described in the following manner: "An admirable and most forcible way to drive up water by fire, not by drawing or sucking it upwards, for that must be, as the philosopher calleth it, *Infra sphæram activitatis*, which is but at such a distance. But this way hath no bounder, if the vessels be strong enough; for I have taken a piece of a whole cannon, whereof the end was burst, and filled it three quarters full of water, stopping and screwing up the broken end, as also the touch-hole; and making a constant fire under it, within twenty-four hours it burst, and made a great crack:—So that having a way to make my vessels, so that they are strengthened by the force within them, and the one to fill after the other, I have seen the water run like a constant fountain stream forty feet high. One vessel of water rarefied by fire driveth up forty of cold water; and a man that tends the work is but to turn two cocks, that, one vessel of water being consumed, another begins to force and refill with cold water, and so successively, the fire being tended and kept constant, which the self same person may likewise abundantly perform in the interim, between the necessity of turning the said cocks."

We here see clearly what was meant by Ramseye and others when they spoke of raising water *by fire*, viz. that it was by steam, which the fire was employed to produce. It will be perceived that Worcester does not here claim to be the first to raise water in large quantities in this manner, thus tacitly admitting that he was aware of previous applications of steam for the purpose. Had he indeed made such a claim, little reliance could have been placed on his statements; but, notwithstanding all that has been said to the contrary, we have seen nothing in the whole tenor of his conduct with regard to his inventions to shake our confidence in his sincerity. In one respect No. 68 differs from the rest, viz. in the detail with which the device is described; but this was most likely designedly done, in order to show its superiority over other "fire water-works," and to point out where it differed from them. Had it been an original idea, there could have been no more inducement to be thus explicit than with the rest; but being of the same nature as others, he would naturally be led to notice the difference. Some writers are incredulous of his having ever put it in practice, notwithstanding his assertions, and the particulars he has specified; and they further contend that his description was not sufficiently perspicuous to enable a person to make such a machine in his own time, and is not now. To neither of these positions can we assent; and the latter, if true, does not affect the character of Worcester, either for veracity or ingenuity, since the avowed design of the Century was rather to enable himself than others to realize the inventions named.

The description appears not only that of a machine in actual use, and from which a similar one might have been made, but, as just intimated, some particulars are mentioned apparently with the sole view of distinguishing it from other devices of the same kind. Had he given a figure we should have learnt more of the details, but not of the general plan. The nature of the force employed (the expansive power of steam) he shows in the clearest light; and its irresistible energy is admirably illustrated by bursting the cannon: indeed, he could not possibly have selected any thing better adapted for the purpose. Few writers however believe the experiment was ever made, from the seeming difficulty of closing the broken end—a circumstance which, perhaps more than any other, has led people to doubt the accuracy of other of his statements: and it must be admitted that if he is not to be believed in this, his assertions in general must be received with great caution. But what great difficulty after all was there in driving a plug tight into the smoothly bored although broken end of a cannon, and securing the plug effectually in its place, by iron straps and screws round the trunnions? Lest the idea of danger should be connected with his apparatus in the public mind, he remarks that he had a way of preventing his vessels from being exploded. He mentions at least three vessels; one, a boiler in which to generate steam, the others, to receive the water previously to its being raised. A separate boiler shows that the apparatus was a modification of Porta's and Kircher's, (Nos. 187, 190)—and lest any one should suppose that the water was required to be heated before it was elevated, he states distinctly that it was not: hence his device bore no resemblance to that of Decaus, (No. 188)— so far from it that the boiler, or "one vessel rarefied by fire," forced up *forty* times its contents "of *cold* water." It appears that the water was raised forty feet only; perhaps being limited to that height by local circumstances, or by the building in which the apparatus was erected. The pressure of steam in his boiler did not therefore much exceed 30lbs. on the inch. As the elevation exceeded that to which the liquid could be raised by atmospheric pressure, he also takes occasion to notice distinctly that it was not done by sucking; and in this he possibly may allude to some such modes of raising water, viz. by using the steam only to produce a vacuum, and to show the difference; for, by employing its elastic force, he could raise water *at one lift* to any height, and his apparatus, instead of a limited application, was adapted to mines and pits of every depth, and hence he appropriately names it "a *most forcible* way." The receiving vessels were charged or filled "one after another," and the stream discharged from them was uninterrupted. One person only was required to attend to the fire under the boiler and "to turn two cocks," i. e. to admit steam alternately into each receiver, so that when one was "consumed" or emptied, the contents of the other began to "force" or be forced up, and the empty one to "fill" or be refilled with cold water, "and so successively." The vessels were large, or it took a long time to fill them, since the man had abundant time to attend the fire in the intervals of turning the cocks.

Notwithstanding the comprehensive sketch that Worcester has given of this machine, a variety of opinions prevails respecting some of its parts, and the arrangement of the whole. In these respects scarcely two writers agree, while some differ widely. Some have supposed it to have consisted of two eolipiles, like those of Heron or Decaus, (Nos. 179 and 188) connected to one ascending pipe, (see Galloway on the Steam-Engine)—an idea, we think, entirely out of the way, since such a plan would possess neither " merit " nor " originality," which the writer just named accords to Worcester's device. It is moreover opposed to the description given, which expressly states that the contents of one vessel rarefied by fire, driveth up forty of *cold* water; whereas, by the supposed construction, *all* the water must have been heated to the boiling point before it could have been elevated at all, and to a temperature still higher before it was raised forty feet.

The principal point undetermined is the mode by which the receivers were charged. Were they so placed that the water flowed into them through a pipe and cock? Or, were they wholly immersed in the tank, well or pond, and furnished with valves opening inwards for the admission of the liquid, and to prevent its return when the steam was turned on? Or, were they placed *above* the water, and charged by atmospheric pressure? The first and second modes have been suggested, because Worcester says he did not raise the liquid by " sucking;" but it does not appear that he meant any thing more than that the contents of the receivers were not expelled from them in this way. As the elevation to which water could be raised at one lift by his machine was only limited by the strength of the vessels, he very naturally observed, to remove an objection which he foresaw might be made to his assertion, that this was not effected by sucking, but by forcing the liquid up. His plan bears the same relation to a forcing pump, as using steam to produce a vacuum in a receiver does to a sucking one; and in distinguishing between the two applications of the vapor to raise water, viz. by its condensation and its expansion, he uses the same terms that we do to show the difference between the two instruments just named. Of a forcing pump we say, it does not raise water by atmospheric pressure, but in opposition to it; and that the elevation is only limited by the strength of the materials and the power employed: now every person acquainted with the subject knows, that it is the expulsion of the water from the cylinder that is referred to, not the mode of filling it; for almost invariably are the vessels or cylinders of forcing pumps charged by sucking, and so they were in Worcester's time.

If the receivers were placed below the reservoir that supplied them, and were fed from it by a pipe, then as there were but two cocks used, they must have been such as are known by the term " three-way,"—one passage to supply steam to each of the receivers, and the other water. There is no difficulty in admitting this, for both three and four way cocks were in use ages before Worcester's days. They are described in the Spiritalia, (problem 31) in Besson's Theatre, Fludd's Simia, (see our 160th illustration, page 354) Ozanam's Recreations, and in several other old authors. One form of them is seen at page 421. Tavernier found, in baths of the east, cocks which at the same mouth supplied " either hot water or cold," (Relation of the Seraglio) and they are described and figured in the Forcible Movements of Decaus: thus prop. xix of Leak's translation is " Of the cock with four vents," and its application is shown in a self-acting " Phneumatique Engine." M. Arago is therefore greatly mistaken, in his

History of the Steam-Engine, in attributing the invention of the four-way cock to Papin. In his zeal to confer honor on the philosopher of Blois, he inadvertently overlooked the old engineer of Normandy.

This plan of supplying Worcester's receivers is certainly far more probable than that of burying them in the water they were to raise. Indeed, we cannot perceive how the latter could answer at all, as the steam would be condensed by the surrounding medium almost as fast as it entered the receivers; so that instead of "one vessel of water rarefied by fire" driving up forty of cold water, it would hardly be able to drive up any. It appears to us impossible for ingenuity to suggest a worse plan, and yet several writers have adopted it.[a] As a proof that Worcester had an engine at work somewhat similar to the one referred to in his 68th proposition, the following extract from the Journal of Cosmo de Medicis, who visited England in the 17th century, has been adduced: "His highness, that he might not lose the day uselessly, went again, after dinner, to the other side of the city, extending his excursion as far as Vauxhall, beyond the palace of the Archbishop of Canterbury, to see an hydraulic machine invented by my Lord Somerset, Marquis of Worcester. It raises water more than forty geometrical feet, by the power of one man only; and in a very short space of time will draw up four vessels of water through a tube or channel not more than a span in width; on which account it is considered to be of greater service to the public than the other machine near Somerset-House." Now if this engine for raising water from the Thames, and which was managed by one man, was moved by steam—and it probably was—we may rest assured that Worcester knew better how to charge his receivers than by immersing them in the river, or in any tank supplied from it. Had he done so, the machine would never have been "considered of greater *service to the public*" than the engine at Somerset-House, which was worked by horses, and distributed water over "a great part of the city." (This last engine most likely consisted of pumps, such as were erected by Bulmer in 1594. See page 296.) As *four* vessels are here mentioned, there were probably that number of receivers employed.

It would be strange if Worcester's receiving vessels were not charged by atmospheric pressure, considering the examples he had before him. To say nothing of this well known mode of charging eolipiles, and other vessels represented in the Spiritalia, (see 173 and 177 of our illustrations) both Porta and Fludd exhibited experiments expressly to show how water is raised into a vacuum formed by the condensation of vapor, (page 407) and Decaus gives such striking applications of it (page 380) that Worcester never could, with a knowledge of these, have plunged his receivers under water. But was he acquainted with the writings of these men? Unquestionably he was. There is evidence in the Century that he examined every source of information, both at home and abroad, and with an eagerness that has perhaps seldom been equalled; and then no person had greater facilities for ascertaining what had been accomplished. He was not a man to set about devising new modes of raising water while ignorant of old ones, or without perusing those writings which treated directly or indirectly upon the subject. Of all his researches, this of raising water was

---

[a] See Millington's Epitome of Philosophy, and Stuart's Descriptive History of the Steam-Engine. The last named writer speaks however very differently in his valuable "Anecdotes of the Steam-Engine." Further reflection convinced him that Worcester was something more than a charlatan, and the machine in question very unlike the one represented in his previous work.

among the earliest and most favorite, as it was the last and most important; and it was impossible for him and Kaltoff to have spent so many years as they did on this and other subjects, without improving old devices and introducing new ones. Then he was most likely acquainted with the machines of Bacon, and with those of Ramseye, and with Ramseye himself, and Bushell too; and also with the engine of motion noticed in the last chapter, of which he was possibly both the inventor and describer. So far therefore from Worcester's machine being imperfect, as some writers have supposed, we are justified in believing it was *superior* in its general plan, and in the arrangement and execution of its several parts, *to any thing then extant*, or previously proposed.

It would be easy to devise a machine corresponding with these remarks and coinciding with the Marquis's account; but the intelligent reader is aware that it would be substantially the same as Savery's. It is surprising that some authors have supposed Worcester could not have filled his vessels by atmospheric pressure, because, say they, the production of a vacuum by the condensation of steam was not then known, "nor even thought of." But such writers were not aware of the experiments of Porta, and they forgot the employment of eolipiles. It has also been said that a machine as perfect as Savery's, and one in which steam acted on a piston, was beyond the state of the arts in Worcester's days. The Century of Inventions is a proof to the contrary, and so is the Collection of Serviere. Every problem in the one, and every device in the other, indicates great excellence of design and ability of execution; and both are replete with proofs of mechanical skill as well as fertility of invention.

To realize Worcester's machine, it is contended that we must depend upon what he has said, and on *nothing more*. But those who prescribe this rule do not themselves adhere to it; and by following it, posterity could hardly comprehend a modern device from its modern description. As Worcester has not mentioned pipes or valves, neither of these essential elements of his apparatus could, by such a rule, be admitted: and if his words are to be construed literally, he employed two score of receivers; and these were also elevated as well as the water within them: "one vessel of water rarefied by fire driveth up forty [vessels] of cold water." By the same rule it was the boiler, not the steam within it, (he never mentions steam) that drove them up. Then there is the *condensation* of the steam in a receiver, after expelling the liquid, which is also not mentioned; and of course the vessel could not again be filled until this had taken place. On the same ground, a cock, tunnel and pipe or pump to feed the boiler, and a furnace door and grate bars, might be considered gratuitous additions, since none of them are mentioned.

Perhaps the most obscure part of the 68th problem, is that which relates to strengthening the vessels "by the force *within* them." Some persons suppose this refers to the figure of the vessels—others, to interior braces. The latter is the most reasonable, but seems hardly reconcilable with the the text, since the same term (*force*) is used as that by which the *active* power which rent the gun is designated.

Notwithstanding the ambiguous manner in which Worcester drew up his Century of Inventions, there are strong indications of his having imparted motion to a piston by steam, and that upon this he depended for being known to posterity. This was the crown of his glory as an inventor —the primary element in the "semi-omnipotent engine," which supported him under the contumely and neglect that he met with. Unfortunately for his fame, the state of the arts was not sufficiently advanced to convince his contemporaries of the importance of "the great machine," and it was left for a future age to adopt. It does not appear equally clear that he was the first thus to use steam. From the description of the engine of motion mentioned at page 423, and the third and fifth devices in Ramseye's patent, it would seem that a working cylinder had been in previous use; nor do we see how the experimenters of the 17th and previous centuries, when seeking for modes of employing steam as a motive agent, could miss it any more than their successors. It is one of those devices that would be detected by such men in every age, just as it has been by the makers of pumps and piston bellows. Fludd, Hoell, Belidor and Westgarth, all employed a piston and cylinder in pressure engines; and some of them were not aware of their having been employed before in such machines. Guerricke, Papin and Newcomen at once adopted them in atmospheric engines, Hautefeuille in explosive engines, and Watt and others in those moved by steam; and why not Garay, Ramseye and Worcester? And even the troublesome neighbor of Zeno also? It required no great sagacity in Worcester to apply steam to move the loaded piston in Fludd's pressure engine, (page 354) and so simple an idea could hardly escape him after he had turned his attention to impart motion by steam. Indeed, he uses an expression which implies that it was a loaded piston to which he gave motion. But even if this idea escaped both Ramseye and Worcester, the apparatus of Guerricke so clearly exhibited the mode of applying steam to move a piston, that the latter could not possibly have remained any longer ignorant of it.

When the three following propositions in the Century are duly considered, every candid mind will, we think, admit that he was really in possession of an engine similar to Leopold's, or to Newcomen's, or to the single acting one of Watt:—

"98. An engine, so contrived, that working the *primum mobile* forward or backward, upward or downward, circularly or cornerwise, to and fro, streight, upright, or downright, yet the pretended operation continueth, and advanceth, none of the motions above-mentioned hindering, much less stopping, the other; but unanimously and with harmony agreeing, they all augment and contribute strength unto the intended work and operation; and, therefore, I call this *a semi-omnipotent engine*, and do intend that a model thereof be buried with me.

"99. How to make one pound weight to raise an hundred as high as one pound falleth, and yet the hundred pound descending doth what nothing less than one hundred pound can effect.

"100. Upon so potent a help as these two last-mentioned inventions, a water-work is, by many years experience and labor, so advantageously by me contrived, that a child's force bringeth up, an hundred foot high, an incredible quantity of water, even two foot diameter, so naturally, that the work will not be heard, even into the next room; and with so great ease and geometrical symmetry, that though it work day and night, from one end of the year to the other, it will not require forty shillings reparation to the whole engine, nor hinder one day's work; and I may boldly call it the most stupendous work in the whole world: not only, with little charge, to drain all sorts of mines, and furnish cities with water, though never so high seated, as well as to keep them sweet, running through several streets, and so performing the work of scavengers, as well as furnishing the inhabitants with sufficient water for their private occasions; but likewise supplying rivers with sufficient to maintain and make them portable from town to town, and for the bettering of lands all the way it runs; with many more advantageous and yet greater effects of profit admirable and consequence. So that deservedly I deem this invention to crown my labours, to reward my expences, and make my thoughts acquiesce in way of further inventions; this making up the whole century, and preventing any further trouble to the reader for the present, meaning to leave to posterity a book, wherein, under each of these heads, the means to put in execution, and visible trial, all and every of these inventions, with the shape and form of all things belonging to them, shall be printed by brass plates."

To an ordinary reader all this appears preposterous, nor without the key can any satisfactory interpretation be given. The first seems incredible, the second impossible, and the third a proof of mental alienation. But in considering them it should be kept in mind, that Worcester's design was to explain the effects and uses of the mechanism he here refers to, and at the same time to *conceal* the moving principle. This he has accomplished in the happiest manner; and in doing it, has furnished a specimen of ingenuity, and of the fertility of his genius, almost equal to the inventions themselves. The three problems certainly refer to a cylindrical steam-engine raising water by means of a pump. In No. 98 he speaks of steam only: this was the *primum mobile* whose effect was the same in whatever direction it was conveyed to the piston; i. e. whether through ascending, descending, curved, angular or straight tubes, or through a number of them meeting in the cylinder from every imaginable direction; the steam from one not interfering with, or being counteracted by, that from others, but the whole "unanimously and with harmony agreeing, they all augment and contribute strength unto the intended work and operation," viz. in pushing the piston along. It seems impossible for Worcester to have selected a feature of äeriform fluids better adapted for his purpose, or to have made use of it more skillfully. In concealing his meaning by riddles, he seems to have equalled the most expert among the ancients.[a] In No. 99 he plays in a similar style upon *the piston*, and has contrived with admirable tact to contradict (apparently) one of the most palpable maxims in mechanics, and thus to divert prying curiosity into a wrong track. The piston was attached by its rod to one end of a working-beam, and a loaded pump-rod to the other, so that when the steam was turned on, the small piston (which he compares to one pound) was pushed down, and consequently the heavy pump-rod, or the water raised by it, (compared to a hundred pounds) elevated "as high as the one pound falleth." In No. 100 he opens his views still further by stating it to be a *water-work*, for draining "all sorts of mines, and furnishing cities with abundance of water, though never so high seated," and that its action depended upon the two last mentioned inventions (Nos. 98 and 99.) In other words, he here contemplates the pump and steam-engine as a whole; but lest the device should be too easily apprehended, he throws in a dash of the enigmatical, declaring it was so contrived "that a *child's* force bringeth up, an hundred foot high, an incredible quantity of water, even [a column] two foot diameter;" that is, a child could by a lever open and close the cocks, or valves, by which steam was admitted into the cylinder. The uniformity of the movements of a steam-engine, and the little noise attending them,—its working incessantly night and day, and the trifling expense required to keep it in repair, are now well understood.

---

[a] Of ancient riddles, that of the Sphynx is one of the neatest. What animal is it that walks on four legs in the morning, on two at noon, and on three in the evening? Œdipus explained it. The animal, he said, was *man*, who in the morning of life (in infancy) crept on his hands and feet, at the noon of life walked erect, and in the evening of his days supported himself with a stick.

Nothing more is necessary to convince us that Worcester here speaks of a steam-engine working a pump. No other solution can be given—no other conclusion arrived at. No one could have written and spoken as he has done without having either seen or possessed a steam engine. Of its value he was fully aware; for in the patent granted by Parliament in 1663 to himself and his heirs for the long term of ninety years, those who pirated the invention were to forfeit five pounds for *every hour* they used it. He tells us that it was the result of "many years experience and labor," and when it was complete, he poured out his feelings in an address to the Deity, a copy of which was found among his papers, entitled "The Lord Marquis of Worcester's ejaculatory and extemporary thanksgiving prayer, when first with his corporeal eyes he did see finished a perfect trial of his water-commanding engine, delightful and useful to whomsoever hath in recommendation either knowledge, profit or pleasure." Can any one suppose he here was mocking his Creator, when, in the privacy of his closet, he prayed that he might not be "puffed up" with the knowledge of this great machine, and returned thanks next to his creation and redemption "for an insight into so great a secret of nature," and finally desired no greater monument than to have one buried with him? Some men have lost their reason by the excitement attending their discoveries. Pythagoras offered a hecatomb to the gods, and Archimedes ran naked through the streets of the city. Worcester acted more like a philosopher and a Christian. Had he imitated the Syracusan, he had probably been more successful in securing attention to his discoveries.

From the latter part of the 99th proposition, we infer that Worcester used a *forcing* pump, as he intimates that the effect was produced by the descent of a weight (on the pump-rod,) not by its ascent; and this agrees with the description and figures of old water-engines. In "Art and Nature," published as before observed in 1633–4, they consist of forcing pumps worked by large tread and other wheels—i. e. the pistons are raised by these but are carried down by their own weight, or that of weights with which they are loaded. These weights were sometimes attached to the rod, at others to the end of the working-beams to which the rods were connected; and hence they were named "beetle-beams," from their resemblance to a large hammer. Loading the piston-rod of pumps did not therefore originate with Moreland or Newcomen, since the practice was older even than Worcester. The piston in Fludd's pressure engine is an example. Such pistons were named "heavie forcers," (a solid piston being named "a forcer," and the upper box of a common pump "a sucker.")

As Worcester is believed to have applied steam to work a pump, it will be asked, did he not perceive its application as a mover of machinery in general—to propel boats, &c.? Yes; and he has left a proof of this also. In a manuscript (see Stuart's Anecdotes, vol. i, 56) he observes, speaking of the device No. 99, "I can make a vessel of as great a burden as the river can bear, to go against the stream, which the more rapid it is, the faster it shall advance, and the moveable part that works it, may be by one man still guided to take advantage of the stream and yet to steer the boat to any point; and this engine is applicable to any vessel or boat whatsoever, without being therefore made on purpose; and it worketh these effects: it roweth, it draweth, it driveth, (if need be) to pass London Bridge against the stream at low water; and a boat laying at anchor, the engine may be used for loading and unloading." Besides the Century, Worcester published what he called "An exact and true Definition of the most stupendous Water-commanding Engine, invented by the Right Honourable, (and deservedly to be praised and admired), Edward Somerset, Lord Marquis of Worcester, and by his lordship himself presented to his most excellent Majesty Charles the Second, our most gracious Sovereign." This was a tract of twenty-two pages, and is supposed to have been printed for the purpose of forming a company to introduce the device. It is written in the same style as the Century, and instead of describing the machine is confined to an enumeration of its properties.

In Worcester's day, patents for useful inventions were often classed with the most unrighteous monopolies, and the holders of them held in general contempt. This may serve to account in some measure for the neglect that Ramseye and Worcester's projects met with. The abominable abuse which Elizabeth, James and Charles made of the power to grant patents, excited general disgust. Courtiers and others obtained monopolies for nearly all the chief branches of trade, and sold rights in them to others, so that prices were raised to an exorbitant height. Had patents been confined to new inventions, the result would have been beneficial; but exclusive grants were obtained to work and sell the commonest articles, as salt, iron, lead, coals, and even bones and rags: with the monopolists of these, (*harpies and horseleeches* as Elizabeth once called them) the authors of discoveries and improvements in the useful arts were confounded. In a masquerade got up for the entertainment of Charles I, in 1633, (among the managers of which were Noy the Attorney General, Sir John Finch and Mr. Selden) were several flings at monopolies, as hints for the king. In the "Antimasque of Projectors," says Maitland, "rode a fellow upon a little horse, with a great bit in his mouth; and upon the man's head was a bit, with head-stall and reins fasten'd, and signified a projector who begged *a patent*, that none in the kingdom might ride their horses, but with such bits as they should buy of him. Then came another fellow with a bunch of carrots upon his head, and a capon upon his fist, describing [representing] a projector who begged a patent of monopoly, as the first inventor of the art to feed capons with carrots; and that none but himself should make use of that invention; and have the privilege for fourteen years, according to the statute."

Putting out of view his political conduct, the fate of Worcester resembled that of great inventors in almost every age. In some respects it was peculiarly severe. The heir of one of the richest and most powerful families of the land, he devoted his wealth and his energies, for more than one third of a century, to useful discoveries; and in his old age he was reduced to borrow small sums to meet his necessities;—and when at last the profligate Charles was restored, although Worcester recovered his demesne, his dwellings, furniture, papers, models and machines had all been destroyed, and he was overwhelmed with debt. Still his energies were stimulated by a consciousness of the importance of his inventions, but which, alas! his contemporaries were unable to appreciate, except by insinuations that they were the fruits of a partial insanity. Finally death stept in and closed his labors forever. Then it was that his widow, who was fully sensible of the value of his great machine, used her exertions to introduce it; but her confessor, a Roman Catholic priest, expostulated with her on the folly and *sin* of her conduct, and solemnly declared to her "on the faith of a priest," that if she did not cease her endeavors, she would not only lose the favor of heaven, but the use of her reason! She died in 1681, and the evil genius of Worcester did not even then cease its persecutions; for posterity, which generally corrects the errors of contemporaries, has not yet done justice to his memory. While a few writers admit the value and originality of his inventions, and account him one of the chief authors of the steam-engine, others condemn the "Century" as a mass of absurdities, and deny his ever having constructed a steam-machine at all. Those persons however who entertain the latter opinion, evince as much credulity as others, for they cannot deny that he has described the peculiar properties of the great *chef d'œuvre* of human ingenuity (a high-pressure steam-engine) with a degree of accuracy of which history affords no parallel; and hence, if he lacked truth he possessed prescience, and while they reject him as an inventor, they must admit him as a prophet.

In the annals of the arts, there is not to be found a more singular example of devotion to their improvement, either as regards the number of years or the amount of treasure spent, the importance of the results or the ardor with which they were pursued, and the efforts made to excite public attention to them. Whatever others may have done before him, they left no account of their labors. Worcester is the first to communicate with the public by means of the press, and to give a tangible description (although an intentionally obscure one) of his discoveries—(for we do not reckon either the device of Branca or Decaus among such.) On this account alone he is entitled to the praise of every modern engineer; and had he but fulfilled his promise of leaving detailed accounts illustrated with engravings, his fame would have endured as long as the steam-engine itself. If he were not the great magician who evoked the mighty spirit that lay dormant in steam—who pointed out its power and the means of employing it—who revived the project of Garay and embodied and extended the apparatus of Porta—it may be asked who was? And although none of his machines are extant, nor any of his immediate successors have had the candor to acknowledge their obligations to him, it is not less the duty of historians to uphold his claims until evidence shall be adduced to establish those of another—until some older and clearer fountain than his Century of Inventions shall be discovered—from which streams equally unacknowledged have been drawn. We cannot but hope that the obloquy and uncertainty under which his name is yet shrouded will eventually be dispersed, when he will be esteemed one of the most remarkable mechanicians of modern times, and be associated with the Dædaluses and Archytases of antiquity.

How similar to Worcester's manner of announcing his discoveries, is the following one from Glauber, an older writer! It appears, at the first glance, as absurd as any thing in the Century. "A certain secret by the help whereof wines are easily transported from mountainous places, remote from rivers and destitute of other conveniences of carriage, so that the carrying of *ten* vessels is of cheaper price than, *otherwise*, the carrying of *one*." This passage, he observes, offended many both learned and unlearned, who "believed the thing impossible, and nothing but dreams and fancies." He was so much quizzed about it as to regret having mentioned it "Many judge this thing incredible because of the want of *winged carts*, that need not horses! confirming one the other in unbeliefe, leading one another after the manner of the blind." His plan was to take the juice of the grape before fermentation commenced, and concentrate it by boiling till it became of "the consistence of honey." The water being thus evaporated reduced the wine to less than one tenth of its former bulk and weight, while it still retained the strength and virtue of the whole; for "new wine decocted and inspissated before its fermentation loseth nothing of its virtues:" hence it could be transported at one tenth the expense. When used, it was to be diluted with the same quantity of water as was evaporated from it. (Treatise on Philos. Furnaces: Lond. 1651, p. 353.)

The adoption of some mode of concentrating wines as above, would produce an immense saving in their freight and carriage over the globe, and would consequently greatly reduce their cost. It would also defeat the enormous frauds that are practiced in the manufacture of artificial wines—mixtures in which not a drop of the juice of the grape is said to enter. Glauber says, "the new wine is not to be inspissated in cauldrons," on account of the taste which it would contract from the metal.

CHAPTER VII.

Hautefeuille, Huyghens and Hooke—Moreland—His table of cylinders—His pumps worked by a cylindrical high-pressure steam-engine—He made no claim to a steam-engine in England—Simple device by which he probably worked his plunger pumps—Inventions of his at Vauxhall—Anecdote of him from Evelyn's Diary—Early steam projectors courtiers—Ridiculous origin of some honors—Edict of Nantes—Papin—Digesters—Safety valve—Papin's plan to transmit power through pipes by means of air—Cause of its failure—Another plan by compressed air—Papin's experiments to move a piston by gunpowder and by steam—The latter abandoned by him—The safety valve improved, not invented by Papin—Mercurial safety valves—Water lute—Steam machine of Papin for raising water and imparting motion to machinery.

Towards the latter part of Worcester's life, a young Frenchman was fast rising into notice. This was John Hautefeuille, the son of a baker at Orleans, and one of the most brilliant mechanicians of the age. He was in his twentieth year when Worcester died. The device for regulating the vibration of the balance in watches by a spring, whence arose the name of pendulum watches, was invented by him, and was subsequently improved by Huyghens. Hautefeuille entered the church and became an abbé. He wrote several tracts on subjects connected with mechanics. In 1678 he proposed steam as a source of power, and applied it to give motion to a piston. Instead of aqueous vapor he also proposed the alternate evolution and condensation of the vapor of alcohol, in such a manner that none should be wasted; and both he and Huyghens gave motion to pistons, by exploding small charges of gunpowder in cylinders. In 1678, Dr. Hooke proposed a steam-engine on the atmospheric principle, but the only information respecting it is in a memorandum to that effect found among the papers of Dr. Robison, the author of the treatise on Mechanical Philosophy.

These examples of imparting motion *to a piston* by aëriform fluids are interesting, inasmuch as they show that the device was not very novel in the middle of the 17th century, and that mechanics in different countries were familiar with it.

We must now refer to another member of the English court, a contemporary of Worcester, and like him actively engaged in the politics of the times, but who on the other hand adhered to the commonwealth until the latter part of Cromwell's administration. We are told that one evening, near midnight, an interview took place between Cromwell and Thurloe his secretary, at the house of the latter, on some state business that required the utmost secrecy. It was not till the matter had been opened that the Protector became aware of a third person being in the room, when he is said to have drawn his dagger, and would have dispatched the supposed intruder, had not Thurloe guaranteed silence on the part of his sleeping attendant. This was Samuel Moreland, the inventor of the plunger pump. He was then employed by and in the confidence of the secretary, and was asleep, or affected to be so, during the interview. On this or some other occasion, he overheard the discussion of a plan to take off the exiled king; to whom he disclosed the whole, and was rewarded with a title at the restoration.

It is not known when Moreland first turned his attention to mechanics: probably not till the restoration. As a favorite of Charles II, and a groom of the privy chamber, he must often have met Worcester at court; while from their congenial habits and pursuits as mechanicians, they were most likely on familiar terms with each other. As master of mechanics to the king, Moreland was no doubt one of those who visited and examined the machine erected by Worcester at Vauxhall, and as a matter of course he often perused the Century of Inventions. He has not however had the ingenuousness to mention any of these things; but notwithstanding this, we cannot believe so far as his applications of steam are concerned, that he was not indebted either to the machine itself, to the Century, or to personal intercourse with Worcester, and probably to them all. The first invention of Moreland that we hear of is the pump that he patented in 1675, and on which, according to one writer, he had previously spent twelve years. This carries the date back to about 1663, the year in which the Century was published. It is not at all unlikely that this famous pamphlet first induced Moreland (as well as many others) to turn his attention to mechanical discoveries, and furnished him with materials to work upon. In the manuscript volume presented by him to the French king in 1683, (see page 273) and now preserved in the British museum, there is a very short chapter on fire or steam engines, of which the following is a translation:—

"*The Principles of the new Force of Fire, invented by the Chevalier Moreland, in the year* 1682, *and presented to his most Christian Majesty,* 1683.

"Water being evaporated by the force of fire, these vapors immediately require a greater space (about two thousand times) than the water occupied before, and too forcible to be always imprisoned, will burst a piece of cannon. But being governed according to the rules of statics, and reduced by science to measure, weight and balance, then they will peaceably carry their burden, (like good horses) and thus become of great use to mankind; particularly to raise water according to the following table, which shows the number of pounds which can be raised 1800 times per hour to six inches in height, by cylinders half filled with water, as well as the different diameters and depths of those cylinders:—

| CYLINDERS. | | Pounds weight to be |
Diameter in Feet.	Length in Feet.	raised.
1	2	15
2	4	120
3	6	405
4	8	960
5	10	1875
6	12	3240

Number of Cylinders, having a diameter of 6 feet and a length of 12 feet,	required to raise the following numbers of pounds weight of water.
1	3,240
2	6,480
3	9,720
4	12,960
5	16,200
6	19,440
7	22,680
8	25,920
9	29,160
10	32,400
20	64,800
30	97,200
40	129,600
50	162,000
60	194,400
70	226,800
80	259,200
90	291,600 "

As this is all that Moreland has left on the subject, it is difficult if not impossible to ascertain the precise construction of his apparatus. He is as silent respecting the manner and details by which the object was accomplished as Worcester himself, and hence the steam-engine of one is quite as much a riddle as that of the other. Were these "cylinders" *generators* of steam—boilers? or were they separate vessels for the reception of water, and from which it was expelled by the vapor, as from the receivers of Savery? or, working cylinders, whose pistons were moved by the expansive force of steam? or, lastly, were they pump chambers, by which the liquid was raised? We suppose they were the last. Had they acted on the principle of Savery's receivers, they could never have been filled and discharged thirty times a minute, or 1800 times an hour. Then as Moreland speaks only of high steam, it can hardly be imagined that he used or thought of using its expansive force to move pistons in the largest cylinders he has named, or made calculations for the employment of ninety of them. Where could he have got a boiler sufficiently strong and capacious to supply a cylinder twelve feet long and six in diameter, to say nothing of the difficulty of making such cylinders? Yet he speaks of them as nothing extraordinary. Now there was no difficulty in making them of all the dimensions named *for his plunger pumps*, (see No. 123 of our illustrations) for the simple reason that they were not required to be bored; as the piston or plunger worked in contact only with the collar of leathers or stuffing box at the top. That it is to these he refers appears also from the terms, "reduced by science to measure, weight and balance," these being the very same that he used when he claimed, by the invention of this pump, to have "reduced the raising of water to *weight* and *measure*," viz. by comparing the weight of the loaded plunger to the quantity of water displaced from the cylinder by its descent, (see page 273)—and thence the number of pounds raised by each cylinder in the preceding table, would be the sum of the weights on each plunger. The term "six inches" probably arose from that being the length of stroke of his experimental plunger; the length of the other cylinders and their effects being calculated from it. The cylinders being only "half filled with water," would then refer to that quantity, or about that, being expelled at each stroke, because the plungers would occupy one half only of the interior capacities of the cylinders. See the figure of one on page 272.

If this view of Moreland's project be correct, then he merely used steam to work his plunger pump; and therefore could not justly claim in 1682 to have invented, but only to have applied, the "force of fire." That he employed a simple form of a high-pressure engine, in other words moved a piston by the elasticity of the vapor, like Hautefeuille and Worcester, we have little doubt. His language intimates that steam was then rendered so manageable as to be applicable to numerous operations "for the benefit of mankind," of which the raising of water was the only one under his consideration. He obviously was in possession of the means of imparting motion *to solids* by steam, and thus making it peaceably to carry burdens, or overcome resistances, "like good horses:"—Indeed, one might almost suppose from his apparent carelessness in not mentioning the mode in which the steam was applied, viz. in giving motion to a piston, that explanation on this point was then no longer necessary.

It is singular that Moreland made no claim for this invention in *England*. Why was this, if he had any? Does it not imply that he did not invent the steam part of the apparatus?—else why not have patented it as well as the pump? for the object deserved it, and the prospects of remuneration were as promising at home as in France. The fact is, he could not claim the piston steam-engine where the labors of Worcester and others were still in remembrance, and where some of their machines were probably extant. As an educated man and an enlightened mechanic, Moreland was not ignorant of the labors of Ramseye, Fludd, Hautefeuille and Worcester. It is pretty clear that he lit his candle at the lamps of these men, and particularly the latter; for in the short chapter on steam quoted above, he has copied both the ideas and the language of the author of the Century of Inventions. One observation is highly creditable to him, if he was the author of the experiments from which it was deduced, viz. the relative volume of steam and water. A quantity of the latter when converted into the former occupies, he observes, 2000 times its former space: modern experiments make it between 1800 and 1900 times.

Of several simple modes by which Moreland may have applied steam to work his pumps, we shall mention one:—Let a small steam-cylinder, open at the top, be placed under the same end of a vibrating beam as the plunger of the pump; the piston rods of both cylinders being connected to the beam: then, by turning a three-way steam-cock, the vapor would rush into the bottom of the steam-cylinder, and pushing up the piston, would raise the beam and the loaded plunger of the pump; and by then turning the cock so as to close the communication between the cylinder and the boiler, and to open one between the former and the external air, the steam would escape, and the weights on the plunger would cause it, with the beam and steam-piston, to descend. By turning the steam-cock as before, the stroke would be repeated. The only objection to such a device is, that it is too crude to be attributed to Moreland; for, from the advantages he possessed in knowing all that had been previously done, there can be little doubt that he was in possession of a self-acting engine, and of the knowledge of increasing its energy according to the different sized pumps required to be worked by it.

Moreland possessed a natural turn for mechanics, and during the latter half of his life devoted himself almost exclusively to the invention and improvement of useful machinery. Were a description of his and Worcester's workshops now extant, it would possess more real interest than any thing which history or tradition has handed down about round-heads and cavaliers. He had a place fitted up at the expense of government, with the requisite apparatus for carrying on his researches, at which Charles sometimes assisted; and he speaks of having moreover expended large sums of his own in experiments, to please the king's fancy. Of the number of curious things here contrived, besides his speaking trumpet, capstan, pumps and steam-engines, we may judge from what is reported of his dwelling house at Vauxhall, every part of which exhibited proofs of his inventive mind; even the side table in his dining room was supplied with a large fountain, and the glasses stood under little streams of water. His coach too contained a portable kitchen with clock-work machinery, by which he could make soup, broil steaks, or roast a joint of meat.

Vauxhall gardens, as a place of public resort, appear to have originated in the curious things constructed there by Moreland. Aubrey, in his History of Surrey, states that in 1667, Sir Samuel "built a fine room at Vauxhall, the inside all of looking-glass, and fountains very pleasant to behold, which is much visited by strangers. It stands in the middle of the garden, covered with Cornish slate, on the point whereof he placed a punchinello very well carved, which held a dial; but the winds have demolished it." "The house [observes Sir John Hawkins] seems to have been rebuilt since the time that Sir Samuel Moreland dwelt in it; and there being a large garden belonging to it planted with a great number of stately trees, and laid out in shady walks, it obtained the name of Spring Gardens, and the house being converted into a tavern or place of entertainment, it was frequented by the votaries of pleasure."

Moreland's attachment to mechanics continued unabated in his old age, and even after his sight was lost. A pleasing proof of this is given in the diary of the celebrated John Evelyn. "October 25, 1695. The archbishop and myselfe went to Hammersmith to visite Sir Sam. Moreland, who was entirely blind; a very mortifying sight. He shewed us his invention of writing, which was very ingenious, also his wooden kalender which instructed him all by feeling, and other pretty and useful inventions of mills, pumps &c. and the pump he had erected that serves water to his garden, and to passengers, with an inscription, and brings, from a filthy part of the Thames neere it, a most perfect and pure water. He had newly buried £200 worth of music books, six feet under ground, being, as he said, love songs and vanity. He plays himselfe psalms and religious hymns on the Theobo."

It is singular that almost all the early English steam machinists and supposed experimenters were *courtiers*. Bacon was Lord Chancellor; Ramseye, groom of the privy chamber to Charles I; Worcester, a marquis and a general; Moreland, a knight, and groom of the privy or bed chamber[a] to Charles II; and Bushell and Savery held offices under the government.

---

[a] It is a natural inquiry, what odd duties were attached to such an office, that *gentlemen* should desire to perform them? and particularly to such beasts as Charles II or George IV? An analysis of the honors which monarchs bestow, would afford amusement and instruction to American readers. It would add to the causes of honest exultation in the founders of our republic, for their excluding such fooleries from our shores. Our example in this respect as well as in others, is destined by Providence to exert a salutary influence on the world. A spirit of enquiry, and ideas of self respect, are already becoming too prevalent for men to be kept much longer in a mere state of pupilage, to be governed like children through the medium of their senses, with pageants and high sounding titles, costumes and ceremonies, tinsel and gewgaws. Those persons who have not reflected on the subject, have need of a large share of faith to believe one half the circumstances connected with the origin and the conferring of titles, so truly pre

The next experiments on steam of which we have any account were made by a Frenchman, but not in France. The reason of this may as well be noticed, since it will serve to show how great the blessings are which we enjoy over the people of the old world. Of all the different species of tyranny under which Europe has groaned and still groans, that by which the inhabitants are compelled to adopt such articles of religious faith as their governors choose to give them, is the most diabolical. This may be considered as the climax of human degradation, and of human oppression. No feeling mind can contemplate without horror the acts of those despots who, not content with consuming the substance and tyrannizing over the bodies of their *subjects*, as they call them, insist on subduing their *minds* and *consciences* also!—despots who, though covered with crime, blasphemously set themselves up as "Heads of the Church!" and "Defenders of the Faith!" and this too by the "grace of God!"—And these Heads of the Church, in order to defend "the Faith," have harassed, plundered, hanged, shot and even burnt alive both men and women who would not acknowledge them as such! Thus it was when the edict of Nantes, which Henry IV established to protect his Protestant subjects in their civil and religious rights, was revoked by Louis XIV, it became the signal for the most violent persecutions of that people. Their children were taken from them and placed under Papist teachers—they were compelled by the penalty of military execution to embrace the Roman faith—a price was set on the heads of those who refused—a twentieth part of their whole number was butchered—half a million fled into other lands (as the Pilgrims did to this) that they might be at liberty to worship God according to their own consciences. In this way the most ingenious and avowedly the most industrious mechanics of France were driven into exile; and by a righteous and retributive Providence, the staple manufactures of that kingdom were transferred to other nations.

Of those who took refuge in England was Papin, a native of Blois; a physician and philosopher, and one of the most talented of the early experimenters on steam and air: a man of whom any country might have been proud, and who, though France *then* cast out as a disgrace, she *now* claims honor to herself for having given him birth; and mourns that the records of his labors are only to be found in foreign archives. What a commentary on religious persecution, that the only claims which Roman Catholic France has or can set up for a share in the invention of the steam-engine, are based on the ingenuity of a Jew and a Protestant!—on Solomon Decaus and Denys Papin.

Through the influence of the celebrated Boyle, Papin was elected a fellow of the British Royal Society in December, 1680. He was an active and useful member, and contributed several interesting papers to the Society's Transactions. In 1681 he invented a method of softening bones, with a view to extract nourishing food from them, viz. by submitting them to the action of steam at high temperatures, in close vessels named *digesters*.

---

posterous are they. The Orders were derived from all sorts of things, as the moon, stars, dogs, horses, swords, flowers, stones, shells, birds, pigs, the Savior, angels, saints, women—and there was even an order of *fools!*—elephants, thistles, mountains, blood, wool, a table—and who does not know that " the *most* honorable" of English Orders at the present day are those of the *bathing tub*, or *bath!* and of the *garter!* The ceremonies attending these can only be equalled for mummeries and childish puerilities by the old interludes, as those of " The Bishop of Fools" and " The Abbot of Unreason." Such are the things that distinguish " the privileged orders " of Europe—that are deemed necessary to maintain " the *dignity* of the crown,"—and the *debasement* of the people.

There seems however to be some mistake respecting the date just mentioned, which is the one generally assigned; for in the second volume of Boyle's Works (by Shaw, Lon. 1725) are details of experiments on boiling beef &c. "in screw'd vessels or digestors," in the beginning of the year 1679—thus: "January 29. Eight days ago I fill'd a screw'd vessel with beef and water together, and when it had continued over a moderate fire for 8 or 9 hours in *balneo mariæ* [a water bath] stopp'd also with a screw, I took the flesh out," &c. "Feb. 10. I boil'd a cow heel after the same manner as I had done the flesh above mention'd, but left it for four hours or more upon a moderate fire; then the vessels being unstopp'd, we found the flesh exceedingly well boiled, and the bones so soft that they might be easily cut with a knife and eaten." "Feb. 12. I repeated the experiment and let the vessel remain exposed to the fire for 12 hours," &c. - - - - - "Hence it appears that many bones and hard tendons, which we daily throw away as unprofitable, may, by the help of a *balneo mariæ* stopp'd with a screw, be converted into good nourishment." pp 550, 551.

Papin's first digesters were as liable to be rent asunder as eolipiles placed on a fire with their orifices stopped. They are figured in detail in Poliniere's *Experiences de Physique*, 2d ed. Paris 1718. Each consisted of a short but very thick tube, of bell-metal, about a foot in length and five inches in diameter, with one end closed. The open end had a collar cast on it, to which the cap or cover was secured by clamps and a screw. The cover and end of the tube were ground together so as to fit air-tight, like a valve to its seat. A few bones and a little water were put in, and the cover screwed down; the vessel was then laid in a horizontal position on a bed of charcoal in a long iron grate. The almost unavoidable rupture of these vessels, led Papin to the invention of the lever safety-valve, which he first applied to them, and afterwards to machines for raising water by steam.

Notwithstanding the practical knowledge of the properties of steam acquired by the employment of digesters, Papin does not appear to have had any idea of using it as a mechanical agent till some years after. His first paper on the subject of raising water is dated July, 1685, (Phil. Trans. vol. xv, page 1093; Abridgment, vol. i, page 539) entitled "A New Way of Raising Water, enigmatically proposed." Three different solutions were sent in, after which he explained. The device was a small fountain, in which the liquid was raised by a piston bellows "put in some secret place, where a body may play the same." The application of the device was then pointed out, viz. to draw water from mines, by means of a running stream located "far distant" from them: in other words, to transmit power to a considerable distance by means of *air*.

His plan was this: a series of air-tight receivers were to be placed, 12 feet above each other, in the shaft; the highest on a level with the ground, and the lowest 12 feet above the bottom of the pit. The water was to be transferred by the pressure of the atmosphere from one receiver to another, till it was discharged above. For this purpose a pipe extended from the water to the bottom of the lowest vessel; another pipe from the lower part of this to the next one, and so on to the top; and to prevent the water from running back, the upper orifice of each pipe was covered by a valve. The mode by which he alternately withdrew air from and admitted it into the receivers, constitutes the main feature of the plan. The upper parts of every two receivers were connected by branch pipes to a long one attached to the bottom of a *separate* air-pump, which was to be placed near a water-wheel impelled by the current; and the piston was to be worked by a crank formed on the shaft of the wheel. The operation of two pumps and four receivers will be sufficient for the purpose of illustration. The pump cylinders were open at the top. They had no valves, and but one small opening at the bottom of each where the long air-pipe was united; and the capacity of each cylinder was equal to that of two receivers. The cranks were so arranged that as one piston ascended the other descended. It was not two adjoining receivers that were connected to the same pump, but the lowest and the next but one above it, i. e. Nos. 1 and 3, while to Nos. 2 and 4 the pipe of the other pump was attached. Then as the piston of the first mentioned pump was raised, the air in Nos. 1 and 3 would be rarefied by rushing into the pump cylinder, and water would be forced into them by the atmosphere: in the mean time the other piston would produce a partial vacuum in Nos. 2 and 4, and they would become filled with the liquid contents of Nos. 1 and 3, in consequence of the air previously in these being driven back by the descent of the piston; so that as the wheel revolved, water would constantly be entering one half of the receivers, and the contents of the other half be discharging. How the water was to be delivered from the highest receiver Papin has not informed us—probably through an orifice covered by a valve opening outwards.

This project was ingenious, but of no practical value; and it failed even in an experiment. In consequence of the extreme elasticity of air, and the great facility with which it dilates and is compressed, little or no effect was produced by the action of the pumps. When a piston descended, the air in the long pipe readily yielded to its impulse without imparting any very sensible compression in the receivers; and on the piston's ascent, the air in the pipe again dilated and no sufficient rarefaction took place, in consequence of the great distance of the receiver from the exhausting apparatus.

On the failure of this he devised another plan. Suppose, for example, that it was required to raise water out of a mine, and that there was no river to turn a wheel to work the pumps nearer than a mile. Papin proposed to place two air-pump cylinders fitted with pistons near the water-wheel, and other two at the mouth of the mine. These were to be connected by a pipe. The action of the pistons moved by the wheel was to *compress* the air in the cylinders, and in the pipe throughout its whole length, under the idea that when the pistons at one end of the pipe were depressed, those at the other would, by the communication of pressure, be elevated; but although the pistons moved by the water-wheel condensed the air, those at the mine stood still. The same cause that led Papin to abandon the first device, also rendered this one useless. If air were incompressible, the plan would have answered: had he employed *water* instead of air, the machine would have performed. Nothing daunted however, he tried again in 1686, and with a somewhat similar apparatus, but one whose action depended upon the *rarefaction* of air. Two large air cylinders, open at top, were placed a short distance apart at the mouth of the mine, and directly over them a cylindrical shaft or axle, supported on journals at each end. Instead of rods being attached to the pistons, a strong rope was fastened to the centre of each, and coiled three or four times round the axle in opposite directions, and fastened to it. Between the cylinders a large drum or wheel was fixed upon the axle, having a long rope wound round it, and the two ends (of the rope) suspended from opposite sides reached half way down the mine. To these ends two large buckets were attached, in which to raise the water. As the drum turned first one way and then the other, one bucket would be raised and the other lowered, like two buckets suspended over a pulley in a well. The design of the air cylinders was therefore to impart an alternate movement to the axle and drum, and consequently to the buckets, by the *descent* of the pistons. (The power that forced these down was the pressure of the atmosphere, and the manner of exciting it will presently be noticed.) Hence as one piston descended, the rope secured to it necessarily drew round the axle, and raised the other; and when this one in its turn was forced down, the movement was reversed and the first one raised.

A communication was made between the under side of the two cylinders by a pipe, and to this another long one was attached at right angles. This last pipe was to be of such a length as to reach to the place where an under or overshot wheel could be applied to work two air-pumps. These were to be furnished with valves and suckers like common sucking pumps, and to the lower part of each the exhausting pipe was to be connected by a branch. These pumps would therefore draw the air out of the cylinders at the mine, and consequently cause the pistons in them to descend. For this purpose, however, some device for alternately opening and closing the communication of each cylinder with the exhausting pipe was required; because if a vacuum were made in both cylinders at the same time, the pressure of the atmosphere on both pistons would be the same, and neither of them would move. To avoid this, Papin introduced at the intersection of the exhausting pipe with the one that connected the cylinders, a four-way cock—three of its passages being joined to the three branches formed by the intersection of the pipes, while the fourth one opened to the air. Thus, supposing the pumps to be constantly at work, and the plug of the cock so turned that the air in one cylinder at the mine might be drawn out, the atmosphere would then push down the piston, provided the external air had access to the under side of the other piston. This was that which the fourth passage of the cock was designed to accomplish; for whenever one cylinder was in communication with the exhausting pipe the other was in communication with this passage, and hence by turning the cock a constant reciprocating motion was imparted to the axle, drum and buckets.

A project something like this, Papin thought, might be applied to work the pumps of the great machine at Marli, (see page 296)—the power of the water-wheels on the Seine being transmitted by air instead of chains &c. The device is creditable to his ingenuity, but he was doomed to experience further disappointment; for, on trial, the air was so slowly drawn from the cylinders, and the difficulties of making the pistons work air-tight were so great, that no practical benefit could be derived by its adoption. He enlarged his pumps and diminished the bore of the pipes in order to accelerate the movement of the pistons, but without success. Had he placed a close vessel, several times larger than his cylinders, in communication with the farther end of the exhausting pipe, and in which a constant vacuum was maintained, then, on turning the cock, the air in the cylinder would have rushed into this vessel, and the piston would immediately begin to descend. This mode of transmitting power is capable of some useful applications. See an account of a proposed pneumatic rail-way in the current journals of the day.

Although these attempts to raise water and transmit power by means of air were unsuccessful, they are interesting for the ingenuity displayed, and also because their failure led Papin to the employment of other agents. Having been invited by the Landgrave of Hesse to accept the professorship of mathematics in the university of Marpurg, in Germany, he left England in 1687; but shortly before his departure, he exhibited to the Royal Society some experiments on the application of gunpowder to produce a vacuum. His apparatus consisted of a small cylinder, in which a piston like that of a common pump-sucker (viz. with an aperture covered by a valve) was fitted to move. The bottom of the cylinder was closed, and when the piston was near the top he exploded a small charge of powder below it, with the hope that the sudden blast of flame would expel all the air through the valve, which instantly closing would prevent its return. A vacuum being thus formed, the pressure of the atmosphere would be excited and might be used as a source of power. He could not however succeed in driving out all the air by the explosion, and the pressure on the piston, (ascertained by attaching weights to a rope passed over a pulley and connected to the piston rod) instead of being 13 or 14 pounds on the square inch, seldom exceeded six or seven. He published an account of these experiments the following year in the *Acta Eruditorum*, a journal published at Leipsic, and which was to Germany what the Journal des Savans was to France and the Philosophical Transactions to England. It was commenced in 1682, and both the latter in 1665.

In 1690, Papin, unable to obtain a sufficient vacuum with gunpowder, turned his attention to steam. In one of his first essays he raised the piston by its expansive force; and then allowing it time to cool and return to its former bulk as a liquid, the pressure of the air forced the piston back. His cylinder was $2\frac{1}{2}$ inches diameter, and closed at the bottom. A small quantity of water was introduced through a hole in the piston, which was pushed down to exclude the air below it, and the hole then stopped by a plug. A brasier of burning coals was now applied to the bottom of the cylinder, and the piston consequently raised by the accumulating vapor. When the piston reached nearly to the top of the cylinder, it was retained there by a latch slipped into a notch in the piston rod: the fire was now removed, and the steam quickly condensed by the lower temperature of the surrounding air: the latch was removed, and the atmosphere pressed the piston down and raised a load of 60 pounds, which was attached by a rope and pulley to the piston rod, being an effective force of $12\frac{1}{4}$ pounds upon every square inch on the upper surface of the piston.[a] A device of this kind Papin thought was applicable to draw water from mines, and to row boats against wind and tide.

It does not appear that Papin made any essential improvement on the apparatus during the four following years; for when he published his "Recueil des diverse Pièces touchant quelques Nouvelles Machines, et autres Sujets Philosophiques, par M. D. Papin, Dr. en Méd. A Casel, 1695," he still contemplated generating the steam in the cylinders; and at every stroke these were either moved from the fire, or the fire from them. It is astonishing that the idea of a fixed and separate boiler did not occur to him. His plan was never tried except as an experiment; and he subsequently abandoned the use of cylinders and pistons, and applied steam to raise water on the plan of Worcester's 68th proposition. This was unfortunate for his fame; for in his experiments with the piston and cylinder he was in possession of every principle of the low-pressure steam-engine, and had he followed up the device he would have borne off the palm from all his contemporaries. Even the high-pressure engine, and all the glory of its development, was then within his reach; but he was no practical mechanic, and his thoughts became diverted into other channels. One of the

---

[a] It is impossible to contemplate the various attempts of Papin to move a piston by *atmospheric pressure*, without noticing the analogy between his contrivances and that of Guericke, and without thinking that the apparatus of this philosopher was present to his mind.

most pleasing and honorable circumstances connected with the history of Papin's labors, is the candid admission of several English writers of his great merits, and their generously expressing regret that his attention should have been diverted when he was so near realizing the most splendid reward. His name is however inseparably connected with the steam-engine, and as long as the safety-valve shall be used the world will be his debtor.

It should not however be supposed that safety-valves were wholly unknown before Papin's time; on the contrary, they were frequently used, although this fact has not been noticed by any writer on the steam-engine. The liability of stills and retorts to be rent asunder led old chemists to apply plugs to openings in those vessels, that the vapor might raise or drive them out and escape ere its tension exceeded the strength of the vessels: such were the plugs in ancient steam deities, see page 399. In some old works on distilling, conical plugs or valves are shown as fitted into cavities on the tops of boilers, and in some cases they were loaded. In the "*Maison Rustique* de Maistres Charles Estienne et Jean Liebault, Docteurs en Medecine," Paris, 1574, folio 196, 197, are figures of two *close* boilers in which the distilling vessels were heated: one formed a water, the other a vapor bath. On the top of each is a conical valve opening upwards. These served both to let out the superfluous steam and to introduce water. Glauber, who contributed several valuable additions to the mechanical department of chemistry, has figured and described, in his Treatise on Philosophical Furnaces, the modes by which he prevented glass retorts or stills from being burst by the vapor. A long stopple or conical valve was fitted to the neck of each, being ground air-tight to its seat, and loaded with a "cap of lead," so that when the steam became too "high" it slightly raised the valve and a portion escaped; the valve then closed again of itself, "being pressed down with the leaden cap and so stopt close." (English Translation, Lond. 1651, p. 306.) The valve on Newcomen's first engine was of this description. In the same work Glauber describes the most philosophical of all safety-valves, viz. a column of mercury enclosed in a bent tube which communicates with the boiler or still, somewhat like the modern mercurial gauge. He also describes that beautiful modification of it known among chemists as the water lute, or quicksilver lute: that is, around the mouth or neck of a vessel a deep cavity is formed and partly filled with water or mercury, as the case may be. A cylindrical vessel, open at top and closed at bottom, forms the cover: it is inverted, the open end being placed in the cavity and dipping as far into the liquid as the internal pressure may require. In "The Art of Distillation, or a Treatise of the choisest Spagyrical Experiments," &c. by John French, Doctor of Physic, Lond. 1651, the author describes the same devices for preventing the explosion of vessels as those mentioned by Glauber. Speaking of the action of such safety-valves he observes, (page 7) "upon the top of a stopple [valve] there may be fastened some lead, that if the spirit be too strong, it will only heave up the stopple and let it fall down again." Papin's claim therefore is not to the valve itself, but to its improvement, or rather to the mode of applying it by means of a lever and moveable weight; thereby not only preventing the valve from being blown entirely out of its place, but regulating the pressure at will, and rendering the device of universal application.

It was not till some years after Savery had introduced his steam machine that Papin proposed the following one, which he announced in a work entitled "Nouvelle maniére pour lever l'eau par la force du feu, mise en lumiére, par M D. Papin, Docteur en Med. Prof. en Mathém. a Casel.

1707." It is inserted here out of chronological order, to keep this notice of his labors unbroken.

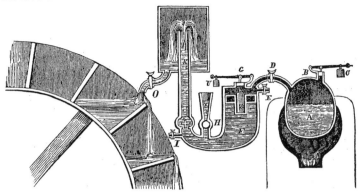

No. 192. Papin. A. D. 1707.

A copper boiler, A, is set in brick work and furnished with a safety-valve, B, whose lever is loaded with the weight C. The steam pipe and cock D connect the boiler with the receiving cylinder F. A hollow float or piston is made to move easily in F, to prevent the steam from coming in contact with the water. A cavity is made in this float for the reception of an iron heater, Z, designed to keep up the temperature of the steam when the latter is admitted into F. The heater is admitted through the opening on the top of F, which is closed by the valve G. X, a funnel through which the water to be raised is introduced, which is kept from returning by closing the cock or valve H. The lower part of F is connected with the rising main K by a curved and tapered tube. The pipe K terminates in a reservoir or air chamber, whence the water is discharged by the pipe O upon an overshot wheel, or conveyed to the place where it may be required. If the receiver be charged from below, a suction pipe (imperfectly represented by the pipe I) was continued to it from the under side of the curved pipe. The steam flowing through the pipe D presses down the piston, and the water beneath it is forced up the pipe K, (the valve at the lower part of K preventing its return.) When the piston has reached the bottom of F, the cock D is shut and the one marked E is opened. H is then opened, and the water rushes in and drives up the piston as before, when the operation is repeated. Water was raised by one of these machines to an elevation of 70 feet, whence it descended and formed a jet d'eau in the court of the Hessian Academy of Arts.

Belidor inserted a figure and description of this machine in the second volume of his Architecture Hydraulique, p. 328.

CHAPTER VIII.

Experimenters contemporary with Papin—Savery—This engineer publishes his inventions—His project for propelling vessels—Ridicules the Surveyor of the Navy for opposing it—His first experiments on steam made in a tavern—Account of them by Desaguliers and Switzer—Savery's first engine—Its operation—Engine with a single receiver—Savery's improved engine described—Gauge cocks—Excellent features of his improved engine—Its various parts connected by coupling screws—Had no safety-valve—Rejected by miners on account of the danger from the boilers exploding—Solder melted by steam—Opinions respecting the origin of Savery's engine—It bears no relation to the piston engine—Modifications of Savery's engine by Desaguliers, Leopold, Blakey and others—Rivatz—Engines by Gensanne—De Moura—De Rigny—Francois and others—Amonton's fire mill—Newcomen and Cawley—Their engine superior to Savery's—Newcomen acquainted with the previous experiments of Papin—Circumstances favorable to the introduction of Newcomen's engine—Description of it—Condensation by injection discovered by chance—Chains and Sectors—Savery's claim to a share in Newcomen's patent an unjust one—Merits of Newcomen and Cawley.

Both philosophers and mechanics were engaged in experiments on air and steam machines about the same time as Papin. Of these, Savery, Amontons, Newcomen and Cawley were the most successful. The two last named have not generally been considered so early in the field; but, from an observation of Switzer, such appears to have been the case. As weekly and monthly 'Journals of Arts' and 'Mechanics' Magazines' had not then been introduced, those who were disposed to communicate their discoveries to the public had no appropriate medium for doing so, except by a separate publication, and this mode but an exceedingly small number of inventors ever adopted: hence it is that not only the dates of several modern inventions are uncertain, but numerous devices and valuable floating thoughts have, with their authors, been constantly passing into utter oblivion. The history of steam as a mechanical agent affords signal proofs of the advantages of inventors recording their ideas: thus the name of Decaus had long been forgotten, when an old tract of his was discovered containing the device we have figured at page 410. This he probably considered the most trifling thing in his book, yet on account of it a place has been claimed for him among the immortal authors of the steam-engine. Moreland, of whose speaking trumpet an account was inserted in the sixth volume of the Philosophical Transactions, and his ideas of the power required to force water to different elevations in the ninth, omitted to publish through the same or any other medium a description of his steam-engine; and by this neglect has lost a large portion of honor that might have been attached to his name. The same may be said of Garay, Ramseye and Worcester. Savery, however, knew better, for he laid his machine before the Royal Society and got it noticed in their Transactions; and when he had subsequently improved it, he published a separate account with illustrations; in consequence of which he has sometimes been considered the author as well as describer of the first working steam-engine.

Of Savery's personal history, less has transpired than of either Moreland's or Worcester's. He evidently was a man of great energy, who raised himself from obscurity by his talents—a self-made man. According to a tradition he commenced life as a working miner, and in process of time became an engineer and thus acquired the title of *Captain*, agreeably to a custom which is said still to prevail among the Cornish miners. He seems to have acquired a competence, if not wealth, previous to the commencement of his experiments on steam, and we shall find that he was as independent in his spirit as in his purse. Switzer, who was intimately acquainted with him, says he was a member of the board of commissioners for the sick and wounded; but this was probably in the latter part of his life, and subsequent to the introduction of his steam machines.

The first invention of Savery that we meet with is in a pamphlet published by him in 1698, on the propulsion of ships in a calm. His plan consisted of paddle-wheels to be worked by the crew. In the first edition of Harris's Lexicon Technicum, A. D. 1704, there is a description, and in the second, 1710, a figure of Savery's "engine for rowing ships." A horizontal shaft passes through the vessel between decks, and to each end a paddle-wheel is attached. On the middle of the shaft is a pinion or trundle wheel, and underneath a capstan upon which a cog wheel is fixed, whose teeth are made to work between those of the pinion. A number of bars are arranged in the capstan, and the crew were to apply their strength to these as in raising an anchor. As the officers of the admiralty after examination declined to adopt it, Savery tells them he had two other important inventions, which he would not disclose until they did him justice in this! He even held up his opponents to ridicule. On the Surveyor of the Navy, who reported against the adoption of his plan as one neither new nor useful, he was very severe. At that time large wigs were commonly worn, and Savery gave a smart rap on that which covered the head of his official adversary. "It is [he observed] as common for lies and nonsense to be disguised by a jingle of words, as for a blockhead to be hid by abundance of peruke." Had Savery been of a timid disposition, we should probably never have heard of him. After enduring one or two rebuffs in attempting to introduce his inventions, he would have retired and sunk unknown into the grave, like thousands of inventors before him.

Of the few incidents preserved respecting his private life, there are two from which it seems that he loved a glass of good wine and a pipe of tobacco; and that, to obtain them, he was in the habit of visiting a tavern. Let not those who eschew such things complain of us for unnecessarily mentioning them, for Savery's first experiments on steam were made in a bar-room, with a wine flask and a tobacco pipe. At such a place and with such implements he is said to have become acquainted with the principles of his famous machine. The circumstance has not been commonly known, or some scientific Boniface would, long ere now, have adopted Savery's head for a sign; and artists would have made him, in the act of experimenting, the subject of a picture. There is a rich but neglected field for historical painters in the facts and incidents connected with the origin and development of useful mechanism.

According to Desaguliers, Savery declared that he found out the power of steam by chance, and in the following manner: "Having drank a flask of Florence [wine] at a tavern, and thrown the empty flask upon the fire, he call'd for a bason of water to wash his hands, and perceiving that the little wine left in the flask had filled up the flask with steam, he took the flask by the neck and plunged the mouth of it under the surface of the water in the bason; and the water of the bason was immediately driven up into the flask by *the pressure of the air*."[a] This illustration of the ascent

---

[a] Exper. Philosophy, edition of 1744, vol. ii, page 466.

Chap. 8.] *His first Steam-Engine.* 455

of water into a vessel from which the air had been expelled by steam, was of course not new in Savery's time, although it appears to have been so to him. Switzer gives a different account. "The first hint from which it is said he took this engine was from a tobacco pipe, which he immers'd to wash or cool it, as is sometimes done: he discover'd by the rarefaction of the air in the tube, by the heat or steam of the water and the gravitation or impulse of the exterior air, that the water was made to spring through the tube of the pipe in a wonderful surprising manner."[a] It was an old practice of veteran smokers, when their (clay) pipes became blackened through use, and more particularly when choked or furred up, to place them in a bright fire till they became red hot, then to remove and allow them to cool. By this operation they were whitened and purified like the incombustible cloth of the ancients, which was cleansed in the same way. But frequently when taken from the fire the mouths of the pipes were plunged slowly into water; steam was thus formed, and rushing through the tubes, was sometimes preceded, often accompanied, and sometimes followed by jets of water. There are however different causes, and far from obvious ones, for the liquid issuing through tobacco pipes under such circumstances, so that it is difficult to perceive what inference Savery drew from the experiment.

But whatever may have led Savery to the subject of steam, he had so far matured his ideas respecting its application to raise water as to erect several engines, and to secure a patent as early as 1698. In June of the following year he submitted a working model to the Royal Society, and made successful experiments with it at the same time. A figure of this engine was published in the Transactions of that year, and may also be found in the first volume of Lowthorp's Abridgment. No. 193 is a copy. It consisted of a close boiler, B, set in a brick furnace A, and two receivers D D supported on a stand, and made of strong copper and air-tight. A suction pipe whose lower end descends into a well, or other place whence water is to be raised, (which may be about 24 feet below D D) and whose upper part, divided into two branches, communicates with the *top* of the receivers. Each branch is furnished with a valve at E E, opening upwards, to prevent the water from returning when once raised. The lower part of the forcing pipe G has

No. 193. Savery's First Engine. A. D. 1698.

also two branches, F F, which communicate with the *bottom* of the receivers, and these branches have also valves, E E, like the others opening upwards. Each receiver has a communication with the upper part of the boiler by steam pipes and cocks C C.

The operation was as follows :—The boiler was two thirds filled with

---

[a] Hydrostatics, edition of 1729, page 325.

water, a fire made under it and the steam raised. One of the cocks was then opened, and the steam passing through filled the receiver by driving the air previously within it into the forcing pipe : the cock was then closed, and the steam within the receiver soon became condensed by the cold air in contact with its exterior surface, or by pouring cold water upon it; hence a vacuum was produced within, and consequently the water in which the lower end of the suction pipe was immersed was driven up by the pressure of the atmosphere so as to fill the void. When this had taken place the same cock was again opened, and the steam rushing in urged by its expansive force the contents of the receiver up the forcing pipe. In this manner water was alternately raised into and expelled from both vessels.

As a practical miner, and consequently conversant with the subject of raising water on a large scale, Savery was better qualified to carry his views into operation than a mere philosopher. His first essay in employing steam was a proof of this. "I have heard him say myself [observes Switzer] that the very first time he play'd, it was in a potter's house at Lambeth, where tho' it was a small engine, yet it [the water] forc'd its way through the roof, and struck up the tiles in a manner that surpris'd all the spectators."

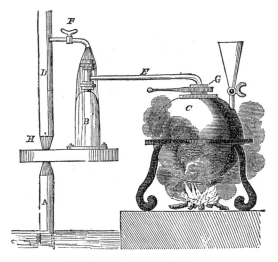

No. 194. Savery's Single Engine.

Sometimes Savery employed but one receiver. No. 194 represents an engine of this kind, erected by him at Kensington. A description of it was first published by Mr. Bradley in his "New Improvements of Planting and Gardening." It is also figured and described by Switzer, who examined it and thought it "the plainest and best proportion'd of any" he had seen. Its effects were considered proportionally greater than those with two receivers. C, a spherical boiler of the capacity of forty gallons, and charged through the tunnel. B, the receiver, which held thirteen gallons A, the suction pipe, sixteen feet long and three inches bore. D, the forcing pipe, of the same bore and forty two feet long. A valve opening upwards was placed in A, and another at the lower part of D, at H. E, the steam pipe, an inch in diameter. G, a sliding valve or cock, furnished with a lever handle. F, a cock in the forcing pipe, to admit cold water to flow upon the receiver. A pipe attached to the tunnel descended into the boiler and served the purpose of a gauge cock.

The operation will be understood from the description of figure No. 193. By turning the handle of G steam is admitted into B, and as soon as the air is expelled from the latter, G is closed and F opened; the affusion of cold water (see the figure) quickly condenses the contained vapor, and hence the receiver becomes charged with water by the pressure of the atmosphere through the suction pipe A. F is then shut and G opened, when the steam issuing from the boiler displaces the water from the receiver, and having no other way to escape the liquid is driven up the pipe D into the reservoir prepared to receive it. As soon as all the water is expelled from the receiver, (which was known by applying the hand to the lower part, for it would be hot) G is shut and F again opened, when the operation is repeated as before.

"When this engine begins to work [says Switzer] you may raise four of the receivers full in one minute, which is fifty two gallons, [less the quantities drawn from F for the purposes of condensation]—and at that rate in an hour's time may be flung up 3120 gallons. The prime cost of such an engine is about fifty pound, as I myself have had it from the ingenious author's own mouth. It must be noted that this engine is but a small one, in comparison of many others of this kind that are made for coal-works; but this is sufficient for any reasonable family, and other uses required for it in watering all middling gardens."

Here is no provision made to replenish the boiler with water except through the tunnel: hence the working of the machine had to be stopped, and the steam within the boiler allowed to escape, before a fresh supply could be admitted. Under such circumstances the boilers were very liable to become injured by the fire when the water became low. They were also exposed to destruction from another cause, the force of the steam; for they had no safety-valves to regulate it, and hence the necessity of the following instructions: "When you have rais'd water enough, and you design to leave off working the engine, take away all the fire from under the boiler, and open the cock [connected to the tunnel] to let out the steam, which would otherwise, was it to remain confin'd, perhaps *burst the engine*."

Savery, from his profession, was aware of the want of an improved mode of draining mines. The influence of the useful arts in enriching a nation was then beginning to be understood. A stimulus was imparted to manufactures, and the demand for coal and the ores of England rapidly increased. As a necessary consequence the depth of the mines increased also; and hence proprietors became anxious to possess some device for clearing them of water, and by which the old, inefficient and excessively expensive horse-gins and buckets might be dispensed with. The cost of drainage was so great in some mines, that their produce hardly equalled the cost of working them: in one mine five hundred horses were constantly employed. Numerous novel projects had been tried and abandoned: what they were we are not informed, but as Ramseye and Worcester and probably others had proposed *fire machines* for the purpose, steam had probably been tried in some way or other and had failed. Having greatly improved his machine, Savery published an account of it, illustrated with engravings, in a pamphlet entitled *The Miner's Friend; or a Description of an Engine for Raising Water by Fire*, with an Answer to the Objections against it. London, printed for S. Crouch, 1702. In his address he begs proprietors not to let the failure of other plans prejudice them against the trial of his. "Its power [he observes] is in a manner infinite and unlimited,

and will draw you water 500 or a 1000 feet high, were any pit so deep.
- - - - - I dare undertake that this engine shall raise you as much water
for eight-pence, as will cost you a shilling to raise the like with your old
engines." The original figures in the Miner's Friend were inserted in
Harris's Lexicon Technicum, in 1704, and copied into Switzer's Hydrostatics in 1729, and by Desaguliers in his Experimental Philosophy in 1744,
(which works are before us) and subsequently into almost every treatise
on the steam-engine. No. 195 is a reduced copy: the figure of the fire
man is an addition.

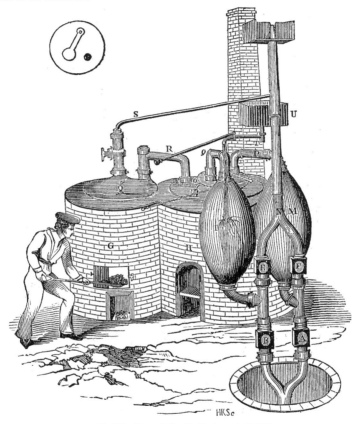

No. 195. Savery's Double Engine. A. D. 1702.

A detailed description of this elegant apparatus is not necessary, since
its operation will be understood from the explanation of the two preceding
machines. It is substantially the same as No. 193, except that this one
has *two* boilers, which are heated by separate furnaces, G H. The additional boiler G was designed merely to supply the other with hot water,
and need not therefore divert the attention of the reader in realizing the
working of the essential parts. The upper end of the suction pipe shown
at the mouth of the pit consists of two branches, which are connected to
similar branches on the lower part of the forcing pipe N. The suction
valves are at B A, and the forcing ones at E F, all opening upwards.

Chap. 8.] *Its excellent features—Coupling Screws.* 459

Between these valves two short curved tubes connect the bottoms of the receivers I M with the branches, as represented, and two other bent tubes, P Q, unite the top of the receivers with the boiler H. On the top of this boiler, and forming a part of it, is a stout round plate, having two openings of the same size as the bore of the tubes last mentioned. In these openings the two steam tubes P Q terminate. Between the openings, and on the under side of the plate, is a moveable disk, which by a short arm is connected to an axle and moved by the long lever shown on the top of the boiler; so that by moving this lever the disk can be made to close either opening, so as to admit or exclude steam from the receivers, and answering every purpose of a three-way cock. It is made somewhat on the plan of the one in No. 189, page 421. The face of the disk is ground smooth, so as to fit close to the under side of the plate, against which it is pressed by the steam. The perpendicular axle by which the disk is turned passes through the plate, and the opening is made tight by a stuffing box. (The plate and moveable disk are represented in the small figure at the top, one of the openings being covered by the disk and the other exposed.) A small cistern, U, is placed over the receivers, and kept supplied with cold water from the forcing pipe by means of a ball cock, viz. a cock that is opened and shut by a ball floating in the cistern. From the bottom of this cistern a short pipe, T, proceeds; and to it is connected, by a swivel joint or stuffing box, another one at right angles. This pipe furnishes water to condense the steam in the receivers, over both of which it can be moved by the rod attached to the plug of the cock as shown in the figure. The upper cistern denotes the place where the water raised by the engine is to be discharged.

A communication is made between the boilers by a siphon or bent tube, R, whose legs extend nearly to the bottom of the boilers. In the leg within the small boiler is a valve opening upwards, which permits the water of G to pass into H, but prevents any returning from the latter. When the attendant wishes to inject into H a fresh supply of water, he increases the little fire kept up under the boiler G, (which is always kept supplied with water by the pipe S,) and as soon as the liquid boils and the force of the steam exceeds that in H, the contents of G, both steam and hot water, are forced through the valve; and thus H is kept supplied without the action of the machine being stopped. The cock on the pipe S is then opened, the small boiler again charged, and the water becomes gradually heated; so that by the time it is wanted in the other boiler, a small addition to the fuel quickly raises its temperature, and it is again forced in as before.

The quantity of water in the boilers was ascertained by *gauge cocks*. These were inserted at the top, (see figure) and pipes soldered to them descended to different depths. The principal boiler had two of these, the other but one.

The general arrangement of this engine and the adaptation of its various parts to each other are admirable, and could hardly be improved. The obviously good workmanship—the improved form of the receivers—and the connection of these with the boilers and pipes, and the latter with each other, *by coupling screws*, thus securing easy access to the valves—are highly creditable to Savery and the workmen he employed. Every part was made of the best materials. The cocks, coupling screws, regulator, valves, and all the pipes immediately connected with them, were of brass; while the boilers, receivers and suction pipes were of " the best hammered copper, of sufficient thickness to sustain the force of the working engine: in short, [continues the inventor] the engine is so naturally adapted to perform what is required, that even those of the most ordinary and meanest capacity may work it for some years without injury, if not hired or employ'd by some base person on purpose to destroy it;"—that is, by inattention or design to permit steam to accumulate within the boilers till they were burst. Some device to prevent this was wanting, viz. a safety-valve or something analogous to it; and it is astonishing that he never thought of such a thing, but permitted his machine for lack of it to fall into disrepute.

The miners could not be induced to adopt it, in consequence of the danger of explosion. "Savery [says Desaguliers] made a great many experiments to bring this machine to perfection, and did erect several, which raised water very well for gentlemen's seats, but could not succeed for mines, or supplying towns where the water was to be raised *very high* and in great quantities; for then the steam required being boiled up to such a strength, as to be ready to tear all the vessels to pieces. - - - - - - I have known Captain Savery at York's Buildings make steam 8 or 10 times stronger than common air; and then its heat was so great, that it would melt common solder, and its strength so great as to blow open several of the joints of his machine; so that he was forced to be at the pains and charge to have all his joints soldered with spelter or hard solder." Ex. Philos. ii, 467.

There has been much discussion respecting the origin of this famous engine; some writers contending that it was wholly Savery's own, others that he derived it from one of Worcester's, or from the Century of Inventions. Desaguliers asserts that Savery, to conceal its origin, " bought up all the Marquis of Worcester's books that he could purchase in Paternoster Row, and elsewhere, and burn'd 'em in the presence of the gentleman his friend, who told me this." But as Savery denied being indebted to any one for it, and as he was certainly a man of great mechanical genius, it is probable that the doctor was imposed upon by his informant. It is not likely that Savery would have committed such an act in the presence of a witness, when there was not only no necessity for one, but every possible inducement for secrecy. Many years before the publication of this charge by Desaguliers (in 1744) the opinion was prevalent that the machine was not original with Savery. In 1729 Switzer remarks, "others say that the learned Marquis of Worcester, in his Century of Inventions, which book I have not seen, gave the first hint for this raising of water." (Hydr. 325.) Dr. Hutton, in his Math. Dictionary, asserts, though on what authority we know not, that Savery knew more of Moreland's experiments than he was willing to acknowledge; and Desaguliers maintains that he invented the story of the experiment with the wine flask " to make people believe that he had not got the idea from Worcester's Century of Inventions."

In reply to the above it may be remarked, that independently of those coincidences of thought that always have and will happen to inventors, there are circumstances which strongly corroborate Savery's own account. In the first place, the experiment with the wine flask was one very likely to occur in the manner he has mentioned, and to a mind like his would naturally lead to a practical application of it. His thoughts, we are told, " were always employed in hydrostatics or hydraulics, or in the improvement of water-works." Then there is no evidence that he was much of a reader: had he been conversant with books, he would not have proposed the propulsion of vessels with paddle-wheels as *new*. These occurred to him as they have done to thousands in every age when devising means to increase the speed of boats; and so it may have been with his steam machine: a device like it would naturally be the result of the experiment with the wine flask, and even without it, when his thoughts were once directed to raising water by steam. Moreover, Savery was ignorant of the safety-valve, the very thing wanted to remove the most formidable objection to his machine; and yet, as we have shown, he might have found it in some popular works on chemistry and distillation,—besides which, Papin's improved application of it had been published several years. (The single machine figured No. 194 was erected by Savery himself as late as 1711 or '12, and it had no safety-valve.)

But whether he derived the hint from Worcester or not, he is entitled to all the honor he has received. He was the first effectually to introduce the device, and the first to publish a description of it in detail. He concealed nothing, but, like a sensible and practical man, explained the whole, and left it to its own merits. No one's claims to a place in the history of the steam-engine were better earned, whether he be considered the reïnventor, or improver only of Worcester's 68th proposition. There are several points of resemblance in the characters of Savery and Oliver Evans. By their energy and indomitable perseverance they *forced* their inventions into public notice in spite of public apathy, and so worked their way into the temple of honorable fame. Both published curious pamphlets, that will preserve their names and inventions from oblivion.

But Savery's steam-engine does not belong to the same family as the modern one, nor can he be said to have contributed to the invention of the latter, except so far as making his contemporaries more familiar with the mechanical properties of aqueous vapor. 'Tis true he employed this fluid in close vessels, and so far he succeeded; but his ideas seem to have been wholly confined to its application to raise water, and in the most direct manner—hence he never thought of *pistons*. Had he turned his attention to impart motion to one of these, he would have left little for his successors to do; but as it was, he did not lead engineers any nearer to the piston engine. He proposed to propel machinery by discharging the water he raised upon an overshot wheel; hence his patent was "for raising water, and occasioning motion to all sorts of mill work." But this was obviously an afterthought, an accidental result, rather than one originally designed or looked for. A piston and cylinder only could have given his machine a permanent place in the arts, either as a hydraulic or a motive one. He accomplished almost all that could be realized without them. The most splendid talents of the present times could have done little more. Papin abandoned the piston and cylinder, and in doing so quenched a halo of glory that would have shone round his name for ever; and Savery, for want of them, notwithstanding his ingenuity, perseverance and partial success, lived to see his device in a great measure laid aside. Savery died about the year 1716.

As Savery's engine became known, several additions to and modifications of it were proposed. A few of these may be noticed:—

Drs. Desaguliers and Gravesande, from some experiments, concluded that single engines were more economical than double ones—a single receiver being "emptied three times whilst two succeeding ones [of a double engine] could be emptied but once a piece." Of single engines Desaguliers erected seven between the years 1717 and 1744. "The first was for the late Czar, Peter I, for his garden at Petersburgh, where it was set up. The boiler of this engine was spherical, (as they must all be in this way, when the steam is much stronger than air) and held between five and six hogsheads; and the receiver held one hogshead, and was filled and emptied four times in a minute. The water was drawn up by *suction* or the pressure of the atmosphere 29 feet high, out of a well, and then pressed up 11 feet higher. Another engine of this sort which I put up for a friend about five and twenty years ago, [1719] drew up the water 29 feet from the well, and then it was forced up by the pressure of the steam 24 feet higher," &c. But these "improved" engines differed in reality but little from Savery's single one, No. 194. Desaguliers furnished his boiler with Papin's steelyard safety-valve; a three-way cock alternately admitted steam into the receiver and water from the forcing pipe to condense it: in other respects the engines were much the same. Savery made no provision to secure his boilers from being exploded; but the safety-valve was not always a preventive in former times, any more than at present. "About three years ago [says Desaguliers] a man who was entirely ignorant of the nature of the engine, and without any instructions, undertook to work it; and having hung the weight at the farther end of the steel-yard, in order to collect more steam to make his work the quicker, he hung also a very heavy plumber's iron upon the end of the steel-yard: the consequence prov'd fatal, for after some time the steam, not being able with the safety-clack to raise up the steel-yard loaded with all this unusual weight, *burst the boiler* with a great explosion, and kill'd the poor man." Exp. Philos. ii, 489.

In a double engine by Leopold, A. D. 1720, the receivers were placed below the water they were to raise: hence the principle of condensation was not required—for as soon as the steam expelled the contents of a receiver, a communication was opened between the upper part of the latter and the atmosphere, so as to allow the steam to escape and a fresh supply of water to enter below. He produced a rotary movement by discharging the water into the buckets of a water-wheel.

When steam is admitted into a receiver, a portion is immediately condensed by the low temperature of the vessel and the cold water within; so that not till a film or thin stratum of hot water is thus formed on the surface, can the full force of the vapor be exerted in expelling the contents. This waste of steam is not however so great as might be imagined, because the water with which it comes in contact still remains on the surface, having become lighter than the mass below by the accession of heat, and consequently preventing the heat from descending: yet various attempts were made to interpose some non-conducting substance between the steam and the water. Papin, as we have seen, used a floating piston. In 1766, Mr. Blakey, an enterprising English mechanic, took out a patent for the application of a stratum of oil or air. To use these he made some corresponding alterations in the receiver; but the advantages were not so great as had been expected. Blakey also introduced a new boiler, consisting of tubes or cylinders completely filled with water and imbedded in the fire. It caused considerable excitement among scientific men, but the danger arising from them, and the explosion of one or more, caused them to be laid aside. He spent several years in France, where he erected some of his engines. He wrote on several subjects connected with the arts. There is a copious and interesting extract from his Dissertation on the Invention and Progress of Fire Machinery, in the Gentlemen's Magazine for 1792, page 502.

Other modifications of Savery's engine were made previous to and about Blakey's time, of which no particular accounts are now extant. In his Comparisons of French and English Arts, (article Horology) Blakey says, "About 1748 another Swiss, named *Rivatz*, appeared in Paris: he understood all the known principles and methods for regulating time in equal parts, to which he added others of his invention. - - - - - - And I can say without pretending to prejudice any one's merit, that I never met with any French or English man who had so much ingenuity and knowledge in mechanical, hydraulic, *fire machinery* principles, &c. as Rivatz." (Gent. Mag. 1702, page 404.)

In 1734 M. Gensanne, a French gentleman, made some improvements on Savery's engine, and by additional mechanism rendered it *self-acting*. The alternate descent of two vessels of water opened and closed the cocks, on much the same principle as that exhibited in Fludd's pressure engine, page 354. (Machines Approuveés, tome vii, 222.) In 1740 M. De Moura, a Portuguese, accomplished the same thing by the ascent and descent of a copper ball or float within the receiver; but the device was too complicated for practical purposes. It is figured and described by Smeaton in the Philosophical Transactions, vol. xlvii, 437, in the Supplement to Harris's Dictionary of the Arts, and in other English works. In 1766, Cambray de Rigny, an Italian, made some additions to Savery's engine so as to make it in a great measure independent of manual assistance. Professor Francois, of Lausanne, having been consulted respecting the draining of an extensive marsh between the lakes Neuchatel, Bienne and Morat, adopted a fire engine on Savery's plan, and which he made self-acting by a more simple device than either of the preceding. A description and good figure of his machine may be seen in the fourth volume of the Repertory of Arts, (1794) page 203. Nuncarrow's improvement on Savery's is described in the American Phil. Transactions, vol. i, 209, in Tilloch's Phil. Mag. vol. ix, 300, and in Galloway's History of the Steam-Engine. An English patent was issued in 1805 to James Boaz, and another in 1819 to Mr. Pontifex, both for improvements on Savery's engine. For further information see the Repertory of Arts, Nicholson's Journal, vol. i, 419, and the Journal of the Franklin Institute.

"A commodious way of substituting the action of fire instead of the force of men and horses to move machines," was proposed in 1699, by M. Amontons, one of the earliest and most useful members of the French Academy of Sciences. He named his machine *a fire mill*. It resembled a large wheel, supported on a horizontal axis, but was composed of two concentric hollow rings, each of which was divided by partitions into a dozen separate cells. The small or interior ring was at a considerable distance from the axis, and the cells communicated with each other through openings made in the partitions and covered by valves or clacks. The cells of the exterior rings had no communication with each other, but a pipe from each connected them with the inner ones. The outer cells contained air, and about one half of the inner ones contained water. The object was to keep this water always on one side, that its weight might act tangentially, and so cause the wheel to revolve, and the machine connected to it. A furnace was built close to a portion of the periphery, and the lower part of the wheel was immersed in water, to a depth equal to that of the exterior cells. When the fire was kindled, the air in the cell against which the flame impinged became rarefied, and, by means of a pipe communicating with an inner cell *below the axle*, forced the water contained in that cell into an *upper* one. This caused that side of the wheel to preponderate, which brought another air cell in contact with the fire, and the fluid becoming expanded by the heat forced up the contents of another of the inner cells into a higher one, as before: in this way every part of the periphery of the wheel was brought in succession in contact with the fire, and the water in the inner cells kept constantly rising on one side of the wheel, thus causing the latter to revolve. The air in the outer cells was cooled as they passed through the water in which the lower part of the wheel dipped.

This device of Amontons is rather an air than a steam machine. It hardly belongs to this part of our subject; but as it may be considered the type of most of the steam wheels subsequently brought forward, we have been induced to notice it here. As a theoretical device, it is highly meritorious, but as a practical one, of little value. There is in Martyn and Chambers's abridged History of the Academy of Sciences at Paris, Lond. 1742, a full account of this wheel, and of the experiments from which it was deduced. (See vol. i, 69.) It was simplified by Leopold.

Towards the close of the 17th century there lived in Dartmouth, a small seaport town on the English channel, two mechanics who combined their energies to devise a machine for raising water by means of steam. Their names were Thomas Newcomen and John Cawley; the first a blacksmith, but sometimes called an ironmonger, the latter a plumber and glazier. The circumstances that led them to the subject have not been recorded, nor have the particular contributions of each been specified. Their efforts were however eminently successful, for to them belongs the honor of having permanently established the employment of steam as a mechanical agent. The date of the commencement of their efforts is unknown, but from the observation of a contemporary writer it seems to have been as early as the first attempts of Savery.

The principal objection of miners to Savery's machine, viz. the enormous force of the steam required, and the consequent frequent explosion of the boilers, &c. was completely avoided by Newcomen and Cawley; for they used steam of little or no greater force than cooks do in common cauldrons—hence it could never explode a boiler or endanger human life. Savery's engine had other disadvantages. It was required to be placed *within* a mine or pit, and in no case farther from the bottom than 25 or 30 feet; whereas Newcomen and Cawley's was erected on the surface, near the mouth of the shaft. Moreover, in those mines which were previously drained by pumps, it could be used to work these as before, without any additional cost for new pipes and pumps; the engine in such cases merely superseding the horses and their attendants. Instead of applying steam like Savery *directly* to the water to be raised, these mechanics made use of it to give motion to a piston and vibrating beam, and through these to common pump rods; hence the device may be considered rather for imparting motion to machines proper for raising water, than as one of the latter.

It is in evidence that Newcomen had some correspondence respecting his machine with Dr. Hooke, and that he was acquainted with what Papin had previously done. This however might very well consist with the idea of giving motion to *a piston* originating with himself or partner; yet as their labors were subsequent to those of the French philosopher, their claims to it, if they ever made any, could not be sustained. Their machine in its essential features is a copy of Guerricke's, and the mode of producing a vacuum under the piston similar to Papin's; but as Papin did not succeed, the reintroduction of a device similar to his, and its successful application to the important purpose of draining mines, belong wholly to them; and the merit of doing this was certainly much greater than can ever be claimed for the abortion of Papin. Fulton did not invent steam boats, but he was the first to demonstrate their utility and to introduce them into use here after they had been tried and abandoned in Europe.

It should not be supposed that the piston engine would not have been realized at the close of the 17th or beginning of the 18th century, if Papin and Newcomen had not lived. The spirit of inquiry that was abroad in their days, and the number of ingenious men engaged in devising means to employ steam as a motive agent, would assuredly have soon brought it into use. Indeed, every improvement in the application of steam seems to have been always perceived by some contemporary projectors, among whom the contest of maturing it was, as in a race, one of speed. " Watt [observes Prof. Renwick] found a competitor in Gainsborough, and but a few weeks would have placed Stevens on the very eminence where Fulton now stands." The circumstances of the times, the increase of English manufactures, and the general want of some substitute for animal labor, were all then favorable to the introduction of the steam-engine. "Had the mines of Cornwall been still wrought near the surface, Savery or Newcomen would hardly have found a vent for their engines. Had not the manufacturers of England been wanting in labor-saving machinery, the double-acting engine of Watt would have been suited to no useful application. A very few years earlier than the voyage of Fulton [to Albany] the Hudson could not have furnished trade or travel to support a steam boat, and the Mississippi was in possession of dispersed hordes of savages."

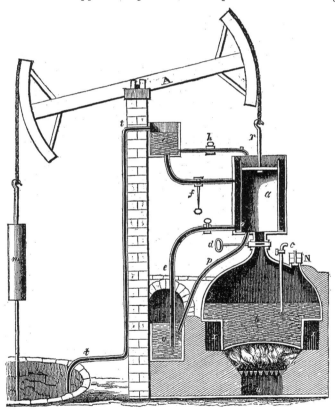

No. 196. Newcomen and Cawley's Engine. A. D. 1705.

The above figure will sufficiently explain the principles and operation of Newcomen and Cawley's first engine; and, when compared with those already noticed, will enable the reader to do justice to all concerned. It will be perceived that although steam is an essential agent, it is not the *primum mobile* of the apparatus: the pressure of the atmosphere is the first mover, and to excite this only was steam employed.

A, in the figure, (No. 196) represents a vibrating beam with arched ends or sectors, from one of which the main pump rod is suspended by a chain. This rod descends into the mine or pit, and is connected to as many other rods as there are pumps to be worked. A counterpoise or heavy weight *m* is fixed to the rod, so as to depress it and raise the other end of the beam in the position represented. *a*, the steam cylinder, open at top, its sides being surrounded by another, and the space between them containing water. *r*, the piston rod and piston, attached to the beam by a chain. *b*, the boiler. *c*, gauge cock. N, safety valve with weights placed directly upon it. *d*, a cock to admit steam into the cylinder. *e*, a pipe and cock to convey the water round *a*, into the well or tank *o*. *f*, a pipe and cock to supply cold water to condense steam in the cylinder. *h*, another pipe and cock to furnish occasionally a little water to the upper side of the piston, to prevent air from passing between the packing and sides of the cylinder: this water was kept at the depth of about two inches. *t t*, a pipe proceeding from one of the pumps in the pit to supply the small cistern with water. *p*, a pipe to convey the steam condensed within *a* into the tank *o*. *w* the ash pit. *x x*, flues round the boiler.

Fire being applied to the boiler and steam generated, the cock *d* is opened and the cylinder filled with steam, provision being made for the escape of the air previously within. *d* is then closed and *f* opened, by which cold water from the cistern is admitted to flow round *a*: this condenses the vapor within, and a vacuum being thus formed under the piston, the latter is pushed down by the atmosphere; consequently the opposite end of the beam is raised, and with it the pump rods and the load of water with which they are burthened. *f* is now closed and *d* again opened, when the counterpoise *m* preponderates, the piston is raised, the cylinder again filled with steam, and the operation repeated. But previous to the admission of vapor the second time into the cylinder, the cock *f* is closed and the one on pipe *e* opened, to allow the water between the cylinders to escape into the tank *o*, this water having become heated by its contact with *a*. As soon as the cylinder is charged anew with steam, a fresh supply of cold water to condense it is admitted by again opening *f*.

The amount of force thus excited depends upon the diameter of the cylinder *a*, or the area of its piston, and the state of the vacuum made under the latter. The dimensions of *a* must therefore be proportioned to the resistance to be overcome—to the quantity of water to be raised from a mine, and the height at which it is to be discharged—and to render an engine of the kind effective, the whole of the steam in *a* should be condensed, and as quickly as possible. These conditions were not very well fulfilled by the apparatus as figured above. Time was required for the cold water between the cylinders to extend its influence from the circumference to the centre of the inner one, in order completely to condense the vapor; hence the movements were extremely slow, the strokes seldom exceeding seven or eight per minute. An accidental circumstance pointed out the remedy, and greatly increased the effect. As the engine was at work, the attendants were one day surprised to see it make several strokes much quicker than usual; and upon searching for the cause, they found, says Desaguliers, "a hole in the piston which let the cold water [kept upon the piston to prevent the entrance of air at the packing] into the *inside* of the cylinder." The water falling through the steam condensed it almost instantaneously, and produced a vacuum with far less water than when applied to the exterior of the cylinder. This led Newcomen to remove the outer cylinder, and to insert the lower end of the pipe f into the bottom of a, so that on opening the cock f a jet of cold water was projected through the vapor. This beautiful device is the origin of the injection pipe still used in low-pressure engines.

Newcomen and Cawley's engine, as figured page 465, was improved in several parts in 1712, and soon after adopted as a hydraulic machine for draining the coal and iron mines in various parts of Europe. Very elaborate engravings of some used in French mines may be seen in the folio edition of Arts et Métiers. See also Desaguliers' Ex. Philos. vol. ii, and Switzer's Hydrostatics.

The application of sectors and chains to pump rods did not originate with Newcomen. They are figured by Moxon, and were probably employed in working pumps in mines previous to the invention of the steam-engine.

We have often thought the heaviest charge against Savery was to be found in his conduct towards Newcomen and Cawley. Their machine was essentially different from his in its principle, construction and mode of action, yet he insisted that it was an infringement upon his patent. He employed the pressure of the atmosphere in charging his receivers, by condensing with cold water the steam within them. So far as regards this mode of forming a vacuum, (he in his receivers and they beneath a piston) there is a resemblance between the two machines, but no farther; and this plan of making a vacuum was not original with him any more than with them. It was no more a new device in his time than his paddle wheels were. The object of Newcomen and Cawley in forming a vacuum was also quite different from his; for they did not raise water into the vacuity, but employed it solely to excite the pressure of the atmosphere upon the upper side of a piston, in order to impart motion to common pump rods. Again, he used the expansive force of high steam: this was the prominent feature in his machine, and the great power that gave efficiency to it; but they did not use this power at all. The weight of the external air, not the expansive force of steam, was the *primum mobile* in their machine, and it was brought into action by the vapor of water at the ordinary boiling point.

But as they formed a vacuum in their cylinder by the condensation of steam, he insisted on having a share in their patent! The fact was his machines had become in a great measure laid aside, and he doubtless perceived that they were destined to be wholly superseded. Desaguliers (in 1744) observes that the progress and improvement of the fire engine were stopped by the difficulties and dangers attending it, till Newcomen and Cawley "brought it to the present form, in which it is now used, and has been near these thirty years." Unless his name was included as a joint patentee, Savery threatened an appeal to the law; and it is said his influence at court, as commissioner for the sick and wounded, gave weight to this ungenerous and unjust demand. Newcomen we are informed was a Quaker, or like Cawley a Baptist, and therefore on principle averse to legal controversy: he was moreover a man of "a great deal of modesty," and so yielded the point. The patent was consequently issued (in 1705) "to Thomas Newcomen and John Cawley of Dartmouth, and Thomas Savery of London."

Another point has been generally overlooked: so far from Newcomen's machine being an infringement or improvement upon Savery's, it was really invented as early if not earlier than the latter. Switzer (Savery's friend) says, "it [Newcomen's engine] is indeed generally said to be an improvement to Savery's engine, but *I am well inform'd* that Mr. Newcomen was as early in his invention, as Mr. Savery was in his, only the latter being nearer the court, had obtain'd his patent before the other knew of it; on which account Mr. Newcomen was glad to come in as a partner to it." (Hydrostatics, 342.) That is, as a partner to his own invention.

To Newcomen and his associate belongs the honor of laying the foundation for the modern engine. The piston engine of Worcester had been forgotten, Papin's was an abortion, and Savery probably never thought of one; hence, whether the Dartmouth mechanicians were aware of its previous employment or not, to them a large share of merit is justly due. They were moreover amiable and unassuming in their manners, and seem to have passed through life without exciting much of that envy that embitters more or less the nights and days of successful inventors. From such men, who can withhold expressions of approbation and esteem? Had they been members of the Roman church, they should have been canonized —could we believe in the efficacy of prayers for the dead, we would have masses performed for the repose of their spirits—and had we the power, every contributor to useful mechanism should be commemorated by an apotheosis.

Cawley died in 1717, but the date of Newcomen's decease has not been ascertained.

CHAPTER IX.

General adoption of Newcomen and Cawley's engine—Leopold's machine—Steam applied as a mover of general machinery—Wooden and granite boilers—Generating steam by the heat of the sun—Floats—Green-houses and dwellings heated by steam—Cooking by steam—Explosive engines—Vapor engines—English, French and American motive engines—Woisard's air machine—Vapor of mercury—Liquefied gases—Decomposition and recomposition of water.

Newcomen and Cawley's engines were found to answer the purpose of raising water so well, that in a few years they were introduced into Russia, Sweden, France and Hungary; and about 1760, one was imported by the proprietors of the old copper mine near Belleville, New Jersey. They in fact imparted a new and very beneficial impulse to mining operations, and quickly raised the value of mining stock. Deluged works were recovered, old mines deepened, and new ones opened, in various districts, both in Great Britain and continental Europe : nor were they confined to draining mines, but were employed to raise water for the use of towns and cities, and even to supply water-wheels of mills. By exciting the attention of ingenious men to their improvement, they became the means of extending manufactures generally, and introduced one which had never before been known in the world, viz. the fabrication of *motive engines*—a manufacture upon which the wealth, power and happiness of nations are destined in a great degree hereafter to depend.

Leopold, to whom we have frequently referred, reflecting on Papin's experiments, suggested the following application of steam to move pistons and to raise water:—

No. 197. Leopold's High Pressure Engine. A. D. 1720.

Two steam cylinders, open at top and provided with pistons $a\ b$, were placed over the boiler c, from the upper part of which a four-way cock d admitted steam alternately into the bottom of each. The pistons were connected by inflexible rods to the ends of two working beams, and to the opposite extremities of the beams were connected, by similar rods, the pistons $f\ g$ of two forcing pumps, whose lower parts were placed in the water to be raised. An attendant turned the plug of the cock to admit steam under one piston, which was pushed up by the expansive force of the fluid, and consequently the piston of the pump connected to the same beam was forced down, and the water in its chamber driven up the rising main i. The cock was then turned to admit steam into the other cylinder, whose piston was raised in like manner; at the same time one passage of the cock opened a communication with the interior of the first cylinder and the external air, so as to allow the steam within to escape.—(See the figure.)

This is the first high-pressure piston engine figured in books. It has been greatly admired, and yet as represented it is useless and impracticable; for when the steam pistons were once raised the whole would remain immoveable, there being no means for causing them to descend. Had Leopold used one beam instead of two, and placed a pump and steam cylinder under each end, the device would have been complete and very effective.

It is singular that the researches of Leopold had not made him acquainted with the fact that four-way cocks were used long before Papin, to whom he attributes them.

With this device of Leopold, we take leave of steam machines. Hitherto they had been employed *only* to raise water, but the period was now approaching when the agency of this fluid as a first mover of machinery in general, was to become indefinitely extended. The engines of Newcomen and Leopold were the links which connected the labors of Heron, Garay, Porta, Worcester, Moreland, Papin and Savery with those of Watt. They opened the way for the introduction of the crank and fly-wheel, which changed completely the character of the old engines. Like Worcester's and Savery's, Newcomen's engine required the constant attention of an attendant to open and close the cocks; but a boy named Potter employed in this service, stimulated by the love of play, ingeniously added cords to the levers by which the cocks were turned, and connecting the other ends of the cords to the moving beam, rendered the machine self-acting, and thus acquired opportunities of joining his sportive companions unknown to his employers. Iron rods were soon after substituted for the cords by Beighton, and finally Watt and Gainsborough, Hornblower, Evans and Trevithick, &c. appeared and made the steam-engine the great prime mover of man.

A few subordinate devices relating to steam and steam-engines may here be noticed. There is in Stuart's Anecdotes an historical note respecting wooden boilers, in which water is heated by furnaces or flues within them. They are traced back to 1663. It may be interesting to some readers to state, that they were in use in the preceding century, and that the device in all probability dates from even a more remote period. They are described in Gesner's "Secrets of Phisicke and Philosophie." In the English translation of 1599, by Baker, to which we have already had recourse, they are twice figured, and thus described: "A wooden bowle or tubbe of a sufficient compasse and largnesse over: in the middes of which tubbe erect and set from the bottom unto the edge or brinke of the same, or rather above it, a great copper vessel, in the forme of a hollow pype. Let a parte of the copper pype descende, in such sort and manner, that the water be contained betweene the outward bored wall of the pype and the parte within of the tubbe: But within that parte of the pype which descendeth by the bottome of the tubbe, let the fire be put and kindled, for the heating of the water." Folio 25. The third part of Glauber's Treatise on Philosophical Furnaces also relates to wooden boilers, in which liquids were heated by a copper retort placed in a fire, and whose neck was inserted in the lower part of the boiler, the liquid circulating through the retort. Eng. Trans. by Dr. French, London, 1652.

We have been informed that an enormous steam boiler for an atmospheric engine was in use many years ago at a copper mine near Redruth, in Cornwall, England, which was composed entirely of large blocks of *granite*, or "moor stone." The water was heated by a furnace, from which iron pipes traversed backwards and forwards in the water.

The old chemists often boiled liquids by the sun's heat, and a writer in the London Magazine for 1750 proposed to substitute the solar rays for common fires in heating steam-engine boilers, viz. by collecting the rays in a focus "by means of a common burning glass, or a large concave reflecting mirror of polished metal, or perhaps more conveniently by the newly revived method of Archimedes, which by throwing the focal point to a greater distance may be capable of many advantages that the others have not." He anticipates three objections:—1. "The focus will vary with the motion of the sun." To obviate this he proposes to make the mirror moveable by machinery attached to the engine itself. 2. "The extreme heat of the focal point." If this should be too intense, it may be moderated by enlarging the focus. 3. "The sun does not constantly shine"—therefore the engine must stop. This objection, he remarks, is common to wind, tide

Chap. 9.] *Explosive Engines.* 471

and other mills; and he thinks in the hot months, at least, a steam-engine might be made to raise by the sun's heat water enough from a well to replenish fish ponds &c. as opportunity served. We have long thought that solar heat will yet supersede artificial fires to a limited extent in raising steam, as well as in numerous other operations in the arts, especially in places where fuel is scarce. It is a more legitimate object of research than one half of the new projects daily brought forward.

The mode adopted by Watt for supplying water to his boilers by means of a *float* attached to a lever, and so arranged as to open and close a valve in an adjoining cistern, was not invented by him. It was employed by Mr. Triewald, the Swedish engineer, in 1745, in his apparatus for communicating heat to green-houses by steam, and is described, with a figure, in the London Magazine for 1755, p. 18—21.

Heating green-houses by steam is mentioned by an English writer in 1660. Rivius, in 1548, speaks of eolipiles being employed to impart an agreeable temperature to apartments in dwellings. Col. Wm. Cook's "Method to warm rooms by the steam of boiling water," is described, with a cut, in the Gentleman's Magazine for 1747, p. 171. A boiler was to be heated by the kitchen fire, and the steam pipe to ascend through one tier of rooms, and descend through another, traversing backwards and forwards in each room according to the temperature required; the escape of the condensed and waste steam being regulated by a cock.

A patent for cooking by steam was taken out in England by Mr. Howard in 1793. He named his apparatus "a pneumatic kitchen." Repertory of Arts, vol. x, 147.

There are two other classes of motive machines that we intended here to notice in some detail; but as they have not come into general use, and this volume having already nearly reached its prescribed limits, a brief sketch may suffice. The origin of most of them may be traced to attempts to supersede steam by more portable fluids, or such as require less fuel to generate. We allude to *explosive* and to *vapor engines*. Of all the devices to which the steam-engine has given birth, none possess greater interest than these. Some were designed to raise water directly, and all of them indirectly. The first class are named from the force by which they act being developed by the firing (generally under pistons) of explosive compounds. These are either concrete or aëriform substances, as gunpowder, a mixture of hydrogen gas and common air, &c. Those of the second class are similar to steam-engines, except that they are worked by elastic fluids evolved from volatile liquids, or such as pass easily and at low temperatures into the aëriform state, as alcohol, ether, &c.

Explosive like steam engines have been made to act in two different ways, according to two opposite properties or effects of the exploded substance—the expansive force developed, and the vacuum or partial vacuum which succeeds. For the purpose of explanation, suppose two large repeating guns or muskets, provided with small charges of powder only, to be secured by a frame in a perpendicular position, with their muzzles upwards, and three or four feet apart. Directly over them let there be adapted a working beam, somewhat as in the last figure, suspended on a fulcrum at an equal distance from each. Suppose the ramrods placed in the barrels with their buttons or plugs so made as to fill the bore, and work air-tight like the piston of a syringe or pump. Let the upper ends of these rods then be connected by a bolt to the ends of the beam, which should be at such a distance above the muzzles that when the plug of one rod is at the bottom of its barrel, that of the other may be just within the muzzle of the other barrel. Now let that musket with whose breech the plug of its ramrod is in contact be first fired, and the rod will instantly be forced like a bullet up the barrel, and by its connection with the beam will cause the other rod to descend. The musket in which this last rod moves is then in its turn to be fired and the rod forced up in the same way. Thus the operation is continued. The reciprocating motion of the beam is converted, if required, into a continuous rotary one by means of a crank or some analogous device.

Engines on this plan have not succeeded, nor is there any probability of their success. There are apparently insuperable objections to them, but which need not here be detailed. The explosion of gunpowder has therefore been more frequently employed to produce a partial vacuum in a cylinder when its piston is raised, in order to excite the pressure of the atmosphere to force it down. Suppose one or more openings, covered by valves or flaps, were made near the upper ends of the muskets mentioned above, i. e. just beneath the pistons or plugs of the ramrods when at the highest point in the barrels, and the powder exploded when they are in that position: the sudden expansion would drive out through the valves most of the air previously in the barrel, the valves would instantly close, and the atmosphere would push down the rod and thus raise the other; which in its turn might be caused to descend by exploding the charge under it, and so on continually. Instead of openings in the cylinders for the escape of the air, some experimenters have made large openings in the pistons and covered them with flaps, (like the suckers of common pumps) so that when the explosion ceased the flaps closed and prevented the air's return. Others have used solid pistons and removed the bottoms of their cylinders, and covered the openings with leather flaps so as to operate as valves and give a freer exit to the air and heated gases. This was the plan adopted by Mr. Morey. Papin used hollow pistons. The vacuum produced in this manner by gunpowder has always been very imperfect. Instead of obtaining a pressure of 14 or 15 pounds on the inch, Papin could not realize more than six or seven.

Gunpowder has also been applied to raise water directly, by exploding it in close vessels like the receivers of Savery, with a view to expel their contents by its expansive force, and also to produce a vacuum in order to charge them—but with no useful result.

Explosive mixtures, formed of certain proportions of an inflammable gas and common air, have been found to produce a better vacuum than gunpowder; for a volume of air equal to that of the gas used is displaced from the cylinder by the entrance of the gas previous to every explosion, and when this takes place nearly the whole of the remaining air is expelled. As yet, however, the best of explosive engines have had but an ephemeral existence. Besides other disadvantages, the heat generated by the flame attending the explosion expands the air that remains, so as to diminish considerably the effect.

Of *vapor* engines, the most promising at one time were those in which the moving force was derived from ether and alcohol. The former boils at about blood heat, or 98° of Fahrenheit's scale, and the latter at 174°, while water requires 212°. The vapor of alcohol, it has been stated, exerts double the force of steam at the same temperature; and if to this it be conceded that the same quantity of fuel produces equal temperatures on both alcohol and water, then the former would seem to be more economical than than the latter. Moreover, in consequence of the different specific gravities of water, alcohol and ether, the cost of vaporizing equal volumes of each varies in a still greater ratio than their boiling points—

this cost being as the numbers 11, 4, 2—thus making the scale preponderate still more in favor of alcohol and ether. Why then, it may be asked, have they not superseded water? Principally because the different volumes of vapor from equal quantities of the three liquids turn back the scale in favor of steam. A cubic inch of water affords 1800 cubic inches of steam, while a cubic inch of alcohol produces about 600 and ether only 300 inches; hence the expense of producing equal *volumes of vapor* (and that is the main point) is actually in favor of steam. It has therefore been deemed more economical to use this fluid than the others, even if they were equally cheap—to say nothing of the danger arising from such an employment of highly inflammable liquids, and the practical difficulties attending their application.

In 1791, Mr. John Barber obtained a patent for an explosive motive engine: he used gas or vapor from "coal, wood, oil, or any other combustible matter," which he distilled in a retort, and "mixed with a proper quantity of atmospheric or common air." See Repertory of Arts, vol. viii, 371. Another patent was issued in 1794 to Robert Street, for an "inflammable vapor force," or explosive engine. He exploded spirits of tar or turpentine mixed with common air under a piston, and forced it entirely out of the cylinder, into which it was again returned (by its own weight) and guided by grooves in the frame work. Repertory of Arts, vol. i, 154. In 1807, a patent was granted in France to M. De Rivaz, for another, in which hydrogen and common air were mixed and exploded. De Rivaz moved a locomotive carriage by the power he thus derived. He also inflamed the gaseous mixture by the electric spark. Dr. Jones, in 1814, made experiments on another. See Journal of the Franklin Institute, vol. i, 2d series, page 18. Mr. Cecil, in 1820, published in the Transactions of the Cambridge Philosophical Society, (Eng.) a description of an explosive engine of considerable merit.

In 1825, Mr. Brown, of London, patented his pneumatic or gas vacuum engine. The very sanguine expectations it excited have now died away. It is figured and described in too many works, both English and American, to require insertion here. In 1826, Mr. Morey, of New Hampshire, patented an explosive engine, and soon after exhibited a large working model in this city, (New York) which we took several opportunities to examine. The piston rods of two vertical and open cylinders were connected to the opposite ends of a vibrating beam. The pistons were made of sheet copper, in the form of plungers, about nine inches diameter, and were made to work air-tight by means of a strip of oiled listing or cloth tied round the upper ends of the cylinders. This was all the packing. Mr. Morey employed the vapor of spirits of turpentine and common air. A small tin dish contained the spirits, and the only heat he used was from a common table lamp. By means of a crank and fly-wheel a rotary movement was obtained, as in the steam-engine.

A singular device for making the atmospheric changes of temperature a means for raising water, was devised by M. Woisard. It consisted of two vessels, one above the other, connected by a tube. The lower one, having a valve in its bottom, was placed in the water to be raised. The upper vessel was exposed to the sun's heat, and within it was a bag or small balloon containing air, and a little ether, or other volatile liquid. "As the atmospheric temperature falls, the balloon will diminish in bulk, the surrounding air will become rarer, and the water will introduce itself into the machine through the valve; and when the temperature again rises, the pressure exerted within the machine by the increasing volume of the balloon, will cause the excess of water to flow out." With the exception of the ether, this device is a modification of the air machines Nos. 174 and 175, figured at page 380.

The vapor of mercury has been tried as a substitute for steam, but without much success. This metal boils at 660°.

Another source of power has been sought in the tremendous force with which the liquefied and solidified gases expand at common temperatures. Liquid carbonic acid, at the low temperature of 32°, has been found to exert a force equivalent to thirty five atmospheres! and every increment of heat adds to its energy. No very practical mode of employing this force as a mechanical agent has yet been matured.

The alternate decomposition and recomposition of water has also been suggested. By decomposing this liquid by galvanic electricity, oxygen and hydrogen gases are produced in the exact proportions in which they combine in water. If these gases be made to occupy the interior of a cylinder when the piston is raised, and the electric spark be then passed through them, they instantly become condensed into a few drops of water, and an almost perfect vacuum is the result, when the atmosphere acts on the piston. The water is then to be reconverted into its constituent gases, and the operation repeated. See "The Chemist," for 1825. For further and more recent information respecting motive engines, consult the Repertory of Arts, Hebert's Register of Arts, London Mechanics' Magazine, and the Journal of the Franklin Institute.

END OF THE FOURTH BOOK.

BOOK V.

NOVEL DEVICES FOR RAISING WATER, WITH AN ACCOUNT OF SIPHONS, COCKS, VALVES, CLEPSYDRÆ, &c. &c.

CHAPTER I.

Subjects treated in the fifth book—Lateral communication of motion—This observed by the ancients—Wind at the Falls of Niagara—The trombe described—Natural trombes—Tasting hot liquids—Waterspouts—Various operations of the human mouth—Currents of water—Gulf stream—Large rivers—Adventures of a bottle—Experiments of Venturi—Expenditure of water from various formed ajutages—Contracted vein—Cause of increased discharge from conical tubes—Sale of a water power—Regulation of the ancient Romans to prevent an excess of water from being drawn by pipes from the aqueducts.

In this book we propose to notice some devices for raising water that are either practically useful, or interesting from their novelty or the principles upon which they act. An account of siphons is added, and also remarks on cocks, pipes, valves, and other devices connected with practical hydraulics.

A fluid moving in contact with another that is comparatively at rest, drags along those particles which it touches, and these by their mutual adhesion carry their neighbors with them; the latter also communicate the impulse to others, and these to more distant ones, until a large mass of the fluid on both sides of the motive current is put in motion. Whatever may be the process by which this is effected, or by whatever name the principle involved may be called, (lateral communication of motion or any other) there is no question of the fact. The operation moreover is not confined to any particular fluid, nor is it necessary that the one moved should be of the same nature as the mover: thus air in motion moves water and other liquids as well as air, and aqueous currents impart motion to aëriform fluids as well as to standing waters. A stream of wind from a bellows bears with it the atmospheric particles which it touches in its passage to the fire—i. e. it sweeps along with it the lining of the aerial tube through which it is urged. Blowing on a letter sheet to dry the ink, or on scalding food to cool it, brings in contact with these substances streams of other air than what issues from the thorax.[a] The operations by which the man in the fable blew hot and cold " out of the same mouth "

---

[a] Does not the same principle perform an important part in respiration ?—the lungs not being wholly inflated by air directly in front of the lips, where particles of that previously exhaled might still linger, but also by currents flowing in from all sides of the mouth or nostrils.

may here be explained: in the first case the hollow hands closely encompassed the mouth and received the warm air from his chest; in the latter, his food was at a distance from his lips, and consequently the heat of his breath was absorbed by the surrounding air and that which was carried along, with it to his soup.

A blast of wind directed over the surface of a placid pond or lake not only creates a current on the latter, but sometimes bears away part of the water with it. A vessel sailing before the wind is aided in her course, though it may be but slightly, by the liquid current produced on the ocean's surface. Storms of wind long continued heap up the sea against the mouths of rivers, and cause them to overflow their banks, while low tides often result from the same agent driving the ocean away in opposite directions. These effects of wind were observed in remote ages. " He raiseth the stormy wind which lifteth up the waves." The river Jordan was " driven back " by wind, so that " all the Israelites passed over on dry ground." By its agency, a passage for the same people was opened through the Red Sea. "And Moses stretched out his hand over the sea, and the Lord caused the sea to go back, *by a strong east wind all that night*, and made the sea dry land, and the waters were divided." Exodus xiv. 21.

On the other hand, rivers and water-falls bear down immense quantities of air with them. Strata of this fluid on the surfaces of rapid streams acquire a velocity equal to that of the latter, and in some places aerial currents thus produced are very sensible. At Niagara they are sufficient to drive mills or supply blasts for a long line of forges. In 1829, while ascending the path on the Canadian side, in order to pass under the grand chute, we entered suddenly into one of those invisible currents under the Table Rock, and were nearly prostrated by it. It is the ascent of this air loaded with minute particles of water, (which are borne up by it in the same manner that it is itself carried down) that contributes to the formation of the solar and lunar rainbows seen at the great North American and other cataracts. Heavy rains bring down oceans of air, and in the shower bellows, or *trombe*, blasts of wind are produced on the same principle. Could we *see* the air brought down by heavy showers, we should behold it rebounding from the earth, something like smoke when driven against a wall or any other plane surface.

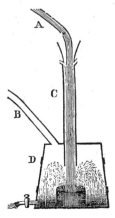

No. 198. The Trombe.

As the trombe illustrates this part of our subject, a figure of one may as well be given. The pipe A discharges water from a reservoir into a funnel placed on the vertical tube C. The end of A terminates in the funnel, and opposite to it is made a number of openings in C, two of which are shown in the cut. The lower end of C enters the close vessel D, and discharges its contents on a stone placed directly under it. As the water from A passes down C, it draws air along with it through the top of the funnel, and also through the holes in the upper part of C. As the liquid dashes against the stone, the air separates and rises to the top of the vessel, whence it is forced by successive volumes through B to the fire, while the water collects at the bottom and is let off by a regulating valve or cock. This machine it will be perceived is a miniature imitation of some of nature's operations; for cascades, water-falls, and also heavy showers of rain, are all natural trombes.

The trombe is of considerable antiquity. It was known to Heron, and is referred to in Pliny's Natural History. Kircher has given several figures of it. See tom. i, 203, of his Mundus Subterraneus, and tom. ii, pp. 310, 347, of his Musurgia Universalis; in which last work he shows its application to supply wind to organs, and by discharging the water from the bottom of the vessel upon a wheel he imparted motion to the keys of those instruments. See also Phil. Trans. Abridg. vol. i, 498.

Liquids raised by currents of air may be illustrated by operations in common life. Whenever water in a well settles to a level with the orifice of the pump pipe, air rushes in (on the ascent of the sucker) and sweeps up with it portions of the liquid in the form of dense rain. On the same principle people are enabled to taste scalding liquids. The next time the reader sips hot soup, or tea, or coffee, he will find himself involuntarily keeping the edge or rim of the spoon or vessel a short distance from his mouth, and protruding his lips till the upper one projects a little over the edge: then drawing in his breath, the entering air *ripples* the surface of the liquid, and by its velocity bears broken portions along, precisely like the pump just mentioned. The liquid particles being thus mixed with comparatively large volumes of cool air, are so reduced in their temperature as to be received without injury and without inconvenience.[a]

Water-spouts appear to be charged in much the same way, whatever may be the active agent in the formation of these singular phenomena; for the sea immediately under their orifice has often been observed to bubble or boil violently, and rise into the spout in disjointed masses.

A stream of water directed into or through a body of the *same liquid*, also communicates motion to those particles of the latter that are in contact with or adjacent to the current. Examples of this are furnished in several of nature's hydraulic operations. That constant oceanic current produced by the trade winds is one. It sweeps round the globe, but is deflected and divided by the varying configuration of the lands that lie in its way. Under the torrid zone, it passses through the Pacific and Indian oceans, whirls round the southern point of Africa, inclines to that continent in again approaching the equator, then stretching across the Atlantic is divided by the South American coast—one part turning northward to the Gulf of Mexico—thence this last division issues as the *Gulf Stream*, and being turned in an easterly direction by the coast of the United States, it bears away past the banks of Newfoundland, and extends its influence to Ireland, Iceland, Norway and the North Sea. This mighty current not only draws with it the liquid channel through which it flows, but the ocean for leagues on each side is carried along with it, or follows in its train;

---

[a] Some of the operations of the mouth are deserving of particular notice. They will be found to elucidate several philosophical principles, and attention to them would certainly have enabled inventors to have anticipated many useful discoveries. We have in a preceding book observed that the mouth is often employed as a forcing pump in ejecting liquids, and as a sucking one when drawing them through siphons, or through simple tubes. We have just seen how it raises hot liquids by drawing a stream of air over them, and machines on the same principle have been made to raise water. It is often used as a bellows to kindle fires, and every body employs it to cool hot victuals by blowing. It even acts as a stove to warm our frozen fingers, by giving out heated air. Many make a condensing air-pump of it, to fill bladders, air-beds and air-pillows; some make an exhausting one of it, and in all it acts continually as both in respiration. How often does it perform the part of a fife, an organ, or a whistle, to produce music? —of an air-gun to shoot bullets and arrows from the sarbacan?—and, not to weary the reader, when employed in smoking a pipe of tobacco, we see in operation the identical principle of increasing the draft of locomotive chimneys by exhaustion—i. e. a sucking apparatus is applied to that extremity of the flue that is the farthest from the fire—a device patented in Europe a few years ago.

and thus it is incessantly transferring to northern latitudes the warm waters of the equinoctial regions.[a]

The volumes of water which shoot from the mouths of the Amazon, Oronoco and Mississippi, continue with almost unabated velocity for leagues into the sea, and impart motion to the contiguous portions of the latter, which are compelled to accompany them in their course.

A current of water not only imparts motion in this manner to a mass of the same liquid when on a level with itself, but it may be applied to *raise* water from a *lower level*. This at first sight does not appear very obvious. A person having a field which he is unable to drain for want of a place of discharge sufficiently low for the purpose, would hardly think his object could be obtained by passing a rapid stream into it from a higher level. To some farmers this would seem the most direct way to deluge the land; yet the thing is not only possible, but in some cases quite easy, as will appear from the following experiments made by M. Venturi in 1797.

From the lower part of the cistern D, No. 199, a horizontal tube proceeded into the vessel A C. The water in D was kept at $32\frac{1}{2}$ inches above the centre of the pipe. Opposite and at a short distance from the pipe was placed the mouth of an inclined rectangular channel or gutter, open at top. The water issuing from the pipe rushed up this channel, and was discharged at B; but as it entered the gutter, the current dragged in with it the contents of A C, until the surface sunk from A to C. From this experiment it is obvious that land on a low level, as at C, might be drained in this manner, and the water discharged above, as at B, wherever a motive current could be obtained. Venturi applied the principle with success to some marshy land belonging to the public.

In the next experiment both air and water are moved by the current, and the pressure of the atmosphere excited to raise water as in the pipe

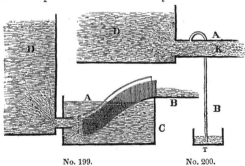

No. 199.   No. 200.

of a pump. The cylindrical tube K (No. 200) was connected to a reservoir of water, D, the surface as before being $32\frac{1}{2}$ inches above its orifice. The pipe K was 18 lines in diameter and 57 long. A glass tube A B was connected to its upper surface at the distance of eight lines from its junction with the reservoir. The other end of the glass tube descended into a vessel, T, containing a colored liquid. When water flowed through

---

[a] Floating substances have often been thrown into the Gulf Stream to ascertain its direction. Upwards of twenty years ago we cast overboard, near the Banks, a common quart bottle carefully corked and sealed, and having a few inches of red bunting tied to the neck. The bottle contained a letter addressed to a gentleman in London, and an open note in English and French, requesting the finder to put the letter into the nearest post-office, American or European, and also a memorandum of the circumstances, date and place of its discovery. Precisely eleven months from the day the bottle was committed to the deep, the letter was delivered by the postman, and accompanied with another from an Irish clergyman. The fragile vessel floated safely ashore near Sligo. Its little pennon excited the attention of a peasant, who broke the bottle, and not knowing what to make of the contents, carried the whole to his priest. This gentleman politely forwarded the letter to its destination, and wrote another containing the particulars just mentioned. Both letters, we believe, were laid before the British Admiralty by the gentleman to whom they were addressed.

K it dragged the air at the mouth of the glass tube with it, the remaining air dilated, and finally the whole was carried out with the effluent water, and the colored liquid rose to the height of 24 inches in A B. The glass tube was then shortened to about 22 inches, when the contents of T rose up and were discharged from K. In another experiment K was placed in nearly a perpendicular position, being inclined a little that the jet might not fall back on itself, but the liquid rose through A B as before. The end of A B where it joined K was flush with the interior surface of the latter. Several small holes were made round K; these diminished the velocity of the issuing current, but no water escaped through the openings.

There is a singular fact relating to the discharge of liquids from different shaped ajutages: for example, more water flows through a short tube than through a simple orifice of the same diameter. A circular opening, of the same diameter as the bore of K in the last figure, was made in a sheet of tin, and the latter attached to a cistern in which the water was kept at a constant altitude of $32\frac{1}{2}$ inches: now while four cubic feet of water escaped through the opening in 41 seconds, an equal quantity passed through K in 31 seconds; and when the length of K was only twice its diameter, the quantity discharged was still greater.

But the quantity discharged may be still further increased if the end of K next the reservoir be made to assume the form of the *contracted vein*. This term is used to designate that contraction which a liquid column undergoes when escaping through an orifice, or when entering a tube. Suppose an aperture, an inch in diameter, made in the bottom of a bucket or a cauldron, and closed by a plug. Then fill the vessel with water, and withdraw the plug. Upon examination the descending column will be found contracted or tapered for a short distance below the orifice, viz. half an inch, or half the diameter of the orifice. The area of the section of the smallest or most contracted part will be to the orifice as 10 to 16 according to Bossut, but when a short cylindrical tube was applied to the orifice, he found the contraction as 10 to 12.3. (The same thing occurs whether the opening be made in the side or bottom of a vessel.) Hence by enlarging the end of K next the reservoir, in the proportions named, the contraction within the cylindrical part of the tube would be avoided, and the discharge consequently increased.

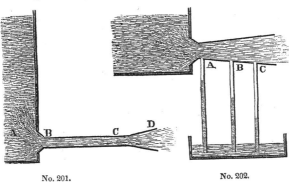

No. 201.   No. 202.

By substituting for K a compound tube of the form and proportions figured at No. 201, the quantity discharged has been ascertained to be more than doubled, being to that delivered by the orifice in the tin plate as 24 to 10! A, the cistern; B, a short conical tube connecting the cylin drical one to the conical frustrum C D. Supposing the diameter of C to be unity or 1, that of D should be 1.8, and the distance between them 9. The increased discharge ceases when the cylindrical part of the tube B C is of considerable length, and of the same bore as the smaller end of C D.

A tube of the form represented at No. 202 was applied by Venturi to the same reservoir, the depth of water in which was also kept at $32\frac{1}{2}$ inches. Three glass tubes, A B C, were connected to the under side of the pipe, and their lower ends inserted into a vessel containing mercury. When water flowed through the pipe the mercury rose 53 lines in A, 20 in B, and 7 in C. These quantities correspond with 62 inches of water in A, 24 inches in B, and 8 in C. The length of the pipe should not exceed four times the diameter of its smaller end, and its sides should not diverge from each other more than what is required to form an angle of from three to four degrees. By this principle it will be perceived, that water may also be raised from a lower level and discharged at an upper one, and in many situations it might doubtless be adopted with advantage. See Nicholson's Journal, vol. ii, and Hachette's Traité Élémentaire des Machines.

Different causes have been assigned for the increased discharge of liquids through conical tubes. One is certainly to be found in the material of which they are made; for when formed of or lined with any substance that repels or refuses to coalesce or be wetted with the effluent water, as wax, tallow, &c. the effect ceases. The phenomenon therefore depends upon the attraction and adhesion of the liquid to the sides of the tubes, which sides exert a capillary force in drawing the particles of the liquid towards them, so as not only to prevent its assuming the figure of the contracted vein when entering the tube No. 202, but also drawing the particles to the diverging sides of the discharging ajutage.

A knowledge of the increased discharge of liquids from conical tubes has led some persons to take advantage of the fact, to the serious injury of others. We have heard of the purchaser of a water power, who according to the covenant was to connect his mill-race with the dam by a trunk of a certain specified bore *at the junction*. This he did, but making the sides of the trunk diverge as in the last figure, the proprietor of the dam was astounded to find the water, as if moved by instinct, giving the new channel the preference, and unaccountably persisting in rushing through it with a velocity that threatened to drain the well supplied reservoir, and leave his own mill to take its rest. This increased discharge is not confined to tubes of a cylindrical or conical form. The walls of the channel may be straight, and its section may be a square, a triangle, &c. as well as a circle.

There is some reason for believing that overreaching in this way is not wholly a modern discovery. No city, ancient or modern, was perhaps ever supplied with water in greater profusion than old Rome; yet the contents of her aqueducts were meted out with economy, and, as in modern times, a revenue was derived from the sale of the water. The superintendence of the aqueducts and of the distribution of the liquid through the streets and houses were always intrusted to a citizen of rank and talents. The celebrated Frontinus held the office under Nerva, by whose directions he wrote two books on the water-works of Rome, the times of their erection, districts of the city supplied by each, the number of public and private fountains, quantities of water discharged from different sized orifices, &c. From him we learn that numerous frauds were practiced in obtaining more than the assigned quantity of the liquid, one of the means for preventing which was this: when a pipe for the supply of a house was to be connected to the castellum or reservoir, (which received the water from one of the aqueducts) a *brass calix*, or short bent tube, (probably the same as the modern ones which connect the lateral pipes to the mains) was delivered by the officer in charge to the workmen, to insert into the castellum; and it was enacted that the bore of the cylindrical leaden pipe should be the same as that of the calix for at least *fifty feet* from the castellum. It is therefore pretty clear that Roman engineers were aware, that the increased discharge through enlarged orifices ceases when a considerable length of pipe of the same bore as the calix intervened.

CHAPTER II.

Water raised by currents of air—Fall of the barometer during storms—Hurricanes commence at the leeward—Damage done by storms not always by the impulse of the wind—Vacuum produced by storms of wind—Draft of chimneys—Currents of wind in houses—Fire grates and parabolic jambs—Experiments with a sheet of paper—Experiments with currents of air through tubes variously connected—Effect of conical ajutages to blowing tubes—Application of these tubes to increase the draft of chimneys and to ventilate wells, mines and ships.

Currents of air and other elastic fluids may be employed to raise water in a manner different from any yet noticed; i. e. not by any modification of the lateral communication of motion, nor by breaking the liquid into minute particles by the motive fluid mixing with them, but by the *removal* or *diminution of atmospheric pressure*. The principle to which we allude is to be found more or less active in nature, and illustrations of it are not infrequent in common life, although for want of reflection they are seldom noticed and are not always understood.

Meteorologists have long observed that storms of wind are accompanied with a diminution of the air's pressure, and that the descent of the mercurial column in the barometer keeps pace generally with the violence of the tempest: thus in hurricanes the depression is much more than during ordinary gales, while in the vortex of a tornado or a whirlwind it is excessive.

Some persons are apt to consider winds as proceeding directly *from* the power that generates them, as a stream of water proceeds from a fire-engine or one of air from a bellows, whereas they as often rush *towards* the source that gives them birth; and hence it is that hurricanes, sometimes if not always, commence at the leeward. Should any mystery appear in this it is easily explained:—if a person blow through a tube, the blast proceeds from him; if he suck air through it, the current is directed to him; when we close a pair of bellows, wind issues from the nozzle; if they are opened while the valve in the lower board is shut, it rushes back through the same channel: so it is with currents in the atmosphere. A partial void is formed in the upper regions, perhaps by electricity, by changes of temperature or humidity, by rarefaction or other causes, and instantly oceans of the fluid matter around rush to restore the equilibrium: then the removal of these oceans necessarily induces others to move also to take their place, and in this way various strata of the atmosphere, for miles and hundreds of miles, are put in motion towards the place where the cause of their movements is located, and in a way not unlike that by which streams of air enter a person's mouth while he sucks an empty tube, or a bellows during the act of opening them.

When the lowering sky and flitting clouds announce the approach of a violent storm, and when, like a demon broke loose, it destroys in its fury nearly every thing in its track, we commonly suppose the mischief is done by the direct *impulse* of the blast—that agitated and groaning forests, trees prostrated, walls and fences leveled, buildings o'erturned and others unroofed, &c. are the results of a tempest sweeping these objects before it, somewhat as we blow dust &c. from a table or from the cover or edge of a book. But this, though sometimes the case, is not always so; for if it were, almost every object blown down by the wind would be found lying in the direction of the blast, whereas they are frequently discovered in the opposite one. The effects enumerated are sometimes caused by winds blowing over a district of country without coming in contact with the earth or the objects upon it, but merely sweeping at some distance above them: at other times similar results are met with at the extreme edge of a storm, and even beyond it. In these cases a partial vacuum produced by the aerial currents often works all the mischief, although it may be, as it frequently is, but of momentary duration. Close buildings have been instantaneously destroyed by the expansion of the air *within them*, their walls being thrown *outwards*, and their roofs projected aloft. The tornado by which the city of Natchez was recently destroyed furnished striking proofs of this removal of atmospheric pressure, and of fearful damages occasioned by the void. The doors and windows of one or two houses left standing amid the general wreck happened to be open, and thus furnished avenues for the dilated air to escape. In some houses the leeward gable ends were pushed out, and the windward ones stood; in others, the leeward walls remained standing while those to the windward were thrown outwards in the face of the storm. Both gable ends were burst out in some, and of others the sudden expansion of the air raised the roofs for a passage, and left more or less of the walls standing.[a]

Persons whose ideas of a vacuum are inseparably associated with air-tight vessels, would hardly suppose that any thing approaching to one could be formed in the open regions above and about us; yet every breath of wind—the gentle zephyr as well as the furious tempest—destroys the equilibrium of the air's pressure, and consequently produces a partial void; and it will be seen in this and the following chapter that a vacuum may be produced and maintained in *open tubes*. It should however be kept in mind, that an *absolute* vacuity is not found in nature nor to be obtained by art: the slightest rarefaction and the best results of the best air-pump are but degrees in the range of a scale, of whose limits we know but little.

A few more familiar illustrations of the removal or diminution of atmospheric pressure by currents of air will not be out of place. And first, who has not, while sitting by a winter's fire, witnessed the coals in the grate brighten suddenly up, and heard the flames and heated air roar in the chimney as if urged for a few moments by some invisible bellows-blower? —phenomena attributed, we believe, in the days of witchcraft, to elves and

---

[a] See an interesting account of this tornado by Dr. Tooley, of Natchez, in the Journal of the Franklin Institute for June, 1840.

fairies, those mischievous imps who, in their wayward moods, sometimes undertook to blow the fires as well as to sweep and sand the floors of the houses they visited, and who, by their screams of delight on leaving their work, were supposed to produce the hollow sounds in the flue as they darted up to join their comrades in the tempest without! It need hardly be observed that it is gusts of wind, sweeping in particular directions over the tops of chimneys, and thereby causing a partial vacuum within them, that thus powerfully increases the draft. But it is not necessary to have fire in the grate, for the effect may be noticed in parlors during the summer months, when those light and ornamental paper aprons with which ladies cover the fronts of their grates are often thus drawn into the flues, and become disfigured and spoiled.

Other examples may be derived from the movements of interior doors, blinds and curtains of windows, &c. While we are writing, the front door of our dwelling is opened, which affords a clear passage from the street to a garden in the rear. The door of the room we occupy opens into the passage, through which a flaw of wind has just passed, and in a twinkling the blinds swing from the windows, and the door is slammed to its frame, by the air in the room rushing to join the passing current, or to fill the slight vacuum produced by it. An open fire-place creates a draft up the chimney, which acts as a pump to draw cold air into the room; hence the complaint, not at all uncommon, of being roasted in front while facing the fire, and at the same time experiencing the unmitigated rigors of winter behind. (In such cases the combustion should be supported by air drawn from without by a pipe terminating beneath the grate—a device patented in modern days, though it was known two centuries ago, and is described by M. Gauger in his treatise on "Fires Improved," a work translated by Desaguliers in 1715.[a]) The motion of every object in nature produces currents of air, and in every possible direction—the movement of the hand in writing or sewing—the trembling of a leaf or of an earthquake—the flight of an eagle or of an insect—the ball whizzing from a cannon's mouth, the creeping of a snail, or a wasp using her forceps.

Artificial illustrations might be quoted without end. Lay two books of the same size, or two pieces of board, six or eight inches apart upon a table, and place a sheet of paper over them; then blow between the books, and the paper, instead of being displaced by the blast, will be pressed down to the table by the atmosphere above it, and with a force proportioned to the intensity of the blast. Instead of the mouth next use a pair of bellows, by inserting the nozzle under one edge of the paper, and the effect will still be the same. The stream of wind may even be directed partly against the under side of the paper, which notwithstanding will retain its place and be pressed down as before. Suspend the books or fix them to the under side of a table, then hold on the paper till the blast is applied, when the sheet will be sustained against gravity. Fold the paper into a tube and blow through it with the mouth, or with bellows—in both cases it will be collapsed. From this experiment we learn that the force which fluids exert against the sides of pipes that contain them, is greatly diminished when they pass rapidly through. We have known a small leak in the pipe that supplied steam to a high-pressure engine, cease to give out vapor every time the communication was opened to the cylinder

---

[a] "Parabolical jambs" (also patented) or backs of grates for reflecting from their polished surfaces the heat into the room, are described in the same interesting little work. At page 140 Desaguliers speaks of bellows invented and patented by Captain Savery—a device of his that is no where else mentioned that we are aware of.

—the particles of the fluid then being hurried along with a velocity too great to allow any of them to change their direction to escape at the leak.

The following abstract of experiments made by us in 1834–5, to illustrate the same principle, may interest some readers:—To ascertain the extent to which atmospheric pressure was removed from under the sheet of paper, we bent a small glass tube at right angles, and placing one end under the paper let it rest on the table, while the other descended into a tumbler containing a little water. Then taking a small pair of bellows, and directing the blast over the pipe, the water rose from one half to three fourths of an inch. The books upon which the sheet laid were then placed within two inches of each other, when the effect was increased, the liquid rising from 1½ to 2 inches. We next laid aside the paper and made use of two tubes, one to blow through and the other to measure the ascent of the liquid.

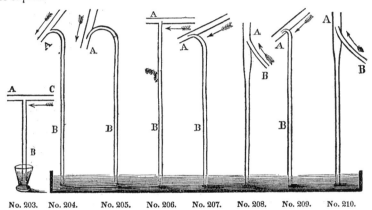

No. 203.　No. 204.　No. 205.　No. 206.　No. 207.　No. 208.　No. 209.　No. 210.

Two leaden or block tin tubes, straight and polished in the inside, were united at right angles. See No. 203. A C the blowing pipe, 8 inches long and half-inch bore. B 12 inches long and three-eighths bore. The upper end of B was joined flush and smooth with the interior of the other, three inches from the end A. Upon applying the mouth to C and blowing in the direction C A, indicated by the arrow, instead of the liquid rising in B, part of the current from the lungs entered that tube and was forced through the water in the tumbler. Various portions of the end A were then cut off without changing the result, until half an inch only remained in front of the joint, when the air no longer descended, but no rarefaction was produced in B. When both tubes were made of the *same bore*, part of the blast descended in B until the whole of A in front of the joint was removed. In numerous trials, the water in the lower end of B was depressed more or less, whether the blast of wind through A was weak or strong. (From these experiments we discover the impropriety of placing cylindrical tubes on chimney tops at right angles to the draft, and especially on locomotive carriages, as was at first proposed. In the Edinburgh Encyclopedia, vol. xvii. p. 457, a carriage by Tredgold is described, and a figure of it given in plate 511. The chimney is represented with a short horizontal tube attached fore and aft to the top, as in No. 203, with a view "to *assist* the draft" by the passage of the air or wind through it. The experiments above show that the reverse would have been the case.)

As part of the air in passing through A, in No. 203, turned off into B, the idea occurred that if the junction of B were made to form an *acute* angle with the longer part of A, then the whole of the aerial current might possibly pass out at A, since to enter B it would have very nearly to *reverse* its direction. The device figured at No. 204 was made to test this. (The part of A in front of the joint was $1\frac{1}{2}$ inches long, which from several experiments we thought produced the best effect, when A was half an inch in the bore—i. e. the length of this part of the blowing tube was three times its diameter.) Upon trial part of the current passed into B and escaped through the liquid, as in the preceding experiment; and even when B was turned up in a vertical direction before entering the water, the same effect took place.

Various modes of uniting the pipes with the view of preventing the blast from entering the vertical one were now tried, and to ascertain the effects produced a glass tube, three feet long and three-eighths of an inch bore, was attached to the vertical or exhausting tube of each. In No. 205 a portion of B protruded into A, so as to form a partition or partial cover to the orifice. Upon blowing through A (in the direction of the arrow) the water sprung up B to the height of 12 inches, and in subsequent trials varied from 10 to 20 inches, according to the strength of the blast. By connecting the glass tube to the blowing end of A and then blowing through B, the liquid rose from 8 to 10 inches; the difference no doubt being caused by the current of air having had greater facilities in one passage than in the other.

We next united two tubes at right angles, but instead of making the joint flush within as No. 203, the upper end of B was cut obliquely, as if to form a mitred or elbow joint. This end was inserted into the under side of A, as represented at No. 206, the open part of B facing A. The object of this device was to ascertain whether the convex part of the vertical tube within A would be sufficient to divert the blast from entering B, while it swept over the upper edge and passed round each side. Previous to connecting the lower end of B with the glass tube we inserted it in water, and upon blowing smartly through A, the liquid rose (10 inches) and was expelled with the air, forming a dense shower. The glass tube was then attached, (by a slip of India rubber) and upon blowing again the water rose, on different trials, from *twenty* to *thirty inches*. The tube A was half-inch bore, and B three-eighths. Various experiments were made to determine the best length of that part of A in advance of the joint: the result was generally in favor of the extent already mentioned.

The end of B cut obliquely, as in the preceding experiment, was now inserted into A at an acute angle. See No. 207. The ascent of the liquid in several trials varied from 20 to 28 inches. A moderate puff raised it 14 inches, but a strong effort of the lungs was required to elevate it over two feet. When the glass tube was connected to A, as in No. 208, and a blast directed through B, the highest range of the liquid was nine inches.

The tubes were next united as in No. 209; that is, the axis of the part of B which entered A coincided with that of the latter, thus leaving an annular space one-eighth of an inch wide for the passage of the blast. The effect of this did not differ so much from No. 208 as was expected. The rise varied from 20 to 30 inches; and not more than half the former amount was produced by reversing the tubes, as in No. 210. The annular passage for the blast in No. 209 was too small, the current was pinched in passing, and its velocity consequently diminished. In another tube in which the space was enlarged, the water rose six inches higher.

We next endeavored to ascertain the effects of varying the form of the discharging ends of the blowing tubes, either by adapting additional ones of a tapering form to them, or by enlarging the ends themselves. Of a number of experiments, the following will be sufficient for our present purpose. In two of the tubes (Nos. 211 and 212) the exhausting pipe did not protrude into the blowing one: in No. 213 it did. As it is difficult to keep up a strong blast from the lungs through a pipe so large in the bore as half an inch, No. 211 was made of quarter-inch tubing, and No. 212 of five-sixteenths. The blowing tube of No. 213 was seven-sixteenths, and the exhausting one three-sixteenths, and all were made of lead. Besides the tubes just named we prepared a dozen conical ones, nine inches long, the small ends one-quarter inch bore, and the large ones varying from three-fourths to 2½ inches. They were made of tin plate, the seams were lapped, and no particular care was taken in their formation. From numerous trials with them in a variety of ways, we obtained the best results with two, one of which was 1⅛ inches at the large end, and the other seven-eighths. But of these the latter, marked C in the cut, generally caused the water to rise highest in the exhausting tube.

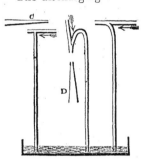

No. 211.   No. 212.   No. 213.

The discharging end of No. 211 extended 1¼ inches from the joint, and the opposite end 2½ inches. When blown through in the direction of the arrow, part of the current descended through the water, but when the conical pipe C was held close to the discharging end the liquid rose in the vertical pipe 9½ inches. A quarter of an inch was next cut off the discharging end and C again applied, when the water rose 12 inches. The end was next remered out with the tapered prong of a file, when the water rose (without C) 11 inches. Another portion was next cut off, leaving only half an inch in front of the joint, and the end swelled out as before, upon which the rise was 7½ inches; but when C was applied the water rose 17½ inches.

In all the trials with C it was necessary, in order to obtain the best results, that its axis should coincide with that of the blowing tube; otherwise the current of air is deflected in its passage. The length of the blowing end of the tube should be no more than what is necessary to give a straight direction to the current. If longer than this, the velocity and strength of the blast is unnecessarily diminished by friction against the prolonged sides. The blowing tube should also be straight and smooth within; for the energy of the blast is less diminished in passing through a straight than through a crooked channel—through a smoothly polished tube than through one whose interior is marked with asperities. Moreover, dints or bruises in a pipe produce counter currents, and materially diminish the ascent of the liquid. In small tubes, the end received into the mouth might be enlarged or cut obliquely to facilitate the entrance of the air; for if the fluid be retarded in its entrance, part of the force exerted by the lungs is uselessly expended. It is immaterial in what position the blowing tube is used.

In No. 212 the blowing tube was jointed to the exhausting one at an angle of 20°. The part in advance of the joint was 1¾ inches. Upon trial, the liquid rose seven inches. The tube D was applied, (its small end being enlarged to five-sixteenths) and the water rose nine inches. The tube was then swelled out by the prong of a file until its orifice was seven-sixteenths of an inch, when the rise was 10½ inches. D was then applied, its end entering the other, and the water rose 18 inches. Previous to this trial D had become slightly bruised in the middle of its length by a fall: the bruises were taken out, and the water rose 24 inches. Various portions were cut from the large end of D, but no diminution of the rise occurred while 3½ inches remained, and this length from several trials gave better results than when the tube was made shorter.

In No. 213, the discharging end of the blowing tube was 1¼ inches long. Without any additional tube, the water rose 16 inches. The end was swelled out, and the liquid rose 19 inches. D was applied, and it rose 29 inches. C was then tried, which made the liquid ascend 31 inches. The discharging end was reduced in length from an inch and a half to half an inch, and the elevation of the liquid was diminished, both with and without the additional tubes C and D.

Two other tubes connected like No. 213 were also tried. From slight variations in the dimensions of the passage way over the end of the exhausting tube, the results varied. Without the additional tube C, one raised the water only seven inches, while with C the rise was 17 inches. The other alone raised the liquid 14 inches, and with C 20½ inches.

It has been seen from preceding experiments, that when two tubes of the same bore are united, as in Nos. 203, 204 and 211, part of the current from the mouth will descend the vertical one, if but half an inch or even less of the discharging end project beyond the joint. To ascertain at what distance from the joint this descent of the current could be counteracted by additional tubes, we connected two pieces of leaden pipe (A and B) five-sixteenths of an inch bore to each other, as in the figure. A was 15 inches long; B four inches, and joined to the other three inches from the blowing end, thus leaving 12 inches in front of the joint. The lower end of B dipped not more than one-tenth of an inch in water.

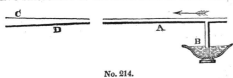

No. 214.

A tapered pipe, C, whose wide end was 1¼ inches and the small one five-sixteenths was attached to A, and upon blowing through A, part of the blast descended through B. Small portions were then successively cut off the discharging end of A, until the air ceased to descend in B. When nine inches remained in front of the joint, but a solitary bubble or two escaped through the water, and after another inch was removed, leaving eight inches in front, the whole current from the lungs passed through A. The conical tube was nine inches long, and after the last result it was divided at D, four inches from the end. Upon removing the part thus cut off, air *again descended* through B.

From this experiment we see that the influence of such terminations as C to cylindrical air tubes, extends to a distance equal to 25 times the tube's diameter. It is however modified by the velocity of the motive current. When high steam is used instead of air, the distance is greatly diminished, and in some cases annihilated. A smoky chimney, or one with a feeble draft, may be cured by enlarging its upper part like the additional tube C in the last figure. The reason why an equal amount of rise in the exhausting tube is not produced by additional ones to such devices as No. 213, arises no doubt from the projection of the exhausting tube into the blowing one, which prevents the blast from sweeping directly into the conical one and *filling* the latter, a condition necessary to the increased ascent.

Some applications of the principle illustrated by the preceding experiments may be noticed :—1. In siphons for decanting corroding or other liquids—for which see remarks on these instruments in a subsequent chapter. 2. Increasing the draft of chimneys, as well as preventing them from smoking. Instead of the old fashioned caps of clay or the moveable ones of iron, let them be made in the form of the annexed figure, and either of sheet iron or copper. A short pipe should be fixed on the chimney, and over it an outer one (shown in the cut) to turn freely, but as close as possible without touching, that the horizontal one to which the latter is attached may veer round with the wind. The vane V keeps the opposite end A to the wind, which enters as indicated by the straight arrow, and in passing through sweeps over the projection and causes a vacuum in the chimney, as in the blowing tubes already described.

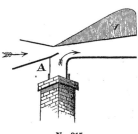

No. 215.

A device of this kind might be made to act in windy weather as a perpetual bellows to blast or refining furnaces, and also to those of steamboats and locomotive carriages. When used on chimneys of the latter, a contrivance to turn and keep the blowing tube fore and aft, as the carriage is turned, would be required. The joint where the perpendicular tube moves over the fixed one might also be made air-tight by an amalgam, on the principle of the water lute. From the experiments with the tubes Nos. 206, '7, '8, '9, '10 and '13, it follows that if the waste steam of a locomotive carriage were discharged over the mouth of the chimney as above, instead of up its centre, the resulting vacuum would be greater.

It is worth while to try whether wells, mines, and the holds of ships, could not be more speedily and effectually ventilated by a similar device than by the common wind sails used in the latter. These displace the noxious vapors by mixing fresh air with them, but by the proposed plan the foul air might be drawn up alone, while the atmosphere would cause a steady and copious supply to stream in at every avenue.

If two or three exhausting tubes, of metal or of any other suitable material, (whose diameter for a ship of the largest class need not exceed three or four inches) were permanently secured in a vessel, their lower ends terminating in or communicating with those parts where noxious effluvia chiefly accumulates, and the upper ends leading to any convenient part of the deck, sides or stern, so that the blowing part could readily be slipped tight into or over them, the interior might be almost as well ventilated, even when the hatches were all down, as the apartments of an ordinary dwelling. It appears to us moreover, that a vessel might by this means be always kept charged with fresh and pure air; for the apparatus might be in operation at all times, day and night, acting as a perpetual pump in drawing off the miasmata. The only attention required would be, to secure the blowing tube in its proper position with regard to the wind during storms. In ordinary weather its movements might be regulated by a vane, as in the figure, when it would require no attention whatever. The upper side of the blowing part of the tube should be cut partly away at the end, so as to facilitate the entrance of descending currents of wind. See the above figure.

CHAPTER III.

Vacuum by currents of steam—Various modes of applying them in blowing tubes—Experiments—Effects of conical ajutages—Results of slight changes in the position of vacuum tubes within blowing ones—Double blowing tube—Experiments with it—Raising water by currents of steam—Ventilation of mines—Experimental apparatus for concentrating sirups in vacuo—Drawing air through liquids to promote their evaporation—Remarks on the origin of obtaining a vacuum by currents of steam.

As the utmost rarefaction which can be produced with blowing tubes by the lungs is exceedingly limited, we next endeavored to ascertain how far it could be carried with currents of *steam*. This fluid presents several advantages. By it a uniform blast can be obtained and kept up, and its intensity can be increased or diminished at pleasure: hence experiments with it can be continued, repeated or varied, till the results can be relied on. As it is inconvenient to measure high degrees of rarefaction by columns of water, mercury was employed for that purpose; and as the blowing tubes &c. if made of lead or block tin would have become soft and bent by the heat, they were all made of copper, while the additional or conical tubes (generally) were of cast brass, and smoothly bored. A detail of all or even half the experiments made would possess no interest to general readers, and would be out of place here; we therefore merely notice such as gave the best results. The force of the highest steam used was equal to a pressure of 90 pounds on the inch. It was measured by the hydrostatic safety-valve described in the Journal of the Franklin Institute, vol. x 2d series, page 2.

While engaged in the prosecution of this subject, we supposed that currents of steam had never been employed to produce a vacuum; but it will be seen towards the close of the chapter, that we were anticipated by a French gentleman, though to what extent we are yet uninformed. We were not aware of the fact until all the following experiments had been matured, and most of them repeatedly performed. The circumstance affords another example of those coincidences of mental and mechanical effort and resource with which the history of the arts is and always will be crowded. The shoemakers' awl was formerly straight, but is now *bent*: the author of the improvement was supposed to have lived in comparatively modern times; but recent researches among the monuments of Egypt have proved, that the artists who made shoes and wrought in leather under the Pharaohs used awls identical in shape with the modern ones.

The expenditure of high steam through open blowing tubes like those figured in Nos. 203 and 204 would obviously be enormous, since there is nothing in them to prevent its passing freely through. They are not therefore so well calculated for practical operations as those in which the end of the exhausting pipe projects into the blowing one and contracts the passage for the vapor, as in Nos. 205—210. These are also better on another account—they produce a better vacuum. Economy in the employment of steam is of the first importance; hence it was desirable to determine if possible that particular construction of the apparatus by which the highest degree of rarefaction may be obtained with the least expenditure of vapor. Fortunately for the solution of this problem, there is one form of the apparatus in which both are eminently combined; for while an increase of the steam's elasticity increased the vacuum, an increased discharge of the vapor was often found to diminish it. This was frequently the case when high steam was employed: for example, if the cock through which steam passed into the blowing tube marked C in No. 217 was wide open, the mercury would sometimes fall two or three inches, but when partially closed, would instantly rise; thus indicating that it is the velocity and not the volume of vapor passing over the orifice of the exhausting pipe, upon which the vacuum depends.

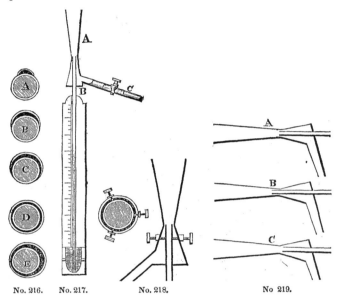

No. 216. No. 217. No. 218. No 219.

We first passed steam through tubes connected like No. 213, both with and without the conical ajutages C D in Nos. 211 and 212. Various proportions of the steam passage over the orifices of the vacuum or exhausting pipes were also employed, as at A, B, C, D, E, No. 216, which represent horizontal sections of the vacuum pipe and steam passage over its orifice. The dark parts show the passage for the steam, and the inner circle the mouth of the vacuum tube. In A the steam channel did not extend over one-fourth of the circumference of the orifice; in B it reached nearly half way round; in C three-fourths; while in D and E it extended entirely round. Upon trial, the vacuum produced by B was greater than that by A; C surpassed B, and D uniformly exceeded them all. We therefore finally arranged the apparatus as shown at No. 217, in which A is a brass tube composed of two conical frustums united at their lesser ends. The longer part, A, was smoothly bored and polished in the direction of its length, to remove any minute ridges left by the borer. The interior diameter of the large end was an inch and an eighth, and of the smallest part nineteen-fortieths, (rather less than half an inch.) The external diameter of the vacuum pipe B was seventeen-fortieths, so that the annular space left round it for the steam was only one-fortieth of an inch in width, being about as small a space as could well be formed without the pipe B touching A. The length of A from the contracted part was $5\frac{3}{4}$ inches. A glass tube three feet long, whose lower end was placed in

Chap. 3.] *Effect of Conical Ajutages.* 491

a vessel of quicksilver, was attached to B, and a scale to measure the ascent of the mercurial column.

When the pressure of the steam in the boiler was equal to 30 pounds on the inch exclusive of atmospheric pressure, and the steam cock C opened, (the hole in its plug was five-sixteenths of an inch in diameter) the mercury rose 9 inches in the vacuum tube B. When the steam was at 40 pounds, the mercury rose 15½ inches. At 50 pounds it reached over 18 inches, at 60 pounds over 19 inches, and at 70 pounds 21 inches. At 80 pounds it was only 21 inches, but on partially closing the cock it sprung up to 22 inches. When the steam was at 90 pounds on the inch, the mercury fell to 20 inches, but on turning the plug of C it rose to 22 inches. These experiments were repeated several times and on different occasions without materially altering the results.

The effect of additional tubes inserted into the open end of A was next observed. Ten or twelve of these were made of tin plate, and of different lengths and taper. The small ends of all were half an inch in diameter, and made very thin, so as to slide into A nearly up to the contracted part, and at the same time to present the least projection possible to the issuing current. The effect of three of these tubes, two of which gave the best results, are stated in the following table. The tube No. 1 was 14 inches long, and its wide end 1¼ inches across. No. 2 was 27 inches long, and 1¾ in diameter at its mouth. No. 3 was five feet long, and its mouth or wide end 2½ inches in diameter.

Pressure of steam in pounds on each square inch.	Vacuum in inches of mercury with apparatus No. 217.	VACUUM WITH ADDITIONAL TUBES.		
		No. 1.	No. 2.	No. 3.
30	9	10	11	—
40	15.5	17	18	10.5
50	18.1	20	20.5	—
60	19.6	20.5	22	—
70	21	21.5	22.8	16.5
80	21	22	23.5	—
	22	22	23.5	—
90	20	20	21	—
	22	22	—	—

In adjusting an additional tube it was moved till its axis coincided with that of A. This was ascertained by the mercury, which oscillated with every movement of the tube, but always rose when it was in the position indicated. On one occasion, when the mercury stood at 15 inches, the additional tube No. 1 was slipped into A and the mercury fell to 12 inches; but this was caused by pushing the tube in too far, i. e. till it touched the vacuum pipe—for on withdrawing it and swelling out the end a little, the mercury rose to 17 inches on the tube being reinserted. A small addition was made to the wide end of No. 1, so that it flared out like a trumpet: on trying it, the mercury stood two inches lower than before.

The fall of the mercury when the steam was raised to 90 pounds, was quite unexpected. It was at first supposed to have been caused by a wrong position of the additional tube, and then to some small object lodged by the steam between the vacuum and the blowing pipes; but on examination nothing of the kind was found. As the mercury still refused to rise, we tried another apparatus similar in all respects to No. 217, except being of rather larger dimensions; but the same thing occurred. When the steam was at 30 pounds the mercury stood at 7¾ inches—at 50 pounds 17 inches—at 60 pounds 20 inches—at 70 pounds 22 inches—at 80 pounds 23½ inches—and at 90 pounds 20 inches! Several experiments seemed to indicate that the length and taper of the additional tubes should vary with the force of the steam, and that the annular passage for the vapor should be contracted as the elastic force of the steam was increased.

Cylindrical pipes applied to the mouth of A, or to those of the additional tubes, caused the mercury to fall; but any plane object held against the current issuing from A, did not affect the vacuum. A piece of board was gradually brought to within *one-fourth of an inch* of the end of A, and of course deflected the steam at right angles; yet the vacuum was not in the least diminished until the board was pushed still nearer. By applying the large ends of the additional tubes to A the vacuum was diminished.

The noise made by the steam issuing from A is indicative of the state of the vacuum. If it be loud and sonorous, the vacuum is not near so high as when the sound is less and *hissing*. In the former case there is generally too much steam escaping—the cock should be partially closed.

No. 218 is a vertical and a horizontal section of the device by means of which the vacuum tube was retained in a position either eccentric or concentric with the blowing one, and by which it could be drawn to one side so as to touch the narrowest part of the latter. Three fine screws with blunt ends were tapped at equal distances from each other into solid projections cast on A, $1\frac{1}{2}$ inches below the contracted part. By these, the exact position of the vacuum tube which gave the best result was accurately ascertained; and it was remarkable how small a change in its position affected the mercurial column. A few examples are annexed:—1. The steam in the boiler being low, and the mercury standing at $3\frac{1}{2}$ inches only, the vacuum tube was drawn by the screws so as barely to touch A, and instantly the mercury fell to 2 inches. 2. When the pipes were clear of each, and the mercury $19\frac{1}{4}$ inches high, as soon as they touched it fell to 16 inches. Similar results took place whatever might be the force of the steam. 3. The mercury fell also when the axis of both tubes did not quite coincide, although a clear passage still remained for the steam, as shown at E, No. 216. In this case, as in the others, the greater flow of vapor on one side probably created cross currents in A, after passing the contracted part. On one occasion, the mercury suddenly fell several inches while the pipes were concentric with each other. Upon examination this was found to be owing to a minute piece of grit, or a film of lead, blown by the steam between the two pipes, where it was wedged in. It did not exceed one-sixteenth of an inch in any direction. It produced the same effect as when the pipes touched. Upon removing it, the steam rose as before.

Another point necessary to be attended to is the position of the orifice of the vacuum pipe with respect to the narrowest part of the blowing one; i. e. whether in a line with it, or in advance, or behind it, as figured at A B C No. 219. To test the effect of these various positions, the vacuum pipe was so arranged by a screw cut on it, as to be pushed in or drawn back at pleasure. In one experiment, the mercury stood at 21 inches when they were on a line, as at A. The vacuum tube was pushed forward three-sixteenths of an inch without any change in the vacuum; but when the pipe protruded three-eighths, as at B, the mercury fell to 18 inches. It was then drawn behind the contracted part of A, and the mercury instantly began to fall. When the orifice was one-fourth of an inch behind, the mercury fell from 21 inches to 4; and when drawn back one-eighth of an inch more, as at C, the steam descended the vacuum tube and blew the mercury out of the vessel that contained it. In another experiment, the vacuum tube was one-fourth of an inch in advance of the contracted part, and the mercury 20 inches high: when the tube was drawn back so that its end was in a line with the contracted part, the mercury rose half an inch. When drawn back one-eighth of an inch, it fell to 17½ inches.

That part of the vacuum tube within the steam chamber, or back end of A, should be straight, and its axis should coincide as nearly as possible with that of A, else the vapor in passing over the orifice will be more or less deflected to one side, and thus diminish the vacuum. Although the blowing tube figured at No. 217 has its mouth opening upwards, in practice we used it in a horizontal position, as at No. 219, or rather inclined downwards, as at No. 221, that the condensed vapor might not fall back and enter the vacuum tube.

No. 220 represents another modification of this mode of removing atmospheric pressure, by which the vacuum may be carried to a greater extent than with No. 217. It consists of two blowing tubes attached to one vacuum pipe. The lower blowing tube in its narrowest part was seven-twentieths of an inch in diameter, and the annular passage for the steam between it and the vacuum pipe was only one-fiftieth of an inch in width. The bore of the steam cock and pipe C was three-tenths of an inch. The upper end of the lower blowing tube was half an inch in diameter, and terminated at the contracted part of a larger one, D, where a space of one-thirtieth of an inch was left for the steam between them. D was six inches long, and its upper end an inch in diameter. It was also furnished with steam by the pipe and cock E. (Both blowing tubes in the accompanying illustration are figured too large for the exhausting one.)

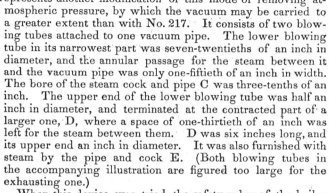

No. 220.

When this device was tried, the safety-valve of the boiler indicated a pressure of 40 pounds on the inch. The cock E was first opened, and as the steam rushed from D the mercury rose 8 inches. E was then closed and C opened, upon which the mercury rose 8.8 inches. Both cocks were then opened, and the mercury rose 16.6 inches.

When the steam was at 60 pounds and E opened, the mercurial column was 9 inches. With E closed and C opened it rose 15 inches. Both cocks were next opened, and the height was increased to 20 inches.

Steam at 80 pounds and E open, the mercury stood at 17.5 inches. C open and E shut, it rose to 21 inches; and when both were opened, it reached to 24½ inches. The addition of another blowing tube over D would most likely have carried it to the full height of the barometer. If D were inserted in a chimney in the direction of the flue, it would not only increase the draft, but the draft would increase the vacuum.

The steam pipe that supplied D was then unscrewed from the cock E, which was left open. The cock C was again opened, and the mercury rose as before to 21 inches, the air rushing through E producing no effect on the column except rendering its surface slightly concave. By often closing and opening the orifice of E with the finger, no sensible change in the vacuum could be perceived.

After removing the vessel of mercury from the bottom of the vacuum pipe, a piece of twine several yards in length which happened to be laying on the ground near by, was drawn into the tube and discharged through D. This was repeated several times. By presenting one end near the end of the glass tube, the whole was almost instantaneously drawn up and thrown out by the steam, although the vacuum tube was continued in a horizontal position nearly two feet before it was connected to the glass one. On applying a vessel of sand, and another of water, to the end of the tube, the contents of both were raised and discharged in the same way.

The vacuum tube of No. 220 was connected to a soda fountain, and an opening one-sixteenth of an inch diameter made in the latter to admit air. The mercury previous to making this opening stood at 16 inches, and it still remained at that height. The opening was next widened to one-eighth of an inch, when the mercury fell to 12 inches. The opening was then made as large as the bore of the vacuum tube, (about five-sixteenths) upon which the mercury fell to six inches.

It is obvious that by connecting one of these blowing instruments to an air-tight vessel, water may be raised into the latter by the atmosphere, and to an elevation corresponding with the vacuum. In one of our earliest experiments, we attached a blowing tube to a soda fountain placed 22 feet above the surface of the water in a well, into which a pipe descended from the upper part of the fountain. But by arranging a series of close vessels at certain distances above each other, (according to the extent of the vacuum obtained by the apparatus) water may be raised in this manner to any elevation—the pressure of the atmosphere transferring it from one vessel to another till it arrive at the place of discharge, as in Papin's plan, described at page 447-8. An English patent was granted in 1839 for a very elaborate French machine of this kind. See Civil Engineer and Architect's Journal, vol. iii, page 51. In December 1840, an American patent was obtained for the same thing by a French merchant of this city. This gentleman has had one constructed from drawings sent from Paris. The receiving vessels were 12 feet apart. The mode of applying the steam is to discharge it at the orifice of the vacuum pipe, over a *small part* of the *periphery*, as at A No. 216. The steam however does not come in contact with the sides of the vacuum tube, as in the preceding figures No. 217 to 220, for this tube does not form one of the walls of the small steam chamber behind its orifice—the chamber being a separate part complete in itself, and having a semicircular recess formed at one side, into which the vacuum pipe is received. There is therefore, between the interior of the vacuum tube and the steam without, not only the thickness of the metal of which that tube is fabricated, but also the thickness of the plate of which the steam chamber is made. Floats are arranged in the interior of the receiving vessels, so that when one of the latter is filled with water from the one below, the float opens a valve to admit the atmosphere to press the contents into the vessel next above it.

There is another mode of raising water to considerable elevations by an apparatus like Nos. 217 and 220, and for which they seem much better adapted than any other, viz. by admitting portions of air to mix with the ascending liquid, as in the examples given at pp. 224, 225. No air-tight receiver would then be required, as both the air and water would be discharged with the steam at the open end of the blowing tube, which, for the reason already stated, should be inclined downwards.

Wherever large volumes of air are required to be withdrawn, as in the ventilation of mines, these instruments we believe would be found as efficient and economical as any device yet tried. A number of vacuum tubes, whose lower ends were made to terminate in different parts of a mine— (they might be of leather or other flexible materials, so as readily to be moved wherever required)—and whose upper ones were connected to one or more blowing tubes through which currents of steam were constantly passing, would effectually withdraw the noxious vapors from below, and induce a more copious supply of fresh air than any forcing apparatus could ever furnish. The waste steam of engines at coal or other mines

Chap. 3.] *Apparatus for evaporating liquids in vacuo.* 495

might be beneficially applied to large blowing tubes, and thus contribute to the same result.

There are other useful applications of these blowing instruments. One of our first attempts was to employ them as substitutes for the expensive air-pumps worked by steam-engines, employed in evaporating sirups and refining sugar by Howard's vacuum plan.

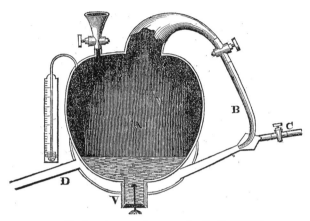

No. 221. Apparatus for evaporating liquids in vacuo.

We fitted up a very strong old still, (No. 221) three feet in diameter and about the same in depth. A jacket of copper was fitted to its lower part so as to form a double bottom. The discharging pipe passed through the jacket, and was closed by a valve V. Steam being conveyed into the jacket heated the liquid within the still, but instead of taking steam from the boiler expressly for this purpose, we made use of that by which the vacuum was produced. The open end of a blowing tube was inserted into the jacket as represented in the cut, and the vacuum tube B connected by a cock to the neck of the still. C the steam cock and pipe leading to the boiler. D a pipe that conveyed the surplus steam from the jacket into the chimney. The orifice of the vacuum pipe within the blowing one was three-eighths of an inch diameter, and the annular space around it for the passage of the steam was the same as in No. 217. At the first trial with this apparatus, 25 gallons of sirup were put into the still through the funnel, and the cock shut. The steam cock C was then opened, and in a few moments the mercury in the gauge rose 15 inches, but in eight minutes fell to 10 inches, the fall being occasioned by the evolution of vapor in the still. The steam in the boiler was raised higher, until the mercury rose to 16 inches; but after the operation had been continued about half an hour it commenced rising, and was at 18 inches when the experiment was closed. On another trial 32 gallons of sirup were poured in, and when C was opened the mercury rose to 22 inches, but in ten minutes fell to 17. In half an hour it began slowly to rise, and in fifteen minutes reached to 20 inches, at which height it remained when the concentrated sirup was withdrawn.

Had a double tube like No. 220 been used, the vacuum might probably have been carried to 28 or 29 inches, and the operation performed in much less time. The experiment however shows how small a tube can withdraw the vapor arising from a surface of seven square feet. It would be an advantage to apply two or perhaps three separate blowing tubes, of different sizes, to each sugar pan—using the largest first, to draw off the the bulk of the vapor, and finishing with the smaller ones. There would be a saving of steam, and the vacuum might be carried higher towards the close of an operation with a very small tube and current.

Another mode of using these tubes to promote evaporation, is to draw air through liquids instead of forcing it through them with pumps, as in the pneumatic processes of concentrating sirups. An open boiler, four feet in diameter, was inverted and placed in another over a fire and containing sirup : a blowing tube, the orifice of whose vacuum pipe was three-fourths of an inch diameter, was connected to the inverted vessel, and it drew so much air under the edges as to cool the liquid to such a degree that the operation of concentration was prolonged to twice the ordinary time.

While engaged in making the experiments described in this chapter, (in 1835) and stimulated by the conviction that we were the first thus to apply currents of steam for the purposes of raising water and promoting the evaporation of liquids at low temperatures, &c. we were exceedingly surprised to learn that something of the kind had been previously done, or proposed to be done, in France. As we had made preparations to secure the invention by a patent here, and by others in Europe, our experiments were discontinued with a view to ascertain the particulars of the French plan, that it might be known whether we were traveling on beaten ground or not; but to the present time we have not obtained any specific description of it, nor do we know whether it consisted of a jet of steam discharged through the centre of a tube, as in Nos. 208, 210, and as applied to increase the draft of chimneys in locomotive carriages, or whether the jet was directed over the outside of a part or the whole of the end of the vacuum tube—nor have we learnt what degree of rarefaction was obtained. We have therefore concluded to insert the preceding notice of our labors, that since we cannot claim priority in the research, we may be allowed the credit, if any be due, for our modes of application, and the extent to which they carried the vacuum and are obviously capable of carrying it, especially by such devices as No. 220.

The whole of the devices, from the blowing tubes described in the last chapter to the apparatus for boiling sugar in vacuo described in this, with the exception of the patented plan of raising water by a series of vessels on different levels, originated entirely with ourselves, nor were we indebted either directly or remotely for so much as a hint in maturing them to any persons or writings whatever; and upon them we have also spent no inconsiderable amount both of time and money. But as we have on several occasions shown that *new* devices, so called, are often *old* ones, it is but just that we should mete to ourselves the same measure which we have given to others. We therefore with pleasure record the fact, that at a meeting of the Paris Academy of Arts and Sciences, held in January, 1833, M. Pellatans read a paper on the dynamic effects of a jet of steam, of which a notice (not a description of the plan) was published in an English journal, and copied into the Journal of the Franklin Institute for March of the same year—vol. x, 2d series, p. 195.

There is also described in the London Mechanics' Magazine, vol. iii, p. 275, an experiment of a current of air from a bellows directed over the orifice of an inverted glass funnel, which was placed in a saucer filled with water. From this (which we did not see till recently) the blowing tubes described in the last chapter might, with a little ingenuity, have been deduced.

CHAPTER IV

Spouting tubes—Water easily disturbed—Force economically transmitted by the oscillation of liquids—Experiments on the ascent of water in differently shaped tubes—Application of one form to siphons—Movement given to spouting tubes—These produce a jet both by their ascent and descent—Experiments with plain conical tubes—Spouting tubes with air pipes attached—Experiments with various sized tubes—Observations respecting their movements—Advantages arising from inertia—Modes of communicating motion to spouting tubes—Purposes for which they are applicable—The Souffleur.

There is a simple mode of raising water which to our knowledge has never been adopted, nor yet suggested—viz. by straight and open pipes, or, as they might be named, *spouting tubes.*

Water is raised in the ram (No. 168) by the force which the liquid acquires in flowing through descending channels, but in the instruments to which we now refer, the same effect is produced by its momentum in passing up vertical ones. So far as respects the force of a liquid in motion, it makes little difference in what direction it moves—whether the liquid rise perpendicularly, or having first descended at one angle it ascend at another. A jet d'eau, deducting all resistances, rises with the velocity with which it would fall through the same space; but in practice, the velocity is diminished by the length, figure and dimensions of the channel through which the liquid flows, and of the ajutage from which it escapes.

Every person's experience teaches him, that a very small force is sufficient to disturb a large body of water, and that the consequent movement of the liquid is long continued after the force is withdrawn. A stone dropt into a tank, or thrown into a pond, causes waves to rise and roll to and fro over their whole surfaces, and some time elapses ere the movements cease. Days and even weeks elapse after a storm is over before the ocean recovers its previous repose. This effect is the result of the great mobility of water; its particles move with such extreme facility among themselves, and so actively impart their motion to each other, that a force once communicated to them is long ere it becomes exhausted. It is the same to a certain extent when waves rise and fall within tubes; for although the friction of liquids against the sides of these channels is considerable, especially in small ones, still the force in the central parts is but slowly consumed. A device therefore by which the oscillation of liquids is employed in transmitting forces, will probably consume as little in the transit as any mechanical device known.

It has already been remarked, that the momentum of a flowing liquid suffers less in passing through a short than through a long tube—through a straight than a crooked one; and we may add that this is more especially true when the figure of the tube is expressly designed to facilitate the passage of the moving liquid, instead of being uniform in its bore throughout. Now in these particulars spouting tubes are eminently superior to others, or they may be made so. They are short, straight, and of a form adapted to the rising wave within them.

Motion is imparted to water in a spouting tube either by depressing the liquid below the orifice and then admitting it to enter, or by excluding it from the tube till the lower orifice of the latter be sufficiently immersed. If a pipe whose lower end is closed be plunged perpendicularly in water, the liquid will rise within it the moment its end is opened; but it will depend upon the length and figure of the tube, and the relative proportion of its two orifices, whether the liquid rush up above the surface without, or slowly reach it and there remain.

The following are selected from a number of experiments made several years ago. Instead of closing the lower orifice, the upper one was closed with the fore finger, the confined air acting the part of a cork, and preventing the liquid from entering until the finger was removed.

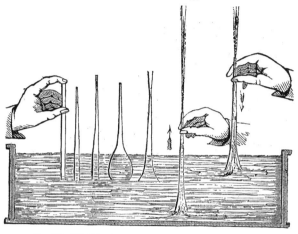

Nos. 222. 223. 224. 225. 226. 227. 228.

EXPER. I.—No. 222, a cylindrical glass tube, 18 inches long and half-inch bore. Its upper orifice was closed air-tight by the finger, and the lower one then held four inches under the surface of the water in the vessel. Upon raising the finger, the liquid rose in the tube six inches; i. e. its momentum carried it two inches higher than the surface in the cistern, and after a few oscillations it settled at the same level. Cylindrical tubes of various sizes were tried at different depths, and the average extent of the rise (above the surface) was equal to half the length of the part of the tube immersed below the surface. If No. 222 dipped four inches, the rise was two—if eight inches, it was four—and if twelve, it was six. By contracting either orifice the effect was diminished.

EXPER. II.—No. 223, a tube slightly conical, 16 inches long, the diameter or bore of the large end half an inch, and that of the small end one-third of an inch. The rise of the liquid in this exceeded that in No. 222. When tried with the large end up, little or no rise took place.

EXPER. III.—No. 224, another tube, 18 inches in length, the diameter of whose upper orifice was three-sixteenths and of the lower seven-eighths of an inch. Four and a half inches of the lower part was cylindrical. When dipped four inches in water and the finger removed, the liquid rose but two inches above the surface. This was owing to the cylindrical form of the lower part of the tube, all the water that entered being required to fill the lower part. When the dip was six inches, the rise was five; when eight, the jet passed out of the tube and ascended sixteen inches. When the tube was lowered to ten inches below the surface, the jet rose thirty inches; and when the end of the tube was twelve inches under the surface, the jet ascended four feet and a half. Fourteen inches dip threw it six feet, and sixteen inches dip caused it to ascend over seven feet. The rise in cylindrical tubes, we have seen, bore the same relation to the dip at various depths; but this experiment shows that the elevation of the jet in conical tubes increases in a much greater ratio.

EXPER. IV.—To include the extreme proportions between the two orifices, we next took a matrass or bolt head (No. 225) and cut a portion from the globe opposite the junction of the neck or pipe. The opening thus made was 3½ inches, and the orifice of the tube three-tenths. When the lower end was thrust two inches below the surface, scarcely any rise took place upon removing the finger; and when half the length of the whole was immersed, say ten inches, the rise did not exceed six or seven. The reason was plain: the large volume of air contained in the lower part could not be expelled instantaneously by the pressure of the liquid column through the small orifice above, but the force of the ascending liquid was consumed in doing this. Various portions were now cut from the lower part, with a view to ascertain the greatest rise that could be obtained with a dip of four inches. This occurred when the diameter of the lower end was reduced to 1¾ inches: the liquid then rose between nine and ten inches above the surface. The upper end was now heated in the flame of a lamp, and the bore enlarged by pressing into it a tapered piece of wood, till the end resembled the conical ajutage C D in No. 201. This caused the liquid to rise an inch higher.

EXPER. V.—A number of conical tubes of the same length, (21 inches) whose wide ends diverged or flared differently, were next procured, with the view of selecting those through which the jet rose the highest, as affording an approximation to the best form. The one represented at No. 226 gave a better result than any other. With a dip of four inches the jet rose thirteen. The diameter of its lower orifice was 1.6 inches, and that of the upper one .4: three inches below the latter, the bore was .2. At seven inches from the small end, the bore was .3—at fourteen inches, .4—and at seventeen inches, .5. The curve given to the flaring part of the lower end should be that which the fluid itself assumes in entering; but that given in the figure is sufficient for all practical purposes to which small instruments of this kind are applicable.

Before proceeding we may observe, that these instruments, simple as they are, and even when charged in the manner indicated above, are susceptible of some useful applications; among which may be named siphons. If the tube No. 226 were bent in the form of one, it might be applied in numerous cases to transfer acids or other liquids; and as it would be charged by the mere act of inserting its short leg into the liquid to be withdrawn, there could be no danger from sucking, &c. as in using the ordinary instrument. It will moreover be perceived from the third experiment, that the extent to which these siphons are applicable increases with the depth to which the short leg can be immersed: but as this chapter is appropriated to the application of spouting tubes to raise water from one level and discharge it at a higher one, their employment as siphons will be illustrated in a subsequent part of this volume.

It will at once occur to every machinist, that to render these tubes of any practical value for raising water, some mode of working them very different from that of alternately opening and closing the upper orifice with the finger, and raising them wholly out of and then plunging them into the liquid, would be required: a mode of regularly and rapidly depressing the liquid within them, that the wave formed by its ascent might rise and fall uniformly.

There is a simple way of doing this:—If the whole of the tube No. 227

be sunk perpendicularly in water, except one or two inches by which it is held, and then raised eight or ten inches, air will enter the small orifice and fill the part previously occupied by the liquid: if the upward movement be very slow, the air will gradually fill the interior without disturbing the surface of the liquid; but if the tube be raised by a rapid movement or slight jerk, the air will then rush into the void with a force that will push down the liquid before it to a considerable depth, so that on the re-ascent of the liquid its momentum will project a portion in the form of a jet, precisely like Nos. 224, 225 and 226. It is surprising how elevated a wave is generated in the tube by the slightest ascent of the latter, provided its movement be made sufficiently quick. The rise of the water, too, follows that of the tube so rapidly that most observers at first suppose them to rise simultaneously. The fact is, the liquid when depressed returns with such velocity as to escape from the tube the instant the stroke is finished, and even before its motion be slackened.

EXPER. VI. A jet may be produced by the *descent* of the tube as well as by its ascent. Let No. 228 be so held that its lower end dip not more than an inch or an inch and a half in the water, and then be pushed quickly down eight or ten inches—a stream will be projected from its upper orifice to an elevation of six or seven feet, and will be instantly followed by another that will reach nearly as high. The same cause operates here as in the upward movement, but it is differently excited. A small part only of the air within is expelled at the end of the stroke, on account of the tube's rapid descent, and consequently the water is prevented from entering; but as soon as this movement of the tube ceases, the liquid rushes in and a portion ascends in the form of a jet. On the subsequent ebb of the wave within, another one rises nearly equal to the first, and causes the second jet. The following experiment will illustrate both movements:— A small glass tube eight inches long, its wide end an inch and five-eighths diameter and its small end one-eighth, was employed. By its upward movement or stroke the extremity of the jet reached to an elevation of nine feet. By the downward stroke a jet rose six feet, which was succeeded by another that reached four feet and a half. Now if both movements are properly combined in a spouting tube of large dimensions, we believe the instrument may be made to raise as much water, in circumstances adapted to its employment, as any other hydro-pneumatic machine.

If the figure given to No. 226 should be found better adapted than any other when the tube is used as a siphon, it does not therefore follow that the same form would be the most suitable to produce *jets* of water. In the former case the instrument acts while at rest, but in the latter a constant and rapid movement is required: hence, to prevent an unnecessary expenditure of the power employed, it should be so formed as to present as little opposing surface to the resistance of the dense fluid in which it works as is consistent with the elevation, or quantity of water to be raised by it. This remark applies particularly to the lower or wide end, for if that part be suddenly expanded or flared like a trumpet, a volume of water of equal diameter has to be displaced in the reservoir every time the tube is pushed down, and also a ring of water whose external diameter is the same (the internal one being bounded by the tube) every time the latter is lifted up. When used as spouting tubes the lower end should therefore flare very little, if any, unless in cases where the outlay of power to work them is of little consequence or of secondary importance. The upper end of a spouting tube, when intended to throw jets from its orifice, should not diverge like that of No. 226, since the elevation of the stream would be thereby diminished: instead of rising in a compact jet, it would

Raising Water with Open Tubes.

sooner become expanded and broken. When, however, one of these instruments is intended to deliver water at a level with its upper orifice only, then the discharging orifice should resemble that of No. 226, or C D in No. 201, as an increased discharge of the liquid would in that case take place: a greater flow of air would enter on the ascent of the tube, and a larger volume of water flow out on its return.

EXPER. VII.—A number of conical tubes, ten inches long, were prepared. The diameter of the small ends of all was $\frac{1}{4}$ inch, while the large ends were respectively 4 inches, $3\frac{1}{2}$, 3, $2\frac{3}{4}$, $2\frac{1}{2}$, $2\frac{1}{4}$, 2, $1\frac{3}{4}$, $1\frac{1}{2}$, $1\frac{1}{4}$ and 1; and besides these, two cylindrical ones of $\frac{1}{2}$ inch and $\frac{1}{4}$ inch bore. With the cylindrical tubes no jet could be produced by any movement given to them, either quick or slow, however deep they were immersed; nor yet when they were inclined. When the conical ones were immersed half their length, and worked without plunging them deeper, no water could be ejected: the cause of this however was not the same in all. In six or seven of the largest, the parts below the surface were too capacious to be filled instantaneously with air through the small orifice above as they were raised. The sound made by the entering fluid (like a person gasping for breath) showed this, especially in the largest. But in the smaller sizes, the air entered as fast as they were raised, and consequently disturbed but slightly the surface of the liquid within.

When any one of them was immersed within an inch of the small end and then moved two or three inches up and down, a jet was thrown out, and from the large ones with considerable force, on account of the greater mass of the liquid put in motion in their lower part. Still, however, the jet did not rise so high from the large as from some of the smaller tubes, because the sides of the former converged so rapidly to the discharging orifice that the liquid particles crossed and counteracted each other as they issued. Short cylindrical ajutages soldered on two of the largest made no sensible improvement. The disadvantages of making the lower parts too wide or spacious for the entering air fully to occupy, was also very apparent when the tubes were raised five or six inches in working them. The water within not being wholly displaced, it hung in them as in an inverted tumbler or bucket, and consequently its weight was added to that of the tube. This not only required an increase of force, but the intended effect was diminished and in a great measure destroyed. The same thing of course occurs if a smaller tube be used, with a large additional part to its lower extremity, as at No. 229. To obviate this by furnishing a larger supply than would enter the smaller orifice, we adapted an air tube whose exterior end was covered by a valve opening upwards, as shown in the cut. The

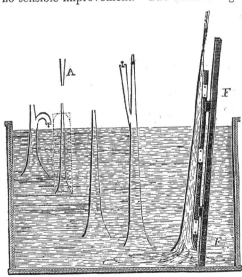

No. 229.   230.   No. 231.   No. 232.   No. 233.

force required to work the larger tubes was very sensible, but with the smaller ones it was scarcely appreciable. Those whose larger ends were 2 inches and 1¾ inches produced the highest jets, but they were obviously too much tapered for practical purposes, and even the sides of the smallest one named, formed too large an angle to be applied with advantage at great depths.

The tube No. 230, two feet one inch in length, was made of tin plate. It consisted of a conical piece 22 inches long, 1¼ inches wide at one end, and ¼ inch at the other. To the wide end a flaring piece, 3 inches long and 4 diameter at the lower edge, was added. This piece was made of sheet lead for the convenience of forming it. When wholly immersed in water, except 2 or 3 inches by which it was held, this tube threw a jet 15 feet high. By the upward stroke the jet rose 12 feet. When the diverging ajutage A (whose contracted part was the same as the orifice of the tube) was slipped on the latter, the jet was dispersed before it rose 8 feet. An inch was cut off the lower end, leaving the diameter 3 inches, upon which the jet rose to about 14 feet. Another inch was then removed, when it rose still lower; yet it might still, by a *quick* back stroke, be thrown nearly as high as at the first. It would therefore seem, that although a large flaring end requires more force to raise it than a small one, yet the increased velocity required to be given to the downward stroke, in order to raise the jet to an equal height, comes to much the same thing. There is a way however by which the resistance which a large flaring end meets with from the water may be avoided in the *upward* stroke, viz. by enclosing the tube in an air-tight *cylindrical* one, of the diameter of the flaring end, as represented by the dotted lines in No. 230: or the instrument might be inserted in a wooden tube, whose specific gravity was about the same or rather less than that of water.

No. 231 was 3½ feet long, formed of copper, and of a regular taper to within four inches of its lower end. Its diameter at the small end was half an inch, and at the lower end 3½ inches, to which a piece flared out to six inches was added. By an upward stroke of 18 inches, the jet rose 17 feet; and by a downward stroke of one foot, it rose to the same height. (These measurements, and the others mentioned, relate to the extreme height to which a small part only of the liquid rose. The main body of the jet seldom reached over two-thirds of the distance.) When the upward stroke was continued 2½ feet, the rushing air pushed all the water out of the tube, and rose up on the outside.

Exper. VIII.—We next prepared a larger tube, and arranged it so as to be worked in a light wooden frame, which was secured in a wine pipe filled with water. (See No. 233. The wine cask is omitted.) This instrument was deemed equal to any that was tried—the quantity of water, and the elevation to which it was raised, being compared with the force employed. It should not, however, be considered as exhibiting anything like the maximum effect which spouting tubes are capable of producing, because the friction of the liquid in passing through so small an orifice as that of No. 233 was very considerable. The reader is therefore requested to bear in mind, that the larger the bore of these tubes, the more favorable would be the result; and that, although jets of water may be thrown very high by them, yet they are better adapted to raise large volumes of water to small heights.

The tube No. 233 was five feet long. It was composed of one piece 4 feet 4 inches in length, .75 of an inch diameter at one end, and 2.9 inches at the other. To this end a piece 5 inches long was added, which made the diameter 5.5 inches; and to this another piece 3 inches long, which made the extreme end of the tube 7.5 inches diameter. The tube as thus formed was secured to a straight strip of wood of nearly the same length, by means of three copper straps, which were soldered to the tube and screwed to the wood. (See the figure.) About a foot from each end, and across the back of the strip, two pieces of wood, 3 inches long and 1½ wide, were secured. They projected half an inch over each side of the strip, and were beveled at the ends, so as to fit into and slide readily up and down in a dovetailed groove formed on the face of the post F F. This post was secured in an inclined position, as represented. When large tubes are used they should always be inclined, that the water once raised above the orifice may not fall into it again and run back. The surface of the water in the cask was 13 inches below the upper end of the tube, and upon working the latter the jet (¾ of an inch diameter) rose 22 feet. A piece of pipe was next slipped on the end, which made the tube a foot longer, and reduced the orifice to half an inch, when the jet rose little if any higher than before. Another tapered piece of pipe was added to the last, making the orifice five-sixteenths of an inch, upon which the jet did not ascend over six or eight feet. An air-pipe, figured at No. 232, was now added, that the water might be fully depressed in the tube on its ascent, but the jet was so pinched at the orifice that no obvious change was perceived.

The upward stroke ought to be so regulated, that the air in rushing down should push nearly all the water out of the tube, that the wave in rising may be urged up with the full pressure of that above it in the reservoir: hence the elevation of the jet produced by the upward stroke of a spouting tube depends chiefly upon the depth of its immersion. But if the upward movement exclude nearly all the water, the downward one if made with due velocity prevents it, or much of it, from entering before the tube itself gets nearly to the end of its stroke, and consequently the effective height of the hydrostatic column is then increased to an extent equal to the length of the stroke. On the other hand, if the upward movement be made so quick that the air has not time to fill the enlarged space below before the stroke is finished, then little or no rise will take place. The operation in this case is the converse of the experiment with the matrass, No. 225.

When the movements of one of these instruments are properly timed, the *inertia* of the descending air and ascending liquid is peculiarly beneficial. In ordinary machines, where the direction of moving masses is reversed, or when they are alternately brought into a state of rest and motion, the inertia is overcome by an outlay of the force employed; but this is not the case with spouting tubes. Thus when a tube is raised, the air descends into the vacuity left by the retiring liquid, and when its momentum is expended, its motion is continued by inertia alone, and consequently the water is pushed down still further. Then again, on the ascent of the liquid the elevation of the jet, or the volume discharged, will be increased if the inertia of the rising wave be suffered to expend itself without interference by an untimely movement of the instrument.

A reciprocating rectilinear movement might be given to spouting tubes by a spring-pole, as in the *canne hydraulique*. The movement, however, should be regulated by that of the wave. This might be accomplished in large tubes by connecting to the moving apparatus a heavy pendulum, whose length could be increased or diminished according to that of the stroke.

If a tank or reservoir be not sufficiently deep for the employment of these tubes, an opening of the proper size and depth might be made at the corner or side in which to work them. When water is to be discharged on a level with the orifice, the upper part of the tube should slide through another fixed to and standing above the bottom of the receiving cistern, that the liquid when once raised may not run back; and, for the same reason, the tube should be inclined. Among other uses to which they are applicable is that of occasionally watering or washing trees and plants. In public gardens and other places, where a jet d'eau cannot otherwise be conveniently obtained, these instruments might be placed in a reservoir and moved by concealed mechanism, so as to produce one; and although it would consist of a succession of jets, the movements might be so regulated that they would appear but one. The motion of the tube itself might also be hid, by making it play in the interior of a fixed one, above whose orifice it need not protrude. In this manner the air in factories, hospitals, and rooms of private dwellings, might be kept cool, and, by perfuming the water, rendered very agreeable and refreshing in sultry weather. In fact, at every place where a fountain is desirable, a vase and spouting tube might be used.

The experiments we have given are very imperfect, but they may serve to excite those persons who have leisure and opportunity to pursue the subject. This mode of raising water is deserving of a rigid investigation, and will amply repay all the labor expended upon it.

There is a natural illustration of spouting tubes in the *Souffleur*, or Blower, on the south side of the Mauritius. The action of the waves has undermined some rocks that run out into the sea from the main land, and has worn two passages that open vertically upwards. They are represented " as smooth and cylindrical [conical ?] as if cut by a chisel." When a heavy sea rolls in, it fills in an instant the caverns underneath, and finding no other egress, a part is forced up the tubes to an elevation of sixty feet. The moment the waves recede, the vacuum left by them causes the wind to rush into the apertures with a noise that is heard at a considerable distance. See a description of this phenomenon in the Saturday Magazine, vol. vi. p. 77.

CHAPTER V.

Nature's devices for raising water—Their influence—More common than other natural operations—The globe a self-moving hydraulic engine—Streams flowing on its surface—Others ejected from its bowels—Subterranean cisterns, tubes and siphons—Intermitting springs—Natural rams and pressure engines—Eruption of water on the coast of Italy—Water raised in vapor—Clouds—Water raised by steam—Geysers—Earthquakes—Vegetation—Advantages of studying it—Erroneous views of future happiness—Circulation of sap—This fluid wonderfully varied in its effects and movements—Pitcher plant and Peruvian canes—Trees of Australia—Endosmosis—Waterspouts—Ascent of liquids by capillary attraction—Tenacity and other properties of liquids—Ascent of liquids up inclined planes—Liquid drops—Their uniform diffusion when not counteracted by gravity—Their form and size—Soft and hard soldering—Ascent of water in capillary tubes limited only by its volume—Cohesion of liquids—Ascent of water through sand and rags—Rise of oil in lamp wicks and through the pores of boxwood

BEFORE taking leave of artificial machines for raising water, a few of the most prominent of those which nature employs may be noticed; for, after all, the best of human contrivances are but imitations of hers.

The extent to which raising of water is carried by nature is wonderful. Persons who have not reflected on the subject would hardly suspect the influence which this operation exerts on our globe; yet it is one which the Creator has adopted to bring about results upon which the happiness of all things living depend. To the elevation of water into the atmosphere, and its return to the earth, the formation of continents and islands, lakes, rivers, fountains, valleys, plains, gravel, sand, mould, &c. are due. The fertility of soil, growth of vegetables, and life of animals, are also to be attributed in a greater or less degree to the same source.

Of nature's machinery, devices to raise, diffuse and collect water are the most *common*. They pervade all her works—the most magnificent and the most minute: and if we turn our thoughts to the world at large and contemplate it as a whole, we find it performing the part of an immense hydraulic engine, one which never stops working, and whose energy never flags. In almost every point of view this feature is obvious. In its exterior our planet is rather aqueous than terrene. Three-fourths of its surface are sunk into basins and scooped into channels for the reception and transmission of water; more than one-half is occupied by the ocean, the principal reservoir; while the other half is intersected in every direction by lakes, rivers, and rivulets innumerable, that convey the dispersed liquid back to the sea. The *motion* imparted to water also exhibits every degree of activity and agitation, from overwhelming torrents and mountainous waves, to the gentle shower that descends as if dropt through the finest cullender, and the placid stream that glides imperceptibly by. Sometimes we behold it running with the speed of a race horse, roaring among rapids, leaping over precipices and darting down cataracts—here dashed into spray, there churned into foam; now winding in eddies and gyrating in whirlpools; passing through channels whose paths are tortuous as those of a serpent, and shooting through others straight as an arrow.

Open channels and reservoirs constitute, however, but a part of nature's hydraulic machinery. In the *interior* of the earth, are close and air-tight reservoirs, and tubes of every imaginable size and figure, and of inconceivable strength. These receive and transmit liquid columns whose hydrostatic pressure would shiver the strongest conduits made by man, while the volumes of water that play within and pass through them render utterly insignificant all the products of artificial engines. We know that rivers sometimes discharge themselves into subterraneous tubes, which, transporting the fluid to a distance, again vomits it up. In this manner water is often conveyed to places where its appearance is difficult to account for, because of the level of all the neighboring regions being far below the aperture of discharge—this being sometimes on the summit of mountains, and often at their sides.

But the transmission of water from one level to another through pipes, is one of the simplest operations in natural as it is in artificial hydraulics. The flexure of the tubes fabricated by nature convert some of them into siphons, and these often decant the contents of caverns in which water slowly accumulates. The liquid rises till it flows over the highest bend in the tube, and the siphon being thus charged continues in operation, like one of ours, until the reservoir that supplies it be emptied, or the contents reduced to a level with the external orifice of the discharging leg. The action then ceases until the cavern be again filled and the operation renewed. Hence intermitting springs, and some of those that ebb and flow.

Natural machines analogous to water-rams, pressure engines, and fountains of compression are doubtless also in operation in the bowels of the earth. In the intricate and infinitely variegated chasms and fissures through which water is falling and gases collecting, the principles of these machines must necessarily be often excited, and on scales of magnitude calculated to strike us with awe. It is not improbable that some of those horrible eruptions mentioned in history and others that have occurred at sea without human witnesses were effected by machinery of this description. The subaqueous eruption which occurred on the south-west coast of Italy, in 1831, was probably an example. A column of water, 800 yards in circumference, was forced to an elevation of sixty feet, and an island formed of the solid materials displaced.

But natural devices are not confined to such as raise liquids by the momentum they acquire in flowing through tubes, or oscillating in waves, nor by the hydrostatic pressure of one volume transmitted by means of airs to another. There are some in which water is raised by solar heat. The liquid is converted into steam or vapor, in which state it is rendered lighter than air, and consequently ascends. This may be considered as nature's favorite plan. It is in operation everywhere, and always. By it water is drawn from every part of the earth's surface—both sea and land, and by it oceans of the liquid are kept suspended above us in the form of clouds, until it again returns in showers of rain and drifts of hail and snow. Of the quantity thus elevated, we may form some rude idea from the calculations of Halley respecting that drawn daily from the surface of the Mediterranean, viz. between five and six millions of tons ! a result which he deduced from experiments. Every person knows that canals require an extra supply of water to meet the expenses of evaporation. By experiments on the canal of Languedoc in France, the annual quantity thus borne off was found to be nearly three feet in depth over its whole area. Clouds of vapor or steam are often observed hanging over marshy ground, until the wind rises and bears them away. In hot seasons copious steams may be seen ascending just after a shower; but in general aqueous vapor thus generated, is invisible as it is impalpable. In clear weather, we are not sensible of its presence or of its movements.

We literally live in it, as in the spray of a fountain, but our perceptions are too gross to detect it.

How simple is this mode of raising water, and yet how effective! How silently does it work, and yet how sure! In its liquid state, water is too heavy to be suspended in the firmament; hence the Creator has made this provision to attenuate its particles by heat. It then rises upwards of its own accord—neither wheels nor cranks, pumps, pistons, pipes, nor even power is required to send them up, or to keep them there; and yet billions of tons are rising every hour, and accumulating in masses so great as to baffle language to describe or thought to grasp. And, what is equally remarkable, neither cisterns are required to contain, nor conduits through which to convey them. The phenomenon teaches us how a heavier fluid may be suspended in a lighter one, and that the proposition of water being 800 times heavier than air, is only conditionally true—depending merely upon the state in which those fluids are ordinarily exhibited to us. To increase our admiration, the salt water of the ocean is during the process of elevation distilled into fresh, thus furnishing among other suggestions that by which navigators have often adopted to sustain life in the extremities of thirst.

Water is also continually being converted into vapor and urged into the atmosphere by subterranean heat. Our planet may be considered, as indeed it was by the ancients, as a cauldron, in which steam is generated by those fires whose flues are volcanos. Oceans of the liquid are incessantly but silently thrown up from this cause. But, as might be expected, from the intricate arrangement of internal chambers and channels of communication, steam must often accumulate in cavities until its elasticity drives up the water that seals the passage to the surface. Hence boiling and thermal springs, and hence also the hot spouting springs of Iceland. According to Olafsen, a Danish traveler, one of the Geysers exhibited a jet at one time 19 feet in diameter and 360 feet high!

Modern authors explain the phenomenon of earthquakes by the accumulation of steam in the bowels of the earth. Plutarch says the Stoic philosophers did the same; but long before Zeno appeared the opinion prevailed, and caused the epithet "shaker of the earth" to be given to *Neptune*. The mechanical as well as chemical operations going on within the earth, are wonderful in their nature and terrible in extent. Well might mythologists locate the workshops of the gods there, and place the forges of Vulcan and the Cyclops at the base of volcanos.

Of contrivances for raising liquids, as developed in the organization of animals, we took some notice in the second and third books. Most if not all of them may be considered modifications of bellows and piston pumps. In the vegetable kingdom, other devices, and such as are based on other principles, are in active operation. This portion of creation exhibits in a striking light the important part which devices for raising water perform in the constitution of our globe. Every tree and every plant, from the towering cedar of Lebanon, to the hyssop that springeth out of the wall, from the wide-spreading banyan to a wheaten straw or melon vine, is a natural pump, through whose tubes water is drawn from the earth or imbibed from the air.

There is something exceedingly pleasing and sublime in the contemplation of the growth of vegetables, the germination of seeds, appearance of sprouts, development of stems, branches, leaves, buds, blossoms, flowers, and fruits—their variegated forms, dimensions, movements, colors, and odors. Some persons who have never turned their attention to this subject till the evening of their days, have been astonished at the wonders which burst on their view. A new state of existence seemed to open upon them. Their perception and estimate of things were changed. Instead of considering the world as calculated only for what man too generally makes it—a scene for the display and gratification of the most groveling and sordid passions, they find it a theatre crowded with enchanting specimens of the Creator's skill, the study of which imparts the sweetest pleasure, and the knowledge of which constitutes the greatest wealth.

Those pious but mistaken people, who incessantly murmur against the world, and long to depart from " this howling wilderness," as they are pleased to term it, reproach their Maker by reviling his work. They are waiting for future displays of his glory, and neglect those ravishing ones by which they are surrounded, forgetting that " the whole earth is full of his glory"—looking for sources of pleasure to come, and closing their eyes on those before them—thirsting for the waters of heaven, and despising the living fountains which the Father of all intellects has opened for them on earth. They seem to think happiness hereafter will not depend upon knowledge, or that knowledge will be acquired without effort—a kind of passive enjoyment, independent of the exercise of their intellectual or spiritual energies. But they have no ground to hope for any such thing. Reasoning from analogy and the nature of mind, the happiness of spirits must consist in being imbued with a love of nature—in contemplating the wisdom and other attributes of the Deity, as they are unfolded in the works of creation. In what else can it consist? It is not probable that human or finite beings of any class can ever know God except through the medium of his works.

It is admitted that the study of nature is a source of exquisite pleasure to intelligent beings, and the most refined one too that the mind can conceive: it is also one that can never be exhausted. Those persons, therefore, who take no pleasure in examining the works of creation here, are little prepared to enter upon more extensive and scrutinizing views of them in other worlds. If they have no relish for an acquaintance with the Creator's works while they live, they have no right to expect new tastes for them after death. The works of God are all *perfect;* those in this world as well as those in others; and he that can look with apathy on a tulip or a rose, a passion flower or a lily, or any other production of a flower garden or a forest, has not begun to live. Besides, we are not sure that other worlds possess more captivating or more ennobling subjects for contemplation and research—more thrilling proofs of the wisdom and beneficence of God.

The circulation of *sap* (sometimes called the *blood* of plants) is one of the most interesting of natural phenomena. It is connected with some of the most delightful feelings of our nature, and with the activity and joys of the brute creation. When in spring its action commences, a sensation of buoyancy pervades all organized beings. The earth begins to put on her richest attire—her inhabitants rejoice in her approaching splendor, and exult in view of the feasts preparing for them. On the other hand, when in autumn her freshness fades and her glory withers, all feel the change. How infinitely varied are the effects of sap and the energy of its movements! Rushing to the summit of the tallest trees, and lingering in the grass of our meadows—shooting up perpendicularly in the poplar and pine, horizontally in the branches of the baobab and oak, and descending in those of the Indian fig-tree and willow. In some plants, accumulating chiefly in their roots, as in the turnip, radish, and potato, and emerging above ground in cucumbers and melons—ascending higher in the bushes of currants and gooseberries, and ranging over those in apple and pear trees. By what wonderful process is sap distilled into liquid honey in the maple, and into wine in the grape? How is it elaborated into fruits of every flavor, and exhaled in perfumes from sweet scented herbs, and in what manner does it contribute to produce every imaginable color and tint in flowers?

By what means does sap form a natural vase in the *pitcher plant*, and then enter it as limpid water, along with rain and dew? This singular production of the vegetable kingdom collects water from the earth and atmosphere in vessels of the same consistence and color as the leaves. Each pitcher is strengthened by a hoop, and furnished with a cover or lid that turns on a fibrous hinge. When dew or rain falls, this cover opens; and as soon as the weather clears, it closes and prevents the water that entered from being wasted by evaporation. There are other plants which store up water much in the same way. Such were the reeds that relieved Alvarado (one of the conquerors of Peru) and his companions from perishing of thirst. Garcilasso, in his Commentaries observes, " The information they had of the water was from the people of the country, who guided them to the canes, some of which contained *six gallons*, and some more."

We know that the juices of plants cannot be raised without force, and that this force must be increased with the elevation to which the liquid is to be lifted. Animals exert a muscular power in working the pumps formed in their bodies, and these machines they put in motion at will. This is not the case with vegetables: yet sap, the pabulum of their life, is elevated to the tops of the highest trees, and apparently with the same facility as it is diffused through microscopic plants. That the force by which this is done is not latent or negative in its nature, is clear, since it may easily be rendered manifest. Cut a branch from a vine in the spring when the sap is rising, and stretch a piece of india rubber over the end of the part that remains, secure it by thread wound round the stump, so as to exclude the air and prevent the wound from healing. In a little while the caoutchouc will be swelled or bulged out by the exuding fluid, and it will continue to swell, however thick it may be, till it burst. A few years ago we treated in this way some branches of an Isabella grape vine, and afterwards applied to one of them a close vessel containing mercury, in which the lower end of a long glass tube was immersed with a view to measure the force excited. In four days the mercury rose 36 inches in the tube, being pushed up by the sap which took its place in the vessel; and but for an accident, by which the apparatus was broken, it would probably have ascended still higher.

But this force, great as it was, is small when compared with that which sends the fluid through trees that grow on the Australian continent and islands. Some of these resemble single tubes, and are filled with a semifluid or soft pith. Tasman, the discoverer of Van Dieman's Land, found trees there whose lowest branches were between 60 and 70 feet above the ground. The French expedition sent in search of the lamented Perouse, found on Cocos island a tree nearly 100 feet high, and only *three inches* in diameter. It was of so hard a texture, that it resisted at first the heaviest blows of an axe; and when the pith was taken out, the thickness of the wood did not exceed $\frac{4}{10}$ of an inch—forming a perfect tube. But this tree was only half the height of some others in the same regions; for several were seen whose diameters were only seven or eight inches, and whose tops towered upwards of 200 feet above the earth! The force that drives sap to such elevations is wonderful indeed; and could it be applied as a mechanical agent, it would be resistless as steam. It might be supposed that a force so energetic—one that would rupture pipes which convey water to our dwellings—would rend asunder most of the delicate pores through which it circulates; and so it would were not their diameter so exceedingly small—for the strength of tubes increases as their bore is diminished.

The ascent of sap has been explained by *Endosmosis*, or transit of bodies through pores. See two interesting papers on this subject in the Journal of the Franklin Institute, vols. xvii and xviii, by J. W. Draper, now Prof. of Chemistry in the New-York University.

Water Spouts constitute a peculiar class of nature's contrivances for raising water. Electricity is supposed to have a controlling influence in their formation; but the mode by which it acts is not clearly understood. More water is *drawn* up by them within the same space of time than by any other natural device. The liquid appears to be borne up the vortex mechanically as solid substances are raised by whirlwinds, except that it is broken by masses of air rushing into and mixing with it. After arriving at the top of the spout, it is dispersed by lateral currents of wind. A drop of water suspended from the conductor of an electrifying machine is supposed to exhibit a miniature water spout. When a vessel of water is placed under it, and the machine put in operation, the drop assumes the various appearances of a spout in its rise, form, and mode of disappearance. Clouds act as cisterns in holding water raised by evaporation; but in water spouts they perform a more singular part, since they are moulded into visible pipes, through which volumes of liquid are conveyed as securely as through those made of solid materials.

Although the rise of sap in trees is attributed to endosmosis, there is reason to believe that *capillary attraction* takes part in the process, as well as in a thousand other operations of nature. When one end of a small glass tube is placed in water, the liquid rises within it; and the height to which it ascends in different tubes, is inversely as their diameters. The phenomenon is more or less common to all liquids when the tubes dipped in them are made of such materials as they readily unite with. This condition is necessary, otherwise the liquid would be depressed. Water rises higher than other liquids in glass tubes; and as these instruments are transparent, they are always adopted in experiments on this subject.

The phenomenon of capillarity has exercised the ingenuity and learning of the most eminent philosophers, and various are the causes to which they have attributed it. Some supposed the atmospheric pressure less within the tubes than without. Others imagined an unknown fluid circulating through them that bore the liquid up; and some ascribed it to moisture on the inside of the tubes. An attractive force existing between the glass and the water is now more generally admitted; and hence in tubes of very small bore, it is said, the glass being nearer the water, attracts it more powerfully, i. e. raises it higher—other writers think the effect is due to electricity. The subject is admitted to be an intricate one, and the manner in which it has been handled by scientific men, has not rendered it very accessible to ordinary readers. Without looking for ultimate causes, the phenomenon, like that of an increased discharge, through diverging ajutages, may be traced to the relative properties of the liquid and the material of the tube, and to the force with which particles of liquids cohere among themselves.

Capillary attraction is exhibited in a great variety of forms. Particles of water, like those of all other liquids, require some force to separate them. A needle or film of lead while dry, will float; and myriads of

Chap. 5.]   *Forms of Drops.*   511

gnats career on the surface of a pond as securely as on land. Some liquids are viscid, and may be drawn into threads; and even water may be stretched into sheets ere its substance be broken: bubbles produced during rains, and those pellicles sometimes formed over the mouths of small vials and the interstices of sieves are examples. Water, moreover, in common with other fluids, unites with some substances more readily than with others. It does not combine with oils, nor adhere to substances impregnated with grease. Hence umbrellas and water-proof dresses are made of oiled silk; and rain rolls off the backs of ducks and other aquatic birds without wetting them, because these fowls dress their feathers with an unctuous fluid which their bodies secrete.

When a vessel contains a liquid that readily unites with it, the liquid stands highest at the edges. Thus in cups of tea or tumblers of water, the fluid climbs up against the sides until it is considerably elevated above the general level. This is observable with milk in a pot, pitch in a cauldron, oil in cans, mercury in vessels lined with an amalgam; melted tin in tinned iron or copper vessels, and fused brass in an iron ladle whose interior has been coated with the alloy, as in the process of hard soldering. If, on the other hand, a liquid has no affinity for, or will not unite with the substance of which the vessel is made, an effect the reverse is produced; that is, the liquid is depressed at the sides, as when mercury is contained in a vessel of glass, wood, or earthen ware; or even in one of metal not lined with an amalgam, or with which the mercury cannot form one. The same thing occurs to fused brass, or lead or tin in crucibles, to water in greasy tubes or dishes, &c.

The same thing, in another form, occurs with *drops* of liquid. When water is sprinkled on a greasy surface, the particles remain separate however near to each other. By blowing against them, they may be rolled over the plate on which they rest without leaving any portion behind; but if the substance on which they are dropped combine readily with moisture their figure is changed; each becomes flattened by spreading, so that two adjacent drops quickly run together. A drop of oil or speck of grease makes a large stain on a lady's dress or a marble table. Quicksilver will not unite with marble, but a small portion dropped on a sheet of tin will spread over it like water on damp paper. A portion of tinmen's solder kept in fusion on clean plates of tin or lead spreads, and is absorbed in like manner. When ink is spilt upon unsized paper, the latter is stained to a considerable extent: round each drop a broad ring of moisture is formed; the darker and grosser particles remaining as a nucleus in the centre.

The different *forms* which drops assume when pendent from solid bodies, are governed by the parts with which they are in contact. When water is sprinkled on a plate partly covered with grease, those particles that fall on the clean parts resemble very flat segments of spheres, while those on the greased parts are larger portions of smaller spheres; the liquid in these swelling out above the base on which they rest, in preference to extending itself like the others upon it. A drop hanging from the point of a wire is elongated vertically—if held between the finger and thumb, it may be stretched out horizontally. If suspended in a ring, its upper surface becomes hollow and its lower one convex, forming a species of liquid cup, and supported somewhat like the dishes which chemists hang over lamps in moveable rings of brass. A drop of liquid in a capillary tube is thus supported; the tube being nothing more than a deep ring.

The *quantity* of liquid contained in pendent drops varies with the extent of surface in contact with the supporting body. When one is ready to fall from an inclined object, as the bottom of a bucket or a tea cup, it may be retained by making the bottom approach nearer to a level; the fluid then spreads and holds by a larger surface. This is illustrated in the case of metals: tin-plate workers commonly take up solder on the face of their irons. The under sides of these instruments are tinned, and being placed upon the metal, a larger or smaller portion is melted and borne off at pleasure. An equal quantity of water may probably be thus suspended from a plane surface, as within a cylinder of the same area.

Numerous facts show, that when not pulled down by gravity, liquids diffuse themselves uniformly on substances with which they combine—as much upwards as downwards. Small drops of water or ink dashed against vertical sheets of paper equally extend themselves from the centre. We are so much in the habit of contemplating fluids in masses, where gravitation greatly preponderates, that we overlook this property in them, or do not suspect its existence. The observation that water never runs "up hill" is proverbial, but it is not correct. Examples might be quoted, in which it prefers to ascend an inclined plane to going down one—to rise in a wet channel, than descend in a dry one. Take a dry piece of glass, or china, the blade of a knife, or the bottom of a saucer, or almost any solid material, and dampen or slightly wet a part of it: place a drop of ink or water near the edge of the wetted part, then incline the saucer so that the drop may be beneath, and make a channel of communication between them, by drawing with a pointed instrument a small streak of fluid from one to the other. The instant this is done, a current will set up with considerable velocity from the drop into the thin sheet above.

This effect takes place on wood and on metals, and even paper. Penmen, who have their paper inclined towards them often witness the experiment in another form, especially when they make the bottom of their strokes thicker than the rest. The ink may then be seen to ascend from the bottom upon the removal of the pen. This takes place if the paper be held vertically. Again, when a large drop of ink falls on a book, it is customary to shake out that which remains in the pen, and to place the latter over the drop as in the act of writing; upon which a large portion of the liquid enters the quill. This is then shaken, and the operation renewed. Here the principle of distribution again appears. There is a surplus below, and a deficiency (or less depth of it) above, and the liquid ascends to produce an equilibrium. Were the pen fully charged with ink before applied to the drop, it could take none from the latter.

Other examples of the ascent of liquids, and even of solids against gravity are familiar to some classes of mechanics, but not to all. When two sheets of tin plate are soldered together in an inclined position, small pieces of solder laid near the lower edge of the joint are drawn up under the face of the iron as soon as the fused mass touches them. Illustrations of this occur in whatever position the joint may be. They are still more common in *hard soldering*, for copper and silversmiths commonly charge their joints on the outside, so that the solder is below or next to the fire when fused.

These experiments are all based on the same principles as the ascent of water in capillary tubes. We see that when a mass of liquid (wholly resting on a plane surface or enclosed in a cylinder) is connected by a short channel to a thin sheet of the same substance above, a part of the mass below will ascend. The channel it should be remembered is a fluid one, for neither water nor any other liquid will thus rise except in channels of the same substance as themselves. The effect does not therefore appear to be due wholly to the material that sustains the liquid, but, to some extent, to that force by which particles of matter congregate with their kind in preference to mingling with others. The aqueous vapor floating in the atmosphere moistens more or less the surfaces of all bodies. Glass tubes are coated with it; but if a capillary tube previous to its use was not thus prepared, it becomes so the instant one end is immersed in water—a stream of vapor (though not obvious to sight) then passes through it: the whole interior is thus coated with aqueous moleculæ accumulating upon it at insensible distances from each other, and those adjacent to the surface of the liquid operate to solicit its ascent through the channel thus prepared for it. The ascent of vapor under these circumstances is unlimited, but that of a liquid column is soon arrested. This however does not prove that the force excited is insufficient to raise liquids to great elevations, but that it is the volume which determines the height. If the quantity be indefinitely small it will be raised indefinitely high. Experiments so far as they have been made prove this; but as the finest of artificial tubes are, when compared to nature's, as a mast is to a needle or a cable to a thread, the ascent of liquids in them must necessarily be very limited. As long as the liquid column can be sustained by adhesion to the sides of a tube it will rise, but when the weight of the central parts (which not being attached to the tube are sustained by cohesion alone) exceeds this force, the ascent ceases.

The force with which particles of some fluids cohere is so energetic that they present the singular spectacle of liquid rods, pendent like icicles or stalactites. When one of these rods is broken an interesting contest between gravitation and cohesion takes place, during which the figure of the pendent changes as one or the other of those forces prevails: it becomes longer while the first predominates, shorter when the latter controls, and stationary when both are balanced. These phenomena may be observed by letting a drop of molasses fall from the point of a knife or a spoon. The globule descends to a considerable distance before it is wholly separated from the portion above, because a rod of the liquid continues to be formed that unites them. When this rod breaks, the part suspended from the mass above is drawn up: a thread over a foot in length is sometimes thus contracted to less than $\frac{1}{4}$ of an inch, strongly reminding one of the elasticity of caoutchouc.

Water rises to considerable heights through sand and other porous bodies—also through rags and threads of cotton, &c. Oil ascends in the wicks of lamps. Capillary siphons formed of cotton wick are employed to supply oil to the journals and working parts of machinery. It is customary with stereotype founders to oil the faces of engraved wooden blocks previous to taking casts from them. These blocks are of box, a species of wood whose texture is exceedingly close. We have often placed some of those used in the illustration of this work on receiving them from the engraver, into a dish containing oil to the depth of $\frac{1}{4}$ inch, and have witnessed the appearance of the liquid at the top within half a minute, and frequently in a quarter of one. Unlike water in glass tubes, the oil here rises entirely out of the tubes in the wood and collects in globules over the orifices.

From the infinite variety and importance of devices for raising liquids that are at work in the animal and vegetable kingdoms and in general nature, the wisdom displayed in their formation and movements, and their wonderful effects, it would seem as if the Creator designed particularly to call man's attention to this department of knowledge, and to induce him to cultivate it. Sources of hydraulic contrivances and of mechanical movements are endless in nature; and if machinists would but study in her school, she would lead them to the adoption of the best principles, and the most suitable modifications of them in every possible contingency.

CHAPTER VI.

SIPHONS—Mode of charging them—Principle on which their action depends—Cohesion of liquids—Siphons act in vacuo—Variety of siphons—Their antiquity—Of Eastern origin—Portrayed in the tombs at Thebes—Mixed wines—Siphons in ancient Egyptian kitchens—Probably used at the feast at Cana—Their application by old jugglers—Siphons from Heron's Spiritalia—Tricks with liquids of different specific gravities—Fresh water dipped from the surface of the sea—Figures of Tantalus' cups—Tricks of old publicans—Magic pitcher—Goblet for unwelcome visiters—Tartar necromancy with cups—Roman baths—Siphons used by the ancients for tasting wine—Siphons, A.D. 1511—Figures of modern siphons—Sucking tube—Valve siphon—Tin plate—Wirtemburg siphon—Argand's siphon—Chemists' siphons—Siphons by the author—Water conveyed over extensive grounds by siphons—Limit of the application of siphons known to ancient Plumbers—Error of Porta and other writers respecting siphons—Decaus-Siphons for discharging liquids at the bend—Ram siphon.

THE *siphon*, or as it is sometimes named the *crane*, is in its simplest form merely a tube bent so as to resemble an inverted letter U or V; and is employed to transfer liquids from one level to a lower one, in circumstances where natural or artificial obstructions prevent a straight pipe from being used; as when rocks or rising grounds intervene between a spring and the place where the water is required, or when the contents of casks and other vessels are to be withdrawn without making openings for the purpose in their bottom or sides. Thus farmers occasionally have water conveyed over hills to supply their barn-yards and dwellings; and portable siphons are in constant requisition with oil and liquor merchants, chemists and distillers. The two branches of a tube that constitute a siphon are commonly of unequal lengths, and named *legs;* the "short" or receiving leg, and the "long" or discharging one. The highest part where the legs are united is known as the apex or bend.

As liquids are raised in siphons by atmospheric pressure, the perpendicular length of the short leg, like the suction pipe of a pump, should never exceed 25 or 28 feet. To put siphons in operation, the air within them must be first expelled. Small ones are sometimes inverted and filled with a portion of the fluid to be decanted, but more frequently the liquid is drawn through the tube by sucking. Other devices for charging them will be noticed farther on.

The action of a siphon does not depend upon any inequality of atmospheric pressure, as some writers on natural philosophy have inadvertently intimated. In one popular work, it is said, " the pressure of the air is more diminished;" and in another, more " weakened or abated" over the discharging than over the receiving orifice; whereas, philosophically speaking, the reverse is the fact: for as the discharging end is nearer the earth, a deeper and consequently heavier column of atmosphere rests over it than over the other. Nor does the effect depend upon any difference in the actual lengths of the legs, for they are often in this respect the same; and sometimes the receiving one is much longer (in an oblique direction) than the other—not yet does their comparative diameters contribute to the results; for the short one may be much more capacious than the long one. It is the difference in the *perpendicular* length of the *liquid columns* within the legs that causes a siphon to act: the column in the discharging leg must exceed in this respect that contained in the receiving one, or no action can take place. By examining the figure in the margin, it will be perceived that the column in the receiving leg extends only from the surface of the liquid in the vessel to the bend, whereas in the other it extends from the bend to the orifice. As the pressure of fluids is as their depth without regard to their volume, the hydrostatic equilibrium of the two columns is destroyed, when the longer one necessarily preponderates, upon which the vacuity left in the upper part of the tube is filled, by the atmosphere driving fresh portions up the other leg.

No. 234.

But siphons could not act at all were it not for that property of fluids by which their particles cling to each other. The tenacity of liquids may be considered like that of solids, only less intense; and thus it is when water flows through a siphon, the descending particles actually drag down those above them, somewhat like a chain or rope unequally suspended over a pulley, when the longer end pulls the shorter one after it. A siphon is in fact a contrivance by which liquid chains or ropes are thus made to act. But for the cohesion of liquids the contents of the discharging leg would drop out like sand, and no further effect would follow—the rope would be broken, and the separated parts fall asunder. The influence of cohesion in the action of siphons is proved by the fact that very short ones continue to operate when removed into a vacuum.[a]

The tenacity, or what might almost be called the malleability of liquids, is beautifully exemplified in soap bubbles. These yield to impressions without breaking. They fall on and rebound from the floor like bladders or balls of india rubber. They shake in the wind, and their figures become altered like that of balloons tossed to and fro in the air: all this they often endure before their shells are broken by evaporation.

Siphons are exceedingly diversified in their forms, materials and uses. They are made of cylindrical and other shaped tubes, and both of uniform and irregular bore. The legs of some are parallel, while in others they meet at every angle—sometimes straight and often crooked—one may be larger than the other, or both may be alike; they also may be separate, one loosely slipping into or over the other. Instead of tubes siphons are sometimes formed by an arrangement of plates, and also by a combination of vases. This plastic property has occasioned their concealment in more various forms than Proteus ever assumed. Siphons are made of tin, copper, iron, silver, glass, lead, earthenware, leather, wood, canes; and (capillary ones) of paper, strips of cloth, threads of cotton, &c. Examples of their various forms and applications will be found noticed in the following historical sketch.

The origin of siphons like that of pumps is lost in antiquity. Some

---

[a] For information on the action of siphons in vacuo, see Boyle's Works, by Shaw, vol. ii. 446. History and Memoirs of the French Academy, translated by Martin and Chambers, vol. iv. 374; and Desaguliers Exper. Philos. vol. ii. 168.

writers of the last century attributed them to Ctesibius, (see page 268,) because they were used in some of his water-clocks, and no earlier application of them was then known. For the same reason the invention of toothed wheels has been erroneously ascribed to him. All the information extant respecting the ancient nations of the East is exceedingly limited, while of their arts and details of their mechanism we know next to nothing. The greater part of our ordinary machines cannot be traced to a higher source than Greece, but Greece itself was colonized by Egyptians; and however much the children of Cadmus may have refined on some departments of the useful arts, the general mechanism of their ancestors is believed to have passed through their hands to those of the Romans, and from the latter to us with little alteration. This was certainly the case with their hydraulic and hydro-pneumatic devices. The siphon is an example. The *name* of this instrument is taken from a Greek word, which signifies simply a tube; but it has been ascertained that the word is of a remoter—of an oriental origin, being derived from *siph* or *sif*, to imbibe or draw up with the breath, and whence comes our expression to *sip*. Now if it can be proved that the siphon was in use, and was charged by sucking before the times of Grecian history, we may safely conclude that a more ancient people furnished the Greeks with both the instrument and its name.

The researches of Rosellini and Wilkinson have settled this point. These gentlemen have brought to light irresistible evidence that siphons were used in Egypt at least as early as 1450 years before Christ. In a tomb at Thebes, which bears the name of Amunoph II, who reigned at the period just named, they are delineated, and in a manner too distinct to admit of any doubts. See No. 235. Several jars are represented upon a frame or stand. Into three of them siphons are inserted; two apparently in operation, and a man is in the act of charging the other by sucking: the contents of the jars being transferred into a large vase supported upon an ornamental stand.

No. 235. Egyptian siphons, 1450, B. C.   No. 236.

Mr. Wilkinson supposes that siphons were invented in Egypt, and were used to decant the Nile water from one vessel to another. He says it is necessary to let this water stand for some time before being used, that the mud suspended in it may settle to the bottom. On this account vases containing it cannot be moved without rendering it again turbid, and the same effect is produced by dipping; hence the use of siphons. The conjecture may be correct, but it does not derive much support from the use of those instruments figured at No. 235; for unless there was some contrivance to prevent the ends of the siphons from going too near the bottom of the jars, scarcely any thing would more effectually disturb and draw off the sediment with the water. The tubes were obviously of some flexible material, and from the manner in which they are held, it would be impossible for the person using them to regulate by hand the depth to which the short legs were immersed. Moreover, another individual (omitted in our illustration) is represented pouring a liquid into one of the jars, an operation that would effectually disturb the sediment.

Instead of water, jars so small probably contained *wines*, and the artist designed to exhibit the mode of *mixing* them; a common practice of old, and one referred to in several parts of the scriptures. The Egyptians were much given to luxurious living, and especially with regard to wine, a fact which the sculptures corroborate, for scenes of gross excess, and in females too, are portrayed. The Jews we know carried with them into Palestine not only the arts but many of the worst habits of the Egyptians, and the excessive indulgence of mixed wines was one. " Woe unto them that are mighty to drink wine, and men of strength to mingle strong drink." Isaiah v, 22. " She hath mingled her wine, she hath also furnished her table." Prov. ix, 2. " They that tarry long at the wine, they that go to seek mixed wine." Ibid xxiii, 30.

Other examples of the early use of siphons are met with. In the tomb of Remeses III. who flourished 1235 B. C., is a representation of an Egyptian kitchen, with the various operations of slaying animals, cutting up the joints and preparing them for cooking—kneading dough with the feet, and paste with the hands—making cakes and confectionary, &c.— Of kitchen furniture, there are tables, jars, plates, cauldrons, bellows, ovens, molds, pestle and mortar, knives, baskets, &c., and suspended on ropes or rods, *a number of siphons;* showing evidently that those instruments were in constant requisition. See No. 236. These were probably adapted for jars of certain depths, unlike those in the preceding figure, which seem to have been appropriated to different sized vessels, and their shape altered as occasions might require.

How singular that these philosophical instruments should have been more common before the siege of Troy than at the present day! And how precious are those monumental records that have preserved this and other facts of the kind!

The circumstance of siphons having been used in Egypt at so early a period may be deemed conclusive that other nations were not ignorant of them. With Egypt, all the famous people of antiquity maintained an intercourse; and enterprising men flocked from all parts to acquire a knowledge of the arts and sciences that were cultivated on the banks of the Nile. Their neighbours, the Jews, as a matter of course, were acquainted with siphons, and there is probably a reference to them in John ii. " Jesus saith unto them, fill the water pots with water. And they filled them to the brim. And he said unto them, *draw* out now and bear unto the governor of the feast, and they bare it." How did they *draw* this liquid ? Certainly not by inclining the jars and *pouring* it out; nor yet does it appear to have been done by *dipping:* for as the large pots were filled to the very brim, this would have caused the liquid to overflow. It is more reasonable to suppose that small siphons were used on the occasion, and that they were charged by sucking, as represented in No. 235. This and this only clearly accounts for the fact that those who drew the liquid were first aware, as they must have been, of the change it had undergone. This change does not seem to have affected the color, for not till he *tasted* of it was the presiding officer himself sensible of its being wine. " When the ruler of the feast had *tasted* the water that was made wine," he " knew not whence it was, but the servants that drew the water knew."

No. 235 probably was designed to represent one of Pharaoh's butlers engaged in that part of his duty which required him to draw and mix the king's drink. Such officers formed part of large establishments among the ancients, and so they do in modern times. Switzer, speaking of small siphons observes, " the insinuation of air is such that wine will not always keep on its regular ascent, without the *butler* puts his mouth sometimes to it, to give it a new suction."

One of the modes by which Ctesibius applied siphons to clepsydræ, will be found figured in a subsequent chapter.

Were the old philosophers of Egypt acquainted with the principle on which the siphon acts? Doubtless they were, else they could never have diversified its form and adapted it with such admirable ingenuity to the great variety of purposes both open and concealed, which we know they did. In connection with hydromancy it was made to play an important part. Magical goblets were often nothing else than modifications of siphons; and from the Spiritalia we learn that they formed the basis of more complex and imposing apparatus. The tricks connected with the glass tomb of Belus, and the miraculous vases in the temple of Bacchus probably depended upon siphons ; and most writers on the vocal statue of Memnon have introduced them as essential parts of the supposed machinery; imitating in this respect the apparatus described by Heron for producing mysterious sounds from the figures of men, birds, &c.

Heron is more diffuse on the subject of siphons than any other writer, Upwards of twenty problems in his Spiritalia relate to or are illustrrated by them; and from him we learn that these instruments were in his time employed on a large scale in draining and irrigating land, viz. by transferring water over hills from one valley to another. This use of the siphon was probably quite as common under the Pharaohs as under the Ptolemies; for Heron does not intimate that it was novel in his time any more than the instrument itself.

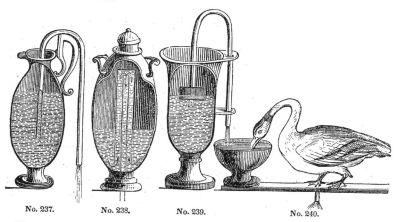

No. 237.   No. 238.   No. 239.   No. 240.

The above figures are illustrations of the first, second, third, and thirtieth problems of Heron's work.

No. 237 (the first figure in the Spiritalia) represents an ordinary siphon resting over the handle of a vase, within which the short leg is inserted. This instrument was charged by sucking, as the more ancient ones in the last cut.

No. 238 exhibits another form of the siphon, consisting of two straight and separate tubes, the smaller one of which is inserted through the bottom of the covered vase, and reaches as high within it as the liquid is required to stand. Over this tube another one is slipped whose upper end is closed air-tight. Hence it is obvious that when the liquid is higher than the orifice of the inner tube which forms the long leg, it will ascend between the tubes and continue to be discharged as in the common siphon, until the surface descends below the lower end of the outer tube, or short leg. Here the liquid is discharged from the bottom of the vessel, not over its rim as in the preceding figure. The siphon admits of a great variety of modifications, some of which in the hands of ancient jugglers contributed not a little to amaze the ignorant. The contents of the large vases, often permanently fixed in temples, could and doubtless often were secretly emptied by contrivances of this kind; the siphons of course being concealed in the ornaments, handles, or other adjuncts. The six vessels of wine placed daily in the temple of Bel, which the priests clandestinely emptied every night, might have been more neatly robbed of their contents by concealed siphons, than by entering through *a secret passage under the altar;* but as the abstraction of the more solid food which the priests pretended was consumed every day by the brazen deity, (forty sheep and twelve measures of floor,) required some contrivance like the latter; the vases were emptied at the same time. [Story of Bel and the Dragon.] The romantic account by Herodotus of the robbery of Rhampsinitus' treasury, shows to what extent the system of secret passages was carried, and the ingenuity with which they were made and concealed.

The velocity with which water flows from an ordinary siphon necessarily diminishes as the surface in the reservoir falls. In some cases a uniform discharge is desirable. No. 239 shows how ancient engineers accomplished this. A float or hollow dish was attached to the end of the short leg, so that the instrument descended with the water. The long leg was passed loosely through two openings in projecting pieces that preserved it in the proper position.

The difference in the specific gravity of liquids was a fruitful source of deception. Many capital tricks were based upon it, especially when the lighter fluids employed were of the same color as those on which they reposed. If for example, a vessel contained oil, wine and water, these liquids could be discharged by a siphon like No. 239 in the same order; and by secretly raising or lowering an ordinary one, or the moveable tube in No. 238, any one liquid could be drawn off. Fresh water being lighter than salt is often found some distance at sea; and sailors, like old jugglers, can draw up either, according to the depth to which their buckets are immersed. Four miles from the mouth of the Mississippi the fresh water is about two feet deep, and at ten miles it may be obtained by careful dipping.

In problem XXX of the Spiritalia, Heron shows how siphons may be concealed within the figures of oxen or other animals in the act of drinking; the orifice of the short leg being at the mouth, and that of the long one in one of the feet. See No. 240. When the bore of the siphon is properly adjusted to the quantity of water flowing into a basin, the animal will appear to drink the whole.

The following represent a number of Tantalus' cups, magic goblets, &c. In No. 241, the long leg of the siphon passes through the bottom of the vessel, and the short one remains above; so that when the liquid rises over the bend, it will be discharged by the siphon into the cavity below.

No. 241.   No. 242.   No. 243.   No. 244.   No. 245

Devices of this kind admit of numerous modifications by which the tube may be concealed. When it is enclosed within the figure of a man, (the water entering at one foot slightly raised, and passing out through the other,) the vessel is named a *Tantalus' cup*, and the liquid instead of entering the mouth, as in No. 240, only rises to the chin, and then runs away—illustrating the classical fable, which represents Tantalus suffering the tortures of thirst in the midst of water that reached to his lips, but which on his attempting to taste sunk below his reach; hence the origin of our word *tantalize*, and its relatives.

> Next, suff'ring grievous torments, I beheld
> Tantalus: in a pool he stood, his chin
> Wash'd by the wave; thirst parch'd he seem'd, but found
> Nought to assuage his thirst; for when he bow'd
> His hoary head, ardent to quaff, the flood
> Vanish'd absorb'd, and at his feet, adust
> The soil appear'd, dried instant, by the gods.
>
> Odys. xi. *Cowper.*

It is supposed the fable was intended to illustrate the influence of avarice, by which misers in the midst of plenty often deny themselves the necessaries and comforts of life.

Sometimes the *sides* and *bottom* of Tantalus' cups are made hollow and the siphon formed within them. No. 242 is one of these. An examination of it will sufficiently explain the construction. A small opening near the bottom (which may easily be concealed) communicates with a passage formed by a partition, above the top of which the liquid must rise before it can pass down the other side into the base of the cup.

In No. 243 the siphon is formed within the handle. The short leg communicates with the lower part of the cup at the swell, so as not easily to be detected, and the long one with the cavity formed below. The figure represents a Tantalus' cup in our possession.

A liquid is retained in one of these as in an ordinary goblet, so long as the surface does not reach above the highest part of the siphon; but if the cup be once inclined so as to set the latter in operation, the contents will gradually be transferred to the hollow base, and this whether the vessel be replaced in an upright position or not. Thus tankards have been so contrived that the act of applying them to the lips charged the siphon, and the liquid instead of entering the mouth then passed through an illegal passage into the cavity formed for its reception below. By making the capacity of the siphon sufficiently large, a person ignorant of the device would find it a difficult matter even to *taste* the contents however thirsty he might be. In the dark ages, simple people would naturally on such occasions give credit to legends respecting mischievous demons loving beer and taking these opportunities to get it. Dishonest publicans whose sign-boards announced "entertainment for man and beast," are said to have occasionally thus despoiled travelers of a portion of their ale or mead, as well as their horses of feed. Oats were put into a perforated manger, and a large part forced through the openings into a receptable below, by the movements of the hungry animal's mouth.

Martial the Roman poet refers to tricks of ancient publicans, and what will surprise some readers, he complains of having had wine foisted on him instead of water. Ravenna was originally built like Venice on piles, and was a sea-port, though now several miles inland. Water has always been extremely scarce at this city, and probably was more so formerly than at present. In the poet's time it seems to have brought a higher price than inferior kinds of wine. Hence his complaint:

> By a Ravenna vintner once betray'd,
> So much for wine and water mix'd I paid;
> But when I thought the purchas'd liquor mine,
> The rascal fobb'd me off with wine.   L. iii, Ep. 57.   *Addison.*

No. 244, a magical pitcher, from the eighth problem of the Spiritalia. The siphon is not employed, but the device is allied to the preceding ones. A horizontal partition or diaphragm perforated with minute holes divides the vessel into two parts. The handle is hollow and air-tight, and at the place where its lower end is connected to the pitcher, a tube proceeds from it and reaches nearly to the bottom. At the upper part of the handle a small hole is drilled, where the thumb or finger can readily cover it. It should be disguised by some neighboring ornament or scroll. If this pitcher be half filled with water and inverted, the liquid would be retained as long as the small hole in the handle was closed—being suspended as in the atmospheric sprinkling pot, No. 69 and 70, and in Tutia's sieve, No. 74. If the lower part be filled with water and the upper with wine, the liquids will not mix as long as the small hole in the handle is closed; the wine can then either be drunk or poured out. If the hole be left some time open, a mixture of both liquors will be discharged. With a vessel of this kind, says an old writer, "You may welcome unbidden guests. Having the lower part already filled with water, call to your servant to fill your pot with wine; then you may drink unto your guest, drinking up all the wine: when he takes the pitcher thinking to pledge you in the same, and finding the contrary, will happily stay away until he be invited, fearing that his next presumption might more sharply be rewarded."

Another old way of getting rid of an unwelcome visiter, was by offering him wine in a cup resembling No. 245. The sides were double, and an air-tight cavity formed between them. When the vessel was filled, some of the liquid entered the cavity and compressed the air within; so that when the cup was inclined to the lips and partly emptied, the pressure being diminished, the air expanded, and drove part of the contents in the face of the drinker. Porta, in his Natural Magic, (Eng. translation, 1658,) mentions several similar devices, but they are all to be found in one form or another in the Spiritalia. One goblet was so contrived that "no man can drink out of it but he who knows the art." The liquid was suspended in cavities and discharged by admitting or excluding air through secret openings. Another one "for making sport with them that sit at table with us," a cup into which wine was poured in the presence of the drinker, but who could derive from it nothing but water, &c.

The necromancers of the Tartars and Cathayans, [Chinese,] says Purchas, "are exceedingly expert in their divellish art. They cause that the bottles in the hall of the great *khan* doe fill the bowls [cups] of their own accord, which also without man's help pass ten paces through the ayre into the hands of the great *khan ;* and when he hath drunke, in like sort they returne to their place." The cups were doubtless filled and moved by some ingenious device; but this being concealed, the whole was of course miraculous.

Among the antiquities of Lunenburg was a magical goblet or ewer, "une aiguiére dans laquelle il y a *un secret hydraulique.*" (Le Curieux Antiquaire, a Leide, 1729, tom ii, 495.)

From the time of Heron up to the 16th and 17th centuries little specific information respecting siphons is to be met with. They were of course known to the Romans. Sir Wm. Gell supposes some modification of them was employed in connecting the large boilers in which water was heated for the public baths. It appears from discoveries made in Pompeii that these vessels were closed on all sides and bore some resemblance to the bodies of modern stills; and that to economize the heat, three of them were placed upon each other. The lowest one in contact with the fire was the largest, and named "the caldarium, that above it the tepidarium, and the uppermost which was supplied with cold water directly from the aqueduct or other reservoir the frigidarium; and they were so contrived, by means of something of the nature of a siphon, that when the water of the lowest was drawn off for the bath, an equal quantity descended simultaneously from the second to the lowest cauldron, and from the uppermost to the second." Julius Pollux, who lived in the second century, informs us that siphons were used for tasting wine. They are also referred to by other ancient writers, but as several instruments were designated by the same name, it is difficult to determine with precision what particular one was, in every case, intended. It is very probable, from the remark of Pollux, which is corroborated by the illustration No. 235, that siphons were employed by ancient *vintners* and private gentlemen for decanting wine, just as the same classes use them at this day.

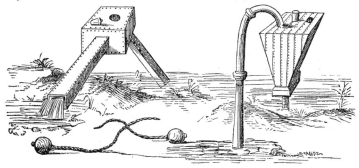

No. 246. A. D. 1511.   No. 247. A. D. 1511.

The earliest modern figures of siphons that we have met with are in the German translation of Vegetius, Erffurt, 1511. The above figures, Nos.

246 and 247 are copies. Both are designed to show the application of these instruments for transferring large quantities of water over rising grounds, as mentioned by Heron. No. 246 is formed entirely of wooden planks strongly nailed together. The upper ends of the two trunks or pipes are united to a square and close box, by means of which they were charged through the opening on the top. The lower orifices were temporarily closed by plugs, figured below with short ropes attached. When the whole was filled, the hole at the top was closed by driving in the stopper, figured near it, and then the two plugs below were withdrawn by means of the ropes.

There is little doubt that *large* siphons made of planks and jointed or lined with pitch would work well, even if they were not perfectly tight, provided the orifice of the discharging leg was considerably lower than the surface of the water in which the short leg was placed.

Heron directed large siphons to be filled through a funnel at the top, and the orifices closed below, as represented in Nos. 246 and 247.

No. 247 was of metal, but charged like the last by means of a wooden box; the opening to admit the water and its stopper being clearly represented. There appears no device for closing the lower ends of this siphon; and as they enter the water perpendicularly, the plugs and ropes used in No. 246 would hardly apply. Probably the short leg was closed by a *valve* opening upwards at the bottom of the box, on which account the latter was made conical to afford room for it to play. This valve would be sufficient for the purpose of charging the siphon, provided the upper part of the box was higher than any other part of the instrument. We therefore suppose that the disproportionate size of the box and its being figured below the bend are errors of the artist.

Of modern improvements, the addition of sucking tubes by which small siphons are now commonly charged was the first. It is uncertain when or by whom they were introduced. They do not appear to have been much used, if at all, before the early part of the last century; for all the siphons described in old treatises on chemistry, distilling, &c. invariably consist of single tubes, which were either charged by immersing them, or by drawing out the air from the orifice of the discharging leg by the mouth. It may contribute to some future history of the siphon to preserve a few of these.

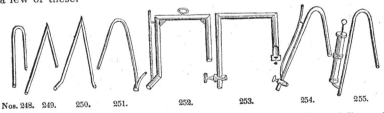

Nos. 248. 249. 250. 251. 252. 253. 254. 255.

The first two are from the English translation of one of Conrad Gesner's works, Lond. 1599. (See page 381 of this volume.) Speaking of drawing off water from the head of a still, the author observes, "You may put certaine *draying pipes* into the cover such as you see here livelie portrayed." Copies of the same are inserted in several other old works. In "Maison Rustique," Paris, 1574, folio 217, they are to be seen, and the instruments are said to have been made of *tin plate*,[a] (*tuyaux de fer blanc*.)

---

[a] This beautiful manufacture (tin plate) which contributes so largely to the furnishing of our kitchens, &c. is supposed to be of ancient date. The Germans were the first makers of it in modern times.

No. 250 is from the "Dictionnaire Œconomique," Paris, 1732, 3d edit. Tome i, 864. It is obviously copied, with the distilling apparatus of which it forms a part, from some older work. No. 248 differs in nothing from those belonging one of the Pharaohs, (No. 236,) while the forms of Nos. 249 and 250 are evidently owing to the material of which they are made, viz. tinned iron; the legs were separate pieces, and their junction formed an acute angle.

The sucking tube is not figured by Decaus, Fludd, Moxon, Boyle, Belidor; nor yet by Rohault, Gravesande, Desaguliers, and the Abbé Nollet, although it was in use before the popular works of the last named authors were published. Switzer, in his Hydrostatics, 1729, has figured a siphon for transferring water over a hill with a short sucking tube attached; but this is placed near the top, and was designed to draw off the air that might accumulate at the bend after the instrument had been some time in use.

In Martin's "Philosophical Grammar," Lond. 1762, sixth edit. No. 251 is represented. The sucking tube appears but as the nucleus of the modern one, being a very short conical piece attached to the extremity of the discharging leg. The figure we suppose was in the previous editions of the work. It was copied into the London Magazine for 1764, p. 584, and is there named "the syphon or crane *in common use.*" But the sucking tube was fully developed before these dates. In "Arts et Metieres," it is not curtailed of its fair proportions. The treatise on the Art of the Cooper, (Art du Tonnelier,) was published in 1763, and in it No. 252 is given as the siphon then used in Paris for emptying wine casks, &c. It was made of tin plate, and for the convenience of hanging it up when not in use, a ring was attached to the upper part. "Ce siphon est connu sous le nom de *pompe.*" (Folio edit. p. 47.)

"L'Art du Distillateur Liquoriste," was published in 1775. In it another valuable modification of the siphon is exhibited. See No. 253. This in its outline resembles the preceding one, being made of the same material. It has no sucking tube, but the discharging leg is closed by a cock, and the receiving one by a light valve opening inwards; hence when once charged, this siphon would always remain so while the cock was kept shut: it could be moved from one vessel to empty another at pleasure, for as soon as the end of the short leg was immersed and the cock opened, it would commence to act. This instrument was named "*siphon à clapet.*" (Folio edit. p. 140.)

The more common form of the siphon as now used is shown at No. 254, a valve in the short leg being dispensed with. *Small* instruments are so easily charged, that little or no advantage is derived from keeping them filled. Liquids confined in them become insipid, and in some cases tainted by the material of the tube; besides, as small siphons are required to decant different liquids, their contents must be discharged every time the liquid is changed. On these accounts the valve has been dropped. The junction of the sucking tube with the discharging leg must always be kept below the surface of the fluid to be drawn, as the virtual length of the leg there terminates. By means of the cock the discharge can always be regulated, and when a receiving vessel is filled—entirely stopped until another vessel is prepared.

Siphons with small syringes attached for the purpose of charging them, are frequently made by silversmiths for decanting wine from ordinary bottles, &c. See No. 255. The capacity of the syringe should equal that of the siphon, as one stroke only (an upward one) of the piston can be used. Atmospheric and forcing *pumps* are often used to charge very long siphons; the former being applied to the discharging, and the latter to the receiving orifice.

Of devices for stopping and renewing the discharge without either cocks or valves, the *Wirtemburg siphon* is the oldest. It was so named from its invention in that city. The legs are of equal length, and to prevent the admission of air when the instrument is not in use, their ends are bent upwards. See No. 256. (For the convenience of discharge, one end is commonly recurved.) The alledged advantages of this siphon over others were more imaginary than real. It was at one time announced as "a very extraordinary machine, performing divers things which the common siphon cannot reach." Thus, when the legs were inserted in different vessels, it was said to preserve the liquid at the same level in both; and although the legs were of equal length, water rose indifferently up one and descended through the other, besides other properties which in fact are *common to all* siphons. Its only peculiarity consists in the ends of the legs being turned upwards, so as to retain the fluid within, and thus be always ready for use: but this retention of the contents, although theoretically true, is in practice hardly attainable, since it requires the orifices to be always preserved on the same horizontal line—a condition extremely difficult to perform, except with very small instruments, and whose ends are turned considerably up. If the ends reach only to a level with the upper side of the flexure, the slightest change of position makes one leg longer than the other; air is admitted, and in a moment the whole contents are expelled. A siphon thus made of inch, or ¾ inch tubing, could not be moved from one vessel to another, or hung against a wall, without the contents being displaced. Disks or stoppers placed over the orifices would prevent this, but they would virtually be valves. The Wirtemburg siphon is consequently seldom seen except in the lecture room. (See Phil. Trans. xv, 846-7, and Lowthorp's Abridg. i, 537-9.)

In 1808, M. Argand, the inventor of lamp burners that go under his name, devised a "valve siphon" precisely similar to No. 253. From remarks made in the journals of the time, he seems to have been considered the introducer of the valve —an erroneous idea. As regards the *construction* of his siphon all that could be claimed by or for him was the mode of connecting the legs to the horizontal part by screws, so that they might easily be separated, either for the purpose of cleaning or more conveniently packing. But Argand's *mode of charging* his siphon was novel. It was effected on the same principle as water is raised by the *canne hydraulique*, (page 372,) viz. by moving the instrument perpendicularly up and down in the liquid, until it became filled. Instead of imparting motion to the whole instrument, which in larger ones would be inconvenient, M. Hachette suggests that the lower part of the receiving leg be connected to the upper part by a flexible tube of leather or cloth impermeable to liquids, so that the part in which the valve is situated need only be moved. See No. 257.

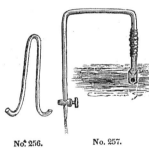

No. 256. No. 257.

Siphons are necessary in numerous manipulations of the laboratory, and modern researches in chemistry have given rise to several beautiful devices for charging them, and also for interrupting and renewing their action. When corrosive liquids or those of high temperatures are to be transferred by siphons, it is often inconvenient, and sometimes dangerous to put them in operation by the lungs. Moreover cocks and valves of metal are acted on by acids, and in some cases would affect or destroy the properties of the fluids themselves.

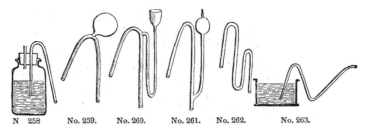

No. 258.   No. 259.   No. 260.   No. 261.   No. 262.   No. 263.

No. 258 shows how hot or corrosive liquids may be drawn off from a wide mouthed bottle or jar. The short leg of a siphon is inserted through the cork; and also a small tube, through which the operator blows, and by the pressure of his breath forces the liquid through the siphon.

No. 259 represents a siphon sometimes employed by chemists. When used, the short leg is first placed in the fluid to be decanted, the flame of a lamp or candle is then applied to the underside of the bulb; the heat rarefies the air, and consequently drives out the greater part of it through the discharging orifice. The finger is applied to this orifice, and as the bulb becomes cool the atmosphere drives up the liquid into the void and puts the instrument in operation.

No. 260 is a siphon by M. Collardeau. It is charged by pouring a quantity of the fluid to be decanted into the funnel; the bent pipe attached to which terminates near the top of the discharging leg. The fluid in descending through this leg bears down the air within it, on the principle of the trombe, and the atmosphere drives up the liquid in the reservoir through the short leg. In experiments with this instrument we invariably found the contents of the charging tube drawn into the siphon whenever the orifice of the discharging leg was not made smaller than the bore of the receiving one. By not attending to this, such siphons will only act as long as water is poured into the funnel.

No. 261. A glass siphon for decanting acids, &c. It is charged by sucking, and to guard against the contents entering the mouth, a bulb is blown on the sucking tube. The accumulation of a liquid in this bulb being visible, the operator can always withdraw his lips in time to prevent his tasting it.

No. 262 is designed to retain its contents when not in use, so that on plunging the short leg deep into a liquid the instrument will operate. This effect however will not follow if the end of the discharging leg descend below the flexure near it, and if its orifice be not contracted nearly to that of a capillary tube.

No. 263 is a siphon by which liquids may be drawn at intervals, viz. by raising and lowering the end of the discharging leg according to the surface of the liquid in the cistern.

Our own labors have developed some novel modifications of the siphon. No. 264 is charged by an apparatus designed as a substitute for the syringe. (See No. 255.) The sucking tube of an ordinary siphon is made to pass through the centre of a much larger pipe. This is closed at the bottom, open at top, and its length equal to that of the short leg. A moveable tube open at bottom and closed above is fitted to slide in the last, and is of such a bore that the space between its sides and the exhausting tube equals the capacity of the siphon. To use this instrument, fill the wide tube with water or some other fluid, and place the short leg into the liquid to be decanted; then close the orifice of the long leg with the finger, and raise the moveable tube (by the ring attached to it) and the siphon will be charged. In using this instrument, the fluid by which it is charged does not mix with that which is decanted, as in No. 260. The apparatus is more simple than a syringe and is not liable to be deranged. By using mercury both the length and bore of the charging tubes may be greatly reduced. As these tubes themselves constitute a siphon, (see No. 238,) the upper end of the small exhausting one should extend a little above that which contains the charging fluid, lest this should occasionally rise over the orifice—in which case the whole would be drawn off. A description of this siphon was published in the Journal of the Franklin Institute for November, 1834.

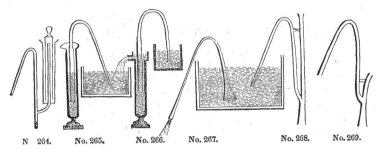

N 264.   No. 265.   No. 266.   No. 267.   No. 268.   No. 269.

Nos. 265-6 represent another mode of charging siphons on the same principle, but the apparatus is more simple and is accompanied with some peculiar advantages. The siphon itself has no exhausting pipe attached to it, but is a bent tube simply. It is put in operation by means of a moveable tube of about the same length as the discharging leg, and having the bottom closed and a lip or spout formed on its upper edge. This tubular vessel is filled with water (or other fluid) and the long leg of the siphon inserted into it. The short leg is then placed into the liquid to be decanted and the moveable tube drawn gently down. The air within becomes rarefied and the instrument charged in consequence of the vacuity left in the long leg by the receding liquid. The moveable tube may then be wholly withdrawn or not as circumstances may dictate. If the liquid is to be decanted at intervals, or the stream increased or diminished, the tube should be used; thus, to lessen or stop the discharge, slide up the tube and as the lip approaches to a level with the surface in the reservoir the stream will become less and less, and by raising it still higher, as in No. 265, entirely stopped. Hence the instrument acts as a perfect cock, by which the liquid may be discharged in single drops or in a full stream, and unlike the ordinary brass taps, it can never leak nor require repairs.

The apparatus also performs the part of a *guage*, viz. by accurately indicating the surface of the fluid within any vessel to which it is attached. Suppose we wish to know the quantity of liquid remaining in a demijohn, or other close vessel, after drawing off part by one of these siphons; all that is required is to slide up the tube till the liquid barely drops from the lip—its surface in the tube will then be on the *same level* as in the demijohn. If the moveable tube be made of glass, the quantity left can always be known at sight, because its surface in the tube would always be visible. A device of this kind might be employed to draw off and to guage the contents of standing casks. It would be better to make the discharging leg of this siphon of rather larger bore than the short one, since the rarefaction would then be more perfect. The discharging leg must always be inserted in the moveable tube before the short one is placed in the liquid to be transferred. (See Journal of the Franklin Institute for July and November, 1834.)

No. 267 is formed of a conical tube, and charged by the act of placing it in the fluid to be transferred. The end of the long leg is first closed tight by the finger, and the short one then immersed as deep as can be conveniently in the liquid. The air being thus confined prevents the liquid from entering, but when the finger is withdrawn, it is urged up the short leg by the hydrostatic pressure of the column over the orifice of the latter, and the *momentum* of the large volume contained in the lower part drives sufficient over the bend to put the instrument in operation. The action of this siphon depends upon the same principle as the spouting tubes described in the last chapter. This siphon is in fact merely one of these bent into a proper form. The bend should be a regular curve in order to present as little obstruction as possible to the liquid in passing over: it should also be *short*, so as to require less of the passing fluid to fill it than a longer one. The proportions of the different parts of these siphons should approach those represented in the cut. Small siphons on this plan are limited in their application to those cases where the short legs can be immersed half their depth or more; but the application of large instruments increases with the depth. (See No. 226 and remarks upon it, page 499.)

Nos. 268–9 are *blowing* siphons, being charged by blowing with the mouth through the tubes connected to the orifices of the discharging legs. This mode of producing a vacuum in one pipe by blowing air through another is sufficiently explained in a previous chapter. In No. 269, the junction of the siphon with the blowing tube is flush or smooth in the interior of the latter, and whenever this is the case a conical ajutage must be added as represented, or the instrument cannot be charged. (See remarks on blowing tubes, pp. 486–7.) The better way is to make the siphon like No. 268, in which a part of the leg projects into the blowing tube and diverts the current of air from the lungs over the orifice, as in Nos. 205–'6, '7, and '13. These are more readily charged than the others, and although they will operate without the conical ajutage, they are much easier charged with it. By such siphons water may be raised one or two feet by a smart puff. They are safe and convenient to transfer acids, &c. as there is not the least danger of receiving any portion into the mouth, as when sucking siphons are used.

Siphons are now used, as they were by the Egyptians in Heron's time, to convey water to considerable distances. When they are laid over ground that is elevated from 20 to 25 feet above the spring, a quantity of air is disengaged from the water at the highest parts of the tube, and accumulating there is very apt to cause the action to cease. To prevent this, a close vessel, furnished with a cock and funnel at the top, should be connected at its bottom to the highest part of the siphon by a stop cock or valve. The air evolved from the water will collect in this vessel and should be occasionally drawn off in the following manner. Shut the lower cock and open the one attached to the funnel; then expel the air by filling the vessel with water and turn the cocks as at first. As fresh portions of air arise from the liquid, they will enter the vessel and drive the water down the discharging leg. When the ground is very uneven at the highest parts, the several eminences of the siphon should be connected by small tubes to the air-chamber.

We have known siphons from a quarter to half a mile in length, and formed of leaden pipes only half an inch in the bore continue running from nine to fifteen months without once stopping, although no air-vessels were attached to them. In one case the pipe was 1200 feet in length, the orifice of the discharging leg was but five or six feet below that of the receiving one, and the highest part of the tube was from 12 to 15 feet above the surface of the spring.

An opinion is current with some writers, that the extreme elevation to which water can be carried by siphons was unknown to the ancients, and that Heron, the most celebrated writer of antiquity upon these instruments, was not himself aware of its limitation to about 30 feet. It is not clear that Heron was thus ignorant; but if he were, it would only show that in this department of the arts he was no practical man. That ancient plumbers and pump-makers, who prepared and laid large siphons were aware of the limitation there can be no doubt, just as the same class of mechanics were in modern times with regard to pumps, before philosophers were informed of the fact or able to account for it. As however siphons for conveying water over hills and to great distances have always been of rare occurrence, (comparatively speaking,) it is not at all surprising that even some hydraulic engineers should have been thus ignorant with regard to them, although familiar with the extent to which water can be raised by atmospheric pumps. If some of these men have talked of conveying water by siphons over mountains, we never hear them speak of raising it to equal elevations through the suction pipes of pumps. Daily experience in applying the latter to various depths prevented them from falling into the error.

Baptist Porta, in the 19th Book of his Natural Magic, speaks of raising water by a siphon to the top of a high tower, and several old writers have the same conceit. This was in accordance with the ancient doctrine of the *plenists*, who denied the possibility of a vacuum. They attributed the ascent of liquids through siphons to nature's abhorrence of a void, and imagined the elevation to be unlimited. But these men were philosophers, whose practical knowledge was confined to portable experiments; had they been working pump-makers they would have known better; they would have become advocates for the opposite doctrine—*vacuists*. So long was the error of the plenists maintained, observes Switzer, "that I have seen a book of Machines, written even in Queen Elizabeth's time, by one *Ward*, an engineer, who ventur'd to give a sketch of a high hill, and a house at the bottom or side, over which, by a vast extended syphon, the water was to be convey'd from one vale to another." The author of the old treatise, entitled, 'Art and Nature,' quoted at pp. 321, 375, was of the same opinion. "*How to convey water over a mountain:* this experiment is as easie to be performed as any of the former, and indeed after the same manner, for you must lay a pipe of lead over the mountain, with one end in the spring or water that you desire to convey, and the other end must lie somewhat lower; then open the pipe at the top of the mountain; stop both ends of the pipe, and with a tunnell fill the pipe full of water; then close it up exactly that neither ayer nor water may come out thereat; then unstop both the ends of the pipe, and the water will run continually," (p. 10.) Decaus appears to have been better informed, if we may judge from his remarks respecting the perpendicular length of pipes of atmospheric pumps. In large engines, he recommends that they be not made over 20 feet; and including the working cylinders, he says, " I am of opinion that it [the water] must not be constrained to rise more than thirty feet in height." The second plate of his " Forcible Movements" represents two atmospheric pumps placed one above the other, and the lowest one raised it from "24 to 30 feet," and the upper one "may raise it from thence 24 or 30 feet" higher. The "Forcible Movements," it will be remembered, was published about thirty years before the discovery of atmospheric pressure.

Contrivances for discharging water from the *highest part* of siphons have often been proposed. They are to be met with in several old authors, and the principle of most of them may be found in the Spiritalia. They are however seldom employed, because circumstances on which they depend rarely occur; and other devices are preferable even under those circumstances. A descriptive account of a few of them may interest some machinists, and be serviceable to others, viz: by preventing them from expending their energies in devising similar things. Indeed in this respect books which contain accounts only of the *best* machines are not always the most useful to *inventors*. In whatever department of the arts these men exercise their talents they are almost certain to fall at one time or another on old devices, which appear to them both new and equal to similar plans in common use. Books therefore which describe rejected and antiquated contrivances are not so worthless as some persons imagine.

One plan to raise water by a siphon consists in enlarging or swelling it out at or near the bend, or what amounts to the same thing, connecting the legs to an air-tight vessel; and when this becomes filled the communication between it and the legs is cut off by valves or cocks, and the contents drawn off. When this is done the vessel remains filled with air, which if admitted into the legs would stop the action of the siphon. It must therefore, in order to expel the air, be filled with some liquid to replace that drawn out. Suppose a siphon of this kind be designed to raise water for the supply of a dwelling, in or near which the vessel is placed, it may then be refilled with refuse or impure water, which on adjusting the cocks will pass down the discharging leg. Then after a short time elapses, the vessel will again be filled with fresh water, which may be again exchanged for the same quantity of impure.

In locations where river, salt, or any other water can thus be exchanged for fresh, and it is desirable to do so, such devices are applicable. (In breweries, distilleries, &c. the descent of one liquid may thus be made to raise another.) It should however be observed that an equal quantity must be given for that received, and it must descend rather more than the latter rises. But when circumstances allow these conditions to be fulfilled, the apparatus is not always to be depended upon; air insinuates itself through the minutest imperfections in the pipes and cocks, and often deranges the whole. One of these siphons is described in Nicholson's Journal, 4to. vol. iv, and in vol. ii, of Gregory's Mechanics. Another in the Bibliotheque Phisico-Economique, which is copied in vol. x, of the Repertory of Arts, 2d series. Another is figured in Art and Nature, A. D. 1633, with two close reservoirs at the top; and Porta, in cap. 3, book xix, of his Magic, describes another, with the close vessel on the top of a tower: the discharging leg is described as terminating in another close vessel of the same size as the one above, and furnished with a cock and funnel through which to fill it, and another cock to discharge the contents: this charging vessel from his description appears to have been placed on the ground a little below the spring and then emptied—if so, the apparatus could not act. He does not appear to have been aware of the necessity of the contents of the lower vessel being discharged from the orifice of a pipe as much below as the receiving vessel on the tower was above the spring. The device (which he probably imperfectly copied from some older author)

Chap. 6.] *Ram Siphon.* 531

would then be tne same as the siphon for raising water which Gravesande has figured in the second volume of his Philosophy, p. 39, plate 74.

The best of these devices are not only subject to derangement by the wear of the cocks and valves, and want of care in opening and closing them at the proper times, but they require almost as much attendance as would suffice to raise the water directly from the spring. On this account various contrivances have been proposed to render them self-acting.

An ingenious device of this kind may be seen in the Gentleman's Magazine for 1747, p. 582. It is named a "*lifting siphon.*" Water from a spring is received into an open cistern, from the bottom of which a pipe descends to a perpendicular depth of 33 feet. The bore of this pipe is closed and opened by two stop-cocks, one at its lower end and the other near the upper, or just below its junction with the cistern. A close vessel to receive the water raised is to be fixed at any required elevation, not exceeding 30 feet above the cistern; and from its bottom a pipe descends to within two inches of the bottom of the cistern. This pipe constitutes the short leg of the siphon, and its upper orifice is covered by a valve to prevent the water that ascends through it from returning. From the top of the close vessel a small or exhausting pipe proceeds down to the one beneath the cistern and is connected to it below the upper cock. Thus united they may be considered as the long leg of the siphon, although water only descends through the lower branch and air through the upper one. The apparatus for alternately opening and closing the cocks (upon which the action of the machine depends) is somewhat similar in principle to that represented at page 354. A bucket containing water is the prime mover; a rope attached to it is passed twice round two rollers, and a counterpoise is suspended from the other end of the rope. When the bucket is partly filled it preponderates, and when it is emptied the counterpoise prevails; hence an alternating movement is imparted to the rollers and to the plugs of the two cocks, as the shanks of these constitute the axles of the rollers.

A plan for making siphons of this description self-acting by means of four vessels placed one over the other, and each provided with a siphon by which its contents may be discharged, was proposed by Mr. Wm. Close, in Nicholson's Journal before referred to.

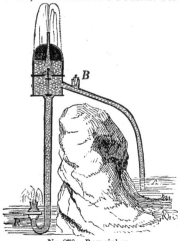

No. 270. Ram siphon.

M. Hachette has combined the ram of Montgolfier with the siphon, in order to discharge water from the apex of the latter: see the annexed figure. A the short leg and R the long or discharging one. The upper end of each terminates in a close chamber within which two valves attached to a perpendicular rod are made to work. The upper valve closes an opening in the horizontal partition that separates the interior of the chamber from the air-vessel and jet pipe above. The seat of the lower valve is at the orifice of R. The distance between the valves is such that when one is closed the other is open. Their movements are produced as in the ram; a coiled spring keeps the upper one closed till the momentum of the fluid in passing through the siphon shuts the lower one. The lower end of R is furnished with a cock, and that of A with a valve opening outwards, for the purpose of charging the siphon through an opening at B. When in operation, the water after running a little while acquires sufficient momentum to shut the lower valve, upon which a portion rushes into the air-vessel and escapes in a jet; the spring then closes the upper valve, and the fluid descends through R till the lower valve is again closed and another jet produced.

CHAPTER VII.

FOUNTAINS: Variety of their forms, ornaments and accompaniments—Landscape gardeners—Curious fountains from Decaus—Fountains in old Rome—Water issuing from statues—Fountains in Pompeii— Automaton trumpeter—Fountains by John of Bologna and M. Angelo—Old fountains in Nuremberg, Augsburg and Brussels—Shakespeare, Drayton and Spencer quoted—Fountains of Alcinous—The younger Pliny's account of fountains in the gardens of his Tuscan villa—Eating in gardens—Alluded to in Solomon's Song—Cato the Censor—Singular fountains in Italy—Fountains described by Marco Paulo and other old writers—Predilection for artificial trees in fountains—Perfumed and musical fountains—Fountains within public and private buildings—Enormous cost of perfumed waters at Roman feasts—Lucan quoted—Introduction of fountains into modern theatres and churches recommended. Fountains in the apartments of Eastern princes—Water conveyed through pipes by the ancients into fields for the use of their cattle—Three and four-way cocks.

ARTIFICIAL fountains and jets d'eau are of extreme antiquity: although they were not (like natural ones) objects of worship among the ancients, they were at least held in great estimation, and unusual care was often taken in designing and decorating them. Indeed no other hydraulic devices have ever been so greatly and so variously enriched with ornament. The pipes of supply were concealed in columns, &c. and their orifices wrought into numerous emblematic figures, (see page 119,) while the basins that received the fluid were generally of polished marble. Sometimes the pipes terminated in statues of men, women, children, animals, birds, fishes, vases, gods, goddesses, &c. From them the fluid spouted high in the air, or was discharged directly into receivers, or broken in its descent by intervening objects: oftentimes it was made to flow over the rim of a vase, to issue from others that seemed to have been accidentally overturned, and not infrequently the figure of a female poured it from a pitcher.

From the facility of applying water as a motive agent another feature was added. Various automata were put in motion by mechanism concealed in the base or pedestal from which the fluid issued—figures of men blew trumpets and played on organs, and automaton birds warbled forth notes on adjacent trees. (Such devices are described by Heron.) All the senses were often gratified at these fountains; the sultry atmosphere was cooled and rendered grateful to the feeling—the sparkling liquid quenched the thirst—sight was gratified in contemplating the design and execution of the whole, and noticing the ever-changing forms assumed by the moving fluid—the pleasure derived from the sound of falling water has ever been noticed by poets—and not to forget the sense of smelling, in those fountains that were designed only to moderate the temperature of the air, the water was often *perfumed*.

The taste of old landscape gardeners for fountains and cascades, serpentine streams, and other "pieces of water-works" although derived from the East, had its origin in nature. "Even as *Paradise* itself (says Switzer) must have been deemed an immodelled and imperfect plan, had it not been watered by the same Omnicient hand which first made it, so our gardens and fields, the nearest epitomy and resemblance to that happy place which is to be met with here below, cannot be said to be any way perfect, or capable of subsisting without it." These men contemplating the world as a garden endeavored to copy it in miniature. They constructed lawns for deer and reared diminutive forests for game—they formed lakes and stocked them with fish—walks were made on the margin of brooks, torrents fell from artificial mountains, and tiny streams wound their way through labyrinths of reeds and of sedge. Springs were seen bursting out of rocks rudely piled up, as if thrown together by nature, while aquatic birds sported in basins below. But they went further, for ascending jets were thrown up so as to resemble bundles of reeds, others were crested like wheat sheafs, or branched out like trees. Sometimes the streams were directed so as to form avenues and alcoves, as of chrystal, which when the sun shone produced a magical effect. Even hedges and borders of gardens were imitated. "The hedge of water (says Evelyn) in forme of lattice-worke which the fontanier caused to ascend out of the earth by degrees exceedingly pleased and surprised me."

Giving the reins still more to their imaginations, these artists were hurried into singular puerilities. They made the fluid to spout from the sides of ships, the mouths of birds, and other incongruous figures. Swarms of heathen deities were also pressed into their service; and not content with a Triton blowing water through his shell, or Neptune pouring it from an urn, figures of the latter were made to rise from the bottom of deep basins, and drawn by spouting dolphins and accompanied with Amphitrite and a legion of sea nymphs, sailed over his fluid domains to allay the tempest that called him up!

Old treatises on water-works are full of such things. In "Art and Nature," Neptune is figured "riding on a whale, out of whose nostrils, as also out of Neptune's trident the water may bee made to spin thorow small pin holes." Other devices consisted of "divers forms and shapes of birds, beasts, or fishes; dragons, swans, whales, flowers, and such like *pretty* conceits, having very small pin holes thorow them for the water to spin out at." The 15th and 16th plates of Decaus' Forcible Movements represent the mechanism of "an engin by which Galatea is drawn upon the water by two dolphins, going in a right line and returning of herself, while a Cyclope plaies upon a flajolet." And the 17th and 18th plates shew Neptune drawn by sea horses, preceded and followed by Tritons, sailing round a rock on which Amphitrite is reposing, and from which water is gushing forth.

Fountains for supplying the inhabitants of *towns* and *cities* are frequently mentioned in scripture, but it is difficult to discriminate between artificial ones and those that were natural. In the early history of Rome some are mentioned. The news of the victory obtained over the Tarquins and the people of Latium was conveyed in an incredibly short time by two young men, said to have been Castor and Pollux, who were met " at *the fountain in the market-place*," at which their horses foaming with sweat were drinking. (Plutarch in Paulus Æmilius.) Statues of Jupiter Pluvius, of the Egyptian god Canopus and others, were erected over fountains, the liquid issuing from different and sometimes from all parts of the bodies. On the day Julius Cæsar was assassinated, he was implored by Calphurnia in consequence of a dream, to remain at home instead of meeting the senators according to appointment, a circumstance to which Shakespeare thus alludes :—

>   *Decius.* Most mighty Cæsar, let me know some cause,
>   Lest I be laughed at when I tell them so.
>   *Cæsar.* Calphurnia here, my wife, stays me at home;
>   She dreamt to-night she saw *my statue,*
>   Which *like a fountain,* with *a hundred spouts,*
>   Did run pure blood, and many lusty Romans
>   Came smiling, and did bathe their hands in it.

Pliny (xxxi, 2.) speaks of a fountain from which water ran "*at many pipes.*" From excavations made at Pompeii, it appears that in almost *every street* there was a fountain, and that bronze statues, through which the water issued were common. Several have been found—four or five are boys of beautiful workmanship; the fluid issued from vases resting on their shoulders or held under their arms, and in some cases from masks. Paintings of elegant fountains, from which the water issued in perpendicular jets from vases, have also been discovered both at Herculaneum and Pompeii.

A circumstance mentioned by Suetonius in his Life of Claudius, the successor of Caligula, although not directly related to this part of our subject, shows that Roman engineers were quite at home in devices analogous to those moving and musical statues, which two centuries ago were so common in European fountains. Previous to drawing off the waters of the lake Fucinus, the emperor exhibited a naval conflict, in which 19,000 criminals were engaged against each other in two fleets. An immense multitude of spectators attended. Claudius presided dressed in a coat of mail, and with him was Agrippina in a mantle of cloth of gold. When the two fleets were ready to engage, *a Triton of silver rose up in the midst of the lake and sounded the charge.*

Of modern street fountains many curious ones are to be seen in Italy, France and Germany, while descriptions of others, no longer extant, may be found in Misson, Blainville, and other writers of the last century. Thirty folding plates, representing some of the most remarkable, are attached to Switzer's Hydrostatics. A colossal statue of Jupiter Pluvius, in a singular stooping position, was designed for a fountain at Tratolino, by John of Bologna. The extremities are of stone, but the trunk is formed of bricks overlaid with cement that has acquired the hardness of marble. A number of apartments are constructed within it—one in the head is lighted through the eye-balls, which serve as windows. To add to the extraordinary effect, a kind of crown is formed by little jetteux that drop on the shoulders and trickle down the figure, shedding a sort of supernatural lustre when irradiated by the sun. One hand of the figure rests on the rock as if to support itself, while the other is placed on the head of a lion, from the mouth of which the principal stream issues.

A fountain designed by Michael Angelo is described by Sir Henry Wotton as 'a matchless pattern,' being 'the figure of a sturdy woman, washing and winding linen clothes; in which act she *wrings out the water* that made the fountain, which was a graceful and natural conceit in the artificer, implying this rule that all designs of this kind should be proper.'

Of remarkable fountains at Nuremberg, Blainville has noticed several. Of one he observes, " Its basin is an octagon in the middle of which stands a large brass pillar; from its chapiters project six muzzles of lions, each of which spurts water into the air out of a twisted pipe. On the cornish are the six cardinal virtues, which squirt water from their breasts. On this pillar stands a less one fluted, upon which are six infants, every one of whom leans on an escutcheon bearing the arms of the empire, those of Nuremberg and other towns; they are all of them sounding trumpets, out of which water jets in plenty. On the top of this second pillar is a fine statue of Justice, with her sword in one hand and her balance in the other; she likewise sends water from her breasts, and supports herself upon a large ostrich which spouts water most bountifully. All this is in brass surrounded with an iron grate carved and gilt." (Travels, i, 197.)

Another at Augsburg he thus describes : " In the middle of the basin is a double pedestal, at the foot of which are several sphinxes and statues jetting water into the basin, some by the mouth, others by their breasts, and three by trumpet-marines. On the four corners of the first pedestal are four fine statues big as life; their feet rest upon four very large shells into which they pour water, some out of vases, *others in another fashion.* Upon the top of the second pedestal is a Hercules combating the Lernean Hydra." (Ibid. 291.)

Old writers represent Brussels as well supplied with water 150 years ago as Rome itself. There were twenty public fountains at the corners of the principal streets, and all adorned with statues. In the herb-market were figures of four beautiful females " squeezing water out of their breasts"—a favorite device, and another equally popular was adopted in a splendid fountain near the Carmelite church in the same city : " Tout près de cette Eglise est le Manneke-pis, c'est la statue d'un garçon, elevée sur une colonne ; du haut, de laquelle il jette de l'eau, comme s'il pissoit, par sa pipe, jour et nuit, dans un bassin qui est au pied de la colonne. C'est une des sept merveilleuses fontaines de la ville." (Le Curieux Antiquaire, tome i, 175.)

Shakespeare often alludes to the figures of old English fountains. In Winter's Tale, Act iv, Scene I, he compares the old shepherd to "a weather bitten conduit of many king's reigns;" that is, to a statue from which the water flowed. Henley in commenting on the passage observes : "Conduits' representing a human figure were heretofore not uncommon. One of this kind, a female form, and *weather beaten* still exists at Hoddesdon in Herts." In As You like It, Rosalind says, she will weep "like Diana in the fountain"—an allusion to that erected at Paul's Cross, where, after the religious images had been destroyed, (see page 106,) "there was set up a curious wrought tabernacle of gray marble, and in the same an alabaster image of *Diana,* and water conveyed from the Thames, prilling from her naked breast,"

Drayton, a poet contemporary with Shakespeare, alludes to fountains and their basins in his *Quest of Cynthia.*

> At length I on a fountain light,
> Whose brim with pinks was platted,
> The banks with daffodilies dight
> With grass, like sleave was matted.

And Spencer in the *Fairy Queen*—

> And in the midst of all a fountaine stood,
> Of richest substance that on earth might bee,
> So pure and shiny, that the silver flood
> Through every channel running one might see.

Fountains have always been indispensable adjuncts in oriental *gardens,* and they doubtless formed conspicuous objects in those of Babylon. The two fountains in the gardens of Alcinous, from their elevated position and the abundance of water they poured forth, must have greatly contributed to the beauty and effect of the surrounding scenery.

> Two plenteous fountains the whole prospect crown'd :
> This through the garden leads its streams around,
> Visits each plant and waters all the ground ;
> While that in pipes beneath the palace flows,
> And thence its current on the town bestows.   Ody. vii. *Pope.*

The younger Pliny's description of his Tuscan villa contains the only detailed account extant of an ancient Roman garden. As might be supposed, fountains and jets d'eau frequently occur. The front of the house faced the south and had several porticos. The terrace was embellished with hedges of box, and the lawn overspread with the soft acanthus. At one end of the front portico a dining room opened on the terrace, and opposite the centre of the portico there was a small area shaded by four plane trees, " in the midst of which a fountain rises, from whence the water running over the edges of a marble basin, gently refreshes the surrounding plane trees and the verdure underneath them." In the same vicinity he describes " a little fountain playing through several small pipes into a vase." Speaking of the view from the front windows of a spacious chamber, he observes, they look " upon a cascade, which entertains at once both the eye and the ear, for the water dashing from a great height foams over the marble basin that receives it below."

After mentioning bathing rooms and other apartments, walks, meadows, groves, trees, &c. Pliny continues—" In one place you have a little meadow, in another the box is cut into a thousand different forms, sometimes into *letters*, expressing the name of the master, sometimes that of the artist ; whilst here and there little obelisks rise, intermixed alternate with fruit trees ; when on a sudden in the midst of this elegant regularity you are surprised with an imitation of the negligent beauties of rural nature. In the centre is a spot surrounded with a knot of dwarf plane trees. Beyond these is a wall planted with the smooth and twining acanthus, where the trees are also cut into a variety of *names* and shapes. At the upper end is an alcove of white marble shaded with vines, supported by four small Carystian pillars. From the bench [or triclinium, a species of couch on which the Romans reclined to eat] the water gushing through several little pipes, *as if it were pressed out* by the weight of the persons who repose themselves upon it, falls into a stone cistern underneath, whence it is received into a fine polished marble basin, so artfully contrived that it is always full without ever overflowing. When I sup here this basin serves for a table, the largest sort of dishes being placed round the margin, while the smaller ones swim about in the form of little vessels and waterfowls. Corresponding to this is a fountain which is incessantly emptying and filling ; for the water which it throws to a great height, falling back into it, is by means of two openings *returned as fast as it is received*." This must have been either a modification of Heron's fountain, (No. 163,) in which the water would *appear* to be returned, or some concealed force pump threw it back.

The practice of eating, and even of sleeping, in gardens during the summer months, has always been more or less common in the East. In Solomon's Song it is obviously alluded to. " Thou that dwellest in the gardens," that " feedeth among the lilies, in a fountain of gardens," or rather a garden of fountains. Indeed a great part of this song seems to refer to that season, (and anxiety for its approach,) when the custom was for the wealthy to remove, like Pliny, to their country villas. It was very common with the rich Greeks and Romans, as well as with the Jews and other Asiatics, "when the winter was past, the rain over and gone; when the flowers appeared on the earth, and the time of the singing of birds was come, and the voice of the turtle heard in the land; when the fig tree put forth her green figs, and the vines with the tender grapes gave a good smell"—to hie away to their villas, and in the figurative language of the East, to dwell in gardens and feed among lilies. The custom is based on some of the finest feelings of our nature, and it is on such occasions only that we can realize some of the most exquisite pleasures which our progenitors in Eden enjoyed. Motezuma, we are informed by Solis, took peculiar pleasure in supping in his gardens, in which were *numerous fountains* and flowers "of delightful variety and fragrance."

That the Jews had fountains in their gardens and often washed in the basins during the heats of summer, we learn from the accounts of Bathsheba and Susannah. The fountains doubtless being shaded with foliage and trees like those mentioned by Pliny.

Cato the censor, that terrible scourge of the luxurious Romans, rendered himself generally obnoxious by the reformations he introduced. Among other measures, "he cut off the pipes by which people conveyed water from the public fountains into their houses *and gardens*," probably on account of its excessive waste in ornamental water-works. Plutarch has quoted an epigram, from which we learn that the *physiognomy* of this celebrated man, like that of Socrates and Phocion, was not very prepossessing.

> With eyes so grey and hair so red,
> With tusks so sharp and keen,
> Thou'l fright the shades when thou art dead,
> And hell wont let thee in. *Langhorne's Trans.*

To give an account of modern street and garden fountains would be an endless task. Descriptions of the most remarkable, as those in the gardens of Frescati and Versailles, are too common to need repetition here. We shall therefore merely notice a few singular ones.

There is no doubt that the general features and essential parts of ancient fancy water-works were preserved in those of modern Italy, whence they were, including water-organs, spread over the rest of Europe. A sketch of those in the gardens at Pratolino will give, says an old writer, a general idea of other Italian works of the kind. "Besides Tritons, Cupids, and other statues which on a sudden cover you with water, other streams issue from between rows of trees, &c. You are led into a grotto, of which the roof alone is said to have cost 30,000 ducats, being all of coral, mother of pearl, and other costly materials; the walls are lined with the same, and the pilasters adorned with an organ, which by means of water plays several tunes. Here your eyes are diverted with a great variety of moving figures: the god *Pan* strikes up a melodious tune with his mouth, at the sight of his mistress standing before him. In another grot, an *angel* carries a trumpet, puts it to his mouth, and gives you a tune upon it. In another, a *clown* carries a dish of water to a *serpent*, which lifts up its head and drinks it. Here you have a mill grinding olives—in another a paper mill with the hammers going. The grotto of *Galatea* shows her coming out of a door in a sea chariot with two nymphs, and having sailed a while upon the water she returns the way she came. In the basin is a large *dolphin* carrying a naked woman on his back, and swimming about with several other figures, all moving as if alive. In another place, you see a curious round table fit to receive fifteen guests, having a fountain playing in the midst, while other streams play between every two persons and supply them with water to cool their wine. The *woman of Samaria* appears next, coming out of her house with two buckets, and having filled them, goes back the same way. Meantime you are diverted with smiths thumping, mills going, and *birds chirping on trees*—all which are set to work by the water."

In Dr. H. Brown's Travels, (Lond. 1685,) are figures of one or two ancient fountains—one, in Carinthia, of the form of a dragon, from whose mouth the water issued.

In the year 916, an embassy proceeded from Constantinople to Bagdad and was received with much pomp by the Caliph Moctader. "In the midst of the great hall in which he gave audience to the ambassadors was *a tree* of massy *gold*, which had (amongst others) eighteen principal branches, whereon were birds of gold and silver, which clapped their wings, and warbled various notes."—(Martigny's History of the Arabians, iii, 323.)

Marco Paulo, in the 13th century, mentions a fountain in the gardens of the "Old man of the Mountain," which gave out wine, milk, and a mixture of honey and water.

Rubriques, in the same century, saw a *silver tree* at the court of the Great Khan, which poured forth milk and wines of different kinds. At the foot were four lions, through each of which passed a tube. On the summit was the figure of an angel holding a trumpet, and which by some interior mechanism was made to sound. It was the work of a French goldsmith.

This predilection for trees as ornaments for fountains and gardens seems to have been of a more ancient date. The *palm tree* of *brass*, which was consecrated to Apollo by Nicias, and placed in a field or garden purchased by him, probably served for a fountain. It must have been of enormous dimensions, since a fragment that was blown off by a storm of wind, "falling upon a large statue demolished it." (Plutarch in Nicias.) The pedestal of this statue has been discovered. A golden statue of Pallas, Plutarch observes, was erected in the temple of Delphi on a *palm tree* of *brass*, which had *golden fruit*. There are two other celebrated trees mentioned in history, but their uses are not indicated. We learn from Herodotus, vii, 27, that Pythius, a native of Lydia, presented Darius with a *plane tree* of *gold*. It was worth $5\frac{1}{2}$ millions sterling according to Montfaucon. The *golden vine* of Aristobolus was valued at 400 talents. It was carried through Rome in Pompey's third triumph, and afterward deposited in the temple of Jupiter Capitolinus. Another one, which Alexander took at the sack of Thebes, was preserved at Rome in the temple of Apollo in Pliny's time. This author has a remark on the decay of the art of working brass, which may here be noticed. He observes, in former times the artists worked to win *fame* and *glory*, "but now as in all things else for gain and *lucre* only," (xxxiv, chap. 2.)

One of the fountains at Versailles was in the form of an *oak tree*, from which the liquid was dispersed in all directions. (It is figured in one of the plates attached to Switzer's work.)

Among the garden water-works at Chatsworth were, "1. Neptune with his nymphs, who seem to sport in the waters, let out by a cock in several columns, and falling upon sea-weeds; 2. a pond where sea-horses continually roll; 3. a *tree of copper*, resembling a willow, and by the turning of a cock every leaf drops water, which represents a shower; 4. a grove of cypress and a cascade with two sea nymphs at top with jars under their arms; 5. at the bottom of the cascade a pond with *an artificial rose*, through which by the turning of a cock the water ascends, and hangs in the air in the figure of that flower; 6. another pond with Mercury pointing at the gods and throwing up water."—Lond. Mag. 1752, p. 7.

Bell, in his account of the Russian embassy to Ispahan, notices those in the gardens of the Scah. In front of the Hall of Audience was "a large fountain of pure water, which springs upward in three pipes and falls into a basin filled with roses, jessamines, and many other fine flowers."

In one of the public gardens of Brussels, among other water-works was an hydraulic organ.—(Le Curieux Antiquaire, tome i, 175.)

The old device of artificial music combined with fountains, is thus mentioned in the 17th Proposition of Worcester's Century of Inventions:—
"How to make upon the Thames a floating garden of pleasure, with trees, flowers, banqueting houses and fountains, stews for all kinds of fishes, a reserve for snow to keep wine in, delicate bathing places and the like; with music made with mills, and all in the midst of the stream where it is most rapid.

Fountains were often placed within ancient public buildings as well as near them. They were common appendages to temples, and the custom, as mentioned in our first book, is still retained by the Turks and other Asiatics. Henry Blount visited Adrianople in 1624, and in describing the mosque, says, there were "*tenne* conduits with cocks on the north side, and as many on the south for people to wash before divine service; to which use also on the west side in the church-yard, are thirty or forty cocks under a fountain, so sumptuous, as excepting one at Palermo, I have not seen a better in Christendome."—(A Voyage into the Levant, Lond. 1638.)

During hot weather, Augustus the Roman emperor slept (observes Suetonius) with his chamber doors open, "and frequently in a portico with waters playing around him."

"In the middle of the square of the Coliseum is a pretty remarkable piece of antiquity, (says Blainville,) though very little minded by most people. Here stood anciently a beautiful fountain, adorned with the finest marbles and columns; and on the top was a bronze statue of Jupiter, from which issued great plenty of water, as may be seen on the reverse of one of Titus' medals. They called it *Meta Sudans*: *meta*, because it was made in the form of a goal; and *sudans*, sweating, because of the water running from the *several parts of the statue*. This fountain was of great use both to the spectators and the gladiators in the *amphitheatre* to refresh themselves. Pope Alexander VII. caused it to be repaired, but since his time it has been entirely neglected."—(Travels, vol. ii, 535.)

The theatres of the Romans were fitted up with numerous concealed pipes that passed in every direction along the walls, and were connected to cisterns of water or to machines for raising the latter. Certain parts of the pipes were very minutely perforated, and were so arranged that by turning one or more cocks, the liquid escaped from them and descended upon the audience in the form of dew or extremely fine rain. This effectually cooled the heated air, and must have been exceedingly refreshing to the immense multitudes, especially in such a climate as Italy. On some occasions the water was *scented* with the richest perfumes. Thus Hadrian, in honor of Trajan his father, commanded water impregnated with saffron and balsam to be sprinkled on the people at the *theatres*. The dining rooms of Nero's golden house were ceiled in such a manner, that the attendants could make it rain either flowers or liquid perfumes. At one feast 100,000 crowns were expended in perfumed waters. Suetonius says they were discharged from "*secret pipes*." The *statues* that adorned the interior of the theatres were made to sweat perfumes on the audience. This was accomplished by making them hollow, drilling in them an infinite number of small holes, and connecting them by secret tubes to reservoirs of scented waters. The practice is alluded to by several authors, and among others by Lucan in the following passage :—

> As when mighty Rome's spectators meet
> In the full theatre's capacious seat;
> At once by secret pipes and channels fed
> Rich tinctures gush from every antique head;
> At once, ten thousand saffron currents flow,
> And rain their odors on the crowd below.

Sometimes rich people left by their wills sums of money to furnish these perfumes and the apparatus for dispersing them. An example is given by Maffei in his 'History of Ancient Amphitheatres.' (Lond. 1730, p. 168.) A Roman lady bequeathed funds to celebrate a hunting of wild beasts in the amphitheatre, and she ordered that *salientes* should be made. This term Maffei understood to mean "those hidden channels or pipes by which with wonderful artifice, [as is twice mentioned by Seneca,] they caused odoriferous liquid to spring up from the bottom to the top of the amphitheatre, which then jetted and spread itself in the air like a very fine shower of rain."

The custom might be adopted with advantage in modern theatres: it would render visits to these crowded places more agreeable and less injurious to health. Why can't the managers announce it in their "bills," among other inducements, just as their predecessors did eighteen centuries since? One of the notices of a public entertainment in Pompeii has been found written on the walls of a bath in that city. It is in these words :— "On occasion of the dedication of the baths, at the expense of Cnæus Alleius Nigidius Maius, there will be the chase of wild beasts, athletic contests, *sprinkling of perfumes*, and an awning."—(Pompeii, vol. i, 148.)

Fountains for cooling the air should constitute part of the ordinary appendages to *churches*, as much as apparatus for heating and lighting them. They should be considered by us, as they were by the ancients, *essential* to the *health* as well as comfort of large assemblies of people. They certainly are as necessary here, especially in the Southern states, as they were in southern Europe. Their construction is so simple, their modifications so various, their application so universal, and their effects so beneficial and *cheap*, that it is surprising they have not been introduced. We don't see why a person might not be as innocently employed in pumping water during worship to supply a fountain or jet d'eau, as in pumping air into the pipes of an organ. But it is unnecessary, for where the fluid would not rise sufficiently high from public reservoirs or pipes that pass through the streets, it might be elevated into a reservoir in the roof the day previous to the sabbath. In this use of fountains ancient architects were clearly in advance of ours.

The custom of cooling the air in private apartments is of great antiquity in Asia, and is still kept up in the dwellings of princes. See a plate in Generale Histoire, tome xiii, p. 311, representing a private apartment in the seraglio of one of the generals of Aurengzebe. An octagon basin with a handsome jet is in the centre of the room, with images of birds floating in the water. On the borders of the basin are trays with refreshments, and the company reclining around on carpets, much in the same manner in which Pliny represents himself and friends feasting around a fountain in his garden.

Henry Blount describing one of the palaces at Cairo in 1624, observes, "In the chiefe dining chamber, according to the capacitie of the roome, is

Chap. 7.] *Ancient Fountains for Cattle.* 541

made one or more richly gilt fountains, which through secret pipes supplies in the middle of the roome a daintie poole, which is so neatly kept, the water so cleare, as makes apparent the exquisite mosaik at the bottome. Herein are preserved fish which have often taken bread out of my hand."

Sometimes the jet is made to fall into basins filled with flowers, the odor of which is dispersed in the spray. Bell describes the hall of audience at Ispahan as a most magnificent room, lined with mirrors of various sizes, the floor covered with carpets of silk interwoven with branches and foliage of gold and silver. In the centre were two basins in which several pipes spouted water that fell among roses and other flowers and produced a fine effect. Another fountain at the entrance threw the water so high that it fell like a thick rain or dew which concealed the Schah from those on the opposite side.

See remarks on the introduction of portable fountains into *private* dwellings at page 361.

No. 271. Ancient fountain for cattle.

That ancient farmers occasionally conveyed water through pipes into fields for the use of their stock, as is now sometimes done, appears from the above cut, from a basso relievo preserved in one of the museums at Rome.—(D'Agincourt's History of the Fine Arts. Sculp. Plate I.)

It has already been remarked, (pp. 163, 170,) that the old Mexicans and Peruvians had fountains, from which the fluid issued through figures of snakes and crocodiles.

There is reason to believe that three and four way-cocks were anciently employed in fountains: they are to be found in the old water-works of Italy and France.—(See L'Art du Plombier in Arts et Mètieres, 4to. edit. p. 560, planche xiii.)

CHAPTER VIII.

CLEPSYDRÆ AND HYDRAULIC ORGANS: Time measured by the sun—Obelisks—Dial in Syracuse.—Time measured in the night by slow matches, candles, &c.—Modes of announcing the hours—" Jack of the clock"—Clepsydræ—Their curious origin in Egypt—Their variety—Used by the Siamese, Hindoos, Chinese, &c.—Ancient hour-glasses—Indexes to water-clocks—Sand clocks in China—Musical clock of Plato—Clock carried in triumph by Pompey—Clepsydra of Ctesibius—Clock presented to Charles V.—Modern Clepsydræ—Hour-glasses in coffins—Dial of the Peruvians. HYDRAULIC ORGANS: Imperfectly described by Heron and Vitruvius—Plato, Archimedes, Plutarch, Pliny, Suetonius, St. Jerome—Organs sent from Constantinople to Pepin—Water organs of Louis Debonnaire—A woman expired in ecstacies while hearing one play—Organs made by monks—Old Regal.

CLEPSYDRÆ and water organs are not strictly included in the general design of this volume; but as they are ancient devices in which water performed an important part, and as they undoubtedly contributed to the improvement of hydraulic machinery, and moreover gave rise to clocks and watches, we were unwilling to omit them.

Sun-dials were the earliest means employed to note the lapse of time. Country people in all ages have marked the passing hours by the shadow of a tree, a post, the corner of a house, or any other permanent object; these were natural gnomons, while the ground upon which their shadows were thrown served as dials. In cities, artificial objects were necessary; hence the obelisks of the Egyptians and other ancient people. These gnomons were placed in open and conspicuous places for public convenience, and many of them from their great elevation threw their shadows to a considerable distance. Sometimes their pedestals formed magnificent buildings. When Dion, after delivering the Syracusans, spake to them on the tyranny of Dionysius, Plutarch says, he stood upon a lofty sun-dial erected by the tyrant: " at first it was considered by the soothsayers a good omen that Dion, when he addressed the people, had under his feet the *stately edifice* which Dionysius had erected; but upon reflecting that this edifice on which he had been declared general, was a sun-dial, they were apprehensive his present power might fall into speedy decline." " The dial of Ahaz" seems to have been a public building of a similar description. The governors of provinces in China assemble on the " time-telling towers" on public occasion. (Atlas Chinensis of Montanus, p. 594.) The Peruvians had pillars erected for measuring time by the sun. Small dials were anciently made of brass or other metals and placed upon columns, or were attached to public buildings. Vitruvius has described several in book ix. of his Architecture, and among them one by Berosus the Chaldean.

But dials are only serviceable while the sun shines. During cloudy weather and after sun-set they are useless; other devices are therefore required to mark the fleeting hours. Of ancient contrivances for this purpose there were two whose action depended one upon fire and the other on water, viz: by burning slow matches, powder, or candles, and by water-clocks. The former were used by the Anglo-Saxons, (see p. 350,) and are still common in Japan, and probably other Asiatic countries. Nieuhoff, in his account of the Dutch embassy to China, says, the Chinese have instruments to show the hour of the day which operate by fire and water. Those that depend upon fire "are made of perfumed ashes." (Ogilby's Trans. 1673, p. 159.) This is too vague to convey an idea of their construction; but from Thunberg's account of those he saw in Japan, we at once learn what they were. For the mensuration of time, observes that enlightened traveler, the Japanese use the bark of the skimmi (anise tree) finely powdered. A box, 12 inches long, being filled with ashes, small furrows are made in the ashes from one end of the box to the other, and so on backwards and forwards to a considerable number. In these furrows is strewed fine powder of skimmi bark, and divisions are made for the hours. The powder is ignited at one end of a groove, it consumes very slowly, and the hours are proclaimed by striking the bells of the temples. (Travels, iii, 228.) Time is also measured in Japan by burning matches, twisted like ropes and divided by knots. When one of these after being lighted has burned down to a knot, and thereby denoted the lapse of a certain portion of time, an attendant announces it by a certain number of strokes on bells near their temples, if in the day time; but in the night, by striking two pieces of wood against each other.—(Ibid. 88.)

In all ancient devices, the passing hours were announced by men appointed for the purpose, a custom still continued over all Asia. Sometimes it was done by the voice. Thus the Turks have an officer (with strong lungs) on the top of every mosque who, stopping his ears with his fingers, proclaims with a loud voice the break of day, noon, three in the afternoon, and twilight. Martial the Roman satirist, refers to a similar practice, and Athenæus mentions "a mercenary hour-teller." Allusions to the same custom are to be found in the Bible—that which ye have spoken in closets "shall be proclaimed upon the house tops." But the more general mode was that which is still so common in the East, viz. by striking a bell, drum, gong, or some other sonorous instrument, and distinguishing the different hours, as in our clocks, by the number of strokes. In modern ages in Europe before the striking parts of town clocks were invented, men struck the hour on a bell, and long after these officers were dispensed with figures of men were made as ornaments to perform the same duty. To these "Jacks of the clock," Shakespeare and other writers of his age often refer. Such clocks are still extant: the one attached to St. Dunstan's church near Temple Bar, London, is often mentioned by writers of the last century, and we believe is still to be seen.

Some authors attribute the invention of *water-clocks* to Ctesibius, and others suppose they were first used under the Ptolemies; but both are mistakes: they were doubtless greatly improved by the Alexandrian mathematician, and probably reached the acme of perfection under the successors of Alexander. In India, Egypt, Chaldea and China, clepsydræ date back beyond all records. They were known at an early period in Greece. Plutarch mentions them in his life of Alcibiades, who flourished in the fifth century B. C. when they were employed in the tribunals at Athens to measure the time to which the orators were limited in their addresses to the judges. Demosthenes and his great rival Æschines allude to this use of them. Plato had water-clocks, and to him was attributed their introduction into Greece. Plutarch in his Philosophy, observes, that Empedocles illustrated the act of respiration by "a clepsidre water hour-glass." (Opin. of Philos.) Julius Cæsar found the Britons in possession of them. Pliny (book vii, 60.) says, men announced with the voice the hours from the shadow of the sun, and that Scipio Nasica set up the first clepsydra "to divide the hours of both day and night equally, by water distilling and dropping out of one vessel into another."

The ancients had various modifications of water-clocks, some were exceedingly simple, and others elaborately constructed, and the forms and decorations wonderfully diversified; but the principle was more or less the same in all, viz. water trickling through a minute channel from one vessel into another. The instruments were made of various materials from glass to gold, and of sizes differing, like modern clocks, from large ones permanently erected for public use to such as were carried in the hand.

Valerianus, who wrote in the 16th century, says the priests of Egypt divided the day into *twelve* hours, because the cynocephalus, a sacred animal, was observed to make a violent noise at those times, and to void urine as often. Cicero mentions a tradition of Trismegistus observing the same thing. The Egyptians, therefore, ornamented their water-clocks with figures of apes, and some were of the form of those animals urinating; hence it would seem that this singular people not only derived enemas from studying the habits of the ibis, but were led to construct clepsydræ from noticing those of monkeys.

As it is impossible to give anything like a history of these machines in this volume, we shall notice a few only, but sufficient to give a general idea of their construction and variety. Sometimes an empty basin with a minute opening through its bottom was placed floating in a cistern of water; the fluid gradually entering filled it in an hour, half an hour, or some other determinate time. It was then emptied and allowed to swim as before; as soon as it became filled, a gong or other instrument was sounded for the information of the public.

" The Siamese measure their time by a sort of water-clock, not like the clepsydra of old, wherein the water descended from above, but by forcing it upwards through a small hole in the bottom of a copper cup placed in a tub of water. When the water has sprung up so long that the cup is full, it sinks down, and those that stand by it, forthwith make a noise with basons, signifying that the hour is expired." (Ovington's Voyage to Surat in 1689, p. 281.)

The *ghurree al*, or clepsydra of the Hindoos, consists of a thin brass cup having a hole in the bottom. " A large vessel is filled with water and this cup placed on the surface; the water rises through the hole, and when it has reached a height marked by a line previously adjusted, the watchman strikes the hour with a wooden mallet on a pan of bell metal."—(Shoberl's Hind. v, 157.)

In other devices, time was measured by *emptying* the vessel. Valerianus observes, that the priests of Acanta, a town beyond the Nile, poured water every day into a vessel, by the dropping of which through a small hole they measured the hours.—(Harris' Lex. Tech.)

Dr. Fryer, who visited India in the 17th century, observed the Hindoos measuring time " by the dropping of water out of a brass basin."— (Travels, 186.)

It is obvious that by adapting the size of an opening in the bottom of a vessel, the entire contents of the latter might be made to flow out in a certain time and with tolerable accuracy; but in refilling it great care was required to introduce precisely the same quantity. To accomplish this, both the vessel and receiver were closed on all sides and connected together, so that when the proper quantity of fluid was once introduced, it could neither escape by leakage or evaporation. Both vessels were shaped *like a pear* and *united at the smaller ends*, through which the passage for the fluid was made; and sometimes sand was used instead of water. Hence the hour-glass of modern days, the only modification of ancient clepsydræ which modern nations continue to use. Nieuhoff observes of Chinese water-clocks, " they bear a resemblance to some great hour-glasses in shape ;" and he says, in several sand was used instead of water. On an ancient bas-relief at Rome, representing the marriage of Thetis and Peleus, Morpheus holds an *hour-glass ;* and from Athenæus we learn that the ancients carried portable ones about with them somewhat as we do watches.

In another variety of clepsydræ, the sides of the vessel from which the fluid escaped were graduated, somewhat like chemists' measuring glasses, and the hours announced as the descending surface of the fluid reached the marks. If the vessel was of a cylindrical or cubical figure the distance between the marks was not uniform, because the water escaped fastest at first, in consequence of the greater pressure of the column over the orifice, which pressure constantly diminished with the efflux ; the surface of the fluid could not therefore descend through equal spaces in equal times. When such formed vessels were used, the relative distances of the marks were probably determined by experiment, although they might have been by calculation. Sometimes the vessels were funnel-shaped, the angle of their sides being so adjusted that an equal distance could be preserved between the marks—unequal quantities of the fluid escaping in equal times. These instruments were generally made of glass, and a cork or some floating image, to which a needle was secured, pointed out the hour as the water sunk. Pancirollus says, the small holes were edged with gold.

In some clepsydræ the fluid was received into a separate vessel to *raise* a floating image that pointed as an index to the hours. Sometimes a boy with a rod, Time with his scythe, and Death with a dart. In this variety of the instrument, it was desirable that the quantity of fluid discharged into the vessel should be uniform at all times ; and to effect this, the floating siphon, No. 239, was sometimes used. Such we presume was the clepsydra of Orontes, which was made " in the form of a small ship floating on the water, and which emptied itself by means of a siphon placed in the middle of it." Dr. Harris, not aware of the property of a floating siphon, could not perceive how the hours were made equal by this contrivance, which, he observes, Orontes devised to remedy the *unequal flow* of water from an open vessel.—(Lex. Tech.)

Nieuhoff noticing the numerous towns in China, upon the greater part of which, he observes, were clepsydræ, says, " upon the clock-house turrets stands an instrument which shows the hour of the day by means of water, which running from one vessel into another *raises a board*, upon which is portrayed a mark for the time of day ; and you are to observe, there is always a person to notice the time, who every hour signifies the same to the people by beating upon a drum, and hanging out a board with the hour writ upon it in large letters." (Ogilby, Trans. 196.) Montanus says these letters were " a foot and a half long." See also Purchas' Pilgrimage, 499.

In another class of ancient clepsydræ, the water dropped upon an overshot wheel, which turned an index in the centre of a circle, round which the hours were marked ; hence our clock and watch dials. " The Chinese have other instruments to know the hour of the day, being somewhat like our clocks with wheels, and they are made to turn with *sand* as millwheels are with water." (Nieuhoff.) At last solid weights were introduced in place of water, and by means of cords gave motion to the index, and thus opened the way still more for the introduction of modern clocks.

It would appear from the description of clepsydræ by Vitruvius and other writers, that the ancients had carried these machines to very great perfection; and as regards ornament, they probably excelled many of our mantel time pieces. They were even combined with music. Thus Plato had one that, during the night, when the index of the dial could not be seen, announced the time by playing upon flutes. Athenæus also constructed one that indicated the hours by sounds, produced by the compression and expulsion of air by water—the same principle as Plato's. Petrarch in enumerating the spoils of Asia which Pompey exhibited at his third triumph, besides cups, chests and beds of gold, a mountain of the same metal, with statues of harts, lions and other beasts; trees, and all kinds of fruits formed of pearls suspended from golden branches, &c. continues, "Of the same substance, there was a *clocke*, so cunningly wrought that the woorkmanshyp excelled the stuffe, and which continually moved and turned about—a right woonderfull and strange sight." —("Phisicke against Fortune," translated by T. Twyne. Lond. 1579, F. 120.)

The want of uniformity in the going of ancient water-clocks was noticed by Seneca, and compared to the differences of opinions entertained by philosophers; and Charles V. after shedding rivers of blood to make men believe the dogmas he wished to impose, amused himself in his retirement in the construction of watches, and was surprised that he could not make two go alike.

No. 272 represents one of the improved clepsydræ of Ctesibius,[a] from Perrault's Translation of Vitruvius.[b] It presents several interesting particulars relating to the state of the useful arts upwards of twenty centuries ago, and is better calculated to impart information to mechanics respecting the ingenuity, and even the workshops and tools, of their ancient brethren, than reams of letter-press. Besides carving, turning, founding, &c. &c.—it shows the practical application of water to move overshot wheels—the art of transmitting motion and of changing its direction by toothed wheels—it exhibits the same principle of measuring time as practised by our clock and watch-makers, viz. by proportioning the number of teeth on wheels to those on the pinions between which they work. The application of the siphon is also interesting, being the same as is used to illustrate the action of intermitting springs. Upon this instrument the renewal of the diurnal movements of the machinery depended: its effect being similar to that of winding up an ordinary clock.

This clepsydra consisted of a cylindrical column placed on a square pedestal, within which the mechanism was concealed. The hours for both day and night were marked upon the column; their inequality at different seasons being measured by unequal distances between the curved lines and by the revolution of the column round its axis once a year. On the pedestal are seen the figures of two boys, one of which was immoveable, but the other rose and pointed out the different hours with his wand. Water (supplied from some reservoir by a concealed pipe) continually dropped from the eyes of the figure on the left, and falling into a dish was conveyed, by a horizontal channel, under the feet of the other figure, where it trickled into a deep vessel, or large vertical tube, whose lower end was closed. In this tube a float was made to rise and fall with the water, and being attached to the feet of the figure with the wand caused it to rise also, and thus to indicate the lapse of time. At the end of 24

---

[a] There was another ancient philosopher of the same name, Ctesibius of Chalcis.
[b] In Barbaro's Vitruvius, Venice, 1567, there are figures of two others equally ingenious, but rather more complex.

hours the tube would be filled, and the figure near the top of the column. It was then that the siphon came into play. Its short leg, as represented in the cut, was connected to the lower part of the tube that contained the float, and its bend reached as high as the upper end of the tube. When the latter therefore was full the siphon would be charged, and the contents of the tube discharged by it into one of the buckets of the wheel. The figure with the wand would then descend, having nothing to support it. The wheel having six buckets only, performed a revolution in six days. To its axis was secured a pinion of six teeth that worked into a wheel with sixty, and on the shaft of this wheel a pinion of ten teeth drove a wheel of sixty-one teeth, which last wheel by its axis turned the column round once in 365 days.

As the accuracy of such a clock depended upon the size of the orifices in the weeping figure, whence the water escaped, to prevent their enlargement by the friction of the liquid, Ctesibius bushed them with jewels.

About the year 807, the Caliph Haroun sent some valuable presents to Charlemagne, and among them a water-clock, which struck the hours by means of twelve little brass balls falling on a bell of the same metal. There were also twelve figures of soldiers, which at the end of each hour opened and shut doors according to the number of the hour.—(Martigny's Hist. Arabians, iii. 92.)

There is a very simple clepsydra in Kircher's Mundus Subterraneus, tom. i, 157. M. Amontons devised another. Mém. Acad. Science, A.D. 1699, p. 51. See also Phil. Trans. vol. xlv, p. 171, and Fludd's Simia. Decaus has given a clepsydra in the fifth plate of his Forcible Movements. A water pendulum is figured in Ozanam's Recreations, p. 388.

N 272. Clepsydra by Ctesibius.

Hour-glasses were formerly placed in coffins and buried with the corpse, probably as symbols of mortality—the sands of life having run out. See Gent. Mag. vol. xvi, 646, and xvii, 264. Lamps found in ancient sepulchres were possibly interred with the same view—to indicate the lamp of life having become extinguished.

Garcilasso mentions a dial by which the Peruvians ascertained the time when the sun entered the equinox : whether these people or the Mexicans had water-clocks we have not been able to ascertain.

Hydraulic Organs do not appear to be of so high antiquity as clepsydræ, but their origin is equally uncertain. Perhaps they were derived from musical water-clocks.

The first organs were probably nothing more than simple combinations of flutes, pipes, and other primitive wind instruments. What the circumstances were that led to the idea of uniting a number of these, and supplying them with wind from bellows instead of the mouth can hardly be conjectured. The first step was probably bag-pipes, and the second the addition of keys or valves. In process of time, the instruments, instead of being made of reeds or other natural tubes, were formed of metal; and their number, variety, and dimensions increased until organs became the most powerful and harmonious, and consequently the most esteemed of all musical machines. The organs mentioned in the Bible were probably portable ones, similar to the modern regal. The ancients divided them into two classes—*pneumatic* and *hydraulic*, or wind and water organs. The difference consisted merely in the modes of supplying the wind—in one it was by means of water, while in the other bellows were worked by men.

Water was employed in various ways in ancient hydraulic organs.

1. By falling through a pipe, it carried down air into a reservoir, as in the trombe or shower bellows, (No. 198.) Here it not only furnished the air but forced it through the pipes. According to Kircher, it was then discharged on a wheel, and gave motion to drums on whose peripheries were projecting pins, which depressed the keys of the instrument, as in the modern barrel organ.

2. It was discharged upon an overshot wheel, and by cranks and levers merely worked common bellows. This may seem strange to some readers, but it must be remembered that these instruments were often of enormous dimensions. Even so rude a people as the Anglo-Saxons, had organs that required " seventy strong men" to work the bellows.

3. Sometimes water was only used in an open tank or cistern, into which a smaller one constituting the air-chamber was inverted. The air was then forced by ordinary or piston bellows into the latter, and displacing the water caused it to rise in the outer vessel, where its constant pressure urged the air through the organ.—See No. 110, and p. 245.

4. The vapor of boiling water or steam was also used, and which of course supplied the place of both wind and bellows. The extent to which steam was used is unknown. It was probably confined chiefly to the temples.

The details of the mechanism of ancient organs that have come down are very imperfect. Their description by Vitruvius and Heron is obscure, and in some parts unintelligible; and they admit that the construction was too complex to be easily comprehended except by those familiar with them.

The earliest distinct notice on record of any thing like a water-organ, is the musical clepsydra of Plato. There is no reason to suppose it was invented by him, but rather the contrary, for he contemned all mechanical speculations. He probably met with it in Egypt, and having introduced it to his countrymen, was (as usual with them) considered its author.

Tertullian, in a Treatise on the Soul, speaks of an organ invented by Archimedes, but of its construction little is known.

From Vitruvius' account of hydraulic organs, and from the last two Problems in Heron's Spiritalia, we learn that they were very elaborate machines. Decaus has amplified some of Heron's devices for producing music by water.

Plutarch in comparing Cato and Phocion, after observing that their severity of manners was equally tempered with humanity, and their valor with caution; that they had the same solicitude for others, and the same disregard for themselves; the same abhorrence of every thing base and dishonorable, &c. observes, that to mark the difference in their characters would require a very delicate expression, like the *finely discriminated sounds of the organ*. This is supposed by Langhorne to have been a water-organ. The elder Pliny refers to them in book ix, cap. 8. Speaking of dolphins, he observes, they are fond of music, especially "the sound of the *water instrument*, or such kind of pipes." We noticed, page 245, a representation of an hydraulic organ on a medal of Valentinian. The silver Triton, mentioned in the chapter on Fountains, that by machinery was made to rise out of the lake and sound a trumpet, may be considered a modification of these instruments, and so may the whistling clock of Athenæus mentioned in the last one.

Suetonius, in his Life of Nero, mentions an hydraulic organ which that emperor took particular pleasure in. It must have been a magnificent affair, since even Nero deemed it of sufficient importance to form the principal object vowed by him, when the empire was in danger from the rebellion of Vindex. Inviting some of the chief Romans to a consultation on public affairs, "he entertained them the rest of the day with an organ of a new kind, and showing them the several parts of the invention, and discoursing about the nature and difficulty of the instrument, he told them he designed to introduce it upon the theatre, if Vindex would permit him." In this passage Suetonius does not state that the machine was a *water-organ*; but in a subsequent one he observes,—" Towards the close of Nero's life, he publicly vowed that if the empire was secured to him (by overcoming the rebels) he would bring out at the games, for his obtaining the victory, *a water-organ*, a chorus of flutes and bag-pipes," &c.

The author of a letter, attributed to St. Jerome, speaks of a large organ at Jerusalem, the sounds of which could be heard at the distance of a thousand paces, or to the Mount of Olives. It consisted of two elephant skins, or rather perhaps resembled two of those animals. There were twelve large bellows and fifteen brass pipes. The two animals were said to represent the Old and New Testaments—the pipes the patriarchs and prophets, and the bellows the twelve apostles. The particulars of its construction are not known.

Organs were used more or less throughout the dark ages, during which several were brought into Europe from the East.

In 757, the Greek emperor Constantine sent two organs to Pepin, king of France. Mezeray says, they were the first seen in that country. Another was sent from Constantinople to Charlemagne in 812; but nothing is known of their construction, except that the last imitated the sounds of thunder, the lyre and cymbal.

In the ninth century, Louis Debonnaire had a water-organ made for his palace at Aix-la-Chapelle, by a Venetian priest named George. Another organ, in which water is supposed not to have been employed, he erected in one of the churches of that city, and its sounds are said to have been so ravishing, that a woman died in ecstacy under their influence.—(See Preface to L'Art du Facteur D'Orgues; Arts et Metières, folio edit. 1778.)

At page 401, we mentioned an organ made by Gerbert, in which steam was employed instead of air.

We find, says Fosbroke, organs with pipes of box-wood, of gold, and organs of alabaster and glass; and some played on with warm water. Brass pipes and bellows are mentioned by William of Malmsbury. "The monks of Italy, of the orders devoted to manual labor, applied themselves to the fabrication of organs; and in the tenth century, a maker was sent into France, whence they insensibly spread over all the western churches."

Of modern hydraulic organs it is unnecessary to enlarge. Several have been noticed in the chapter on Fountains. They have become nearly extinct. See Kircher's Musurgia Universalis, Fludd's Simia, Decaus' Forcible Movements; Misson, Blainville, Breval, and Keysler's Travels.

The old *Regal*, a diminutive species of organ, still used in some parts of Europe, was sometimes acted on by water; at least so it would seem from a remark of Lord Bacon in his *Sylva*. Speaking of music, he particularizes the tones from the percussion of *metals*, as in bells—of *air*, as in the voice while singing, in whistles, organs, and stringed instruments—" and of *water*, as in the nightingale pipes of *regalls*, or organs, and other hydraulics, which the ancients had, and Nero did so much esteeme, but are now lost."—Cent. ii, 102.

CHAPTER IX.

SHEET LEAD: Lead early known—Roman pig lead—Ancient uses of lead—Leaden and iron coffins—Casting sheet lead—Solder—Leaden books—Roofs covered with lead—Invention of rolled lead—Lead sheathing. LEADEN PIPES: Of great antiquity—Made from sheet lead by the Romans—Ordinance of Justinian—Leaden pipes in Spain in the ninth century—Damascus—Leather pipes—Modern iron pipes —Invention of cast leaden pipes—Another plan in France—Joints united without solder—Invention of drawn leaden pipes—Burr's mode of making leaden pipes—Antiquity of window lead—Water injured by passing through leaden pipes—Tinned pipes. VALVES: Their antiquity and variety—Nuremberg engineers. COCKS: Of great variety and materials in ancient times—Horapollo—Cocks attached to the laver of brass and the brazen sea—Also to golden and silver cisterns in the temple at Delphi—Found in Japanese baths—Figure of an ancient bronze cock—Superior in its construction to modern ones—Cock from a Roman fountain—Numbers found at Pompeii—Silver pipes and cocks in Roman baths—Golden and silver pipes and cocks in Peruvian baths—Sliding cocks by the author. WATER-CLOSETS. Of ancient date—Common in the East. TRAPS for drains, &c.

A FEW subordinate inventions, but such as are of some importance in practical hydraulics have been reserved for this chapter, viz : sheet lead, pipes, valves and cocks, water-closets and traps.

Lead was probably worked before any other metal. Its ores abound in most countries, and frequently reach to the surface; they are easily reduced; the metal fuses at a low temperature; it is soft and exceedingly plastic. Lead is mentioned as common at the time of the Exodus. It was among the spoils taken by the Israelites from the Midianites, and articles made of it were ordered to be melted up. The Phenicians exported tin and lead from Britain. Both are enumerated in the graphic account of the commerce of Tyre, in the 27th chapter of Ezekiel. The Romans worked lead mines in France, Spain and Britain; Pliny says, those in the former countries were deep and the metal procured with difficulty; but in Britain it was abundant, and " runneth ebb in the uppermost coat of the ground." Several Roman mining tools and pigs of lead have been found in England. In 1741, two pigs were dug up in Yorkshire. Their form was similar to that in which the Missouri lead is cast, but more than twice the weight. Each weighed 150 lb. and was inscribed in raised letters with the name of the reigning emperor, Domitian.—(Phil. Trans. Abrid. ix, 420.)

The uses to which lead was put by the ancients were much the same as at present. The fishermen of Egypt sunk their nets with it just as ours do. A portion of a net with " sinkers" attached is preserved in the Berlin Museum. Leaden statues are ancient. There was one of Mamurius at Rome. They probably preceded those of bronze, and perhaps formed part of the spoil of the Midianites mentioned above. The Romans had leaden coffins; a device adopted more or less in all ages. Double leaden coffins (observes Fosbroke) occur in the Anglo-Saxon era, not made of plain lead, but folded in a very curious and handsome manner. For the mode of making coffins and their singular forms, consult L'Art du Plombier, Arts et Mètieres, tome xiii, a Neuchatel, 1781.[a]

The art of casting lead into sheets on beds of sand, as now practised by plumbers, is of immense antiquity. The terraces of Nebuchadnezzar's hanging gardens were covered with sheets of lead soldered together, to retain moisture in the soil. The composition of ancient solder for lead, we know from Pliny, was the same as ours. It is uncertain whether the art of uniting lead by "burning," that is, by fusing two edges together (without solder) was known. Pliny says, "two pieces of lead cannot possibly be soldered without tin glass." Either therefore the ancients had not the art of "burning" pieces of lead together, or Pliny was not aware of it.

Tablets of lead were anciently used to write on. Job alludes to them. Books composed of leaden leaves are figured by Montfaucon. To such tablets, we presume, Pliny refers, when he speaks of lead "driven with the hammer into thin plates and leaves."—(Nat. Hist. xxxiv, 17.)

The employment of sheet lead as a covering for roofs ascends to the earliest ages in the East. It is still extensively used there. Tavernier, in speaking of the mosques at Aleppo, says their domes were covered with lead, and so was the roof of the great hall of the Divan at Constantinople. He mentions an inn or caravansary, the roof of which was covered with the same metal. Henry Blount, who traveled in Egypt and Turkey in 1634, found the roofs of the mosques and seraglios at Adrianople covered with lead. Count Caylus mentions ancient sheet lead half a line thick taken from the inner dome of the Pantheon. Gregory of Tours describes an old temple of the Gauls, which was extant in the time of the Emperor Valerian, and had a leaden roof. (Montfauc. Supp.) Paulinus built a church at Catarick, Eng. which was burnt by the Pagans; he built another of wood, which was the mother church of British Christianity, " and enclosed the whole building with a covering of lead." The churches and castles of Europe in the middle ages were almost uniformly covered with this metal. In a statute passed in the fourth year of Edward I. of England, (A. D. 1276,) to ascertain the value of real estate, commissioners were appointed to visit "castles, and also other buildings compassed about with ditches [to determine] what the walls, buildings, timber, stone, *lead*, and other manner of covering is worth."—(Statutes at large.)

Leaden seals on woolen cloth were used in Henry IV.'s reign.

It is uncertain whether the ancients were acquainted with the process of forming lead into sheets by passing it between rollers. If they were, the art, like many others, became lost, and was not revived till the 17th century. A close examination of specimens of ancient sheet lead might determine the question.

Rolling or "milling" of lead was invented by Mr. Thos. Hale, in 1670, about which time the first mill was erected at Deptford. The inventor met with violent opposition from shipwrights, because the lead, from its smooth surface, uniformity of thickness, and low price, began to be generally adopted for sheathing vessels, in place of the old wooden and leather sheathing. And as it was used also for gutters and roofs of houses, "the

---

[a] About twenty years ago iron coffins were introduced in England and secured by patent; but they were not then by any means a new thing under the sun : for the Parsees of India for ages buried their dead in them. " Ces idolatres adorent le feu comme une divinité, considerant le bois comme sa viande ; d'où il vient qu'ils ne mettent pas leurs morts dans les cercueils de bois, *mais de fer.*"—(C. Antiquaire, iii. 846.)

plumbers were as industrious as the shipwrights to decry the lead;" but finding their opposition in a great measure fruitless, " some of them now began to cry it up, and have set up engines to mill it themselves."— (Collier's Dict. Art. England.)

A paper in the Phil. Transactions, 1674, erroneously attributes the invention to Sir Philip Howard and Major Watson. These gentlemen were associated with Hale in the patent, and merely contributed their influence to introduce the new manufacture, especially to sheathe the public ships. (Abrid. i, 596.) The large ship built by Archimedes was sheathed with lead.

PIPES for the conveyance of water have been made of earthenware, stone, wood and leather, but more generally of lead and copper. *Leaden pipes* extend back to the dawn of history. They were more or less common in all the celebrated nations of old. In the old cities of Asia, Egypt, Greece, Syria, &c. they were employed to convey water wherever the pressure was too great to be sustained by those of earthenware. The same practice is still followed : thus in Aleppo, both leaden pipes, and those of stoneware are used, and in all probability just as they were when this city was known to the Greeks as Bercœa, and to the Jews in David's time as Zobah. Archimedes used pipes of lead, to distribute water by engines in the large ship built for Hiero ; and the same kind were doubtless employed in conveying water to the different terraces of the famous gardens of Babylon. The great elevation to which the fluid was raised would render earthenware or wooden pipes entirely inapplicable.

We have no information respecting the mode of making leaden pipes previous to the Roman era; but as that people adopted the arts and customs of older nations, we may be assured that their tubes, as well as their pumps and other engines, were mere copies of those made by the plumbers of Babylon and Athens, Egypt and Tyre. All ancient pipes yet discovered are said to have been made from sheet lead ; viz : strips of sufficient width folded into tubes and the edges united by solder. We learn from Vitruvius that Roman plumbers generally made them in ten feet lengths, the thickness of the metal being proportioned to their bore, according to a rule which he gives in book viii, cap. 7. of his Architecture. Large quantities of Roman leaden pipes have been found in different parts of Europe, varying in their bore from one to twelve inches. Some of them are very irregularly formed, their section being rather egg-shaped than circular. Montfaucon has engraved several specimens. On large ones belonging to the public, the name of the consul under whom they were laid was cast upon them. Others that supplied the baths of wealthy individuals have the owners' names; and sometimes the maker's name was cast on them. Of small leaden pipes, Frontinus mentions 13,594 of one inch bore that drew water from one of the aqueducts. Pompeii was but a small provincial town, of which not more than one-third has been explored, and yet a great many tons of pipes have been found. The consumption of lead for pipes must have been enormous in old Rome, not only from their great number, but on account of the large dimensions of the principal ones. Pliny might well observe, " Lead is much used with us for sheets to make conduit pipes."—(xxxiv, cap. 17.)

An ordinance of Justinian respecting a bagnio erected at Constantinople by one of the dignitaries of the empire is extant : " Our imperial will and pleasure is, that the *leaden pipes* conducting the water to the Achillean bagnio, contrived by your wisdom, and purchased by your munificence, be under the same regulations and management as have been appointed 'n the like cases ; and that the said pipes shall only supply such bagnios

Chap. 9.] *Cast Leaden Pipes—First Articles of Cast-Iron in England.* 553

and nymphæa as you shall think fit," &c. Constantinople has for ages been supplied with water through leaden pipes. The Sou-terazi or water towers, are mere contrivances to facilitate the ascent and descent of the fluid through pipes.

Leaden pipes have been uninterruptedly employed in some or other of European cities since the fall of the empire. In the middle of the ninth century water was conveyed by them to supply Cordorva, in Spain, under the Caliph Abdulrahman II. who also caused that city to be paved. This is the oldest pavement on record in modern cities. Benjamin of Tudela, who visited Damascus in the 12th century, says, the river *Pharpar* (see 2 Kings, chap. v,) slideth by and watereth the gardens; " but *Abana* is more familiar and entereth the citie, yea, by helpe of art in conduits [pipes] visiteth their private houses."—(Purchas' Pilgrim.)

The ancient inhabitants of the island Arados ingeniously obtained fresh water from the bottom of the sea. They sunk down over the spring a large bell of lead, to the upper part of which was attached a pipe of leather that conveyed the fluid to the surface.—(Pliny, v, 31.)

Some of the Roman earthenware pipes were made to screw into each other. Old leaden pipes laid, A. D. 1236, to supply London, are mentioned at page 294. Most modern pipes of large bore are now made of cast iron. The largest sizes now laying to supply this city, are nine feet in length, three feet internal diameter, and weigh from 3500 to 3800 pounds.

The first improvement on the ancient mode of making leaden pipes was matured in England in 1539. It consisted in *casting* them complete in short lengths, in molds placed in a perpendicular position. After a number were cast, they were united to each other in a separate mold, by pouring hot metal over the ends until they run together. The device for " burning" or melting the ends was exceedingly ingenious, and such pipes are still made to some extent in Europe. In the 30th year of Henry VIII. (observes Baker in his Chronicles of the Kings of England,) " the manner of casting pipes of lead for conveyance of water under ground without using of soder, was first invented by Robert Brocke, clerk, one of the king's chaplains, a profitable invention; for by this, two men and a boy will do more in one day, then could have been done before by many men in many days. Robert Cooper, goldsmith, was the first that made the instruments and put this invention in practice."—(Edit. of 1665, p. 317.)

Five years afterwards, Ralph Hage and Peter Bawde made the first articles of *cast iron* in England.—(Ibid.)

In the reign of Henry IV. of France, a native of St. Germain, devised another mode of casting pipes and burning them together. The mold was used in a horizontal position, and the metal poured in at one end. When a pipe was cast, it was not drawn entirely out of the mold, but one or two inches were left near the spout where the metal entered, so that when another length was cast, the hot metal running over the end of the previous pipe fused it, and both became as one. The tube was then drawn nearly out and another one cast and united to it in like manner, and so on till any required length was attained.—(See Planche, vii. L'Art du Plombier in Arts et Mètieres.)

Sometimes pipes formed of sheet lead have their seams united by "burning." A strip of pasteboard is packed against the inside of the seam, and the tube (if small) filled with sand; the edges are then melted with a soldering iron, and the deficiency made up with a bar of lead, in the same way as when a bar of solder is used. The old mode of burning these seams was by pouring hot lead upon them, and generally a projection of metal was left along the seam. The ancient pipes figured by Montfaucon have a similar projection. The plan of drawing leaden pipes through dies like hollow wire was first proposed by M. Dalesme, in the Transactions of the French Academy of Sciences, in 1705. It was subsequently brought forward by M. Fayolle in 1728; but it was not till 1790 that such pipes were made. In that year, Mr. Wilkinson, the celebrated English iron master, took out a patent for drawing them, since which period they have become general in England, France, and the U. States.—(See Reper. of Arts, 1st series, vol. xvi.)

In 1820, a singular mode of making leaden pipes was patented in England by Mr. T. Burr. A large and very strong cast iron cylinder, in which a metallic piston is made to work, is secured in a vertical position. To the underside of the piston a strong iron rod is fixed, its lower end being cut into a screw or formed into a rack for the convenience of forcing the piston up, either by means of a steam engine or any other suitable first mover. To the upper side of the piston is secured a polished cylindrical rod, rather longer than the cylinder, and of the same diameter as the bore of the pipe. The cylinder forms a mold in which the pipe is first cast, and this rod is the core. The bottom of the cylinder may be open; but the top is strongly closed, with the exception of a circular and polished opening at the centre, of a size equal to the external diameter of the pipe. Suppose the piston now drawn down to near the bottom of the cylinder, the upper end of the polished rod will stand a little above the circular opening, and an annular space will be left between them equal to the required thickness of the tube. The cylinder is then to be filled with fused lead through an opening at the top, (which is to be stopped up by a screw-plug or any other device,) and as soon as the metal begins to assume the solid state, the piston is slowly raised; this necessarily forces the lead through the annular space in the form of a tube, which is then wound on a reel as fast as formed.

Various cylinders are employed according to the different sized tubes. For half inch pipe, one 18 inches long, six or seven inches internal diameter, and the sides three or four inches thick would be required. Plates with openings of different sizes may be adapted to one cylinder. They may be made to slide in recesses cast in the top.

This mode of forming leaden tubes is the same in principle as that by which some of earthenware have been made: the clay being put into a square and close trunk, is forced by a piston through an annular space, adapted to the thickness and bore of the tubes required. At first sight the process appears difficult. It also seems strange that *solid* lead can thus be squeezed through an aperture into the form of a tube; but it should be remembered that this metal is extremely soft when heated to near the fusing point; and that the mode only differs from that of making clay pipes in requiring a greater force. Tubes made in this way are in general more solid than others. This arises from the large body of metal of which they are formed being poured while very hot into the cylinder, so that there is little danger of flaws or fissures. These pipes may also be made in much greater lengths than by any other plan. A manufactory of them has recently been established in Philadelphia.—(See Repertory of Arts, for 1820, vol. xli, p. 267.)

From the quantities of pipes used of old, it appears singular that the art of drawing them was not discovered, especially as the *Tire-Plomb* or glazier's vise for drawing "window lead" is of ancient date—a most beautiful machine, and one far more ingenious and interesting than the drawbench; one too by which lead is worked at a single operation into very difficult forms, and such as require the metal of different thickness in the same piece.[a]—(See L'Art du Vitrier, pl. v, Arts et Mètieres.)

It has long been known that water conveyed through leaden pipes becomes more or less impregnated with a poisonous solution of that metal; a fact of which the ancients were fully aware, and which made them very scrupulous in using it for purposes of domestic economy. Hippocrates and Galen denounced its employment both for cisterns to contain, and tubes through which to conduct water. Ancient architects were of the same opinion; thus Vitruvius observes, that water drawn from leaden tubes is very pernicious, and adds, "we should not, therefore, conduct water in pipes of lead if we would have it wholesome." The Medical Transactions of modern times, and works on mineral poisons abound with examples of the fatal effects of drinking water from reservoirs and pipes of this metal.

Several modes have been devised to render leaden pipes innoxious. In 1804, an English patent was obtained for coating their interior surfaces with tin. This was effected in the following manner :—Suppose a workman engaged in making tubes of half an inch bore; he first pours lead into an iron mold and forms a pipe two feet long, an inch thick, and nearly an inch in the bore : as soon as the lead poured in becomes solid, he withdraws the steel mandril which formed the interior of the tube, throws in a little rosin dust, and inserts a half inch mandril, between which and the inside of the tube a certain space is left. Into this space he then pours melted tin, which as it collects below, causes the rosin to float on its surface, as it rises to the top, and lubricates the hot sides of the leaden tube. Both metals thus become united, and when the tin becomes solid the mandril is taken out; and the tube, thus plated with tin, is passed to the drawbench, and drawn out to the required length like an ordinary leaden tube. There is some difficulty in making the tin unite uniformly to the lead, and when this does not take place the pipes are apt to be broken in drawing; for as the two metals do not stretch equally, the thin lining of tin is pulled apart; and if the lead does not separate at the same place, its surface is exposed, and the strength of the tube greatly diminished at such places.— (Repertory of Arts, 2d series, vol. v.)

In 1820 another English patent was issued for a similar plan, the difference consisting chiefly in a mode of better securing the union of the tin with the lead.—Ibid. vol. xxxviii.

In 1832, the author of this volume took out a patent for coating leaden pipes with tin, by passing them, after being drawn and otherwise finished, through a bath of the fluid metal. As there is a difference in the fusing points of tin and lead of about 200° Fahrenheit, there is no difficulty in the process. By this plan tubes are effectually tinned both inside and out, and any imperfections or fissures are soldered up. The operation is exceedingly simple and the expense trifling. The process is patented in England, where the tubes are, we believe, more extensively used than in this country.—(See Journal of the Franklin Institute for November, 1832, and May, 1835.)

Valves and *Cocks* are too essential to hydraulic engines to be omitted in this work. The principle of the valve has always been in use for a variety of purposes. Doors are valves, and were so named by the ancients. Those of the private apartments of Juno were contrived by Vulcan to close of themselves. Thus Homer sings :

---

[a] In one of the apartments of a villa at Pompeii, there was a large glazed bow-window. The glass was thick, tinged with green, and " set in *lead* like a modern casement." —(See Encyc. Antiq. pp. 57, 398.)

> Touched with the secret key, the doors unfold;
> Self-closed behind her shut the valves of gold.[a] *Iliad*, xiv.

It is probable that all valves were originally in the form of doors; that is, mere flaps or clacks moving on a hinge, and either laying horizontally like a trap-door, inclined like some of our cellar doors, opening vertically as an ordinary door, or suspended by hinges from the upper edge; and sometimes they consisted of two leaves like folding doors. Examples of all these are still common. Isis was represented by the ancient Egyptians with "the key of the sluices of the Nile" in her hand; the instrument by which the doors or valves, like the locks in our canals were opened and closed.

The most ancient musical wind instruments known in the Eastern world are provided with valves, as the primitive bag-pipes, and the Chinese variation of this instrument, which Toreen describes as consisting of " a hemisphere to which thirteen or fourteen pipes are applied, and catching the air blown into it by valves." The pastoral flute of Pan, from its expressing thirty-two parts, he supposes to have been of a similar construction. (Osbeck's Voyage, ii, 248.) Valves were of course employed in the organs of Jubal, as well as in the bellows belonging to his celebrated brother and other antediluvian blacksmiths. The ninth problem of the Spiritalia relates to valves. Conical metallic valves were used by Ctesibius in the construction of clepsydræ. In most of the old representations of pumps, flaps of leather, loaded and stiffened with pieces of wood or lead are figured. Agricola has given figures of no other. These *clacks*, as they are named, are in most cases preferable to the most perfect spherical or conical valves of metal: the smallest particle of sand adhering to these makes them leak; besides which, they are liable to stick. We have known them replaced with common clacks. Amontons, in experimenting with a forcing pump, found the valves, which were of highly polished metal and well fitted, adhere so strongly to their seats, that he had to substitute leather clacks for them.

The spindle valve, or such as have a long shank to prevent their rising too high, and guiding them when descending, is said to be of French origin.

We have sometimes used a simple valve on the lower box of a pump. It consisted of a short pipe of thin and very soft leather secured to the upper side of the box. When the sucker was raised, the water rushed through this pipe, and when the stroke ceased, it was instantly collapsed by the pressure of the fluid above it, and then fell down on one side of the box.

Cocks are a species of valve, but not self-acting like the latter. In pumps and bellows the momentum of the entering fluids opens the valves,

---

[a] Doors opened and closed by secret machinery were formerly much in vogue. Heron made those of a temple thus to act. Vitruvius speaks of doors that closed by themselves, (and when opened, rose sufficiently high to clear the carpet.) In the old cities of Europe, the gates were moved by concealed mechanism to prevent a surprise. Those at Augsburg were famous. A single person only could enter at a time, and he was inclosed between two gates till the object of his visit was ascertained. As soon as he approached the first one, it opened of itself, he entered, "and it closed upon his heels." On reaching the second it acted in like manner. During these operations, the visitor saw no person, although he was exposed to the scrutiny of officers within. The magistrates of Nuremburg, desiring to have a gate of the same kind for the security of their city, sent some engineers to take a model; but after several examinations, they returned home and reported "that without pulling down the walls, and all the masonry, it was not in the power of *Beelzebub* himself to find out how it was contrived, or to make one like it in a thousand years."—(Blainville's Travels, i, 250.)

while their own weight serves to close them; but in ordinary cocks, the plugs must be turned by some external force. Cocks of wood, brass, and other metals, and made on the principle of those now in use are extremely ancient. There is reason to believe that ancient modifications both of valve and plug cocks were quite as numerous as modern ones. It is certain that the Greeks, Romans (and most probably the Babylonians and Egyptians also) had far richer specimens of these instruments, both as regards the material and workmanship than any thing of the kind in modern days.

Horus Apollo, or Horapollo, an Egyptian of the fourth century, wrote a work " Concerning the Hieroglyphics of the Egyptians," and he informs us that the priests gave the form of a lion to "the mouths *and stops* [cocks] of consecrated fountains," because the inundation of the Nile occurred when the sun was in Leo.—(Encyc. Anti. i, 185, note.)

The contents of those enormous metallic vases mentioned in both sacred and profane history, were undoubtedly discharged through cocks, although these are not always indicated: as the *laver of brass* made by Bezaleel out of the mirrors of the Israelitish women: the *brazen sea* also, which was cast by a Tyrian brass-founder for Solomon. This unrivaled vase was, according to Josephus, of an hemispherical form. It was sixteen feet in diameter and between eight and nine in depth; " an hand-breadth" in thickness, and contained about 15,000 gallons. The brim was wrought like the brim of a cup, with flowers of lilies; " and under the brim of it round about, there were knops cast in two rows when it was cast." It was supported on a pedestal which rested on twelve brazen statues of oxen, from whose mouth the liquid is supposed to have been drawn. This splendid vessel was removed from off the statues by Ahaz—" he took down the sea from off the brazen oxen that were under it, and put it upon a pavement of stones." It was subsequently carried to Babylon by Nebuchadnezzar.

When Sylla pillaged the temple of Delphi, he found a vase of silver so large and heavy that no ordinary carriage could support it. He therefore had it cut up. (Plutarch in Sylla.) Herodotus, i, 51, in enumerating the gifts of Crœsus to the same temple, mentions a cistern of gold, and one of silver of immense dimensions, (perhaps the same taken by Sylla,) also silver *casks* and basins—that these had cocks is certain, for he observes that a statue of a boy was attached to one of them, and the water was discharged through *one of his hands.* This shows how variegated were the figures and orifices of ancient cocks. The Japanese indulge a similar taste, and have doubtless inherited it from their remote progenitors. Some of their bronze idols are made to serve as *fountains*, and the water issues from *the fingers* of some, while others hold a vase from which it flows, as in the Greek and Roman designs of Oceanus and Neptune. The Dutch on first visiting the Japanese found the baths of these people supplied with cold and warm water by means of pipes " and copper cocks."—(Montanus' Japan, translated by Ogilby, pp. 94, 279, 449, and Thunberg's Travels, iii, 102.)

Bronze or brass cocks were as common in old Rome and probably other ancient cities, as they are in any modern one. The immense number of pipes that conveyed water to the houses, baths, fountains, &c. must have kept a great number of founders constantly at work in making and repairing them. We learn from Vitruvius that every *main* pipe that passed through the streets, had a large cock, by which the water was let in or excluded, and that these cocks were turned as similar ones now are, with an iron key. Several specimens of ancient cocks are extant. Among these, a very large one discovered in the ruins of a temple built by

Tiberius at Capri, and preserved in the Museum at Naples, is not the least interesting. No. 273 is a figure of it. The plug has become by time immoveable, and having been shut when last used, the water within it is still confined. This is made evident, for when two men raise the cock, the splashing sound of the fluid is distinctly heard.

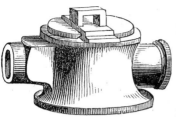

No. 273. Ancient bronze cock.

This cock was found attached to a reservoir, but in what manner it was connected we know not—by solder? screws?—particulars that cannot be determined by the sketch. Had we an opportunity of examining it we would endeavor to ascertain its weight, dimensions, &c.—whether the substance of the plug and chamber are the same, and if the former is secured in the latter by slightly riveting its lower edge, as in our small cocks, or by a washer and screw as in large ones.

The mode of forming the *handle*, or that part by which the plug is turned, in a separate piece from the latter, is decidedly superior to the common practice of casting both in one piece. It is a common occurrence to throw aside a cock and replace it with a new one, simply because this part has been broken from the plug, and can only be remedied by replacing the latter with it. Now this would never occur if cocks were made like this ancient one, for the part alluded to might be renewed with the same facility as the key of a door or the handle of a hammer. The mode of attaching this part to the plug by sliding it between two *dove-tailed grooves*, is ingenious, simple, and very effective.

In a great portion of modern cocks the area of the opening through the plug seldom exceeds one half of that through the chamber; but in the above one, the chamber is sufficiently large to allow a uniform passage-way throughout.

The modern name of these instruments is supposed to have arisen from their having been made in the form of the male of the domestic fowl; hence weather-cock, the cock of a gun, &c.

The luxury of the Romans under the empire led them to monstrous excesses, particularly with regard to baths; the water to supply which was often conveyed through pipes of *pure silver*, and of course through *cocks of the same*. Seneca, in a letter to Lucilius, describing the humble villa of the great Scipio, deplores this degeneracy of his countrymen. " I write to you [he says] from the villa of Scipio Africanus, where I at present am, and have worshipped his manes and his altars ... I surveyed this villa, which is built with square stone and surrounded with a wall. I viewed the groves and towers planted and erected on each side: a capacious cistern and basin for water is below the house and gardens, large enough to supply a whole army; next a small bath, and that something dusky. It was a sensible pleasure to compare the manners of Scipio with ours. In this little hole, this corner, did that terror of Carthage, to whom alone Rome owed her not being taken a second time, wash and refresh himself, after being tired with his country toils; for he used the country exercises and ploughed his ground himself, as the ancients were wont to do. Beneath this humble roof he stood, and this plain unartful floor supported him. Who now, in our days, would endure so mean a bath? Every man now thinks himself poor if the walls of his bath shine not with large orbs of precious stones—unless the Alexandrian marble be embossed, crusted over and varied with Numidian borderings—unless they are covered with Mosaic—if the vaulted roof be not of glass—unless the Thusian stone, formerly so rare and only to be found in some particular temple or public building, line the cistern, into which he descends after sweating, without soul or life, if the water pours not on him from *silver conduits*. I speak now only of the pipes and baths of the vulgar; but what shall I say when I come to those of the freed-men? How many statues! How many rows of pillars supporting no weight, but placed there merely for the sake of expense and ornament!" &c. &c.

No. 274. Ancient Roman cock.

No. 274 is another ancient cock from the third volume of Montfaucon's Antiquities. It will serve as a specimen of the richness and variety of ornament with which these instruments were sometimes decorated. The figure standing on the head of a dolphin, and which formed the handle by which the cock was opened and closed, is supposed to have represented the Genius of the garden, in which the fountain was placed. Another highly ornamented cock, or rather part of one, is also engraved in the same work; but as it appears to be merely that part by which the plug was turned, it is omitted. There are several bronze jet pipes for fountains extant, and in great variety of shapes. They were sometimes plated with gold, as appears from traces of it left on some of them.

Much additional information respecting the use of cocks among the Romans has been obtained from the ruins of Herculaneum and Pompeii. Several have been found in the houses and baths. Some were attached to pipes, fountains, and to boilers on large moveable tripods, or braziers, and also to urns or vases, similar to our tea and coffee urns. Most of them are ornamented with lions' heads, &c. In one brazier, the cock is quite plain, and resembles those which are known to plumbers as *stop-cocks*. In some of the braziers, the grate bars are *hollow*, that the water might circulate through them, and the cocks are inserted just above the bottom of the boilers, that a little water might always be retained to prevent the fire from destroying them.

In the baths of *Claudius* the water ran through *pipes of silver*. At Lanuvium, in the ruins of a villa of *Antoninus Pius*, a SILVER COCK was found which served for a fountain. It weighed thirty-five Roman pounds, and was inscribed " Faustinæ Nostræ."—(Encyc. Antiq. vol. i, 456.) I was shown, says BREVAL, in his " Remarks on Europe," several curious fragments that had been dug out of the gardens of Mæcenas. Among these, were some huge leaden pipes that conveyed the water from the Claudian aqueducts into a subterraneous bathing-room. The magnificence of the place must have been suitable, no doubt, to the immense wealth and delicacy of a Roman of his rank; especially, if what I was assured was the fact, that some lesser tubes discovered among the same rubbish were of *solid silver*.

Nothing, says Blainville, could equal in richness the apartments of Caracalla's baths. Columns, statues, rarest marbles and jaspers, and pictures of an immense value were lavished on every one of them. The very pipes, both large and small, which conveyed water into the bathing apartments, were all of the *finest silver*. This particular is recorded by several [ancient] authors, and among others by Statius.

Otho, in a feast given to Nero, almost deluged his guests with a most precious liquid perfume, which, " by opening certain *cocks*," gushed out of silver and golden tubes that were placed in different parts of the room.

As water was conveyed by pipes into the houses and temples of ancient Mexican and Peruvian cities, it might thence be concluded, in the absence of direct testimony, that *cocks*, at least wooden spigots, were in use also; but there is evidence of the fact. We are informed, that in a palace of Atabalipa, there was a bath or " golden cisterne, whereto were by two pipes from contrary passages, brought both cold water and hot, to use them mingled or asunder at pleasure." (Purchas' Pilgrim, 1073.) Now that these pipes were furnished with cocks, is expressly asserted by Garcilasso, in a passage we have already quoted. (See page 170.) Cisterns and pipes, both of silver and gold were used in the temple at Cusco.

" Golden pipes" are mentioned by the prophet Zechariah, iv, 2 and 12.

We gave a figure of a siphon cock at Nos. 265–6, and shall here describe a sliding one, contrived and used by us several years ago. A, No. 275 represents a short brass or copper tube, with a stuffing-box fitted to its upper end: the lower end is soldered to a pipe proceeding from a reservoir, or from a main in the street. B a smooth and smaller tube, having its lower end closed, works through the stuffing-box: to its upper end, which is also closed, a knob or handle is fixed, and just below, there is a spout for discharging the water. At the middle of B, a number of holes are drilled through its sides, or they may be in the form of slits. Now while these openings are kept above the stuffing-box (as shown in the cut) no water can be discharged; but as soon as B is pushed down, so as to bring them below the stuffing-box, the fluid rushes through them and escapes at the spout. To stop the discharge B is then raised, as in the figure. There should be one or two small projecting pieces near the lower end of B to prevent its being pulled entirely out of A. The pressure of the water tends to keep B from sliding down, when the instrument is not in use, even if the friction of the stuffing-box were not sufficient. The external edges of the slits should be smooth to prevent them from catching hold of the packing while passing through it. Of this, there is however but little danger in small cocks, and in those of larger size, that part of B through which they are made might be slightly contracted.

No. 276 represents one of these cocks attached to a cistern, with the openings within the stuffing-box, and consequently the fluid escaping. The length of the slits should always be less than the depth of the packing.

No. 277 exhibits a stop-cock, or one whose ends are straight and alike, (such as plumbers solder in the middle of pipes.) A straight tube C D is closed by a partition or disk in the middle of its length: as the water which flows from the reservoir always remains in the end C, the object is to open a communication for it to pass into D. To accomplish this, slits or other shaped openings are made through the pipe on both sides of the disk, and a shorter but wider tube E, with a stuffing-box at each end is fitted to slide over C D. Thus, to allow water to pass into D, all that is required is to move E (by the two projecting handles) till both series of openings are inclosed by it; while to stop the flow through D, E must be moved back towards C as in the figure. The upper figure in the cut is another form of the same thing. The sliding tube H is the smallest, and has one end closed like Nos. 275 and 276, while F and G are separate pieces. Its action will be sufficiently obvious from the preceding remarks.

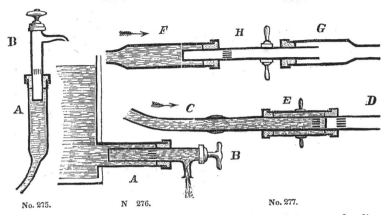

No. 275.   N 276.   No. 277.

Large cocks on this principle may be made for half the cost of ordinary ones, while the expense of keeping them in order is too trifling to be noticed—occasionally to renew the packing is all that could be required.

Water-Closets have been greatly improved by modern artists, but they are an ancient and probably Asiatic device. The summer chamber of Eglon, king of Moab, (Judges, iii, 20–25) is supposed to have been one. They were introduced into Rome during the republic, and are noticed by several ancient writers. Those constructed in the palace of the Cæsars were adorned with marbles, arabesque and mosaics. At the back of one still extant, there is a cistern, the water of which is distributed by cocks to different seats. The pipe and basin of another has been discovered near the theatre at Pompeii, where it still remains. Heliogabalus concealed himself in one, and whence he was dragged by his soldiers and slain.

Water-closets seem to have been always used in the East, and for reasons which Tavernier and other oriental travelers have assigned. Numbers are erected near the mosques and temples. A similar custom prevailed in old Rome, Constantinople, Smyrna, and probably all ancient cities. In the city of Fez, "round about the mosques, are 150 common houses of ease, each furnished with a cock and marble cistern, which scoureth and keepeth all neat and clean, as if these places were intended for some sweeter employment."—(Ogilby's Africa, 1670, p. 88.) In his "Relation of the Seraglio," Tavernier describes a gallery, in which were several water-closets. "Every seat [he observes] has a little cock." He mentions others, in which the openings were covered by a plate, which by means of a spring "turned one way or the other at the falling of the least weight upon it."

Sir John Harrington is said to have introduced water-closets into England in Elizabeth's reign, and some writers have erroneously ascribed their invention to him. They are described in the great French work on Arts and Manufactures, by M. Roubo, who says, they were long used in

France before being known in England. Those which he has figured are however on the ancient plan, without traps, and such are still to be found in oriental cities. They are not to be compared with the modern ones. (See L'Art du Menuisier, folio edit. 1770, Pl. 69; Gell's Pompeiana; A Dissertation on Places of Retirement, Lond. 1751; Fryer's Travels in India and Persia, Lond. 1698.)

Devices for preventing the ascent of offensive vapors from sinks, sewers, drains, &c. are named *traps*. As these are simple in construction, and applicable under all circumstances, and yet are little known, we have inserted a sketch of a few of the most common. They are all modifications of the same principle.

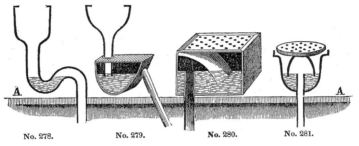

No. 278.   No. 279.   No. 280.   No. 281.

A A represent a floor or covering of a sink or sewer, and the object is to discharge refuse water or slops of any kind into the latter without allowing currents of air to rise through the passage. No. 278 is a leaden pipe bent at one part into the form of a letter S, which part constitutes the trap. One extremity enters the sink, and to the other, which is turned up perpendicularly, the basin of a water-closet, or a common funnel is attached. The flexures of the tube must be such, that whatever liquid is thrown down the basin, a portion will always remain in the bent part below *so as to seal the passage completely*, as shown in the cut. The basin and trap may be placed in a room at any distance above the sink or sewer, provided both are connected by an air-tight tube.

No. 279 is named a D trap, from its resemblance to that letter. It is of the kind generally used in water-closets, for which purpose it is always made of lead, and about twelve inches long, five wide, and ten or eleven deep. The pipe that enters the sink is soldered to one end and near the top. The other one to which the basin is attached descends six or seven inches through the top at the opposite extremity of the trap. By this arrangement water is retained within to a level with the lower edge of the pipe that enters the sink, while the perpendicular pipe dips between one and two inches below the surface. Hence although impure air in the sink can readily ascend into the trap, it cannot enter the tube on which the basin is placed; for to do so, it would have to descend through two inches of the water to reach the orifice of the tube; and then to ascend through an equal column within the latter before it could rise into the basin.

No. 280 is a form of trap used over the openings of street sewers, for which purpose they are commonly constructed of stone or brick and lined with cement. The figure is that of a square box open at top. A pipe is inserted through the bottom at one side to connect it with the sewer. This pipe stands about half way up the inside of the box, and above it there is a bent rectangular partition attached at three of its sides to the box, while the fourth side extends into the middle and dips two inches below the orifice of the pipe, and consequently that depth in water, thus cutting off all external communication with the air in the sewer. A loose grate fits into a recess on the edge of the box, and is occasionally removed to take out the dirt that passes the grate. Small traps of the kind, and made of cast iron, are sometimes used in the drains of private houses.

No. 281 is named a *bell* trap from its figure. Such are generally of small dimensions, and are mostly used in kitchens, over the channels or tubes through which refuse fluids are discharged into sinks or drains. The end of the pipe projects two or three inches into the trap, consequently a quantity of water must always remain within at the same elevation. Over the pipe a bell or inverted cup dips about half an inch into the water, and is of such a size as to leave sufficient room for the fluid to descend between it and the sides of the trap, and also to pass under its edge and rise into the pipe, and so escape into the drain. The cup or bell is connected to a brass grate that drops into a recess cast round the inner edge of the trap.

The origin of traps is, we believe, unknown. The principle is precisely the same as in the water-lute of old chemists. Glauber used contrivances identical with Nos. 278 and 281, instead of cocks to close retorts, &c. Instead of water he sometimes used mercury, when the contents were of a corrosive nature.

END OF THE FIFTH BOOK.

APPENDIX.

John Bate—Phocion—Well worship—Wells with stairs—Tourne-broche—Raising water by a screw—Perpetual motions—Chain pumps in ships—Sprinkling pots—Old frictionless pump—Water power—Vulcan's trip-hammers—Eolipiles—Blow-pipe—Philosophical bellows—Charging eolipiles—Eolipilic idols referred to in the Bible—Palladium—Laban's images—Expansive force of steam—Steam and air—Wind-mills—Imprisoning chairs.

SOME facts and observations having occurred to us during the progress of this work which could not be inserted in their proper places, a few are added by way of appendix. While engaged on the last chapter, a large collection of old books was imported into this city from Europe, in which we fortunately found a perfect copy of "Nature and Art," mentioned at pp. 321, 421. From the title, which is annexed, it will be seen that our conjectures respecting its author and date of publication were correct. "The Mysteries of Nature and Art in foure severall parts. The first of water-works : the second of fire-works : the third of drawing, washing, limning, painting and engraving : the fourth of sundry experiments. The *second edition*, with many additions unto every part. By John Bate, Lond. 1635."

At page 19, we quoted an example of frugality in Dentatus cooking his simple food while he swayed the destinies of Rome. There is a parallel case in one of the most virtuous of the Greeks, viz. *Phocion*. Alexander esteemed him, but could never induce him to accept of gifts, although he was always poor. At one time the Macedonian warrior sent him out of Asia a hundred talents as a mark of his regard; but when the envoys arrived with the treasure at Athens, Phocion was inflexible—he would not touch it. They then followed him to his house, and were astonished beyond measure to find the wife of this truly great man making bread, and himself *drawing water*.

Worship of Wells, pp. 33–37. "The worship of this well of St. Edward was particularly forbid by Oliver Sutton, bishop of Lincoln, in the time of Edward I. This well worship is strictly forbidden in King Edgar's canons, and K. Cnute's laws, as 'twas in a council at London under Archbishop Anselm, in the year 1102; and some of our best criticks observe that what is translated *will*-worship in Colossians, ii, 23, should be *well*-worship."—Hearne's Preface to Robert of Gloucester's Chronicle.)

Wells with Stairs, p. 53. An extraordinary well of this kind was built by Pope Clement VII. in 1528.—(See Lond. Mechanics' Mag. vol. ii, 208.)

Tourne-broche, p. 75. In the 33d year of Henry VI. A. D. 1454, an ordinance was established for reducing the expenses of the king's household. Instead of a larger number, only "vj children of ye kechyn tourne-broches" were appointed, i. e. to turn the spits.—(Proceedings and Ordinances of the Privy Council of England, edited by Nichols, vol. vi, 229.)

Raising Water through a Screw, p. 140. Some persons deceived by the apparent facility of working a water screw, especially when its journals are delicately fitted to their bearings and they turn with little friction, imagine that it not only elevates the liquid with a less expense of force than any other machine, but with less than is due to the quantity raised; hence it has often been adopted in projects for the perpetual motion. When arranged so as to be turned by an *overshot wheel*, it constitutes one half of the first attempts at a solution of that impossible problem, under the impression that it would raise and discharge upon the wheel all the water expended in moving it! The inclined position of a water-screw is supposed to contribute to this imaginary result, for, say these reasoners, the water then arrives at the top by naturally flowing along each convolution, while the force consumed is little if any more than would be required to turn the tube if empty!—the fluid being thus raised in a different manner and with much less force, than when lifted directly and perpendicularly by the piston of an atmospheric pump, or driven up by that of a forcing one!

In these projects, the action of the wheel depends of course as much upon the screw, as that of the latter does upon the wheel; in other words, each is designed to turn the other: but the very idea of two machines reciprocally moving each other at the same time is palpably absurd. The two forces will either be equal or unequal. If they are alike both would be in equilibrio, and the machines would remain at rest; and if at any time one force exceeded the other, the same result would necessarily take place, for the smaller could not then overcome the greater. If the wheel could transmit its entire force to the screw, (undiminished by resistance from the air, the friction of its bearings and that of the intermediate mechanism,) it would still be impossible for the latter to return it, because to do so a greater force than that derived from the wheel would be required; a machine cannot be moved and at the same time move its mover. When moved, its force is less than that by which it is moved; and if it becomes the mover, its force must exceed that of the machine to which it imparts motion.

The effect of any machinery composed of levers, cranks, wheels, &c. and moved by water, animals, or men, can never exceed the power that moves it, for there is nothing in wood, iron and brass, or in any combination of them, by which they can *create force*, or, what is the same thing, give out more than is imparted to them. As well might we expect to see a carriage returning of itself from a long journey, and laden with the horses that drew it from home.

Wilkins has given a chapter in his Mathematical Magic on "composing a perpetual motion by fluid weights." His prominent plan was raising water by a screw, and discharging it on float boards attached to the screw itself. He quotes older authors who indulged the same whim. Visions of great mechanical discoveries often burst upon the ingenious prelate, as well as on lay inventors: in such seasons he was in ecstacies. When he first thought of obtaining power by means of a water-screw, he says, "I could scarce forbear with Archimedes to cry out, *Eureka! Eureka!* it seeming so infallible a way for the effecting of a perpetual motion, that nothing could be so much as probably objected against it: but upon trial and experience I find it altogether insufficient for any such purpose."

In the Gentleman's Magazine for 1747, p. 459, there is a description and figure of a similar device—either water or balls were to be raised through a screw and dropped upon an overshot wheel. It was devised by a Col. Kranach, of Hamburgh, who, in a pamphlet, declared he ha spent thirty years in perfecting it. He proposed it as a substitute for wind and water-mills, and particularly for raising water and ore from mines. In the same work for 1751, p. 448, there is "a self-moving wheel." And at p. 391, "a self-moving machine;" the latter by a Polish Jesuit; it consisted of a wheel, ropes, pulleys, a pump, weights, &c. and of course, like Kranach's, could no more move of itself than a lamp-post, nor increase any force imparted to it than could a collection of paving stones.

If a perpetual motion could be obtained by a water-wheel and screw as above, then it would follow that a bricklayer's laborer could convey a hod of mortar or a bucket of water to the top of a building with a much less expenditure of force by traveling along a circular stair-way, than by ascending directly up a ladder, and whether he carried the load on his shoulder or dragged it after him by a cord. But the fact is, a 100 lb. of water cannot by *any* contrivance whatever be conveyed to the top of a building with a force less than would be required to pull up the same weight of stone or mortar in a bucket: it can no more be wheedled out of its gravity by passing it up an inclined plane than a vertical one—through a helical tube than through a straight one.

Chain-Pumps in Ships, p. 154. John Bate, describing a chain-pump in 1633, says, a short brass chamber smoothly bored was inserted in the lower end. The pistons were fitted to this, and the rest of the pipe was of larger bore. The chain was of iron and carried round by a sprocket wheel. Each piston consisted of a disk of *horn* between two of leather. Such a pump, he observes, "goeth very strongly, and therefore had need be made with wheels and wrought by horses, for so the water is brought up at Broken Wharfe in London." He names the chain-pump "an engin whereby you may draw water out of a deep well, or mount any river water . . . . . Also it is used in *great ships*, which I have seen."—(Mysteries of Nature and Art.)

Atmospheric Sprinkling Pots, p. 194. When Louis, duke of Orleans and Milan, brother of Charles VII. was murdered, (A. D. 1407,) his widow, as a symbol of her distress and an indication that the rest of her life would be spent in tears, adopted the *chantepleure* or garden pot as an heraldic device; and which, with the motto, *plus ne m'est riens*, she had engraved upon almost every thing in her house. No. 282 is a figure of the instrument. (Devises Heroiques, par M. C. Paradin, A Lyon, 1557.)

No. 282. No. 283.

No. 283 is another old form of the atmospheric sprinkler, from a Latin Collection of Emblems of the early part of the 17th century. The motto on a flying scroll was *Modo Spiritus Adsit*. Air was admitted through a small opening near the top, which was closed with the point of the finger.

The sixth and seventh problems of Heron's Spiritalia relate to these instruments. The two figures there given are hollow spheres; a small circle round the bottom being perforated, and a minute orifice near a ring or handle on the top. In one there is a partition, so that two different liquids could be contained within; and wine, or hot and cold water, discharged as one or the other of the orifices at the top was uncovered.

Fire-Engines and Bellows Pumps, pp. 241, 321. No. 284 is a bellows or frictionless pump, from the first edition of Bate's Mysteries of Nature and Art. It is identical with the fire engine referred to in our third book, except being placed within an open frame instead of a cistern fixed upon wheels. For its description, see pp. 321–2. (The leathern bag which connected the two brass vessels is not figured by the old artist.)

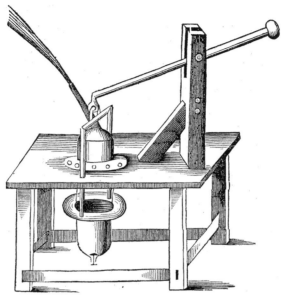

No. 284.   Old frictionless or bellows pump, A. D. 1633.

Water-Wheels, p. 282. There are indications in the Iliad that Vulcan used water power, and that it was by the dextrous concealment of it and the mechanism by which it was transmitted that enabled him to excite in so high a degree the astonishment of his contemporaries, and to give rise to those wonderful stories of his skill that are even yet extant. When engaged at the anvil Homer represents him, like a modern smith, with a single pair of bellows. Thus Thetis found him "sweating at his bellows huge;" but in other scenes, he is exhibited rather as manager of extensive forges for the *reduction* of metals; the fires being urged by a large number of bellows moved either by water or some other inorganic force. Like a superintendent of modern iron or copper works, ordering the bellows to be thrown into geer, and the blasts increased or diminished as circumstances require : so Vulcan "turning to the fires, he bade the bellows heave;" then

      Full twenty bellows working all at once,
      Breathed on the furnace, blowing easy and free.

Of Vulcan's numerous works none were more celebrated by the ancients than the two *androids* which assisted him at the anvil. They were obviously nothing more than ingenious devices for concealing the mechanism by which motion was communicated to the sledges they held in their hands :—in other words, mere *trip hammers*, and worked most likely by a distant water-wheel. The rods or levers which communicated the motion were probably concealed under the floor, and terminated at the feet of the figures, while Vulcan could easily throw them in and out of geer unperceived. It can readily be imagined what the effect of two well executed working images of this kind must have been in early times.

Eolipiles for Fusing Metals, p. 397. The surprising effects produced in modern days by steam and those more important ones which it is destined hereafter to accomplish, will always render examples of its early employment in the arts interesting. The use of eolipiles as bellows, like that of atmospheric sprinklers for watering pots, has long been discontinued, and both have almost passed into oblivion. We shall therefore offer no apology for inserting the following additional illustrations of the use of the former in bygone times. No. 285 is a steam blow-pipe from the 2d edit. of John Bate's work. His description forms an admirable comment on Wilkins's observation, (p. 396,) that eolipiles were used in melting glass and metals. This remark of the bishop has been quoted by several writers, but not one has, to our knowledge, endeavored to show how steam was thus applied, although every mechanic on perusing Wilkins's book would, like ourselves, feel anxious for information on the subject.

No. 285. Eolipile for glass blowing.   No. 286. Eolipile for fusing metals.

The first figure consists of a lamp and a copper ball or eolipile, placed on and heated by a furnace or brazier. The apparatus is named "a device to bend glasse canes, [tubes,] or to make any small work in glasse." "Let there be a vessel of copper about the bignesse of a common foot-ball : let it have a long pipe at the top, which must be made so that you may upon occasion screw on lesser or bigger vents made for the purpose. Fill this one third part with water, and set it over a furnace of coals, as B; and when the water beginneth to heat, there will come a strong breath out of the nose of the vessel that will force the flame of a lamp placed at a convenient distance, as A; if you hold your glasse in the extension of the flame, it will melt suddenly; so you may work what you will thereof." Bate observes, that some persons instead of this apparatus used a pipe (the common mouth pipe) fastened on a bench between a crotched stick, as figured at C. He himself occasionally employed this, but considered it not so convenient as the eolipile.—(Mysteries of Nature and Art, Lond. 1635.)

In 1650, Dr. John French published "The Art of Distillation, or a Treatise on the Choisest Spagyrical Preparations .... with descriptions of the chiefest furnaces and vessels used by *ancient* and modern chemists."

Of old devices three eolipiles are figured: one is precisely the same as above described by Bate. French observes, that it "blows a candle to make the flame thereof strong for the melting of glasses and nipping them up." No. 286 is another for fusing metals. A large eolipile is permanently connected to a furnace, the blast being conveyed through a brick wall. The following is all that he says respecting it: D "signifies that which blows a fire for the melting of any metall or such like operation, and it blows most forcibly with a terrible noise." The water was introduced through an opening at the top. E is a portable eolipile to be held in the hand, and the blast applied to fixed objects. It appears from French, and also from Ercker's work on Metallurgy, that eolipiles when used for blowing fires and fusing metals, were formerly known as *The Philosophical Bellows*, a circumstance that renders their disappearance from modern writings still more singular.

Since the insertion of illustration No. 185, we have met with an English translation of Ercker's work, by Sir John Pettus, " of the Society of the Mines Royal," under Charles II. but who appears to have derived little wealth from mining speculations, since he rendered Ercker's book into English while confined in prison for debt. The translation is illustrated with fine copperplate engravings, and a dictionary of technical terms is subjoined. Under the word *bellows*, Pettus mentions the " philosophical bellows;" the common smith's bellows, and very large ones that were worked by water-wheels, and which, he observes, were made " in imitation of the nature of a *cow beast*, which in drawing in and forcing out her breath, is said to *bellow*"—a quaint definition of bellows, but one which, we believe, gives the true etymology of the word. Of the antiquity of " philosophical bellows" there can be little doubt. They were probably used by the fancy glass-blowers of Egypt, Greece and Rome, as well as by other artists in the reduction of metals. The transition from blowing ordinary fires with eolipiles to such operations was obvious and easy. There is a passage in the book of Joshua which seems to refer to the early use of them. In one of the contests of that warrior with the Canaanites, it is said he chased them to *" Mizrephoth-maim"*—a word signifying "*burnings of waters,*" and "*furnaces where metals are melted.*" A place that probably derived its name from extensive forges that were urged by blasts from eolipiles.

Charging Eolipiles by Atmospheric Pressure, pp. 395, 407. Dr. French observes, " You must heat them very hot, then put the noses thereof (which must have a very small hole in them, no bigger than a pin's head may go in) into a vessel of cold water, and they will presently suck in the water." Roman eolipiles were charged in the same way, as is clear from their description by Vitruvius, for they had but *one* opening, through which, he says, they were filled with water, and out of which the blast issued.

Eolipilic Fire-Blowers and Idols, pp. 398–400. In addition to those passages of Scripture which we have supposed alluded to eolipiles, a few others may be named. The sacred writers, it is well known, often contrast the power and other attributes of God with the impotency of idols: to adapt their instructions to idolaters, they represent the Almighty as excited with anger, wrath, fury, &c. apparently in reference to such passions being exhibited (as we know they were) by idols, and particularly eolipilic idols. Why should God be represented as blasting or consuming men with streams of fire from his *mouth,* and with smoke from his *nostrils?* kindling coals by his *breath?* Why is his anger said to *smoke,* to *burn,* to *wax hot,* &c. if it be not in reference to such idols as Pusterich, or those images described by Carpini? " By the blast of God," says

Job, "the wicked perish, and by the breath of his nostrils are they consumed," i. e. as fuel on the hearth is consumed by the blast of an eolipile. The Psalmist, describing God, says, "there went up a smoke out of his nostrils, and fire out of his mouth; coals were kindled by it." "Behold [says Isaiah] the name of the Lord cometh from far burning with his anger, [or the grievousness of flame as the margin has it,] his lips are full of indignation, and his tongue as a devouring fire; and his breath as an overflowing stream shall reach to the midst of the neck." Again, "Tophet is ordained of old, yea for the king it is prepared: he hath made it deep and large: the pile thereof is fire and much wood, the breath of the Lord like a stream of brimstone doth kindle it."

It appears to us that here and in similar passages are allusions to eolipiles of the human form, and to such images as Pusterich, from whose eyes, mouths and nostrils issued streams of flame, smoke, steam, &c. Perhaps it will be said the expressions are *figurative*: true they are so; but then there is in them an allusion to the *things* from which the figures are derived. When God is said to *melt* his people, to *refine*, to take away the *dross* from them, every one perceives the allusions to metallurgical operations, because such operations are known to all; and equally clear would the passages quoted above appear had eolipilic blowers and idols continued in use to our times. We should then have perceived that such expressions as *the sword of his mouth, swords of fire, flaming swords*, &c. were neither of figurative origin nor application only; for from the variety of eolipilic images, there is little doubt that inflammable fluids were made to issue from different parts of them, and in various shapes—from their mouths as *tongues* of fire, and from the hands as *flaming swords*, &c. We know that ancient priests were exceedingly expert in working prodigies by inflammable fluids, of which numerous examples might be quoted. When Octavius was in Thrace, he consulted the oracle of Bacchus, and the ministers of the temple finding it their interest to gratify him, contrived that when the wine was poured on the altar, a body of flame should burst out and ascend above the roof of the temple; a portent, observes Suetonius, "that had never happened to any but Alexander the Great, when he was sacrificing at the same altar." They could, of course, as easily have made the flame dart from the mouth and eyes of an idol as from the altar, if their views had so required it.

But if it should be contended that the passages quoted, rather gave rise to idols like Pusterich, i. e. were hints which heathen priests worked from in order to produce or imitate the same effects, it will not affect the inference we wish to draw from them, viz. the antiquity of steam and vapor images. In connection with this subject, it may be observed, that the famous Palladium of Troy was probably an eolipilic idol, in which inflammable fluids were used; for on certain occasions *flashes of fire* darted from *its eyes*, as from the mouth and forehead of Pusterich.

If biblical critics would pardon our temerity, we would also suggest that the *Lares* or images which Rachel stole from her father's dwelling were, like the small Saxon idol, (p. 398,) and those referred to in Isaiah, (p. 400) eolipilic fire blowers. They have exceedingly perplexed commentators, who after suggesting numerous explanations, generally conclude by observing that their nature and uses are unknown; but had these writers called to mind the ancient employment on the domestic hearth of brazen eolipiles of the human form, they would have perceived that the name of Laban's images gave an indication of what they were. In all ancient languages proper names were invariably expressive of some prominent feature, attribute, or design of the objects named: so of these images—they were named "*teraphim*," a word signifying "*blowers*," from *teraph*, " *to blow*." So also the eolipilic idol *Pusterich* was named from *pusten*, " to blow." (See p. 399.) Eolipiles, like the *Lares*, were located on the hearth, and as they were avowedly made and named after a god, (Eolus,) and were designed to imitate him in producing blasts of wind, (Varro makes the lares gods of the air,) it was natural enough to adopt them as household deities. Rachel was evidently an intelligent and very shrewd woman; and as we have no reason to suppose she was an idolater after having lived twenty years in the same house with Jacob, (if indeed she ever was,) it is not at all likely that she coveted the images as *idols*, but only as domestic utensils of real utility—utensils which she had long been in the habit of using, and such as were highly desirable in setting up housekeeping for herself.

Expansive Force of Steam, p. 409. The Stoics, says Plutarch, attributed earthquakes to aqueous vapor generated within the earth by subterranean heat. (Opin. Philos.) No stronger proof that the ancients were *familiar* with the force of steam could be desired : the idea could never have occurred except to men practically acquainted with the irresistible energy of this fluid when confined. If by no other means, we may be sure they had frequent proofs of this energy in the rupture of eolipiles when their vents were closed. The hypothesis of Plato respecting the conversion of water into *air* and *fire*, (mentioned below,) shows him to have been a close experimenter on steam at different temperatures. The old theory of boiling springs being forced from the interior by steam, implies also an acquaintance with devices for raising water by it.

Identity of Steam and Air, pp. 395-400, 418-421. This erroneous opinion doubtless dates back to the early ages, during which it led to the invention of eolipiles, and to the *first* mechanical application of aqueous vapor, viz. to *blow fires*, instead of wind from bellows. It is singular, however, that such an opinion should have been maintained at so late a period as the close of the 17th century—that modern as well as ancient philosophers should have taught that water rarefied by heat was converted into air, and that air condensed by cold was returned into water. Besides the examples already given, we add a few more. Of the elements into which philosophers formerly resolved all things material, viz : *earth, water, air* and *fire*, Plato suspected the last three were but modifications of one ; at any rate, he supposed they were convertible into each other—that water attenuated by heat was dilated into air, (steam,) and that this by a higher temperature became an invisible and glowing fluid or fire. (Plutarch, Opin. Philos.) Plutarch himself, in his Treatise on Cold, observes, " aire when it doth gather and thicken is converted into water, but when it is more subtile it resolveth into fire ; as also in the like case, water by rarefaction is resolved into aire." Pliny, in speaking of winds says, " aire is gathered into a waterie liquor." The sweating of walls, breathing on glass, moisture on the outside of a tumbler of water, &c. were considered proofs that cold condensed air into water. Lord Bacon, in his Sylva, Expers. 27 and 76, speaks of " the means of turning aire into water," and Exp. 91, relates to " the version of water into aire." Norton, (a contemporary of Bacon,) in his " Rehearsal of Alchemy," versifies the old doctrine thus :—

> But ayre condens'd is turn'd to raine,
> And water rarefied comes ayre again.

Wind-Mills, p. 418. These were known in England in the 13th century. At the battle of Lewes, A. D. 1264, " there was many a modre sonne broght to grounde and the kynge of Almayne was taken in a *wynde mylle.*"—(Hearne's Glossary to Peter Langtoft's Chronicle.)

Intelligence of Animals exemplified in Raising Water, p. 74. Plutarch in his comparison of land and water animals, says, oxen were employed in raising water for the king of Persia's gardens at Susa, " by a device of wheels which they turned about in manner of a windlass." Each ox was required to raise one hundred buckets daily, and as soon as that number was completed, no efforts of the attendants could induce him to add another. Attempts were made to deceive the animals but without effect, so accurately " did they keep the reckoning."

Imprisoning Chairs, p. 429. Such devices are very ancient. The *first* proof of Vulcan's mechanical ingenuity is said to have been a throne or chair of gold, with secret springs. This he presented to his mother, and no sooner was Juno seated in it than she felt herself pinioned and unable to move. The gods interfered, and endeavored to release her, but without effect; and it was not till the artist had sufficiently punished her for her want of affection towards him that he consented to let her go.

Nabis, the tyrant of Lacedemon, had a device for extorting money from the wealthy. It was a statue of a female clothed in rich apparel. When any one refused to part with his wealth, the tyrant introduced him to the image, which by means of springs, seized him in its arms, and put him to the most excruciating torments, by forcing numerous bearded points into his body.

Rotary Pumps, Eolipiles, Steam-Guns, &c. In " Mathematical Recreations, or a collection of sundrie excellent problems, out of ancient and modern philosophers; written first in Greek and Latin, lately compil'd in French by H. Van Etten, and now in English, Lon. 1674," is a rotary pump similar to the one we have figured at p. 285 : it is named " a most soveraign engine to cast water high and far off to quench fires." A gooseneck like those now used is also figured—also an atmospheric garden pot —magic cups—three-way cocks—ear trumpets, and eolipiles. Of the last, the author says, " some make them like a ball, some like a head painted, representing the wind—some put within an eolipile a crooked tube of many foldings to the end that the wind impetuously rolling to and fro within, may imitate the voice of thunder—some apply near to the hole small windmills, or such like, which easily turn by reason of the vapors." One problem relates to the " charging of a cannon without powder." This was done, 1st, by *air* as in the air-gun ; and 2d, by *steam,* the latter fluid to be generated from water confined in the breech.

Olaus Magnus mentions eolipilic war machines, apparently similar to those described by Carpini, (see page 400.) They are distinguished from every species of guns : he calls them " *brazen horses that spit fire :* they were placed upon turning wheels, and carried about with versatile engines into the thickest body of the enemy : they prevailed so far to dissolve the enemy's forces, that there seemed more hopes of victory in them than in the souldiers."—(History of the Goths, book ix, chap. 3, Eng. Trans. Lond. 1658.)

SUPPLEMENT

ON

ORACULAR AND FIGHTING EOLIPILES

In consequence of a suggestion that a little additional matter on Eolipilic automata would add interest to this volume, a few specimens accompanied with cursory observations are subjoined. The figures themselves constitute, perhaps, a better exposition than any thing which can now be written on the devices which they represent—devices once wielded with terrible effects by both sacerdotal and military engineers.

Like extinct natural monsters, oracular and warring Eolipiles have disappeared from the earth and left scarcely any authentic vestiges behind. They belonged to certain states or conditions of society which they could not survive. Indigenous to ages of darkness, they flourished only in the absence of light. Receding, as civilization advanced, it may be said of them, as of spectres, they flutter at dawn and vanish as soon as the sun (of science) has risen. But they are not the less interesting subjects of research because of the evils they inflicted on our species, any more than are geological remains of mammoth beings which preyed on inferior tribes. Antique Eolipiles are in some respects the richest of artificial, as fossil bones are of natural, relics. Both are unique memorials of past times—vivid remembrancers of strange beings and dark deeds. The former afford proofs of stupendous animals reigning as monarchs over the woods and waters of the old world; and the latter remind us of moral monsters, preying with surprising facility upon all classes of men.

Pictorial representations of idolatrous and fighting eolipiles are exceedingly rare; and these, few as we find them, if not transferred to modern pages will soon be irrecoverably lost. Those which follow, though deplorably imperfect and obscure, will be acceptable to most readers, if not to all. Examples of the employment of elastic and inflammable fluids under singular circumstances, they can hardly fail to elicit the attention of inquirers into the origin and history of motive mechanism. They may afford hints on old and lost arts. Nor do they lack interest to general, or even learned readers; for, besides illustrating ancient society and manners, they reflect light on the darkest passages of poetry and romance: they add strength to the conviction that much which ancient literature has failed to explain, a close examination of ancient arts may yet render clear. Even the Eolipile, simple as it seems, promises to conduct inquirers, like the clew of Ariadne, through labyrinths as perplexing as those which puzzled old travellers to Egypt and Crete.

Of all the freaks of poor human nature, idolatry is the strangest; and, taken in connection with evils springing from it, the most infectious and fatal of maladies. Hitherto ineradicable, inexpugnable, it has tainted all epochs, polluted all people. Its ravages have been more destructive than war, more distressing than famine. It has been the fertile source of both. Superstition, the parent of idolatry, is peculiar to man, unless demons be tormented by it, which is not unlikely; for, besides its associations being truly diabolical, (it has every where erected altars to Baals and furnished victims to Molochs,) it seems the natural, and may be the universal punishment of mental debasement. It is to the mind what premature decrepitude is to the body—a horrible penalty for violating a fundamental law of our nature, for stunting the soul's growth, for not cultivating the intellectual with the physical faculties, that both might expand and improve together; that infant puerilities might be succeeded by youthful intelligence and masculine knowledge. Instead of this, superstition unites dwarfed and crippled minds to grown up bodies—stocks the world with souls blind to their destinies and duties, and consequently to the great purposes of existence lost. Where else, then, can such abortions be more appropriately consigned than to the hades of ignorance—of sottish delusions—to murky regions, where the sickly imagination sits an incubus on the prostrate judgment, and visions of insanity are reckoned as realities; where the occupants wander among shades, and mutter the gibberish of phantoms.

A stranger to natural causes, startling phenomena have ever filled the barbarian with dread. To account for such things he peoples the elements with imaginary beings, who control, as he supposes, all mundane affairs at their will. Meteorological commotions, pain, sickness, death, and every public and private calamity, were held as manifestations of their power or their wrath; hence the idea of propitiating beings so mighty and malignant; hence idolatry with its direful progeny, magic, divination, necromancy, and their congeners; and hence too the rise of those astute spirits who, from the beginning, have subdued the million by working on their fancies and fears—who have raised themselves into gods and sunk the rest of mankind into brutes.

Idols were almost invariably modeled after hideous forms, because designed to excite terror. This was in accordance with the principles on which demonolatry was founded. As fear was to be awakened it was essential to make them correspond, as nearly as could be, with the evils they had power to inflict or emotions they were designed to inflame. To have made them more attractive than repulsive would have been preposterous, since it would have been neglecting the cultivation of that passion upon which their efficiency rested. Their makers knew their business better. In nothing is the versatility of ancient genius more apparent than in representations of the horrible—in conjuring up images to cause the timid to tremble and the bold to recoil—the most hideous of hybrids, in which were combined features derived from every thing on the earth and in the waters under the earth calculated to excite abhorrence and dread. Perhaps it is not too much to say that here also little is left for professors of the *fine* arts to do, except to imitate works of old masters. Invention seems out of the question. Our best and worst specimens of diablerie and the monstrous are but copies and caricatures of originals in old galleries of furies, minotaurs, hydras, chimæras, centaurs, sphinxes, fauns, dragons, griffins, gorgons, satyrs, harpies, hippogriffs, and other unearthly combinations of human bodies with those of beasts, birds, fish, reptiles and demons.

But ghastly, terrific or fiendish features were not always deemed sufficient. It was expedient to communicate active qualities, such as might influence other senses than the sight, and which, being appropriate to the character an idol was intended to sustain, might serve still further to establish or increase its fame. Thus, some moved their heads, arms hands, eyes; others spoke, groaned, smiled, perspired, laughed, &c. &c.

A few, like the image of Nabis, squeezed unbelievers to death in their arms, and others, like the gods of the Zidonians, in their fury swallowed offenders alive. The repeated declarations in the Bible that gods of stone, wood and metal, neither saw, heard, ate nor "spake through their throats," &c. imply that by priestly artifice these and other functions were imitated. Had all been dumb, motionless statues, this constant denial of such powers to them would have been nugatory.

The date of androidal idols is unknown: they appear to have been co-eval with the use of metals—are perhaps of a still earlier date, for modern savages have attempted them. They were found so effectual as to have become important instruments in the hands of rulers in ante-historic eras; while to devise and work them became the profession of priests. As society advanced the treasures of states and temples were expended in their production, and the influence of both was exercised in establishing their reputation: a union of wealth and intelligence which accounts for the perfection and celebrity of many ancient androids.

Ever on the look out for novel and imposing devices, the founders and fosterers of idolatry were too close observers to overlook the most appalling of nature's displays, and too keenly alive to their interests to remain ignorant of the means of imitating them. At an early day those gods were counted the greatest that had power over fire and controlled atmospherical tempests—that spake in thunder and whose darts were the electric fluid. On this belief Eolipilic idols arose, a class certainly among the most productive if not among the most ancient. They were necessarily the work of the founder, not of the carver, and, as already intimated, not a few of the "brazen" or "molten" images of the Old Testament were more or less allied to them—an inference justified by numerous allusions to blasts of flame, smoke and wind issuing from their mouths and eyes, &c. There was probably less difficulty in the apotheosis of Eolipilic images than of others. When idolatry was universal few could refuse subjection to deities that rivalled Neptune in shaking the ground—Jupiter in his character of the thunderer; and Pluto—the grim and inexorable—the sulphur-enthroned god—in the worst of his functions. To none were apotrophic hymns so fervently addressed, for none looked more threatening and fierce, or gave out such awful manifestations of wrath.

Of their authors or inventors there is no room to doubt. They were men whose intelligence was far in advance of their times, who monopolized knowledge for the sole interest of their class. Claiming kindred with heaven, freed from worldly cares, clothed in reverend vestments, they lived apart from other people; holy and artless in appearance, yet adepts in artifice and very devils in craft. Hierophantic magicians sojourned in temples, feasted on tythes and got rich by means of idols. They moved gods to compassion by wires, and roused them to anger by explosive compounds. Their professional attainments are indisputable. In the roguish departments of physics they were never surpassed. What resources and talents did those of Egypt display in competing with Moses, even to the development of lower forms of life! The laboratory was their study, natural science the volume over which they pored, the knowledge of latent phenomena their wealth. It is impossible to think on the variety, magnitude and difficulties of some of their impostures without conceding to them excelling ingenuity and impudence sublime. In chemistry and mechanics they were profound: of their contrivances few were more successful than those to which both

Pusterich, an Eolipilic Idol.

sciences contributed; but of all their chemico-mechanical productions perhaps none performed greater deeds of renown than the Eolipile. To accomplish its purposes this instrument put on a strange diversity of shapes, and was endowed with such attributes as its adroit managers required; but, purposely disguised as it was, and its movements ingeniously masked, its former tricks are not entirely concealed by the veil which time has dropped over the stirring dramas of ancient life. It may be detected, though too remote to be distinct. In the deepest obscurity its performances are too peculiar to be mistaken. It appears to have flourished in mythologic and heroic ages, and, naturally enough, these were the times of its greatest achievements. Besides a few minor engagements, it was principally employed in personating three remarkable characters:—a god, a warrior, and a guardian of treasure. In the temple it descended with neophytes into the sacred chambers and took part in the lesser and sublimer mysteries, while at the altar it confirmed the faith of its worshipers by miracles wrought in their presence.

In war its effects were once equally decisive. Its appearance alone sufficed, like the head of Medusa, to petrify opponents with horror. Superstitious troops (in early times all were superstitious) were astounded at the sight of an enemy, supernatural in form, borne along in chariots of clouds and whirlwinds of fire; no stronger proof of the gods being against them could be adduced. Like the affrighted Philistines under a similar persuasion, their hearts would melt within them, and ere they fled they exclaimed with the warriors of Canaan, "Wo unto us! Who shall deliver us out of the hands of these mighty gods?"

As a serpent or dragon, it couched by the portals of palaces or lay at the entrance of caverns to protect the plunder its owners had gotten together.

The annexed figures and subsequent remarks may serve to elucidate in a feeble degree a few of its performances under each of these characters.

Idols, especially Eolipilic ones, belong to a department of ecclesiastical history hitherto little examined and less understood. True, they recall no very pleasing associations, yet they make us acquainted with many curious transactions. This figure is a representation of Pusterich, a bronze Eolipilic god of the ancient Germans, described at page 399, to which the reader is referred. The burning fluids and flame issued from the mouth and the eye or orifice in the middle of the forehead.

This is not near so repulsive as many ancient and modern idols: compared with some it might almost be deemed engaging. Perhaps its admirers were too far advanced to relish a mongrel deity, or one with an extra number of heads or limbs. It is but one among many of its kind which might be adduced, had we the history of numerous bronze images extant, or of others noticed in antiquarian works. Several have openings behind and fitted for plugs, as if designed for charging them with liquids.

No. 287. Ancient Eolipilic Idol.

There is an impressive resemblance between this figure and that of a Cyclop, and there may be a real similitude between idols of this kind and the three fabled sons of Neptune and Amphitrite. As remarked further on, *fire-breathing* and other mythic monsters were not all mere visions, mystic emblems, or hieroglyphical pictures, but actual brazen beings, of the forms and with many of the functions described—in other words, Eolipilic idols, personified as all idols were. The reader need not be reminded of the relation of the Cyclops to fire, since they were aids to Vulcan, and were destroyed by Apollo for manufacturing or ejecting, like Pusterich, thunderbolts. They are sometimes described as having but one eye, at other times represented with three—two in the ordinary places, and a third in the forehead, as in the preceding figure. [See plate page 141, vol. 1, Fosbroke's Encyc. Antiq.] This idol is supposed to have belonged originally to a high antiquity, and may possibly be a genuine Cyclop.

Two or three more metallic deities, which appear to be Eolipilic, might here be introduced; but as the fact is uncertain, and nothing but conjectures could accompany them, we forbear. Had more data been accessible the subject would needs be a thrilling one. No work of imagination could be richer in interest or more fertile in intrigue and plots than accounts of idolatrous androids of the more advanced nations of old, of the puppet-machinery in each famous temple, and the by-play by which the reverend showmen set them off to advantage, lulled suspicion and kept their audiences in the right humor. We may descant as we please on epic poets, on tragic and comic authors and actors, but what were the best of them compared to those proto-fathers of fiction and histrionic professions? Men whose theatres were temples, whose stages were altars : master players on the passions, who excited what emotions they pleased, and impressed on their congregations an abiding sense of the realities of the illusions they exhibited. The subject reaches down to the nonage of society and comes up with it to our own days; has relation to the most stupendous system of deception ever conceived, and the most successful one ever practised by man upon man; affords the most deplorable and durable examples of human credulity and cunning; involves the early history of all races and of nearly all arts. Its exposition of principles of ancient science would be highly instructive, and their villanous applications often amusing. The mystery that envelopes it irresistibly whets curiosity. The little that is known makes us anxious to push aside the skreen that hides from our view the ingenious and elaborate mechanism by which pagan monks emasculated the species and kept an awe-stricken world at their feet.

The following figures illustrate the fighting qualities of the Eolipile. As a war-instrument it became better known than as an oracle confined in temples. In the field it was exposed to the scrutiny of the curious as well as of its immediate managers, so that, whether captured or not, the secret of its construction could not long remain one, or the device be confined, if much employed, to one people. Nor did it cast off its pretensions to divinity with this change of occupation, but rather sustained them, for it was as a god that it first became terrible in battle—as such its military achievements shook neighboring nations with alarm and acquired for it a celebrity that has reached to our times. The nature of its performances remained the same as at the altar, except that it now did not hesitate to destroy those whom it could not convince.

Every people, no matter how barbarous, esteemed their own gods superior to others. It was indispensable to the interests of priests to keep this conviction alive under all exigencies; hence while victories served to establish it, defeats did not overthrow it. These, it was artfully sug gested, were only proofs of a deity having become temporary offended, either for not being properly invoked or on account of indignities offered to his ministers. It was only to make his protégés sensible of his displeasure that on such occasions he left them a prey to their foes! Pagan history is full of examples, they abound in the Iliad, which opens with one. Thus the character of an oracle or idol, and the influence of its officials were ingeniously preserved whether those who trusted in it became conquerors or conquered, victors or victims. Such was the practice under ordinary circumstances, the god remaining the while undisturbed in his fane; but when extraordinary calamities threatened, when an invading army approached and his worshippers were menaced with captivity or famine, corresponding efforts were made to appease and even to compel him to be propitious. Bribes were held out, votive gifts, hecatombs and new temples promised—processions in his honor were got up, with sacred banners, relics, &c. borne aloft, (an European practice through the middle ages, and an Asiatic one yet.) Then to make sure of success by connecting his fate with that of his followers, the latter took him down from his shrine and carried him to the battle-ground, under a belief that he would not suffer himself to be taken if he were disposed to leave them in the lurch. On the same principle idolaters of every age have acted. The early Jews were not free from the strange infatuation, nor is it easy to see how they could have been better informed previous to or at the period of the Exodus. They were as much attached to idols as the Egyptians, and took the first opportunity that the absence of Moses presented for making an image of Apis. After the severe defeat at Aphek, some of the ignorant got up a cry to bring the ark to the camp and renew the contest under its auspices. "When it cometh among us it may save us out of the hands of our enemies." To this the better informed probably acceded with the hope that Jehovah would protect it, and the people for its sake, but they were mistaken—they were routed, thirty thousand were slain, "the ark of God was taken," and exhibited in the principal cities of the captors for a period of seven months, during which Phenician priests and artists were probably not very scrupulous in examining its contents, its designs and decorations, the cherubim of hammered gold, their forms, features, wings, &c.

In this same manner warring Eolipiles became known to others than their designers: as gods and demi-gods they made their debut in battle. As such they were victorious, and as such were eventually captured. Exaggerated accounts of some of the earliest are preserved in mythological annals. So awful were their attributes and so terrific their appearance, that their very looks overcame their opponents. Of this Briareus was an example; but when their artificial nature became known they put on less formidable shapes, their efficacy then depending more on what they did than how they looked. In comparatively modern epochs they never, however, attained much beauty, if we might judge of the one on the following page.

The age to which the specimen figured in the next cut belonged is unknown. It and No. 289 are from a Latin folio published in Paris in 1535, containing Vegetius on Military Machinery and Institutions, Elian on Tactics, Frontinus on Strategems, and the Book of Modestus on Military Affairs:—collated from *Ancient* codices by BUDEUS, the celebrated French critic. Attached to, and paged with Vegetius, are one hundred and twenty folio illustrations, rudely executed on wood. They are copies of those of the old German translation to which we have frequently referred, with the exception of a couple of reduced fac-similes which are now before the reader. (a)

No. 288. Ancient Fighting Eolipile.

As not a word of explanation accompanies this singular figure, (nor any other in the book,) and little or nothing is to be found in Vegetius or other Roman authors to aid us, all that we can offer must be received as conjecture. If the magnitude of the machine be judged from other illustrations in the collection, it was colossal. No object is portrayed near it by which to infer its relative dimensions. The general outline represents the human bust, and the whole seems to have been an enormous Pusterich on wheels. It probably combined the god with the warrior, assuming the character of each as occasion required. It is no bad representative of both; and the powers it possessed of punishing its enemies are as obvious as they were awful. The ignited jet issued from the conical tube whose wide end is riveted to the forehead—(a small pipe descending from it to the bottom of the bust, as in the air-vessels of fire-engines,) and possibly, also, out of its eyes and mouth. The prolongation of the nose, and the daggers projecting from the mouth, were intended to ward off blows during assaults, and to prevent access to it, lest the orifice or orifices should be spiked or otherwise closed. Pointed projections of this kind are quite common adjuncts in old war engines.

As this Eolipile is figured at rest and not in use, neither fire, fire-place, nor the mode of charging it is delineated. The fuel was probably applied in the lower part of the bust behind, though it may have been kindled

---

(a) " Fl. Vegetii Renati viri illustris de re militari libri quatuor. Sextiivlii Frontini viri consularis de strategematis libri totidem. Æliani de instruendis aciebus liber unus. Modesti de vocabulis rei militaris liber unus. Item picturæ bellicæ cxx. passim Vegetio adjectæ. Collata sunt omnia ad antiquos codices, maxime Budæi, quod testabitt r Ælianus. Parisiis, mdxxxv."

externally, the head being for that purpose inclined backwards and resting on the cornigerous and auricular prolongations, which would, like the feet of a caldron, form a tripod to support it. But much allowance must be made for old illustrations. Scarcely ever is an attempt made to delineate interior parts or external details. One object of the horn and ears was obviously to vary the direction of the jet, to incline the tube to the right or left, up or down, somewhat in the manner of the syringe engine of Besson. The wheels are solid, and as there are but two, some mechanism for preserving the image in an upright position was necessary: as they moved on separate axles the tube could as readily be turned in a lateral direction as it could be elevated or depressed. The manner of conveying this machine to considerable distances is not indicated, probably because it was rather intended as a stationary means of defence, than, like the next, a moveable one for attack.

No. 289. Eolipilic War-Dragon.

Here is a variety of the griffin, hippogriff, or dragon genus, placed on four wheels, and evidently designed to break the ranks of an opposing army, by being driven through them. The burning liquids rushed out of two rows of small holes on the upper jaw or lip: the effect forcibly reminding one of mythic monsters from whose nostrils went forth smoke, and from whose mouths issued flame. No provision is shown for raising or lowering the jets, nor was any necessary, for from the elevation and position of the orifices, troops among whom this engine forced its way could not avoid either right or left its fluid and scorching missive. The rod held by the captain or leader is enlarged and pierced or cloven at its upper end, where it is joined to the head: it is apparently a lever by which the plug of a cock was turned to open and shut off the discharge. We may suppose the passage was closed in the present position of the lever, and that to open it the manager pulled back the end he grasps, until, like a modern artillerist, he became sufficiently in the rear to be out of harm's way when the jets found vent; he then could join his associates in directing the monster's movements. The wheels, as in the last figure, are represented solid, a feature undoubtedly genuine; for it was the uniform practice to attempt to stop the progress of such war-chariots as had wheels with spokes, by throwing spears, &c. between the latter;

and hence such wheels were sometimes covered with boards or plates of iron previous to entering into battle.

The sword or dagger-like tongue kept an enemy from approaching too near in front, while the flames protected both sides. It would not have answered the purposes of this war-engine to have made its sides horrent with bayonets, for they would have retarded its progress by contact with every obstacle within their reach. Its efficiency depended chiefly on the velocity and precision of its movements, it would therefore be divested of every thing calculated to interfere with these. The inclination of the tongue was designed to remove obstacles from the path. Had the spike been horizontal it would have transfixed objects it met with, and the progress of the machine would soon have been stopped. This machine is apparently represented as in times of peace, for, unlike most others in the collection, no signs of war are delineated in the landscape. The fire was perhaps applied externally, as in the case of Pusterich, the brazen monster belonging to the Tyrant of Agrigentum, and other ancient devices of the kind: but this part of the subject is very obscure. Like chariots with swords and scythes fixed to them, and others with similar weapons revolving in their fronts, this machine when in active service was most likely urged forward by horses yoked behind; or by a number of men applying their force to bars attached to and radiating from the rear—both ancient and very common war devices.

An enormous Eolipile, formed after the above pattern, charged with inflammable liquids, and driven furiously and unexpectedly upon a superstitious foe, must not only have borne all before it, like a modern locomotive, but must have rendered opposition hopeless until its contents were expended.

The dimensions of this war dragon cannot safely be inferred from those of the men attached to it, for in most of the plates in the work whence it is taken, no kind of proportion is preserved. Soldiers raising ladders to scale the walls of high towers are often drawn sufficiently tall to reach the roof with their hands.

As the name of a war machine, the term *dragon* was continued to modern times. It was early given to pieces of ordnance, to devices resembling in their attributes ancient Eolipilic monsters. Culverines were originally called fiery-dragons. The Draconarii of the Romans bore dragons on their standards; the Parthians, Indians, Persians, Scythians, Assyrians, Normans, Saxons, Welsh, and all the Celtic and Gothic nations painted the same thing upon their banners and pennons, as the Chinese, Russians, Tartars, &c. do now. Modern dragoons have probably also derived their designation from soldiers who formerly managed Eolipilic dragons, as in the preceding figure; the name being preserved in war's vocabulary after the office and instrument were forgotten. Orders of chivalry were named after the dragon, and heraldry abounds with its figures.

Let us now turn to the history of the Goths, by Olaus Magnus. (Basil ed. 1567.) The fourth chapter of the ninth book is headed, " *De æreis equis ignivomis* "—" Of brazen horses that vomit fire." The materials of the chapter are condensed from the History of the Danes, by Saxo Grammaticus, a writer who flourished A. D. 1140. The principal incident relates to the stratagetic skill of an old king, *Regnerus*, who was eventually put to death by his sons, Daxon and Dian. On one occasion the two rebellious brothers invaded their father's kingdom, having been furnished for the purpose with a large army by king Ruthenus, whose

Brazen Horses that vomited Fire. 583

daughters they had married. Alarmed at the mighty forces brought against him, Regnerus ordered a number of brazen, fire-breathing horses to be secured on chariots, and whirled suddenly into the densest body of his enemies. The manœuvre succeeded, and his unnatural sons were put to flight. It appears that the chariots and their burdens were exceedingly massive, since they overwhelmed whatever opposed them We add the passage at large from Saxo. It will be perceived that *he is silent* respecting the fire-vomiting faculty of the metallic chargers, though that was clearly implied in the opinion of the Gothic historian; an opinion that can hardly be questioned.

Post hæc Regnerus, expeditionem in Hellesponticos parans, vocataque Danorum concione, saluberrimas se populo leges laturum promittens, ut unusquisque paterfamilias, secut ante, quem minimi inter liberos duxerat, militaturum exhiberet, ita tunc valentioris operæ filium aut probatioris fidei servum armaret, edixit. Quo facto omnibus, quos ex Thora procreverat, filiis, præter ubbonen, assumptis, Hellespontum ejusque regem Dian variis contusum bellis lacessendo perdomuit. Ad ultimum eundem creberrimis discriminibus implicatum extinxit. Cujus filii Dian et Daxon, olim Ruteni regis filias maritali sorte complexi, impetratis a socero copiis, ardentissimo spiritu paternæ vindictæ negotium rapuerunt. Quorum Regnerus immensum animadvertens exercitum, diffidentia copiarum habita, *equos æneos* ductilibus rotalis superpositos ac versatilibus curriculis circumductos in confertissimos hostes maxima vi exagitari præcepit. Quæ res tantum ad laxandam adversariorum aciem valuit, ut vincendi spes magis in machinamento quam milite reposita videretur, cujus intolerabilis moles, quicquid impulit obruit. Altero ergo ducum interfecto altero fuga sublapso, universus Hellesponticorum cessit exercitus. Scithæ quoque, Daxon arctissimo materni sanguinis vinculo contingentes, eodem obstriti discrimine refuruntur. Quorum provincia Witserco attributa, Rutenorum reg. parum viribus fidens, formidolosa Regneri arma fuga præcurrere maturavit.

[Saxo Grammatici Historia Dania. Edited by P. E. Muller. Copenhagen, 1839. Liber ix. p. 452.]

In a note on the Equos Æneos, the editor, not knowing that such things had ever been, observes, "commentum nescio unde petitum."

No. 290. Eolipilic War-Engines.

The cut No. 290 is copied from the rude illustrations of the fourth and fifth chapters, Book ix, of Olaus Magnus. A figure of one of the brazen horses is in the foreground, but as usual it is a mere outline, and was perhaps designed by the illustrator of the Gothic historian's work from the meagre description its pages or those of Saxo afford. Nothing definite can be derived from it which the text does not furnish. Neither the carriage nor its load comes up to the description : the words imply that the images had some elastic and revolving mechanism of their own, and versatile chariots meant something more than common carts.

The fifth chapter (Book ix) is on the same subject, and to this effect. 'Vincentius in Spec. Histo. L. xxxi. Cap. 10, asserts that the king of the Indians, commonly called Prester John, being attacked by a powerful army of Ethiopian Saracens, enemies of the christian faith, delivered himself by a stratagem not unlike that of Regnerus, for he made *copper images of men* and mounted each upon a horse. Behind every image was a man to govern it, and to blow with a bellows, through holes made for the purpose, on fumid materials inserted beforehand into the body of the image. Provided with a large number of these he proceeded vigorously against his enemies, whom Vincentius calls Mongols or Tartars. The mounted images being ranged side by side in front of the hostile army, their managers were directed to advance, and when arrived within a short distance of the foe to commence blowing with their bellows the smoking fire within, and with a continual blast to fill the air with darkness—the consequences of which were that many of the invaders were slain and others took to sudden flight. Large numbers of horsemen and horses were burnt to death and some reduced to ashes *by Greek-fire*, composed of the following ingredients, by the artificers of Prester John:

> Aspaltum, nepta, dragantum, pix quoque Greca,
> Sulphur, vernicis, de petrolio quoque vitro,
> Mercurii, sal gemmæ Græci dicitur ignis.
> ITEM: Sulphur, petrolium, colopho, resi, terebinthi,
> Aspaltum, camphora, nepta, armo, benedictum.'

Magnus could make nothing out of these old poetic recipes. He thought it would be a vain task to attempt their explanation, and wicked to revive the invention. He seems to have been of an opinion—once heartily entertained—that the souls of the authors of Greek-fire and gunpowder were reaping their appropriate rewards in perdition, doomed for ever to taste of torments which their "devilish devices" inflicted on others. Vincentius, or Vincent De Beauvais, was a learned monk of the 13th century, and one of the most voluminous writers whose works furnished employment to the first race of printers. He died about 1260. His "Speculum Historiale" was printed in 1473. The most striking incident drawn from it by the Gothic writer we quoted at page 400, from Carpini, a contemporary monk, who began his travels in 1245, and to whom he of Beauvais was most likely indebted for it.

If the reader will now look again at the last cut he will find on the back ground a miniature of one of the brazen horsemen in the act of attacking the Mongols, and with a living soldier on the crupper performing his part of the business with bellows. There is certainly an air of romance about these figures; but accounts of them reaching us through ages and hot-beds of legends, might be expected to be loaded with apocryphal matters. Of the main feature, that of ejecting flame and smoke, there is no room to question, since it is corroborated by old writers on Greek-fire, by the brazen horses of Saxo, and the preceding figures in this supplement. But Carpini's relation does not savor so much of poetry as may be supposed. The principal difficulty is in mounting the images on *natural* horses; but this is not a necessary inference. They may have been artificial as well as their automaton riders—and we believe were so—were secured, like those mentioned by Saxo, on carriages, and behind them the bellows-blowers were located. If this is not what Carpini meant, we should say he misunderstood his informant. Living horses, with flames roaring and rushing from orifices close to their eyes and ears, would be as likely to be affrighted as those they attacked: however drilled, they could not in such circumstances be managed without difficulty and without requiring the whole attention of their riders, but the latter were entirely engaged in urging the fires at the most critical periods of the charge, leaving the animals to pursue the right course of themselves. We presume the metalline images were a species of Hippocentaurs, the flames issuing from the human bust, and the fluid and other materials contained in the spacious abdomen below.

It is said these equestrian images cast forth Greek-fire; were they then Eolipiles? mounted Pusterichs? i. e. were they charged with liquids, or with dry substances, which once ignited continued of themselves to burn until the whole became expended? From the want of specific information it is difficult to arrive at a definite conclusion on this point. The evidence, however, preponderates in favor of their Eolipilic character. Had the contents been a composition similar to any thing used in modern pyrotechnics, what need of fire to heat them and of bellows to urge the fire? How did the flaming stream continue to issue from its orifice with unabated force as the material diminished within, as it sank far below the place of exit? Would not the image be liable to explode ere its contents were half emptied? If not, why have metallic images? Those of fragile materials would have done. Again, the reaction of the jet, like that of a rocket, would require no small force to be overcome: it would be very apt to shoot the brazen warriors back among their friends, instead of their carrying destruction among their foes. But not one of these objections, and others which might be named, apply to Eolipiles—to a liquid discharged by the elasticity of its own vapor, or the vapor itself thus shot forth. With these instruments the employment of fuel was necesary and the application of a blast in time of action important if not indispensable. But, what is more to the point, Greek-fire *was* a liquid. See p. 307, 8. Meyrick, in his account of ancient armor, gives its composition from an author of the time of Edward III. Several ingredients enumerated are mentioned in the preceding recipes from Vincentius:—An equal quantity of pulverized rosin, sulphur and pitch; one fourth of opopanax and of pigeons' dung well dried, were dissolved in turpentine water, or oil of sulphur: then put into a close and strong glass vessel and heated for fifteen days in an oven, after which the whole was distilled in the manner of spirit of wine, and kept for use. Another account makes it to consist chiefly of turpentine water (spirits of turpentine) slowly distilled with turpentine gum. It was said to ignite by coming in contact with water.

Two distinct modes of dispersing the horrible fluid are mentioned; one by forcing-pumps, the other by "blowing" it through tubes and from the mouths, &c. of metallic monsters. The former is noticed in connection with naval warfare, and the latter, if we mistake not, was chiefly employed in conflicts on land. Any one can see how difficult it would be for soldiers promptly to apply pumps in the confusion of battle. Apparatus equal to our fire-engines would have been of little effect, for the jets could but feebly be sustained, and worse directed while the reservoirs, engines and men were in motion, whirling hither and thither, now advancing and anon retreating. We read also of portable "*siphones*" being also used, but these and the necessary vessels to hold the liquid were still less likely to be effective except on ships in close combat; where to keep up conflagrations, the fluid could be ejected, cold and unignited, on parts already kindled—as if our engines were to be employed to lanch oil or turpentine on objects already in flames. On ship-board, the reservoirs were always at hand, and both men and the fixed pumps they worked relatively at rest, and moreover protected either between decks or in equally secure locations, so that one or two individuals alone sufficed to direct the fiery streams over a galley's bow or sides, and through flexible or jointed ajutages.

The expression "*blown* through tubes," &c. could, of course, have no reference to any thing like the sarbacan, nor to any employment of human lungs. No adequate and no continuous force could have been obtained except by artificial means, and of those by none so readily as by the Eolipile. That this instrument was intended, the figures in the cut strongly indicate. If the *vapor* of the fiery liquid was ejected, we know that nothing else could have answered. But both the idea and expression are used at this day with respect to modern Eolipiles: engineers "blow off" steam by opening a safety valve or other aperture of a boiler; and when one of these explodes, on shore or afloat, how often is it said of missing individuals and objects, they were "blown overboard"—or "blown to such and such distances." On a review then of the particulars that have reached us respecting the famous Greek-fire, it seems that the machinery for ejecting it on shipboard was a species of pump; and on land by large boilers, suspended on wheels and driven by horses or men, made in fantastic forms of men and animals, from whose mouths the flaming torrents were ejected. This, ancient writers have asserted, and the figures we have given confirm.

That Greek-fire was rather the revival of an old thing than the discovery of a new one, and that both the fire and the machines for dispersing it—Eolipilic devices infinitely more grotesque than any figured on these pages—were known in extremely remote times, is, we think, pretty clear. Under this impression some further remarks are submitted with the view of eliciting attention to a curious and interesting subject of archeological research—one which, it will be conceded, *appears* to reflect light on old legends as well as on old Eolipiles.

The history of idolatrous and other Eolipilic automata is lost or perhaps never was written, and now the opportunity, the materials and men for preparing it are gone; the requisite knowledge did not sufficiently transpire beyond the walls of temples, and even there was confined to a privileged few. Such a record could only have been furnished by those who had every earthly inducement to suppress it—by men whose private labors were devoted to disguise the elements of deceptive devices they employed, and whose public administrations still further concealed them. It may therefore be concluded that such an exposé was never made, or, if made, religiously reserved for the perusal of heads of colleges or the eyes of arch-magicians alone. It is to be regretted that so valuable a fund of hidden knowledge, of mechanical and chemical combinations, of singular discoveries and inventions; a bibliotheca for philosophers and artisans, illustrating, probably, every branch of ancient science and ex posing the secret workings of some of the shrewdest spirits of antiquity—should be lost. It would have enabled us to repeat staple tricks of Babylonian sorcerers and soothsayers, and would have placed us in a more favorable position for observation than was Pharaoh when he commanded "the magicians of Egypt and the wise men thereof" to exhibit their skill in his presence.

It is with Eolipiles as with other *materiel* of old jugglers. The few broken specimens and straggling notices which have come down are interesting but unsatisfactory; they tantalize with a sip, and make the mouth water for more, provoking a thirst which they cannot allay. That these instruments are of a very high antiquity is undeniable, and that they were occasionally used to eject inflammable fluids for deceptive and destructive purposes is equally certain. The resemblance in the forms and functions of those we have figured to mythological fire-spouting monsters, is too striking to escape observation. And is there any absurdity in supposing both were artificial; that the latter were literally what they are described; and that stories of dragon-killing heroes are not quite so romantic as they appear? A literal interpretation of such matters may appear preposterous, but a slight view of the subject will convince unprejudiced minds that it is not half so absurd as many received metaphorical solutions, nor is it, like them, embarrassed with insurmountable difficulties; on the contrary, it renders things intelligible which paleologists have not ventured to explain, and which, without reference to Eolipilic automata, we presume they never can explain—things so bizarre they know not what to make of them. But once admit they were what they pretend to be, and there is little difficulty in receiving them; interpret them by some other rule, and we are at once cast adrift on the ocean of conjecture.

Admit that mythic characters obtained celebrity from battling with Eolipilic opponents; that some, at least, of the dragons and many-headed monsters of antiquity performed actions ascribed to them—belched out smoke and flame, shrieked and growled, and on the approach of strangers or "curious impertinents" shook themselves, sprung from their caves, (they were commonly and for good reasons located in dark places) often destroyed those who attacked them, and sometimes disappeared in sudden bursts of thunder and amidst showers of thunderbolts—very much as their descendants, the steam-dragons of the present day, unfortunately now and then do. Admit this, and passages in history, poetry and tradition, hitherto inexplicable, become recitals of facts; embarrassing enigmas are unriddled, and the supposed offspring of fancy are found sober children of truth. That Greek and Roman writers did not perceive this is little to the point, since they do not appear to have been acquainted with fighting Eolipiles; they were therefore necessarily at a loss to explain, except by metaphor, conflicts between these machines and heroes of ancient days. But the presiding spirits at Eleusis and Delphos could have furnished the clew, and, had it suited their views, could have illustrated the entire series of fire-breathing monsters, by reference to their own collections; for, as before remarked, Eolipiles went from the altar to the field.

In those remote times, when superstition reigned paramount, when common objects and events were construed into omens and uncommon ones were looked on as prodigies, the defeat of an army by fire-breathing warriors would form an epoch in barbarian annals; exaggerated descriptions of flaming chariots, of giants, dragons, hippogriffs and hybrids of every horrid form, and possessing supernatural powers, would be blazoned abroad and become permanently preserved in tradition. It could not be otherwise; and that such was really the case is evident, for mythology and remote history is replete with these very things; with battles between Gods, Cyclops and Titans. But in process of time the artificial nature of warring Eolipiles would sooner or later be suspected and ascertained. Intrepid individuals took courage to attack and had the good fortune to destroy one. Success made them heroes, if not something more. To swell their fame the form and faculties of their strange opponents were distorted, and the story repeated, with every addition that a love of the marvellous could invent or credulity receive, till, as ages rolled away, it became just what such stories yet extant are—stories of monster-killing gallants from Jason to Saint George.

WARS OF THE GIANTS.

In the wars of the giants, fire, thunder and thunderbolts were the chief destructive agents, and these, we are told, were produced by and ejected from monsters, apparently precisely in the manner of Pusterich. Some had more heads and arms than have Hindoo deities, with bodies terminating, like that of Dagon, in legs resembling fish or serpents. When brought into battle their terrible aspects and the volumes of flame they poured forth filled their enemies, the *gods*, with consternation. Defeated, these fled into Egypt, where they learned the nature of their ardent foes. Jupiter, Hercules, and their associate refugees having thus ascertained that their victors were not invincible, recovered courage, returned, and were at last victorious. Now what, when stripped of oriental ornament, does this amount to, but a conflict similar to that between Prester John and his Mongolian invaders; between Regnerus and his unnatural sons, and others in which fire-spouting images, figured in this supplement, were employed? The most ingenious conquering, whether gods or mortals were combatants. The names of the mythic parties were misnomers, for the deities were ignorant braggarts—they could not withstand their " earth-born " enemies, but fled for refuge and instruction into other lands. The accounts remarkably resemble Chinese bulletins of fights with Europeans—contests between modern " Celestials " and " outside barbarians." For, ancient like, existing "sons of heaven" seem to have placed at first as much dependence upon their divine pretensions and their comminations as in their weapons, and therefore were defeated. The giants were probably ingenious or scientific men—the Roger Bacons of their day—in advance of the age and consequently denounced, as such have ever been, by self-styled heirs of heaven, as infidel dogs or children of Tartarus.

The circumstance of the divinities flying to Egypt when they could not cope with the fire-breathing monsters, or rather with the cunning monster-makers, is remarkable. There they, like less pretenders, improved themselves in knowledge. That it was an early Pharaonic policy to encourage the discontented of neighboring nations, is abundantly proved in the Old Testament. " Wo to them that go to Egypt for help—that strengthen themselves in the strength of Pharaoh!" [See *Isa.* chaps. 30 and 31; *Jerem.* 42 and 43.] How deep and general must have been the impression of the power of the Pharaohs to call forth the declaration—" Now the Egyptians are *men* and not God; their horses flesh and not spirit."

TYPHON.

Here is a description of Typhon, the most famous of fighting giants—can it be doubted that he was a genuine Pusterich? "He had numerous heads resembling those of serpents or dragons. Flames of devouring fire rushed hissing from his mouth and eyes; he uttered horrid yells like the dissonant shrieks of different animals. He was no sooner born than he warred with the gods and put them to flight." Not a circumstance is here mentioned that does not accord with his alleged artificial character, and there are few others which do not harmonize with it. He went to battle as soon as born, that is, as soon as he was made. The whole family was said to be "earth-born"—the members rising out of the ground completely formed, &c.; indications of their gross not ideal nature, of their secret construction in subterranean workshops—the latter a precaution essential to the recognition of, and belief in their supernatural origin.

> She sings, from earth's *dark womb* how Typhon rose,
> And struck with mortal fear his heavenly foes.—[*Ovid, Met.* v.]

The name, *Typhon*, is derived from a word signifying, "*to smoke.*" The goddess of night was the mother of monsters; an enigma beautifully expressive of the secret fabrication of Eolipilic imagery. Typhon and his brethren were moreover sons of Tartarus as well as of Terra—were brought forth of earth by the assistance of hell—a trait still further significative, and particularly of the element by which they were animated, that from which their terrors were derived. Demons they were in shape, occupations and attributes; in the torments they inflicted and the victims they slew; tangible, and the most perfect representations of evil principles and passions. The paternity of these monsters is the same as that given to modern ordnance, so true it is that similar things ever produce the same ideas. A thousand times have guns and gunpowder been described as infernal inventions, as conceptions injected by demons and matured by their influence.

Does the idea seem too gross for contending gods and demi-gods to fight with Eolipiles? Let it be remembered that Milton could find no warring engines so appropriate for Satan and his hosts as artillery. In fact, poets can only arm mortal or immortal warriors with weapons and agents that are *known*, although they may exaggerate them. All symbolic imagery must be derived, directly or remotely, from earthly types. The author of Paradise Lost necessarily followed, in this respect also, the old mythologists he copied, and as "fiery monsters," whether guns or Eolipiles, are not in their nature and effects much unlike, we find little difference in ancient poetic descriptions of one, and modern poetic descriptions of the other. Indeed they might often be interchanged without detection. The monsters described by Milton as mounted upon wheels, whose mouths with hideous orifices gaped, and which, with impetuous fury, belched from their deep throats chained-thunderbolts and iron hail, are therefore no stronger proofs of guns and gunpowder being known during the English Commonwealth, than are fire-breathing hybrids of mythology, of the early use of Eolipilic engines.

THE COLCHIAN BULLS AND DRAGON.

If we turn to later examples we shall find circumstances leaking out which betray the artificial character of mythic monsters. The Argonautic, like all early expeditions, was of a piratical nature. Its object the Colchian treasury, or the "golden fleece," a term in ancient Syriac implying treasures of gold. These were protected by a dragon, and by two brazen-horned and hoofed bulls, which flashed from their mouths and nostrils flames and smoke. As usual, they were located at the entrance of a cave.

> "Thick smoke their *subterraneous* home proclaims;
> "From their broad nostrils pour the rolling flames."
> [*Apollonius*, L. iii.]

The daughter of Æetes (the Colchian king) becomes enamored of Jason. The lovers swear eternal fidelity to each other; and to save the adventurer's life, Medea explains to him the secret of the monster's powers. Thus informed, and furnished with an ointment to protect his face and hands from the singeing blast at the onset, he approached with a smiling countenance, as well he might, and quickly, to the chagrin of the monarch, subdued the "brazen" monsters. If any doubt remains respecting the true character of this transaction, it is greatly if not wholly removed by the subsequent conduct of Medea. She every where evinces familiarity with the principles of the Eolipile—with secret applications of fire, steam, sulphur, inflammable fluids and explosive compositions. (See page 120.) By the adroit use of these, which she introduced into Greece, she became celebrated as the most expert enchantress of antiquity. It was by a clever but diabolical trick in Pyrotechnics she destroyed Creusa, while, further to be revenged on her unfaithful husband, she contrived to set his palace in flames and then disappeared in a chariot drawn by winged dragons!—probably some startling pyrotechnic device learned from the magicians at her father's court, and under the cover of which she withdrew; unless we are to suppose she was blown up by the explosion of one of her own caldrons or compounds.

There is no improbability in the supposition that attempts at flying were somewhat frequent in remote ages, and that jugglers and artists, like Dædalus, did then, as in subsequent times, get up exhibitions of the kind; but, be this as it might, it may be taken for granted that so expert a pyrotechnist as Medea, was at no loss in sending up a chariot with an artificial representation of herself, on the same principle as such things have been done from time immemorial in India and among the Chinese. They were common a few centuries ago in Europe. Like most old writers on fire-works, John Bate gives directions how to make "fire-drakes" and "flying-dragons." The latter were to be constructed of ribs of light and dry wood, or with whalebone "covered with muscovie glasse and painted." They were to be filled with "petrars,"—fiery serpents were attached to their wings, which were arranged to shake when the monster moved. A sparkling composition was to burn at the mouths and tails, and one or two large *rockets* were to be attached, "according to the bignesse and weight of each dragon." The trick of Simon Magus, in presence of Claudian or Nero, was perhaps allied to that by which the Colchian enchantress astounded her adopted countrymen. Giving out that he would prove his divinity, or his alliance with the gods, by flying, he appeared at the appointed time, as the story says, on the top of a high tower, whence he flung himself, (or an artificial substitute,) and floated for some time in the air, supported by demons or *dragons*. The latter no doubt as real as the huge scarabeus which Dr. John Dee, state-conjurer to Elizabeth, made, and which flew off with a man on its back, and took a basket of provisions for the journey.

Oriental literature is laden with aerial exploits of this nature—of enchanters, who like Medea, or Urganda in Amadis de Gaul, transported men through the air on artificial serpents and dragons, and of conflicts between knights and monsters. But for the loss of those volumes on "curious arts,"—the pile of magical books burnt at Ephesus—(*Acts*, xix. 19.) many an ancient and modern prodigy might have been explained. We know with what ardor marvellous tricks and stories were devised and concocted in the middle ages, and with what avidity gaping multitudes received them. Even at this very day similar tricks are played off successfully by monks to unsuspicious congregations. Is it any wonder, then, to find pagan boors in Roman times, and others in the darkest of mythic epochs, dupes to expert jugglers ? We may regret the infatuation of remote ages, but we should not forget how, in comparatively late days, traditions arose and swelled in wonder as years rolled over them, and how mechanical devices, simple in themselves, but not comprehended by the public, were metamorphosed into supernatural productions, which increased in mystery and magnitude as the times when they were contemplated receded from those of their birth. Had printing not been introduced we might have competed with the ancients in prodigies, and prodigies as fully believed; for there are few old examples derived from tangible mechanism, or pure phantasma, that have not been imitated by modern manufacturers. But alas for these! the revival of letters is the bane of their fame. Stripped of their borrowed garments they stand before us as ordinary mortals—a predicament most of their predecessors would be in, had we equal facilities to disrobe them.

The manner of taming the dragon at Colchis is characteristic. It was the work of Medea rather than of Jason, accomplished privily, and at midnight. Instead of instructing the leader of the Grecian adventurers to attack it as he attacked the bovine monsters, armed with his faulchion and club—a species of combat that might have alarmed the palace, she adopted a process more quiet and equally effective; in fact, just such an one as might have been expected from her.

> "To make the dragon sleep that never slept,
> Whose crest shoots dreadful lustre; from his jaws
> A triple tire of forked stings he draws,
> With fangs and wings of a prodigious size:
> Such was the guardian of the golden prize.
> Yet him, *besprinkled with Lethæan dew*,
> The fair enchantress *into slumbers threw*." [*Met.* vii.]

That is, in unadorned prose, she turned or threw on the concealed boiler and furnace a shower of *cold water;* and thus, without injuring the dragon, sent him as effectually to sleep as a steam-engine is without steam—the very device which has been recommended to render harmless a boiler when ready to explode.

The incident mentioned by Apollonius of the dragon hissing so horribly and loud, when the two lovers approached, as to cause neighboring forests to echo back the sound and make distant people start in their dreams, is pure hyperbole : if modified to an ordinary growl it is hardly reconcileable with what he just before narrates of the lady being so cautious of awakening the numerous palace-guards as to escape through by-paths barefoot. Sensible of the solecism he in the next breath ascribes the undisturbed repose of Æetes and his family to magic. It would however be futile to attempt to extract unadulterated truth in *every* particular from labored fiction, and particularly in dragon history, to make out where truth and fable meet, where one begins or the other ends. Facts woven up in old poetry were like woollen threads in Babylonian garments—valued in proportion as they were embellished. The poet's like the sculptor's or embroiderer's skill was measured by the art with which ordinary materials were lost in forms and ornament. Few think of aluminous earth while viewing the splendid vase, and none look for truth unadorned in works of classic artists.[a]

THE CHIMÆRA.

The Chimæra destroyed by Bellerophon looks very like another specimen of Eolipilic ingenuity, though represented of course as a living animal, agreeably to legendary tradition and poetic license. Homer describes it as

> .... Lion faced,
> With dragon tail, shag bodied as the goat,
> And from his jaws ejecting streams of fire. [*Il.* vi.]

The most popular of ancient explanations supposes this monster signified a burning mountain, whose top, on account of its *desolate* nature, was the resort of lions, [an obvious contradiction] the middle being fruitful, abounded with goats, the marshy ground at the bottom swarmed with serpents, and Bellerophon by cultivating the mountain subdued it! Such is one of the best specimens of classical guessing, and yet both mountain and its inhabitants were suppositious—assumed for want of better grounds of conjecture. It is observable that old fire-breathing monsters are represented as akin to each other: thus the Chimæra, the dragon which guarded the golden fruit in the garden of the Hesperides, Cerberus and others, were related to Typhon and the rest of the giants—as if to intimate their common nature, so that, according to mythology itself, if one was an automaton, all, or nearly all, partook of the same character. If the mountain supplied the true solution of the Chimæra, it should furnish a key to unriddle the rest, but it would be impossible to locate volcanoes where fiery dragons were—in gardens, cellars, palaces, &c. and still more so to make them travel abroad and rush hither and thither in battle.

How much more reasonable to admit the Chimæra to have been an Eolipilic dragon; its description is then natural, its appearance and performances credible, and its demolition by the great captain consistent. Old demi-gods did not acquire their titles by wielding the mattock.

If the figure No. 289 had a couple more heads and were furnished with the caudal terminus of a lizard or cayman, it would form no bad representation of the Chimæra.

CACUS.

As like causes produce like effects, so in early as in later times disbanded soldiers turned often robbers. Too idle to work, numbers of these ruffians lived by private plunder when opportunities ceased for sharing public spoils. Not a few of the old heroes belonged to this class, and among them was *Cacus*. The story of this famous thief is an admi-

---

[a] There is a striking likeness in the manners, customs and superstitions of the Colchians, as portrayed by Apollonius Rhodius, and those of the people described by Saxo and Olaus Magnus. It would be a curious fact if fighting and juggling Eolipiles, or the knowledge of them, lingered in the regions of the Euxine and Caspian from the adventure of the Argonauts to the battles in which the automatons represented in figs, 289 and 290 are said to have been employed. It was from Scythia the arts of brass-founding and working in metals descended to lower latitudes, according to Pliny.

rable comment on the state of society in his day, besides furnishing another specimen of fraud preying on credulity by means of Eolipiles. A son of Vulcan, he knew something of machinery and of the wonders, honest and dishonest, his father wrought by it. As usual, he occupied a cave favorably located for his purposes.

> . . . . See yon rock that mates the sky,
> About whose feet such heaps of rubbish lie;
> Such indigested ruin; bleak and bare,
> How desert now it stands, exposed in air!
> 'Twas once a robber's den, enclosed around
> With living stone, and deep beneath the ground
> The monster Cacus, more than half a beast,
> This hold, impervious to the sun, possess'd. [*En.* viii. *Dryden.*]

At the cavern's mouth he had a triple-headed image, which (not its owner) belched black clouds and livid fire. It was at length destroyed by Hercules, who we have seen had some experience in such matters. The success of Cacus in levying contributions from the fields and folds of the simple inhabitants of the neighborhood, and on drovers passing through it, appears to have been due to the tact by which he made it generally believed that he and the monster were one and the same individual:—a common ruse this in such cases, and one by no means peculiar to mythic epochs. He made his forays in the night, and lay concealed during the day.

The personification of Eolipilic and other images was in keeping with their design, and necessary to preserve their influence over the ignorant. As they sustained the characters of gods and demigods, they were addressed as such. The practice differs but little from what is now in vogue; fire-engines, mills, ships, guns, &c. have male and female designations, are often spoken of as if endowed with spontaneity and passions; but with not half the propriety as androids representing and performing functions of living beings. Sometimes these are so delineated in their appearance, feelings, employments, &c. that no doubt of men being intended could arise, were they not at other times associated with attributes and deformities unknown to humanity. The solution is however easy:—The ancients like the moderns gave their names to certain classes of devices, and it is descriptions of these which we confound with the persons after whom they were named—the artificial dragon of Cacus with that individual. The same cause of misapprehension may take place with regard to men and things of our day. What, for example, must people think, some thousands of years hence, of Washington and Franklin, if all memorials of them should then be lost except a few statements, of which one described them as floating monsters, 300 feet in length, with scores of brazen mouths through which they vomited floods of fire and roared so loud as to make mountains quake:—or according to another they were of less majestic size, but showering volumes of smoke from iron throats, trembling with passion when obstructed in their progress, and then starting forward, gasping and galloping over the ground with almost lightning speed, and leaving trains of fire behind! Land and water dragons! What *could* such people think unless informed that 74 gunships and locomotive carriages often bore the christian names and surnames of those celebrated men.

GERYON.

Geryon, another demigod, resembled Cacus in appearance but not in circumstances and condition, for he was a prince, and rich in flocks and herds, and to guard them had a dog with two heads and a dragon with seven; both of which were overcome by Hercules, who also slew their owner and seized the cattle as his rightful spoil. This Quixote of mythology travelled in quest of strange adventures, and enriched himself, as all heroes did and do, by rapine. In his time, as in Job's, wealth consisted principally in cattle; and cattle stealing was, as in subsequent times, not held dishonorable—except when unsuccessful. Gods and demigods followed and acquired fame by the profession. Of primitive moss-troopers none equalled Mercury and Hercules in cunning; it was therefore a sad mistake in Cacus to seize eight of Geryon's kine while in the possession of such a bold and knowing drover as Alcides. Though he succeeded in getting them unperceived into his den, his fire-spitting image had no fears for the enraged loser, who was too familiar with such things to dread them.

This primitive prevalence of robbery sufficiently accounts for the adoption of secret and extraordinary devices to scare night thieves from folds and dwellings of the rich; and sure we are that modern ingenuity might be taxed in vain to produce one better adapted to terrify the ignorant and keep the dishonest at bay, in dark and grossly-superstitious times, than flame-ejecting Eolipiles. On the approach of a thief, the concealed attendant had only to open a cock to send a scorching blast on the offender, or the latter might himself unconsciously be made to open it by his weight—a species of contrivance perfectly in character with the genius and acknowledged productions of ancient artists. Vulcan was full of such conceits. Even now a grim-looking image of the kind would excite no little horror among stupid burglars, while it would strike savages dumb.

The word Geryon, according to some paleologists, signified thunderbolts, and was allusive to the hissing, piercing, overwhelming and scorching blasts which issued from the dog and dragon, or from a triple-bodied monster called Geryon: not a slight intimation this of their Eolipilic nature. In fact, to consider them as figurative creations, and the rest of the characters and objects real, is inconsistent; unless it be conceded that Geryon's cows were kept from thieves by metaphors, and that these were hacked and shattered by material clubs and faulchions. It would have required some flaming similes to frighten experienced cattle-lifters like Cacus and Autolycus from their destined prey, or to induce them to yield up acquired spoils.

To resolve these "brazen" monsters into mere creations of the brain, appears to us as reasonable as to explain away in like manner metalline automata of the Bible—representing them as having had no connection with the crucible, but simple abstractions: the serpent, for example, as emblematical of the cunning of Moses, and the calf of stupidity in the people. By the same process, we might interpret the "bronze" vessel or statue in which Eurystheus concealed himself from Hercules into an imaginary symbol of excessive fear; and so with the brazen bull of Phalaris and horse of Aruntius, in which human victims were consumed, and their shrieks made to resemble the bellowing of oxen, by reverberating through interior tubes: a device probably as old as Amalekitish artists, and even older. The calf or heifer cast by the Israelites in the wilderness "lowed," according to the Koran. (Chap. vii.)

No one can doubt the ability of workmen ancient as Vulcan and the Cyclops to produce machinery of the kind. If one fact be more prominent than another in the earliest records, sacred and profane, it is the perfection to which brass-founding had arrived, and the amazing extent to which metallic imagery was carried. This was a natural result of idolatry. Superstition was the nurse of these arts; the keenest intellects and finest workmen were engaged in them. The grand distinction between the useful professions of past and present times, is not due to any difference in capacity or skill, but to the estimation in which the arts were and are held. The ancients were ignorant of their destined influence on human happiness and glory, and therefore only such branches were patronized as strengthened the hold of chief priests and rulers on the multitude.

CERBERUS.

It is said of Hercules that he went about subduing the powerful, relieving the oppressed, and exposing fraud; but when occasions required he obviously acted the juggler himself. The last and greatest of his twelve labors—his Cerberean adventure—bears on every feature traces of trick. He here employs the very device which Cacus, Geryon, Æetes and others had found so successful. To play it off well would establish his fame over all competitors. Having destroyed every earthly dragon he had heard of, he undertakes to wind up his achievements in that line by proving his prowess upon the one which guarded the gates of hell. It was therefore given out that he was about to bring up Cerberus to light and exhibit him to mortal view. This would eclipse all other dragon transactions, and this he accomplished! Is it asked how? Why, by entering a "*dark cavern*" on Mount Tænarus, and after a while dragging to its mouth a three-headed dog—an Eolipilic automaton! As the exhibition was of course made in the night, the affrighted spectators, and all not in the secret, could not doubt, at the distance they stood, the presence of the canine guardian of Tartarus; its eyes glaring with living fire, smoke pouring from its jaws, its movements and the noise it made, would more than ensure conviction. The public part of the performance being over, the exhibiter, agreeably to promise, instantly set about (no doubt to the gratification of the audience and particularly of Eurystheus) to remove the monster to its own domicile. There is no room to doubt this—he certainly pulled it back to the place whence he drew it forth, and none were so bold as to follow and see how he succeeded. Probably not one of the beholders but would rather his hands and feet had changed places than have ventured within the cave on this occasion.

We can form a pretty accurate idea of the sonorous "roarings," the "hissings," and "variegated yells" of mythic monsters, by similar sounds produced when steam is blown off, through various formed orifices, from modern Eolipiles.

A distinction is observable in the characters and applications of fire-vomiting images. Those which represented gods or warriors partook more or less of the human figure, while such as guarded enclosures for cattle, habitations, and places where riches were kept, put on forms compounded of dogs, serpents, lizards, bats, &c. *i. e.* were *dragons*—an idea derived from the employment of household mastiffs and shepherd curs. (A beautiful illustration of the practice of protecting houses is seen on entering the vestibule of "The house of the Tragic poet" at Pompeii. On the mosaic pavement is lively represented a fierce and full-sized dog, collared and chained, in the act of barking, and ready to spring upon the intruder. At his feet is the caution, in legible letters, *cave canem*, beware of the dog.) Griffins, or dragons, says Pliny, formerly guarded gold mines, and in old illustrated works some queer-looking nondescripts are seen performing that duty. The sentiment was once universally received; it still has believers in benighted parts of Europe, and over a great part of the East. It was encouraged by interested individuals to keep timid thieves at a distance. Ridiculous as it appears, it accords with every other occupation of dragons. Why not protect rich mines as well as a few pounds of metal? The story or the fact gave rise to the fable of Cerberus; for Tartarus, its occupants and their occupations were all derived from earthly, tangible types.

Pluto was an extensive mining proprietor, Tartarus his subterranean domains; its fires his furnaces. Demons were felons condemned "to the mines," where, naked and in "chains," some toiled in darkness, and were urged to unnatural exertions by the lashes of inexorable overseers; others, ghastly from inhaling the poisonous fumes, appeared still more so in the glare of sulphurous fires, in which they roasted and smelted the ores. Their punishment was endless, their sentence irrevocable; they had no hopes of pardon and no chance of escape. Cerberus freely permitted all to enter the gate, but not one to pass out. There were no periods of cessation from labor; their fires never went out; both night and day the smoke of their torments ascended; groans never ceased to be heard, nor the rattling of chains and shrieks of despair. Acheron, Cocytus and Styx were subterranean streams, each possessing some peculiar feature or property, while near Phlegethon arose a stream of carburetted hydrogen, a phenomenon not uncommon on the earth's surface, but often occurring in mines. Such is the most probable exposition of the origin of Tartarus. From what else, indeed, could the heathen have derived the idea at epochs anterior to Scripture descriptions of hell, and before prophets or apostles flourished? We know that the ancients sent their worst felons to the mines, and that these places presented the most vivid representations of severe and *ceaseless* punishment which the earth affords. The greater part of the convicts ere they entered these dreary regions took their last look of the sun. With shuddering horror, pale, and eyes aghast, they viewed their lamentable fate. Milton's description of hell was literally true of ancient mines and subterranean smelting furnaces.

> "A dungeon horrible, on all sides around
> As one great furnace flam'd, yet from those flames
> No light, but rather darkness visible
> Serv'd only to discover sights of wo;
> Regions of sorrow, doleful shades, where peace
> And rest can never dwell, hope never comes
> That comes to all; but torture without end
> Still urges, and a fiery deluge, fed
> With ever-burning sulphur unconsumed."

Does the reader think the picture too highly colored for mortal perdition? Why, it lacks a modern trait, one more revolting than the ancients ever imagined. Boys and girls from six to ten years and upwards, *born* and *bred* in coal-pits, less knowing than brutes, and incomparably worse cared for, are, or were recently, wholly employed in dragging and pushing on all fours, and perfectly denuded, laden sledges through dark, broken, wet and tortuous passages or sewers to the pit's mouth! And this too in a christian and enlightened land, where no small part of the people's earnings are consumed by an opulent hierarchy! Is it possible for hell itself so effectually to efface God's image, or to heap such accumulated woes on infant and unoffending victims? Pluto and his myrmidons would have quaked with passion at the bare proposal of such a scheme; yet it, and other evils scarcely less sickening and vile, have their defenders among those who worship the molochs of monarchy and mammon. Heaven help the oppressed of this earth—the creators but not partakers of its wealth—who industriously toil, and through excessive penury prematurely die—urged to produce a maximum amount of work with a minimum of rest and food—who with their offspring groan in hopeless misery here, and are threatened with endless torments in another life if they remain not *satisfied* "in that station into which," some reverend and blaspheming despots say "it hath *pleased God* to call them!"

The reason why sulphur figured so largely in descriptions of Tartarus must be apparent to all conversant with mining and metallurgical operations. It is the earth's internal fuel, the most profuse of subterranean inflammable substances. It pervades most mineral bodies; and not minerals alone, but in metalliferous ores it wonderfully abounds. All the principal ores of commerce are sulphurets; iron, silver, copper, tin, lead, zinc, &c. Of these some contain 15, and others 50 per cent. and upwards of sulphur, to get rid of which constitutes the chief difficulty in their reduction. In order to this they are "roasted" at a low red heat for six, twelve, twenty, and some for thirty hours, that the sulphur may be volatilized, and not till its blue flames cease is the signal realized to increase the heat and fuse the metal. Thus, for every ton of the latter, half a ton, and often a whole ton of the former has to be driven off in flames and vapor; so that it was with strict propriety said that Pluto's fires were fed with it. Comparatively speaking, they consisted of little else, and little else was felt or seen. It impregnated every object, while from its offensive odor and suffocating fumes none could escape. Immense quantities of common brimstone are obtained by collecting and condensing the vapors that ascend from smelting furnaces; and it may have been this, or a native mass, which formed the throne or usual seat of the lord of the lower regions. As long as the earth endures, volcanos burn, and minerals are reduced, there will be, as in Pluto's time, artificial as well as natural fires of ever-burning sulphur.

There are passages in Maundeville's Travels corroborative of Carpini's images and Pliny's Griffins. He speaks of artists in northern Asia as wonderfully expert in automatical contrivances—"fulle of cauteles and *sotylle disceytes*," making "bestes and bryddes, that songen full delectabely, and meveden be craft, that it semede thei weren quyke." In his 28 Cap. he describes a valley rich in gold and silver, (in the "lordchippe of Prester John,") but it abounded with devils, and few men who ventured there for treasure returned. This was the story, and we need not say how like a primitive artifice to scare people from intruding. "And in mydde place of that vale, *undir a roche*, is an hed and the visage of a devyl bodyliche, fulle horrible and dreadfulle to see, and it schewethe not but the hed to the schuldres. But there is no man in the world so hardy, Cristene man ne other, but that he wold ben a drad for to behold it; and that it wolde semen him to dye for drede, so hideouse is it for to beholde. For he beholdethe every man so scharpley, with dredfulle eyen that ben evere more movynge and sparklynge as fuyr, and chaungethe and sterethe so often in dyverse manere, with so horrible countenance, that no man dar not neighen [approach] towardes him. *And fro him comethe smoke and stynk and fuyr*, and so much abhomynacioun, that unethe no man may there endure." This was one of the tricks which the traveller could not tell, whether it was done "by craft or by negromancye." From automata he saw in the country belonging to Prester John's father-in-law, [China] he was led to conclude that artists there surpassed all men under heaven for deceptive inventions. That devices like the one just described, or similar to the brazen horses of Regnerus, were in vogue in the East, in Maundeville's time, appears from Marco Paulo, who mentions magic contrivances for darkening the air with clouds of smoke, &c. in use by the military, and under cover of which many were slain. Marco himself was once in danger of his life on such an occasion: he escaped, but several of his associates were cut off.

DRAGONS.

Had not ideas of fire-spouting nondescripts been exceedingly ancient they had never become so intimately and universally mixed up with human affairs. Throughout the old world the *dragon* was the *ne plus ultra* of impersonations of the horrible—the king of monsters. It is so now, and a more appalling one, or one invested with more terrific qualities cannot be devised. So deeply was its image impressed on ancient minds that it pervaded history, song, and all religions. We meet with it in the Scriptures as well as in the classics. The devil, from his reputed connection with smoke and liquid fire, is named "the great dragon." In old religious processions, and in the "mysteries" or dramatic representations of the church, Satan was symbolized by an image of a dragon spitting fire. The author of the apocalypse seems to allude to mythic fire-breathing images in the following passage. " If any man will hurt them, fire proceedeth out of their mouths and devoureth their enemies." [a]

The universal custom of exhibiting figures of dragons in ecclesiastical and civic pomps was a mythic relic—a practice continued from times when captured idols and warring Eolipiles were led in triumph. Then, objects of superstitious dread, they now amused spectators: at the coronation of Anne Bulleyn, a "foyste" or galley preceded the lord mayor's barge; "in which foyste was a great red dragon, continually moving and casting forth wild fire: and round about the said foyste stood terrible, monstrous and wilde men, casting fire and making a hideous noise." If the truth could be known, there would be found little difference between this modern monster and some of its ancient namesakes.

No chimerical being was ever so celebrated as the dragon. To it temples were dedicated, of which some remained in classical eras. The practice is continued in China. Of an official dignitary it is said, ere he entered on the duties of his office (at Canton,) he one morning paid his devotions at eight temples, of which one was consecrated to the god of fire, another to the god of wind, and a third to the dragon or dragon-king. "The festival of the dragon-boats" is another relic of times when these artificial monsters were in vogue. [*Chinese Rep.* iii. 95, 47.] The legends of China and Japan teem with dragon allegories and apologues. The figure is an imperial emblem, and as such is wrought on robes, painted on porcelain, carved on dwellings, ships, furniture and other

---

[a] In Scandinavian and ancient British history, and throughout northern Asia and Europe, the dragon was the universal minister of vengeance. It was eventually made typical of all destructive agents—of water as well as fire. It became a symbol of the deluge, on which account figures of it pouring water from the mouth were adopted in ancient fountains. Some of these have been noticed in this volume. May not St. John have had one in view when he wrote "And the dragon cast out of his mouth water as a flood." [*Rev.* xii. 13-27.]

works of art. No people retain so many characteristics of times when Eolipilic monsters flourished. They act on the same principle as old warriors did, by trying to frighten their enemies with warlike scarecrows, with pompous orders—assuming the language of gods and addressing other people as devils, dogs and reptiles. Their taste for the horrible extends to civil life; things of the wildest forms which imagination can furnish or nature reveal are most highly prized.

As a guardian of temples, sacred groves and treasures, the celebrity of the dragon has continued to present times. Enforcing a principle in ancient ethics, it kept the ignorant honest by frightening them. But when it lost this magic power, and enchanted chambers could no longer be relied on, eastern monarchs sought out natural monsters to guard their precious stones and living jewels. Deformed negroes; the most hideous of nature's abortions, are now the sentinels of eastern treasuries and seraglios.

Mythic dragons had commonly a multiplicity of heads. This was in keeping with their design and with the taste of the times. Each additional member adding horror to their appearance and furnishing in the mouth and eyes additional orifices for the issuing flames; like fire-engines that eject several streams. The device is very analagous to others common in old war-engines. The idea was adopted by the author of the most figurative book of the Scriptures. He speaks of " a great red dragon with seven heads and ten horns." The figure No. 288 it will be seen has one horn. Most of the idols of the Hindoos, and of the orientals generally, have numerous heads, and some have horns. By dragons in the Bible, crocodiles, or large serpents, are commonly intended, but chimerical or mythic beings are obviously intended in such passages as the one above quoted.

Another characteristic in dragon biography, attributed to rather modern individuals, was an undoubted trait in the patriarchs of the species. When one was overcome without being demolished, it was generally led in triumph, in the manner of Theseus showing off the Marathonian bull in the streets of Athens—or of Saint Romain leading with his stole a fierce dragon to the market-place at Rouen—the victor receiving the congratulations of his countrymen on his prowess, and the prisoner behaving the while, as well behaved prisoners should—i. e. silently submitting to the will of the captors. Suppose the dragon figured at No. 289, exhausted of its contents, (in battle it would often require fresh charging,) its movements put a stop to, and in that condition captured; what follows, but that the victors put one end of a rope round its neck and the other in their hands; and have we not then a perfect representation of a fiery monster becoming harmless as a lamb and tamely submitting to be led about, as ancient chronicles have it, "like a meke beaste and debonayre."

But the dragon was dedicated to Minerva; and to whom else could it have been so appropriately devoted? One might almost fancy she mounted this popular form of the Eolipile on her cap as a compliment to old artists. Certainly if the patroness of the useful arts had now to select an expressive symbol of her best gift to mortals, she would adopt the same thing in its modern shape—a miniature engine and boiler. This she would consider, like Worcester, her "crowning" device. But it is perhaps said, the ornament on her crest was an emblem of war. Well, was not that the chief use to which Eolipilic dragons were put? Then was she not so familiar with artificial lightning and thunder as to nave rivalled her father in hurling them at will on her foes. She took part in the wars of the giants, and destroyed not the least of the kindred of Typhon herself. Another circumstance indicative of her acquaintance with Eolipilic contrivances is the fact, (noticed on a previous page,) of her image at Troy having the faculty of sending flames from its eyes.

It were easy thus to proceed and point out the artificial character of most of the imaginary monsters of antiquity—to render in a high degree probable, that, like acknowledged androidal and automatal productions of Vulcan, Dædalus, Icarus, Perillus, and other artists named by Pliny in his 34th Book, they were originally mechanical, pyrotechnical or eolipilic images; sometimes combining two or more and occasionally other elements in their functions and movements; that the faculty of locomotion attributed to some accorded not only with applications of modern mechanism, but with avowed artificial contrivances of ancient artists, and that their material natures were, in after times, construed into the ideal, either from ignorance or by the imagination of poets—but this is unnecessary. Enough has been said to induce the reader to pursue the subject, or to reject the hypothesis as untenable. The antiquity of Eolipiles is unquestionable. Their origin is lost in remote time. We know they were made in fantastic and frightful forms, were used as idols, designed to spout fluids and eject fire—the very attributes ascribed to mythic monsters. Is it unreasonable then to suppose the latter had no existence except as Eolipiles? But if it be contended they were wholly figurative, from what were the conceptions derived, if Eolipiles were not the *things* they symbolized; and how account for coincidences which nothing else in nature or in art can produce? One observation more, and we conclude:—

Early applications of Eolipiles and their present employment as steam boilers, suggest some interesting analogies. Emblems of half civilized times and races, they connect the remote past with the present. Ordained as it were to move in advance of the arts and astonish mankind, they have lost none of their virtue. If their ancient vagaries shook communities with alarm, their current deeds are eliciting the world's admiration. They furnished tradition with marvellous stories, and modern history is engaged in recording their wonders. They supplied materials for the earliest and worst chapters in the earth's annals; to them and their effects will be devoted some of the latest and best. Formerly they feebly personated Gods; now, the sole animators of our grand motive engines, they annihilate time and space by their movements and laugh at all physical resistance. Children watch their operations with ecstacy and old men hardly believe what they see. Once an instrument of the worst of tyrannies, the Eolipile is becoming the most effectual agent in the extinction of tyrants. Instead of acting, as of yore, on human fears; debasing the mind and furthering the views of oppressors, it captivates the judgment of the wisest, elevates nations in morals, and confers on them wealth and extended domain. The gem of old miracle-mongers, it is the staple device of living magicians, for its present improvers and users are the genuine representatives of Pharaonic Savans and mythologic Magi.

New-York, July, 1845.

THE END.

INDEX.

ABACUS, 430
Achelous and Hercules, fable of explained, 120
Adam, traditions of, 12
Adventures of a bottle, 478
Æschylus, singular death of, 365
Agricola, quoted, 69, 72, 117, 126, 151, 219, 240, 278, 315
Agriculture, 79, 118. Implements of, 12, 132
Air, its properties, 176—186. Ancient experiments on, 192. Rarefaction by heat, 374—380. Liquids raised by currents of, 224-5, 447, 473
Air-barometer, 188, 375
—— beds, 177. Guns, 181, 192, 270, 425
—— chambers to pumps, 265, 269, 270, 306, 307, 326, 337, 338, 371
—— machines, 374—380, 447, 473
—— pumps, 179, 180, 190, 403, 426
Air and steam, their supposed identity, 395—400, 418, 420, 421, 572
Ajutages, conical, effect of, 479, 480, 486—488, 490—496
Albertus, 104
Alchemists, 395, 407, 408
Alcithæ, taken off by priests, 385
Alcoholic engines, 441, 472
Aleppo, water-wheels at, 115
Alexandria, library at, 415. Wells at, 54, 55
Alfred the Great, measured time by candles, 350
Algerines, their superstitions, 36
Altars, 107. Tricks at, 383—385
Alum, used to make wood incombustible, 304
America, ancient arts in, 159—172
American water-works, 298—301. Fire-engines, 339—348. Wells, 50, 160, 164, 298
Amontons, his fire-mill, 463
Androids and automata, 104, 183, 294, 534, 568, 573
Anecdotes, of Mahomet, 10. Dentatus, 19. Darius, 22. Egyptian priests, 22. Two elephants, 39. A boy and goose 39. Alexander, 39. A caliph, 42. Cleanthes, 56. An ass, 74. An Indian Cacique, 107. Ctesibius, 121, 122. Valentinian, 196. A raven, 203. A Spanish pump-maker, 224. A Basha, 316. A Dutch Burgher, 366. Lord Bacon, 375. Marquis of Worcester, 392. Zeno, 393. Cromwell, 442. Savery, 454. Phocion, 537, 565. Oxen, 573
Angelo, M. 534

Anglo-Saxons, worshiped wells, 36. Homily, ibid. Buckets, 67. Mirrors, 121. Swape, 99. Windlass, 72. Steam idols, 398
Animals, employed to raise water, 74, 117, 573. Devices of, 365. Their physiology illustrated, 180, 181, 209, 210, 256—258
Anthemius and Zeno, 393
Antipater, 282
Antlia of the Greeks, 213
Anvil, blacksmith's, 12, 43, 240, 241
Aquarius, 85
Arabs, 41
Arago, 145, 411, 433
Archimedes, 141, 360, 438
Archytas, 7, 268
Argand's siphon, 525
Arkwright, 359
Arts, useful, their origin, &c. 2, 6, 11, 12, 81, 83, 232, 282
Artificial hands and feet, 4
Atabalipa, 169
Astronomy, 85
Aqueducts, 165—169, 212
Auto da Fe, 351
Awls, 87, 489
Atmosphere, its properties, 176—189. Discovery of its pressure, 187, 425, 426. Its pressure diminished during storms, 481. By currents of air, 482-8. By currents of steam, 489—496. By currents of water, 479
Atmospheric pumps, 173, 175, 187—191, 206—230
—————— sprinkling pots, 194, 195, 567, 573

B.

Babylon, 79. Hydraulic engine at, 133, 303
Bacchus, tricks at his temples, 200, 385
Bacon, Roger, 403
—— Lord, 416—17, 550
Balls supported on jets of air, steam, and water, 270, 395
Barbers, 121, 162. See preface.
Barometer, 190, 375, 481
Basket, swinging, for raising water, 85, 86
Bate, John, 321, 375, 421, 565, 568, 569
Baths, 120, 147, 169, 393, 552, 558
Bears employed in tread-wheels, 74
Beer, 87
Bees, 257, 276
Beds, air and water, 177, 178. Bedcloths, 178
Bedsteads, 87, 178, brazen feet. See preface

Bel and the Dragon, 519
Bellona, 308
Bells used as fire-engines, 313, 314. Bellmen, 315
Bellows, 87, 90, 177, 180, 232—243, 261, 268, 483. Origin of the word, 570. Vulcan's 233, 240, 268, 568. African, 235, 246, 252. Asiatic, 236. Egyptian, 237, 238. Madagascar, 246, 252. Lantern, 240, 241. Rotary, 252, 255
———— piston, 244—252. Philosophical, 396, 397, 569, 570
———— pumps, 205, 210, 235, 241, 243, 568
Belt, hydraulic, 137
Belzoni, 124
Berenice's hair, 143
Besson's Theatre, quoted, 69, 114, 126, 152, 218, 280, 317, 410
Bethlehem, well at, 48
Bible quoted or illustrated, 2, 10, 11, 19, 22, 24, 26, 30, 31, 33, 35, 38, 40, 44, 51, 52, 84, 87, 90, 95, 117, 132, 195, 292, 283, 291, 303, 366, 391, 393, 400, 476, 517, 536, 560, 565, 571
Birmah, raising water in, 73
Blacksmiths, Adam one, 12. Vulcan, 240. Grecian, 241. *See preface.*
Blakey, 462
Blood, circulation of, 132, 257
Blow-pipes, 19, 234, 569. Eolipilic, 569
Blowing tubes, Ewbank's, 484—496
———— siphons, Ewbank's, 527, 528
Boa Constrictor, 180
Boats, steam and others, 258, 403, 419, 423, 438
Boilers, of wood and of granite, 470. Heated by the sun, 471
———— ancient, 392, 393, 522. Of coiled tubes, 394
Branca, 418
Buckets, 32, 52, 54, 63, 73, 83, 167, 171, 302, 314, 315, 340, 341. War caused by one, 67, Ancient metallic ones, 230
Bucket engines, 64—66, 128. Ancient, 573
Burckhardt at Petra, 42. At Hamath, 116

C

Calabash, 14, 16
Cambyses, 377
Camera Obscura, 379, 403, 430
Candles, time measured by, 350
Canne hydraulique, 372, 373
Canopus, origin of these vessels, 23
Capillary attraction, 510, 513
Cardan, 396, 407, 410
Carpenter's tools and work, 87
Carthagenian wells, 27, 38
Cast iron, 268, 553
Cato the Censor, 536
Cauldrons, 19—21, 120, 162, 171, 237, 238, 391, 394
Caus. See Decaus.
Cawley, 464, 468
Cecil's motive engine, 473
Cement, 58
Centrifugal pumps, 229, 230, 290, 291
Chains, golden, 162. Watch, 323
Chain of pots machine, 122—132
Chain-pumps, 148—158, 567. In ships, 154—157, 567
Chain and sector, 467
Chairs, imprisoning, 29, 573. Cane, 323
Chemists' siphons, 526
Chemnitz pressure engine, 362
Child's rattle, inventor of, 268
Chili, aqueducts in, 165
Chimneys increasing the draft of, 395, 482, 483. 488

Chinese tinker, 20, 248. Juggler, 199. Wells, 28, 30, 35, 83. Proverbs, 30, 31. Printing, Windlass, &c. 69, 70. Irrigation, 82, 83, 86. Noria, 112. Chain-pump, 150. Ships, 158. Bellows, 248. Digesters, 393. Clocks, 545
Chisels of gold, and faced with iron, 5
Chuck, eccentric, 285
Churches, fountains in, recommended, 540
Circulation of the blood, 132, 257
Cisterns, 48, 58, 169, 170, 557
Cistern pole, 57
Cleanthes, 56
Clepsydræ, 95, 542—7
Clocks and watches, 122, 285, 323. Substitutes for, 350
Clysters, 260
Coach with portable kitchen, 445
Coal-pits, raising water from, by fire, 419
Cocks, of gold, silver, &c. 170, 559, 560. Three and four way, 354, 355, 421, 433, 449, 462, 523. Ancient, 394, 557, 559. Guage, 459. Siphon, 527. Sliding, 561
Coffins of iron and lead, 551
Cog-wheels, 71, 72, 114, 121. Engine for cutting teeth, 323
Cohesion of liquids, 513, 515
Coining, Roman mode of, 6
Combs, 87. *See preface.*
Condensation of steam by injection discovered by chance, 466
Conon, 143
Constantinople, water-works at, 43, 553. Ancient baths at, 552. Fire-engines at, 316
Constellations, 143
Cooking by steam, 471
Cord and bucket for raising water, 53—56
Cornucopia, 119, 120
Cortez, 159—163
Coryatt, 78
Cotton gin, 359
Coupling screws, 326, 459
Couvre feu, 350
Crassus, bad practice of his, 311
Crates, saying of his about war, 309
Creusa, 383
Crucibles, Egyptian, 87, 234
Crusaders, 32, 372
Cupelo furnaces, 397
Cupping, 202, 203
Curfew bell, 350
Ctesibius, 121, 122, 192, 213, 259, 266—270, 547

D

Dædalus, inventions ascribed to him, 268
Damasking linen, 323
Danaus, carried pumps to Greece, 130
Darius, anecdote of, 22
Decaus, 319, 380, 410—13, 529, 533
Delphic oracle, 241, 392
Democritus, saying of his, 57
Demons, superstitions respecting, 313,482,521
Demosthenes, his father a cutler, 6
De Moura's steam-engine, 463
Dentatus, anecdote of, 19
Desaguliers, his bucket-engines, 63, 66. Remarks on Savery, 460. Steam-engine, 461
Dials, 542
Digesters, 392, 393, 446, 447
Discoveries in the arts, way to make, 359
Diogenes, buried heels up, 36
Distaff and spindle, 283
Distilling, 381, 393, 407
Divination, by water, 34, 36. With cups, 200, 201. With fire, 383, 571. Steam, 392, 399
Diving ships and apparatus, 430, 431
Dogs in tread wheels, 74, 75

INDEX. 603

Dolls, 87, 268
Doors and gates, self-moving, 384, 555, 556
Dove-tailing, 87, 268
Dowry of Scipio's daughter, 121
Drawing water, imposed as a punishment, 84, 131, 533
Drebble, C. 188, 323, 381, 430
Drinking vessels, 4, 11, 14—16, 162, 195, 205, 520—2
Drops of liquid, 511
Dropping tubes, 199
Dunstan, St. 104, 105. Adroit trick of his, 107
Dutch scoop, 93. Fire-engines and hose, 328, 329. Inventions. See preface.
Dwellings, heated by steam, 471. Fountains in, recommended, 361, 540

E
Ear-trumpets, 379, 573
Ecclesiastes, a fine passage in, illustrated, 132
Ecclesiastics, devices of, 103—108, 383—387, 392, 398—400
Eddystone light-house, 258, 366, 367
Egypt, labor of, what it was, 86
Egyptian wells, 26—28. Customs, 34, 78, 81, 83, 87. Noria, 113. Shadoof, 94, 95. Mental, 85. Chain of pots, 123, 131. Screw, 142. Siphons, 516. Clepsydræ, 544, 547. Goldsmiths, 234. Fire-engine, 307
Emblematic devices, 32, 194, 203, 261, 314, 557
Endosmosis, 510
Engines of motion, 419, 423
Engines to extinguish fires, of great antiquity, 303. Employed in ancient wars, 303, 305. Referred to by Apollodorus, 235, 304. Described and figured by Heron, 305. Portable engines, 311. Syringe engine, 315, 317, 321. German, 318, 319, 324, 331. English, 320, 321, 322, 332—335, 568. French, 324, 325, 327, 329, 336. Dutch, 328, 331. American, 339, 344, 345. Rotary, 285, 573. Steam fire-engines, 338, 346—349
English, inventions of, 323. See preface. Water-works. 294—296 Fire-engines, 320, 332—335. Steam-engines, 420, 437, 455, 465
Eolian harp, 104
Eolipiles, an emblematic device, 261. From Heron's Spiritalia, 394. For blowing fires, 395—400, 573. Increasing draft of chimneys, 395, 401. Diffusing perfumes, 401. Producing music, 401. Fusing metals, 397, 569, 570. In the human form, 398, 399. Used in war, 400. Charging, 395, 407, 570
Eolipilic idols, 398, 400, 570
Eolus, god of winds, 400
Ephesus, fountains at, 49
Epigrams, ancient, 282, 537
Erckers, blowing eolipiles from, 307, 570
Evans, Oliver, 424
Evaporation of water from the earth, 506
Ewbank's experiments on raising water, 225. Mode of propelling vessels, 406. Blowing tubes, 484—496. Spouting tubes, 497—504. Mode of evaporating liquids in vacuo, 495. Experiments on the force of sap, 509. Increasing the draft of chimneys, 488. Ven-
̇lating ships and mines, 488. Siphons, 527, 528. Siphon cocks, 527. Tubular valve, 556. Sliding cocks, 560, 561. Tinned leaden pipes, 555
Explosion of boilers, 392
Explosive motive engines, 441, 450, 471—473

F
Faye's La, improved tympanum, 111
Feast of Cana, siphons used at, 517
Females, employment of ancient, 283
Fetters, Lacedemonians bound with their own, 84. Found on skeletons in Pompeii, 29
Fire, modes of obtaining it, 197. Sacred, 196, 197. Superstitions respecting, 312—314. Protecting buildings from, 304, 349. Greek fire, 307. Laws respecting, 351. Raising water by, 374—384, 418, 419, 431, 442. Kindling on altars, 383, 384
Fire-escapes, 350
Fire-engines, 302—349, 573
Firemen, Roman, 309. American, 340—345
Fire-places, 483
Fish, fishing, 86, 87, 185. Nets, 550. Salting fish, 86
Fitch, John, 424
Flatterers, among men of science, 143, 145
Flies, curious mechanism of, 182, 183
Flying, 103, 104, 324, 430
Fly-wheels, 278, 283
Floats for steam-boilers, 471
Fludd, Robert, 65, 194, 219, 354, 407
Forcing pumps, 262
Forks derived from China, 70. Their use in Europe, 76—78
Fortification, a moveable one, 430
Fortune, wheel of, 119
Fountains, 27, 33, 34, 35, 41, 43, 49, 119, 163, 170. Artificial, 30, 361, 379, 445, 532—541
Fountain lamps, &c. 193
Forcing pumps, 262—281
Francini's bucket machine, 128
Francois' steam machine, 463
French inventions. See preface. Water-works, 277, 296—298. Fire-engines, 317, 324—331, 336
Frictionless pumps, 208, 209, 321, 274, 568. See bellows pumps.
Frogs, climb by atmospheric pressure, 183
Fuel, in steam idols, 400. Lord Bacon on, 417
Fulton, Robert, 359, 464
Fusee, 71. Fusee windlass, 69—71
Future happiness, erroneous views of, 508

G
Gaining and losing buckets, 64—66, 128
Galileo, 104, 187, 188
Games, ancient, 81
Garcilasso, 167, 170, 509
Gardens, Egyptian, 101. Babylonian, 134. Mexican, 163, 537. Peruvian, 171. Roman, 536. Italian, 537. Persian, 539, 573
Floating, 539
Garden watering pots, 194, 195, 567, 573.
Garden syringes, 261
Gates and doors, closed by machinery, 556
Gauls, 36. Induced to invade Rome by the report of a smith, 19
Geese, in tread wheels, 75
Genevieve, St. 37
Gensane's engine, 463
Gerbert, 104, 401
German snail, 138. Bellows pump, 207. Fire-engines, 319, 323—326. German inventions. See preface.
Gesner, 381
Geysers, 411, 507
Glass, painting on, 4. Engine for working, 323. Mirrors, 121. Glass tomb of Belus, 200
Glass tubes, curious motion of, 429
Glazier's vise, antiquity of, 554

INDEX.

Glauber, device of his, 440. Safety valves used by him, 451. Wooden boilers, 470
Glue, ancient, 87,
Gnat, the, a boat builder, 258
Goats employed in tread wheels, 152. Battles between, 366
Goblets for unwelcome guests, 521. Magical, 520
Gold-beating, 87
Golden legend, extract from, 313, 314
Goose-neck joint, 307, 322, 327, 573. Substitute for, 324—326
Gosset's frictionless pump, 208
Goths employed bears in tread wheels, 74
Gravity, suspension of objects against, 142
Greece, wells in, 27, 36. Antiquities found in them, 50. Water raised from them by the swape, 96
Greeks, 268
Green-houses heated by steam, 471
Guage cocks, 459
Guage, mercurial, 451
Guerricke, Otto, 181, 190, 426
Gulf stream, 477, 478
Guns, repeating, 430. Air, 181, 192, 270, 379, 573. Steam, 395, 423, 573
Gunpowder, 143, 383. Known to Roger Bacon, 403. Engines moved by, 441, 450, 472
Gutters, for raising water, 88, 91, 92. Spouts of, ornamented, 119

H

Hair, coloring it practised of old, 120. Berenice's hair, 143. Pulling up trees by one of Sampson's hairs, 418
Hamath, water-works at, 115, 116
Hammer, its origin and history, 5, 6
Hand, used as a cup, 11, 40, 52. Artificial hand, 4
Haskin's quicksilver pump, 274 275
Hautefille, 441
Heart, the, a pump, 258
Hegisostratus, wooden foot of, 4
Helepoles, 304
Heliopolis, fountain at, 43, 49
Heliogabalus, 177, 561
Heraldic devices, 261, 314, 396
Herculaneum, wells at, 28, 29, 55. Fountains at, 534
Hereward, the Saxon, 36
Herodotus, quoted, 4, 11, 12, 20, 22, 27, 58, 79, 80, 81, 84, 96, 133, 241, 260
Heroes, old mechanics the true, 5
Heron, 65. His fountain, 361. Air-machines, 378. Account of his Spiritalia, 385, 386. Eolipiles from, 394
Hieroglyphics, American, 164
Hindoos, their mode of drinking, 11. Wells, 30, 33, 35, 38, 52. Carrying water, 84. Picotah, 97. Swinging basket, 85. Jantu, 89, 90. Syringes, 260, 261. Water-clocks, 544
Hire La, his double acting pump, 271
Holy water, derived from the heathen, 166, 196, 386. Ancient vase for selling it, 387. Used in consecrating bells, 313, 314, and various other articles, 196
Homer, quoted, 19, 21, 22, 33, 233, 240, 250, 536. Kept a school at Scio, 131
Honors, titles of, absurd origin of some, 144, 145, 445, 446
Hookah, 270
Hooke, Dr. 441
Horn of abundance, 119, 120
Horn, drinking, saying respecting it, explained, 205

Hour-glasses, 545, 547
Hose pipes, 304, 326—328, 345
House warming, 37
Hudibras, 265
Hurricanes, commence at the leeward, 481
Huyghens, 441
Hydraulic belt, 137
Hydraulic ram, 367—372
Hydraulic machines, ancient, 7, 10, 81, 131 132—135, 267. Used as first movers of machinery, 128, 140, 158
Hydrostatic press, 276

I

Idols, 82, 106—108. Eolipilic, 398, 399, 570—572
Impostures, 23, 106—108, 376—378. See *juggling*.
Imprisoning chairs, 429, 573
Incas of Peru. Aqueducts erected by them, 165—168
Incendiaries, 308, 350. Punishment of, 351
India ink, 70
Indians, American, 50, 107, 180
Inertia, 373, 503
Intermitting springs, 506
Inventions, how realized, 359. Few recorded, 416. Cause of this, 427. Advantages of recording them, 453. Century of, 64, 140, 362, 428—438
Inventors, old, concealed their discoveries, and why, 427. Caricatured, 439
Iodine, discovery of, 414
Iron cauldrons soldered, 20. Iron statues, 142. Planing iron, 283
Iron first cast in England, 553
Irrigation, 28, 79, 80, 83, 84, 95, 118, 119, 126, 131, 132, 163. Aquarius, an emblem of, 119
Italian mode of raising water to upper floors 63. Fountains, 534, 537

J

Jack, old name of a man-servant, 75. Smoke-jack, ibid.
Jack of Hilton, a Saxon eolipile, 398
Jacks of the clock, 543
Jacob's well, 38, 42, 44
Jaculator fish, 257
Jantu, 89. Alluded to by Moses, 90
Japanese water-works, 125, 557. Clocks, 543
Jehoahaz, portrait of at Thebes, 116
Jets d'eau, 163, 532—541
Jeweled holes for pivots of watches, 122, 547
Jews, their wells, 25, 33, 34. Watering land, 86. Their arts, 133
Joseph's well, 38, 45—47. Divining cup, 200
Josephus, quoted, 38, 40, 54
Juggling, jugglers, magicians, &c. 23, 106—108, 198—201, 376—385, 519, 521—523
Juvenal, quoted, 19, 43, 121, 310, 311, 312, 377. Banished, 48

K

Kircher, on the speaking statue of Memnon, 377. Bellows-pump from, 243. Turned a spit by an eolipile, 396. His mode of raising water by steam, 422
Kitchens, 75. Egyptian, 237, 238, 517
Kites, boys', 422
Knives of gold and edged with iron, 5. Portable knives, 205
Koran, quoted, 10, 54, 117

L

Laban's images, 571
Ladders, portable, 350, 431

INDEX. 605

Lakes Mœris and Mareotis, 80
Lamps, 242, 243, 421
Lantern bellows, 237—240. Pump, 241, 242
Lares. See Idols.
Lateral communication of motion, 475—480
Laver of brass, 557
Law, a bar to the progress of the arts, 427
Lead, pigs of: leaden roofs, coffins, rolled lead, pipes, &c. 163, 211, 550—554
Leaden pipes tinned, 555
Leather pipes, 304, 326
Lenses, concave and convex, 380, 381
Leopold's Fire-engines, 329—331. Steam-engines, 462, 469
Level, 268
Library, in an ancient ship, 147. Alexandrian, destroyed, 415
Lions' heads on cocks, gutters, &c. 119, 557
Liquor tasters, 195, 199
Llama of Peru, 257
Load-stone for suspending an iron statue, &c, 142
Lobster's tail, mechanism of, 258
Locomotive carriages, 403, 423, 424, 473. Increasing draft of chimneys of, 397, 488
London water-works, 294—296, 321, 434, 567
Looking-glasses, 121
Lucan, quoted, 108, 125, 540
Lustral vase and water, 387

M
Macaroni, kneading, 91
Machines, worked by the feet, 90, 237—239. War machines, 305
Machines of Ctesibius, 122, 192, 213, 259,266—270, 547
Madagascar, bellows of, 246, 252
Magic goblets, 518, 520
Magnet, ancient one, 142
Mahomedans, traditions and customs of, 12, 35, 36
Mahomet, 10, 54. His coffin, 142
Man, his body a living pump, 257. His past and future condition, 388—390, 508
Manco Capac, 168, 172
Mangle, Chinese, 90
Manuscripts, 108
Mariner's compass, 143
Marli, water-works at, 296—298
Mars, represented, 308
Martial, quoted, 521
Mastodon, tradition of, 165. No extinct animals of the ox kind thirty feet high, 210
Mathesius, 410
Mechanic powers, origin of some, 1. Implements, 5, 6
Mechanics, ancient, little known of them, 3, 4. An account of their works and workshops would have been invaluable, 4. The true heroes of old, 4, 5. Formerly seated when at work, 139, 240. Advantages of studying the mechanism of animals, 258. Old priests first-rate mechanics, 104, 401, 441
Mechanism, revolving, 282—284
Medea, inventress of warm and vapor baths, 120
Medicines, quack, 120
Memnon, statue of, 377, 401
Mercurial guage and safety valves, 451
Metals, hammered into plates, 2, 283, 551. Drawn into wire, 2. And into pipes, 554. Ancient works in, 6, 87, 162, 171, 557. See preface.
Metallic mirrors, 121
Mexicans, 34, 159—162
Mills, 282, 419, 423

Mines, ventilation of, 488. Raising water from. See Agricola, Ramseye, Savery, New comen, Worcester.
Mirrors, 87, 121, 172
Moclach, a vizier, 55
Momentum, 883, 366, 367, 373. Animals have a knowledge of it, 365
Monks, their ingenuity and professions, 104—108, 386
Montgolfier's ram, 369—372
Moon, Wilkin's project to reach it, 103
Moreland, Samuel, his pump and speaking-trumpet, 273. Steam-engine, 441—445
Morey's motive engine, 473
Motion, transmitted by air, 448. Rotary, 282—284
Motive engines, 423, 472—474
Mouth, various operations of, 477
Musical machines, 17, 381. See Eolipiles, Memnon.
Mythology, Egyptian, 82—85. Peruvian, 167

N
Naamah, the supposed inventress of spinning, 283
Nabis, his cruelty, 573
National vauntings, 402
Natural pumps and devices for raising liquids, 209, 210, 256—258, 505—513
Neptune, 507
Nero, his golden house, 539. Water-clock, 549
Nets, fishing, 86, 87
New-Amsterdam, wells in, 299. Fires and fire-wardens in, 339, 340
New-York, minutes of common council, 299 300. Old treasury note, 300. Fire-engines, 341—345
Newcomen and Cawley's engine, 296, 464 —468
Niagara falls, currents of air at, 476
Nineveh, well at, 26, 36
Noria, Chinese, 112. Egyptian, 113. Spanish, 114. Roman, 113. Syrian 116. Mexican, 163
Nuremburg, in the 16th century, 324. Curious report of Engineers, 556

O
Omar, logic of, 415
Oracle at Delphi, prediction of, 241
Organs, 547—550
Organ-makers, priests, 401
Oscillation of liquids, 497
Osiris, 82. Made his own plough, 83, 132
Ovid, quoted, 11, 13, 52, 76, 120
Oysters, swallowing, 181. Their movements, 257

P
Paddle-wheels, 454. Their antiquity, 406. Substitutes for, 291, 406
Palladium of Troy, 12, 571
Panama chains, 162
Paper, Chinese mill, 90. Made by steam, 388. Experiments with a sheet of, 483. Marbling, 323
Papin, 1. His air-gun, 181. Driven from France by religious persecution, 446. His digesters, 447. Safety valve, 447, 451. Air-machine, 447—450. Explosive engine, 450. Steam machines, 450—452
Parabolic jambs of fire-places, 483
Paris, water-works of, 296—298. Fire-engines, 327—331, 336
Pascal, his experiments on atmospheric pressure, 189
Patents and patentees, old, 439

606 INDEX.

Paving cities, 553
Pedal for pounding rice, 90, 91
Pegu, customs in, respecting water, 35
Pelanque, 164
Pendulum machine to raise water, 92, 93. Pendulum for watches, 441
Penelope and Ulysses, 283
Perfumed fountains, 539, 540
Perfumes dispersed by eolipiles, 401
Perpetual motions, 566, 567
Peruvians, their Asiatic origin doubtful, 172. Whistling bottles of, 17. Mirrors, 121, 172. Wells and irrigation, 165—167. Common utensils of gold, 171. Ancient city disinterred, 17. Sucking tubes, 204. Not ignorant of the bellows, 253—256. Dials, 542, 547
Persians, worshiped wells, 36. Ambassadors thrown into wells, 27. Raising water, 96, 573. Fountains, 539, 541
Persian wheel for raising water, 115
Peter Martyr, quoted, 106, 314
Pewter and pewterers, 162, 260
Philadelphia, water-works of, 300. Fire-engines, 344
Phocion, 537, 565
Piasa, a bird that devoured men, 165
Picotah, a machine to raise water, 97
Pins and needles, 87, 121. First made in England, 323
Pipes, water, flexible, 258. In Mexico, 163. In Peru, 170. Asia, 211. Pompeii, 211, 552. Rome, 213, 552. Of earthenware, 58. Of leather, 304. Of lead, 552, 553. Drawn, 554. Tinned, 555
Pipkins, 18, 19
Pistons, 206, 214, 215, 307, 438
Piston bellows, 244—253
Piston and cylinder, various applications of, 358, 359, 425
Piston steam-engine, of Worcester, 435—437. Of Hautefille and Huyghens, 441. Of Papin, 450. Of Newcomen, 465. Of Leopold, 469
Plato, his views of mechanics, 3. His musical clocks, 543, 548
Play-bills, ancient one, 540
Pliny the elder, quoted, 9, 15, 19, 35, 43, 68, 79, 81, 96, 130, 192, 194, 203, 212, 213, 265, 270, 549, 551. His death, 28
Pliny the younger, his letter to Trajan, 309. Account of his gardens, 536
Plough, 82, 83, 132. Engine for drawing, 423
Plutarch, quoted, 3, 12, 81, 118, 311, 366, 537, 542, 548
Poison, in wells, 40
Pompeii, its discovery, 29. Antiquities found in, 29, 30, 43, 55, 211, 552
Porta, Baptist, 1, 413. Quoted, 379, 381, 430. his Digester, 393. Raised water by heat, 379. By steam, 407—409. By a siphon, 529
Potter, a boy, who made the steam-engine self-acting, 470
Pressure engines, 352—362. Natural, 506
Prester, John, fights the Mongals with eolipiles, 400
Printing, 2, 70, 388
Printers' devices, 194
Projectors, ridiculed in a public procession, 439
Propelling vessels on water. See Paddle-wheels.
Pulley, its origin, used by the Egyptians, 59. Used for raising water, 58—63
Pumps, atmospheric: of uncertain origin, 212. Mentioned by Pliny, 96, 213. See also 211—230. Limits to which water rises in, 190, 223. These limits known to old pump-makers, 191. Deceptions with, 224, 225. Bag-pump, 209. Bellows, do. 205—210. Burr, do. 214. Centrifugal, do. 229, 230. German, do. 138, 207, 218, 219. Natural, do. 209, 210. Liquor, do. 215. Spanish, do. 217, 224
Pumps, forcing, 262—293. Common pump, 263. Enema, 263. Bellows, 241, 257, 207, 568. Double acting, 271. Mercurial, 275. Natural, 209, 210, 256—258. Stomach, 264. Plunger, 272, 444. Perkins' 281. Rotary, 284—291, 373. Reciprocating rotary, 292, 293
Pumps, lifting, 277—279
Pusterich, a steam idol, 399
Pythagoras, 438

Q

Quadrant, the, invented by Godfrey, 143
Quern, hand-mill, 282
Quippus, historical cords of the Peruvians, 168, 172

R

Rain at Thebes, a prodigy, 81
Rams, battering, 366. Siphon ram, 531. Water rams, 366—372. Natural water rams 506
Ramseye, his patent for raising water by fire, 419
Razors, bronze, 121. Mexican, 162
Reciprocating rotatory pumps, 292, 293
Regal, 550
Religious persecutions, 446
Remora, sucking fish, 185
Respiration, 475, 477
Richard III. his coffin, a watering trough, 49
Riddles, 437
Rigny, De, his steam machine, 463
Rivatz, a Swiss machinist, 462
Rivaz, his motive engine, 473
Rivius, eolipiles from, 396
Rocking machine for raising water, 93
Rolling press, 323
Rome, invaded by the Gauls from the report of a smith, 19. Houses in, 310
Roman wells, 28, 34, 40, 41, 50. Chain of pots, 124, Water screw, 138. Fire-engines, 310, 311. Firemen, 309. Fountains, 533, 539, 540. Mirrors, 121
Roode of Grace, an English idol, 106
Rope pump, 136
Rotatory movements, 282—284
Rotatory pumps, 281—291. Defects of, 291
Russia, pumps in, 220

S

Safety valves, 387, 391, 447, 451
Sails of ships, 268
Saladin, 47, 372
Salting fish, in Egypt, 86. Its revival in Europe, 86
Sanguisuchello, 203
Sap, ascent of, 507—509
Sarbacans, 256
Sarcophagii, used as watering troughs, 49, 99
Sauce pans, 21
Savery, his experiments and engines, 453—460. His bellows, 483
Saw, 268
Scipio, his baths, 558. Dowry of his daughter, 121
Scoop, to raise water, 93. Scoop wheel, 111
Scots, worshiped wells, 37
Screws for raising water, 137—142, 566
Scythian tradition, 12

Seneca, quoted, 26, 310, 394, 558
Serviere, his inventions, 63, 91, 285, 429
Shadoof, Egyptian, 94, 95
Shakespeare, quoted, or illustrated, 195, 350
  401, 534, 535
Sheet-lead and other metals, 551
Ships, steam, of Garay, 403. Of Ramseye,
  419. Ventilation of, 488
Ship-building, ancient, 146, 147. Chinese, 158
Ship pumps, 143, 147, 154—157, 214—217, 227,
  567
Shoes, ancient, 50. Motezuma's, 161
Shrine of Becket, 106
Siamese water-clocks, 544
Sieve, Tutia carrying water in one, 198. The
  trick explained, ibid.
Silk, watering of, 323
Silver pipes, cocks and cisterns, 170, 557, 560
Siphons, 192, 193, 212, 268, 514—532. Capillary, 513. Natural, 506. Ram, 531. Act
  in vacuo, 515. Other devices so named,
  212, 213, 304, 307, 311, 315
Smoke jacks, 75
Smoking tobacco, 270, 454, 477
Soap-making, raising ley by steam, 413, 414.
  Great sums expended on soap, ibid. Soap
  factory in Pompeii, ibid.
Socrates, 537
Soldering, 551. Cast iron, 20. Phenomenon
  attending, 512
Solomon, cisterns of, 48
Souffleur, 504
Spaniards, their conquest of America, 159
Spanish pump-maker, anecdote of, 224. Spanish steam-ship, 403—406. Chain of pots,
  126. Bells, 314
Speaking tubes, 106, 107. Trumpets, 273,
  342. Heads, 106, 108, 377
Spectacles, 70
Speculums, 121
Sphinx, 119, 437
Spindle, spinning, &c. 283, 284
Spiral pump, 363
Spiritalia, a work written by Heron, 270, 306,
  312, 376, 386, 415, 518—520
Spouting tubes, 497—504
Sprinkling vessels, atmospheric, 194—196, 567
Spurting snake, 257
Statues, 43. Iron one of Arsinoe, 142. Of
  Memnon, 377, 401. Leaden, 535, 550
Steam, its effects, 359, 388—391. Its mechanical properties, 391, 392, 407. 409.
  Supposed identity with air, 395—400, 418
  —421, 572
Steam-boats, 403, 419, 423, 424, 438
Steam-boilers of coiled tubes, 394. Of wood
  and granite, 470
Steam-engines, 284, 359, 425. Heron's, 394.
  Garay's, 404. Branca's, 418. Classification
  of, 425. Worcester's, 437. Moreland's,
  442—444. Papin's, 450—452. Savery's,
  455—460. Other engines, 462—464. Newcomen's, 465. Leopold's, 469. Made self-acting by a boy, 251, 470
Steam-guns, 395, 573
Steam-idols, 395, 398, 399, 570—572
Steam-machinists, courtiers, 445
Stings of bees, 257
Stomach pump, 264
Stoves, Chinese, 70. Fire-places, 483
Strabo's account of Memnon, 377
Stuyvesant, Peter, proclamations by him,
  339, 340
Suckers, boys', 181. Natural, 182, 183, 184,185
Sucking tubes, 203, 204. Sucking wounds, 202
Suction, 201, 202
Sugar boiling in vacuo, 495

Sulphur baths, ancient, 120
Sun, raising water by the heat of, 378—380
  Distilling by, 381. Raising steam by, 470
Surgical instruments, found at Pompeii, 282
Swape, 94—103
Syene, well at, 48
Syracuse, dial at, 542. The name of Archimedes' ship, 146
Syringe, 259—261. An emblematic device,
  261. Used as fire-engines, 312, 315—317

T

Tacitus, quoted, 214
Tanks, water, 83
Tantalus, city of, 98. Cups, 520
Tartar necromancy, 522
Tenures, 398
Teraphim, 571, 572
Tezcuco, 161. Supplied with water, 163
Theatres, fountains in, recommended, 540
Thebes, a wonder at, 81
Themistocles, 43
Theodorus, of Samos, 5
Thirst, modes of quenching, 11. Sufferings
  from, 31, 32
Tinkers, Chinese, 20, 248
Tlascala, 159. Its water-works, 160
Tobacco smoking, 270, 454, 477. Engine for
  cutting, 323
Toledo, old water-works at, 294
Toltecs, 160
Tools, 5, 87, 132, 172, 268
Toothed wheels, 71. 72, 114, 121
Tornados, 482
Torricelli, 187, 188
Tourne-broche, 75, 76, 565. Eolipilic, 396.
  398, 429
Towers, war, 304
Toys, 87, 268
Traditions of the Mahomedans and Scythians, 12. Arabs, 95. Peruvians, 167
Trajan, his directions respecting fires, 309
Traps for drains, 562, 563
Tread-wheels, 73, 74, 76, 116, 117, 152
Treasury note, copy of an old one, 300
Trees of Australia, 509. Of silver, brass, &c.
  538
Trevithick's pump, 280
Tricks. See Juggling.
Triton, musical, of silver, 534
Trombe, or shower bellows, 476
Troy, fountains at, 49
Trumpets, speaking, 273, 342. Ear, 379, 573
Tubal Cain, his bellows, 232. See Vulcan.
Turkish fountains, 31. Fire-engines, 316
Tutia, carrying water in a sieve, 198, 386
Tympanum for raising water, 110, 114
Tyre, a well at, 38. Glass mirrors made at,
  121

U

Union joints, 326, 459

V

Vacuum, B. Porta on, 379. Produced by
  steam, 407, 489. In open tubes, 482—496
  Boiling sugar in, 495
Valentinian, anecdote of, 196
Valves, 232, 235, 268, 307, 555, 556. Safety,
  387, 391, 447, 451
Vapor engines, 441, 472, 473
Vases, ancient, 16, 17. For lustral water, 387
Vauxhall, engines at, 434. Gardens, 445
Vegetius, old translation of, 177, 207, 217
  430, 522
Veneering, 87
Ventilation of mines, ships, &c. 488
Venturi, experiments by, 478—480

INDEX.

Vestals, 195—197, 383
Virgil, quoted, 3, 11, 12, 13, 34, 117, 238, 302
Vision of Mahomet, 98
Vitruvius, quoted, 9, 34, 109, 113, 117, 124, 139, 192, 244, 266, 269, 392, 395
Vulcan, bellows of, 232, 239, 240. Trip hammers, 568. Imprisoning chair, 573

W

Wagons, steam, 423, 424
Walrus, climbs by atmospheric pressure, 183
Wars, warriors, 3, 308, 359. *See preface.*
Watches and clocks, 71, 122, 323, 350, 441, 542—547
Watch chains, 323
Water, its importance in nature and the arts, 9, 302, 358. Supposed identity with air, 395, 418, 420, 421, 572. Worshiped, 33—37, 565. Penalty for stealing, 43. Fresh dipped from the sea, 519
Water beds, 178. Bombs, 349. Canes, 372. Carriers, 83, 84. Closets, 561. Hammer, 367. Lute, 451. Power, 480. Rams, 368—371. Spouts, 477, 510. Wheels, 125, 282, 568
Water-works at Hamath, 116. In Japan, 125. At Babylon, 133—135. Of the Peruvians, 165—172. Mexicans, 160—162. At Augsburgh, Bremen, Toledo, Paris, and London, 294, 295, 296. Old ones described, 438. In America, 298. Roman, 267, 480
Watt, 145, 258. *See preface.*
Waves, 366, 367
Wedge, 268
Weather-glasses, 375
Weeping images, 108
Wells, 24—49, 52, 71, 241, 565. Solon's laws respecting, 27. Reflections on, 48. Ventilation of, 488. Worshiped, 33—37, 565. Ancient American wells, 50, 160—167. Their examination desirable, 50
Wheels, flash, 94. Other wheels to raise water 109—116. Wheel of fortune, 118. Scoop 111. Persian, 115. Capstan 77.

Wheels, tread, 73—76, 116, 117, 152. **Cog,** 71, 72, 114, 121
Whistles, in Peruvian bottles, 17. Chaldean cups, 201
Whitney, Eli, 359
Whitehurst's water ram, 368
Wilkins, Bishop, quoted, 103, 104, 396, 418, 429—431, 566
Windlass, 68—72, 573
Wind-mills, 125, 139, 151, 158, 418, 572
Wines, concentration of, 440. Mixed, 517. Siphons for tasting, 195, 516, 522
Wine flask, Savery's experiment with, 454—455
Winifred's well, miracle at, 37
Winnowing machine, 70
Wire, ancient, 2, 87, 121, 162, 323. Mill for drawing, 423
Wirtz's spiral pump, 363
Woden's well, 35
Woisard's air machine, 473
Women, early experimenters on steam, 391
Wooden hams, 70
Worcester, Marquis of, his century of inventions, 64, 140, 362, 428—440, 539. His steam-boat, 438. Steam-engine, 437. His character and death, 439
World, the, an hydraulic machine, 505
Wynken de Worde, quoted, 314

X

Xerxes, his chair, 4. Sends a distaff to his general, 283

Y

Yoke, 83, 117. Description of an Egyptian one, 84
Yucatan, ancient wells in, 164, 165

Z

Zem Zem, the holy well of Mecca, 35, 43, 44, 45
Zeno, his quarrel with Anthemius, 393
Zodiac, signs of. 85

Made in the USA
Coppell, TX
27 July 2020